Logistics Engineering Handbook

Logistics Engineering Handbook

Edited by **G. Don Taylor**

CRC Press
Taylor & Francis Group
Boca Raton London New York

CRC Press is an imprint of the
Taylor & Francis Group, an **informa** business

CRC Press
Taylor & Francis Group
6000 Broken Sound Parkway NW, Suite 300
Boca Raton, FL 33487-2742

First issued in paperback 2019

© 2008 by Taylor & Francis Group, LLC
CRC Press is an imprint of Taylor & Francis Group, an Informa business

No claim to original U.S. Government works

ISBN-13: 978-0-8493-3053-7 (hbk)
ISBN-13: 978-0-367-38797-6 (pbk)

Library of Congress Card Number 2006004732

Library of Congress Cataloging-in-Publication Data

Logistics engineering handbook / editor, G. Don Taylor.
    p. cm.
  Includes bibliographical references and index.
   ISBN 978-0-8493-3053-7 (alk. paper)
   1. Systems engineering--Handbooks, manuals, etc. 2. Logistics--Handbooks, manuals, etc. I. Taylor, G. Don. II. Title.

TA168.L64 2008
658.5--dc22
    2007022017

**Visit the Taylor & Francis Web site at
http://www.taylorandfrancis.com**

**and the CRC Press Web site at
http://www.crcpress.com**

This handbook is dedicated to my children, Alex and Caroline.

Alex always makes me laugh and he is the best pal I've ever had.
We think so much alike it seems that we are almost the same guy!
My time with him is treasured.

Caroline is the sweetest little person I've ever known.
She has stolen my heart forever and has made the word "Daddy"
my most cherished title.

Contents

Section I   Introduction to Logistics Engineering

Section II   Logistics Activities

Section III   Topics in Transportation Management

Section IV   Enabling Technologies

Section V   Emerging and Growing Trends

Preface

Logistics activities are critical integrating functions in any type of business. Annual expenditures on logistics in the United States alone are equivalent to approximately 10% of the U.S. gross domestic product. Logistics expenditures represent an even larger percentage of the world economy. Thus, achieving state-of-the-art excellence in logistics functions, and attaining the inherent cost reductions associated with outstanding logistics efforts, is very important in terms of competitiveness and profitability. As logistics tools evolve in comprehensiveness and complexity and as the use of such tools becomes more pervasive in industry, it is increasingly difficult to maintain a position of leadership in logistics functions. In spite of the importance of the topic, logistics education often lags industry requirements, especially in terms of engineering-based needs. This handbook seeks to fill this void by providing a comprehensive reference tool that could be effectively used as an engineering textbook or as a complete and versatile professional reference.

This handbook provides comprehensive coverage of both traditional methods and contemporary topics in engineering logistics. It introduces the reader to basic concepts and practices in logistics, provides a tutorial for common logistics problems and solution techniques, and discusses current topics that define the state of the logistics market. The book is comprised of 30 chapters divided into 5 major sections. In each section, the reader will likely note that many of the chapters are written by leading experts in their field.

Although each major section of the book can be considered a stand-alone segment, the handbook is perhaps strongest when read or studied in the order presented. The first section, *Introduction to Logistics Engineering*, focuses on providing basic background information that defines the topic of engineering logistics. Chapters in this section discuss logistics from a historical perspective, discuss the economic impact of logistics functions, and introduce the reader to general logistics tools. Common metrics are discussed so that progress relative to logistics goals can be measured, and logistics is discussed from a system's perspective.

The second section on *Logistics Activities* delves into activities that commonly fill the workdays of logisticians. The section begins with chapters discussing important business-oriented issues like customer service, purchasing and sourcing. The section then provides chapters dealing with demand forecasting, facility layout and location, inventory management, material handling, warehousing, distribution networks and transportation systems management. The reader should find that the important chapter on facility layout and location is particularly comprehensive.

The third section is entitled *Topics in Transportation Management*, and goes into detail on issues related specifically to freight transport. Chapters discuss specific issues such as dispatching and pricing/rating in the trucking industry, but also provide information of more general interest, such as classic transportation problems, the management of freight imbalance, and yield management/capacity planning.

The *Enabling Technologies* discussed in Section IV of the book discuss those enabling technologies that are currently being exploited to great benefit in the logistics industry. Chapters include discussions of logistics tracking technologies, electronic connectivity techniques and software systems, and use of the Internet. Also included are a chapter on reliability, maintainability, and supportability in logistics systems, and a chapter discussing how logistics activities can be funded and justified.

Finally, the fifth section of the book deals with *Emerging and Growing Trends*. Chapters in this section deal with green logistics, reverse logistics and associated packaging needs, global logistics concerns, outsourcing, the use of third-party logistics providers, and the increasing reliance on intermodal transportation. Other chapters discuss the very timely topics of logistics in the service industry and the growing importance of securing the supply chain. This section makes the handbook particularly useful to savvy logistics professionals wishing to exploit possible future trends in logistics practice.

In spite of the growing importance of logistics as a necessary condition for business success, no comprehensive engineering-oriented handbook exists to support educational and reference needs for this topic. Although colleges and universities are starting to pay greater attention to logistics, business schools seem to be well ahead of engineering schools in terms of the development of educational materials, degree programs, and continuing education for logisticians. It is notable and telling that several of the contributing authors for this engineering-based handbook are business school professors. While business schools produce very capable logisticians, there is certainly also a great need for more technical logisticians, whether they come from industrial, systems or even civil engineering or related programs. This comprehensive *Logistics Engineering Handbook* is therefore needed to support education and reference needs for the more technically oriented logisticians. Although contributing authors do not, in the editor's view, make their chapters overly analytical, a more rigorous and mathematics-based treatment of many important topics has been encouraged.

If the engineering/technical orientation of the handbook is the key difference in comparison to other handbooks on the market, another distinguishing feature is that it provides an entire section dedicated more or less to freight transit. Even though transportation is the largest component of logistics expenses, the best engineering references seem to focus more on traditional issues such as plant layout and location, material handling, and classical transportation problems. This handbook covers those vital topics also, but offers an additional focus on transportation management and on freight transit in particular.

A final distinguishing factor for the handbook is that each chapter includes either a brief "case study" overview of an industrially motivated problem or a tutorial using fabricated data designed to highlight important issues. In most cases, this is a discussion that focuses on applications of one or more topics discussed in the chapter, in the form of either a separate section or as a "breakout" at the end of the chapter. In some cases, the case study environment is imbedded within the chapter so that key points can be illustrated with actual case data throughout the chapter. This feature of the handbook helps to ensure that the topics are relevant and timely in terms of industry needs. It also enables the reader to see direct application of the techniques presented in the chapters. Furthermore, having a required case study in every chapter served as a reminder to the contributing authors that the handbook has been designed to be a useful teaching and reference tool, not a forum for theoretical work.

The book should be equally useful as either a textbook or as part of a professional reference library. Beginning with the initial chapters, the handbook can be useful as either a course introduction or as a professional refresher. The comprehensive coverage of logistics activities and topics presented subsequently is likewise useful in either a classroom or business setting. Hopefully, the reader will agree that the chapters in this handbook have been written, in many cases, by the world's leading experts in their field and that the handbook provides a "one-stop shopping" location for logistics engineering reference materials ranging from basics, to traditional problems, to state-of-the-market concerns and opportunities.

About the Editor

G. Don Taylor, Jr. is the Charles O. Gordon Professor and Department Head of the Grado Department of Industrial and Systems Engineering at Virginia Polytechnic Institute and State University in Blacksburg, Virginia. In addition to leading this distinguished department, he has broad-based research interests in several aspects of logistics systems. He has particular interest in seeking state-of-the-art solutions to large-scale, applied logistics problems using simulation and optimization techniques. His recent work has been primarily in the truckload trucking and barge transportation industries.

Prior to joining Virginia Polytechnic Institute and State University, Professor Taylor held the Mary Lee and George F. Duthie Endowed Chair in Engineering Logistics at the University of Louisville where he was co-founder of a multi-university center, the Center for Engineering Logistics and Distribution. He has also held the rank of Full Professor at the University of Arkansas, where he was also the Arkansas Director of The Logistics Institute. He has held a visiting position at Rensselaer Polytechnic Institute and industrial positions at Texas Instruments and Digital Equipment Corporation.

He has a PhD in Industrial Engineering and Operations Research from the University of Massachusetts and MSIE and BSIE degrees from the University of Texas at Arlington. He has served as Principal Investigator (PI) or Co-PI on approximately 70 funded projects and has written more than 200 technical papers. This handbook is his eighth edited book or proceedings. He is a registered Professional Engineer in Arkansas and an active leader in the field of industrial and systems engineering.

Contributors

Tolga Bektaş
Department of Management and Technology
University of Québec in Montréal
Montréal, Québec, Canada

David Bennett
Aston Business School
Aston University
Birmingham, United Kingdom

Douglas R. Bish
Grado Department of Industrial and Systems
  Engineering
Virginia Polytechnic Institute and State University
Blacksburg, Virginia, U.S.A.

Ebru K. Bish
Grado Department of Industrial and Systems
  Engineering
Virginia Polytechnic Institute and State
  University
Blacksburg, Virginia, U.S.A.

Benjamin S. Blanchard
Virginia Polytechnic Institute and State University
Blacksburg, Virginia, U.S.A.

K. Bulbul
Manufacturing Systems Engineering Program
Sabanci University
Istanbul, Turkey

C. Richard Cassady
Department of Industrial Engineering
University of Arkansas
Fayetteville, Arkansas, U.S.A.

Lap Mui Ann Chan
Grado Department of Industrial and Systems
  Engineering
Virginia Polytechnic Institute and State University
Blacksburg, Virginia, U.S.A.

Teodor Gabriel Crainic
Department of Management and Technology
University of Québec in Montréal
Montréal, Québec, Canada

John R. English
College of Engineering
Kansas State University
Manhattan, Kansas, U.S.A.

Richard Germain
College of Business
University of Louisville
Louisville, Kentucky, U.S.A.

Joseph Geunes
Department of Industrial and Systems Engineering
University of Florida
Gainesville, Florida, U.S.A.

Marc Goetschalckx
School of Industrial and Systems Engineering
Georgia Institute of Technology
Atlanta, Georgia, U.S.A.

Sunderesh S. Heragu
Department of Industrial Engineering
University of Louisville
Louisville, Kentucky, U.S.A.

William Ho
Operations and Information Management Group
Aston Business School
Aston University
Birmingham, United Kingdom

Ricki G. Ingalls
School of Industrial Engineering and
    Management
Oklahoma State University
Stillwater, Oklahoma, U.S.A.

M. Eric Johnson
Center for Digital Strategies
Tuck School of Business
Dartmouth College
Hanover, New Hampshire, U.S.A.

Mustafa Karakul
School of Administrative Studies
York University
Toronto, Ontario, Canada

Thomas L. Landers
College of Engineering
University of Oklahoma
Norman, Oklahoma, U.S.A.

Carman Ka Man Lee
Division of Systems and Engineering Management
School of Mechanical and Aerospace Engineering
Nanyang Technological University
Singapore

Chi-Guhn Lee
Department of Mechanical and
    Industrial Engineering
University of Toronto
Toronto, Ontario, Canada

Bacel Maddah
Engineering Management Program
Faculty of Engineering and Architecture
American University of Beirut
Beirut, Lebanon

Ryan E. Maner
J. B. Hunt Transport
Lowell, Arkansas, U.S.A.

Alejandro Mendoza
Department of Industrial Engineering
University of Arkansas
Fayetteville, Arkansas, U.S.A.

Qiang Meng
Transportation Engineering
National University of Singapore
Singapore

Benoit Montreuil
Department of Operations and Systems of Decision
University of Laval
Montréal, Québec, Canada

Edward A. Pohl
Department of Industrial Engineering
University of Arkansas
Fayetteville, Arkansas, U.S.A.

Warren B. Powell
Department of Operations Research and Financial
    Engineering
Princeton University
Princeton, New Jersey, U.S.A.

Luke Ritter
Trident Global Partners
Annapolis, Maryland, U.S.A.

Manuel D. Rossetti
Department of Industrial Engineering
University of Arkansas
Fayetteville, Arkansas, U.S.A.

Darren M. Scott
Center for Spatial Analysis
School of Geography and Earth Sciences
McMaster University
Hamilton, Ontario, Canada

Loay Sehwail
College of Business
University of Wisconsin Oshkosh
Oshkosh, Wisconsin, U.S.A.

Yen-Ping Leow Sehwail
School of Industrial Engineering and Management
Oklahoma State University
Stillwater, Oklahoma, U.S.A.

A. Sen
Manufacturing Systems Engineering Program
Sabanci University
Istanbul, Turkey

Gunter P. Sharp
Georgia Institute of Technology
Atlanta, Georgia, U.S.A.

M. Grazia Speranza
Department of Quantitative Methods
University of Brescia
Brescia, Italy

James R. Stock
Department of Marketing
University of South Florida
Tampa, Florida, U.S.A.

Joel L. Sutherland
Center for Value Chain Research
Lehigh University
Bethlehem, Pennsylvania, U.S.A.

Kevin Taaffe
Department of Industrial Engineering
Clemson University
Clemson, South Carolina, U.S.A.

Tarek T. Taha
J. B. Hunt Transport
Lowell, Arkansas, U.S.A.

G. Don Taylor, Jr.
Grado Department of Industrial and Systems
  Engineering
Virginia Polytechnic Institute and State University
Blacksburg, Virginia, U.S.A.

Dušan Teodorović
University of Belgrade
Belgrade, Serbia
and
Virginia Polytechnic Institute and State University
Blacksburg, Virginia, U.S.A.

Gunduz Ulusoy
Manufacturing Systems Engineering Program
Sabanci University
Istanbul, Turkey

Katarina Vukadinović
University of Belgrade
Belgrade, Serbia

Xiubin Wang
Wisconsin Transportation Center
University of Wisconsin at Madison
Madison, Wisconsin, U.S.A.

Gary L. Whicker
J. B. Hunt Transport
Lowell, Arkansas, U.S.A.

Wayne Whitworth
College of Business
University of Louisville
Louisville, Kentucky, U.S.A.

Rosalyn A. Wilson
R. Wilson, Inc.
Laurel, Maryland, U.S.A.

Thomas G. Yeung
Department of Industrial and Automatic Control
Ecole des Mines de Nantes/IRCCyN
Nantes, France

I

Introduction to Logistics Engineering

1

Logistics from a Historical Perspective

Joel L. Sutherland
Lehigh University

1.1 Defining Logistics

Logistics is a word that seems to be little understood, if at all, by nearly anyone not directly associated with this professional and very important discipline. Many, when hearing someone say they work in the logistics field, associate it with some quantitative, technological, or mathematical practice. Some even confuse *logistics* with the study of language (i.e., *linguistics*). The fact is, logistics is a very old discipline that has been, currently is, and always will be, critical to our everyday lives.

The origin of the term logistics comes from the French word "logistique," which is derived from "loger" meaning quarters (as in quartering troops). It entered the English language in the nineteenth century.

The practice of logistics in the military sector has been in existence for as long as there have been organized armed forces and the term describes a very old practice: the supply, movement, and maintenance of an armed force both in peacetime and in battle conditions. Logistics considerations are generally built into battle plans at an early stage, for it is logistics that determine the forces that can be delivered to the theater of operations, what forces can be supported once there, and what will then be the tempo of operations. Logistics is not only about the supply of materiel to an army in times of war, it also includes the ability of the national infrastructure and manufacturing base to equip, support and supply the armed forces, the national transportation system to move the forces to be deployed, and its ability to resupply that force once they are deployed.

The practice of logistics in the business sector, starting in the later half of the twentieth century, has been increasingly recognized as a critical discipline. The first professional association of logisticians was formed in 1963, when a group of practitioners and academicians formed the National Council of Physical Distribution Management, which in 1985 became the Council of Logistics Management, and then in 2004 the Council of Supply Chain Management Professionals ("The Council"). Today, this organization has thousands of members around the world. A sister organization, The International Society of Logistics (or SOLE), was founded in 1966 as the Society of Logistics Engineers. Today, there are numerous professional associations throughout the world with essentially the same objectives: to conduct research, provide education, and disseminate knowledge for the advancement of the logistics discipline worldwide.

The Council, early on, recognized that there was confusion in the industry regarding the meaning of the term logistics. Over the years, they have provided, and adjusted to changing needs, a definition of logistics that is the most widely accepted definition worldwide. Just as important, they recognized that the relationship between logistics and supply chain management was not clearly understood by those who used these terms—often interchangeably. The Council struggled with the development of a broader definition of logistics and its' relationship to supply chain management that would be widely accepted by practitioners around the world. In 2003, the Council published the following definitions, and boundaries and relationships, for logistics and supply chain management:

1.1.1 Definition of Logistics Management

Logistics management is that part of supply chain management that plans, implements, and controls the efficient, effective forward and reverse flow and storage of goods, services, and related information between the point of origin and the point of consumption in order to meet customers' requirements.

1.1.1.1 Logistics Management—Boundaries and Relationships

Logistics management activities typically include inbound and outbound transportation management, fleet management, warehousing, materials handling, order fulfillment, logistics network design, inventory management, supply–demand planning, and management of third-party logistics services providers. To varying degrees, the logistics function also includes sourcing and procurement, production planning and scheduling, packaging and assembly, and customer service. It is involved in all levels of planning and execution—strategic, operational, and tactical. Logistics management is an integrating function, which coordinates and optimizes all logistics activities, as well as integrates logistics activities with other functions including marketing, sales manufacturing, finance, and information technology.

1.1.2 Definition of Supply Chain Management

Supply chain management encompasses the planning and management of all activities involved in sourcing and procurement, conversion, and all logistics management activities. Importantly, it also includes coordination and collaboration with channel partners, which can be suppliers, intermediaries, third-party service providers, and customers. In essence, supply chain management integrates supply and demand management within and across companies.

1.1.2.1 Supply Chain Management—Boundaries and Relationships

Supply chain management is an integrating function with primary responsibility for linking major business functions and business processes within and across companies into a cohesive and high-performing business model. It includes all of the logistics management activities stated earlier, as well as manufacturing operations, and it drives coordination of processes and activities with and across marketing, sales, product design, finance, and information technology.

1.2 Business Logistics and Engineering Logistics

Before moving on, it is probably helpful to understand the differences that exist between business logistics and engineering logistics. The fact is, there are few, if any, significant differences between the two except that logistics engineers are often charged with handling the more "mathematical" or "scientific" applications in logistics. For example, the business logistician might be concerned with building information systems to support supply chain management, whereas the logistics engineer might be looking for an optimal solution to a vehicle routing problem within defined time windows. This is important to understand as examples are provided throughout the remainder of this chapter.

1.3   Historical Examples of Military Logistics

Without supplies, no army is brave—Frederick II of Prussia, in his *Instruction for his Generals* 1747

Business logistics is essentially an offshoot of military logistics. So it behooves us to look at the military side of the logistical coin first. For war is not just about tactics and strategy. War is very often about logistics.

Looking at most wars throughout history, a point can be identified at which the victory of one side could no longer be prevented except by a miracle—a point after which the pendulum was tipped heavily to one side and spending less and less time on the other. Logistics is absolutely the main factor that tends to tip the pendulum. The following examples illustrate the importance of logistics in military campaigns of the past.

1.3.1   Alexander the Great

Alexander the Great and his father Philip recognized the importance and improved upon the art of logistics in their time. Philip realized that the vast baggage train that traditionally followed an army limited the mobility of his forces. In order to compensate he made the troops carry their own weapons, armor, and some provisions while marching, minimizing the need for a transportation infrastructure. Oxen and oxcarts were not used as they were in many other campaigns during earlier "ancient" times. Oxen could achieve a speed of only 2 miles per hour, their hooves were unsuitable for carrying goods for long distances, and they could not keep up with the army's daily marches, which averaged 15 miles per day. The army did not use carts or servants to carry supplies, as was the practice of contemporary Greek and Roman armies; horses, camels, and donkeys were used in Alexander's baggage train because of their speed and endurance. As necessary, road builders preceded the army on its march to keep the planned route passable.

Alexander also made extensive use of shipping, with a reasonable sized merchant ship able to carry around 400 tons, while a horse could carry 200 lbs (but needed to eat 20 lbs of fodder a day, thus consuming its own load every 10 days). He never spent a winter or more than a few weeks with his army on campaign away from a sea port or navigable river. He even used his enemy's logistics weaknesses against them, as many ships were mainly configured for fighting but not for endurance, and so Alexander would blockade the ports and rivers the Persian ships would use for supplies, thus forcing them back to base. He planned to use his merchant fleet to support his campaign in India, with the fleet keeping pace with the army, while the army would provide the fleet with fresh water. However, the monsoons were heavier than usual, and prevented the fleet from sailing. Alexander lost two-thirds of his force, but managed to get to a nearby port where he reprovisioned. The importance of logistics was central to Alexander's plans, indeed his mastery of it allowed him to conduct the longest military campaign in history. At the farthest point reached by his army, the river Beas in India, his soldiers had marched 11,250 miles in eight years. Their success depended on his army's ability to move fast by depending on comparatively few animals, by using the sea wherever possible, and on good logistic intelligence.

1.3.2   The Romans

The Roman legions used techniques broadly similar to the old methods (large supply trains, etc.), however, some did use those techniques pioneered by Philip and Alexander, most notably the Roman consul Marius. The Romans' logistics were helped, of course, by the superb infrastructure, including the roads they built as they expanded their empire. However, with the decline in the Western Roman Empire in AD fifth century, the art of warfare degenerated, and with it, logistics was reduced to the level of pillage and plunder. It was with the coming of Charlemagne in AD eighth century, that provided the basis for feudalism, and his use of large supply trains and fortified supply posts called "burgs," enabled him to campaign up to 1000 miles away, for extended periods.

The Eastern Roman (Byzantine) Empire did not suffer from the same decay as its western counterpart. It adopted a defensive strategy that, in many ways, simplified their logistics operations. They had interior lines of communication, and could shift base far easier in response to an attack, than if they were in conquered territory—an important consideration due to their fear of a two-front war. They used shipping and considered it vital to keep control of the Dardanelles, Bosphorous, and Sea of Marmara; and on campaign made extensive use of permanent magazines (i.e., warehouses) to supply troops. Hence, supply was still an important consideration, and thus logistics were fundamentally tied up with the feudal system—the granting of patronage over an area of land, in exchange for military service. A peacetime army could be maintained at minimal cost by essentially living off the land, useful for Princes with little hard currency, and allowed the man-at-arms to feed himself, his family, and retainers from what he grew on his own land and given to him by the peasants.

1.3.3   Napoleon in Russia

As the centuries passed, the problems facing an army remained the same: sustaining itself while campaigning, despite the advent of new tactics, of gunpowder and the railway. Any large army would be accompanied by a large number of horses, and dry fodder could only really be carried by ship in large amounts. So campaigning would either wait while the grass had grown again, or pause every so often. Napoleon was able to take advantage of the better road system of the early 19th century, and the increasing population density, but ultimately still relied upon a combination of magazines and foraging. While many Napoleonic armies abandoned tents to increase speed and lighten the logistics load, the numbers of cavalry and artillery pieces (pulled by horses) grew as well, thus defeating the objective. The lack of tents actually increased the instance of illness and disease, putting greater pressure on the medical system, and thereby increasing pressure on the logistics system because of the larger medical facilities required and the need to expand the reinforcement system.

There were a number of reasons that contributed to Napoleon's failed attempt to conquer Russia in 1812. Faulty logistics is considered a primary one. Napoleon's method of warfare was based on rapid concentration of his forces at a key place to destroy his enemy. This boiled down to moving his men as fast as possible to the place they were needed the most. To do this, Napoleon would advance his army along several routes, merging them only when necessary. The slowest part of any army at the time was the supply trains. While a soldier could march 15–20 miles a day, a supply wagon was generally limited to about 10–12 miles a day. To avoid being slowed down by the supply trains, Napoleon insisted that his troops live as much as possible off the land. The success of Napoleon time after time in Central Europe against the Prussians and the Austrians proved that his method of warfare worked. However for it to work, the terrain must cooperate. There must be a good road network for his army to advance along several axes and an agricultural base capable of supporting the foraging soldiers.

When Napoleon crossed the Nieman River into Russia in June 1812, he had with him about 600,000 men and over 50,000 horses. His plan was to bring the war to a conclusion within 20 days by forcing the Russians to fight a major battle. Just in case his plans were off, he had his supply wagons carry 30 days of food. Reality was a bit different. Napoleon found that Russia had a very poor road network. Thus he was forced to advance along a very narrow front. Even though he allowed for a larger supply train than usual, food was to be supplemented by whatever the soldiers could forage along the way. But this was a faulty plan. In addition to poor roads, the agricultural base was extremely poor and could not support the numbers of soldiers that would be living off the land. Since these 600,000 men were basically using the same roads, the first troops to pass by got the best food that could easily be foraged. The second troops to go by got less, and so forth. If you were at the rear, of course there would be little available. The Russians made the problem worse by adopting a scorched earth policy of destroying everything possible as they retreated before the French. As time went by, soldiers began to straggle, due to having to forage further away from the roads for food and weakness from lack of food.

The situation was just as bad for the horses. Grazing along the road or in a meadow was not adequate to maintain a healthy horse. Their food had to be supplemented with fodder. The further the army went into Russia, the less fodder was available. Even the grass began to be thinned out, for like food the first horses had the best grazing, and those bringing up the rear had it the worse. By the end of the first month, over 10,000 horses had died!

Poor logistics, leading to inadequate food supplies and increasingly sick soldiers, decimated Napoleon's army. By the time Napoleon had reached Moscow in September, over 200,000 of his soldiers were dead and when the army crossed into Poland in early December, less than 100,000 exhausted, tattered soldiers remained of the 600,000 proud soldiers who had crossed into Russia only five months before.

1.3.4   World War I

World War I was unlike anything that had happened before. Not only did the armies initially outstrip their logistics systems with the amount of men, equipments, and horses moving at a fast pace, but they totally underestimated the ammunition requirements, particularly for artillery. On an average, ammunition was consumed at ten times the prewar estimates, and the shortage of ammunition posed a serious issue, forcing governments to vastly increase ammunition production. But rather than the government of the day being to blame, it was faulty prewar planning, for a campaign on the mainland of Europe, for which the British were logistically unprepared. Once the war became trench bound, supplies were needed to build fortifications that stretched across the whole of the Western Front. Given the scale of the casualties involved, the difficulty in building up for an attack (husbanding supplies), and then sustaining the attack once it had started (if any progress was made, supplies had to be carried over the morass of "no-man's land"), it was no wonder that the war in the west was conducted at a snail's pace, given the logistical problems.

It was not until 1918, that the British, learning the lessons of the previous four years, finally showed how an offensive should be carried out, with tanks and motorized gun sleds helping to maintain the pace of the advance, and maintain supply well away from the railheads and ports. World War I was a milestone for military logistics. It was no longer true to say that supply was easier when armies kept on the move due to the fact that when they stopped they consumed the food, fuel, and fodder needed by the army. From 1914, the reverse applied, because of the huge expenditure of ammunition, and the consequent expansion of transport to lift it forward to the consumers. It was now far more difficult to resupply an army on the move. While the Industrial nations could produce huge amounts of war materiel, the difficulty was in keeping the supplies moving forward to the consumer.

1.3.5   World War II

World War II was global in size and scale. Not only did combatants have to supply forces at ever greater distances from the home base, but these forces tended to be fast moving and voracious in their consumption of fuel, food, water, and ammunition. Railways proved indispensable, and sealift and airlift made ever greater contributions as the war dragged on (especially with the use of amphibious and airborne forces, as well as underway replenishment for naval task forces). The large-scale use of motorized transport for tactical resupply helped maintain the momentum of offensive operations, and most armies became more motorized as the war progressed. After the fighting had ceased, the operations staffs could relax to some extent, whereas the logisticians had to supply not only the occupation forces, but also relocate those forces that were demobilizing, repatriate Prisoners of War, and feed civil populations of often decimated countries.

World War II was, logistically, as in every other sense, the most testing war in history. The cost of technology had not yet become an inhibiting factor, and only a country's industrial potential and access to raw materials limited the amount of equipment, spares, and consumables a nation could produce. In this regard, the United States outstripped all others. Consumption of war material was never a problem for the United States and its allies. Neither was the fighting power of the Germans diminished by their huge expenditure of war material, nor the strategic bomber offensives of the Allies. They conducted a stubborn, often brilliant defensive strategy for two-and-a-half years, and even at the end, industrial production was still rising. The principal logistic legacy of World War II was the expertise in supplying far-off operations and a sound lesson in what is, and what is not, administratively possible.

During World War II, America won control of the Atlantic and Pacific oceans from the German and Japanese navies, and used its vast wartime manufacturing base to produce, in 1944, about 50 ships, 10 tanks, and 5 trained soldiers for every one ship, tank, and soldier the Axis powers put out. German soldiers captured by Americans in North Africa expressed surprise at the enormous stockpiles of food, clothing, arms, tools, and medicine their captors had managed to bring over an ocean to Africa in just a few months. Their own army, though much closer to Germany than the American army was to America, had chronic shortages of all vital military inventory, and often relied on captured materiel.

Across the world, America's wartime ally, the Soviet Union, was also outproducing Germany every single year. Access to petroleum was important—while America, Britain, and the Soviet Union had safe and ready access to sources of petroleum, Germany and Japan obtained their own from territories they had conquered or pressed into alliance, and this greatly hurt the Axis powers when these territories were attacked by the Allies later in the war. The 1941 Soviet decision to physically move their manufacturing capacity east of the Ural mountains and far from the battlefront took the heart of their logistical support out of the reach of German aircraft and tanks, while the Germans struggled all through the war with having to convert Soviet railroads to a gauge their own trains could roll on, and with protecting the vital converted railroads, which carried the bulk of the supplies German soldiers in Russia needed, from Soviet irregulars and bombing attacks.

1.3.6   The Korean War

The Korean War fought between the U.S.-led coalition forces against the communists offered several lessons on the importance of logistics. When the North Korean Army invaded South Korea on June 25, 1950, South Korea, including the United States, was caught by surprise. Although there were signs of an impending North Korean military move, these were discounted as the prevailing belief was that North Korea would continue to employ guerrilla warfare rather than military forces.

Compared to the seven well-trained and well-equipped North Korean divisions, the Republic of Korea (ROK) armed forces were not in a good state to repel the invasion. The U.S. 8th Army, stationed as occupation troops in Japan, was subsequently given permission to be deployed in South Korea together with the naval and air forces already there, covering the evacuation of Americans from Seoul and Inchon. The U.S. troops were later joined by the UN troops and the forces put under U.S. command.

In the initial phase of the war, the four divisions forming the U.S. 8th Army were not in a state of full combat readiness. Logistics was also in a bad shape: for example, out of the 226 recoilless rifles in the U.S. 8th Army establishment, only 21 were available. Of the 18,000 jeeps and 4 × 4 trucks, 55% were unserviceable. In addition, only 32% of the 13,800 6 × 6 trucks available were functional.

In the area of supplies, the stock at hand was only sufficient to sustain troops in peacetime activities for about 60 days. Although materiel support from deactivated units was available, they were mostly unserviceable. The lack of preparedness of the American troops was due to the assumptions made by the military planners that after 1945 that the next war would be a repeat of World War II. However, thanks to the availability of immense air and sea transport resources to move large quantities of supplies, they recovered quickly.

As the war stretched on and the lines of communication extended, the ability to supply the frontline troops became more crucial. By August 4, 1950, the U.S. 8th Army and the ROK Army were behind the Nakton River, having established the Pusan perimeter. While there were several attempts by the North Koreans to break through the defense line, the line held. Stopping the North Koreans was a major milestone in the war. By holding on to the Pusan perimeter, the U.S. Army was able to recuperate, consolidate, and grow stronger.

This was achieved with ample logistics supplies received by the U.S. Army through the port at Pusan. The successful logistics operation played a key role in allowing the U.S. Army to consolidate, grow, and carry on with the subsequent counteroffensive. Between July 2, 1950 and July 13, 1950 a daily average of 10,666 tons of supplies and equipment were shipped and unloaded at Pusan.

The Korean War highlights the need to maintain a high level of logistics readiness at all times. Although the U.S. 8th Army was able to recover swiftly thanks to the availability of vast U.S. resources, the same cannot be said for other smaller armies. On hindsight, if the U.S. 8th Army had been properly trained and logistically supported, they would have been able to hold and even defeat the invading North Koreans in the opening phase of the war. The war also indicates the power and flexibility of having good logistics support as well as the pitfalls and constraints due to their shortage.

1.3.7 Vietnam

In the world of logistics, there are few brand names to match that of the Ho Chi Minh Trail, the secret, shifting, piecemeal network of jungle roadways that helped the North win the Vietnam War.

Without this well-thought-out and powerful logistics network, regular North Vietnamese forces would have been almost eliminated from South Vietnam by the American Army within one or two years of American intervention. The Ho Chi Minh Trail enabled communist troops to travel from North Vietnam to areas close to Saigon. It has been estimated that the North Vietnamese troops received 60 tons of aid per day from this route. Most of this was carried by porters. Occasionally bicycles and horses would also be used.

In the early days of the war it took six months to travel from North Vietnam to Saigon on the Ho Chi Minh Trail. But the more people who traveled along the route the easier it became. By 1970, fit and experienced soldiers could make the journey in six weeks. At regular intervals along the route, the North Vietnamese troops built base camps. As well as providing a place for them to rest, the base camps provided medical treatment for those who had been injured or had fallen ill on the journey.

From the air the Ho Chi Minh Trail was impossible to be identified and although the United States Air Force tried to destroy this vital supply line by heavy bombing, they were unable to stop the constant flow of men and logistical supplies.

The North Vietnamese also used the Ho Chi Minh Trail to send soldiers to the south. At times, as many as 20,000 soldiers a month came from Hanoi through this way. In an attempt to stop this traffic, it was suggested that a barrier of barbed wire and minefields called the McNamara Line should be built. This plan was abandoned in 1967 after repeated attacks by the North Vietnamese on those involved in constructing this barrier.

The miracle of the Ho Chi Minh Trail "logistics highway" was that it enabled the "impossible" to be accomplished. A military victory is not determined by how many nuclear weapons can be built, but by how much necessary materiel can be manufactured and delivered to the battlefront. The Ho Chi Minh Trail enabled the steady, and almost uninterrupted, flow of logistics supplies to be moved to where it was needed to ultimately defeat the enemy.

1.3.8 Today

Immediately after World War II, the United States provided considerable assistance to Japan. In the event, the Japanese have become world leaders in management philosophies that has brought about the greatest efficiency in production and service. From organizations such as Toyota came the then revolutionary philosophies of Just in Time (JIT) and Total Quality Management (TQM). From these philosophies have arisen and developed the competitive strategies that world class organizations now practice. Aspects of these that are now considered normal approaches to management include kaizen (or continuous improvement), improved customer–supplier relationships, supplier management, vendor managed inventory, collaborative relationships between multiple trading partners, and above all recognition that there is a supply chain along which all efforts can be optimized to enable effective delivery of the required goods and services. This means a move away from emphasizing functional performance and a consideration of the whole supply chain as a total process. It means a move away from the silo mentality to thinking and managing outside the functional box. In both commercial and academic senses the recognition that supply chain management is an enabler of competitive advantage is increasingly accepted. This has resulted in key elements being seen as best practice in their own right, and includes value for money, partnering, strategic procurement policies, integrated supply chain/network management, total cost of ownership, business process reengineering, and outsourcing.

The total process view of the supply chain necessary to support commercial business is now being adopted by, and adapted within, the military environment. Hence, initiatives such as "Lean Logistics" and "Focused Logistics" as developed the U.S. Department of Defense recognize the importance of logistics within a "cradle-to-grave" perspective. This means relying less on the total integral stockholding and transportation systems, and increasing the extent to which logistics support to military operations is outsourced to civilian contractors—as it was in the 18th century. From ancient days to modern times, tactics and strategies have received the most attention from amateurs, but wars have been won by logistics.

1.4   Emergence of Logistics as a Science

In 1954, Paul Converse, a leading business and educational authority, pointed out the need for academicians and practitioners to examine the physical distribution side of marketing. In 1962, Peter Drucker indicated that distribution was the "last frontier" and was akin to the "dark continent" (i.e., it was an area that was virtually unexplored and, hence, unknown). These and other individuals were early advocates of logistics being recognized as a science. For the purpose of this section we define the science of logistics as, the study of the physical movement of product and services through the supply chain, supported by a body of observed facts and demonstrated measurements systematically documented and reported in recognized academic journals and publications.

In the years following the comments of Converse and Drucker, those involved in logistics worked hard to enlighten the world regarding the importance of this field. At the end of the twentieth century, the science of logistics was firmly in place. Works by Porter and others were major contributors in elevating the value of logistics in strategic planning and strategic management. Other well-known writers, such as Hoskett, Shapiro, and Sharman, also helped elevate the importance of logistics through their writings in the most widely read and respected business publications. Because these pioneers were, for the most part, outsiders (i.e., not logistics practitioners) they were better able to view logistics from a strategic and unbiased perspective.

The emergence of logistics as a science has been steady and at times even spectacular. Before the advent of transportation deregulation in the 1980s, particularly in the 1960s and 1970s, "traffic managers" and then "distribution managers" had the primary responsibility for moving finished goods from warehouses to customers on behalf of their companies. Little, if any, attention was given to managing the inbound flows. Though many of these managers no doubt had the capacity to add significant value to their organization, their contribution was constrained by the strict regulatory environment in which they operated. That environment only served to intensify a silo mentality that prevailed within many traffic, and other logistics related, departments.

The advent of transport deregulation in the 1980s complemented, and in many cases accelerated, a parallel trend taking place—the emergence of logistics as a recognized science. The rationale behind this was that transportation and distribution could no longer work in isolation of those other functional areas involved in the flow of goods to market. They needed to work more closely with other departments such as purchasing, production planning, materials management, and customer service as well as supporting functions such as information systems and logistics engineering. The goal of logistics management, a goal that to this day still eludes many organizations, was to integrate these related activities in a way that would add value to the customer and profit to the bottom line.

In the 1990s, many leading companies sought to extend this integration end-to-end within the organization—that is, from the acquisition of raw materials to delivery to the end customer. Technology would be a great enabler in this effort, particularly the enterprise resource planning (ERP) systems and supply chain planning and execution systems that connect the internal supply chain processes. The more ambitious of the leaders sought to extend the connectivity outward to their trading partners both upstream and downstream. They began to leverage Internet-enabled solutions that allowed them to extend connectivity and provide comprehensive visibility over product flow.

As we turned the corner into the 21st century, the rapid evolution of business practices has changed the nature and scope of the job. Logistics professionals today are interacting and collaborating in new ways within their functional area, with other parts of the organization, and with extended partners. As the traditional roles and responsibilities change, the science of logistics is also changing. Logistics contributions in the future will be measured within the context of the broader supply chain.

1.5 Case Study: The Gulf War

1.5.1 Background

The Gulf War was undoubtedly one of the largest military campaigns seen in recent history. The unprecedented scale and complexity of the war presented logisticians with a formidable logistics challenge.

On July 17, 1990, Saddam Hussein accused Kuwait and the United Arab Emirates of overproduction of oil, thereby flooding the world market and decreasing its income from its sole export. Talks between Iraq and Kuwait collapsed on August 1, 1990. On August 2, Iraq, with a population of 21 million, invaded its little neighbor Kuwait, which had a population of less than two million. A few days later, Iraqi troops massed along the Saudi Arabian border in position for attack. Saudi Arabia asked the United States for help. In response, severe economic sanctions were implemented, countless United Nations resolutions passed, and numerous diplomatic measures initiated. In spite of these efforts Iraq refused to withdraw from Kuwait. On January 16, 1991, the day after the United Nations deadline for Iraqi withdrawal from Kuwait expired, the air campaign against Iraq was launched. The combat phase of the Gulf War had started.

There were three phases in the Gulf War worthy of discussion: deployment (Operation Desert Shield), combat (Operation Desert Storm); and redeployment (Operation Desert Farewell). Logistics played a significant role throughout all three phases.

1.5.1.1 Operation Desert Shield

The Coalition's challenge was to quickly rush enough troops and equipment into the theater to deter and resist the anticipated Iraqi attack against Saudi Arabia. The logistical system was straining to quickly receive and settle the forces pouring in at an hourly rate. This build-up phase, Operation Desert Shield, lasted six months. Why the six-month delay? A large part of the answer is supply.

Every general knows that tactics and logistics are intertwined in planning a military campaign. Hannibal used elephants to carry his supplies across the Alps during his invasion of the Roman Empire. George Washington's colonial militias had only nine rounds of gunpowder per man at the start of the

Revolution, but American privateers brought in two million pounds of gunpowder and saltpeter in just one year. Dwight Eisenhower's plans for the June 1944 invasion of Normandy hinged on a massive buildup of war materiel in England. The most brilliant tactics are doomed without the ability to get the necessary manpower and supplies in the right place at the right time.

During the six-month build up to the Gulf War, the United States moved more tonnage of supplies—including 1.8 million tons of cargo, 126,000 vehicles, and 350,000 tons of ordnance—over a greater distance than during the two-year build up to the Normandy invasions in World War II.

Besides the massive amount of supplies and military hardware, the logistics personnel also had to deal with basic issues such as sanitation, transport, and accommodation. A number of these requirements were resolved by local outsourcing. For example, Bedouin tents were bought and put up by contracted locals to house the troops; and refrigerated trucks were hired to provide cold drinks to the troops.

Despite the short timeframe given for preparation, the resourceful logistics team was up to the given tasks. The effective logistics support demonstrated in Operation Desert Shield allowed the quick deployment of the troops in the initial phase of the operation. It also provided the troops a positive start before the commencement of the offensive operation.

1.5.1.2   Operation Desert Storm

It began on January 16, 1991 when the U.S. planes bombed targets in Kuwait and Iraq. The month of intensive bombing that followed badly crippled the Iraqi command and control systems. Coalition forces took full advantage of this and on February 24, 1991 the ground campaign was kicked off with a thrust into the heart of the Iraqi forces in central Kuwait. The plan involved a wide flanking maneuver around the right side of the Iraqi line of battle while more mobile units encircled the enemy on the left, effectively cutting lines of supply and avenues of retreat. These initial attacks quickly rolled over Iraqi positions and on February 25, 1991 were followed up with support from various infantry and armored Divisions.

To the logisticians, this maneuver posed another huge challenge. To support such a maneuver, two Army Corps worth of personnel and equipment had to be transported westward and northward to their respective jumping off points for the assault. Nearly 4000 heavy vehicles were used. The amount of coordination, transport means, and hence the movement control required within the theater, was enormous.

One reason Iraq's army was routed in just 100 hours, with few U.S. casualties, was that American forces had the supplies they needed, where they needed them, when they needed them, and in the necessary quantities.

1.5.1.3   Operation Desert Farewell

It was recognized that the logistical requirements to support the initial build up phase and the subsequent air and land offensive operations were difficult tasks to achieve. However, the sheer scope of overall redeployment task at the end of the war was beyond easy comprehension. To illustrate, the King Khalid Military City (KKMC) main depot was probably the largest collection of military equipment ever assembled in one place. A Blackhawk helicopter flying around the perimeter of the depot would take over an hour. While the fighting troops were heading home, the logisticians, who were among the first to arrive at the start of the war, were again entrusted with a less glamorous but important "clean up job." Despite the massive amount of supplies and hardware to be shipped back, the logisticians who remained behind completed the redeployment almost six months ahead of schedule.

Throughout the war, the Commanding General, Norman Schwarzkopf, had accorded great importance to logistics. Major General William G. (Gus) Pagonis was appointed as the Deputy Commanding General for logistics and subsequently given a promotion to a three-star general during the war. This promotion symbolized the importance of a single and authoritative logistical point of contact in the Gulf War. Under the able leadership of General Pagonis, the efficient and effective logistical support system set up in the Gulf War, from deployment phase to the pull-out phase, enabled the U.S.-led coalition forces to achieve a swift and decisive victory over the Iraqi.

Both at his famous press conferences as well as later in his memoirs, Stormin' Norman called Desert Storm a "logistician's war," handing much of the credit for the Coalition's lightning-swift victory to his chief logistician, General Gus Pagonis. Pagonis, Schwarzkopf declared, was an "Einstein who could make anything happen," and, in the Gulf War, did. Likewise, media pundits from NBC's John Chancellor on down also attributed the successful result of the war to logistics.

1.5.2 Lessons Learned from the Gulf War

1.5.2.1 "Precision Guided" Logistics

In early attempts inside and outside of the Pentagon to assess the lessons learned from the Gulf War, attention has turned to such areas as the demonstrated quality of the joint operations, the extraordinary caliber of the fighting men and women, the incredible efficacy of heavy armor, the impact of Special Forces as part of joint operations on the battlefield, and the success of precision-guided weapons of all kinds. Predictably lost in the buzz over celebrating such successes was the emergence and near-seamless execution of what some have termed "precision-guided" logistics.

Perhaps, this is as it should be. Logistics in war, when truly working, should be transparent to those fighting. Logistics is not glamorous, but it is critical to military success. Logisticians and commanders need to know "what is where" as well as what is on the way and when they will have it. Such visibility, across the military services, should be given in military operations.

1.5.2.2 "Brute Force" Logistics

In 1991, the United States did not have the tools or the procedures to make it efficient. The Gulf War was really the epitome of "brute force" logistics. The notion of having asset visibility—in transit, from factory to foxhole—was a dream. During the Gulf War, the Unites States did not have reliable information on almost anything. Materiel would enter the logistics pipeline based on fuzzy requirements, and then it could not be readily tracked in the system.

There were situations where supply sergeants up front were really working without a logistics plan to back up the war plan. They lacked the necessary priority flows to understand where and when things were moving. It was all done on the fly, on a daily basis, and the U.S. Central Command would decide, given the lift they had, what the priorities were. Although progress was eventually made, often whatever got into the aircraft first was what was loaded and shipped to the theater. It truly was brute force.

Even when air shipments were prioritized there was still no visibility. Although it is difficult to grasp today, consider a load being shipped and then a floppy disk mailed to the receiving unit in the theater. Whether that floppy disk got where it was going before the ship got there was in question. Ships were arriving without the recipients in the theater knowing what was on them.

Generally speaking, if front-line commanders were not sure of what they had or when it would get there, they ordered more. There were not enough people to handle this flow, and, in the end, far more materiel was sent to the theater than was needed. This was definitely an example of "just-in-case" logistics. When the war ended, the logistics pipeline was so highly spiked that there were still 101 munitions ships on the high seas. Again, it was brute-force logistics.

The result was the off-referenced "iron mountains" of shipping containers. There was too much, and, worse yet, little, if any, knowledge of what was where. This led, inevitably, to being forced to open something like two-thirds of all of the containers simply to see what was inside. Imagine the difficulty in finding things if you shipped your household goods to your new house using identical unmarked boxes. Since there were a great number of individual users, imagine that the household goods of all of your neighbors also were arriving at your new address, and in the same identical boxes.

That there was this brute force dilemma in the Gulf War was no secret. There just wasn't any other way around it. The technology used was the best available. Desert Storm was conducted using 286-processor technology with very slow transfer rates, without the Internet, without the Web, and without encrypted satellite information. Telexes and faxes represented the available communication technology.

1.5.2.3 "Flying Blind" Logistics

This was an era of green computer screens, when it took 18 keystrokes just to get to the main screen. When the right screen was brought up, the data were missing or highly suspect (i.e., "not actionable"). In contrast to today, there were no data coming in from networked databases, and there was no software to reconcile things. There were also no radio frequency identification tags. In effect, this was like "flying blind."

In fact, nothing shipped was tagged. Every shipment basically had a Government bill of lading attached to it, or there were five or six different items that together had one bill of lading. When those items inevitably got separated, the materiel was essentially lost from the system. Faced with this logistics nightmare, and knowing that there was often a critical need to get particular things to a particular place at a particular time, workarounds were developed.

As a result of our experience in the Gulf War, the Department of Defense (DOD) has subsequently been refining its technologies and testing them through military joint exercises and deployments and contingencies in such places as Bosnia, Kosovo, and Rwanda. Specifically, the DOD has focused on the issue of logistics management and tracking and on how technology can enable improvements in this mission critical area. The DOD has improved its logistics management and tracking through policy directives and by engaging with innovative technology companies in the development and leveraging of technical solutions.

The DOD now has clear knowledge of when things are actually moving—the planes, the ships, what is going to be on them, and what needs to be moved. Communication is now digital and that represents a quantum leap in capability and efficiency from the first war in Iraq. Operators now get accurate information, instantaneously, and where needed. The technology exists to absorb, manage, and precisely guide materiel.

1.5.3 Applying Lessons Learned from the Gulf War

1.5.3.1 Operation Enduring Freedom

While troops raced toward Baghdad in the spring of 2003, digital maps hanging from a wall inside the Joint Mobility Operations Center at Scott Air Force Base, Ill, blinked updates every four minutes to show the path cargo planes and ships were taking to the Middle East. During the height of the war in Iraq, every one of the military's 450 daily cargo flights and more than 120 cargo ships at sea were tracked on the screen, as was everything stowed aboard them—from Joint Direct Attack Munitions to meals for soldiers.

In rows of cubicles beneath the digital displays, dozens of military and civilian workers from the U.S. Transportation Command (TRANSCOM) looked at the same maps on their computer screens. The maps, along with an extensive database with details on more than five million items and troops in transit, came in handy as telephone calls and e-mail queries poured in from logisticians at ports and airfields in the Persian Gulf: How soon would a spare part arrive? When would the next shipment of meals arrive? When was the next batch of troops due? With just a few mouse clicks, TRANSCOM workers not only could report where a ship or plane was and when it was due to arrive, but also could determine which pallet or shipping container carried what. In many cases, logisticians in the field also could go online, pull up the map and data and answer their own questions.

Vice Admiral Keith Lippert, director of the Defense Logistics Agency (DLA) says the war in Iraq validated a new business model that moves away from "stuffing items in warehouses" to relying on technology and contractors to provide inventory as needed. The agency, which operates separately from TRANSCOM, is responsible for ordering, stocking, and shipping supplies shared across the services. In addition, the Army, Navy, Air Force, and Marines have their own supply operations to ship items unique to each service. The DLA supplied several billion dollars worth of spare parts, pharmaceuticals, clothing and 72 million ready-to-eat meals to troops during the war.

Military logisticians have won high marks for quickly assembling the forces and supplies needed in Iraq. Advances in logistics tracking technology, investments in a new fleet of cargo airplanes and larger ships, and the prepositioning of military equipment in the region allowed troops to move halfway around the world with unprecedented speed. Troops were not digging through containers looking for supplies they had ordered weeks earlier, nor were they placing double and triple orders in hopes that one of their requests would be acted upon, as they did during the Gulf War in 1991. While the military transportation and distribution system may never be as fast or efficient as FedEx or UPS, its reliability has increased over the past decade.

Nonetheless, challenges remain. Several changes to the way troops and supplies are sent to war are under consideration, including:

- Further improvement of logistics information technology systems
- Development of a faster way to plan troop deployments
- Consolidated management of the Defense supply chain

While TRANSCOM has gotten positive reviews for moving troops and supplies to the Middle East, concerns have been raised about how the services moved supplies after they arrived in the field.

Perhaps the most valuable logistics investment during the war was not in expensive cargo aircraft or advanced tracking systems, but in thousands of plastic radio frequency identification labels that cost $150 apiece. The tags, which measure eight inches long by about two inches wide, contain memory chips full of information about when a shipment departed, when it is scheduled to arrive and what it contains. They are equipped with small radio transponders that broadcast information about the cargo's status as it moves around the world. The tags enable the Global Transportation Network to almost immediately update logistics planners on the location of items in the supply chain.

These tags were a key factor in avoiding the equipment pileups in warehouses and at desert outposts that came to symbolize logistics failings during the first Gulf War. The tags also saved hundreds of millions of dollars in shipping costs, logisticians say. For example, British soldiers spent almost a full day of the war searching cargo containers for $3 million in gear needed to repair vehicles. Just as they were about to place a second order for the gear, a U.S. logistician tapped into a logistics tracking system and was able to locate the supplies in the American supply network.

Rapid response to shifting requirements is clearly the fundamental challenge facing all logisticians, as relevant in the commercial sector as it is in the military environment. The commercial logistician requires the same thing that the combatant commander requires: situational awareness. We all need an in-depth, real-time knowledge of the location and disposition of assets.

Indeed, Wal-Mart, arguably the channel master for the world's largest, most globally integrated commercial supply chain, has embarked on a passive RFID initiative that is very similar to the Department of Defense's plans. The retailer mandated that suppliers tag inbound materiel with passive RFID tags beginning at the case and pallet level. Wal-Mart established a self-imposed January 2005 deadline to RFID-enable its North Texas operation, along with 100 of its suppliers. The first full-scale operational test began on April 30, 2005. Based on the success of this initial test Wal-Mart expanded its supplier scope and deployment plan for RFID and by early 2007 reported that some 600 suppliers were RFID-enabled.

While there have been some solid successes early on, there are now many suppliers (in particular the smaller ones) that are dragging their feet on RFID adoption due to an elusive return on investment (ROI). Current generation RFID tags cost about 15 cents, while bar codes cost a fraction of a cent. Suppliers have also had to absorb the cost of buying hardware—readers, transponders, antennas—and software to track and analyze the data. The tags also have increased labor. Bar codes are printed on cases at the factory, but because most manufacturers have yet to adopt RFID, tags have to be put on by hand at the warehouse. The retail giant also experienced difficulties rolling out RFID in their distribution network. Wal-Mart had hoped to have up to 12 of its roughly 137 distribution centers using RFID technology by the end of 2006, but had installed the technology at just five. Now Wal-Mart has shifted gears from their distribution centers to their stores where they believe they will be better able to drive sales for their suppliers and to get product on the shelf, where it needs to be for their customers to buy. By early 2007 there were roughly 1000 stores RFID-enabled with another 400 stores planned by the end of the year.

Regardless of where Wal-Mart places their priorities, with this retail giant leading the charge, and driving industry compliance, it is expected that this initiative will have a greater, and more far-reaching, impact on just the retail supply chain. Virtually every industry, in every corner of the planet, will be fundamentally impacted sometime in the not-too-distant future. Clearly the lessons learned in military logistics are being applied to business logistics and as a result engineering logistics.

2

Economic Impact of Logistics

Rosalyn A. Wilson
R. Wilson, Inc.

2.1 Expenditures in the United States and Worldwide

As the world continues to develop into a homogenized global marketplace the growth in world merchandise trade has outpaced the growth in both global production and the worldwide economy. In 2006, world merchandise trade increased 8%, while the global economy rose only 3.7%.* Globalization has dramatically shifted where logistics dollars are spent as developing countries now account for over one-third of world merchandise exports. Increased world trade means higher demand for logistics services to deliver the goods. Expenditures for logistics worldwide are estimated at well over $4 trillion in 2006 and now account for about 15% to 20% of finished goods cost.† Growth in world merchandise trade, measured as export volume, has exceeded the growth in the worldwide economy, as measured by Gross Domestic Product (GDP), for close to two decades. Although the worldwide economy slowed to some extent in late 2006 and early 2007, trade volumes are predicted to continue to rise well into the next decade.

This phenomenal growth in world trade has profound implications for logistics. In the past five years the demand for shipping has outstripped the capacity in many markets, altering the supply demand equilibrium and pushing up prices. It now costs from 15% to 20% more to move products than it did in 2002. Shifts in global manufacturing as the United States continues to move manufacturing facilities to other global markets with lower labor costs, such as China, India, and South Korea, are redrawing the landscape for transportation strategies. The growth was led by Asia and the so-called transition economies (Central and Eastern Europe and the Russian Trade Federation). In real terms these regions experienced 10–12% growth rates in merchandise exports and imports. China, for instance, has seen the most dramatic trade growth, with a 27% jump in 2006. The World Trade Organization (WTO) recently

---

* World Trade Organization Press Release, "World Trade 2006, Prospects for 2007," April 12, 2007.
† Estimated from a 2003 figure for global logistics of $3.43 trillion. Report from the Ad Hoc Expert Meeting on Logistics Services by the United Nations Conference on Trade and Development's (UNCTAD) Trade and Development Board, Commission on Trade in Goods and Services, and Commodities, Geneva, July 13, 2006.

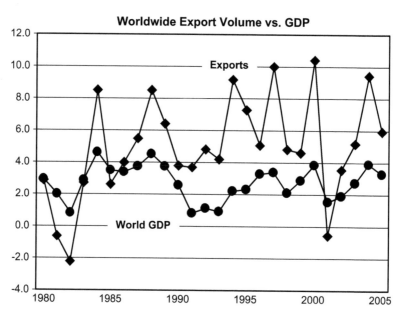

FIGURE 2.1 Worldwide export volume vs. GDP. (From World Trade Organization, International Trade Statistics, 2006.)

reported that China's merchandise exports actually exceeded those of the United States, the market leader, for the second half of 2006. Worldwide export volumes as a percentage of world GDP appear in Figure 2.1.

Studies have shown that total expenditures as a percentage of GDP are generally lower in more efficient industrialized countries, usually 10% or less. Conversely less-developed countries expend a much greater portion of their GDP, 10–20%, on logistics. Where a country falls on the spectrum depends on factors such as the size and disbursement of the population, the level of import and export activity, and the type and amount of infrastructure development. The relative weights for the components of total logistic costs vary significantly by country, with carrying costs accounting for 15–30%, transportation expenditures for another 60–80%, and administrative costs for the remaining 5–10%. Logistics cost in the United States have been holding steady at just under 10% of GDP. The breakout for the components of U.S. logistics costs are 33% for carrying costs, 62% for transportation costs, and about 4% for administrative costs. Additional detail is provided in Figure 2.2.

During 2005, the cost of the U.S. business logistics system increased to $1.18 trillion, or the equivalent of 9.5% of nominal GDP. Logistics costs have gone up over 50% during the last decade. The year 2005 was a year of record highs for many of the components of the model, especially transportation costs, mostly trucking. Transportation costs jumped 14.1% over 2004 levels, and 77.1% during the past decade. Yet, total logistics costs remained below 10% of GDP.

2.2 Breakdown of Expenditures by Category

The cost to move goods encompasses a vast array of activities including supply and demand planning, materials handling, order fulfillment, management of transportation and third-party logistics (3PLs) providers, fleet management, and inventory warehouse management. To simplify, logistics can be defined as the management of inventory in motion or at rest. Transportation costs are those incurred when the inventory is in motion, and inventory carrying costs are those from inventory at rest awaiting

2005 U.S. Business Logistics System Cost

				$ Billions
Carrying Costs - $1.763 Trillion All Business Inventory				
Interest				58
Taxes, Obsolescence, Depreciation, Insurance				245
Warehousing				90
			Subtotal	393
Transportation Costs				
Motor Carriers:				
Truck - Intercity				394
Truck - Local				189
			Subtotal	583
Other Carriers:				
Railroads				48
Water	I 29	D 5		34
Oil Pipelines				9
Air	I 15	D 25		40
Forwarders				22
			Subtotal	153
Shipper Related Costs				8
Logistics Administration				46
TOTAL LOGISTICS COST				1183

FIGURE 2.2 Breakdown of U.S. business logistics system costs. (From 17th Annual State of Logistics Report, Rosalyn Wilson, CSCMP, 2006.)

the production process or in storage awaiting consumption. The third broad category of logistics cost is administrative costs, which encompass the other costs of carrying out business logistics that is not directly attributable to the first two categories. The cost of the U.S. business logistics system as measured by these three categories was $1183 billion in 2005.*

2.2.1 Carrying Costs

Carrying costs are the expenses associated with holding goods in storage, whether that be in a warehouse or, as is increasingly done today, in a shipping container, trailer, or railcar. There are three subcomponents that comprise carrying cost. The first is interest and that represents the opportunity cost of money invested in holding inventory. This expense will vary greatly depending on the level of inventory held and the interest rate used. The second subcomponent covers inventory risk costs and inventory service costs and comprises about 62% of carrying cost expense. These are measured by using expenses for obsolescence, depreciation, taxes, and insurance. Obsolescence includes damages to inventory and shrinkage or pilferage, as well as losses from inventory which cannot be sold at value because it was not moved through the system fast enough. In today's fast paced economy with quick inventory turns, obsolescence represents a significant cost to inventory managers. The taxes are the *ad valorem* taxes collected

* Logistics expenditures for the United States have been measured consistently and continuously for the "Annual State of Logistics Report" developed by Robert V. Delaney of Cass Logistics in the mid-1980s and continued today by Rosalyn Wilson. The methodology used by Mr. Delaney was based on a model developed by Nicholas A. Glaskowsky, Jr., James L. Heskett, Robert M. Ivie in *Business Logistics*, 2nd edition, New York, Ronald Press, 1973. The Council for Supply Chain Management Professionals (CSCMP) has sponsored the report since 2004.

on inventory and will vary with inventory levels. Insurance costs are the premiums paid to protect inventory and mitigate losses. The final subcomponent is warehousing. Warehousing is the cost of storing goods and has traditionally included both public and private warehouses, including those in manufacturing plants. The market today includes a wide variety of storage possibilities from large mega-distribution centers, to smaller leased facilities, to container and trailer-storage yards.

In 2005, inventory carrying costs rose 17%—the highest level since 1971. The increase was due to both significantly higher interest rates than in 2004 and a rise in inventories. The average investment in all business inventories was $1.74 trillion, which surpassed 2004's record high by $101 billion. Both the inventory-to-sales ratio and the inventory-to-factory shipments ratio have been rising steadily in recent years. Inventories have been slowly creeping up since 2000, reversing the trend to leaner inventories from the previous decade. The globalization of production has driven the economy away from the lean just-in-time inventory management model of the 1990s. Stocks are increasingly maintained at a higher level in response to longer and sometimes unpredictable delivery times, as well as changes in distribution patterns. Manufacturers and retailers have struggled to achieve optimum inventory levels as they refine their supply chains to mitigate uncertain delivery times, add new sources of supply, and become more adept at shifting existing inventories to where they are most advantageous. On an annualized basis, the value of all business inventory has risen every year since 2001, as depicted graphically in Figure 2.3.

2.2.2   Transportation Costs

Transportation costs are the expenditures to move goods in various states of production. This could include the movement of raw materials to manufacturing facilities, movement of components to be included in the final product, to the movement of final goods to market. Transportation costs are measured by carriers' revenues collected for providing freight services. All modes of transportation are included: trucking, intercity and local; freight rail; water, international and domestic; oil pipeline; both international and domestic airfreight transport; and freight forwarding costs, not included in carrier

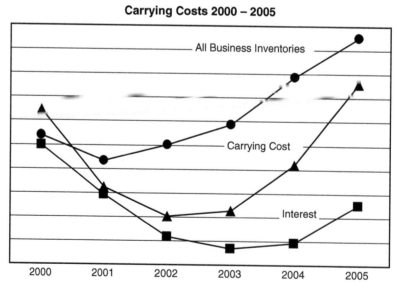

FIGURE 2.3   Costs associated with inventories. (From 17th Annual State of Logistics Report, Rosalyn Wilson, CSCMP, 2006.)

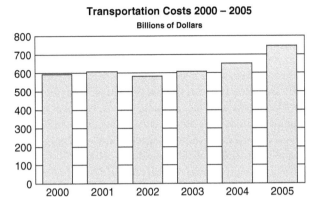

FIGURE 2.4 Transportation costs. (From 17th Annual State of Logistics Report, Rosalyn Wilson, CSCMP, 2006.)

revenue. Transportation includes movement of goods by both public and private, or company-owned, carriers. The freight forwarder expenditures are for other value-added services provided by outside providers exclusive of actual transportation revenue which is included in the modal numbers. Transportation costs are the single largest contributor to total logistics costs, with trucking being the most significant subcomponent. Figure 2.4 shows recent values for these costs.

Trucking costs account for roughly 50% of total logistics expenditures and 80% of the transportation component. Truck revenues are up 21% since 2000, but that does not tell the whole story. In 2002, trucking revenues declined for the first time since the 1974–1975 recession. During this period demand was soft and rates were dropping, fuel prices were soaring, insurance rates were skyrocketing. The trucking industry was forced to undergo a dramatic reconfiguration. About 10,000 motor carriers went bankrupt between 2000 and 2002, and many more were shedding their terminal and other real estate and non-core business units to survive.* While the major impact was the elimination of many smaller companies with revenues in the $5–$20 million range, there were some notable large carriers including Consolidated Freightways. Increased demand and tight capacity enabled trucking to rebound in 2003 and it has risen steadily since.

Trucking revenues in 2005 increased by $74 billion over 2004, but carrier expenses rose faster than rates, eroding some of the gain. The hours-of-service rules for drivers have had a slightly negative impact by reducing the "capacity" of an individual driver, at the same time a critical driver shortage is further straining capacity. The American Trucking Association (ATA) has estimated that the driver shortage will grow to 111,000 by 2014. Fuel ranks as a top priority at trucking firms as substantially higher fuel prices have cut margins. However, for many the focus has shifted from the higher price level to the volatility of prices. The U.S. trucking industry consumes more than 650 million gallons of diesel per week, making it the second largest expense after labor. The trucking industry spent $87.7 billion for diesel in 2005, a big jump over the $65.9 billion spent in 2004.

Rail transportation has enjoyed a resurgence as it successfully put capacity and service issues behind. Freight ton-mile volumes have reached record levels for nine years in a row. Despite a growth of 33% since 2000, rail freight revenue accounts for only 6.5% of total transportation cost. Intermodal shipping has given new life to the rail industry, with rail intermodal shipments more than tripling since 1980, up from 3.1 to 9.3 million trailers and containers. Sustained higher fuel prices have made shipping by rail a more cost-effective mode than an all truck move. High demand kept the railroad industry operating at

---

* Donald Broughton tracks bankruptcies in a proprietary database for A.G. Edwards and Sons.

near capacity throughout 2005, bumping revenue 14.3%. The expansion of rail capacity has become a paramount issue. The Association of American Railroads (AAR) has reported that railroads will spend record amounts of private capital to add new rail lines to double and triple track existing corridors where needed. In addition, freight railroads are expected to hire 80,000 new workers by 2012.

Water transportation is comprised of two major segments—domestic and international or oceangoing. The international segment has been the fastest growing segment leaping over 60% since 2000, from $18 billion to $29 billion. This tracks with the dramatic growth in global trade. Domestic water traffic, by comparison, has actually declined 30% since 2000, falling from $8 billion to $5 billion in 2005. The United States continues to struggle with port capacity problems, both in terms of available berths for unloading and throughput constraints which slow down delivery.

Water transportation faces many obstacles to its continued health. Given the expected growth in international trade U.S. ports are rapidly becoming inadequate. Many ports are over fifty years old and are showing signs of neglect and obsolescence and many have narrow navigation channels and shallow harbors that do not permit access by deep draft vessels which are becoming predominant in the world-wide fleet. The U.S. ports system is close to reaching the saturation point. The World Shipping Council estimates that over 800 ocean freight vessels make over 22,000 calls at U.S. ports every year, or over 60 vessels a day at the nation's 145 ports. Even worse, while the U.S. has done little more than maintain our ports, ports throughout Asia and Europe have become more modern and efficient, giving them an edge in the global economy. As global trading partners build port facilities to handle the larger ships the U.S. places itself at an even greater competitive disadvantage.

The domestic waterway system, the inland waterways, and Great Lakes, has also been the victim of underinvestment. For too many years there has been a lack of resources aimed at maintaining and improving this segment of our transportation network and it is beginning to have dramatic impacts on the capacity of the system. Dredging has fallen behind and the silt built up is hampering navigation and the nation's lock systems are aged and crumbling, with 50% of them obsolete today. Revitalizing this important transportation segment and increasing its use could have a significant impact on reducing congestions and meeting demand for capacity. Although it is not very prevalent now, waterways could even handle containers. A single barge can move the same amount of cargo as 58 semi-trucks at one-tenth the cost.

The air cargo industry has both a domestic and an international side. It is primarily composed of time-sensitive shipments for which customers are willing to pay a premium. Both markets are strong with international revenue up almost 88% since 2000 and domestic revenues up 32% during the same period. Although the air cargo market is thriving and growing, it is still a relatively small share of the whole, representing only about 5% of transport costs. Airfreight revenues increased by $6 billion during 2005, which was an increase of 17.6% over 2004. Along with the growth in revenue came skyrocketing expenses, especially for fuel. In 2003, fuel represented about 14% of operating expenses and in 2005 the percentage had grown to 22%.

The next segment, oil pipeline transportation, accounts for slightly over 1% of total transportation costs. It includes the revenue for the movement of crude and refined oil. We have not added much capacity in the last decade and costs have remained stable, so revenues have been largely constant since 2000.

The final segment, forwarders, has increased over two and half times since 2000, rising from $6 billion to $22 billion. It is important to note that this segment does not include actual transportation expenses, those are picked up in the figures for each mode. Freight forwarders provide and ever increasing array of services as they adapt to meet the changing needs of shippers who chose to outsource their freight needs. The most basic function of a forwarder is to procure carrier resources and facilitate the freight movement. Globalization was a boon to such third-party providers as they specialized in the processes and documentation necessary to engage in international trade. Today forwarders offer such services as preparation of export and import documentation, consolidation and inspection services, and supply chain optimization consulting.

2.2.3 Administrative Costs

The final component of logistics cost is administrative costs and it has two subcomponents: shipper-related costs and logistics administration costs. Shipper-related costs are expenses for logistics-related functions performed by the shipper that are in addition to the actual transportation charges, such as the loading and unloading of equipment, and the operation of traffic departments. Shipper costs actually amounts to less than 1% of total logistics costs.

Logistics administration costs represent about 4% of total logistics costs. It includes corporate management and support staff who provide logistics support, such as supply chain planning and analysis staff and physical distribution staff. Computer software and hardware costs attributable to logistics are included in this category if they cannot be amortized directly elsewhere.

2.3 Logistics Productivity over the Past 25 Years

There has been a dramatic improvement in the U.S. business logistics system in the past 20 years. Inventory carrying costs as a percentage of GDP has declined about 40%. Transportation costs as a percentage of GDP dropped by 8% and total logistics costs declined by 23%. Logistics costs as a percentage of nominal GDP has been below 10% since 2000, despite a 25% increase in the last two years. Imports into the United States, as measured by TEUs, has jumped from under 50 million units to over 400 million in the past 26 years, despite the fact that the capacity growth rate of the nation's transportation infrastructure has been static.

Logistics costs in the United States, and to some extent Europe, have dropped significantly since the deregulation of the transportation modes in the 1980s. Much of the gain was due to reductions in inventory costs. The improved performance of the U.S. logistics sector can be traced to the regulatory reforms in the 1980s. All modes were substantially deregulated, including trucking, rail and air, and after a period of six to eight years of adjustment the economy began to reap the benefits of enhanced productivity, rationalized rail lines, and expanded use of rate contracts. Investments in public infrastructure, particularly the interstate highway system and airports, initially contributed to improved performance in the industry. For the last decade the United States has seriously lagged behind in the necessary investment to sustain the growth however. Much of the gain has come from private innovations and companies agile enough to change rapidly with the times. Examples are the appearance and then explosive growth of the express shipping market, just in time and lean inventory practices which are now being replaced with carefully managed inventories that can be redirected instantaneously, mega retail stores like Wal-Mart and Target with clout to influence logistics practices, and logistics outsourcing.

Over the last 15 years, there has not been a dramatic shift in the relative weights for each of the components that make up total logistics costs. Carrying costs represented 39% of total logistics costs in 1989 and account for 33% today, while transportation costs have climbed from a 56% share to a 62% share of the total. With the exception of carrying costs, each of the other components have risen over 60% since 1989, with both transportation and shipper-related costs jumping 75%. (See Fig. 2.5 for a graphical depiction of trends.)

The nation's railroads move over 50% of all international cargo entering the United States for some portion of the move. International freight is expected to double its current level by 2025. Although the railroads have made heavy investments in recent years in equipment and additional labor, average train speed is falling. Truck vehicle-miles traveled on U.S. highways have nearly doubled in the last 25 years. According to the Federal Highway Administration (FHWA), the volume of freight traffic on the U.S. road system will increase 70% by 2020. Also by 2020, the highway system will have to carry an additional 6.6 billion tons of freight—an increase of 62%. Slower trains mean higher costs and more congestion. Statistics published by the AAR show that average train speed for the entire United States declined from 23 miles per hour in 2000 to less than 22 miles per hour in 2005. The rail freight network was rationalized shortly after the passage of the Staggers Act in 1980 and is now about one-half the size it was, prior

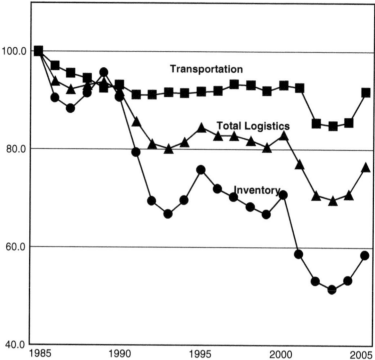

FIGURE 2.5 Logistics series as a percentage of GDP. (From 17th Annual State of Logistics Report, Rosalyn Wilson, CSCMP, 2006.)

to 1980. The leaner system is more productive however, and carries almost double the number of ton-miles the old system carried. Yet, shippers are pushing for even more efficiencies in this area. Will the old strategies applied so successfully in the past work in the rapidly changing global environment? Perhaps, the evidence will show that to maintain the gains we have made and to improve the U.S. world competitiveness will require innovation and a re-engineering of supply chain management. Leading the pack in this arena is the contract logistics market.

Market location has become one of the most important drivers of logistics cost. The push by the United States to locate manufacturing facilities offshore to take advantage of less expensive labor and abundant resources has caused a shift in trade patterns. Logistics services that were traditionally performed largely by developed nations are now increasingly being carried out by emerging economies. Now developing countries move finished goods, in addition to raw materials.

The growth and market clout of mega-retailers like Wal-Mart increased the pressure to reduce costs and increase efficiency, forcing many companies to outsource pieces of their supply chain, often to offshore resources. However, global manufacturing is driving many companies to devise innovative strategies for ensuring reliable sources of goods. The ongoing shift of manufacturing to Asia has added stress to an already congested and overburdened domestic transportation system, particularly on shipping in the Pacific. The region has already been operating at full capacity.

Another interesting demographic is the number of small companies now participating in global trade, which had been the purview of large multinational companies until the late 1990s. Over 80% of corporations surveyed in 2002, ranging from small businesses to global giants, indicated that they operated on a global scale. Most operate distribution, sales or marketing centers outside of their home markets.

The globalization of trade and logistics operations has led to the development of international opera-tors based in the regional hubs of developing regions, with Hong Kong, Singapore, United Arab Emirates, and the Philippines. These entities have refined their processes and often employed state-of-the-art equipment to enhance their productivity. The infrastructure has often been built from the ground up with today's global climate in mind. These companies now account for over 30% of global terminal operations.

Many U.S. shippers are contracting their logistics out to non-U.S.-based providers. The estimated value globally for contract logistics services has exceeded $325 billion, with the U.S. portion estimated to be about $150 billion. Shippers are now outsourcing one or more of their supply chain management activities to 3PLs service suppliers. These providers specialize in providing integrated logistics services that meet the needs of today's highly containerized freight system. These companies have proven to be particularly adaptable to the changing global environment including the use of larger and faster ships, containerization of freight, increased security requirements, new technologies to track and monitor shipments, and the rise in air transport for time-sensitive shipments. The global marketplace seemed to emerge overnight and most companies were not prepared or agile enough to respond to the changes. A new knowledge-based needed to be acquired and the rules were constantly changing. Third-part providers provided the answers to these problems. These companies filled the niche and became experts, enabling even the smallest firms to operate multinationally. The most successful of these companies control a major share of the market and they play a key role in our ability to expand our supply chains into international markets.

3

Logistics
Engineering Tool Chest

Dušan Teodorović
University of Belgrade and Virginia Polytechnic Institute and State University

Katarina Vukadinović
University of Belgrade

3.1 Introduction

Logistic systems are systems of big dimensions that are geographically dispersed in space. Their complexity is caused by many factors. Interactions between decision-makers, drivers, workers and clients; vehicles, transportation and warehousing processes, communication systems and modern computer technologies which are very complex. Logistics has been defined by the Council of Logistics Management as "... the process of planning, implementing, and controlling the efficient, effective flow and storage of goods, services, and related information from point of origin to point of consumption for the purpose of conforming to customer requirements." This definition includes inbound, outbound, internal, and external movements, and return of materials for environmental purposes.

Many aspects of logistic systems are stochastic, dynamic, and nonlinear causing logistic systems to be highly sensitive even to small perturbations. Management and control of modern logistic systems is based on many distributed, hierarchically organized levels. Decision-makers, dispatchers, drivers, workers, and clients have different interests and goals, different educational levels, and diverse work experience. They perceive situations in different ways, and make a lot of decisions based on subjective perceptions and subjectively evaluated parameters.

Management and control of modern logistic systems are based on Management Science (MS), Operations Research (OR), and Artificial Intelligence (AI) techniques. Implementation of specific control actions is possible because of a variety of classical and modern electronic, communication, and information technologies that are vital parts of logistic infrastructure. These technologies significantly contribute to the efficient distribution, lower travel times and traffic congestion, lower production and transportation costs, and higher level of service.

Observation, analysis, prediction of future development, control of complex systems, and optimization of these systems represent some of the main research tasks within OR. Analysis of system behavior assumes development of specific theoretical models capable of accurately describing various system processes. The developed mathematical models are used to predict system behavior in the future, to plan future system development, and to define various control strategies and actions. Logistic systems characterized by complex and expensive infrastructure and equipment, great number of various users, and uncertain value of many parameters, have been one of the most important and most challenging OR areas.

Artificial Intelligence is the study and research in computer programs with the ability to display "intelligent" behavior. (AI is defined as a branch of computer science that studies how to endow computers with capabilities of human intelligence.) In essence, AI tries to mimic human intelligent behavior. AI techniques represent convenient tools that can reasonably describe behavior and decision-making of various decision-makers in production, transportation, and warehousing. Distributed AI and multi-agent systems are especially convenient tools for the analysis of various logistic phenomena.

During the last decade, significant progress has been made in merging various OR and AI techniques.

3.2   Operations Research: Basic Concepts

The basic OR concepts can be better described with the help of an example. Let us consider the problem of milk distribution in one city. Different participants in milk distribution are facing various decision problems. We assume that the distributor has a fleet composed of a few vehicles. These vehicles should deliver milk and dairy products to 50 different stores. The whole distribution process could be organized in many different ways. There are number of feasible vehicle routes. The dispatcher in charge of distribution will always try to discover vehicle routes that facilitate lowest transportation costs.

Store managers are constantly facing the problem of calculating the proper quantity of milk and dairy products that should be ordered from the distributor. Unsold milk and other products significantly increase the costs. On the other hand, potential revenue could be lost in a case of shortage of products.

Both decision problems (faced by distributor dispatcher and store managers) are characterized by limited resources (the number of vehicles that can participate in the milk distribution, the amount of money that could be invested in milk products), and by the necessity to discover optimum course of action (the best set of vehicle routes, the optimal quantities of milk and dairy products to be ordered).

Operations Research could be defined as a set of scientific techniques searching for the best source of action under limited resources. The beginning of OR is related to the British Air Ministry activities in 1936, and the name Operations Research (Operational Research) has its roots in research of military operations. The real OR boom started after World War II when OR courses were established at many American Universities, together with extensive use of OR methods in industry and public sector. The development of modern computers further contributed to the success of OR techniques.

Formulation of the problem (in words) represents the first step in the usual problem solving scheme. In the next step, verbal description of the problem should be replaced by corresponding mathematical formulation. Mathematical formulation describes the problem mathematically. Variables, objective function, and constraints are the main components of the mathematical model. To build a mathematical model, analysts try to establish various logical and mathematical relationships between specific variables. The analysts define the objective function, as well as the set of constraints that must be satisfied. Depending on the problem context, the constraints could be by their nature physical, institutional, or financial resources. The generated feasible solutions are evaluated by corresponding objective function values. The set of feasible solutions is composed of all problem solutions that satisfy a given set of constraints. It is very difficult (and in majority of cases impossible) to produce mathematical model that will capture all different aspects of the problem considered. Consequently, mathematical models represent simplified description of the real problem. Practically, all mathematical models represent the compromise between the wish to accurately describe the real-life problem and the capability to solve the mathematical model.

3.2.1 Problem Solving Steps

Many real-life logistic and transportation problems can be relatively easily formulated in words (Fig. 3.1). After such formulation of the problem, in the next step, engineers usually translate problem's verbal description into a mathematical description.

Main components of the mathematical description of the problem are variables, constraints, and the objective. Variables are sometimes called *unknowns*. While some of the variables are under the control of the analyst, some are not. Constraints could be physical resources, caused by some engineering rules, laws, guidelines, or due to various financial reasons. One cannot accept more than 100 passengers for the planned flight, if the capacity of the aircraft equals 100 seats. This is a typical example of physical constraint. Financial constraints are usually related to various investment decisions. For example, one cannot invest more than $10,000,000 in road improvement if the available budget equals $10,000,000. Solutions could be feasible or infeasible. Solutions are feasible when they satisfy all the defined constraints. An objective represents the end result that the decision-maker wants to accomplish by selecting a specific program or action. Revenue maximization, cost minimization, or profit maximization are typical objectives of profit-oriented organizations. Providing the highest level of service to the customers represents the usual objective of a nonprofit organization.

Mathematical description of a real-world problem is called a mathematical model of the real-world problem. An algorithm represents some quantitative method used by an analyst to solve the defined

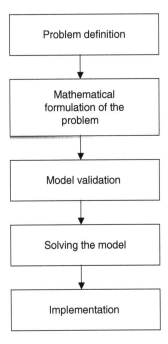

FIGURE 3.1 Problem solving steps.

mathematical model. Algorithms are composed of a set of instructions, which are usually followed in a defined step-by-step procedure. An algorithm produces a feasible solution to a defined model with the goal to find an optimal solution. Optimal solution to the defined problem is the best possible solution among all feasible solutions. Depending on a defined objective function, optimal solution corresponds to maximum revenue, minimum cost, maximum profit, and so on.

3.3 Mathematical Programming

In the past three decades, linear, nonlinear, dynamic, integer, and multiobjective programming have been successfully used to solve various engineering, management, and control problems. Mathematical programming techniques have been used to address problems dealing with the most efficient allocation of limited resources (supplies, capital, labor, etc.) to meet the defined objectives. Typical problems include market share maximization, production scheduling, personnel scheduling and rostering, vehicle routing and scheduling, locating facilities in a network, planning fleet development, etc. Their solutions can be found using one of the mathematical programming methods.

3.3.1 Linear Programming

Let us consider a rent-a-car company operations. The total number of vehicles that the company owns equals 100. The potential clients are offered 2 tariff classes at $150 per week and $100 per week. The potential client pays $100 per week if he or she makes the reservation at least 3 days in advance. We assume that we are able to predict exactly the total number of requests in both client-tariff classes. We expect 70 client requests in the first class and 80 client requests in the second class during the considered time period. We decide to keep at least 10 vehicles for the clients paying higher tariffs. We have to determine the total numbers of vehicles rented in different client tariff classes to reach the maximum company revenue.

Solution:

As we wish to determine the total numbers of vehicles rented in different client tariff classes, the variables of the model can be defined as:

x_1—the total number of vehicles planned to be rented in the first client-tariff class
x_2—the total number of vehicles planned to be rented in the second client-tariff class

Because each vehicle from the first class rents for $150, the total revenue from renting x_1 vehicles is $150x_1$. In the same way, the total company revenue from renting the x_2 vehicles equals $100x_2$. The total company revenue equals the sum of the two revenues, $150x_1 + 100x_2$.

From the problem formulation we conclude that there are specific restrictions on vehicle renting and demand. The vehicle renting restrictions may be expressed verbally in the following way:

- Total number of vehicles rented in both classes together must be less than or equal to the total number of vehicles.
- Total number of vehicles rented in any class must be less than or equal to the total number of client requests.
- Total number of vehicles rented in the first class must be at least 10.
- Total number of vehicles rented in the second class cannot be less than zero (non-negativity restriction).

The following is the mathematical model for rent-a-car revenue management problem:
Maximize

$$F(X) = 150x_1 + 100x_2$$

subject to:

$$x_1 + x_2 \leq 100$$
$$x_1 \leq 70$$
$$x_2 \leq 80$$
$$x_1 \geq 10$$
$$x_2 \geq 0$$

In our problem, we allow variables to take the fractional values (we can always round the fractional value to the closest feasible integer value). In other words, all our variables are continuous variables. We also have only *one* objective function. We try to maximize the total company's revenue. Our objective function and all our constraints are linear, meaning that any term is either a constant or a constant multiplied by a variable. Any mathematical model that has one objective function, all continuous variables, linear objective function and all linear constraints is called a linear program (LP). It has been seen through many years that many real-life problems can be formulated as linear programs. Linear programs are usually solved using widely spread Simplex algorithm (there is also an alternative algorithm called Interior Point Method).

As we have only two variables, we can also solve our problem graphically. Graphical method is impractical for mathematical models with more than two variables. To solve the earlier-stated problem graphically, we plot the feasible solutions (solution space) that satisfy all constraints simultaneously. Figure 3.2 shows our solution space.

All feasible values of the variables are located in the first quadrant. This is caused by the following constraints: $x_1 \geq 10$, and $x_2 \geq 0$. The straight-line equations $x_1 = 10, x_1 = 70, x_2 = 80, x_2 = 0$, and $x_1 + x_2 = 100$ are obtained by substituting "≤" by "=" for each constraint. Then, each straight-line is plotted. The region in which each constraint is satisfied when the inequality is put in power is indicated by the direction of the arrow on the corresponding straight line. The resulting solution space of the rent-a-car problem is shown in the Figure 3.3. Feasible points for the problem considered are all points within the boundary or on the boundary of the solution space. The optimal solution is discovered by studying the direction in which the objective function $F = 150 x_1 + 100 x_2$ rises. The optimal solution is shown in the Figure 3.3.

The parallel lines in the Figure 3.3 represent the objective function $F = 150 x_1 + 100 x_2$. They are plotted by arbitrarily assigning increasing values to F. In this way, it is possible to make conclusions about the slope and the direction in which the total company revenue increases.

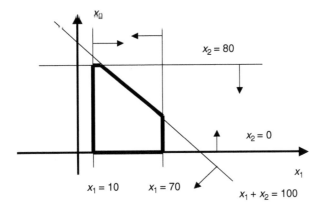

FIGURE 3.2   Solution space of the rent-a-car revenue management problem.

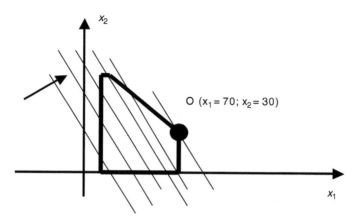

FIGURE 3.3 The optimal solution of the rent-a-car problem.

To discover the optimal solution, we move the revenue line in the direction indicated in Figure 3.3 to the point "O" where any further increase in company revenue would create an infeasible solution. The optimal solution happens at the intersection of the following lines:

$$x_1 + x_2 = 100$$
$$x_1 = 70$$

After solving the system of equations we get:

$$x_1 = 70$$
$$x_2 = 30$$

The corresponding rent-a-car company revenue equals:

$$F = 150 \, x_1 + 100 \, x_2 = 150(70) + 100(30) = 13,500$$

The problem considered is a typical resource allocation problem. Linear Programming helps us to discover the best allocation of limited resources. The following is a Linear Programming Model:

Maximize

$$F(X) = c_1 x_1 + c_2 x_2 + c_3 x_3 + \cdots + c_n x_n$$

subject to:

$$a_{11}x_1 + a_{12}x_2 + a_{13}x_3 + \cdots + a_{1n}x_n \leq b_1$$
$$a_{21}x_1 + a_{22}x_2 + a_{23}x_3 + \cdots + a_{2n}x_n \leq b_2$$

$$\tag{3.1}$$

$$a_{m1}x_1 + a_{m2}x_2 + a_{m3}x_3 + \cdots + a_{mn}x_n \leq b_m$$
$$x_1, x_2, \ldots, x_n \geq 0$$

The variables $x_1, x_2, \ldots, x_n$ describe level of various economic activities (number of cars rented to the first class of clients, number of items to be kept in the stock, number of trips per day on specific route, number of vehicles assigned to a particular route, etc.).

3.3.2   Integer Programming

Analysts frequently realize that some or all of the variables in the formulated linear program must be integers. This means that some variables or all take exclusively integer values. To make the formulated problem easier, analysts often allow these variables to take fractional values. For example, analysts know that the number of first class clients must be in the range between 30 and 40. Linear program could produce the "optimal solution" that tells us that the number of first class clients equals 37.8. In this case, we can neglect the fractional part, and we can decide to protect 37 (or 38) cars for the first class clients. In this way, we are making small numerical error, but we are capable to easily solve the problem.

In some other situations, it is not possible for analysts to behave in this way. Imagine that we have to decide about a new warehouse layout. You must choose one out of numerous generated alternatives. This is kind of "yes/no" ("1/0") decision: "Yes" if the alternative is chosen, "No," otherwise. In other words, we can introduce binary variables into the analysis. The variable has value 1 if the i-th alternative is chosen and value 0 otherwise. The value 0.7 of the variable means nothing to us. We are not able to decide about the best warehouse layout if the variables take fractional values. When we solve problems similar to the warehouse layout problem we work exclusively with integer variables. These kinds of problems are known as integer programs, and corresponding area is known as Integer Programming. Integer programs usually describe the problems in which one, or more, alternatives must be selected from a finite set of generated alternatives. Problems of determining the best schedule of activities, finding the optimal set of vehicle routes, or discovering the shortest path in a transportation network are typical problems that are formulated as integer programs. There are also problems in which some variables can take only integer values, while some other variables can take fractional values. These problems are known as mixed-integer programs. It is much harder to solve Integer Programming problems than Linear Programming problems.

The following is the Integer Programming Model formulation:

Maximize

$$F(X) = \sum_{j=1}^{n} c_j x_j$$

subject to:

$$\sum_{j=1}^{n} a_{ij} x_j \leq b_i \qquad \text{for } i = 1, 2, ..., m \tag{3.2}$$

$$0 \leq x_j \leq u_j \qquad \text{integer for } j = 1, 2, ..., n$$

There are numerous software systems that solve linear, integer, and mixed-integer linear programs (CPLEX, Excel and Quattro Pro Solvers, FortMP, LAMPS, LINDO, LINGO, MILP88, MINTO, MIPIII, MPSIII, OML, OSL).

A combinatorial explosion of possible solutions characterizes many of the Integer Programming problems. In cases when the number of integer variables in a considered problem is very large, finding optimal solution becomes very difficult, if not impossible. In such cases, various heuristic algorithms are used to discover "good" solutions. These algorithms do not guarantee the optimal solution discovery.

3.4   Heuristic Algorithms

Many logistic problems are combinatorial by nature. Combinatorial optimization problems could be solved by exact or by heuristic algorithms.

The exact algorithms always find the optimal solution(s). The wide usage of the exact algorithms is limited by the computer time needed to discover the optimal solution(s). In some cases, this computer time is enormously large.

The word "heuristic" has its roots in Greek word "ευρισκω" that means "to discover," or "to find." Heuristic algorithm could be described as a combination of science, invention, and problem solving skills. In essence, a heuristic algorithm represents procedure invented and used by the analyst(s) in order to "travel" (search) through the space of feasible solutions. Good heuristics algorithm should generate quality solutions in an acceptable computer time. Complex logistic problems of big dimensions are usually solved with the help of various heuristic algorithms. Good heuristic algorithms are capable of discovering optimal solutions for some problem instances, but heuristic algorithms do not guarantee optimal solution discovery.

There are few reasons why heuristic algorithms are widely used. Heuristic algorithms are used to solve the problems in situations in which exact algorithm would require solution time that increases exponentially with a size of a problem. For example, in case of a problem that is characterized by 3000 binary variables (that can take values 0 or 1), the number of potential solutions is equal to 2^{3000}.

In some cases, the costs of using the exact algorithm are much higher than the potential benefits of discovering the optimal solution. Consequently, in such situations analysts usually use various heuristic algorithms.

It could frequently happen that the problem considered is not well "structured." This means that all relevant information is not known by the analyst, and that the objective function(s) and constraints are not precisely defined. An attempt to find the "optimal" solution for the ill-defined problem could generate the "optimal" solution that is in reality poor solution to the real problem.

The decision-makers are frequently interested in discovering "satisfying" solution of real-life problems. Obtaining adequate information about considered alternatives is usually very costly. At the same time, the consequences of many possible decisions are not known precisely causing decision-makers to come across with a course of action that is acceptable, sufficient, and logical. In other words, "satisfying" solution represents the solution that is satisfactory to the decision-makers. Satisfactory solution(s) could be generated by various heuristic algorithms, after limited search of the solution space.

Great number of real-life logistic problems could be solved only by heuristic algorithms. Large number of heuristic algorithms are based on relatively simple ideas, and many of them have been developed without previous mathematical formulation of the problem.

3.4.1 "Classical" Heuristic Algorithms

The greedy and interchange heuristics are the widely used heuristic algorithms. Let us clarify the basic principles of these algorithms by analyzing the traveling salesman problem (TSP). The TSP is one of the most well known problems in OR and computer science. This problem can be defined as follows: Find the shortest itinerary which starts in a specific node, goes through all other nodes exactly once, and finishes in the starting node. In different traffic, transportation, and logistic problems, the traveling salesman can represent airplanes, boats, trucks, buses, crews, etc. Vehicles visiting nodes can deliver or pick up goods, or simultaneously perform pick up and delivery.

A typical solution process of the TSP is stepwise as in the following: (a) First an initial tour is constructed; (b) Any remaining unvisited nodes are inserted; (c) The created tour is improved. There are many developed algorithms for each step.

Before discussing various heuristic algorithms, let us define the "scenario" of the TSP. A traveling salesman starting and finishing its tour at one fixed point must visit $(n - 1)$ points. The transportation network connecting these n points is completely connected. This means that it is possible to reach any node from any other node, directly, without going through the other nodes (an air transportation network is a typical example of this type of network). The shortest distance between any two nodes equals the length of the branches between these nodes.

FIGURE 3.4 "Triangular inequality."

From this, it is certain that the following inequality is satisfied:

$$d(a,b) < d(a,c) + d(c,b) \tag{3.3}$$

for any three nodes *a*, *b*, and *c*.

We also assume that the matrix of shortest distances between the nodes is symmetrical. The nodes *a*, *b*, and *c* are shown in Figure 3.4.

3.4.2 Heuristic Algorithm Based on Random Choice

The TSP could be easily solved by the following simple heuristic algorithm:

Step 1:   Arbitrarily choose starting node.
Step 2:   Randomly choose the next node to be included in the traveling salesman tour.
Step 3:   Repeat Step 2 until all nodes are chosen. Connect the first and the last node of the tour.

This algorithm is based on the idea of random choice. The next node to be included in the partial traveling salesman tour is chosen at random. In other words, the sequence of nodes to be visited is generated at random. It is intuitively clear that one cannot expect that this algorithm would give very good results, as it does not use any relevant information when choosing the next node that is to be included in the tour. On the other hand generating sequences of nodes at random can be repeated two, three, …, or ten thousand times. The repetition of generating various solutions represents the main power of this kind of an algorithm. Obviously, the decision-maker can choose the best solution among all solutions generated at random. The greater the number of solutions generated, the higher the probability that one can discover a "good" solution.

3.4.3 "Greedy" Heuristic Algorithms

"Greedy" heuristic algorithms build the solution of the studied problem in a step-by-step procedure. In every step of the procedure the value is assigned to one of the variables in order to maximally improve the objective function value. In every step, the greedy algorithm is looking for the best current solution with no look upon future cost or consequences. Greedy algorithms use local information available in every step. The fundamental concept of greedy algorithms is similar to the "Hill-climbing" technique. In case of "Hill-climbing" technique the current solution is continuously replaced by the new solution until it is not possible to produce further improvements in the objective function value. "Greedy" algorithms and the "Hill-climbing" technique are similar to the hiker who is trying to come to the mountaintop by never going downwards (Fig. 3.5).

As it can be seen from Figure 3.5, hiker's wish to never move down while climbing, can trap him or her at some of the local peaks (local maximums), and prevent him or her from reaching the mountaintop (global maximum). "Greedy" algorithms and the "Hill-climbing" technique consider only local improvements.

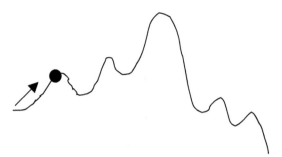

FIGURE 3.5   Hiker who is trying to come to the mountaintop by going up exclusively.

The Nearest Neighbor (NN) heuristic algorithm is a typical representative of "Greedy" algorithms. This algorithm, which is used to generate the traveling salesman tour, is composed of the following algorithmic steps:

Step 1:   Arbitrarily (or randomly) choose a starting node in the traveling salesman tour.
Step 2:   Find the nearest neighbor of the last node that was included in the tour. Include this nearest neighbor in the tour.
Step 3:   Repeat Step 2 until all nodes are not included in the traveling salesman tour. Connect the first and the last node of the tour.

The NN algorithm finds better solutions than the algorithm based on random choice, as it uses the information related to the distances between nodes.

Let us find the traveling salesman tour starting and finishing in node 1, using NN heuristic algorithm (Fig. 3.6). The distances between all pairs of nodes are given in the Table 3.1.

The route must start in node 1. The node 2 is the NN of node 1. We include this NN in the tour. The current tour reads: (1, 2). Node 3 is the NN of node 2. We include this NN in the tour. The updated tour reads: (1, 2, 3). Continuing in this way, we obtain the final tour that reads: (1, 2, 3, 4, 5, 6, 7, 1). The final tour is shown in Figure 3.7.

Both algorithms shown ("random choice" and "greedy") repeat the specific procedure a certain number of times unless a solution has been generated. Many of the heuristic algorithms are based on a specific procedure that is repeated until solution is generated.

When applying "greedy" approach, the analyst is forced, after a certain number of steps, to start to connect the nodes (in case of TSP) quite away from each other. Connecting the nodes distant from each other is forced by previous connections that significantly decrease the number of possible connections left.

3.4.4   Exchange Heuristic Algorithms

Exchange heuristic algorithms are based on the idea of interchange and they are widely used. The idea of interchange is the idea to start with the existing solution and check if this solution could be improved.

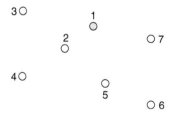

FIGURE 3.6   Network in which a traveling salesman tour should be created using NN heuristic algorithm.

TABLE 3.1 The Distances between All Pairs of Nodes

	1	2	3	4	5	6	7
1	0	75	135	165	135	180	90
2	75	0	90	105	135	210	150
3	135	90	0	150	210	300	210
4	165	105	150	0	135	210	210
5	135	135	210	135	0	90	105
6	180	210	300	210	90	0	120
7	90	150	210	210	105	120	0

Exchange heuristic algorithm first creates or selects an initial feasible solution in some arbitrary way (randomly or using any other heuristic algorithm), and then tries to improve the current solution by specific exchanges within the solution.

The good illustration of this concept is two-optimal tour (2-OPT) heuristic algorithms for the TSP [3-OPT and *k*-optimal tour (*k*-OPT) algorithms are based on the same idea]. Within the first step of the 2-OPT algorithm, an initial tour is created in some arbitrary way (randomly or using any other heuristic algorithm). The two links are then broken (Fig. 3.8). The paths that are left are joined so as to form a new tour. The length of the new tour is compared with the length of the old tour. If the new tour length is less than the old tour length, the new tour is retained. In a systematic way, two links are broken at a time, paths are joined, and comparison is made. Eventually, a tour is found whose total length cannot be decreased by the interchange of any two links. Such a tour is known as two-optimal tour (2-OPT).

After breaking links (*a*, *j*) and (*d*, *e*), the node *a* has to be connected with node *e*. The node *d* should be connected with node *j*. The connection between node *a* and node *d*, as well as the connection between node *j* and node *e* would prevent creating the traveling salesman tour. In case of 3-OPT algorithm in a systematic way three links are broken, new tour is created, tour lengths are compared, and so on.

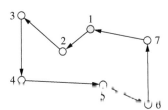

FIGURE 3.7 Traveling salesman tour obtained by the *NN* heuristic algorithm.

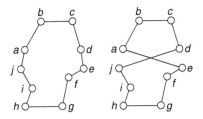

FIGURE 3.8 Interchange of two links during 2-OPT algorithm.

2-OPT algorithm is composed of the following algorithmic steps:

Step 1: Create an initial traveling salesman tour.
Step 2: The initial tour is the following tour: $(a_1, a_2, ..., a_n, a_1)$. The total length of this tour is equal to D. Set $i = 1$.
Step 3: $j = i + 2$.
Step 4: Break the links (a_i, a_{i+1}) and (a_j, a_{j+1}) and create the new traveling salesman tour. This tour is the following tour: $(a_1, a_2, ..., a_i, a_j, ..., a_{i+1}, a_{j+1}, a_{j+2}, ..., a_1)$. If the length of the new tour is less than D, than keep this tour and return to Step 2. Otherwise go to Step 5.
Step 5: Set $j = j + 1$. If $j \leq n$ go to Step 4. In the opposite case, increase i by 1 $(i = i + 1)$. If $i \leq n - 2$ go to Step 3. Otherwise, finish with the algorithm.

By using the 2-OPT algorithm, we will try to create the traveling salesman tour for the network shown in Figure 3.6. The distances between nodes are given in Table 3.1. The traveling salesman should start his trip from node 1. The initial tour shown in Figure 3.7 is generated by the NN algorithm. It was not possible to decrease the total length of the initial tour by interchanging of any two links (Table 3.2). Our initial tour is 2-OPT.

The k-opt algorithm for the TSP assumes breaking k links in a systematic way, joining the paths, and performing the comparison. Eventually a tour is found whose total length cannot be decreased by the interchange of any k links. Such a tour is known as k-OPT.

3.4.5 Decomposition Based Heuristic Algorithms

In some cases it is desirable to decompose the problem considered into smaller problems (subproblems). In the following step every subproblem is solved separately. Final solution of the original problem is then obtained by "assembling" the subproblem solutions. We illustrate this solution approach in case of the standard vehicle routing problem (VRP).

There are n nodes to be served by homogeneous fleet (every vehicle has identical capacity equal to V). Let us denote by v_i $(i = 1, 2, ..., n)$ demand at node i. We also denote by D vehicle depot (all vehicles start their trip from D, serve certain number of nodes and finish route in node D).

Vehicle capacity V is greater than or equal to demand at any node. In other words, every node could be served by one vehicle, that is, vehicle routes are composed of one or more nodes.

TABLE 3.2 Steps in the 2-OPT Algorithm

Broken Links	New Traveling Salesman Tour	Tour Length
(1, 2), (3, 4)	(1, 3, 2, 4, 5, 6, 7, 1)	765
(1, 2), (4, 5)	(1, 4, 3, 2, 5, 6, 7, 1)	840
(1, 2), (5, 6)	(1, 5, 3, 4, 2, 6, 7, 1)	1020
(1, 2), (6, 7)	(1, 6, 3, 4, 5, 2, 7, 1)	1140
(1, 2), (7, 1)	(1, 7, 3, 4, 5, 6, 2, 1)	960
(2, 3), (4, 5)	(1, 2, 4, 3, 5, 6, 7, 1)	840
(2, 3), (5, 6)	(1, 2, 5, 4, 3, 6, 7, 1)	1005
(2, 3), (6, 7)	(1, 2, 6, 4, 5, 3, 7, 1)	1140
(2, 3), (7, 1)	(1, 2, 7, 4, 5, 6, 3, 1)	1095
(3, 4), (5, 6)	(1, 2, 3, 5, 4, 6, 7, 1)	930
(3, 4), (6, 7)	(1, 2, 3, 6, 5, 4, 7, 1)	990
(3, 4), (7, 1)	(1, 2, 3, 7, 5, 6, 4, 1)	945
(4, 5), (6, 7)	(1, 2, 3, 4, 6, 5, 7, 1)	810
(4, 5), (7, 1)	(1, 2, 3, 4, 7, 6, 5, 1)	870
(5, 6), (7, 1)	(1, 2, 3, 4, 5, 7, 6, 1)	855

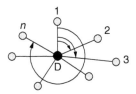

FIGURE 3.9 Sweep algorithm.

Problem to be solved could be described in the following way: Create set of vehicle routes in such a way as to minimize the total distance traveled by all vehicles.

Real-life VRP could be very complex. One or more of the following characteristics could appear when solving some of the real-life VRP: (a) Some nodes must be served within prescribed time intervals (time windows); (b) Service is performed by heterogeneous fleet of vehicles (vehicles have different capacities); (c) Demand at nodes is not known in advance; (d) There are few depots in the network.

The Sweep algorithm is one of the classical heuristic algorithms for the VRP. This algorithm is applied to polar coordinates, and the depot is considered to be the origin of the coordinate system. Then the depot is joined with an arbitrarily chosen point that is called the *seed point*. All other points are joined to the depot and then aligned by increasing angles that are formed by the segments that connect the points to the depot and the segment that connects the depot to the seed point. The route starts with the seed point, and then the points aligned by increasing angles are included, respecting given constraints. When a point cannot be included in the route as this would violate a certain constraint, this point becomes the seed point of a new route, and so on. The process is completed when all points are included in the routes (Fig. 3.9).

In case when a large number of nodes need to be served, the Sweep algorithm should be used within the "clustering-routing" approach. In this case, considering clockwise direction, the ratio of cumulative demand and vehicle capacity should be checked (including all other constraints). The node that cannot be included because of the violation of vehicle capacity or other constraints becomes the first node in another cluster. In this way, the whole region is divided into clusters (zones). In the following step, VRP is solved within each cluster separately. Clustering is completed when all nodes are assigned to clusters (Fig. 3.10). It is certain that one vehicle can serve all nodes within one cluster. In this way, the VRP is transformed into few TSP.

The final solution depends on a choice of the seed point. By changing locations of the seed point it is possible to generate various sets of vehicle routes. For the final solution the set of routes with minimal total length should be chosen.

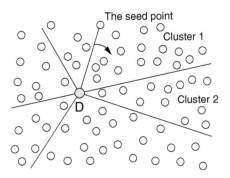

FIGURE 3.10 Clustering by Sweep algorithm.

3.5   Algorithms' Complexity

Various heuristic algorithms could be used to solve a specific problem. Decision-makers prefer to use algorithms that have relatively short CPU time (execution time) and provide reasonably good solutions. One might ask, which one of the developed algorithms is better for solving the TSP? The execution time highly depends on the CPU time, programming language, speed of a computer, etc. To objectively compare various algorithms, a measurement of algorithms' complexity has been proposed that is independent of all computer types and programming languages. The "goodness" of the algorithm is highly influenced by the algorithm's complexity. The complexity of the algorithm is usually measured through the total number of elementary operations (additions, subtractions, comparisons, etc.) that the algorithm requires to solve the problem under the worst case conditions.

Let us assume that we have to solve the TSP. We denote by n the total number of nodes. We also denote by E the total number of elementary operations. Let us assume that E equals:

$$E = 4n^4 + 5n^3 + 2n + 7 \tag{3.4}$$

As n increases, the E value is largely determined by the term n^4. We can describe this fact by using the "O-notation." The "O-notation" is used to describe the algorithms' complexity. In the considered example, we write that the algorithm's complexity is $O(n^4)$, or that solution time is of the order $O(n^4)$. The "O-notation" neglects smaller terms, as well as proportional factors. It could happen that for small input sizes an inefficient algorithm may be faster than an efficient algorithm. Practically, the comparison of the algorithms based on "O-notation" is practical only for large input sizes. For example, the algorithm whose complexity is $O(n^2)$ is better than the algorithm whose complexity is $O(n^3)$.

Many real-life problems can be solved by the algorithms whose solution time grows as a polynomial function of the problem size. We call such algorithms polynomial algorithms. The problems that can be solved by polynomial algorithms are considered as *easy problems*. Large instances of easy problems can be solved in "reasonable" computer times using an adequate algorithm and a "fast" computer.

All optimization problems can be classified into two sets. By P we denote the set of problems that can be solved by polynomial algorithms. All other problems, whose solution is difficult or impossible, belong to the set that is called *NP-Complete*. No polynomial time algorithms have been created for the problems that belong to the set *NP-Complete*.

Polynomial algorithms are "good" algorithms [e.g., the algorithms whose complexity is $O(n^2)$, $O(n^5)$, or $O(n^6)$]. The algorithm whose complexity is $O(n \log n)$ also belongs to the class of polynomial algorithms, as ($n \log n$) is bounded by (n^2). Developing appropriate polynomial algorithm could be, in some cases difficult, time consuming, or costly.

Non-polynomial algorithms [e.g., the algorithms whose complexity is $O(3^n)$ or $O(n!)$] are not "good" algorithms. When the algorithms' complexity is, for example, $O(3^n)$, we see, that the function in the parentheses is *exponential* in n. One might ask, "Could a faster computer help us to successfully solve "difficult" problems?" The development of faster computers in the future will enable us to solve larger sizes of these problems; however, there is no indication that we will be able to find optimal solutions in these cases. Every specific problem should be carefully studied. In some cases, it is not an easy task to recognize an "easy" problem and to make the decision regarding the solution approach (optimization vs. heuristic). All heuristic algorithms are evaluated according to the quality of the solutions generated, as well as computer time needed to reach the solution. In other words, good heuristics algorithm should generate quality solutions in an acceptable computer time. Simplicity and easiness to implement these algorithms are the additional criteria that should be taken into account when evaluating a specific heuristic algorithm.

Heuristic algorithms do not guarantee the optimal solution discovery. The closer the solution produced is to the optimal solution, the better the algorithm. It is an usual practice to perform "Worst Case Analysis," as well as "Average Case Analysis" for every considered heuristic algorithm. Worst Case

Analysis assumes generating special numerical examples (that appear rarely in real life) that can show the worst results generated by the proposed heuristic algorithm. For example, we can conclude that the worst solution generated by the proposed heuristic algorithm is 5% far from the optimal solution. Within the Average Case Analysis, a great number of typical examples are usually generated and analyzed. By performing statistical analysis related to the solutions generated, the conclusions are derived about the quality of the solutions generated in the "average case." The more real-life examples are tested, the easier it is to evaluate specific heuristic algorithm.

3.6 Randomized Optimization Techniques

Many heuristic techniques that have been developed are capable of solving only a specific problem, whereas metaheuristics can be defined as general combinatorial optimization techniques. These techniques are designed to solve many different combinatorial optimization problems. The developed metaheuristics are based on local search techniques, or on population search techniques. Local search-based metaheuristics (Simulated Annealing, Tabu Search, etc.) are characterized by an investigation of the solution space in the neighborhood of the current solution. Each step in these metaheuristics represents a move from the current solution to another potentially good solution in the current solution's neighborhood. In case of a population search, as opposed to traditional search techniques, the search is run in parallel from a population of solutions. These solutions are combined and the new generation of solutions is generated. Each new generation of solutions is expected to be "better" than the previous one.

3.6.1 Simulated Annealing Technique

The simulated annealing technique is one of the methods frequently used in solving complex combinatorial problems. This method is based on the analogy with certain problems in the field of statistical mechanics. The term, simulated annealing, comes from the analogy with physical processes. The process of annealing consists in decreasing the temperature of a material, which in the beginning of the process is in the molten state, until the lowest state of energy is attained. At certain points during the process the so-called thermal equilibrium is reached. In case of physical systems we seek to establish the order of particles that has the lowest state of energy. This process requires that the temperatures at which the material remains for a while are previously specified.

The basic idea of simulated annealing consists in performing small perturbations (small alterations in the positions of particles) in a random fashion and computing the energy changes between the new and the old configurations of particles, ΔE. In case when $\Delta E < 0$, it can be concluded that the new configuration of particles has lower energy. The new configuration then becomes a new initial configuration for performing small perturbations. The case when $\Delta E > 0$ it means that the new configuration has higher energy. However, in this case the new configuration should not be automatically excluded from the possibility of becoming a new initial configuration. In physical systems, "jumps" from lower to higher energy levels are possible. The system has higher probability to "jump" to a higher energy state when the temperature is higher. As the temperature decreases, the probability that such a "jump" will occur diminishes. Probability P that at temperature T the energy will increase by ΔE equals:

$$P = e^{-\frac{\Delta E}{T}} \tag{3.5}$$

The decision whether a new configuration of particles for which $\Delta E > 0$ should be accepted as a new initial configuration is made upon the generation of a random number r from the interval $[0, 1]$. Generated random number is uniformly distributed. If $r < P$, the new configuration is accepted as a new initial configuration. In the opposite case, the generated configuration of particles is excluded from consideration.

In this manner, a successful simulation of attaining thermal equilibrium at a particular temperature is accomplished. Thermal equilibrium is considered to be attained when, after a number of random perturbations, a significant decrease in energy is not possible. Once thermal equilibrium has been attained, the temperature is decreased, and the described process is repeated at a new temperature.

The described procedure can also be used in solving combinatorial optimization problems. A particular configuration of particles can be interpreted as one feasible solution. Likewise, the energy of a physical system can be interpreted as the objective function value, while temperature assumes the role of a control parameter. The following is a pseudo-code for simulated annealing algorithm:

Select an initial state $i \in S$;
Select an initial temperature $T > 0$;
Set temperature change counter $t := 0$;
Repeat
    Set repetition counter $n := 0$;
    Repeat
        Generate state j, a neighbor of i;
        Calculate $\Delta E := f(j) - f(i)$
        if $\Delta E < 0$ then $i := j$
        else if random $(0, 1) < \exp(-\Delta E/T)$ then $i := j$;
        Inc(n);
    Until $n = N(t)$;
    Inc(t);
    $T := T(t)$;
Until stopping criterion true.
where:
S—finite solution set,
i—previous solution,
j—next solution,
$f(x)$—criteria value for solution x, and
$N(t)$—number of perturbations at the same temperature.

It has been a usual practice that during the execution of the simulated annealing algorithm, the best solution obtained thus far is always remembered. The simulated annealing algorithm differs from general local search techniques as it allows the acceptance of improving as well as nonimproving moves. The benefit of accepting nonimproving moves is that the search does not prematurely converge to a local optimum and it can explore different regions of the feasible space.

3.6.2 Genetic Algorithms

Genetic algorithms represent search techniques based on the mechanics of nature selection used in solving complex combinatorial optimization problems. These algorithms were developed by analogy with Darwin's theory of evolution and the basic principle of the "survival of the fittest." In case of genetic algorithms, as opposed to traditional search techniques, the search is run in parallel from a population of solutions. In the first step, various solutions to the considered maximization (or minimization) problem are generated. In the following step, the evaluation of these solutions, that is, the estimation of the objective (cost) function is made. Some of the "good" solutions yielding a better "fitness" (objective function value) are further considered. The remaining solutions are eliminated from consideration. The chosen solutions undergo the phases of *reproduction*, *crossover*, and *mutation*. After that, a new generation of solutions is produced to be followed by a new one, and so on. Each new generation is expected to be "better" than the previous one. The production of new generations is stopped when a prespecified stopping condition is satisfied. The final solution of the considered problem is the best solution generated

TABLE 3.3   Encoded Values of Variable x

String	Value of Variable x	String	Value of Variable x
0000	$0 = 0*2^3 + 0*2^2 + 0*2^1 + 0*2^0$	1000	$8 = 1*2^3 + 0*2^2 + 0*2^1 + 0*2^0$
0001	$1 = 0*2^3 + 0*2^2 + 0*2^1 + 1*2^0$	1001	$9 = 1*2^3 + 0*2^2 + 0*2^1 + 1*2^0$
0010	$2 = 0*2^3 + 0*2^2 + 1*2^1 + 0*2^0$	1010	$10 = 1*2^3 + 0*2^2 + 1*2^1 + 0*2^0$
0011	$3 = 0*2^3 + 0*2^2 + 1*2^1 + 1*2^0$	1011	$11 = 1*2^3 + 0*2^2 + 1*2^1 + 1*2^0$
0100	$4 = 0*2^3 + 1*2^2 + 0*2^1 + 0*2^0$	1100	$12 = 1*2^3 + 1*2^2 + 0*2^1 + 0*2^0$
0101	$5 = 0*2^3 + 1*2^2 + 0*2^1 + 1*2^0$	1101	$13 = 1*2^3 + 1*2^2 + 0*2^1 + 1*2^0$
0110	$6 = 0*2^3 + 1*2^2 + 1*2^1 + 0*2^0$	1110	$14 = 1*2^3 + 1*2^2 + 1*2^1 + 0*2^0$
0111	$7 = 0*2^3 + 1*2^2 + 1*2^1 + 1*2^0$	1111	$15 = 1*2^3 + 1*2^2 + 1*2^1 + 1*2^0$

during the search. In case of genetic algorithms an encoded parameter set is used. Most frequently, binary coding is used. The set of decision variables for a given problem is encoded into a bit string (chromosome, individual).

Let us explain the concept of encoding in case of finding the maximum value of function $f(x) = x^3$ in the domain interval of x ranging from 0 to 15. By means of binary coding, the observed values of variable x can be presented in strings of the length 4 (as $2^4 = 16$). Table 3.3 shows 16 strings with corresponding decoded values.

We assume that in the first step the following four strings were randomly generated: 0011, 0110, 1010, and 1100. These four strings form the initial population P(0). In order to make an estimation of the generated strings, it is necessary to decode them. After decoding, we actually obtain the following four values of variable x: 3, 6, 10, and 12. The corresponding values of function $f(x) = x^3$ are equal to $f(3) = 27$, $f(6) = 216$, $f(10) = 1000$ and $f(12) = 1728$. As can be seen, string 1100 has the best fitness value.

Genetic algorithms is a procedure where the strings with better fitness values are more likely to be selected for mating. Let us denote by f_i the value of the objective function (fitness) of string i. The probability p_i for string i to be selected for mating is equal to the ratio of f_i to the sum of all strings' objective function values in the population:

$$p_i = \frac{f_i}{\sum_j f_j} \tag{3.6}$$

This type of reproduction, that is, selection for mating represents a proportional selection known as the "roulette wheel selection." (The sections of roulette are in proportion to probabilities p_i.) In addition to the "roulette wheel selection," several other ways of selection for mating have been suggested in the literature.

In order to generate the next population P(1), we proceed to apply the other two genetic operators to the strings selected for mating. Crossover operator is used to combine the genetic material. At the beginning, pairs of strings (parents) are randomly chosen from a set of previously selected strings. Later, for each selected pair the location for crossover is randomly chosen. Each pair of parents creates two offsprings (Fig. 3.11).

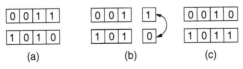

FIGURE 3.11   A single-point crossover operator: (a) two parents (b) randomly chosen location is before the last bit (c) two offsprings.

After completing crossover, the genetic operator mutation is used. In case of binary coding, mutation of a certain number of genes refers to the change in value from 1 to 0 or vice versa. It should be noted that the probability of mutation is very small (of order of magnitude 1/1000). The purpose of mutation is to prevent an irretrievable loss of the genetic material at some point along the string. For example, in the overall population a particularly significant bit of information might be missing (e.g., none of the strings have 0 at the seventh location), which can considerably influence the determination of the optimal or near-optimal solution. Without mutation, none of the strings in all future populations could have 0 at the seventh location. Nor could the other two genetic operators help to overcome the given problem. Having generated population P(1) [which has the same number of members as population P(0)], we proceed to use the operators reproduction, crossover, and mutation to generate a sequence of populations P(2), P(3), and so on.

In spite of modifications that may occur in some genetic algorithms (regarding the manner in which the strings for reproduction are selected, the manner of doing crossover, the size of population that depends on the problem being optimized, and so on), the following steps can be defined within any genetic algorithm:

Step 1: Encode the problem and set the values of parameters (decision variables).

Step 2: Form the initial population P(0) consisting of n strings. (The value of n depends on the problem being optimized.) Make an evaluation of the fitness of each string.

Step 3: Considering the fact that the selection probability is proportional to the fitness, select n parents from the current population.

Step 4: Randomly select a pair of parents for mating. Create two offsprings by exchanging strings with the one-point crossover. To each of the created offsprings, apply mutation. Apply crossover and mutation operators until n offsprings (new population) are created.

Step 5: Substitute the old population of strings with the new population. Evaluate the fitness of all members in the new population.

Step 6: If the number of generations (populations) is smaller than the maximal prespecified number of generations, go back to Step 3. Otherwise, stop the algorithm. For the final solution choose the best string discovered during the search.

3.7 Fuzzy Logic Approach to Dispatching in Truckload Trucking

3.7.1 Basic Elements of Fuzzy Sets and Systems

In the classic theory of sets, very precise bounds separate the elements that belong to a certain set from the elements outside the set. For example, if we denote by A the set of signalized intersections in a city, we conclude that every intersection under observation belongs to set A if it has a signal. Element x's membership in set A is described in the classic theory of sets by the membership function $\mu_A(x)$, as follows:

$$\mu_A(x) = \begin{cases} 1, \text{if and only if } x \text{ is member of A} \\ 0, \text{if and only if } x \text{ is not member of A} \end{cases} \tag{3.7}$$

Many sets encountered in reality do not have precisely defined bounds that separate the elements in the set from those outside the set. Thus, it might be said that waiting time of a vessel at a certain port is "long." If we denote by A the set of "long waiting time at a port," the question logically arises as to the bounds of such a defined set. In other words, we must establish which element belongs to this set. Does a waiting time of 25 hours belong to this set? What about 15 hours or 90 hours?

The membership function of fuzzy set can take any value from the closed interval [0, 1]. Fuzzy set **A** is defined as the set of ordered pairs **A** = {x, $\mu_A(x)$}, where $\mu_A(x)$ is the grade of membership of element x in set **A**. The greater $\mu_A(x)$, the greater the truth of the statement that element x belongs to set **A**.

Fuzzy sets are often defined through membership functions to the effect that every element is allotted a corresponding grade of membership in the fuzzy set. Let us note fuzzy set **C**. The membership function that determines the grades of membership of individual elements x in fuzzy set **C** must satisfy the following inequality:

$$0 \leq \mu_C(x) \leq 1 \quad \forall x \in X \tag{3.8}$$

Let us note fuzzy set **A**, which is defined as "travel time is approximately 30 hours." Membership function $\mu_A(t)$, which is subjectively determined is shown in Figure 3.12.

A travel time of 30 hours has a grade of membership of 1 and belongs to the set "travel time is approximately 30 hours." All travel times within the interval of 25–35 h are also members of this set because their grades of membership are greater than zero. Travel times outside this interval have grades of membership equal to zero.

Let us note fuzzy sets **A** and **B** defined over set X. Fuzzy sets **A** and **B** are equal (**A** = **B**) if and only if $\mu_A(x) = \mu_B(x)$ for all elements of set X.

Fuzzy set **A** is a subset of fuzzy set **B** if and only if $\mu_A(x) \leq \mu_B(x)$ for all elements x of set X. In other words, **A** $\subset$ **B** if, for every x, the grade of membership in fuzzy set **A** is less than or equal to the grade of membership in fuzzy set **B**.

The intersection of fuzzy sets **A** and **B** is denoted by **A** $\cap$ **B** and is defined as the largest fuzzy set contained in both fuzzy sets **A** and **B**. The intersection corresponds to the operation "and." Membership function $\mu_{A \cap B}(x)$ of the intersection **A** $\cap$ **B** is defined as follows:

$$\mu_{A \cap B}(x) = \min \left\{ \mu_A(x), \mu_B(x) \right\} \tag{3.9}$$

The union of fuzzy sets **A** and **B** is denoted by **A** $\cup$ **B** and is defined as the smallest fuzzy set that contains both fuzzy set **A** and fuzzy set **B**. The membership function $\mu_{A \cup B}(x)$ of the union **A** $\cup$ **B** of fuzzy sets **A** and **B** is defined as follows:

$$\mu_{A \cup B}(x) = \max \left\{ \mu_A(x), \mu_B(x) \right\} \tag{3.10}$$

Fuzzy logic systems arise from the desire to model human experience, intuition, and behavior in decision-making. Fuzzy logic (approximate reasoning, fuzzy reasoning) is based on the idea of the possibility of a decison-making based on imprecise, qualitative data by combining descriptive linguistic rules. Fuzzy rules include descriptive expressions such as small, medium, or large used to categorize the linguistic (fuzzy) input and output variables. A set of fuzzy rules, describing the control strategy of the operator (decision-maker) forms a fuzzy control algorithm, that is, approximate reasoning algorithm, whereas the linguistic expressions are represented and quantified by fuzzy sets.

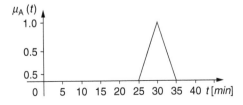

FIGURE 3.12 Membership function $\mu_A(t)$ of fuzzy set **A**.

The basic elements of each fuzzy logic system are rules, fuzzifier, inference engine, and defuzzifier. The input data are most commonly crisp values. The task of a fuzzifier is to map crisp numbers into fuzzy sets. Fuzzy rules can conveniently represent the knowledge of experienced operators used in control. The rules can be also formulated by using the observed decisions (input/output numerical data) of the operator. Fuzzy rule (fuzzy implication) takes the following form:

If x is **A**, then y is **B**

where **A** and **B** represent linguistic values quantified by fuzzy sets defined over universes of discourse X and Y. The first part of the rule "x is **A**" is the premise or the condition preceding the second part of the rule "y is **B**" which constitutes the consequence or conclusion.

Let us consider a set of fuzzy rules containing three input variables x_1, x_2, and x_3 and one output variable y.

Rule 1: If x_1 is $\mathbf{P}_{11}$ and x_2 is $\mathbf{P}_{12}$ and x_3 is $\mathbf{P}_{13}$, then y is $\mathbf{Q}_1$,
or
Rule 2: If x_1 is $\mathbf{P}_{21}$ and x_2 is $\mathbf{P}_{22}$ and x_3 is $\mathbf{P}_{23}$, then y is $\mathbf{Q}_2$,
or
Rule k: If x_1 is $\mathbf{P}_{k1}$ and x_2 is $\mathbf{P}_{k2}$ and x_3 is $\mathbf{P}_{k3}$, then y is $\mathbf{Q}_k$.

The given rules are interrelated by the conjunction *or*. Such a set of rules is called a disjunctive system of rules and assumes the satisfaction of at least one rule. It is assumed that membership functions of fuzzy sets $\mathbf{P}_{k1}$ and $\mathbf{P}_{k3}$ ($k = 1, 2, ..., K$) are of a triangular shape, whereas membership functions of fuzzy sets $\mathbf{P}_{k2}$ and $\mathbf{Q}_k$ ($k = 1, 2, ..., K$) are of a trapezoidal shape. Let us note Figure 3.13 in which our disjunctive system of rules is presented.

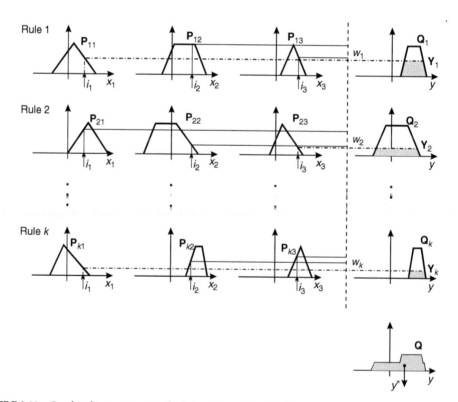

FIGURE 3.13 Graphical interpretation of a disjunctive system of rules.

Let the values i_1, i_2, and i_3, respectively, taken by input variables x_1, x_2, and x_3, be known. In the considered case, the values i_1, i_2, and i_3 are crisp. Figure 3.13 also represents the membership function of output **Q**. This membership function takes the following form:

$$\mu_Q(y) = \max_k \left\{ \min \left[\mu_{P_{k1}}(i_1), \mu_{P_{k2}}(i_2), \mu_{P_{k3}}(i_3) \right] \right\}, \quad k = 1, 2, \ldots, K \tag{3.10}$$

whereas fuzzy set **Q** representing the output is actually a fuzzy union of all the rule contributions Y_1, Y_2, ..., Y_k, that is:

$$Q = Y_1 \cup Y_2 \cup \ldots \cup Y_k \tag{3.11}$$

It is clear that

$$\mu_Q(y) = \max \left\{ \mu_{Y_1}(y), \mu_{Y_2}(y), \ldots, \mu_{Y_k}(y) \right\} \tag{3.12}$$

Consider rule 1, which reads as follows:

If x_1 is P_{11} and x_2 is P_{12} and x_3 is P_{13}, then y is Q_1.

The value $\mu_{P_{11}}(i_1)$ indicates how much truth is contained in the claim that i_1 equals P_{11}. Similarly, values $\mu_{P_{12}}(i_2)$ and $\mu_{P_{13}}(i_3)$, respectively, indicate the truth value of the claim that i_2 equals P_{12} and i_3 equals P_{13}. Value w_1, which is equal to

$$w_1 = \min \left\{ \mu_{P_{11}}(i_1), \mu_{P_{12}}(i_2), \mu_{P_{13}}(i_3) \right\} \tag{3.13}$$

indicates the truth value of the claims that, simultaneously, i_1 equals P_{11}, i_2 equals P_{12} and i_3 equals P_{13}.

As the conclusion contains as much truth as the premise, after calculating value w_1, the membership function of fuzzy set Q_1 should be transformed. In this way, fuzzy set Q_1 is transformed into fuzzy set Y_1 (Fig. 3.13). Values w_2, w_3, ..., w_k are calculated in the same manner leading to the transformation of fuzzy sets Q_2, Q_3, ..., Q_k into fuzzy sets Y_2, Y_3, ..., Y_k.

As this is a disjunctive system of rules, assuming the satisfaction of at least one rule, the membership function $\mu_Q(y)$ of the output represents the outer envelope of the membership functions of fuzzy sets Y_1, Y_2, ..., Y_k. The final value y^* of the output variable is arrived at upon defuzzification, that is, choosing one value for the output variable. In most applications an analyst or decision-maker looks at the grades of membership of individual output variable values, and chooses one of them according to the following criteria: "the smallest maximal value," "the largest maximal value," "center of gravity," "mean of the range of maximal values," and so on (Fig. 3.14).

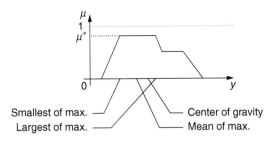

Smallest of max. Center of gravity
Largest of max. Mean of max.

FIGURE 3.14 Defuzzification methods.

3.7.2   Trucks Dispatching by Fuzzy Logic

Transportation companies receive a great number of requests every day from clients wanting to send goods to different destinations. Each transportation request is characterized by a large number of attributes, including the most important: type of freight, amount of freight (weight and volume), loading and unloading sites, preferred time of loading and/or unloading, and the distance the freight is to be transported. Transportation companies usually have fleets of vehicles consisting of several different types of vehicle. In addition to the characteristics of the transportation request, when assigning a specific type of vehicle to a specific transportation request, the dispatcher must also bear in mind the total number of available vehicles, the available number of vehicles by vehicle type, the number of vehicles temporarily out of working order, and vehicles undergoing technical examinations or preventive maintenance work. When meeting transportation requests, one or more of the same type of vehicle might be used. In other cases, several different types of vehicles might be used. Depending on the characteristics of the transportation request and the manner in which the transportation company operates, vehicle assignments to transportation requests can be made several times a day, once a day, once a week, and so on. Without loss of generality, we considered the case when dispatching is carried out every day based on the principle "today for tomorrow." In other words, dispatchers have a set amount of time (one day) to match available vehicles to transportation tasks that are to begin the following day.

Assigning vehicles to planned transportation tasks is a daily problem in every transportation company. In most cases, dispatchers responsible for assigning the vehicles rely primarily on their experience and intuition in the course of decision-making. Experienced dispatchers usually have built-in criteria ("rules") which they use to assign a given amount of freight to be sent a given distance to a given vehicle with given structural and technical-operational characteristics (capacity, ability to carry freight certain distances, and so on).

A good dispatcher must have suitable abilities and skills, and his training usually requires a long period. The problem we consider is not one requiring "real-time" dispatching (which is needed to dispatch ambulances, fire department vehicles, police patrol units, taxis, dial-a-ride systems, and so on). However, the large number of different input data and limited time to solve the problem of assigning vehicles to requests can certainly create stressful situations for the dispatcher. These reasons support the need to develop a system that will help the dispatcher to make decisions.

3.7.2.1   Statement of the Problem

Let us consider the vehicle assignment problem within the scope of the following scenario. We assume that a transportation company has several different types of vehicle at its disposal. The number of different vehicle types is denoted by n. Individual vehicle types differ from each other in terms of structural and technical-operational characteristics. We also assume that the transportation company has a depot from which the vehicles depart and to which they return after completing their trip

Let us consider a delivery system in which different types of freight are delivered to different nodes. We also assume that after serving a node, the vehicle returns to the depot. The reasons for such a delivery tactic are often because of the fact that different types of freight cannot be legally delivered in the same vehicle, and that different types of freight belong to different clients of the transportation company. As the vehicle returns to the depot after serving a node, we note that the routes the vehicle is to take are known. As shown in Figure 3.15, we are dealing with a set of routes in the form of a star, with each route containing a node to be served. Let us denote by m the total number of transportation requests to be undertaken the following day. Let us also denote by T_i the i-th transportation request, $(i = 1, 2, ..., m)$. Every transportation request T_i is characterized by four parameters (v_i, Q_i, D_i, n_i), where v_i is the node where freight is to be delivered when executing transportation request T_i, Q_i is the amount of freight to be transported by request T_i, D_i is the distance freight is to be transported in request T_i (the distance between depot D and node v_i), and n_i is the number of trips along route $\{D, v_i, D\}$ that can be made by one vehicle during the time period under consideration (one day).

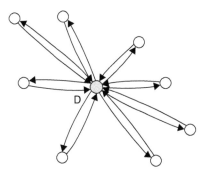

FIGURE 3.15 Depot D and nodes to be served.

In order to simplify the problem, we will assume that the number of possible trips n_i that can be made along route {D, v_i, D} is independent of the vehicle type.

We shall denote by C_j the capacity of vehicle type j taking part in the service ($j = 1, 2, ..., n$). The number of available type j vehicles is denoted by N_j. We also assume that

$$n_i \geq 1 \qquad i = 1, 2, ..., m \qquad (3.14)$$

Based on the discussed relation, we conclude that the vehicle can serve any node within the geographical region under consideration at least once a day and return to the depot.

Depending on the values of D_i and Q_i and the capacity C_j of the vehicle serving node v_i, one or more trips will be made along route {D, v_i, D} during the day being considered. One type or a variety of vehicle types can take part in the delivery to node v_i. Let us first consider the case when only one type of vehicle takes part in serving any node. The more complicated case when several different types of vehicle serve a node is considered later. We would also note that in some cases there is the possibility of the transportation company not being able to serve all nodes with its available transportation capacities.

The standard VRP consists of designing a route to be taken by the vehicles when serving the nodes. In most articles devoted to the classical routing problem, it is assumed that the capacity of the serving vehicle is greater than or equal to demand in any node. In our case, the routes to be taken by the vehicles are known (Fig. 3.15). We shall denote by f_{ij} the number of trips (frequency) to be made by a type j vehicle when executing transport request T_i. It is clear that $f_{ij} \geq 0$ ($i = 1, 2, ..., m, j = 1, 2, ..., n$).

The problem we considered is to determine the value of f_{ij} ($i = 1, 2, ..., m, j = 1, 2, ..., n$) so that the available vehicles are assigned to planned transportation tasks in the best possible way.

3.7.2.2 Proposed Solution to the Problem

The total number of vehicles N available to the dispatcher at the moment he assigns vehicles is

$$N = \sum_{j=1}^{n} N_j \qquad (3.15)$$

As already mentioned, the problem considered is the assignment of N available vehicles to m transportation requests. This belongs to the category of OR problems known as assignment problems.

Some transportation requests are "more important" than others. In other words, some clients have signed long-term transportation contracts, and others randomly request transportation that will engage transportation capacities for longer or shorter periods of time. In some cases there is no absolutely precise information about the number of individual types of vehicle that will be ready for operation the following day. Bearing in mind the number of operating vehicles and the number of vehicles expected to be operational the following day, the dispatcher subjectively estimates the total number of available vehicles by type. Some vehicle types are more "suitable" for certain types of transportation tasks than others. Naturally, vehicles with a 5 t capacity are more suitable to deliver good within a city area than those with a 25 t capacity. On the other hand, 25 t vehicles are considerably more suitable than 5 t or 7 t vehicles for long-distance freighting.

As we can see, the vehicle assignment problem is often characterized by uncertainty regarding input data necessary to make certain decisions. It should be emphasized that the subjective estimation of individual parameters differs from dispatcher to dispatcher, or from decision-maker to decision-maker. The number of available vehicles of a specific type might be "sufficient" for one dispatcher, while another dispatcher might think this number "insufficient" or "approximately sufficient." Also, one dispatcher might consider a certain type of vehicle "highly suitable" regarding a certain distance, while other dispatchers might consider this type of vehicle "suitable" or "relatively suitable." Clearly, a number of parameters that appear in the vehicle assignment problem are characterized by uncertainty, subjectivity, imprecision, and ambiguity. This raises the need in the mathematically modeling phase of the problem to use methods that can satisfactorily treat uncertainty, ambiguity, imprecision, and subjectivity. The approximate reasoning model presented in the following section is an attempt to formalize the dispatcher's knowledge, that is, to determine the rules used by dispatchers in assigning vehicles to transportation requests.

Approximate reasoning model for calculating the dispatcher's preference when only one type of vehicle is used to meet every transportation request

It can be stated that every dispatcher has a pronounced subjective feeling about which type of vehicle corresponds to which transportation request. This subjective feeling concerns both the suitability of the vehicle in terms of the distance to be traveled and vehicle capacity in terms of the amount of freight to be transported.

Dispatchers consider the suitability of different types of vehicles as being "low" (LS), "medium" (MS), and "high" (HS) in terms of the given distance the freight is to be transported. Also, capacity utilization (the relationship between the amount of freight and the vehicle's declared capacity, expressed as a percentage) is often estimated by the decision-maker as "low" (LCU), "medium" (MCU), or "high" (HCU).

The suitability of a certain type of vehicle to transport freight different distances, and its capacity utilization can be treated or represented as fuzzy sets (Figs. 3.16 and 3.17).

Vehicle capacity utilization is the ratio of the amount of freight transported by a vehicle to the vehicle's capacity. The membership functions of the fuzzy sets shown in Figures 3.16 and 3.17 must be defined individually for every type of vehicle.

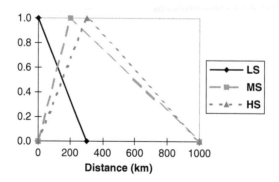

FIGURE 3.16   Membership functions of fuzzy sets: LS is low, MS is medium, and HS is high suitability in terms of distance.

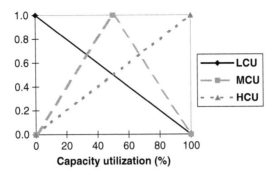

FIGURE 3.17 Membership functions of fuzzy sets: LCU is low, MCU is medium, HCU is high vehicle capacity utilization.

The decision-maker assigns transportation requests to individual types of vehicle bearing in mind above all the distance to be traveled and the capacity utilization of the specific type of vehicle. When dispatching, the decision-maker–dispatcher operates with certain rules. Based on conversations with dispatchers who deal with the vehicle assignment problem every day, it is concluded that the decision-maker has certain preferences:

"Very strong" preference is given to a decision that will meet the request with a vehicle type having "high" suitability in terms of distance and "high" capacity utilization.

Or

"Very weak" preference is given to a decision that will meet the request with a vehicle that has "low" suitability regarding distance and "low" capacity utilization.

The strength of the dispatcher's preference can be "very strong," "strong," "medium," "weak," and "very weak." Dispatchers most often use five terms to express the strength of their preference regarding the meeting of a specific transportation request with a specific type of vehicle. These five preference categories can be presented as corresponding fuzzy sets **P1**, **P2**, **P3**, **P4**, and **P5**. The membership functions of the fuzzy sets used to describe preference strength are shown in Figure 3.18. Preference strength will be indicated by a preference index, **PI**, which lies between 0 and 1, where a decrease in the preference index means a decrease in the "strength" of the dispatcher's decision to assign a certain transportation request to a certain type of vehicle.

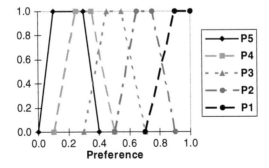

FIGURE 3.18 Membership functions of fuzzy sets: **P1** is very strong, **P2** is strong, **P3** is medium, **P4** is weak, **P5** is very weak preference.

TABLE 3.4 Approximate Reasoning Algorithm for a Vehicle with a Capacity of 14 t

		Capacity Utilization		
		LCU	MCU	HCU
Suitability	LS	P5	P4	P3
	MS	P3	P2	P2
	HS	P2	P1	P1

For every type of vehicle, a corresponding approximate reasoning algorithm is developed to determine the dispatcher's preference strength in terms of meeting a specific transportation request with the type of vehicle in question. The approximate reasoning algorithms for each type of vehicle differ from each other in terms of the number of rules they contain and the shapes of the membership functions of individual fuzzy sets. For example, for a vehicle with a capacity of 14t, the approximate reasoning algorithm reads as shown in Table 3.4.

Using the approximate reasoning by max–min composition, every preference index value is assigned a corresponding grade of membership. Let us denote this value by P_{ij}. This value expresses the "strength" of the dispatcher's preference that the i-th transportation request be met by vehicle type j. Similar approximate reasoning algorithms were developed for the other types of vehicle.

Calculating the dispatcher's preference when several types of vehicle are involved in meeting requests

Up until now, we have only considered the vehicle assignment problem when one type of vehicle is used to meet every transportation request. Some transportation companies often use several different types of vehicle to meet a specific transportation request. When meeting requests with several different types of vehicle, every request can be met in one or several different ways. For example, if the amount of freight in the i-th request equals $Q_i = 18$ t and if we have two types of vehicle whose capacities are 5 t and 7 t, respectively, there are four possible alternatives to meeting the i-th request shown in Table 3.5.

The first of the possible alternatives to meet any transportation request is the one in which only one type of vehicle is used, the vehicle with the greatest capacity. Every other alternative differs from the previous to the effect that there is a smaller share of vehicles with a higher capacity and a greater share of vehicles with a smaller capacity. The last possible alternative uses vehicles with the smallest capacity.

Let us denote the following:

Q_{ijk} is the amount of freight from the i-th request transported by vehicle type j when request T_i is met using alternative k.

N_{ijk} is the number of type j vehicles that participate in meeting request T_i when request T_i uses alternative k.

TABLE 3.5 Comparison of the Total Number of Ton-Kilometers Realized for the Four Different Ways of Assigning Vehicles to Transportation Requests

Possible Ways of Assigning Vehicles to Transportation Requests	Amount of Time Needed to Assign Vehicles to Planned Transportation Requests	Total Number of Realized Ton-kilometers	Percentage of Realized Ton-kilometers
I	2 hr 30 min	163,821	92.26%
II	2 hr 15 min	154,866	87.15%
III	40	152,727	86.01%
IV	40	170,157	95.83%

It is clear that the total freight Q_i from transportation request T_i that is met using transportation alternative k equals the sum of the amount of freight of request T_i transported by individual types of vehicles that is,

$$\sum_{j=1}^{n} Q_{ijk} = Q_i \qquad (3.16)$$

The capacity utilization (expressed as a percentage) λ_{ijk} of vehicle type j that takes part in meeting transportation request T_i using alternative k can be defined as,

$$\lambda_{ijk} = \frac{Q_{ijk}}{C_j \, N_{ijk} \, n_i} 100 \, [\%] \qquad (3.17)$$

Let us denote by P_k the dispatcher's preference to use service alternative k to meet transportation request T_i. It is clear that,

$$P_k = \frac{\displaystyle\sum_{j=1}^{n} N_{ijk} \, C_j \, n_i \, P_{ij}}{\displaystyle\sum_{j=1}^{n} N_{ijk} \, C_j \, n_i} \qquad (3.18)$$

Corresponding dispatcher's preference P_{ij} must be calculated for every type of vehicle j taking part in meeting transportation request T_i. Preference values P_{ij} are calculated based on approximate reasoning algorithms.

Based on relation 3.18, dispatcher preference to meet transportation request T_i with any of the possible service alternatives k can be calculated.

Heuristic algorithm to assign vehicles to transportation requests

The basic characteristics of every transportation request are the amount of freight that is to be transported and the distance to be traveled. Therefore, requests differ in terms of the volume of transportation work (expressed in ton-kilometers) to be executed, and in terms of the revenues and profits that every transportation request brings to the transportation company. It was also emphasized in our previous remarks that a company might have long-term cooperation with some clients, while other clients request the transportation company's services from time to time. Therefore, some transportation requests can be treated as being "more important," or "especially important requests," having "absolute priority in being carried out," and so on. All of this indicates that before assigning vehicles to transportation requests, the requests must first be sorted. The requests can be sorted in descending order by number of ton-kilometers that would be realized if the request were carried out, in descending order of the amount of freight in each request, in descending order of the requests' "importance" or in some other way. The manner in which the requests are sorted depends on the company's overall transportation policy. It is assumed that sorting of the transportation requests is made before vehicles are assigned to transportation requests.

The heuristic algorithm of assigning vehicles to transportation requests consists of the following steps:

Step 1: Denote by i the index of transportation requests. Let $i = 1$.

Step 2: Generate all possible alternatives to meet transportation request T_i.

Step 3: Denote by $k(i)$ the index of possible alternatives to meet transportation request T_i. Let $k(i) = 1$.

Step 4: Analyze alternative $k(i)$. If available resources (number of available vehicles of a specific type) allow for alternative $k(i)$, go to Step 5. Otherwise go to Step 7.

Step 5: Determine the preference for every type of vehicle that takes part in implementing alternative $k(i)$ using an approximate reasoning by max–min composition.

Step 6: Calculate the dispatcher's preference to use alternative $k(i)$ to meet transportation request T_i. Use relation 3.18 to calculate this preference.

Step 7: Should there be any uninvestigated alternatives, increase the index alternative value by 1 ($k(i) = k(i) + 1$) and go to Step 4. Otherwise, go to Step 8.

Step 8: Should none of the potential alternatives be possible owing to a lack of resources, transportation request T_i cannot be met. The final value of the dispatcher's preference (when there is at least one alternative possible) equals the maximum value of the calculated preferences of the considered alternatives. In this case, transportation request T_i is met by the alternative that corresponds to the maximum preference value.

Step 9: Decrease the number of available vehicles for the types of vehicle that took part in meeting transportation request T_i by the number of vehicles engaged in meeting the request.

Step 10: If any transportation requests have not been considered, increase the index by $i(i = i + 1)$ and return to Step 2.

3.7.2.3 Numerical Example

The developed algorithm was tested on a fleet of vehicles containing three different types of vehicle. Capacity per type of vehicle and their respective number in the fleet are: $Q_1 = 4.4\,t$ ($N_1 = 48$ vehicles), $Q_2 = 7.0\,t$ ($N_2 = 49$ vehicles), $Q_3 = 14\,t$ ($N_3 = 42$ vehicles).

Table 3.6 presents the characteristics of the set of 78 transport requests to be met. As can be seen from Table 3.6, each of the 78 transportation requests is characterized by amount of freight Q_i and distance D_i. The transportation work undertaken by the transportation company could be expressed in ton-kilometers (tkm). Based on the characteristics of the transportation requests, it is easy to calculate that the total number of the ton-kilometers to be carried out by the transportation company equals

$$\sum_{i=1}^{78} Q_i\, D_i = 177{,}570.3 \text{ tkm} \tag{3.19}$$

The quality of the solution obtained can be measured as the percentage of realized transportation requests and the percentage of realized ton-kilometers. As the transportation company's profit directly depends on the number of effected ton-kilometers, it was decided that the quality of the solution obtained should be judged on the basis of the total number of realized ton-kilometers. The solutions obtained from the developed model were compared with those obtained by an experienced dispatcher. Let us consider the following four ways of assigning vehicles to transportation requests:

1. An experienced dispatcher assigned vehicles to the transportation requests. The dispatcher was not given any instructions regarding the manner in which the assignments should be made.
2. An experienced dispatcher assigned vehicles to the transportation requests. The dispatcher was asked to assign only one type of vehicle to each transportation request.
3. Vehicles were assigned to transportation requests based on the developed algorithm, with only one type of vehicle being assigned to each transportation request.
4. Before assigning vehicles, the transportation requests were sorted by descending order of ton-kilometers. Vehicles were assigned to transportation requests using the developed algorithm, to the effect that one or several different types of vehicle took part in meeting each request.

The results obtained are shown in Table 3.7.

TABLE 3.6   Characteristics of 78 Transport Requests to Be Met

Request Number	Request Amount of Freight (Tons)	Distance (km)	Daily Number of Trips by One Vehicle	Request Number	Request Amount of Freight (Tons)	Distance (km)	Daily Number of Trips by One Vehicle
1	22.0	42.0	2	40	11.0	180.0	1
2	3.0	25.0	4	41	13.0	12.0	5
3	7.0	138.0	1	42	28.0	198.0	1
4	39.0	280.0	1	43	34.0	265.0	1
5	6.0	75.0	2	44	52.0	140.0	1
6	17.0	189.0	1	45	2.0	180.0	1
7	5.0	45.0	2	46	1.5	17.0	5
8	21.0	110.0	1	47	3.0	29.0	3
9	8.0	180.0	1	48	67.0	270.0	1
10	27.0	42.0	2	49	1.0	87.0	2
11	43.0	197.0	1	50	1.7	195.0	1
12	2.0	317.0	1	51	5.0	49.0	2
13	6.0	180.0	1	52	8.0	165.0	1
14	16.0	78.0	2	53	12.0	87.0	2
15	25.0	78.0	2	54	28.0	65.0	2
16	34.0	57.0	2	55	24.0	29.0	3
17	23.0	57.0	2	56	21.0	12.0	5
18	12.0	129.0	1	57	17.0	369.0	1
19	9.0	32.0	3	58	19.0	100.0	2
20	21.0	21.0	4	59	17.0	120.0	1
21	7.0	180.0	1	60	18.0	140.0	1
22	7.0	87.0	3	61	31.0	190.0	1
23	4.0	49.0	2	62	3.0	120.0	1
24	26.0	127.0	1	63	8.0	108.0	2
25	22.0	240.0	1	64	4.0	140.0	1
26	19.0	220.0	1	65	3.0	17.0	5
27	14.0	100.0	2	66	9.0	98.0	2
28	15.0	121.0	1	67	4.4	78.0	2
29	38.0	27.0	4	68	4.4	78.0	2
30	41.0	129.0	1	69	4.2	112.0	1
31	8.0	160.0	1	70	3.5	5.0	6
32	9.0	180.0	1	71	27.0	15.0	5
33	16.0	70.0	2	72	12.0	5.0	6
34	21.0	161.0	1	73	7.5	98.0	2
35	32.0	180.0	1	74	18.7	210.0	1
36	42.0	120.0	1	75	6.5	180.0	1
37	16.0	132.0	1	76	21.0	600.0	1
38	12.0	12.0	5	77	13.5	120.0	1
39	9.0	27.0	4	78	4.9	120.0	1

TABLE 3.7   Alternatives to Meeting the *i*-th Request

Alternative Number	Number of Vehicles in Services	
	7 t	5 t
1	3	0
2	2	1
3	1	3
4	0	4

The developed model shows indisputable advantages compared to the dispatcher, particularly concerning the amount of time needed to assign vehicles to planned transportation requests. It might also be noted that the model sufficiently imitates the work of an experienced dispatcher. Using the model, it is possible to achieve results that are equal to or greater than the results achieved by an experienced dispatcher. Testing a large number of dispatchers and testing the model on a large number of different examples would confirm whether the model gives better results than the dispatcher in every situation.

Bibliography

Bodin, L. et al., Routing and Scheduling of Vehicles and Crews: The State of the Art, *Comput. Oper. Res.*, 10, 63–211, 1983.

Council of Logistics Management, http://www.clm1.org/mission.html, 12 February 98.

Gillett, B. and Miller, L., A Heuristic Algorithm for the Vehicle Dispatch Problem, *Oper. Res.*, 22, 340–352, 1974.

Hillier, F.S. and Lieberman, G.J., *Introduction to Operations Research*, McGraw-Hill Science, Columbus, OH, 2002.

Holland, J., *Adaptation in Natural and Artificial Systems*, University of Michigan Press, Ann Arbor, 1975.

Klir, G. and Folger, T., *Fuzzy Sets, Uncertainty and Information*, Prentice-Hall, Englewood Cliffs, NJ, 1988.

Larson, R. and Odoni, A., *Urban Operations Research*, Prentice-Hall, Englefood Cliffs, NJ, 1981.

Taha, H., *Operations Research: An Introduction* (8th Edition), Pearson Prentice Hall, Upper Saddle River, NJ, 2006.

Teodorović, D. and Vukadinović, K., *Traffic Control and Transport Planning: A Fuzzy Sets and Neural Networks Approach*, Kluwer Academic Publishers, Boston-Dordrecht-London, 1998.

Winston, W., *Operations Research*, Duxbury Press, Belmont, CA, 1994.

Zadeh, L., Fuzzy Sets, *Inform. Con.*, 8, 338–353, 1965.

Zimmermann, H.-J., *Fuzzy Set Theory and Its Applications*, Kluwer, Boston, 1991.

4

Logistics Metrics

Thomas L. Landers
University of Oklahoma

Alejandro Mendoza
University of Arkansas

John R. English
Kansas State University

4.1 Introduction

With the growth of logistics and supply chain management (SCM), there is an urgent need for performance monitoring and evaluation frameworks that are balanced, integrated, and quantitative. Gerards et al. define logistics as "the organization, planning, implementation and control of the acquisition, transport and storage activities from the purchase of raw materials up to the delivery of finished products to the customers." SCM is defined by the Council of Supply Chain Management Professionals (CSCMP) [2] as follows: "(SCM) encompasses the planning and management of all activities involved in sourcing and procurement, conversion, and all logistics management activities. Importantly, it also includes coordination and collaboration with channel partners which can be suppliers, intermediaries, third-party service providers, and customers. SCM integrates supply and demand management within and across companies." Current frameworks of performance evaluation within most organizations are sets of known performance measures or metrics (PMs) that have evolved over time. CSCMP [2] defines performance measures as "indicators of the work performed and the results achieved in an activity, process or organizational unit. Performance measures should be both nonfinancial and financial."

Monitoring the performance of a given process requires a well-defined set of metrics to help us establish goals within organizations. Managers need guidance in identifying useful performance metrics, their associated units, unique data characteristics, monitoring techniques, and benchmarks against which such metrics can be compared.

A metric is a standard measure that assesses an organization's ability to meet customers' needs or business objectives. Many performance metrics are ratios relating inputs and outputs, thus permitting assessment of both effectiveness (the degree to which a goal is achieved) and efficiency (the ratio of the resources utilized against the results derived) in accomplishing a given task [3].

Metrics generally fall into two categories: (*i*) performance metrics and (*ii*) diagnostics metrics [4]. PMs are external in nature and closely tied to outputs, customer requirements, and business needs for the process. A diagnostic metric reveals the reasons why a process is not performing in accordance to expectations and is internal in nature. The CSCMP standards for delivery processes [5] stress key performance indicators (KPI) to be monitored by summary tools, such as scorecards or dashboards.

There is a growing body of knowledge and publications on topics of performance measurement and benchmarking for logistics operations. Frazelle and Hackman [6] developed a warehouse performance index using data envelopment analysis. Frazelle [7] has continued to report warehouse metrics and best practices. In 1999, the Council of Logistics Management (now the CSCMP) published a business reference book [8] on the topic. Several articles provide good reviews of performance measurement in logistics [9–11]. Other articles have proposed performance measurement frameworks, including identification and clustering of metrics [12–20].

Two major themes have emerged in the field of performance measurements for business processes in general and logistics processes in particular. The first is to maintain breadth of measurement across functions and objectives. Kaplan and Norton [21] proposed the Balanced Scorecard approach, with metrics in multiple categories (e.g., financial, operational efficiency, service quality, and capability enhancement). The Warehouse Education and Research Council (WERC) periodically reports on performance measurement in distribution centers [22]. Secondly, performance measurement should span the full supply chain. The Perfect Order Index (POI) [23] has emerged as a preferred best practice for measuring full-stream logistics and includes as a minimum the following attributes: on-time, complete, damage-free, and properly invoiced. POI requires discipline and integration of information systems across supply chain partners [23].

A typology measuring relative sophistication of logistics management approaches has been developed by AT Kearney [15]. This typology divides companies into four different stages. In Stage I, companies use very simple measures that are expressed in terms of dollars, where information usually comes from the financial organization using very few accounting ratios. In Stage II, companies begin to use simple measures of distribution in terms of productivity to evaluate performance. The use of measures is normally in response to a given problem. In Stage III, companies are proactive and have set meaningful goals for operations. The sophistication of performance measurements is very high. In Stage IV, companies integrate performance data with financial data and are thus able to integrate functional goals.

Comparability of measures, errors in the measurement systems, and human behavior are some of the issues in establishing and monitoring PMs. The marginal benefit of information gathered must exceed marginal costs. Trimble [4] points out that the PMs must be "SMART": Specifically targeted to the area you are measuring, the data must be Measurable (accurate and complete), Actionable (easy to understand), Relevant, and the information inferred from the data must be Timely.

Euske [24] provides a five-step process for developing a measurement system:

1. Establish the problem or goal and its context.
2. Identify the attributes, inputs, and outputs to be evaluated.
3. Analyze the way the measures are obtained.
4. Replace unsatisfactory measures with ones that fulfill the requirements.
5. Perform a cost-benefit analysis to assess the benefit of using a given measurement system.

Lockamy and Cox [18] establish three primary categories of performance measurements: customer, resource, and finance. Within each category, functions are identified. The customer category contains the marketing, sales, and field services functions; the resources category is made up of production, purchasing, design engineering, and transportation functions; and the finance category includes cost, revenue, and investment functions. The PMs for each of these three categories and their associated functions are typically assumed to be independent of one another. As discussed in Byers et al. [25] and implemented by Harp et al. [26], it is necessary to construct performance metrics that monitor performance vertically throughout an organization as well as integrating performance horizontally across the organization giving rise to balanced, full-stream logistics measurements [25].

Boyd and Cox [27] apply this integrating requirement to a case study. Specifically, they implement a negative branch approach (a cause and effect approach developed by Goldratt [28]) to analyze the value-added impact of existing PMs within a pressboard manufacturing process. Through the case study, they clearly demonstrate that performance metrics should not be blindly selected. Specifically, an effective performance metric framework facilitates continuous improvement for the organization.

In summary, performance metrics are data collected from a process of transformation from inputs into outputs to evaluate the existing status of a process. Performance metrics are systematically related to norms and other data. Transformations may include production processes, decision processes, development process, logistics processes, and so on.

4.2 Logistics Data

The monitoring of logistics systems is critical to measuring the quality of service. Data for logistics performance metrics are similar to traditional categories of data in other quality control applications. Quality control data are categorized into two types: attribute and variable. Variable data are measurements that are made on a continuous spectrum. For example, cycle times for receiving materials and issuance of stock are variable data as used in most organizations for a given service type. Alternatively, attribute data are classifications of type. For example, a package either meets or fails to meet packaging standards. Extending this concept further, if 100 packages are selected at random, the proportion of packages meeting inventory accuracy would also be considered as attribute data.

4.2.1 Attribute Data in the Logistics Area

Table 4.1 presents a set of logistics performance metrics of attribute type. Each metric has been either used or recommended for use within a given organization as described in the third column of Table 4.1. In the subsequent discussion, a framework suitable for mostly all logistics systems is presented that more completely enumerates logistics metrics.

4.2.2 Variable Data in the Logistics Area

Table 4.2 provides examples of the current and planned use of variable data within logistics environments. The logistics function is a complex process in which sub-operations are intertwined and may be

TABLE 4.1 Logistics Attribute Data

PM#	Performance Metric	Source
1	Data entry accuracy (total track frequency)	United Parcel Service [29]
2	Preservation and packaging	Defense Logistics Agency [26]
3	Inventory accuracy	Defense Logistics Agency [26]
4	Resolutions complete	Defense Logistics Agency [26]
5	Customer complaints	Defense Logistics Agency [30]
6	Damage freight claims	J.B. Hunt [30]
7	Carrier on-time pickup	Lucent Technologies [30]
8	% Location accuracy	Whirlpool [30]
9	% Empty miles	J.B. Hunt [30]
10	Picks from forward areas	Lucent Technologies [30]
11	Pick rate	Global Concepts [30]
12	% Perfect orders	Global Concepts [30]

TABLE 4.2 Logistics Variable Data

PM#	Performance Metric	Source
1	Cycle time for receipt of material	Defense Logistics Agency [26]
2	Cycle time for issuance of stock	Defense Logistics Agency [26]
3	Cost of nonconformance	Arkansas Best Freight [30]
4	Cost of maintenance	Lucent Technologies [30]
5	Transportation cost	J.B. Hunt [30]
6	Inventory on hand	Lucent Technologies [30]
7	Customer inquiry time	Defense Logistics Agency [30]

confounded. The performance must be considered in view of the process natural variation. For example, consider cycle time for the receipt of material as used by the Defense Logistics Agency (DLA). The DLA records periodic cycle times and reports average cycle times to the appropriate management. The cycle time varies from one study period to the next for a given service (e.g., binable, high-priority items). The cycle time may occasionally exceed requirements; whereas at other times, it may fall significantly below the requirement. Personnel directly involved with the process know to expect variation from one period to the next, but management usually becomes concerned when requirements are either missed or exceeded. Performance requirements must be considered in the context of expected variation. Consequently, logistics functions should be monitored such that the process is controlled and evaluated in accordance to its natural variation.

In any process, whether it be manufacturing, logistics, or other service, natural variation is present, and must be properly addressed. The sub-processes should be controlled to within the range of their natural variation. Only when nonrandom patterns exist should operators adjust the process, because reaction to random behavior inevitably increases process variation. Patterns should be judged as non-random only based upon sound statistical inference. Statistical process control (SPC) provides the framework for statistical inference. SPC builds an environment in which it is the desire of all employees and supply chain partners associated with the process to strive for continuous improvements. Without top-level support, SPC will fail. The following section presents the tools suitable for logistics processes.

4.3 Statistical Methods of Process Monitoring

Statistical process control is a powerful collection of problem-solving tools useful in achieving process stability and improving process capability through the reduction of variability. The natural variability in a process is the effect of many small unavoidable causes. This natural variability is also called a "stable system of chance causes" [31]. A process is said to be in statistical control when it operates under only chance causes of variation. On the other hand, unnatural variation may be observed and assigned to a root cause. These unnatural sources of variability are referred to as assignable causes. Assignable causes can range from improperly adjusted machines to human error. A process or service operating under assignable causes is said to be out of SPC.

4.3.1 Seven Tools of SPC

Statistical process control can be applied to any process and relies on seven major tools, sometimes called the magnificent seven [31]:

1. Histogram
2. Check sheet
3. Pareto chart

   4. Cause and effect diagram
   5. Defect concentration diagram
   6. Scatter diagram
   7. Control chart

Histogram: A histogram represents a visual display of data in which three properties can be seen (shape, location or central tendency, and scatter or spread). The typical histogram is a type of bar chart with the vertical bars ordered horizontally by value of a variable. The vertical scale measures frequencies.

Check sheet: A check sheet is a very useful tool in the collection and interpretation of data. For example, a check sheet may capture data for a histogram. Events are tallied in categories. A check sheet should clearly specify the type of data to be collected as well as any other information useful in diagnosing the cause of poor performance.

Pareto chart: The Pareto chart is simply a frequency distribution (or histogram) of attribute data arranged by category. The Pareto chart is a very useful tool in identifying the problems or defects that occur most frequently. It does not identify the most important defects; it only identifies those that occur most frequently. Pareto charts are widely used for identifying quality-improvement opportunities.

Cause and effect diagram: The cause and effect diagram is a tool frequently used to analyze potential causes of undesirable problems or defects. Montgomery [31] suggests a list of seven steps to be followed when constructing a cause and effect diagram: (*i*) define the problem, (*ii*) form the team to perform the analysis, (*iii*) draw the effect box and the center line, (*iv*) specify the major potential cause categories and join them as boxes connected to the center line, (*v*) identify the possible causes and classify them, (*vi*) rank the causes to identify those that impact the problem the most, and (*vii*) take corrective action.

Defect concentration diagram: The defect concentration diagram is a picture of the process or product. The different types of defects or problems are drawn on the picture, and the diagram is analyzed to determine the location of the problems or defects.

Scatter diagram: The scatter diagram is used to identify the potential relationship between two variables. Data are plotted on an x-y coordinate system. The shape of the scatter diagram indicates the possible relationship existing between the two variables.

Control chart: The control chart is a graphical display of a quality characteristic that has been measured or computed from a sample versus the sample number or time.

4.3.2   Control Charts in the Logistics Area

To separate assignable causes from the natural process variation, we make use of control charts. Control charts are the simplest procedure of on line SPC (Fig. 4.1). These charts make possible the diagnosis and correction of many problems, and help to improve the quality of the service provided. Control charts also help in preventing frequent process adjustments that can increase variability. Through process improvements, control charts often provide assurance of better quality at a lower cost. Therefore, a control chart is a device for describing in a precise manner exactly what is meant by statistical control [27].

A control chart contains a centerline that represents the in-control average of the quality characteristic. It also contains two other horizontal lines called the upper control limit (UCL) and lower control limit (LCL). If a process is in control, most sample points should fall within the control limits. These limits are typically called "3-sigma (3σ) control limits." Sigma represents the standard deviation (a measure of variability, or scatter) of the statistic plotted on the chart. The width of the control limits is inversely proportional to the sample size n.

Control charts permit the early detection of a process that is unstable or out of control. However, a control chart only describes how a process is behaving, not how it should behave. A particular control chart might suggest that a process is stable, yet the process may not actually be satisfying customer requirements.

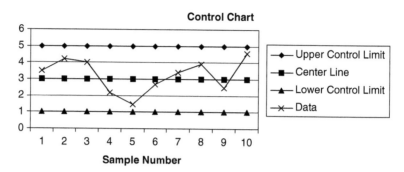

FIGURE 4.1 Control chart.

4.3.3 Error Types

In a control chart, the distance between the centerline and the limits control decision-making based on error. There are two types of statistical error. A type I error, also known as a false alarm or producer's risk, results from wrongly concluding that the process is out of control when in fact it is in control. A type II error, also known as the consumer's risk, results from concluding that the process is in control when it is not. Widening the control limits in a control chart decreases the risk of a type I error, but at the same time increases the risk of a type II error. On the other hand, if the control limits are moved closer to the center line, the risk of having a type I error increases while decreasing the risk of a type II error.

4.3.4 AT&T Runs Rules

The identification of nonrandom patterns is done using a set of rules known as run rules. The classic Western Electric (AT & T) handbook [32] suggests a set of commonly used decision tools for detecting nonrandom patterns on control charts. A process is out of control if any one of the following applies:

1. One data point plots outside the 3-sigma limits (UCL, LCL).
2. Two out of three consecutive points plot outside the 2-sigma limits.
3. Four out of five consecutive points plot outside the 1-sigma limits.
4. Eight consecutive points plot on one side of the center line.

The rules apply to one side of the center line at a time. For example, in the case of rule 2, the process is judged out of control when two out of three consecutive points falling beyond the 2-sigma limits are on the same side of the center line.

4.3.5 Types of Control Charts

Quality is said to be expressed by variables when a record is made of an actual measured quality characteristic. The $\bar{x}$, R, and S control charts are examples of variables control charts. When samples are of size one, individual (I) charts are suggested for monitoring the mean, and moving range (MR) charts are suggested for monitoring the variance. On the other hand, when a record shows only the number of articles conforming or nonconforming to certain specified requirements, it is said to be a record by attributes. The p, c, and u charts are examples of control charts for attribute data. One other important control chart is the moving centerline exponentially weighted moving average (EWMA), which is very effective in monitoring data that are not independent.

4.3.5.1 Control Charts for Variable Type Data

When dealing with variable data, it is usually necessary to control both the mean value of the quality characteristic and its variability. To monitor the mean value of the product, the $\bar{x}$ control chart is often used. Process variability can be monitored with a control chart for the standard deviation called the S chart, or a control chart for the range, called the R chart. The R chart is the more widely used. The $\bar{x}$ and R (or S) charts are among the most important and useful on-line SPC techniques. When the sample size, n, is large, $n > 12$, or the sample size is variable, the S chart is preferred to the R chart for monitoring variability.

4.3.5.2 Control Charts for Attribute Data

It is known that many quality characteristics cannot be represented numerically. Items inspected are usually classified as conforming or nonconforming to the specifications of that quality characteristic. This type of quality characteristic is called an attribute. Attribute charts are very useful in most industries. For example, from the logistics perspective, it is often necessary to monitor the percentage of units delivered on-time, on-budget, and in compliance with specifications.

The p-chart is used to monitor the fraction nonconforming from a manufacturing process or a service. It is based on the binomial distribution (number of successes in n trials) and assumes that each sample is independent. The fraction nonconforming is defined as the ratio of nonconforming items in a population to the total number of items in that population. Each item may have a number of quality characteristics that are examined simultaneously. If any one of the items being scrutinized does not satisfy the requirements, then the item is classified as nonconforming. The fraction nonconforming is usually expressed as a decimal, although it is occasionally expressed as the percent nonconforming.

There are many practical situations in which working directly with the total number of defects or nonconformities per unit or the average number of nonconformities per unit is preferred over the fraction nonconforming. The c-chart assumes that the occurrence of nonconformities in samples of constant size is rare. As a result, the occurrence of nonconformity is assumed to follow the Poisson probability distribution. The inspection unit must be the same for each sample.

4.3.5.3 Control Chart for Moving Centerline Exponentially Weighted Moving Average

The use of variable control charts implies the assumption of normal and independent observations. If the assumption of normality is violated to a moderate degree, the $\bar{x}$ control chart used to monitor the process average will work reasonably well due to the central limit theorem (law of large numbers). However, if the assumption of independence is violated, conventional control charts do not work well. Too many false alarms disrupt operations and produce misleading results. The moving centerline exponentially weighted moving average (EWMA) is effectively a one-step-ahead predictor to monitor processes when data are correlated. The moving centerline EWMA chart is also recommended for use in the logistics arena for a performance metric that is subject to seasonal variation.

4.3.6 Construction of Control Charts

Table 4.3 (variable type) and Table 4.4 (attribute, or fraction nonconforming type) summarize the parameters and equations for commonly used control charts applicable to logistics performance measurement.

4.4 Logistics Performance Metrics

The authors have developed a logistics performance measurement methodology through centers in the National Science Foundation Industry/University Cooperative Research Center program: Material Handling Research Center (MHRC), The Logistics Institute (TLI), and Center for Engineering Logistics

TABLE 4.3 Construction of Control Charts (Variable Type)

	Estimators of Mean	Estimator of Variation		Control Limits
$\bar{x}$ and R charts	Fixed sample size $$\bar{\bar{x}} = \frac{\bar{x}_1 + \bar{x}_2 + \cdots + \bar{x}_m}{m}$$	$$\bar{R} = \frac{R_1 + R_2 + \cdots + R_m}{m}$$	$\bar{x}$ chart	$UCL = \bar{\bar{x}} + A_2\bar{R}$ $Center\ Line = \bar{\bar{x}}$ $LCL = \bar{\bar{x}} - A_2\bar{R}$
		$R = x_{max} - x_{min}$	$\bar{R}$ chart	$UCL = D_4\bar{R}$ $Center\ Line = \bar{R}$ $LCL = D_3\bar{R}$
$\bar{x}$ and S charts	Fixed sample size $$\bar{\bar{x}} = \frac{\bar{x}_1 + \bar{x}_2 + \cdots + \bar{x}_m}{m}$$	$$\bar{S} = \frac{1}{m}\sum_{i=1}^{m} S_i$$	$\bar{x}$ chart	$UCL = \bar{\bar{x}} + A_3\bar{S}$ $Center\ Line = \bar{\bar{x}}$ $LCL = \bar{\bar{x}} - A_3\bar{S}$
	Variable sample size $$\bar{\bar{x}} = \frac{\sum_{i=1}^{m} n_i\bar{x}_i}{\sum_{i=1}^{m} n_i}$$	$$\bar{S} = \left[\frac{\sum_{i=1}^{m}(n_i - 1)S_i^2}{\sum_{i=1}^{m} n_i - m}\right]^{\frac{1}{2}}$$	S-chart	$UCL = B_4\bar{S}$ $Center\ Line = \bar{S}$ $LCL = B_3\bar{S}$
Individual measurements (sample size 1)	$$\bar{x} = \frac{x_1 + x_2 + \cdots + x_m}{m}$$	$$\overline{MR} = \frac{\sum_{i=1}^{m}|x_i - x_{i-1}|}{m - 1}$$	$\bar{x}$ chart	$UCL = \bar{x} + 3\dfrac{\overline{MR}}{d_2}$ $Center\ Line = \bar{x}$ $LCL = \bar{x} - 3\dfrac{\overline{MR}}{d_2}$
			$\overline{MR}$ chart	$UCL = D_4\overline{MR}$ $Center\ Line = \overline{MR}$ $LCL = D_3\overline{MR}$

and Distribution (CELDi). A workshop of invited industry leaders in logistics produced the initial framework [30].

Figure 4.2 presents the framework for the generic design of performance measures necessary for monitoring logistics support functions within most organizations. Clearly, there is overlap among each of the four groups, and as observed in Boyd and Cox [27], it is suggested that each PM be heavily scrutinized for its added value.

There are four primary groups of PMs presented in the framework: financial, quality, cycle time, and resource. In the design of a metrics framework, it is necessary to maintain balance and integration across each of these groups. These four primary groups represent a holistic view of the design of PMs necessary to evaluate and monitor the performance of most logistics support functions. The financial group represents the necessary dimension of evaluating short- and long-term profits to ensure the strong financial position of an organization. The quality group represents the dimension of evaluating an organization's quality of meeting customer expectations (external and internal). The cycle time group represents the necessity of evaluating process velocity and consistency. Finally, the resource dimension accounts for the necessary provision of process resources and the utilization and efficiency of processes.

TABLE 4.4 Construction of Control Charts (Fraction Nonconforming Type)

	Estimators of Central Tendency	Estimator of Variation	Control Limits
1. p-charts	Fixed sample size		Fixed sample size
	$$\bar{p} = \frac{\sum_{i=1}^{m} D_i}{mn} = \frac{\sum_{i=1}^{m} \hat{p}_i}{m}$$		$UCL = \bar{p} + 3\sqrt{\dfrac{\bar{p}(1-\bar{p})}{n}}$ $CL = \bar{p}$ $LCL = \bar{p} - 3\sqrt{\dfrac{\bar{p}(1-\bar{p})}{n}}$
		$\sigma_p^2 = \dfrac{p(1-p)}{n}$	*p*-chart
	Variable sample size	$\sigma_{\hat{p}} = \sqrt{\dfrac{p(1-p)}{n}}$	Variable sample size
	$$\bar{p} = \frac{\sum_{i=1}^{m} D_i}{\sum_{i=1}^{m} n_i}$$		$UCL_i = \bar{p} + 3\sqrt{\dfrac{\bar{p}(1-\bar{p})}{n_i}}$ $CL_i = \bar{p}$ $LCL_i = \bar{p} - 3\sqrt{\dfrac{\bar{p}(1-\bar{p})}{n_i}}$
			p-chart
2. Number nonconforming type	$\bar{c} = \dfrac{\sum D_i}{\sum n_i}$	$\sqrt{\bar{c}}$	*c*-chart	$UCL = \bar{c} + 3\sqrt{\bar{c}}$ $Center\ Line = \bar{c}$ $LCL = \bar{c} - 3\sqrt{\bar{c}}$
	$\bar{u} = \dfrac{\bar{c}}{m}$	$\sqrt{\bar{u}/n}$	*u*-chart	$UCL_i = \bar{u} + 3\sqrt{\dfrac{\bar{u}}{n}}$ $CL_i = \bar{u}$ $LCL_i = \bar{u} - 3\sqrt{\dfrac{\bar{u}}{n}}$

The framework provides a high-level, balanced, and integrated approach. Table 4.5 categorizes and describes the subgroups of performance metrics in each of the four major groups. Tables 4.6 through Table 4.9 summarize recommended performance metrics for the framework in Figure 4.2 and Table 4.5: financial metrics (Table 4.6), quality metrics (Table 4.7), cycle time metrics (Table 4.8), and resource metrics (Table 4.9).

4.5 Case Study

The following case study demonstrates SPC applied to logistics performance metrics. The application is cycle time and quality metrics for material flow in a point-of-use pull system.

In the ideal application for just-in-time (JIT) manufacturing, there is a single product in high-volume continuous demand. Synchronized JIT production results in components being delivered directly from the supplier to the point of use, just at the time of need. Components are received and handled in standard, reusable containers. One stage in assembly is completed just as the resulting work-in-process (WIP) is needed in the next stage. If there is a time delay between two successive operations, a small temporary buffer storage area is provided between the operations. These buffers are called kanbans and

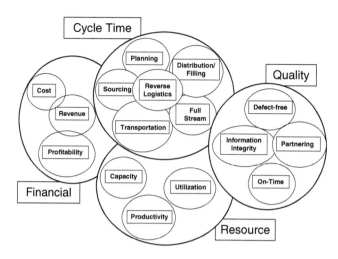

FIGURE 4.2　Performance metrics framework.

serve three purposes in JIT doctrine: (*i*) limit WIP inventory, (*ii*) maintain shop discipline and house-keeping, and (*iii*) provide process visibility. Removal of a workpiece from the kanban empties the buffer and serves as a pull signal for another unit to be produced and placed in the kanban.

In the more typical case of high-mix, low-volume production, some WIP inventory, including compo-nent stocks, may be required. However, it is preferred to minimize the WIP by using the principles of JIT to the extent possible. An engineered storage area (ESA) supports point-of-use pull logic for material flow in the high-mix, low-volume shop. The two-bin system is perhaps the most simple and visible method of deploying component stocks into the workcenter ESA. The stock for a component is split into two storage bins. In the simplest form of two-bin system the quantities are equal. When the first bin is emptied, a replenishment order is initiated. If the stock level is reviewed continuously, then the second bin must contain a sufficient amount of material to meet production needs during the replenishment lead time. If the stock is reviewed periodically, then the second bin should contain sufficient stock to meet demand

TABLE 4.5　Description of Performance Metric Subgroups

Group	Subgroup	Description
Financial	Cost	Focus on cost elements
	Profitability	Consideration of both cost and revenue elements
	Revenue	Focus on revenue elements
Quality	Defect-free	Encompasses all elements of a perfect order outside of on-time and information integrity
	Information integrity	Measurement of information accuracy in the system
	On-time	Meeting partner/customer on-time commitments
	Partnering	Teaming of logistics players, including employees, in order to accomplish value-driven goals
Cycle Time	Distribution/filling	Focuses on distribution and filling time
	Full stream	Spans the entire supply chain
	Planning	Elapsed time related to planning and design
	Reverse logistics	Measurement of elapsed time related to returns
	Sourcing	Focuses on sourcing elapsed time
	Transportation	Measurement of transit time
Resource	Capacity	Related to the output capability of a system
	Productivity	Comparison of actual output to the actual input of resources
	Utilization	Comparison of actual time used to the available time

TABLE 4.6   Financial Metrics Subgroup

Metric	Data	Units	SPC
Annual cost of maintenance by operator	Total amount of money spent on maintenance in a fiscal year	$/operator-year	Pareto chart
Cost per operation	(Total cost)/(total number of operations)	$ per activity	x-bar and R charts or moving centerline EWMA and MR if the data are seasonal
Cost per piece	(Total cost)/(total number of pieces)	$ per piece	x-bar and R charts or moving centerline EWMA and MR if the data are seasonal
Cost per transaction	Cost of a transaction	$ per transactions	x-bar and R charts or moving centerline EWMA and MR if the data are seasonal
Cost per unit of throughput	Cost of a specific facility	$ per unit of throughput for a facility	x-bar and R charts or moving centerline EWMA and MR if the data are seasonal
Cost variance	Cost variance is the ratio between actual costs and standard costs. This ratio most likely varies with particular seasons	Percentage	Moving centerline EWMA and MR charts (assuming the data are seasonal)
Economic value added	(Net operating profit after taxes) – (capital charges)	$	x-bar and R charts
Gross profit margin	[(Sales) – (cost of good sold)]/(sales)	Percentage	Individual and MR charts
Increase in profile adjusted revenues per CWT*	[Positive trend in profile (segment of the market or a particular customer) adjusted revenues)]/[CWT]	$ per CWT	u chart with variable sample size
Inventory carrying Cost	Costs related to warehousing, taxes, obsolescence and insurance (total cost of warehousing, taxes, obsolescence, insurance)/(total cost)	Percentage	Individual and MR charts
Inventory on hand	Total cost of inventory on hand	Percentage	Individual and MR charts
Inventory shrinkage	Total money lost from scrap, deterioration, pilferage, etc.	$	Pareto chart
Logistics operating expenses	Inventory carrying cost + transportation + shipper expenses + distribution + administrative	$	Pareto chart
Material handling rate	(Material handling expense)/(material handling asset value)	Percentage	Individual and MR charts
Net profit margin	(Net profit after taxes)/(sales)	Percentage	Individual and MR charts or Trend charts
Operating expenses before interest and taxes	Cost of goods sold	$ per unit time	x-bar and R charts
Operating ratio	(Cost of goods sold + selling costs + general and administration cost)/(sales)	Percentage	SPC: Individual and MR charts or Trend charts
Payables outstanding past credit term	(Accounts payable past credit term)/(# accounts)	Percentage	P chart, variable sample size
Receivable days outstanding	(Accounts receivable)/(sales per day). Refers to the average collection period	Percentage	x-bar and R charts
Return on assets	(Net profits after taxes)/(total assets)	Percentage	Individual and MR charts or Trend charts
Return on investment	(Income)/(investment capital)	Percentage	Individual and MR charts
Revenue growth percentage	(Revenue at the end of a period) – (Revenue at the end of previous period). Change in revenue over time	Percentage	Individual and MR charts

continued

TABLE 4.6 Financial Metrics Subgroup (continued)

Metric	Data	Units	SPC
Transportation cost per unit (piece, CWT, mile)	(Total transportation costs)/(total number of units)	$ per unit	u chart, variable sample size
Trend	Growth or shrinking market share. Change in market share over time	Percentage	Individual and MR charts

* CWT = $ per 100 pounds of weight.

during lead time plus the periodic review interval. If there is variability in the supply or demand processes, then safety stock may be required, which increases the standard quantity in both bins.

Bar code labeling and radio frequency communications promote the efficiency of an ESA. Each bin location is labeled with a barcode. When bin 1 is emptied, the user scans the barcode to trigger a replenishment cycle. As the replenishment is put away, the location barcode is again scanned to close out the cycle.

A printed wiring board (PWB) assembly operation in manufacturing of telecommunications switching equipment served as a case study for point-of-use pull material flow through an ESA. Figure 4.3 depicts the layout and flow.

The general framework in Figure 4.2 is utilized to design performance metrics specific to the point-of-use pull system in the PWB assembly shop. This custom system is compatible with available

TABLE 4.7 Quality Metrics Subgroup

Metric	Data	Units	SPC
Data entry accuracy	(# Errors)/(# transactions)	Percentage	p-Chart with variable sample size
Document accuracy	(# of orders with accurate documentation)/(total # orders)	Percentage nonconforming	p-Chart with variable sample size
Forecast accuracy	Mean absolute deviation or mean square error. This metric refers to the difference (error) between forecasted and actual	Percentage	x-bar and R charts
Inventory accuracy	(Parts in stock)/(parts supposed to be in stock). Refers to the total number of parts reported by the system of being in stock versus the actual number of parts present in stock	Percentage	p-Charts with variable sample size
Record accuracy	(Number of erroneous records)/(total number of records)	Percentage nonconforming	p-Chart with variable sample size
Tracking accuracy	(Entities in known status)/(total entities). This metric measures the accuracy of tracking job orders e.g., by lot control.	Percentage	p-Chart with variable sample size
On-time delivery	(On-time deliveries)/(total deliveries)	Percentage nonconforming	p-Chart with variable sample size
On-time entry into the system	(Orders with timely system entry)/(total orders)	Percentage nonconforming	p-Chart with variable sample size
On-time loading	(On-time loaded orders)/(total orders)	Percentage nonconforming	p-Chart with variable sample size
On-time marshalling	(Orders ready on time)/(total orders)	Percentage nonconforming	p-Chart with variable sample size
On-time pick up	(On-time pick-ups)/(total pick ups)	Percentage nonconforming	p-Chart with variable sample size
On-time put away	(Orders with timely put away)/(total orders)	Percentage nonconforming	p-Chart with variable sample size

TABLE 4.8 Cycle Time Metrics Subgroup

Metric	Data	Unit	SPC
Cycle sub-time, distribution/filling	Cycle sub-time. Cycle time at the distribution/filling segment	Time units	x-bar and R charts
Fill rate	(Number of lines filled)/(number of lines requested in order)	Percentage	p-Chart with variable sample size
Stock-to-non-stock ratio	(Material shipped)/(total material in stock). Percentage of material shipped by regular stock	Percentage	p-Chart with variable sample size
Cycle time (full stream)	Cycle time. Total or full stream cycle time. Elapsed time between order entry until cycle completion is visible in the computer system	Time units	x-bar and R charts
Days in inventory by item	(Units in inventory)/(average daily usage)	Days	x-bar and R charts
Cycle sub-time, planning/design	Cycle sub-time. Cycle time at the planning and design segment	Time units	x-bar and R charts
Cycle sub-time, reverse logistics	Cycle sub-time. Cycle time for the returns segment	Time units	x-bar and R charts
Cycle sub-time, sourcing	Cycle sub-time. Cycle time for the sourcing segment	Time units	x-bar and R charts
Point of use deliveries	(Number of deliveries)/(total deliveries)	Percentage nonconforming units	p-Chart with variable sample size
Supplier direct deliveries	(Total number of supplier deliveries)/(total number of deliveries)	Percentage	p-Chart with variable sample size
Throughput rate	(WIP)/(cycle time). WIP is the inventory between start and end points of a product routing	Units/time	x-bar, R charts
Cycle sub-time, transportation	Cycle sub-time. Cycle time for the transit segment	Time units	x-bar and R charts
Expedite ratio	(Number of shipments expedited)/(total number of shipments)	Percentage	p-Chart with variable sample size
Off-line shipments	(Number of off-line shipments)/(total number of shipments). Represents the percentage of off-line shipments	Percentage nonconforming units	p-Chart with variable sample size

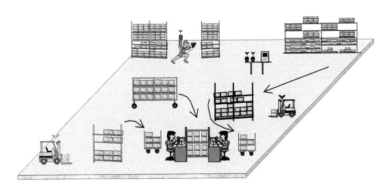

FIGURE 4.3 Case study in electronics manufacturing logistics.

TABLE 4.9 Resource Metrics Subgroup

Metric	Data	Units	SPC
Asset turnover	($ Sales)/($ assets)	Percentage	Individual and MR charts
Asset utilization	(Capacity used)/(capacity available)	Percentage	x-bar and R charts
Cube utilization (load factor)	(Cubic space used)/(cubic space available)	Percentage	x-bar and R charts
Downtime	(Total downtime)/(total available time)	Percentage	x-bar and R charts
Empty miles	(Total empty miles)/(total miles)	Percentage	u-Chart with variable sample size
Empty trailers/containers	(Total empty trailers/containers)/(total trailers/containers)	Percentage	p-Chart with variable sample size
Idleness	(Idle time)/(total available time)	Percentage	x-bar and R charts
Inventory turns	(Sales @ cost)/(average inventory @ cost)	Percentage	Individual and MR charts
Labor utilization	(Total labor used)/(total labor planned to use)	Percentage	Moving centerline EWMA
Material burden	(Good material)/(total material consumed)	Percentage	x-bar and R charts
Network efficiency	(Full enroute miles)/(total miles)	Percentage	u-Chart with variable sample size
Pack rate	(Orders packed)/(employee). Refers to the number of orders packed by a person in a given period of time (minutes, hours, days, etc.)	Packages per employee per unit of time	u-Chart with variable sample size
Pick rate	(Pieces)/(employee), (lines)/(employee), (orders)/(employee). Refers to the number of pieces or orders or lines picked by an employee in a given period of time	Pieces per employee per unit of time, lines per employee per unit of time, orders per employee per period of time	u-Chart with variable sample size
Productivity-on road	(Miles traveled by truck)/(number of days, weeks, etc.). Refers to the number of miles traveled by a truck in a period of time (day, week, etc.)	Miles per period of time	Individual and MR charts
Ratio of inbound to outbound	(Inbound transactions)/(outbound transactions)	Percentage	p-Chart with variable sample size
Receiving rates	(Number of pieces/orders/lines in a given time)/(# employees). Receiving of pieces or orders or lines per employee in a given period of time	Pieces or lines or orders per time unit per employee	u-Chart with variable sample size
Revenue or profit per square foot	(Revenue or profit)/(total space in square feet)	$ per square foot	u-Chart with variable sample size
Revenue per associate	(Total revenues)/(total number of associates)	$ per associate	u-Chart with variable sample size
Shipments per associate	(Number of shipments)/(number of associates)	Shipments per associate	u-Chart with variable sample size
Shipping rate	(Number of pieces/orders/lines in a given time)/(# employees). Shipping of pieces or order or lines per employee in a given period of time	Pieces or lines or orders per time unit per employee	u-Chart with variable sample size
Trailer turns	(Trips)/(period of time). Refers to the total number of trips by a trailer in a given period of time (day, week, month, etc.)	Turns per period of time	Individual and MR charts
Trailer/tractor ratio	(Number of trailers)/(number of tractors)	Trailers per tractor	u-Chart with variable sample size

information systems at the company and can be used to move the organization to an environment that views logistics performance with respect to natural process variation. The facility is positioned to identify areas of excellence in current performance as well as opportunities for improvement.

4.5.1 Point-of-Use/Pull System

One of the processes at the facility is using point-of-use material presentation and pull-logic material flow. Management desires to monitor on-time delivery and accuracy of order filling for components supplied from the stockroom (or supplier) to the shop floor. The layout of the process and the available data resources are identified as follows.

Components are kept in a stock room that fills demand (in varying quantities) to seven different shops. Within each shop, there are associated delivery zones (dz). Each dz has an ESA, consisting of carton flow-rack stock points for components set up as a two bin system with working bins and reserve bins. Both bins have equal quantities of the same product as identified by their product number, or stock keeping unit (SKU). If the SKU from the working bin is depleted, an order is filled from the reserve bin. The box that has been used as a reserve is then moved forward to become the working bin.

When a reserve bin is moved forward to the working bin, a worker scans the barcode of the SKU and places a magnetic sticker on the bin beside the SKU barcode. The magnetic sticker indicates that the part needs to be restocked. The scan triggers a signal to the stockroom computer that notifies the stockroom personnel that the SKU should be restocked. The maximum desired timeframe for the SKU to be restocked is 4 h. When the SKU is delivered to the dz by the stockroom, a second scan is performed. This second scan triggers a signal to the stockroom indicating that the part has been delivered, and this signal marks the time of delivery. If the SKU is delivered within 4 h of the first scan, the action of delivery is considered on-time or good performance. On the other hand, if the SKU is not delivered within the timeframe, the delivery is considered past due. The elapsed time between scans is the total delivery time and is an important performance metric for this process.

If an SKU cannot be delivered from the stock room after the first scan, because it is out of stock, the system automatically sends an order signal to the outside supplier. When an order for the SKU is placed with a supplier, the part is expected to be delivered within five days. Deliveries within this timeframe are considered successful. If the delivery time exceeds five days, the performance is poor and the order is considered short and remains an open transaction. Every Monday the total number of parts short (i.e., open transactions) is collected. On Friday, the system is checked for the number of transactions that have been closed during the week. The difference between the two numbers (open transactions on Monday, closed transactions on Friday) is considered the total number of shortages. Number of shortages is a metric indicating the quality of the logistics system. The number of shortages divided by the number of SKUs within a dz is the performance metric of choice. Due to limitations of the data system, shortages occurring between Tuesday and Sunday are not reported until the following Monday. Therefore, there is a time lag in reporting shortages, and the reported shortages do not necessarily match orders being filled.

A planner is responsible for an assigned set of SKUs. There are 11 planners in the facility. If an SKU order should be placed with a supplier, the associated planner is responsible for the placement of the order and the final delivery of the part to the stock room. The performance of each planner is based on the number of open transactions. This performance metric should be monitored at the planner, shop, and facility levels.

Additionally, the size of the bins is related to the SKU volumetrics. If the packaging is modified by the vendor, the new package may not fit in its designated bin. Therefore, there is a need for monitoring the exceptions to standard packaging.

4.5.2 Performance Metrics

Three key metrics are identified, in view of available data resources, to monitor performance within the facility: delivery time, shortages, and standard packaging.

4.5.2.1 Delivery Time

Delivery time is the metric for monitoring time required to move SKUs from the stock room to the different dz. The target lead time for this operation is 4 h. The shop monitors the number of orders exceeding the 4 h requirement on a per-shift basis. Delivery time is transformed to attribute form at the facility. Attribute data implies that there are two possible events: success or failure. Failure in this case means that the lapsed time between the first scan (need for a SKU to be restocked) and the second scan (SKU restocked) exceeds 4 h. A p-chart with variable sample size is the preferred SPC method for tracking this attribute data. In the case of delivery time performance, the p-chart is used to monitor the percentage of deliveries made within 4 h. These data are collected automatically from the company's database. The formulas used to calculate a p-chart with variable sample size are given in Table 4.4. Figure 4.4 shows the performance of shop 1 for a particular month, and Figure 4.5 shows the performance of the first shift of shop 1 for the same month.

The centerline (CL) in Figure 4.4 is calculated as follows: the total number of nonconforming deliveries (173) is divided by the total number of samples (396) or 0.437. The data being plotted represents the fraction p of nonconforming deliveries. For the first sample the fraction nonconforming is equal to the total nonconforming deliveries for the sample (7) divided by the total number of deliveries for that sample (25). The 3-sigma control limits are calculated by placing control limits at three standard deviations beyond the average fraction nonconforming. For example, the control limits for sample one are:

$$LCL_1 = 0.437 - 3 * \sqrt{\frac{0.437 * (1 - 0.437)}{25}} = 0.437 - 0.297 = 0.139$$

If the LCL for any given sample is smaller than zero, then the value of the LCL is truncated to zero.

$$UCL_1 = 0.437 + 3 * \sqrt{\frac{0.437 * (1 - 0.437)}{25}} = 0.437 + 0.297 = 0.734$$

The resulting control limits and raw data are plotted in Figure 4.4.

The data are also used to monitor the performance of the shop on a per-shift basis. The calculation of the 3-sigma control limits is done in the same way as those for the shop performance. For example, for sample 15 on the first shift, the fraction nonconforming p is $3/15 = 0.2$. The fraction nonconforming, as well as the 3-sigma control limits, for each sample of the data on a per-shift basis are presented in Figure 4.5.

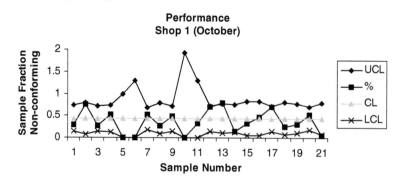

FIGURE 4.4 p-chart for attributes (shop 1).

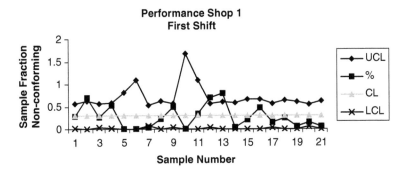

FIGURE 4.5   p-chart for attributes (shop 1, first shift).

As can be seen in Figures 4.4 and 4.5, the points exceeding the UCL indicate a lack of stability in the process. The source of the nonrandom pattern should be determined and eliminated. The points plotting outside control limits require investigation, with the cause assigned and eliminated. Once the cause is eliminated, the associated points are no longer considered in the calculations, revised limits are calculated and the new plot is inspected for points plotting outside limits. Only extended in-control performance can be used to judge the capability of the process.

4.5.2.2   Shortages

The number of shortages is used to monitor the number of open transactions. The number of shortages is a performance metric that indicates the quality of the supply side of logistics systems and is readily available from data sources. The performance should be evaluated on planner, shop and facility levels. The later performance metric provides an aggregate view of all the combined shops, implying both the necessary horizontal as well as vertical dimensions of a balance PM system as suggested in Harp et al. [26].

The number of shortages, like the delivery time, is transformed to the attribute form for the facility. An open order must be closed within five days. Failure in this case is the failure to close an open order within the five-day time frame. A p-chart is recommended to track the percentage of open transactions. Since each SKU is assigned to different planners, a p-chart is allocated to each planner. Planner performance is based upon the percentage of open transactions to the total number of SKUs assigned to the planner. Figure 4.6 shows a p-chart used to monitor the number of shortages of planner 5. Figure 4.7 shows the p-chart used to monitor open transactions at the aggregate level.

Since each shop has a specific number of SKUs, a p-chart is also used to track the percentage of open transactions within a shop (ratio of open transactions to the total number of SKUs in a shop). In Figure 4.0,

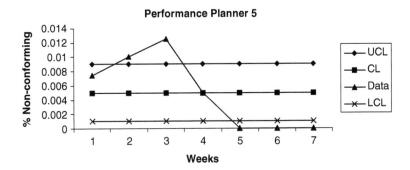

FIGURE 4.6   p-chart for attributes (planner 5).

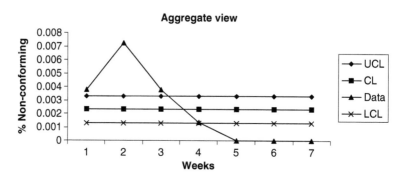

FIGURE 4.7 p-chart for attributes (aggregate).

the p-chart is used to monitor the number of open transactions for shop 2. Furthermore, open transactions per shop should be monitored on an aggregate view as shown in Figure 4.9.

The points plotting outside the UCL indicate lack of stability in the process. Those points must be investigated and assigned to a cause that should be eliminated. After the points associated with this cause are eliminated from the calculation, new revised limits are calculated and plotted. The new plot is inspected for stability.

4.5.2.3 Exceptions to Standard Packaging

Exceptions to standard packaging are also monitored for the process. This data presents the proportion of exceptions to standard packaging by shop.

Since management is interested only in the number of incorrect packaging incidents in relation to the total number of packages, a Pareto chart is recommended to monitor standard packaging. The Pareto chart for nonconforming packaging across all seven shops is shown in Figure 4.10. The data are categorized and ranked showing the cumulative percentage of incorrect packaging incidents by shop. The percentages are obtained by dividing the number of incorrect packaging incidents per shop by the total number of incidents. As a histogram showing the frequency of root causes, the Pareto chart is helpful in prioritizing corrective action efforts. The Pareto chart is used to identify major causes of phenomena like failures, defects, delays, etc. If a Pareto diagram is used to present a ranking of defects over time, the information is useful for assessing the trend of individual defects, frequency of occurrence, and the effect of corrective actions.

Intuitively, the shops with more SKUs will have a greater percentage of incorrect packaging incidents. However, as can be seen in Figure 4.10, shop 5 has the second greatest percentage of wrong packages even though it has the second smallest number of SKUs. The combination of Pareto charts and trend

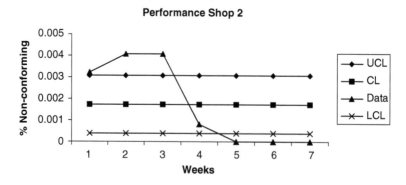

FIGURE 4.8 p-chart for attributes (shop 2).

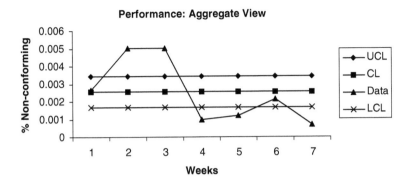

FIGURE 4.9 p-chart for attributes (aggregate).

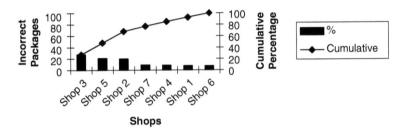

FIGURE 4.10 Pareto chart for nonconforming packaging (shops 1–7).

charts will provide the benefit of a better analysis tool, because the trend chart provides a tool for monitoring the process in view of its natural variation.

References

1. Gerards, G., ten Broeke, A.M., Kwaaitaal, A., vander Muelen, P.R.H., Spijkerman G., Vegter, K.J., Willemsen, J.Th.M. Performance Indicators in Logistics. *Approach and Coherence*, IFS Publications: UK, 1989.
2. CSCMP and Supply Chain Visions. *Supply Chain Management Process Standards: Enable Processes*, Council of Supply Chain Management Professionals: Oak Brook, IL, 2004.
3. Mentzer, J.T., Ponsford, B. An Efficiency/Effectiveness Approach to Logistics Performance Analysis. *Journal of Business Logistics*, vol. 12, no. 1, 1991, pp. 33–61.
4. Trimble, D. How to Measure Success: Uncovering the Secrets of Effective Metrics. Online Learning Center, Sponsored by ProSci, 1996, available at: http://www.prosci.com/metrics.htm. (accessed March 22, 2006).
5. CSCMP and Supply Chain Visions. *Supply Chain Management Process Standards: Deliver Processes*, Council of Supply Chain Management Professionals: Oak Brook, IL, 2004.
6. Frazelle, E.H., Hackman, S.T. The Warehouse Performance Index: A Single-Point Metric for Benchmarking Warehouse Performance, MHRC Final Report, #MHRC-TR-93-14, 1993.
7. Frazelle, E. *World Class Warehousing and Material Handling*, McGraw-Hill: New York, 2002.
8. Keebler, J.E., Durtsche, D.A. *Keeping Score: Measuring the Business Value of Logistics in the Supply Chain*, Council of Logistics Management: Oak Brook, IL, 1999.
9. Caplice, C., Sheffi, Y. Review and Evaluation of Logistics Metrics. *The International Journal of Logistics Management*, vol. 5, no. 2, 1994, pp. 11–28.

10. Caplice, C., Sheffi, Y. Review and Evaluation of Logistics Performance Measurement Systems. *The International Journal of Logistics Management*, vol. 6, no. 1, 1995, pp. 61–74.

11. Chow, G., Heaver, T.D., Henriksson, L.E. Logistics Performance: Definition and Measurement. *International Journal of Physical Distribution and Logistics Management*, vol. 24, no. 1, 1994, pp. 17–28.

12. Andersson, P., Aronsson, H., Storhagen, N.G. Measuring Logistics Performance. *Engineering Costs and Production Economics*, vol. 17, 1989, pp. 253–262.

13. Chan, F.T.S. Performance Measurement in a Supply Chain. *International Journal of Advanced Manufacturing Technology*, vol. 21, 2003, pp. 534–548.

14. Donsellar, K.V., Kokke, K, Allessie, M. Performance Measurement in the Transportation and Distribution Sector. *International Journal of Physical Distribution and Logistics Management*, vol. 28, no. 6, 1998, pp. 434–450.

15. A.T. Kearney, *Measuring and Improving Productivity in Physical Distribution*: The Successful Companies Physical Distribution Management: Chicago, IL, 1984.

16. Keebler, J.E., Manrodt, K.B. The State of Logistics Performance Measurement. *Proceedings of the Annual Conference,* Council of Logistics Management: Oak Brook, IL, 2000, pp. 273–281.

17. Legeza, E. Measurement of Logistics-Quality. *Periodica Polytechnica Series on Transportation Engineering*, vol. 31, no. 1–2, pp. 89–95.

18. Lockamy, A., Cox, J. *Reengineering Performance Measurement*, Irwin Professional Publishing: New York, 1994.

19. Mentzer, J.T., Konrad, B.P. An Efficiency/Effectiveness Approach to Logistics Performance Analysis. *Journal of Business Logistics*, vol. 12, no. 1, 1999, pp. 33–61.

20. Rafele, C. Logistics Service Measurement: A Reference Framework. *Journal of Manufacturing Technology Management*, vol. 15, no. 3, 2004, pp. 280–290.

21. Kaplan, R.S., Norton, D.P. *The Balanced Scorecard: Transforming Strategy into Action,* Harvard University Press: Cambridge, MA, 1996.

22. Hill, J. (Ed.) Using the Balanced Scorecard in the DC. WERC Sheet, Warehousing Education and Research Council, July 2005, pp. 3–4.

23. Novak, R.A., Thomas, D.J. The Challenges of Implementing the Perfect Order Concept. *Transportation Journal*, vol. 43, no. 1, 2004, pp. 5–16.

24. Euske, K.J. *Management Control: Planning, Control, Measurement, and Evaluation*, Addison-Wesley Publishing Co., Menlo Park, CA, 1984.

25. Byers, J.E., Landers, T.L., Cole, M.H. A Framework for Logistics Systems Metrics. TLI Final Report, #TLI-MHRC-96-6, 1996.

26. Harp, C., Buchanan, J., English, J.R., Malstrom, E.M. Design of Group Metrics for the Evaluation of Logistics Performance. TLI Final Report, #TLI-MHRC-97-6, 1997.

27. Boyd, L.H., Cox, J.F. A Cause and Effect Approach to Analyzing Performance Measures, *Production and Inventory Management Journal,* Third Quarter, 1997, pp. 25–32.

28. Goldratt, E. *It's Not Luck*, North River Press: Great Barrington, MA, 1994.

29. Harp, C., Alsein, M., English, J.R., Malstrom, E.M. Quality Monitoring at UPS, TLI Final Report, #TLI-MHRC-97-6, 1997.

30. Mendoza, A., English, J.R., Cole, M.H. Monitoring and Evaluation of Performance Metrics. TLI October Workshop, Summary Report, Arkansas, 1997.

31. Montgomery, D.C. *Introduction to Statistical Quality Control*, 3rd Edition, John Wiley & Sons, Inc.: New York, 1996.

32. *Statistical Quality Control Handbook*, Western Electric Corporation: Indianapolis Ind., 1956.

5

Logistics as an Integrating System's Function

Benjamin S. Blanchard
Virginia Polytechnic Institute and
State University

5.1   Introduction

The objective of this chapter is to view logistics from a total system's perspective (i.e., the "total enterprise") and within the context of its entire life cycle, commencing with the initial identification of a "need" and extending through system design and development, production and/or construction, system utilization and sustaining support, and system retirement and material recycling and/or disposal.

Historically, logistics has been viewed in terms of activities associated with physical supply, materials flow, and physical distribution, primarily associated with the acquisition and processing of products through manufacturing and the follow-on distribution of such to a consumer (customer). The emphasis has been on relatively small consumable components and not on "systems" as an entity. More recently, the field of logistics has been expanding to greater proportions through the development of supply chains (SCs) and implementation of the principles and concepts of supply chain management (SCM), with logistics being a major component thereof. Even with such growth and redefinition, the emphasis has continued to be on the processing of relatively small components in relation to manufacturing and production processes and the establishment of associated supplier networks. The issues dealing with initial system and/or product design, system utilization and sustaining life-cycle support, and system retirement and material recycling and/or disposal have not been adequately addressed within the current spectrum of logistics.

An objective and challenge for the future is to address logistics in a much broader context, reflecting a total system's approach. The interfaces and interaction effects between the various elements of logistics and the many other functional elements of a system are numerous and their interrelationships could have a great impact on whether or not a given system will be able to ultimately accomplish its intended mission successfully. In this context, logistics and its supporting infrastructure, considered as a major element of a total system, can provide an effective and efficient integrating function. Further, there are logistics requirements in all phases of a typical system life cycle, and this integrating function must be life-cycle oriented as design and management decisions made in any one phase of the life cycle can have a significant impact on the activities in any other phase. Thus, it is important to address this logistics integrating function within the context of the "whole" in order to be life-cycle complete; that is, the implementation of a system's life-cycle approach to logistics.

5.2   Logistics—Total "System's Approach"

In defining a *system*, one needs to consider all of the products, processes, and activities that are associated with the initial development, production, distribution, operation and sustaining support, and ultimate retirement and phase-out of the system and its elements. This includes not only those procurement and acquisition functions that provide the system initially, but those subsequent maintenance and support activities that enable the system to operate successfully throughout its planned period of utilization. Thus, the make-up of a "system" should include both the prime elements directly related to the actual implementation and completion of a specific mission scenario (or series of operational scenarios) and those sustaining logistics and maintenance support functions that are necessary to ensure that the specified system operational requirements are fulfilled successfully and in response to some specified customer (consumer) need. Accordingly, the "logistics support infrastructure" should be considered (from the beginning) as a major "subsystem" and addressed as such throughout the entire system life cycle.

Referring to Figure 5.1, the various blocks reflect some of the major activities within the system life cycle. Initially, there is the identification of a specific customer/consumer need, the development of system requirements, and the accomplishment of some early marketing and planning activity (block 1). This leads to design and development, involving both the overall system developer and one or more major suppliers (blocks 2 and 3, respectively). Given an assumed design configuration, the production process commences, involving a prime manufacturer and a number of different suppliers (blocks 4 and 3, respectively). Subsequently, the system is transported and installed at the appropriate customer/user operational site(s), and different components of the system are either distributed to some warehouse or directly to the operational site (blocks 5 and 7, respectively). In essence, there is a forward (or "outward") flow of activities; that is, the flow of activities from the initial identification of a need to the point when the system first becomes operational at the user's site, which is reflected by the shaded areas in Figure 5.1.

In addition to the forward flow of activities as indicated in Figure 5.1, there is also a reverse (or "backward") flow, which covers the follow-on maintenance and support of the system after it has been initially installed and operational at the customer's (user's) site. Referring to Figure 5.1, this includes all activities associated with the accomplishment of on-site or organizational maintenance (block 7), intermediate-level maintenance (block 8), factory and/or depot-level maintenance (blocks 4 and 6), supplier maintenance (block 3), and replenishment of the necessary items to support required maintenance actions at all levels; for example, special modification kits, spares and repair parts and associated inventories, test and support equipment, personnel, facilities, data, information, etc. System "maintenance" in this instance refers to both the incorporation of system modifications for the purposes of improvement or enhancement (i.e., the incorporation of new "technology insertions" throughout the system life cycle), as well as the accomplishment of any scheduled (preventive) and/or unscheduled (corrective) maintenance required to ensure continued system operation. Associated with a number of the blocks as seen in Figure 5.1 are the activities pertaining to the recycling of materials for other applications and/or disposal of such, and the supporting logistics activities as required (e.g., blocks 3, 4, 6, 7, and 8).

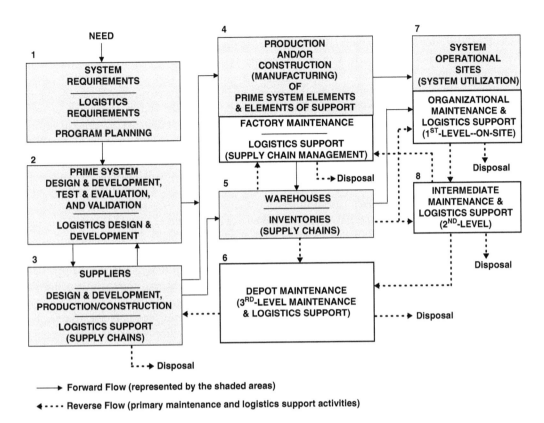

FIGURE 5.1   System operational and logistics support activities.

In the past, the various facets of logistics have been oriented primarily to the "forward" flow of activities shown in Figure 5.1 (i.e., the shaded blocks), and have not addressed the entire spectrum to include the "reverse" flow as well. This, of course, has included the different aspects of "business logistics," emphasized throughout the commercial sector and, more recently, the wide spectrum of activities pertaining to SCs and SCM. More specifically, emphasis has been on (*i*) the initial physical supply of components from the various applicable sources of supply to the manufacturer, (*ii*) the materials handling, associated inventories, and flow of items throughout the production process, and (*iii*) the transportation and physical distribution of finished goods from the manufacturer to the customer's operational site(s). With the advent of SCs and SCM, the physical aspects of logistics have been expanded to include the application of modern business processes, contracting and money flow, information transfer, and related enhancements using the latest electronic commerce (EC), electronic data interchange (EDI), information technology (IT), and associated methods and models.*

In the defense sector, the field of logistics has, for the most part, included a majority of the activities identified within both the forward and reverse activity flows presented in Figure 5.1; that is, the various

---

* An excellent source for material dealing with the various aspects of business logistics, supply chains, and supply chain management is the Council of Supply Chain Management Professionals (CSCMP), 2805 Butterfield Road, Suite 200, Oak Brook, IL 60523 (web site: http://www.cscmp.org). Some good references include: (a) *Journal of Business Logistics (JBL)*, published by CSCMP; (b) Coyle, JJ, E.J. Bardi, and C.J. Langley, *The Management of Business Logistics*, 7th Edition, South-Western, Mason, OH, 2003; and (c) Frazelle, E.H., *Supply Chain Strategy: The Logistics of Supply Chain Management*, McGraw-Hill, New York, NY, 2002.

aspects of business logistics and sustaining system maintenance and support. The principles and concepts of integrated logistic support (ILS), introduced in the mid-1960s, emphasized a total integrated system-oriented life-cycle approach, with such objectives as (*i*) integrating support considerations into system and equipment design, (*ii*) developing support requirements that are related consistently to readiness objectives, to design, and to each other, (*iii*) acquiring the required support in an effective and efficient manner, and (*iv*) providing the required support during the system utilization phase at minimum overall cost. The implementation of ILS requirements greatly expanded the scope of logistics in terms of the entire system life cycle. In recent years, the advent and establishment of SCs and SCM, along with the development and application of appropriate technologies, has expanded the field even further. However, while logistics requirements, as currently being practiced in the acquisition and operation of systems, reflect some definite overall improvement, these requirements have and continue to be addressed primarily "after-the-fact," as an independent entity, and downstream in the life cycle. In other words, logistics requirements have not been treated as a major element of a given system, nor have they been adequately addressed in the design process at a time when the day-to-day technical and management decisions being made have the greatest impact on the resulting logistics and maintenance support infrastructure later on. More recently, this deficiency has been recognized and the principles and concepts of acquisition logistics have been initiated to provide additional emphasis on addressing logistics early in the system design and development process.*

At this point, there is a need to progress to the next step by integrating and implementing the best practices of each; that is, the commercial and defense sectors. More specifically, this can be facilitated by (*i*) addressing logistics from a total system's perspective, (*ii*) considering the logistics support infrastructure as a major element of that system, (*iii*) viewing logistics in the context of the entire system's life cycle, and (*iv*) by properly integrating logistics requirements into the system design process from the beginning.

In responding to the first item, it should be noted that there is both a vertical and horizontal integration process that applies here. First, one must consider a system as being included in somewhat of a "hierarchical structure." For example, there may be a need for an airplane, within the context of a higher-level airline, and as part of an overall regional air transportation capability. Logistics requirements must be properly integrated both upward and downward, as well as horizontally across the spectrum at any level. Further, and in response to second item, the logistics requirements for any given system should be directly supportive of the mission requirements for that system and should evolve from this, and not the reverse. In this context, it is necessary to consider the logistics requirements, at any given level, as a major subsystem and in support of the system-level requirements at that level. Additionally, logistics requirements should be based on the entire life cycle of the system being addressed and, to be meaningful, should be included as an inherent part of the system design process from the beginning. These requirements should be specified from a top-down and/or bottom-up perspective and not just from an after-the-fact bottom-up approach.

5.3   Logistics in the System Life Cycle

While there may be some slight variations relative to specific wording and organization of material, it is assumed that the basic elements of logistics are as shown in Figure 5.2. The intent is the view these overall logistics requirements, from both the commercial and defense sectors, and to integrate such into major categories providing a "generic" approach. Referring to Figure 5.2, logistics requirements stem from higher level system-oriented requirements, and can be properly integrated into what may be referred to as the logistics support infrastructure. Inherent within this integration process is the

---

* A broad spectrum of logistics from an "engineering" orientation is included in: Blanchard, B.S., *Logistics Engineering and Management*, 6th Edition, Pearson Prentice Hall, Upper Saddle River, NJ, 2004.

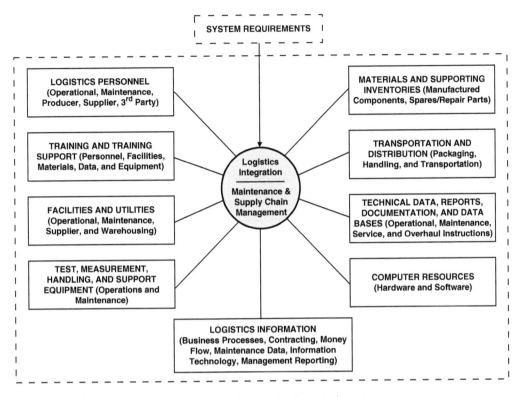

FIGURE 5.2 The "logistics support infrastructure"—a major element of a system.

implementation SC and SCM concepts and principles and the application of analytical techniques and models, EC/EDI/IT methods, and so on as appropriate. Thus, the configuration reflected in Figure 5.2 includes the integration and application of products, processes, personnel, organizations, data and information, and the like, with the objective of ensuring that the system(s) in question can be effectively and efficiently supported throughout its planned life cycle.

The system "life cycle" involves different phases of activity evolving from the initial identification of a need and continuing through system design and development, production and/or construction, system operation and sustaining support, and system retirement and material recycling and/or disposal. While these phases are often considered as being strictly sequential in their relationship to each other, there is actually some concurrency required as illustrated in Figure 5.3.

As indicated in Figure 5.3, the system life cycle goes beyond that pertaining to a specific product. It must simultaneously embrace the life cycle of the production and construction process, the life cycle of the logistics and system support capability, and the life cycle of the retirement and material recycling and disposal process. In this instance (and for the purposes of illustration), there are four concurrent life cycles progressing in parallel, and the top-down and bottom-up interfaces and interaction effects among these are numerous.

The need for the system comes into focus first. This recognition results in the initiation of a formalized design activity in response to the need; that is, conceptual design, preliminary system design, detail design and development, and so on. Then, during early system design, consideration should simultaneously be given to its production. This gives rise to a parallel life cycle for bringing a manufacturing capability into being. As shown in Figure 5.3, and of great importance, is the life cycle of the "logistics support infrastructure" needed to service the system, its production process, the associated material recycling and/or disposal process, and itself. These individual life cycles must be addressed as an

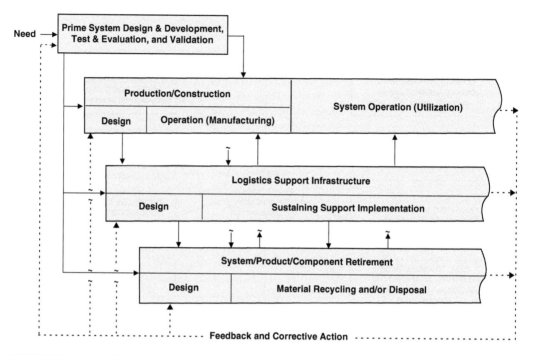

FIGURE 5.3 System life-cycle applications—a concurrent approach.

integrated entity, from a top-down (and then bottom-up) perspective, and each time that a new system need is identified one should evolve through this process. This is not to infer an overall lengthy, redundant, and costly activity, but to an overall process or "way of thinking." The objective is to address all of the producer, supplier, customer, and related activities (and associated resources) necessary in response to an identified need, whenever, wherever, and for as long as required. The "logistics support infrastructure" is an integral part of this requirement, and there are critical logistics activities in each phase of the life cycle.

5.3.1 Logistics in the System Design and Development Phase

Activities in this early phase of the life cycle pertain to design and development of the entire system and all of its elements, and not just limited to design of the prime mission-related components only. This phase commences with the identification of a "need" and evolves through conceptual design, preliminary system design, detail design and development, test and evaluation, and leads to the production and/or construction phase. Inherent within these activities is the accomplishment of a feasibility analysis to determine the best "technical" approach in responding to the stated need, definition of system operational requirements and the maintenance and support concept, accomplishment of functional analysis and requirements allocation, conductance of trade-off studies and design optimization, system test and evaluation, and so on.

An important part of this early system requirements definition process is the establishment of specific quantitative and qualitative technical performance measures (TPMs), to include appropriate performance-based logistics (PBL) factors, as "design-to" requirements; that is, an input to the overall design process in the form of criteria which lead to the selection of components, equipment packaging approaches, diagnostic schemes, and so on. It is at this point when specific system design requirements are initially defined from the top down, providing design guidelines for major subsystems and lower-level elements of the system (including the logistics support infrastructure). These basic early front-end activities, which constitute an iterative process overall, are illustrated in Figure 5.4, with logistics requirements for the entire system life cycle being noted. These activities constitute an integrated composite of design functions for the individual life cycles presented in Figure 5.3.

A prime objective is to design and develop a system that will not only fulfill all of the required "operational" needs, but one that can be supported both effectively and efficiently throughout its planned life cycle. Inherent within the ultimate design configuration are the appropriate attributes (or characteristics) necessary to ensure that the desired functionality, reliability, maintainability, supportability (serviceability), quality, safety, producibility, disposability, and related features, are incorporated, and that the system will "perform" as required. In additional to the system effectiveness side of the spectrum (i.e., the technical characteristics), one must deal with the economic factors as well. These, in turn, must be viewed in terms of the overall system life cycle; that is, life-cycle revenues and cost. If one is to properly assess the risks associated with the day-to-day engineering and management decision-making process throughout system design and development, the issue of "cost" must also be addressed from a total life-cycle perspective; that is, life-cycle cost (LCC). Although individual decisions may be based on some smaller aspect of cost (e.g., item procurement price), the individual(s) involved is remiss unless he or she views the consequences of those decisions in terms of total cost. Decisions made in any one phase of the life cycle will likely have an impact on the activities in each of the other phases.

In addressing the issue of "cost-effectiveness," one often finds a lack of total cost visibility; for example, the unknown factors represented by bottom part of the traditional "iceberg." For many systems, the costs associated with design and development, construction, initial procurement and installation of capital equipment, production, and so on, are relatively well known. We deal with, and make decisions on the basis of these costs on a regular basis. However, the costs associated with the operation (utilization) and sustaining maintenance and support of a system throughout its life cycle are often hidden. This includes not only the initial acquisition and implementation of the "logistics support infrastructure" for a given system, but the sustaining maintenance and support of that infrastructure throughout the system life cycle. The lack of total cost visibility has been particularly notable through the past decade or so when systems have become more complex and have been modified to include the "latest and greatest technology" without consideration of the cost impact downstream. In essence, we have been relatively successful when addressing the short-term aspects of cost but have not been very responsive to the long-term effects.

At the same time, the past is replete with instances where a large percentage of the total LCC for a given system is attributed to the downstream activities associated with system operation and sustaining

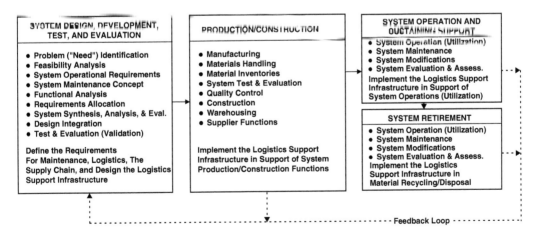

FIGURE 5.4 Logistics in the system life cycle—a system engineering emphasis.

support (e.g., up to 75% for some systems). When addressing "cause-and-effect" relationships, one often finds that a significant portion of this cost stems from the consequences of decisions made during the early phases of planning and design (i.e., conceptual and preliminary system design). Decisions pertaining to the selection of technologies and materials, the design of a manufacturing process, equipment packaging schemes and diagnostic routines, the performance of functions manually versus using automation, the design of maintenance and support equipment, and so forth, can have a great impact on the downstream costs and, hence, LCC. Additionally, the ultimate logistics support infrastructure selected for a system during its period of utilization can significantly affect the cost-effectiveness of that system overall. Thus, including life-cycle considerations in the decision-making process from the beginning is critical. From Figure 5.5, it can be seen that the greatest opportunities for impacting total system cost are realized during the early phases of system design. Implementing changes and system modifications later on can be quite costly.

Historically, logistics has been considered "after-the-fact," and activities associated with the "logistics support infrastructure" have not been very popular in the engineering design community, have been implemented downstream in the life cycle, and have not received the appropriate level of management attention. Although much has been done to provide an effective and efficient system support capability (with the advent of new technologies, the implementation of SC and SCM practices, the application of sophisticated analytical models and methods for analysis and evaluation purposes, etc.), accomplishing all of this after-the-fact can be an expensive approach, as system-level design boundaries have already been established without the benefits of allowing for accomplishment of the proper design-support trade-offs. Logistics requirements have been established as a consequence of design and not an integral part of the process from the beginning. Thus, and with future growth in mind, it is imperative that logistics requirements be (*i*) addressed from inception, (*ii*) established—as top-level system requirements are initially determined during conceptual design, (*iii*) developed through the establishment of design criteria ("design-to" factors) as an input to the overall system design process, (*iv*) properly integrated with the other elements of the system on an iterative basis, and (*v*) considered as an integral part of the engineering process in system design and development. This can best be facilitated through

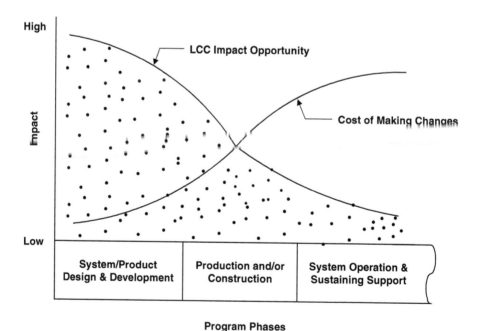

FIGURE 5.5   Activities affecting life-cycle cost.

development and implementation of a logistics engineering function as an integral part of the overall system engineering process.*

"System engineering" constitutes an interdisciplinary and integrated approach for bringing a system into being. In essence, system engineering is "good engineering" with special areas of emphasis: (*i*) a top-down approach that views the system as a whole, versus a bottom-up-only process characteristic of many of the more traditional engineering functions; (*ii*) a life-cycle orientation that addresses all of the phases identified in Figures 5.1, 5.3, and 5.4; (*iii*) the establishment of a good comprehensive system requirement baseline from the beginning; and (*iv*) the implementation of an interdisciplinary and integrated (or team) approach throughout the system design and development process to ensure that all design objectives are addressed in an effective and efficient manner. The system engineering process is iterative and applies across all phases of the life cycle.†

Referring to Figure 5.4, application of the concepts and principles of systems engineering is particularly important throughout the early stages of system design and development (reflected in the first block), with special emphasis on the establishment of system-level requirements. It is during the conceptual design phase when the basic requirements for logistics and system support are first established, one way or another. It is at this stage during the initial determination of system-level requirements when the design criteria (i.e., "design-to" requirements) for the "logistics support infrastructure" are developed, and when the greatest impact on the downstream activities and LCC can be realized. It is at this early stage when logistics engineering activities should be initiated and inherent within implementation of the system engineering process. Referring to the Figure 5.4, a few key system engineering activities, including the development of logistics requirements, are described through the following steps:

1. Problem ("need") identification and feasibility analysis

The system engineering process commences with the identification of a want or desire for something and is based on a real (or perceived) deficiency. For example, the current system capability is not adequate in terms of meeting certain performance goals, is not available when needed, cannot be logistically supported, or is too costly in terms of operation. As a result, a new system requirement is defined along with its priority for introduction, the date when the new system capability is required by the customer (user), and the anticipated resources necessary for acquiring the new system. Through a needs analysis, the basic functions that the system must perform are identified (i.e., primary and secondary), along with the geographical location(s) where these functions are to be performed and the anticipated period of performance. In essence, one must define the "what" requirements (versus the "how"). A complete description of need, expressed in quantitative performance and effectiveness parameters where possible, is essential.

A feasibility analysis is then accomplished with the objective of evaluating the different technological approaches that may be considered in responding to the specified need (i.e., correcting the deficiency). For instance, in the design of a communication system, should one incorporate a fiber-optic, cellular, wireless, or a conventional hard-wired approach? In designing an aircraft, to what extent should one incorporate composite material? In designing a new transportation capability, to what degree should the operation of the various passenger vehicles be automated or accomplished through the use of human operators? In the development of new equipment, should packaging considerations favor "logistics

---

* Logistics and the design for supportability (serviceability), implemented as an integral part of the systems engineering process, are discussed in detail in: Blanchard, B.S. and W.J. Fabrycky, *Systems Engineering and Analysis*, 4th Edition, Pearson Prentice Hall, Upper Saddle River, NJ, 2006.

† There are different definitions and approaches to "system engineering" being implemented today depending on one's background and experience. However, there is a common top-down, life-cycle oriented, interdisciplinary, and iterative theme throughout. A good source for definitions and activities in the field is the *International Council On Systems Engineering* (*INCOSE*), 2150 N. 107th St., Suite 205, Seattle, WA, 98133 (web site: http://www.incose.org).

transport" by air, by waterway, or by ground vehicle? At this point, it is necessary to (*i*) identify the various design approaches that can be pursued to meet the requirements, (*ii*) evaluate the most likely candidates in terms of performance, effectiveness, logistics requirements, and life-cycle economic criteria, and (*iii*) recommend a preferred approach for application. The objective here is to select an overall technical approach, and not to select specific hardware, software, and related system components.

It is at this early point of program inception (reflected by block 1, Fig. 5.4) when logistics engineering involvement in the design process must commence. The questions are, (*i*) What type of a logistics support infrastructure is envisioned? (*ii*) Have the logistics requirements been identified and justified through the appropriate system-level trade-off analysis? (*iii*) Is the approach feasible? The objective is to determine top-level system goals, approach, and general plan for acquisition, and the logistics support infrastructure constitutes a major element of the system in question.

2. *System operational requirements and the maintenance concept*

Once a system need and a technical approach have been identified, it is necessary to develop the anticipated operational requirements further in order to proceed with system design as planned. At this point, the following questions should be asked: What specific mission and associated operational scenarios must the system perform? Where (geographically) and when are these scenarios to be accomplished and for how long? What are the anticipated quantities of equipment, software, people, facilities, etc., required and where are they to be located? How is the system to be utilized in terms of on-off cycles, hours of operation per designated time period, etc.? What are the expected effectiveness goals for the system (e.g., availability, reliability, design-to-LCC, etc.)? What are the expected environmental, ecological, social, cultural, and related conditions to which the system will be subjected throughout its operational life?

The establishment of a comprehensive description of operational requirements from the beginning is necessary to provide a good foundation, or baseline, from which all subsequent system requirements evolve. If one is to design and develop a system to meet a given customer (user) requirement, it is important that the various responsible members of the design team know the mission objectives and just how the system will be utilized to meet these objectives. Of particular interest are the anticipated geographical deployment and the type of operational scenarios to be accomplished. While one certainly cannot be expected to cover all future areas of operation, some initial assumptions pertaining to operational scenarios, anticipated utilization, the stresses that the system is expected to experience, etc., must be made. The question is, how can one accomplish design without having a pretty fair idea as to just how the system will be utilized? This question is particularly relevant when determining the design requirements for reliability, maintainability, supportability (serviceability), and for the logistics support infrastructure. Thus, it is appropriate to develop a few of the more rigorous operational profiles and to design with these in mind. Figure 5.6 provides a partial visualization of what might be included in defining operational requirements.

While all of this may appear to be rather obvious, it is not uncommon for the design community to identify a few of the more easily defined operational requirements, proceed with the design, modify such requirements later on, redesign to meet a changing set of requirements, and so on, which (in turn) can often result in a rather costly process with much time and resources wasted. The objective here is initiate a more thorough and comprehensive approach from the beginning, to provide increased visibility early and identify potential problem areas, to allow for completion of the appropriate trade-offs facilitating an effective and efficient system capability output, and to reduce the risks often inherent throughout the design process. The logistics support infrastructure must be an inherent consideration in this early establishment of system-level requirements.

The system maintenance concept, developed during the conceptual design phase, constitutes a "before-the-fact" series of illustrations and statements pertaining to the anticipated requirements for

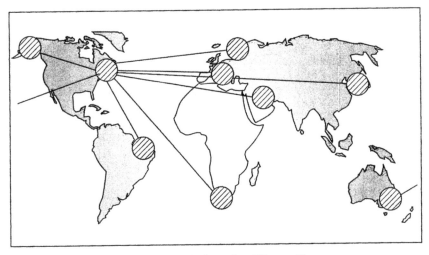

Number of Units in Operational Use per Year

Geographical Operational Areas	Year Number										Total Units
	1	2	3	4	5	6	7	8	9	10	
1. North & South America	–	–	10	20	40	60	60	60	35	25	310
2. Europe (2)	–	–	12	24	24	24	24	24	24	24	180
3. Middle East	–	–	12	12	12	24	24	24	24	24	156
4. South Africa	–	–	12	24	24	24	24	24	24	24	180
5. Pacific Rim 1	–	–	12	12	12	24	24	24	12	12	132
6. Pacific Rim 2	–	–	12	12	12	12	12	12	12	12	96
Total	–	–	70	104	124	168	168	168	131	121	1,054

Average Utilization: 4 Hours per Day, 365 Days per Year

FIGURE 5.6 System operational requirements (overall profile).

system support throughout the life cycle. The objective is to address the following questions: What logistics and maintenance support requirements are anticipated for the system throughout its life cycle? Where (geographically) and when must these support activities for the system be accomplished? To what depth (in the design of the system and its hierarchical structure) should maintenance and support be accomplished? To what level(s) should maintenance and support be accomplished (organizational, intermediate, depot, manufacturer, supplier, third-party, etc.)? Who (what organizations) will be responsible for maintenance and support at each level? What are the "design-to" effectiveness requirements for the logistics support infrastructure (e.g., availability, logistics response time, material processing time, reliability of transportation, total logistics cost, etc.)? What are the expected environmental conditions to which the system will be subjected during the performance of logistics and maintenance support functions?

Referring to Figure 5.2, the objective is to address all of the major logistics and maintenance support activities associated with both the forward and reverse flows as illustrated. These activities need to be projected further and in the context of the operational requirements for the system in question. Figure 5.7 is included as an extension to the operational requirements illustrated in Figure 5.6.

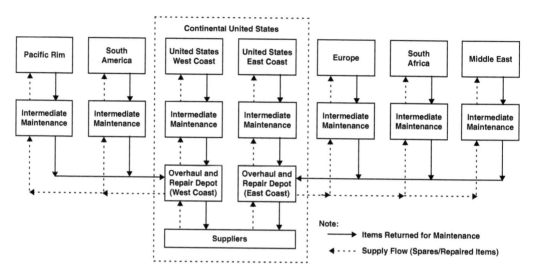

FIGURE 5.7   Top-level system maintenance and support infrastructure.

Whereas in the past these activities were primarily considered after-the-fact and further downstream in the life cycle, the objective here is to attempt to respond to the above questions at an early stage, promote life-cycle thinking early as indicated in Figure 5.5, identify potential high-risk areas that may require special attention, and to build in the logistics support infrastructure into the system design process in a timely manner. The objective is to foster early front-end "visibility" even though it may be difficult (if not impossible) to define all of the basic requirements at this time.

3. System TPMs

Evolving from the definition of system, operational requirements and the maintenance concept is the identification and prioritization of key quantitative performance ("outcome") factors. The objective is to establish some specific "design-to" quantitative requirements as an input to the design and development process, as opposed to waiting to see how well the system will perform after the basic design has been completed. Historically, such requirements for specific equipment items, software packages, etc., have been covered partially through the specification of selected performance factors such as speed, throughput, range, weight, size, power output, accuracy, frequency, and so on. However, in most cases, the specification of higher level performance requirements for the system overall have not been specified. For example, to what level of operational availability should the system be designed to meet? To what level of effectiveness must the logistics support infrastructure be designed in order to meet the required availability requirement(s) for the system? To what level of LCC should the system be designed to meet?

By addressing only lower level requirements for any given system, there often is the tendency to optimize design at the element level in a given system hierarchy, while at the same time suboptimizing the requirements for the system overall. Thus, it is imperative that commencing with the definition of requirements at the system level, considering the various applicable mission scenarios, constitutes a critical early step in accomplishing the activities shown in the first block, Figure 5.4. Further, these early requirements for the system form the basis for establishing lower level requirements for design of the logistics support infrastructure.

4. System functional analysis and requirements allocation

The functional analysis constitutes a complete description of the system in "functional" terms. This includes an expansion of all of the activities and processes accomplished through the forward and reverse flows illustrated in Figure 5.1. A function refers to a specific or discrete action (or series of actions) that is necessary to achieve a given objective; that is, an operation that the system must perform to accomplish its mission, a logistics activity that is required for the transportation of material, or a maintenance action that is necessary to restore the system for operational use. Such actions will ultimately be accomplished through the use of equipment, software, people, facilities, data, or various combinations thereof. However, at this point, the objective is to specify the "Whats" and not the "Hows"; that is, what needs to be accomplished versus how it is to be done. The functional analysis is an iterative process, commencing with the initial identification of a consumer need, of breaking requirements down from the system-level, to the subsystem, and as far down the hierarchical structure as necessary to identify input design criteria and/or constraints for the various elements of the system.*

Referring to Figure 5.4, the functional analysis may be initiated in the early stages of conceptual design as part of the problem (need) identification and feasibility analysis task, and can be expanded as required in the preliminary system design phase. Through the development of system operational requirements, operational functions are identified and expanded as shown in Figure 5.8. These operating functions lead to the identification of maintenance and support functions as illustrated at the bottom of the figure. The identified maintenance and support functions also constitute an expansion of the established maintenance concept. Development of the functional analysis can best be facilitated through the use of functional flow block diagrams (FFBDs), as illustrated through the expanded integrated flow presented in Figure 5.9.

Referring to Figure 5.1, logistics requirements can initially be identified by describing the specific functions to be accomplished in progressing from block 3 to block 4, from block 4 to blocks 5 and 7, and from block 7 backward to blocks 8, 6, 4, and 3, respectively. This may include a procurement function, material processing function, packaging and handling function, transportation function, warehouse storage function, maintenance function, communication function, data transmission function, and so on. The objective is to identify all of the basic functions that must be accomplished by the logistics support infrastructure for the system being addressed. Accomplishing such at this point in the life cycle enables early "visibility" which will allow for the incorporation of any necessary design changes easily and economically.

Given a good comprehensive functional description of the system, the next step is to commence with the identification of the specific requirements for hardware, software, people, facilities, data, and/or various combinations thereof. The process is to analyze each of the major blocks in the appropriate FFBD to determine the resource requirements necessary for the performance of the function in question. There are input factors, expected output requirements, controls and/or constraints, and mechanisms, which must be determined. Through the accomplishment of design trade-offs, the best mix of resource requirements (e.g., hardware, software, people, etc.) for each function can be determined. These resources can then be combined, integrated, and lead to the identification of the various lower level elements of the system, as illustrated in Figure 5.10.

The process of breaking the system down into its elements is accomplished by partitioning. Common functions are grouped, or combined, so as to provide a system packaging scheme with the following objectives in mind: (*i*) system elements may be grouped by geographical location, a common environment, or by similar types of equipment or software; (*ii*) individual system "packages" should be as independent as possible with a minimum of "interaction effects" with other packages; and (*iii*) in breaking down a system into subsystems, select a configuration in which the "communication" (i.e., negative interaction effects) between the different subsystems is minimized. An overall design objective is to divide the system into elements such that only a very few (if any) critical events can influence or change the inner workings of the various packages that make up the overall system architecture. This leads to an open-architecture approach to

---

* In applying the principles of system engineering, one should not identify or initiate the purchase of one piece of equipment, or module of software, or data item, or element of logistics support without first having justified the need for such through the functional analysis. On many projects, items are often purchased early based on what is perceived as a "requirement" but later determined as not being necessary. This practice can turn out to be quite costly.

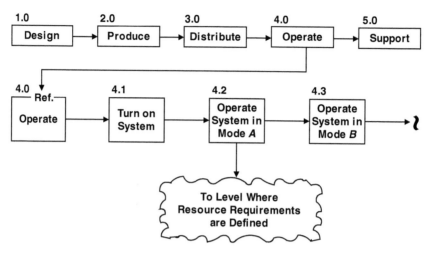

Functional Flow Block Diagram (Partial)

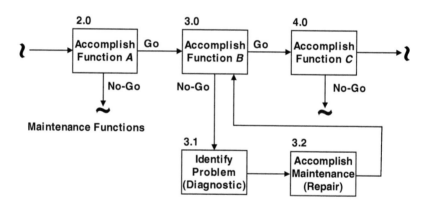

Transition from Operational Functions to Maintenance Functions

FIGURE 5.8   Functional flow diagrams (example)

design which, in turn, should facilitate the incorporation of system changes, technology insertions, and future improvements later on in the life cycle without causing a major configuration redesign.*

Referring to Figure 5.10, the question is, given the requirements for the system (stated in quantitative terms), what specific design-to requirements should be specified for Unit A, Unit B, logistics support infrastructure, transportation and distribution, facilities, and so on? For instance, if there is an Operational Availability (Ao) requirement of 0.90 for the system as an entity, what should be specified for the logistics support infrastructure in order to meet the system-level requirement? If, on the other hand, the system availability requirement is 0.998, then the requirements for the logistics support infrastructure may be different.

---

* Refer to Blanchard, B.S., *Logistics Engineering And Management*. 6th Edition, Pearson Prentice Hall, Upper Saddle River, NJ, 2006, Chapter 4, pp 150–172.

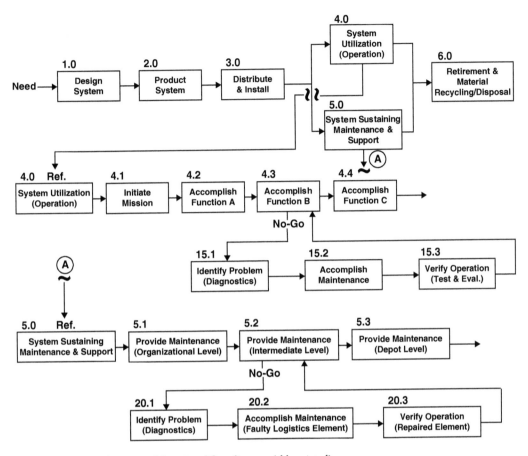

FIGURE 5.9   Partial integrated functional flow diagram (abbreviated).

With regard to the logistics support infrastructure, the objective is to establish some specific design-to goals early (before the fact) and develop a balanced configuration that will best respond to the overall system-level requirements, rather than wait until the design is relatively "fixed" and then have to live with the results. One key performance measure of concern is the overall availability of the logistics support capability, another is logistics response time, a third is total logistics cost (TLC) or the cost per logistics support action, and so on. Top level requirements must then be allocated (or apportioned) downward to the level necessary for providing a good and meaningful input for the design. An example of a few design to goals are noted here:

1. The response time for the logistics support infrastructure shall not exceed four hours.
2. The procurement lead time for the acquisition of any given component shall not exceed 48 hours.
3. The reliability of the overall transportation capability shall be 0.995, or greater.
4. The transportation time between the location where on-site (organizational) maintenance is accomplished and the intermediate-level maintenance shop shall not exceed eight hours.
5. The probability of spares availability at the organizational level of maintenance shall be at least 95%.
6. The warehouse utilization rate shall be at least 75%.
7. The mean time between maintenance (MTBM) for the logistics support infrastructure shall be 1000 or greater.
8. The time for processing logistics information shall not exceed 10 min.

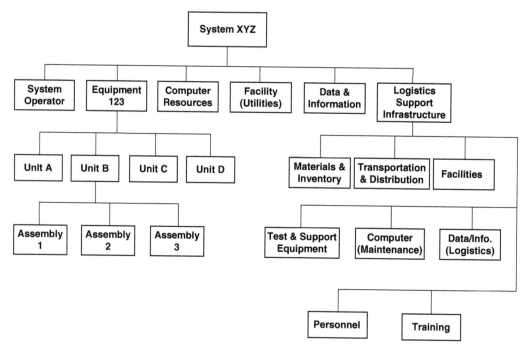

FIGURE 5.10   Hierarchy of system elements.

9. The processing time for removing an obsolete item from the operational inventory shall not exceed 12 hours, and the cost per item processed shall be less than "x" dollars.
10. The TLC for the logistics support infrastructure shall not exceed "y" dollars per support action.

Referring to Figure 5.11, one can visualize the traceability of requirements from the top-down in order to meet such for the overall system as an entity, and performing this function at an early stage in the life cycle will facilitate the accomplishment of the necessary trade-offs and analyses, hopefully leading to an effective and efficient logistics support infrastructure capability. The specific quantitative "design-to" requirements must, of course, be tailored to the overall system-level requirements.

5. System synthesis, analysis, and evaluation

Referring to Figure 5.1, given a set of input requirements from the beginning, there is an iterative and continuous process of synthesis, analysis, and evaluation, which ultimately leads to the development of an effective and efficient logistics support infrastructure configuration. For instance, at this point decisions are made pertaining to specific procurement policies, outsourcing requirements, material handling methods, selection of packaging and transportation modes, determination of inventory levels and warehousing locations, establishment of SCs, application of automation techniques, development of information processing and database requirements, determining maintenance levels of repair, and so on. Accomplishing these design-related analyses is facilitated through the selective application of the many and various operations research (OR) models or tools discussed throughout the other chapters of this handbook and in the literature.

6. System design integration

System design begins with the identification of a customer (consumer) need and extends through a series of steps as noted in Figure 5.4. Design is an evolutionary top-down process leading to the definition of a

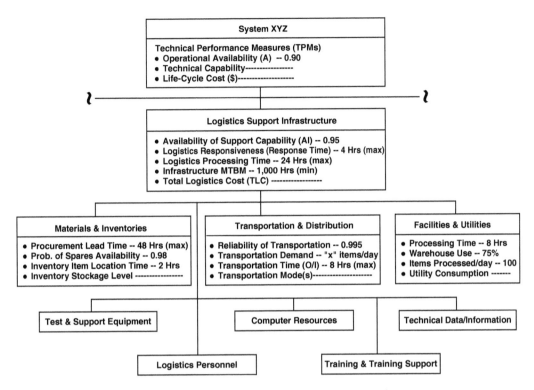

FIGURE 5.11 Allocation of technical performance measures for logistics (example).

functional entity that can be produced, or constructed, with the ultimate objective of delivering a system that responds to a customer requirement in an effective and efficient manner. Inherent within this process is the integration of many different design disciplines, as well as the proper application of various design methods, tools, and technologies. Figure 5.12 provides an example showing many of the different design characteristics that must be considered and properly integrated in order to meet the specified requirements at the system level.

Effective design can best be realized through implementation of the system engineering process. Logistics engineering must be an integral part of this process, along with other design disciplines as applicable (e.g., electrical engineering, industrial engineering, mechanical engineering, reliability engineering, etc.). The role of logistics engineering is twofold: (*i*) to ensure that the prime mission-related elements of the system are supportable (serviceable) through the incorporation of the proper design characteristics or attributes; and (*ii*) to design the logistics support infrastructure to provide the life-cycle support required. In this capacity, logistics engineering can serve as a design integration function across the broad spectrum of the system and throughout its development.

5.3.2 Logistics in the Production and/or Construction Phase

Referring to the four life cycles in Figure 5.3, system-level design requirements (including those for the logistics support infrastructure) evolve from the first life cycle which, in turn, provides an input for the three lower level life cycles. The more "traditional" logistics requirements and associated SCs, particularly those in the commercial sector, have evolved primarily around the second life cycle. There are logistics engineering functions associated with the design and evaluation of the production or construction process, the development of supplier requirements and supply chains, the development of distribution and

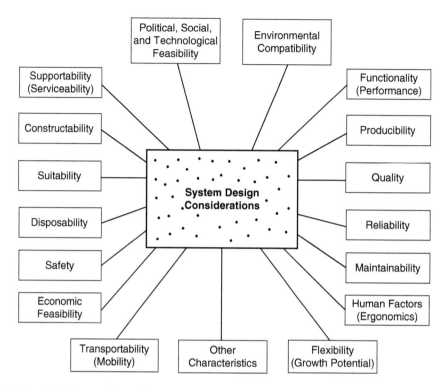

FIGURE 5.12   Typical system design characteristics.

warehousing requirements, and so on. However, these requirements must be properly integrated within the context of the whole; that is, the entire spectrum of activity illustrated in Figure 5.1 and the four life cycles shown in Figure 5.3.*

5.3.3   Logistics in the System Operation and Sustaining Support Phase

Throughout the system operational or utilization phase (refer to Fig. 5.4), logistics functions will include providing the necessary support in response to:

1. Changes in system-level requirements and/or when new technologies are inserted for the purposes of enhancement. Each time when a new **requirement** evolves, the system engineering process is implemented as appropriate; that is, there will be some redesign effort, synthesis and analysis, test and evaluation, etc. Such system-level changes will usually result in changes not only involving the prime mission-related elements of the system but the logistics support infrastructure as well.
2. Scheduled and unscheduled maintenance activities for the system and its elements as required. This will involve procurement functions, material handling tasks, transportation and distribution activities, maintenance personnel and facilities, etc. Logistics activities in this area are reflected by the reverse flow in Figure 5.1.

---

* Logistics in the more "traditional" sense refers to the wide spectrum of activities described in the literature and taught primarily in "business-oriented" programs in the academic community. Such coverage is also described in the other chapters of this handbook. The emphasis herein is to integrate these activities from a system's perspective and within the context of its entire life cycle.

While logistics activities for system life-cycle support are often not properly addressed from the beginning, the need for such is indeed essential if the system is to ultimately accomplish its planned mission, both effectively and efficiently.

5.3.4 Logistics in the System Retirement and Material Recycling/Disposal Phase

Referring to the fourth life cycle in Figure 5.3, logistics requirements for this phase pertain to the:

1. Retirement and phase-out of system components from the inventory throughout the system operational phase, and the subsequent recycling of these items for other uses and/or for disposal. This function is supplemental to those activities presented in Section 5.3.3.
2. Support required when the system (and all of its elements) is no longer needed and is ultimately retired from the operational inventory. This function relates to the recycling and/or disposal of components, the refurbishment of land and facilities for other uses, related data and documentation, and so on.

While this phase of the life cycle is often ignored altogether, the logistics requirements can be rather extensive here, particularly if new facilities are required (for the purposes of material decomposition), new ground handling equipment is needed, special environmental controls are necessary, and so on. Again, the anticipated logistics requirements here must be addressed from the beginning; that is, in conceptual design along with the many other requirements pertaining to the system overall.

5.4 Summary and Conclusions

As a prerequisite to determining the specific logistics requirements for any given system, a good understanding of the overall environment is necessary; that is, the geographical location where the system is likely to be deployed and utilized, nature and culture of the operating agency or organization (the "user"), availability of appropriate technologies and associated resources, system procurement and acquisition processes, political structure, and so on. Additionally, it should be recognized that systems today are operating in a highly "dynamic" world and the need for agility and flexibility is predominant.

While individual perceptions on today's challenges will differ depending on personal experiences and observations, there are a number of trends that appear to be significant. For example, there is more emphasis today on total systems versus the components of systems; the requirements for systems are constantly changing; systems are becoming more complex with the continuous introduction of new technologies; the life cycles of many current systems are being extended for one reason or another while, at the same time, the life cycles of most technologies are becoming relatively shorter (due to obsolescence); there is a greater degree of outsourcing than practiced in the past; and there is more globalization and greater international competition today.

In response to some of these challenges, one needs to view logistics and the various elements of the supply chain (SC) in a much broader context than in the past. More specifically:

1. A total top-down systems approach must be assumed, with the "logistics support infrastructure" included as a major subsystem and oriented to a specific set of mission objectives. Viewing the components of such on an individual-by-individual basis is no longer feasible.
2. A total life-cycle approach to logistics must be implemented. There are logistics requirements and activities in each and every phase of the system life cycle, and these requirements must be treated as an integrated entity since the activities in any one phase could have a significant impact on those in the other phases. If one is to minimize the technical and management risks in the day-to-day decision-making process, then such decisions must be made in the context of the whole.

3. The ultimate logistics support infrastructure configuration must be agile and highly flexible, incorporating an open-architecture approach in design. System-level requirements are constantly changing, and the integration of these requirements (both horizontally and vertically) with other systems are becoming more complex. A new approach to design is necessary to facilitate the incorporation of future changes at minimum total life-cycle cost LCC.
4. Logistics requirements must be established early in the life cycle and in conjunction with the development of system-level requirements from the beginning during the conceptual design phase. This is essential if one is to influence and "optimize" the design for maximum supportability and economic feasibility.
5. The accomplishment of logistics objectives for any type of system can best be realized through implementation of the system engineering process. "Logistics engineering" must be an inherent and active part of this process from inception.

To summarize, the nature of logistics is life-cycle oriented and involves the integration of many different elements, both internally and externally. The elements of logistics must be properly integrated within (as illustrated in Fig. 5.2), integrated with the prime mission-related elements of the system in question, and integrated externally with comparable components of other systems operating in an overall higher level hierarchy. Thus, one might consider logistics as an integrating system's function.

5.5   Case Study—Life-Cycle Cost Analysis

One of the key TPMs for a system is its projected LCC, which is an indicator of the overall economic value of the system in question. Past experience is replete with instances where a large percentage of the total cost of a given system can be attributed to downstream activities pertaining to logistics and system maintenance and support; that is, the logistics support infrastructure as described throughout this section of the handbook. Further, as illustrated in Figure 5.5, the LCC for a system is highly dependent on design and management decisions made early in the life cycle, and that the greatest opportunity for influencing LCC occurs early in the conceptual and preliminary system design phases. Thus, it is at this early stage in the system life cycle when it is essential that the logistics support infrastructure be introduced and addressed within the context of the overall systems design and development process. Further, it is at this early stage when the implementation of life-cycle cost analysis (LCCA) methods can be applied to properly assess various potential system design alternatives and their impact on logistics and system support. Given the significance of LCC as a measure of system economic value and, in particular, logistics support, it was decided to include an abbreviated LCCA case study in this section of the handbook.

In accomplishing a LCCA, there are certain steps that the analyst should perform to acquire the desired result. For the purposes of illustration, the following represents a generic approach:

1. *Define system requirements.* Define system operational requirements and the maintenance concept. Identify applicable TPMs and describe the system in functional terms, utilizing the functional analysis at the system level as required (refer to Figs. 5.6 through 5.11).
2. *Describe the system life cycle.* Establish a baseline for the development of a cost breakdown structure (CBS) and for the estimation of costs for each year of the projected life cycle. Show all phases of the system life cycle and identify the major activities in each phase (refer to Figs. 5.1, 5.3, and 5.4).
3. *Develop a CBS.* Provide a top-down and/or bottom-up cost structure to include all cost categories for the initial allocation of costs (top-down) and the subsequent collection and summary of costs (bottom-up). Develop the appropriate cost-estimating relationships (CERs), estimate the costs for each activity in the life cycle and for each category in the CBS, develop a typical cost profile, and summarize the costs through the CBS network.
4. *Select a cost model for analysis and evaluation.* Select (or develop) a mathematical or computer-based model to facilitate the life-cycle costing process. The model, developed around the applicable CBS,

must be valid for and sensitive to the specific system configuration being evaluated. Accomplish a sensitivity analysis by evaluating input–output data relationships and to verify model application.

5. *Evaluate the applicable baseline system design configuration being considered.* Apply the computerized model in evaluating the baseline design configuration being considered for adoption. Develop a cost profile and a CBS summary, identify the high-cost contributors, establish the critical cause-and-effect relationships, highlight those system elements that should be investigated for possible opportunities leading to design improvement and potential cost reduction, and recommend design changes as feasible. It is at this stage in the LCCA process when the analyst can pinpoint the costs associated with the proposed logistics support infrastructure, its elements, and their respective percent contribution to the total.

6. *Identify feasible design alternatives and select a preferred approach.* After accomplishing a LCC evaluation for the given baseline configuration, it is then appropriate to extend the LCCA to cover the evaluation of multiple design alternatives (as applicable). Develop a cost profile and CBS summary for each feasible design alternative, compare the alternatives equivalently, perform a break-even analysis, and select a preferred design approach.

When accomplishing a complete LCCA for a large system, the detailed steps and the data requirements can be rather extensive and beyond the limits of coverage in this handbook. However, through the information presented herein, derived from an actual case study of a large communications system, it is hoped that the process and results are complete enough to demonstrate the importance of a life-cycle costing application to logistics.

5.5.1   Description of the Problem

A large metropolitan area has a need for a new communication system network capability (i.e., identified as System XYZ herein) that will enable day-to-day active communication between each and all of the following nodes: (*i*) a centralized city operational terminal located in the city center; (*ii*) three remote ground district operational facilities located in the city's suburban areas; (*iii*) 50 ground vehicles patrolling the city and within a 30-mile range; (*iv*) five helicopters flying at low altitude and within a 50-mile range; (*v*) three low-flying aircraft within a 200-mile range; and (*vi*) a centralized maintenance facility located in the city's outskirts. The proposed network needs to enable live two-way voice and data communication, 24-hours per day, and throughout all of its branches and to anyone of the stated nodes as required.

In response to this new system requirement, a need and feasibility analysis was accomplished, a solicitation for proposal was distributed to all known qualified potential sources of supply, and two prospective suppliers responded, each with a different design approach. The objective at this point is to evaluate each of the two supplier proposals, on the basis of system life cycle cost, and to select a preferred approach; that is, Configuration A or Configuration B.

1.   System operational requirements and the maintenance concept

Referring to Section 5.3.1, the first major step in accomplishing a LCCA is to establish a good "baseline" description of system operational requirements, maintenance concept, primary operational TPM requirements, and a top-level system functional analysis. Replicating the material presented in Section 5.3.1, paragraphs 1, 2, 3, 4, and in Figures 5.6 and 5.7, for the proposed new communication system network capability is required. While the specific requirements may change, establishing a good initial foundation, upon which to build the LCCA, is essential. The level of detail will, of course, vary with the goals and depth of required analysis.

2.   The system life cycle

Having described the basic operational and maintenance support requirements for System XZY, the next step is to present these requirements in the context of a proposed life-cycle framework. The objective is to

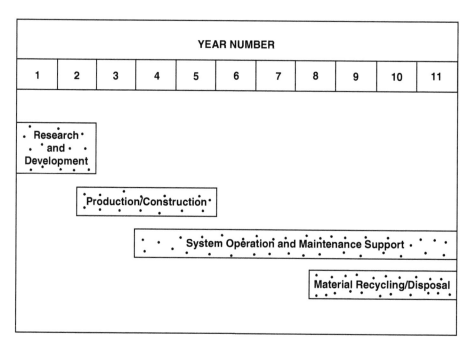

FIGURE 5.13   System XYZ life-cycle plan.

identify the applicable phases of the life cycle and all of the activities within each phase. Figure 5.13, which constitutes a simplified abstraction taken from Figure 5.3, provides an illustration of the framework for the LCCA. This, in turn, forms the basis for collecting and categorizing costs for the analysis; that is, research and development cost, production and/or construction cost, operation and maintenance cost, and system retirement cost.

3.   The CBS

Given the planned program phases and the anticipated activities in each phase (shown in Fig. 5.13), the next step is to develop a CBS, or a top-down and/or bottom-up structure for the purposes of cost estimation and the collection of costs by category. The proposed CBS for System XYZ is presented in Figure 5.14, and must include all of the costs pertaining to the system; that is, direct and indirect costs, contractor and supplier costs, customer (user) costs, design and development costs, production costs, hardware costs, software costs, data costs, logistics costs, and so on.

Referring to Figure 5.14, the objective is to estimate the applicable costs for each of the categories indicated. In estimating LCC, this becomes a bottom-up effort, employing the application of various CERs, activity-based costing (ABC) methods, and utilizing the appropriate analytical models and/or tools to help facilitate the process. In developing a CBS, the analyst needs to know what is included (or left out), and how the various costs are developed. While a detailed description of what is included in each category of a CBS is required to provide the visibility desired, the summary structure in Figure 5.14 is considered to be sufficient for the purposes herein.

4.   *Cost estimation and the development of cost profiles for the proposed design configurations being evaluated—Configuration A and Configuration B*

Within the context of the System XYZ life-cycle plan (Fig. 5.13) and the CBS (Fig. 5.14), the costs for each of the two proposed design configurations being evaluated were determined and are presented as shown in Figure 5.15. The costs for each of the four major categories (i.e., research and development cost,

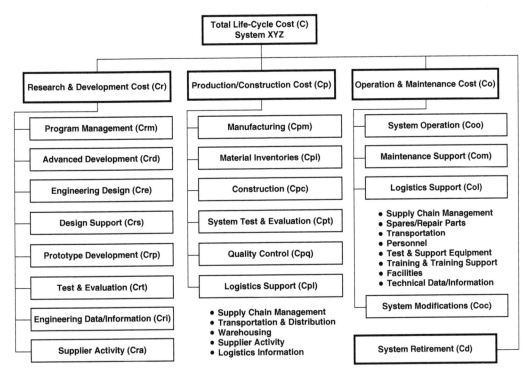

FIGURE 5.14 Cost breakdown structure for system XYZ.

production and/or construction cost, etc.) were determined for each configuration, utilizing bottom-up estimating methods, and are summarized in the Figure 5.15.

Referring to Figure 5.15, the costs are summarized in terms of the estimated inflated budgetary costs for the planned 11-year life cycle, which is reflected by the top profile, or identified as the total cost; that is, \$7,978,451 for Configuration A and \$8,396,999 for Configuration B. A second summary profile is

Cost Category	Life Cycle Year											Total ($)
	1	2	3	4	5	6	7	8	9	10	11	
Configuration A												
Research & Development (Cr)	615,725	621,112										1,236,837
Production/Construction (Cp)		364,871	935,441	985,911	986,211							3,272,434
Operation & Maintenance (Co)				179,203	207,098	448,248	465,660	483,945	503,122	523,297	544,466	3,355,039
System Retirement (Cd)									27,121	41,234	45,786	114,141
Total Cost ($)	615,725	985,983	935,441	1,165,114	1,193,309	448,248	465,660	483,945	530,243	564,531	590,252	7,978,451
Present Value Cost – 6% ($)	580,875	877,525	785,396	922,887	891,760	316,015	309,770	303,627	313,851	315,234	310,945	5,927,885
Configuration B												
Research & Development (Cr)	545,040	561,223										1,106,263
Production/Construction (Cp)		379,119	961,226	982,817	987,979							3,311,141
Operation & Maintenance (Co)				192,199	225,268	456,648	472,236	592,717	613,005	625,428	650,342	3,827,843
System Retirement (Cd)								20,145	35,336	45,455	50,816	151,752
Total Cost ($)	545,040	940,342	961,226	1,175,016	1,213,247	456,648	472,236	612,862	648,341	670,883	701,158	8,396,999
Present Value Cost – 6% ($)	514,191	836,904	807,045	930,730	906,659	321,937	314,089	384,510	383,753	374,621	369,370	6,143,809

FIGURE 5.15 Life-cycle cost profile for system XYZ.

included in terms of present value (PV) cost, required for the evaluation of comparable alternatives on the basis of economic equivalence. A 6% cost of capital was assumed for this LCCA effort.

5. Evaluation of alternative design configurations and the selection of a preferred approach

On the basis of the results shown in Figure 5.15, it appears that Configuration A is the preferred approach, because the present value (PV) cost of $5,927,885 is less than that for the other configuration. The question is, How much better is Configuration A, and at what point in time does this configuration assume a position of preference? It should be noted that, on the basis of acquisition costs only (i.e., Categories Cr and Cp), it appears as though Configuration B would be preferred ($4,417,404 for B and $4,509,271 for A). However, based on the overall LCC, Configuration A is preferred. Relative to the time of preference (i.e., when A assumes the point of preference), the analyst conducted a breakeven analysis as illustrated in Figure 5.16. From the figure, it can be seen that Configuration A assumes a favorable position at about the 7 year to 7 month point in the projected life cycle. It was decided in this instance that this was early enough for the selection of Configuration A.

6. Further analysis and enhancement of the selected configuration

Having initially selected Configuration A as being preferred over the alternative, the next step is to further evaluate the costs that make-up the $7,978,451 for this configuration, identify the high-cost contributors, determine cause-and-effect relationships, and re-evaluate System XYZ design to determine whether improvements can be implemented which will result in an overall reduction in LCC. A breakout of the costs for this configuration is presented in Figure 5.17.

Referring to Figure 5.17, for example, it should be noted that the costs associated with logistics activities (i.e., *Cpl* and *Col*) make up about 21.38% of the total. Within this spectrum, the categories of spares and/or repair parts and transportation represent high-cost contributors (4.57% and 3.73%, respectively)

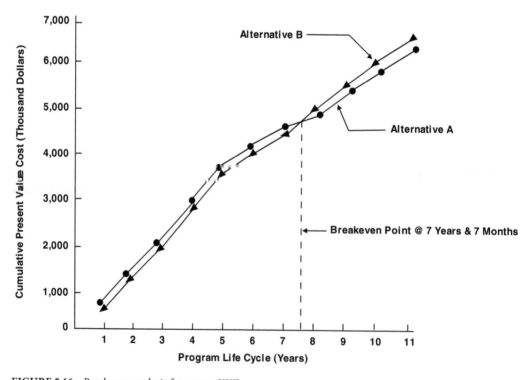

FIGURE 5.16   Breakeven analysis for system XYZ.

Configuration A

Cost Category	Cost ($) (Undiscounted)	Percent (%)
1. Research & Development (Cr)	1,236,660	15.50
(a) Program Management (Crm)	79,785	1.00
(b) Advanced Development (Crd)	99,731	1.25
(c) Engineering Design (Cre)	276,852	3.47
(d) Design Support (Crs)	193,876	2.43
(e) Prototype Development (Crp)	89,359	1.12
(f) Test & Evaluation (Crt)	116,485	1.46
(g) Engineering Data/Information (Cri)	75,795	0.95
(h) Supplier Activity (Cra)	304,777	3.82
2. Production/Construction (Cp)	3,272,762	41.02
(a) Manufacturing (Cpm)	1,716,166	21.51
(b) Material Inventories (Cpi)	453,176	5.68
(c) Construction (Cpc)	95,741	1.20
(d) System Test & Evaluation (Cpt)	228,184	2.86
(e) Quality Control (Cpq)	76,593	0.96
(f) Logistics Support (Cpl)	702,902	8.81
(1) Supply Chain Management	39,892	0.50
(2) Transportation & Distribution	219,408	2.75
(3) Warehousing	168,345	2.11
(4) Supplier Activity	263,289	3.30
(5) Logistics Information	11,968	0.15
3. Operation & Maintenance (Co)	3,354,939	42.05
(a) System Operation (Coo)	1,458,461	18.28
(b) Maintenance Support (Com)	768,325	9.63
(c) Logistics Support (Col)	1,002,891	12.57
(1) Supply Chain Management	79,785	1.00
(2) Spares/Repair Parts	364,615	4.57
(3) Transportation	297,596	3.73
(4) Personnel	153,984	1.93
(5) Test & Support Equipment	46,275	0.58
(6) Training & Training Support	24,733	0.31
(7) Facilities	20,744	0.26
(8) Technical Data/Information	15,159	0.19
(d) System Modifications (Coc)	125,262	1.57
4. System Retirement (Cd)	114,092	1.43
GRAND TOTAL	7,978,451	100.00

FIGURE 5.17 Cost breakdown structure summary.

under Category *Col*. Additionally, transportation and distribution costs within Category *Cpl* are also relatively high (2.75%). Through a re-evaluation of the basic design configuration, the extensive requirements for spares and/or repair parts could perhaps be reduced through some form of reliability improvement, particularly for critical items with relatively high failure rates. For transportation, it may be possible to repackage elements of the system such that internal transportability attributes in the design can be improved, or to select alternative modes of transportation that will still meet the TPM requirements for the system overall, but at a lesser overall cost.

Through implementation of this process on an iterative basis, experience has indicated that significant system design improvements can often be realized. It should be noted that by improving one area of concern, the result could lead to an improvement in another area. For example, if improvement can be made in the spares and/or repair parts area (within Category *Col*), this also may result in a reduction of the maintenance support cost (Category *Com*) as well. There are numerous interactions that could occur throughout the analysis process, and care must be exercised to ensure that improvement in any given area will not result in a significant degradation in another.

5.5.2 Summary

The implementation of the LCCA process, particularly during the early stages of system design and development, can provide numerous benefits to include: (*i*) influencing the overall system design for maximum effectiveness and efficiency from a total life-cycle perspective; (*ii*) facilitating the design of the logistics support infrastructure capability from the beginning when the incorporation of any required changes can be accomplished easily and at minimum cost; and (*iii*) providing early front-end visibility by identifying potential high-cost areas and the risks associated with such. Additionally, LCCA can be applied at any stage in the system life cycle for the purposes of assessment, and for the identification of high-cost areas and the major contributors for such. This case-study approach addresses the steps and process for accomplishing a good LCCA effort.

II

Logistics Activities

6

Customer Service

Richard Germain
Wayne Whitworth
University of Louisville

6.1   Introduction

Customer service has long been of interest to the logistics field. The listing of activities provided by the National Council of Physical Distribution Management's 1963 original definition of physical distribution included freight transportation, warehousing, material handling, protective packaging, inventory control, plant and warehouse site selection, market forecasting, and customer service (Bowersox et al. 1968). Customer service has not lost its central role in logistics and supply chain management. According to Lambert et al. (2006), managing and integrating business processes across the supply chain requires attention to eight basic processes centered on the management of customer relationships, customer services, demand management, order fulfillment, manufacturing flow, supplier relationships, product development, and returns. While current definitions of logistics and supply chain management generally do not list activities, customer service is a core concept in how firms compete to gain a competitive advantage. In its broadest sense, customer service refers to the points of contact orchestrated by sellers in the exchange process with customers. At this broad level, a sales representative at Nordstrom driving to a competitor to acquire out-of-stock merchandise for a valued customer is a customer service activity. While logisticians may scoff at such excessive service, it is really very much like a firm seeking inventory from a distant warehouse when one more proximate to a valued customer is out of stock. This chapter discusses customer service from a logistics perspective. First, we provide a more detailed definition of customer service. This is followed by a discussion of the economic rationale for how firms select the level of service to provide and of ABC analysis. Next, we present recent thought on the connection between integration and logistics customer service. A short case illustrating some of the issues raised in this chapter concludes our contribution.

6.2   Perspectives on Customer Service

During the 1970s, LaLonde and Zinszer (1976) asked managers what customer service meant in their firm. Managers responded in one of the following three ways: (*i*) as a set of activities, including order processing, tracing, and invoicing; (*ii*) as set of measurable system outputs, including fill rates, lead-times, and percent of orders shipped complete; or (*iii*) as a business philosophy. Later definitions refined and developed these perspectives to varying degrees, but this classification schema retains its original usefulness. This is because customer service means different things not only to different firms, but may mean different things to different functions within the same firm. In addition, while a firm may view customer service from a particular perspective, all three are interrelated. The following sections delve more deeply into the three perspectives.

6.2.1 Customer Service as Activities and Measurable System Outputs

From a logistics perspective, the activities of customer service often focus on the seller's order cycle; that is, all the activities that occur between a customer placing an order and the customer receiving the order. As a matter of convention, we shall refer to this as lead-time from the buyer's perspective. At first glance, this may seem to exclude after-sales support, but this is not the case if we treat after-sales support as a distinct order cycle that services an extant product or need. Figure 6.1 shows an abbreviated order cycle for Procter & Gamble after their 1994 reengineering and redesign (McKenney and Clark 1995). Here we see the major elements of the order cycle: order acquisition; order processing; shipment control and billing; and delivery execution. Some of the activities involved include order receipt, order entry, credit check, order transmission to plant or warehouse, product availability check, warehouse load planning, advance shipping notification (ASN), invoicing, truck loading, and order delivery.

Figure 6.1 also provides a listing of measures accumulated by Proctor & Gamble at various stages during the order cycle. The logistics customer service mix, as defined by Bowersox et al. (2002), consists of three major components: (*i*) product availability which includes fill rates, stockouts, and percent of orders shipped complete; (*ii*) operational performance which includes speed, consistency, flexibility, and recovery; and (*iii*) service reliability which includes mis-shipments (goods delivered to the wrong location) and damage. Fill rates and stockout rates are measures of whether the firm possessed inventory when demanded.* Inventory may be defined at the stock keeping unit (SKU) level or SKUs may be aggregated to generate availability measures across product categories or brands. Orders shipped complete, refers to inventory availability for specific orders: for example, the order shipped complete rate would drop if the seller were short just one case across a multiple SKU order. Two major elements of operational performance, speed and consistency, are derived from the concept that the amount of time required to complete each activity in the order cycle possesses a mean and a standard deviation. The average for all activities of the order cycle across a specified time period equals speed. The buyer may state something like, "The lead-time from Acme Corp. is 14 days." The standard deviation across orders over a specified time period represents consistency. The buyer may then say, "While the lead-time from Acme Corp. is 14 days, it may take as long as 24 days or as little as 10 days." Variability in the order cycle is critically important as the seller is forced to hold extra inventory "just in case" the vendor delivers product later than expected or if demand during lead-time is higher than expected. If we return to the Proctor & Gamble order processing system (Fig. 6.1), we can see that most (but not all) of their performance measures fall into product availability, operational performance, or service reliability. These include order cycle time, percent of orders shipped on-time (i.e., within an expected length of time),

---

* A stockout may be defined as the percentage of time that demand was not satisfied over a specified time period. The fill rate may be defined as the depth of stockouts. The stockout frequency is often measured as the 1-(number of times demand was satisfied/number of times a good was demanded) while the fill rate is often measured as units available/units demanded. As a simple example, suppose 30 units of a good are on the retail shelf at the start of a day. Thirty customers enter demanding one unit each, followed by 5 customers seeking 4 units apiece. The stockout frequency equals 1-(30/[30 + 5]) = 14% or an in-stock rate of 86%. The fill rate, however, equals 30/(30 + 5 × 4) = 60%. The fill rate is generally a more rigorous measure of product availability than the in-stock rate.

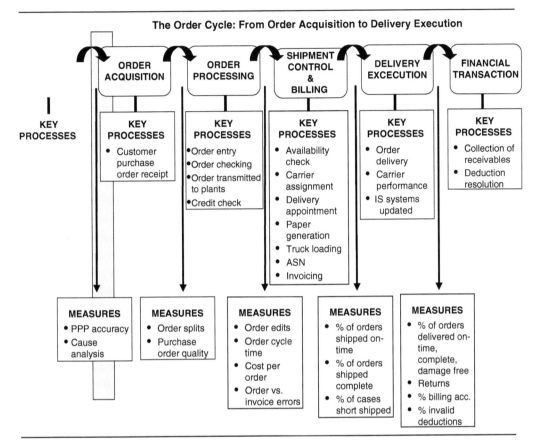

FIGURE 6.1 Abbreviated order processing system at Proctor & Gamble after 1994 re-engineering. (Adapted from McKenney J.L. and Clark, T.H., *Procter and Gamble: Improving Consumer Value Through Process Redesign* 1995.)

percent of cases short shipped (which is a measure of availability), and returns (which may be the result of poor service reliability).

Those measures of performance not captured by product availability, operational performance, or service reliability are artifacts of the order cycle being embedded within a much larger organizational process. This is consistent with the concept that activities, policies, and procedures occurring before and after exchange are critical to successful customer service operations (Ballou 1992, Lambert and Stock 1993). Figure 6.2 provides a classification of customer service elements based upon activities conducted before, during, and after a transaction.

Returning to Figure 6.1, to the left of the order acquisition element is order generation: this represents processes involved in ensuring that software systems have correct pricing and customer information. The former is especially important as trade promotions vary by SKU, time, and region. To the left of order generation would reside the formal policies and procedures that identify service targets (e.g., fill rates, on-time deliveries, etc.), which are comparable to the pretransaction elements shown in Figure 6.2. To the right of the delivery execution element (Fig. 6.1) resides financial transactions. This concerns the collection of monies. Thus, the order cycle is embedded within a larger order to cash cycle. This, in turn, is embedded within a larger manufacture-to-cash cycle, which itself is embedded within the procure-to-cash cycle, and which is finally embedded into the cash-to-cash cycle. The order splits measure of performance (between order acquisition and order processing) is a measure of the percentage of orders

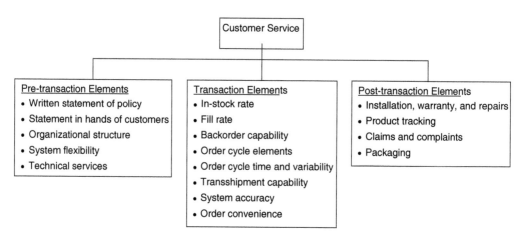

FIGURE 6.2 Temporal classification of customer service elements. (Adapted from Ballou, R.H., *Business Logistics Management*, Prentice Hall, New Jersey, 1992.)

received by fax, phone, and electronically. The connection to customer service is that an electronic order removes a manual step (entering an order into Procter & Gamble's system). But a concern just as important is that electronic orders cost less. The financial consideration is significant when one considers the number of orders that Procter & Gamble processes each year. A performance measure under financial transactions would be day sales outstanding; that is, the average number of days required between shipments leaving the seller's dock (which is when the company counts an order as a sale on its income statement and is able to bill the customer) and receipt of monies. The results of the 1994 redesign of the order processing system along with the introduction of new performance measures, cause analysis, and value pricing [their term for everyday low pricing (EDLP)] were quite remarkable. Billing accuracy rose from 83% to 93%; perfect orders rose from 55% to 75%, deductions and allowances resolved in favor of Procter & Gamble rose from 15% to 65%. Day sales outstanding for accounts receivable fell considerably as well. Thus the firm's focus on customer service processes and their ability to effectively redesign the system and institute new performance measures resulted in two critical effects: (*i*) customer service levels rose—this ultimately leading to top-line growth; and (*ii*) cash flow improved and costs fell—this leading to bottom-line growth.

6.2.2   Customer Service as a Philosophy

Customer service as a philosophy is a mission encouraged by the firm through long-term investment in people. When we say that a firm has superior customer service, we usually think of how customers are treated by the staff of the selling firm at all levels of contact. This is not a matter of technique; that is, it is not a matter of designing effective order processing systems or of implementing sophisticated inventory management systems that determine the appropriate amount of cycle and safety stock (Miller and Le Brenton-Miller 2005). Rather, it is a matter of a deeply embedded mission patiently cultivated by selective recruitment and relevant training. Firms that focus on customer service as a philosophy often do not serve low-end markets. Indeed, they often sacrifice low-cost operations for a core capability of quality that focuses on customer satisfaction. For example, Nordstrom's training budget is about four times and wages are about three times the industry average. Sales quotas are strict and low performers are pruned from the workforce. At the same time, inventory per square foot at the store is several times higher relative to department store competitors. This is sharply contrasted against Wal-Mart, a retailer that focuses on low-cost operations. Service is important, but the target market consists of customers that are economical. Wal-Mart's service is therefore commensurate with the price-conscious shopping behavior of the target market (Miller and Le Brenton-Miller 2005).

6.2.3 Economic Rationale for Determining the Level of Customer Service

A basic microeconomic concept is that the firm should operate at the point where marginal cost equals marginal revenue (Ayers and Collinge 2004). For a company selling a differentiated product, each additional unit of output increases supply. As a result, the firm must lower unit price to sell the additional units. Total revenue will therefore increase as output increases, but at a decreasing rate. The total-cost curve is comprised of two components: total-fixed cost and total-variable cost. As a relatively simplified explanation, the total-cost curve increases at an increasing rate because of the nonlinear requirements of labor across various output levels. Marginal cost and marginal revenue may be estimated as the respective slopes of these curves: profit is maximized at the point where the slopes of the curves are equal.

Panel A of Figure 6.3 illustrates these concepts as applied to the case of logistics customer service. For the sake of argument, suppose that the service is the firm's fill rate for a specific SKU. The market responds to an increasing fill rate, but the response is not linear. A percentage change in the rate when the fill rate is 90% yields a greater sales response than does a percentage change when the fill rate is 97%. The cost associated with the fill rate is primarily a function of the inventory carrying cost. The carrying cost reflects the variable cost associated with holding inventory: for example, financing charges, insurance, shrinkage, etc. Total inventory is comprised of cycle stock, to meet expected demand, and safety stock to meet unexpected demand. The higher the fill rate, the greater the amount of safety stock required. The cost curve is asymptotic at a 100% fill rate. This is because the firm must hold an infinite amount of inventory to be 100% certain of never running out of stock.

The point at which the slope of the total-revenue curve in Panel A equals the slope of the total-cost curve is where profit is maximized. Panel B provides a case where the firm holds too much inventory— the fill rate is too high. The firm should reduce the level of inventory. If such is undertaken, revenue will fall, but costs will fall faster and profit will increase. Ballou (1992) provides a case example of a firm with a probability of being out of stock during lead-time at 99%. Applying the model in Figure 6.3 led to the conclusion that excessive inventory was on hand. The result was that the firm lowered the percentage to 93 and profits were increased.

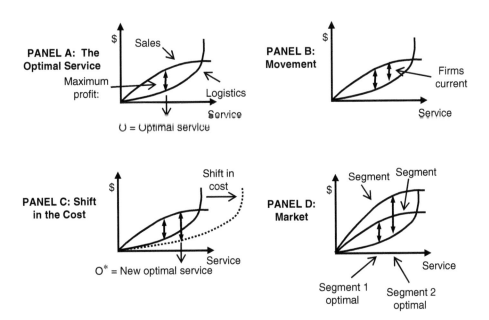

FIGURE 6.3 Economic representation of optimal customer service level.

Panel C of Figure 6.3 illustrates a shift in a cost curve. Any technology, innovation, or business process that lowers cost at any given service level results in a downward shift in the cost curve. There are numerous examples: bar codes and UPC systems, warehouse management systems, demand planning systems, six sigma tools (e.g., flowcharting), etc. Specifically, suppose (*i*) a firm uses the Internet for order processing purpose and (*ii*) the length of the order cycle is the considered element of customer service. The cost curve shifts downward because of efficiencies in processing orders, order cycle time decreases, and sales revenue and profit increase. A less-fleet competitor suddenly finds themselves in a position of having to reduce their lead-time, but their costs increase because they simply experience a movement along their curves (as in Panel B) and not a shift in their cost curve (as in Panel C). The profit of the less-fleet competitor falls. Across time, the innovative firm that focuses on operational improvements may accrue a significant competitive advantage. At the broadest level, this describes the competition between Wal-Mart (the process innovator) and K-Mart (the process laggard).

6.2.4   ABC Analysis

ABC analysis, sometimes referred to as Pareto analysis, is a method that classifies items under consideration in terms of similar relative value, rank orders the identified classes, and then develops policies distinct to the classes. ABC analysis obtained it name because when used on products based on product value or inventory turnover rates, the three classes that are often formed are referred to Category A, B, and C products. In terms of logistics customer service, companies often classify both products and customers. For example, three customer types are identified and ranked in terms of their relative importance to the seller and three product types are identified and ranked according to their profitability. The nine product-by-customer cross-classification types could then be arranged into four priority groups. The most important customer-product priority group may be provided a 99% fill rate and delivery within 48 h, whereas the lowest priority customer-product group may be served at a 90% fill rate and delivery within 120 h (Lambert and Stock 1993).

The ABC approach to determining logistics customer service levels is connected to the economic approach. Panel D of Figure 6.2 shows two sales response curves, one for each of two segments the firm serves. Segment 1 may use the product in critical processes. For instance, an out-of-stock situation by a vendor would result in a plant shutdown. Firms in segment 1 are willing to pay a premium for some combination of higher order shipped complete rates, lower lead-times, and tightly consistent lead-times: that is, just in time (JIT) type service. Firms in segment 2 are less sensitive to service levels and hence are not willing to pay a premium for better service. One curve defines the firm's cost structure, but notice that firms in segment 1 should receive higher service at a premium price because this maximizes the segment's profit. ABC analysis is consistent with this approach if the classes of identified firms possess distinct and unique sales response curves.

6.2.5   Role of Integration

Integration refers to lateral coordination and communication. The hierarchical nature of the typical triangular organizational chart presents barriers to effective internal integration. One anecdote relates that marketing discounted the price of a product variant because of excessive inventory. The production function increased output of the variant as they thought the increase in demand was the result of a shift in market preference. The firm then made variants of the products that had to be sold at a discount rather than being in the more enviable position of making variants that could be sold at a premium. Seminal logistics textbooks advocated the importance of integration from a system's perspective (Bowersox et al. 1968). Specifically, they were interested in the trade-offs between pairs of functions: for example, transportation and inventory management. Technological developments or the diffusion of extant technologies since the 1960s, especially enterprise resource planning systems (ERP), the Internet and electronic ordering, and bar codes, have provided tremendous opportunities not only within the firm (for more effective logistics), but also across firms (for more effective supply chain management).

Internal integration refers to lateral communication within the firm whereas external integration refers to lateral communication with other firms and may occur upstream or downstream. Internal communication within the firm is cultivated through a number of sources: cross-functional teams (permanent, project or *ad hoc* based); and specialized software including ERP systems. Cross-firm integration is often software driven and includes collaborative planning, forecasting, and replenishment (CPFR) and centralized multi-echelon inventory systems.

Although theoretical approaches differ, research has demonstrated the importance of integration in cultivating logistical performance, which includes logistical customer service. The kind of logistical customer service referred to here is measurable system outputs such as order cycle times, order cycle consistency, and fill rates: it does not refer to the extent to which the firm has adopted customer service as a philosophy. Figure 6.4 illustrates the manner in which various research studies have conceptualized the connection of integration to logistical performance. Each of the four approaches enjoys empirical support: (*i*) the individual effects model (Closs and Savitskie 2003); the causal chain model (Stank, Keller et al. 2001; Sanders and Premus 2005); (*ii*) the unified integration model (Rodrigues et al. 2004); and (*iii*) the interactive model (Germain and Iyer 2006). Scaling and sampling differences make it difficult to assess which theory best represents the underlying "truth" of the world. Regardless we can state that greater levels of internal and external integration associate with better logistical performance and hence a higher level of logistics customer service. Furthermore, the two interact in predicting logistical performance—that is, the higher the internal integration, the greater the association of external integration with logistical service performance (Germain and Iyer 2006). The latter researchers also demonstrated that logistical service performance correlates with higher levels of financial performance including return on investment and profit growth. This research stream supports the idea that pretransactional elements are critical to logistics customer service. Internal integration when treated as cross-functional teams is a dimension of organizational structure and when treated as software is a dimension of the firm's information technology resource base. External integration in measure reflects an operational

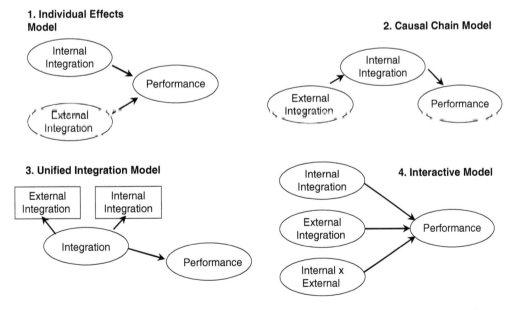

FIGURE 6.4 Competing models of integration's effect on logistics performance. (From Germain, R. and Iyer K., *Journal of Business Logistics*, 3, 3, 2006.)

efficiency orientation—that is, the firm is focusing on operations as a core competency. When external integration takes the form of CPFR, the integration of functions (i.e., promotions, product category management, and supply chain management) across firms (i.e., manufacturers and retailers) provides the policy, infrastructure, communication channels, information, and data for firms to increase measurable systems' outputs including fill rates, delivery speed and consistency, and percent of orders shipped complete. And this is accomplished at lower cost. Essentially, both internal and external integration seem to shift the firm's cost curve downward (Panel C, Fig. 6.3). When both internal and external integration increase, the shift in the curve is greater than the sum of the individual shifts (the result of the significant interaction). Firms that take advantage of both internal and external integration, neither of which is easy to imitate, are generating a sustainable competitive edge that offers a long-run edge advantage.

6.3   Case Example: Beverage Industry

6.3.1   Background

Located in the Mid-Western U.S., Acme is a producer of distilled spirits, wines, and related products. Branding is a core competency of the firm. Quality, customer service, and growth through acquisition complement this strength. Until the early 1980s, Acme focused its growth on brand acquisition, either through acquiring distribution rights or by outright purchase of the brand itself. Although research and development efforts did produce several new products in the mid-1980s, only one product survived the test of time. If possible, new products that required capital investments were first produced by contract bottlers until market penetration warranted production in-house.

Products were marketed under the three-tiered distribution system using distributors, wholesalers, and retailers. Acme maintained close relationships with all levels of the distribution channel. However, obtaining sales and market information on a timely basis was a major challenge because of lack of supply chain integration to move sales data downstream in a timely fashion. The market driven philosophy of the company focused heavily on the customer. This was a major challenge to maintain low finished goods inventory while obtaining a 100% order fill rate within three working days.

6.3.2   New Product Category

During the mid-1980s, a new product category was introduced. Products within this category combined wine or distilled spirits with fruit-based drinks, respectively referred to as wine coolers or mixed cocktails. Products in this category were generally sweet and targeted young adults. Major competitors quickly introduced brands. Acme soon began to experiment with a product that combined lemonade with distilled spirits. Success in selected test markets suggested that the company could compete against wine coolers. During the early 1990s, four flavors of the mixed cocktails were developed referred to here as Orange, Fruit, Green, and Teal. Sales were forecasted in the range of one million cases nationally. Sales of Orange and Fruit far exceeded initial expectations, while sales of Green were disappointing and production scheduling became a constant issue.

Finished goods inventory averaged 48.9 days across all four flavors. Inventory of product Green exceeded 105 days, while Fruit inventory averaged nine days. To complicate production scheduling, bottle deposit laws and labeling requirements varied from state to state. This increased the required amount of raw materials inventory. In addition to proprietary ingredients inside the bottle, each of the four brands typically had a minimum of seven ingredients excluding the base distilled spirit, as well as the additional components for packaging shown in Table 6.1.

Production of the mixed cocktails was outsourced to a small contract bottler. The contract bottler required a minimum production lot size of 10,000 cases per flavor. Additionally, production schedules were updated weekly with a premium charge for changes made during the weekly planning horizon. A rolling monthly

TABLE 6.1 Lot Size Requirements

Component	Quantity Required Per Case
Outer shipper	1
Inner divider	1
Bottles	24
Front labels (state specific)	24
Back labels	24
Closures (caps)	24
Neck wraps	24

schedule sent to the contract bottler and materials suppliers served as a general planning tool, but this was unreliable because of lack of integration within the supply chain. Production materials such as glass, labels, and closures had lead-times in excess of 14 weeks, thus complicating sudden changes in demand because of promotions or shifts in consumer preference. Change orders to suppliers became a constant struggle with frequent variations in reported sales and raw materials inventory. All suppliers required minimum order quantities and frequent changes to label laws on a state-by-state basis were commonplace.

Order fill rates declined after the initial introduction. Fill rates declined below 80% while finished goods inventory increased overall. Production scheduling was tasked with reducing inventory, improving fill rates, and reducing costs to maintain brand position and profitability. The constant dilemma faced by planning was how to accomplish this when the minimum lot size was 10,000 cases per flavor.

Orders were shipped across the United States from the single contract bottler, often in LTL quantities to distributors requiring minimal case quantities of a particular flavor to round out inventory. Typical monthly sales by flavor were Fruit, 12,000 cases, Green, 950 cases, Orange, 9500 cases, and Teal, 5300 cases.

Imbalance of finished goods inventory was only part of the problem. Many of the ingredients were perishable and minimum order quantities frequently meant raw materials spoilage increased with slower moving flavors such as Green. Planners saw customer service levels decline with fill rates falling below 80%. Additionally, planners were given the edict to reduce total inventory and improve fill rates and customer service. Compounding these problems was depletion reports from distributors which were 30 days old and seasonality of some products led to inaccurate forecasting and building finished goods inventory after peak demand had occurred.

6.3.3 Solution

Several remedial activities were taken that eventually led to improved service levels and lower operating costs. Among other actions, Acme:

1. Hired a second contract bottler with smaller lot size requirements to handle flavors with lower customer demand.
2. Re-engineered the production process at the initial contract bottler to improve lot size flexibility and short-run production capabilities.
3. Developed a break-bulk packaging facility for input materials. This facility received materials in bulk, repackaged them in discreet production lot size units, and made deliveries to the contract bottlers on a JIT basis.
4. Improved supply chain communications, especially on distributor depletion reports. Information lag time reduction target from 30 days to 5 days were established and monitored.
5. Created a simple, yet effective, electronic database with production requirements broken down by flavor and material components. These were updated daily and sent to suppliers and bottling operations on a weekly basis.

At the end of three months, customer fill rates exceeded 93% and finished goods inventory were reduced to less than six days. The overall cost reductions allowed Acme to lower the retail price to further pursue sales and market share objectives.

References

Ayers, R.M. and Collinge, R.A., *Microeconomics: Explore and Apply*, Pearson Prentice Hall, Upper Saddle River, NJ, 2004.

Ballou, R.H., *Business Logistics Management*, Prentice Hall, Upper Saddle River, NJ, 1992.

Bowersox, D.J., Closs, D.J., and Cooper, M.B., *Supply Chain Logistics Management*, McGraw-Hill, Boston, MA, 2002.

Bowersox, D.J., Smykay, E.W., and LaLonde, B.J., *Physical Distribution Management*, Macmillan, New York, 1968.

Closs, D.J. and Savitskie, K., Internal and external logistics information technology integration, *International Journal of Logistics Management*, Vol. 14, No. 1, 63–76, 2003.

Germain, R. and Iyer, K., The interaction of internal and downstream integration and its association with performance, *Journal of Business Logistics*, Vol. 3, No. 3, 302–320, 2006.

LaLonde, B.J. and Zinszer, P., *Customer Service: Meaning and Measurement*, National Council of Physical Distribution Management, Chicago, IL, 1976.

Lambert, D.M., Croxton, K.L., Garcia-Dastugue, S.J., Knemeyer, A.M., Rogers, D.S., Bolumole, Y., Gardner, J.T., Goldsby, T.J., and Pohlen, T.L., *Supply Chain Management: Processes, Partnerships, Performance*, Supply Chain Management Institute, Sarasota, FL, 2006.

Lambert, D.M. and Stock, J.R., *Strategic Logistics Management*, Irwin: Chicago, IL, 1993.

McKenney, J.L. and Clark, T.H., *Procter and Gamble: Improving Consumer Value Through Process Redesign*, HBS Case, Harvard Business School Publishing, Boston, MA, 1995.

Miller, D. and Le Breton-Miller, I., *Managing for the Long Run: Lessons in Competitive Advantage from Great Family Businesses*, Harvard Business School Press, Boston, MA, 2005.

Rodrigues, A.M., Stank, T.P., and Lynch, D.F., Linking strategy, structure, process, and performance in integrated logistics, *Journal of Business Logistics,* Vol. 25, No. 2, 65–94, 2004.

Sanders, N.R. and Premus, R., Modeling the relationship between firm IT capability, collaboration, and performance, *Journal of Business Logistics*, Vol. 26, No. 1, 1–23, 2005.

Stank, T.P., Keller, S.B., and Daugherty, P.J., Supply chain collaboration and logistical service performance, *Journal of Business Logistics*, Vol. 22, No. 1, 29–48, 2001.

Wisner, J.D., Leong, G.K., and Tan, K.C., *Principles of Supply Chain Management: A Balanced Approach*, Thomson Southwestern, Mason, OH, 2005.

7
Purchasing and Sourcing

Chi-Guhn Lee
University of Toronto

7.1 Introduction

Purchasing and procurement is the process of procuring materials, supplies, and services. Recently, the term "supply management" has been increasingly adopted to describe this process as it is related to many functions in a firm and pertains to a professional capacity. Employees who serve in this function are known as buyers, purchasing agents, or supply managers. Depending on the size of the organization and the importance of the purchasing function in the firm, buyers may further be ranked as senior management.

7.2 History and Economic Importance

Studies on purchasing date back as far as 1832 when Charles Babbage wrote a book titled *On the Economy of Machinery and Manufacturing*. Prior to 1900, purchasing was recognized as an independent function by many railroad organizations. The first book specifically addressing institutionalized purchasing was *The Handling of Railway Supplies—Their Purchase and Disposition*, written by Marshall M. Kirkman in 1887. Early in the twentieth century, purchasing has drawn more attention as The National Association of Purchasing Agents was founded in 1915. This organization eventually became known as the National Association of Purchasing Management (NAPM) and is still active today under the name The Institute for Supply Management (ISM). Purchasing became an academic discipline as Harvard University offered the first college course on purchasing and the first college textbook on the subject came to light, authored by Howard T. Lewis in 1933 [1]. Early buyers were responsible for ensuring a reasonable purchase price and maintaining operations (avoiding shutdowns due to stock outs). Both World Wars brought more attention to the profession because of the shortage of materials and the alterations in the market. Still, up until the 1960s, purchasing agents were basically order-placing clerical personnel serving in a staff-support position.

In the 1960s, purchasing managers were first seen as professionals, not clerks. This is the time when purchasing became more integrated with a materials system. As materials became a part of strategic planning, the importance of the purchasing department increased. In the 1970s, the oil embargo and the shortage of almost all basic raw materials brought much of business world's focus to the purchasing arena. By the late 1980s, the cost of buying materials represented about 60% of the cost of goods sold, which fueled the conceptual shift from purchasing to supply management [2]. The advent of just-in-time (JIT) purchasing techniques in the 1980s, with its emphasis on inventory control and supplier quality, quantity, timing, and dependability, made purchasing a cornerstone of competitive strategy. By the 1990s, the term "supply chain management" had replaced the terms "purchasing," "transportation," and "operations," and purchasing had assumed a position in organizational development and management.

Organizations must receive parts and materials to make goods for sale, equipment for production and operations, and expendable supplies like pen and paper. Supplies range from office supplies to crude oil and to manufacturing equipment. Purchasing derives its importance to an organization from two sources: cost efficiency and operational effectiveness. From a pure cost standpoint, the importance of purchasing is clear. Table 7.1 shows the cost of purchased materials as percentage of sales revenue in several industries.

Without effective purchasing practices, operations in a firm may be disrupted, customer service levels may fall, and long-term customer relationships may be damaged. Before any product can be manufactured, supplies must be available—and the availability must meet certain conditions. Meeting these conditions may be considered the goal of purchasing.

The importance of purchasing in any firm is largely determined by four factors: availability of materials, absolute dollar volume of purchases, percent of product cost represented by materials, and the types of materials purchased. Purchasing must concern itself with whether or not the materials used by the firm are readily available in a competitive market or whether some are bought in volatile markets that are subject to shortages and price instability. If the latter condition prevails, creative analysis by top-level purchasing professionals is required.

If a firm spends a large percentage of its available capital on materials, the sheer magnitude of expense means that efficient purchasing can produce a significant savings. Even small unit savings add up quickly when purchased in large volumes. When a firm's materials costs are 40% or more of its product cost

TABLE 7.1 The Importance of Purchasing in U.S. Manufacturing Industry

Industry	Percent of Sales Dollar
Food and kindred products	56.4
Textile	59.0
Wood product	58.3
Petroleum	81.0
Machinery	50.7
Transportation equipment	64.1
Beverage and Tobacco	36.2
Average U.S. manufacturing firm	52.7

Source: Adapted from U.S. Bureau of Census, Economics and Statistics Administration, U.S. Department of Commerce, December 2005, Annual Survey of Manufacturers [http://www.census.gov/prod/2005pubs/am0431gs1.pdf, Statistics for Industry Groups and Industries: 2004, M04 (AS)-1].

(or its total operating budget), small reductions in material costs can increase profit margins significantly. In this situation, efficient purchasing and purchasing management again can make or break a business.

Perhaps the most important of the four factors is the amount of control purchasing and supply personnel actually have over materials availability, quality, costs, and services. Large companies tend to use a wide range of materials, yielding a greater chance that price and service arrangements can be influenced significantly by creative purchasing performance. Some firms, on the other hand, use a fairly small number of standard production and supply materials, from which even the most seasoned purchasing personnel produce little profit, despite creative management, pricing, and supplier selection activities.

7.3 Purchasing Process and Performance Checklist

While individual purchases may appear quite different, there is a general, underlying purchasing process. The process may be described as identifying a need, understanding market and identifying potential suppliers, generating a request for quote (RFQ) and negotiating, awarding a contract and implementation, and evaluating the purchase and the supplier.

7.3.1 Need Identification and Analysis

There are many ways to identify a need in an organization. A department may submit a request to purchasing for supplies such as pencils, paper, and production equipment. An order can be initiated by an automated system or manually and submitted to purchasing through an electronic data interchange (EDI) system or simply through a phone call. Once an order is accepted by purchasing, all stake-holders should be identified, and inputs should be collected from all stake-holders before generating a comprehensive and detailed need.

7.3.2 Market Analysis and Identification of Potential Suppliers

Market and industry need to be analyzed to identify opportunities before identifying potential suppliers. Identifying potential suppliers can be as simple as verifying contact information of suppliers or as complex as asking for a preproposals and supplier meetings. To some extent, this depends on the type of purchase and on the product or service being purchased. Once the potential suppliers have been identified, a RFQ or a request for proposal (RFP) will be prepared.

7.3.3 RFQ Generation and Negotiation

Once a set of potential suppliers has been identified, a RFQ must be generated and posted to invite bids. Based on the bids submitted, potential partners will be identified and post-bid negotiation should be conducted with identified suppliers. Through a careful analysis of suppliers and their offers, purchase decision must be made.

7.3.4 Contract Award and Implementation

A contract is awarded to the identified supplier. From this point, the major responsibility of purchasing is to make sure the correct goods are delivered in the correct quantity at the right place. If not, purchasing takes some action to fill the gaps.

7.3.5 Evaluating the Purchase and the Supplier

Most purchasing organizations summarize the accumulated experience with a supplier through many transactions and many purchases. When one transaction goes awry, purchasing may contact the supplier

TABLE 7.2   Procurement Performance Checklist

Leadership Value-Creation Characteristics	Not Yet on Radar	On the "to-do" List	Now Underway	Getting Results
Our supply management organization is involved in setting, not just executing, company strategy.				
Our key suppliers provide innovation throughout the new product or service development process that helps fuel our growth.				
Procurement is involved in identifying and managing alliance and outsourcing opportunities (not just in negotiating and contracting).				
Our company systematically applies advanced cost-management strategies across our spend base.				
We understand our supply risks and have mitigation strategies in place for all major spend categories.				
Our supply management organization fosters cross-functional teaming throughout the company.				
Our procurement processes reflect best practices and are applied company-wide.				
Our procurements tools (such as e-sourcing, requisition-to-pay and contract management) function together as a system to allow for efficient execution of procurement processes.				
Our organization is actively developing and strengthening the employee skills required to successfully apply advanced techniques.				
We have a comprehensive management plan to attract and retain the best talent for supply management.				
Total				

Scoring (sum of your "now underway" and "getting results" check marks)
9–10 = Well prepared —must continue to evolve in a rapidly changing environment
5–8 = Approaching readiness—must fill critical gaps to reach leadership level
0–4 = Falling behind—must build foundation capabilities and launch remedial actions to avoid a competitive disadvantage

Source: From A.T. Kearney, *2004 Assessment of Excellence in Procurement: Creating Value Through Strategic Supply Management*, 2005. Available at: http://atkearney.com/shared_res/pdf/AEP_2004_S.pdf

to avoid future problems. When many transactions fail to meet standards, purchasing then seeks new suppliers. Finally, practitioners can learn what purchasing leaders are doing to improve their purchasing process as given in Table 7.2.

7 4   Sourcing and Supply Management

Kraljic (1983) notes that "purchasing has evolved into supply management," and companies have developed an array of strategies to improve the performance of their primary supply channels [3]. Some of the approaches used include tight partnership with the suppliers, JIT deliveries, and implementation of a sophisticated information system for smoother transactions [3–7]. These approaches generally require a reduced supplier base, which in fact has been a trend in recent years and constitutes the major principle of the JIT purchasing strategy. It is not rare that supplier based is shrunk down to one—a single sourcing strategy.

7.4.1   Single versus Multiple Sourcing

The two major sourcing strategy options available to firms are single and multiple sourcing. Single sourcing can be broadly defined as fulfillment of all of organization's needs for a particular purchased item from one supplier [8]. On the other hand, multiple sourcing refers to a purchasing strategy of an identical part being purchased from two or more suppliers. Single sourcing has typically been regarded as an integral part of a JIT system as multiple sourcing is considered a violation of the JIT principle of elimination of waste. However, practitioners have feared the risks associated with the single sourcing strategy.

The risk and benefits associated with sourcing strategies can be grouped into five categories: disruption of supply, price escalation, inventory and scheduling, technology access, and quality [8]. The category of disruption of supply includes the risk that the vendor, for whatever reason, will decide to terminate the sourcing relationship, thereby voluntarily cutting off the supply. The risk of a vendor escalating price can be realized when an only vendor tries to take advantage of being the only source of supply. These two categories of risks represent the two most commonly cited reasons for not following a single sourcing strategy. The other categories, which tend to be less frequently cited but may be of even greater significance, are related to inventories and schedules, technology, and quality. The last category has been the main drive for a single sourcing strategy.

Many firms are preparing contingency plans for mitigating the risk associated with the supply of critical materials. Commonly employed strategies are (*i*) maintaining multiple sourcing or at least listing or database of alternative suppliers for the critical parts, (*ii*) enhancing integration with suppliers by periodically evaluating supplier's process control and financial strength, listening to rumors advising of potential concerns, arranging frequent meetings with suppliers with senior management attending, and sharing updates on capacity, quarterly forecasts, and production scheduling with key suppliers, (*iii*) ensuring that suppliers themselves have their own contingency plans, (*iv*) preparing emergency product reformulation plans and (*v*) keeping higher safety stocks for critical materials.

HP successfully reduces risk by portfolio approach, which is an instance of multiple sourcing [9] by making mix of different types of contracts with different vendors with an objective to improve the overall risk or return characteristics. In the traditional approach variability was passed along to suppliers as past deals are not necessarily connected to future deals so relationship building is not critical. HP overhauled its procurement function by using a combination of long-term structured contracts, short-term unstructured contract, and spot market purchases to make the consolidated be more efficient. HP has used this successfully for electricity at a San Diego plant [9].

While multiple sourcing can be used to mitigate the associated risk, it violates the main principle of the JIT system and has its own drawbacks. Multiple sourcing strategies usually increase the administrative work at buyer's side, increase fixed cost associated with purchasing, and make it difficult to involve suppliers in the business plan. In conclusion, neither single sourcing nor multiple sourcing is always the best sourcing arrangement and the best strategy depends on the individual industry or market and the specific purchasing situation. A general comparison of these sourcing strategies is presented in Table 7.3.

TABLE 7.3 Comparison of Single and Multiple Sourcing Strategies

Strategy	Advantages	Disadvantages
Single sourcing	The supplier can reduce the price due to economies of scale. Setup costs associated with purchasing can be reduced. These costs include transportation cost, tooling and fixture cost, and administration cost. Long-term relationships can result in mutual cost reduction. Quality control and scheduling are easier.	The level of risk is higher. Dependency on a supplier can be more than optimal level. Small-size part suppliers have difficulties in entering the market. Buyer may have to maintain a high level of inventory. Purchase price is typically higher in a single sourcing situation.
Multiple sourcing	There is insurance against failure at one plant as a result of fire, strikes, quality, delivery problems, and so on. Competitive situation will prevent one supplier from becoming complacent. The buyer is protected against a monopoly.	Technical knowledge is shared among many suppliers, who can potentially help competitors. Without standardization, tooling and fixture cost can be significant. It is hard to build a long-lasting partnership. Increased administration work at buyer's side.

7.4.2 Supplier Selection

The decision of which suppliers to work with is as important as that of how many suppliers to work with. In today's competitive operating environment it is impossible to successfully produce low-cost, high-quality products without satisfactory vendors. This is why the vendor selection problem has been studied extensively and today we have a wide range of vendor selection models available in the purchasing literature [10].

Frequently, the relevant objectives in purchasing are in conflict. The vendor with the lowest per unit price may not have the best quality or delivery rating of the various vendors under consideration. Consequently, the firm must analyze the tradeoffs among the relevant criteria when making its vendor decisions. The analysis of these trade-offs is particularly important in modern manufacturing. For example, in JIT environments the trade-offs among price, quality, and delivery reliability are particularly important [11].

Multi-objective analysis has several advantages over single-objective analysis in that it allows the various criteria to be evaluated in their natural units of measurement and therefore eliminates the necessity of transforming them to a common unit of measurement such as dollars. In addition, such techniques present the decision-maker with a set of noninferior (or nondominated) solutions. Another advantage of multi-objective techniques is that they provide a methodology to analyze the impacts of strategic policy decisions. Such decisions frequently entail a reordering of the priorities on a firm's objectives. For example, the adoption of a JIT manufacturing strategy increases the emphasis on the quality and timeliness of delivery of components purchased from external sources. In addition, firms employing JIT strategies often attempt to reduce the number of vendors which supply material inputs. Changes in emphasis such as these often affect the cost that firms must pay for items purchased from vendors. The multi-objective approach to vendor selection provides decision-makers with a method to systematically analyze the effects of policy decisions on the relevant criteria in their vendor selection decisions.

The literature has identified several dimensions that are important for the multiple objective vendor selection decision [12,13]. These include net price, delivery, performance, history, capacity, communication system, service, geographical location, and so on. Dickson (1966) identified 23 different criteria evaluated in the vendor selection process. In that article, quality was seen as being of extreme importance while delivery, performance history, warranties and claim policies, production facilities and capacity, price, technical capability, and financial position were viewed as being of considerable importance in the vendor selection process.

The operations research community has also addressed the vendor selection problem in many different ways. Weber et al. (1991) discuss the complexity of the supplier selection decision from an operations research perspective and review the literature on this subject [11,13]. Degraeve et al. (2000) and Ghodsypoura and O'Brien (2001) address the vendor selection decision in the framework of the cost criterion, while Roethlein and Mangiameli (1999) address the problem from the supplier's perspective [14]. Rosenthal et al. (1995) study a vendor selection problem in which purchase decisions are to be made when some of suppliers offer bundling of their products [15]. Tajbakhsh et al. (2005) study a multi-supplier inventory management problem with suppliers offering random discounts and apply the multi-supplier inventory model to the problem of identifying a profit-maximizing set of suppliers [16].

7.4.3 Supply Contracts

Buyer–supplier relationships can take many forms and some can be made formal through a binding contract. Several different contract types have been used to ensure adequate supplies and timely deliveries. In a supply contract, the buyer and supplier may agree on:

- Pricing
- Supply quantity
- Return policy
- Delivery lead times

The precise design of supply contract may vastly influence the performance of supply chain—both buyer and supplier—as well as the performance of individual members in the supply chain. Sometimes, a supply contract may not be in the best interest of all the members, which gives rise to the importance of contracts enabling coordination among independent firms in a supply chain.

In the last few years many academic researchers and industry practitioners have investigated various forms of supply contracts and recognized supply contracts not merely as a tool to ensure adequate supply and deliveries but also as a leverage to improve supply chain performance. Suppose a typical supply chain consisting of a single supplier and a single retailer, in which the retailer assumes all the risk of having excessive inventory after sales, whereas the supplier takes no risk; without any risk, the supplier would like the retailer to increase the order quantity so as to minimize the profit loss due to shortage in the retailer's location. If the supplier wishes to share some of the risk with the retailer, the order size may increase and both the supplier and the retailer can enjoy the increased profit.

Recently, a number of supply contracts have been studied and they allow this risk sharing between supplier and buyer, thereby increase profit for individual members of a supply chain [17].

7.4.3.1 Buy-Back Contract

With a buy-back contract the supplier charges the retailer a fixed wholesale price per unit and agrees to buy-back unsold goods for a prespecified price. This will shift risk from the retailer to the supplier and results in an increased order size, which then reduces the likelihood of shortage at retailer's location and increases supplier's profit.

7.4.3.2 Revenue-Sharing Contract

In a revenue-sharing contract, the retailer shares a predetermined share of its revenue with the supplier in return for a reduced wholesale price. Again, this will result in an increased order size placed by the retailer and both parties will get increased profits.

7.4.3.3 Quantity-Flexibility Contract

Under a quantity–flexibility contract the supplier guarantees full refund for unsold goods up to a specified quantity in the contract. Notice that a quantity–flexibility contract gives a full refund for a portion of the unsold goods, whereas a buy-back contract provides partial refund for all unsold goods.

7.4.3.4 Sales-Rebate Contract

A supplier with a sales-rebate contract should provide a rebate for each item sold above a certain quantity, which gives the retailer an incentive to increase the order quantity.

These contracts achieve the coordination in a supply chain by inducing the retailer to order more than he or she would with a standard wholesale-price contract, under which a unit price is paid by retailer for goods ordered and no other payments are made between the retailer and the supplier. Though these types of contracts perform superbly in theory, they have various implementation drawbacks in practice. For example, buy-back contracts require the supplier to have an effective reverse logistics system and revenue sharing contracts incur a significant administrative cost.

7.4.4 JIT and Economic Order Quantity Purchasing

Manufacturing companies that use economic order quantity (EOQ) purchasing, either classical EOQ model or a variation thereof, increasingly are faced with the decision of whether or not to switch to the JIT purchasing policy. This is a complex decision, requiring careful examination of each system and its possible impact on a variety of factors, such as cost, quality, and flexibility of the operations [18].

Just in time is one of the most celebrated modern manufacturing techniques and its use has helped many firms in becoming more productive and competitive. JIT is designed to virtually eliminate the need to hold items in inventory and is defined as "to produce and deliver finished goods just in time to be sold, sub-assemblies just in time to be assembled into goods, and purchased materials just in time to be transformed into fabricated parts" [19]. Most JIT companies view JIT purchasing as a significant component of their JIT implementation and a major factor in their success.

Despite the impressive success of JIT production systems, many companies still use the traditional EOQ-based approach to determine their purchase orders. This is particularly true for small manufacturing firms which cannot effectively implement JIT purchasing [20]. The traditional inventory management practices center around the EOQ model which focuses on minimizing the inventory costs rather than minimizing the inventory [21].

There are a large number of studies comparing EOQ and JIT systems [22–25]. Most of these studies advocate the use of JIT over EOQ. Johnson and Stice (1993) conclude that traditional inventory management techniques may under-emphasize the cost of maintaining large inventories and that JIT may under-emphasize the costs on not maintaining inventories, particularly since such costs are often difficult to identify and measure [21].

7.4.4.1 EOQ Purchasing Plans

According to the EOQ model, a manufacturer places several orders to its suppliers every year, with the size of each order being enough to satisfy the production demand for a certain period of time. For this model, the most EOQ that minimizes the total annual costs can be obtained mathematically.

One of the basic assumptions of the EOQ model is that no shortages are allowed. However, in this chapter, we consider a variation of the EOQ model and allow shortages at cost. As a result the cost function will incorporate a backlog penalty cost for those units on backorder and therefore the cost function consists of inventory holding cost, backorder cost, setup cost, and order cost.

The following notation will be used for the EOQ model with backlog:

Q = maximum inventory level
B = maximum backorder level
h = unit inventory holding cost per unit time
b = unit backlog cost per unit time
D = constant and deterministic demand per unit time
K = replenishment setup cost per nonzero order
$Q + B$ = order quantity
P_E = unit cost under EOQ
T_E = total cost under EOQ

From the inventory profile shown in Figure 7.1 we can easily identify that area 1 corresponds to the holding cost and area 2 corresponds to the backlog cost. The total cost is defined by

$$T_E = \frac{hQ^2}{2(Q+B)} + \frac{bB^2}{2(Q+B)} + \frac{K+D}{Q+B} + P_E D \qquad (7.1)$$

The first term is the holding cost, the second the backlog cost, the third the setup cost and the last the total purchase cost. The first-order condition leads to the following optimal solutions.

$$B^* = \sqrt{\frac{2DKh}{b^2 + bh}},$$

$$Q^* = \sqrt{\frac{2DKb}{bh + h^2}}$$

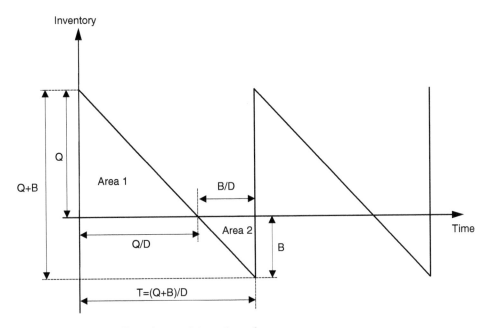

FIGURE 7.1   Inventory profile under an EOQ purchase plan.

Therefore, the optimal order quantity is simply given by

$$EOQ^* = Q^* + B^* = \sqrt{\frac{2KD}{h+b}}\left(\sqrt{\frac{b}{h}} + \sqrt{\frac{h}{b}}\right) \tag{7.2}$$

It is easy to see that Q^* and B^* approach to the standard EOQ and 0, respectively as b goes to infinity and hence the model reduces to the standard EOQ model. Under an optimal EOQ purchase plan, the total cost will be

$$T_E = \sqrt{\frac{2DKhb}{h+b}} + P_E D \tag{7.3}$$

7.4.4.2   JIT Purchasing Plans

Under the JIT system, much of the holding costs and some components of the ordering costs (e.g., preparation of purchase orders for each delivery) can be significantly reduced or eliminated. On the other hand, since JIT requires timely deliveries of a small quantity, costs such as transportation and inspection costs can increase. In an effort to reduce such cost increase, buyers are trying to find suppliers in buyer's vicinity and to improve the quality at the supplier's facility. Nonetheless, the unit purchase cost under a JIT purchase plan is typically higher than that under an EOQ plan. Fazel (1997) argues that the unit price captures these cost increases (e.g., storage, inspection, transportation, preparation of purchasing orders, etc.), whereas Schniederjans and Cao (2001) incorporate the savings from storage space requirement reduction. Aligned with the latter view, the total JIT cost function can be defined by

$$T_J = P_J D - FN \tag{7.4}$$

where P_J is the unit purchase price under the JIT purchase plan, F is the annual cost to own and maintain a square foot of storage space, and N is the number of square feet saved by adopting a JIT purchase plan.

Maximum JIT Purchase Price

For an item with a given demand, D, we can also find the highest price, P_{max}, that the manufacturer can pay to purchase the item on a JIT basis and still be economically better off than using EOQ purchasing. The highest JIT purchase price, allowing backlog at penalty of b, the maximum unit price under a JIT purchase plan that still justifies the adoption of a JIT purchase plan is given by [26].

$$P_{max} = P_E + \frac{\sqrt{2bhK}}{\sqrt{D(b+h)}} + \frac{FN}{D} \tag{7.5}$$

If the unit purchase price under the JIT plan is higher than P_{max}, the EOQ purchase plan offers a lower cost.

Alternatively, given a unit purchase price P_J under the JIT plan and P_E under the EOQ plan, the maximum annual demand that justifies the merit of using EOQ can be derived from Equations 7.3 and 7.4 and given by

$$D_{max} = \frac{bhK + FN(b+h)(P_J - P_E) + \sqrt{bhK\left[bhK + 2FN(b+h)(P_J - P_E)\right]}}{(b+h)(P_J - P_E)^2} \tag{7.6}$$

7.5 Auctions and e-Procurement

The increasing use of computers has enabled e-marketplaces as a major way to improve procurement efficiency. An e-marketplace is a many-to-many market mechanism which allows an enterprise to be connected to its business partners using its application and network infrastructure [27]. Via e-marketplaces, both buyers and suppliers can reach larger markets and gain vital information about the market situation to improve their performance related to procurement process.

Prior to the widespread of the Internet, EDI has been employed to process transactions associated with procurement. The primary objective of EDI in procurement is the efficient processing of transactions in a very secure and reliable way. Common uses of EDI include invoicing, purchase orders, pricing, advanced shipment notices, electronic funds transfer, and bill payment. Benefits of EDI includes reductions in document preparation and processing time, inventory carrying costs, personnel costs, information flow, chipping errors, returned goods, lead times, order cycle times and ordering cost, billing accuracy, customer satisfaction, and so on. However, potential drawbacks to EDI, such as high setup cost, lack of standard formats, and incompatibility of computer hardware and software, have hampered the wide acceptance, especially amongst small- to medium-sized firms.

Recently, the advent of the Internet has changed the prospect of EDI. The Internet seems to make EDI an obsolete technology as well as play a complementary role to EDI. The future of EDI is likely to involve a combination of the more traditional value-added networks and the Internet—as opposed to the Internet substituting EDI transactions.

7.5.1 Auctions in Action

Auctions have been used as pricing tool when an item has no commonly accepted price. Auctions, an interactive means of matching buyers and sellers, are expected to account for more than half of online business transactions in the near future. In April 1999, Sprint Corp. began a typical procurement effort by e-mailing RFP to 150 suppliers, 91 of whom responded to the call. eBreviate, a San Francisco-based Internet auction service provider, and A.T. Kearney, a procurement consultant firm, set up an Internet auction and coordinated the logistics of the suppliers from Canada, Mexico, and the United States. During a 4-h online auction, 800 bids were submitted, the market price was established, and two-thirds of contenders were eliminated from competition. Sprint then resumed a conventional procurement process with face-to-face negotiations, and finally settled on 20 suppliers. The total cost of this buy was $450 million over three-year life of the purchase contracts, and saved the total procurement cost by 18% as well as six weeks' reduction in the buying process.

The use of auctions for transportation service procurement has been prevalent as the medium of obtaining carrier services for shippers. Typically, single lane contracts are bid for by carriers where a single lane represents a commitment to move a specified volume from an origin to a destination. However, cost structures for carriers exhibit economies of scope [28]. That is, carriers would like to obtain a set of contracts that collectively represents traveling as few empty kilometers or miles as possible where an empty distance is the situation where a vehicle is traveling without any load. In the truck-load (TL) procurement context carriers would place single bids on sets of distinct lanes. Allowing single bids on sets of distinct lanes allow carriers to express synergies that exist for certain lanes.

A chemical industry giant DuPont is one of those who realize the advantage of auctions in transportation service procurement. It adopted computerized bidding using Pittsburgh-based CombineNet's Decision Guidance System for its large, complex transportation sourcing decisions. This new system allows vendors to submit so-called "expressive bids" to tailor bids to their strengths and needs. Using this new approach to procurement, DuPont awarded 12,000 ocean lanes and conducted a procurement process for transportation services.

7.5.2 Bundling in Procurement and Combinatorial Auctions

Bundling in purchasing is gaining more and more importance, mainly due to advances in purchasing practices, globalization, and availability and speed of information association with electronic purchasing tools and capabilities. A bundle may contain any combination of products and/or services and the concept of bundling has the potential of improving the efficiency of the procurement process. Bundling can occur (*i*) for a one-time purchase, for example production machinery, (*ii*) for continuously or regularly purchased items such as office products, that may be combined in a blanket purchase order (PO) procured from an aggregator, (*iii*) and for both short- and long-term contracts.

When an auction of multiple items is performed, it is often desirable to allow bids on bundles or combination of items (e.g., transportation service procurement auctions as described earlier), as opposed to only on single items. Such an auction is called combinatorial, and has been applied in a variety of environments involving economic transactions, and they have the potential to play a critical role in improving the efficiency of supply management. Examples are numerous: Logistics.com has conducted B2B procurement combinatorial auctions in the transportation industry; Home Depot successfully has used combinatorial auctions to procure transportation services and reported a significant savings over its traditional procurement process [29]; IBM, on behalf of Mars Incorporated, has performed combinatorial auctions for procurement [30]; and Net Exchange (www.nex.com) procures transportation services for Sears Logistics [31].

However, the use of a combinatorial auction as a procurement tool is facing with major challenges. It is not until these issues are successfully addressed before the widespread of combinatorial auctions for procurement. Among many important practical design issues and challenges [32], there is the requirement by the auctioneer (or buyer) to solve an NP-hard integer program to determine the bidders that are to supply requested items or services [33,34]. This processed is referred to as winner determination.

Bidders also have complex decision problem of evaluating a number of possible bundles so as to identify the one with the maximum utility. This optimization problem is application specific and most likely to be NP-hard because of the sheer number of potential bundles. There have been only a few academic studies done on this aspect of combinatorial auctions and they are mostly in the context of transportation service procurement [25,35,36]. Mathematical modeling to aid bidders in determining the appropriate bundle of items in combinatorial auctions should enable more practical use of combinatorial auctions.

7.6 Case Studies

7.6.1 JIT versus EOQ Purchase Plans

Candle and Fragrance, a Canadian specialty candle maker, has recently restructured their procurement practice from EOQ-based to JIT system. The firm has a facility near the city of Toronto, where 145 employees are producing a variety of candles and other related products. The restructuring project has focused on medium jar candle production, as it is the major product of the firm. This product is manufactured in more than 80 different fragrances and equal number of colors, generating 6400 combinations. A single type container (medium jar) is used for all these products and the annual demand for this jar is close to 1 million. The company currently purchases medium jars using an EOQ model.

An industrial engineer has led the project and carefully compared costs and benefits of the two purchasing plans. Under the current EOQ plan, the medium jars are ordered in a six-week cycle, the usage rate is 20,000 units per week and a 3% waste should also be included in the calculation. Jars are delivered on pallets, each of which can hold 768 jars. Therefore, an order in a six-week cycle involves 161 pallets or 123,648 jars. Because jars are made of heavy glass, the storage has two-storey rack. The dimension of a pallet is 45 inches by 45 inches, or 15 square feet, requiring a warehouse space of 1215 square feet to store 161 pallets. The firm estimates that each square foot of warehouse costs $5.95 per month to maintain, being broken down to $5.75 for rent and $0.20 for insurance. Therefore, annual cost per square foot is $71.4.

The company approached its glass supplier and proposed a JIT purchase system on a weekly basis. This means that the firm will buy 27 pallets per week totaling 20,736 pieces of medium jar per week and hence the space savings will be 1013 square feet (the firm can remove 134 pallets). The supplier agreed to a JIT delivery on every Friday to the specified quantity but at a higher selling price. The firm also estimated that the shortage in inventory would incur $0.20 per piece. Summarizing the cost analysis to get the following:

P_E = $1.52 (unit price under the EOQ plan)
P_J = $1.56 (unit price under the JIT plan)
h = $0.38 (inventory holding cost per unit time)
K = $1185 (ordering cost including transportation cost)
F = $71.4 (annual cost to own and maintain a square foot of facility)
N = 1013 (number of square feet saved by JIT plan)
b = $0.10 (backlog cost per unit time)

The unit purchase cost under a JIT plan is higher than that under the EOQ plan. By comparing the optimal cost under two purchasing policies—EOQ purchase and JIT purchase—the unit purchase cost at which the purchasing manager is indifferent between the two policies can be computed using Equation 7.5 as follows:

$$P_{max} = \$1.52 + \frac{\sqrt{2 \times \$0.1 \times \$0.38 \times \$1185}}{\sqrt{1,000,000 \times (\$0.1 + \$0.38)}} + \frac{\$71.4 \times 1013}{1,000,000} = \$1.61$$

Therefore, the firm concludes that it is beneficial to adopt a JIT purchase plan as long as the supplier agrees to accept unit price less than $1.61. Alternatively, the manager is indifferent between two purchase strategies when the annual demand is given as below:

$$D_{max} = 2,330,029 \text{ (units/year)}$$

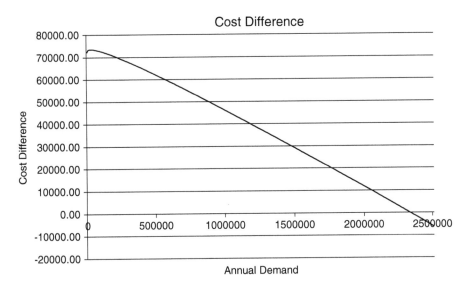

FIGURE 7.2 Total cost difference.

which implies that the JIT purchase plan is better for annual demand less than or equal to 2,330,029 medium jars. Because the current annual demand is roughly a million, the JIT purchase plan outperforms the EOQ plan in this case. Moreover, the difference in optimal cost under the two policies can be graphed as below:

As of May 2005, the firm has an annual demand less than 1 million pieces and therefore the project concluded that the transition from an EOQ plan to a JIT plan was beneficial to the firm.

References

1. Lewis, H.T., *Industrial Purchasing*, Prentice-Hall, Inc., New York, 1933.
2. Bloomberg, D.J., LeMay, S., and Hanna, J.B., *Logistics*, Prentice-Hall, Inc., Upper Saddle River, New Jersey, 2002.
3. Kraljic, P., Purchasing must become supply management, *Harvard Business Review*, 1983, 109–117.
4. Fisher, M.L., What is the right supply chain for your product?, *Harvard Business Review*, 1997, pp. 105–116.
5. Krajewski, L.J. and Ritzman, L.P., *Operations Management: Strategy and Analysis*, 5th ed., Addison-Wesley, MA, 1999.
6. Ross, D.F., *Competing through Supply Chain Management*, Kluwer Academic Publishers, Dordrecht, 2000.
7. Stadtler, H. and Kilger, C., *Supply Chain Management and Advanced Planning*, 2nd ed., Springer-Verlag, Berlin, 2002.
8. Treleven, M. and Schweikhart, S.B., A risk/benefit analysis of sourcing strategies: single vs. multiple sourcing, *Journal of Operations Management*, 7, 1988, pp. 93–114.
9. Billington, C., HP cuts risk with portfolio approach, *Purchasing*, 131, 2002, pp. 43–45.
10. Degraeve, Z., Labro, E., and Roodhooft, F., An evaluation of vendor selection models from a total cost of ownership perspective, *European Journal of Operational Research*, 125, 2000, pp. 34–58.
11. Weber, C.A., and Current, J.R., A multiobjective approach to vendor selection, *European Journal of Operational Research*, 68, 1993, pp. 173–184.

12. Dickson, G.W., An analysis of vendor selection systems and decisions, *Journal of Purchasing*, 2, 1966, pp. 5–17.

13. Weber, C.A., Current, J.R., and Benton, W.C., Vendor selection criteria and methods, *European Journal of Operational Research*, 50, 1991, pp. 2–18.

14. Roethlein, C.J. and Mangiameli, P.M., The realities of becoming a long-term supplier to a large total quality management customer, *Interfaces*, 29, 1999, pp. 71–81.

15. Rosenthal, E.C., Zydiak, J.L., and Chaudhry, S.S., Vendor selection with bundling, *Decision Sciences*, 26, 1995, pp. 35–48.

16. Tajbakhsh, M.M., Lee, C.-G., and Zolfaghari, S., An inventory model with random discount offerings, Working paper, Department of Mechanical and Industrial Engineering, University of Toronto, 2005.

17. Cachon, G.P., Supply chain coordination with contracts, in *Handbooks in OR and MS*, Vol. 11, A.G. de Kok and S.C. Graves, Eds., Elsevier, Amsterdam, The Netherlands, 2003, Chapter 6.

18. Fazel, F., A comparative analysis of inventory costs of JIT and EOQ purchasing, *International Journal of Physical Distribution and Logistics Management*, 27, 1997, pp. 496–504.

19. Schonberger, R.J., *Japanese Manufacturing Techniques*, The Free Press, New York, NY, 1982.

20. Temponi, C., Implementation of two JIT elements in small-sized manufacturing firms, *Production and Inventory Management Journal*, 36, 1996, pp. 23–29.

21. Johnson, G.H. and Stice, J.D., Not quite just-in-time inventories, *The National Public Accountant*, 38, 1993, pp. 26–29.

22. Chyr, F., Lin, T.M., and Ho, C.F., Comparison between just-in-time and EOQ system, *Engineering Costs and Production Economics*, 18, 1990, pp. 233–240.

23. Jones, D.J., JIT and the EOQ model: odd couples no more, *Management Accounting*, 72, 1991, pp. 54–57.

24. Lee, S.M. and Ansari, A., Comparative analysis of Japanese just-in-time purchasing and traditional US purchasing systems, *International Journal of Operations and Production Management*, 5, 1985, pp. 5–14.

25. Lee, C.-G., Kwon, R.H., and Ma, Z., A carrier's optimal bid generation problem in combinatorial auctions for transportation procurement, *Transportation Research Part-E*, 43, 2007, pp. 173–191.

26. Schniederjans, M.J. and Cao, Q., An alternative analysis of inventory costs of JIT and EOQ purchasing, *International Journal of Physical Distribution and Logistics Management*, Vol. 31, No. 2, 2001, pp. 109–123.

27. Raju, C.V.L., Narahari, Y., and Shah, S., Procurement auctions using actor-critic type learning algorithm, *Proceedings of the IEEE International Conference on Systems, Man, and Cybernetics*, 5, 2003, pp. 4588–4594.

28. Caplice, C. and Sheffi, Y., Optimization-based procurement for transportation services, *Journal of Business Logistics*, 24, 2003, pp. 109–128.

29. Elmaghraby, W. and Keskinocak, P., Combinatorial auctions in procurement, in *The Practice of Supply Chain Management*, C. Billington, T. Harrison, H. Lee, and J. Neale, Eds., Kluwer Academic Publishers, 2003, Chapter 15.

30. Hohner, G., Rich, J., Ng, E., Reid, G., Davenport, A.J., Kalagnanam, J.R., Lee, H.S., and An, C., Combinatorial and quantity discount procurement auctions benefit Mars, *Incorporated and its suppliers Interfaces*, 33, 2003, pp. 23–35.

31. Ledyard, J., Olson, M., Porter, D., Swanson, J., and Torman, D., The first use of a combined value auction for transportation services, *Interfaces*, 32, 2002, pp. 4–12.

32. Pekec, A. and Rothkopf, M., Combinatorial auction design, *Management Science*, 49, 2003, pp. 185–1503.

33. de Vries, S. and Vohra, R., Combinatorial auctoins: a survey, *INFORMS Journal on Computing*, 15, 2003, pp. 84–309.

34. Sadeh, N. and Song, J., Multi-attribute supply chain negotiation: coordinating reverse auctions subject to finite capacity considerations, *Proceedings of the ACM Conference on Electronic Commerce*, 5, 2003, pp. 53–60.

35. Kwon, R.H., Lee, C.-G., and Ma, Z., An integrated combinatorial auction mechanism for truck-load transportation procurement, Working paper, Department of Mechanical and Industrial Engineering, University of Toronto, 2006.

36. Song, J. and Regan, A., Combinatorial auctions for transportation service procurement: the carrier perspective, *Transportation Research Record*, in press, 2005.

8

Demand Forecasting in Logistics: Analytic Hierarchy Process and Genetic Algorithm-Based Multiple Regression Analysis

William Ho
Aston University

Carman Ka Man Lee
Nanyang Technological University

Demand forecasting plays an important role in today's integrated logistics system. It provides valuable information for several logistics activities including purchasing, inventory management, and transportation. To minimize the total logistics cost, an accurate and reliable forecasting approach should be developed and adopted. In real-world situations, both quantitative and qualitative factors affecting the demand should be taken into consideration simultaneously. Since analytical hierarchy process (AHP) and genetic algorithm (GA) have emerged as the promising methodologies for dealing with a wide variety of decision-making problems, this chapter presents a AHP-based approach to analyze the priority rankings of all relevant factors first, and then a GA-based multiple regression analysis approach to formulate a forecasting mathematical equation. This chapter provides a novel approach by combining both quantitative and qualitative approach for demand forecasting and this approach is implemented in a leading electronic company in Hong Kong.

8.1 Introduction

Logistics management is the process of planning, implementing, and controlling the efficient and cost-effective flow and storage of raw materials, work-in-process inventories, finished products, and related information from the point of origin to the point of consumption for the purpose of conforming to customers' requirements [1]. Logistics management is sophisticated because it involves numerous complicated activities including customer service, demand forecasting, distribution management, information maintenance, inventory management, materials handling, order processing, packaging, purchasing, reverse logistics, transportation, warehousing, and so on. It is undoubted that these activities are inter-related. For instance, reducing the inventory of finished products will reduce the inventory carrying costs and warehousing costs, but may lead to stock-out as a result of reduced levels of customer service. Because of this relationship, logistics management can also be regarded as the administration of various activities in an integrated system [2].

Inventory management is an important element in the integrated logistics system because it occupies the largest proportion of the total cost. If the amount of inventory held is much higher than the actual demand, there is a chance of not being sold out, and obsolescence cost is incurred because the products cannot be sold in original price or cost. In case the inventory level is kept too low, stock-out costs are incurred as customers may choose the substitutes rather than waiting. In the other words, keeping optimal inventory level is utmost important. Due to the presence of a wide variety of uncertainties in the real-world situations, however, the optimal inventory level is difficult to determine. One of these uncertainties is demand uncertainty, that is, the amount of finished products or services that customers will require at some point in the future is unknown. Demand forecasting, therefore, is a dominant attribute of the inventory management. Besides inventory management, demand forecasting provides valuable information for the purchasing and transportation problems, as illustrated in Figure 8.1. On the basis of forecasts, the decision-makers of the logistics companies can decide on the amount of raw materials to be purchased from the suppliers (i.e., the purchasing problem) to meet the production requirement, decide on the amount of work-in-process and finished products to be stored in the warehouses (i.e., the inventory problem), and decide on the amount of finished products to be transported to the customers (i.e., the

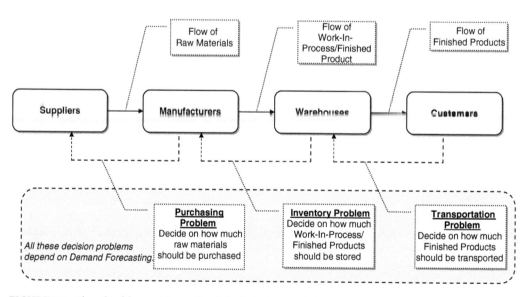

FIGURE 8.1   The role of demand forecasting in logistics/supply chain management.

transportation problem) so as to meet the customers' demands. It is, therefore, believed that the demand forecasting plays a crucial role in logistics and supply chain management.

There are extensive forecasting techniques available for anticipating the future. The classification of the techniques is shown in Figure 8.2 in which the approaches are either qualitative or quantitative. Firstly, the qualitative approach is normally applied to the cases where the historical data are not applicable or available. In such cases, decision-makers forecast the demand on the basis of their experience and judgment. Generally, the demand forecasting not only depends on the quantifiable factors such as the sales in the past, but also the nonquantifiable information including the corporation's policies, competitors' strategies, customers' preferences, and so on. Because the qualitative approach takes the nonquantifiable information into consideration, it can generate a clearer picture for the decision-makers to forecast the demand. Secondly, the quantitative approach can be adopted when the historical data of the variable to be forecasted is available. The basic assumptions are that the information can be quantified, and the future demand will follow or coincide with the trend of the past. Time-series and casual methods are the two commonly used quantitative approaches. The former method applies statistical techniques to discover a pattern in the historical data such as trend, cyclical, seasonal, and irregular, and then extrapolate this pattern into future. Examples of time-series method are moving average, exponential smoothing, and so on. Rather than identifying the trend, casual method aims at developing the casual relationships between the demand and its input factors. A typical example of this method is regression analysis. For example, customer demand is influenced by four factors such as product price, advertising expenditures, promotion policy, and seasonality. Regression analysis is then used to develop an equation showing how these factors are related to the customer demand.

Heizer and Render [3] stated that the casual method is more practical and powerful than the time-series method because it considers other factors relating to the demand to be forecasted instead of merely historical sales records. The casual method, therefore, is selected as one of the approaches in this chapter to deal with the demand forecasting. Besides the quantitative approach, the qualitative approach based on the AHP is also adopted. This chapter is organized as follows: Section 8.2 reviews the relevant literature studying the demand forecasting in logistics. Section 8.3 describes the principles of the qualitative and quantitative approaches. Section 8.4 provides a case study for illustrating how the approaches work. Finally, the conclusion is made in Section 8.5.

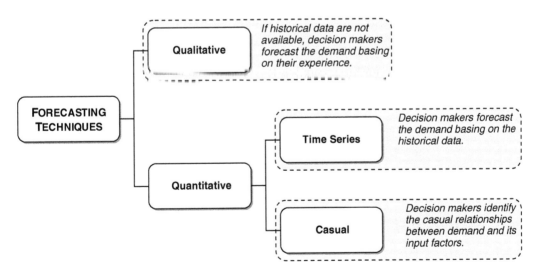

FIGURE 8.2 The classification of forecasting techniques.

8.2   Literature Review

Although demand forecasting is a crucial activity in logistics management, it has attracted less attention. Korpela and Tuominen [4] applied an AHP-based approach to demand forecasting. Both quantifiable and nonquantifiable factors that are relevant to the decision are considered in the approach. There are mainly three steps in the forecasting process such as construction of the hierarchy, assignment of priorities to the elements in the hierarchy, and calculation of the demand forecasting. The outputs of the approach are the overall probabilities for the alternative demand growth ranges. Spedding and Chan [5] noticed that the future event may be different from the structure of historical time series in a volatile business environment. The authors, therefore, proposed a Bayesian dynamic linear time-series model to forecast the demand in a dynamically changing environment. The routine forecasts can be updated by subjective intervention, for example, manager's experience. Kandil et al. [6] compared different types of the time-series method such as exponential smoothing for forecasting the demand of fast developing utility. According to the results, it was noticed that no single forecasting method had shown a constant and stable performance over the forecasting period. To overcome this drawback, it was suggested that the relevant factors affecting the demand should be taken into consideration as much as possible. In addition, some related qualitative information should be incorporated into the forecasting model. Jeong et al. [7] built a generic casual forecasting model, which is applicable to various areas of supply chain management. The coefficients of the model were determined using a heuristic method called the GA. The model was implemented to forecast the products' quality in a glass manufacturing company, and the demand of residential construction. Snyder [8] used the Croston method, which is an adaptation of exponential smoothing, to forecast the sales of slow and fast moving inventories. The method incorporates a Bernoulli process to capture the sporadic nature of the demand, and allows the average variability to change over time. Ghobbar and Friend [9] evaluated 13 forecasting methods including time series and casual for demand of intermittent parts in aviation industry. Since it was found that most of the methods produced poor forecasting performance, a predictive error forecasting model was developed. Chang et al. [10] presented a forecasting model for the sales of the printed circuit board. In the model, the correlations among the factors were identified using the gray relation analysis, whereas the effects of seasonality and trend are considered using the Winter's exponential smoothing method. Heuristic methods called the artificial neural networks and GA were adopted to solve the model. Liang and Huang [11] agreed that the logistics activities are interrelated in the supply chain management. The authors, therefore, developed a demand forecasting method with information sharing among different stakeholders of a supply chain to minimize the total cost for the entire supply chain. The method, which belongs to the class of time series, was tackled using the GA.

Based on the detailed discussion of the literature, two observations have been made. Firstly, for the quantitative approach, the casual method is superior to the time-series method in terms of the adaptability to the real-world situations. Since the demand forecasting is normally influenced by many factors, identifying the relationship between the factors is more effective than just focusing on the historical sales data. Due to this reason, the casual method instead of the time-series method is adopted and discussed in this chapter. The second observation is that the qualitative approach was paid less attention to when compared with the quantitative one. However, as mentioned earlier, some of the factors affecting the demand are nonquantifiable. In order to have a more accurate and reliable prediction, these factors should be taken into consideration, too. In the following section, the methodologies of an AHP-based qualitative approach and a GA-based quantitative approach are described.

8.3   Methodology

8.3.1   Analytic Hierarchy Process

The AHP, developed by Saaty [12], is a theory of measurement for dealing with quantifiable and non-quantifiable criteria. It can be applied to numerous areas such as performance measurement in higher education [13], and demand forecasting in logistics [4]. Since the AHP can provide a systematic framework for the decision-makers to interact and discuss about every factors relating to the decisions, it is selected as a tool to analyze the criteria affecting the demand, and most importantly, determine the demand growth in the future. The approach is similar to that in Korpela and Tuominen [4], except the field of application. The criteria affecting the demand are different when the types of the finished products are not the same. The implementation of our approach on a real case is carried out in Section 8.4.1.

The AHP consists of three main operations including hierarchy construction, priority analysis, and consistency verification. First of all, the decision-makers need to break down a complex multiple criteria decision problem into its component parts of which every possible attributes are arranged into multiple hierarchical levels. For example, overall goal, criteria, attributes of each criterion are in the first, the second, and the third levels, respectively. After that, the decision-makers have to compare each cluster in the same level in a pairwise fashion basing on their own experience and knowledge. For instance, every two criteria in the second level are compared at each time whereas every two attributes of the same criteria in the third level are compared at a time. Since the comparisons are carried out through personal or say subjective judgments, some degree of inconsistency may be occurred. To guarantee that the judgments are consistent, the final operation called consistency verification, which is regarded as one of the most advantages of the AHP, is incorporated in order to measure the degree of consistency among the pairwise comparisons by computing the consistency ratio [14]. If it is found that the consistency ratio exceeds the limit, the decision-makers should review and revise the pairwise comparisons. Once all pairwise comparisons are carried out in every level, and are proved to be consistent, the judgments can then be synthesized to find out the priority ranking of each criterion and its attributes. The overall procedure of the AHP is shown in Figure 8.3.

The first step of the AHP is to develop hierarchy of problem, that is, the demand forecasting in this chapter in a graphical representation which helps to illustrate every factor. For example, the performance of competitors is one of the criteria influencing the demand of which the attributes include product feature, pricing policy, and so on.

Constructing a pairwise comparison matrix is intended to derive the accurate ratio scale priorities. The relative importance of two criteria is examined at a time. A judgment is made about which is more important and by how much. Besides criteria, every two attributes of each criterion are compared at a time. The priorities can be represented by numerical, verbal, and graphical judgments. Subjective judgment can be depicted using quantitative scales which are usually divided into nine-point scale in order to enhance the transparency of decision-making process. In verbal judgment, preference of "equally preferred" is given a numerical rating of 1, whereas preference of "extremely preferred" is given a numerical rating of 9.

Synthesization is carried out after all the judgments have been determined together with all the comparisons being made. The most popular AHP software, Expert Choice, includes two synthesis modes: ideal and distributive. The ideal synthesis mode assigns the full priority of each criterion to its corresponding best (highest priority) attribute. The other attributes of the same criterion receive priorities proportionate to their priorities relative to the best attribute. The priorities for all the attributes are then normalized so that they sum to one. When using this mode, the addition or removal of "not best" attributes will not affect the relative priorities of other attributes under the same criterion. The distributive synthesis mode distributes the priority of each criterion to its corresponding attributes in direct proportion to the attributes' priorities. When using this mode, the addition or removal of an attribute results in a readjustment of the priorities of the other attributes such that their ratios and ranks can change and affect the priorities of the other attributes.

Consistency test will be conducted to ensure that the result is accurate and reliable, and all judgments are tested and evaluated so as to have a satisfactory result. The principal eigenvalue, which is used to calculate the consistency of judgments, captures the rank inherent in the judgments within a tolerable range. In general, the judgments are considered reasonably consistent provided that the consistency ratio is less than 0.1. Based on each attribute's priority and its corresponding criterion priority, the individual priority is summed to calculate the overall priority ranking.

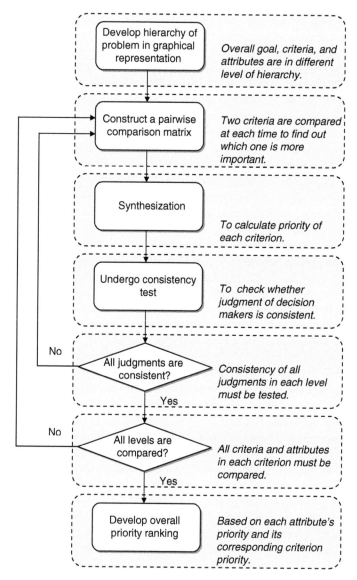

FIGURE 8.3 The flowchart of the AHP.

8.3.2 GA-Based Multiple Regression Analysis

One of the advantages for using the AHP is that both quantifiable and nonquantifiable factors are considered in the demand forecasting process. However, the outputs of the AHP are mainly the priority rankings of the criteria and attributes, and the probabilities of each range of demand growth in future. The exact value of forecasted demand cannot be generated using this approach. To compensate for this, a GA-based multiple regression analysis is adopted. In brief, the multiple regression analysis is to develop a forecasting mathematical equation, whereas the GA is to determine the coefficients of the equation so that the accuracy or performance of the equation is maximized or the error of the equation is minimized. The overall procedure of this quantitative approach is illustrated in Figure 8.4.

Regression analysis is used as a casual forecasting method in this chapter. This method develops a mathematical equation to identify the casual relationships between a dependent variable, that is, the

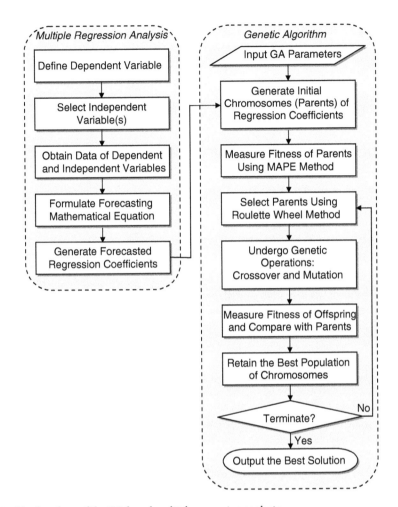

FIGURE 8.4 The flowchart of the GA-based multiple regression analysis.

variable being predicted, and one or more than one independent variable(s), that is, the variable(s) being used to predict the exact value of the dependent variable. There are two types of regression analysis: simple linear and multiple. The simple linear regression analysis consists of one dependent variable together with one independent variable, whereas the multiple regression analysis involves two or even more independent variables. Since the factors affecting the demand are not limited to one, the multiple regression analysis is studied and applied. The selection of the independent variables is dominated by the availability of the data in terms of quantitative, and the strong relationship with the dependent variable. Herein, the dependent variable is the demand forecasted, whereas the criteria and attributes defined in the AHP can be regarded as the independent variables. These independent variables can be included in the multiple regression analysis if they are quantifiable, and with high priority rankings. The AHP therefore provides valuable information for the multiple regression analysis, while at the same time the multiple regression analysis compensates for the AHP as mentioned earlier.

As can be observed from Figure 8.4, after the dependent and independent variables are defined, and the historical data of all variables are obtained, a forecasting mathematical equation can be formulated. It is noted that the use of regression analysis for trend projection is a time-series method rather than a casual method. Consider the quantitative data of the m independent variables in period t ($t = 1, 2, \ldots, n$) is collected. The proposed forecasting mathematical equation can be constructed in Equation 8.1. Besides, the mean absolute percentage error (MAPE) method, formulated in Equation 8.2, is used to evaluate the accuracy and performance of the multiple regression analysis.

$$F_t = c_0 + c_1 x_{1t} + c_2 x_{2t} + \cdots + c_m x_{mt} \tag{8.1}$$

$$\text{MAPE} = \frac{1}{n} \sum_{t=1}^{n} \frac{|F_t - A_t|}{A_t} \tag{8.2}$$

where

A_t = actual value of demand in period t
F_t = forecasted value of demand in period t
x_{it} = value of independent variable i in period t ($i = 1, 2, \ldots, m$)
c_i = forecasted regression coefficients of independent variable i
c_0 = constant coefficient

The regression coefficients indicate the relative importance of the corresponding independent variable in forecasting the value of the dependent variable, and dominate the accuracy of the equation. To minimize the error of forecasted demand, the best sets of coefficients must be generated. To accomplish this goal, the GA is utilized, and thus the approach is so-called GA-based multiple regression analysis. Before describing the procedure of the GA in the approach shown in Figure 8.4, the background of the general GA is presented in the following.

GA, developed by John Holland in the 1960s, is a stochastic optimization technique. Similar to other heuristic methods like simulated annealing (SA) and tabu search (TS), GA can avoid getting trapped in a local optimum by the aid of one of the genetic operations called mutation. Actually, the basic idea of GA is to maintain a population of candidate solutions that evolves under a selective pressure. Hence, it can be viewed as a class of local search based on a solution-generation mechanism operating on attributes of a set of solutions rather than attributes of a single solution by the move-generation mechanism of the local search methods, like SA and TS [15]. In the recent years, it has been successfully applied to a wide variety of hard optimization problems such as the traveling salesman and quadratic assignment problems [16,17]. The success is critical due to GA's simplicity, easy operation, and great flexibility. These are the major reasons why GA is used to optimize the coefficients of Equation 8.1.

Genetic algorithm starts with an initial set of random solutions, called population. Each solution in the population is called a chromosome, which represents a point in the search space. The chromosomes evolve through successive iterations, called generations. During each generation, the chromosomes are evaluated using some measures of fitness. The fitter the chromosomes, the higher the probabilities of being selected to perform the genetic operations: crossover and mutation. In the crossover phase, the GA attempts to exchange portions of two parents (i.e., two chromosomes in the population) to generate an offspring. The crossover operation speeds up the process to reach better solutions. In the mutation phase, the mutation operation maintains the diversity in the population to avoid being trapped in a local optimum. A new generation is formed by selecting some parents and some offspring according to their fitness values, and by rejecting others to keep the population size constant. After the predetermined number of generations is performed, the algorithm converges to the best chromosome, which hopefully represents the optimal solution or may be a near-optimal solution to the problem.

The procedure of the GA for optimizing the coefficients is described as follows. First of all, the GA parameters are set by the decision-makers. The parameters include:

- Population size: number of chromosomes in the population.
- Iteration number: number of generations performed.
- Crossover rate: ratio determining the number of chromosomes to undergo crossover.
- Mutation rate: ratio determining the number of chromosomes to undergo mutation.

After that, initial chromosomes represented in continuous value are generated randomly. There are $(m + 1)$ genes in each chromosome if m independent variables are selected in the phase of multiple regression analysis. Each chromosome is then measured by Equation 8.2. The roulette wheel selection method is adopted to select some chromosomes for performing one crossover and two mutation operations. The fitness of the offspring (i.e., new chromosome) will be measured and may become a member of the population if it possesses a relatively good quality. These steps form one iteration, and then the roulette wheel selection method is performed again to start the next iteration. The GA will not stop unless the predetermined number of iterations is conducted. The detailed procedure of the roulette wheel selection method, the crossover operation, and the mutation operations is discussed in the following subsections. Besides, the implementation of this quantitative approach on a real case is carried out in Section 8.4.2.

8.3.2.1 Selection

The roulette wheel selection method [16] is adopted in order to choose some chromosomes to undergo genetic operations. The approach is based on an observation that a roulette wheel has a section allocated for each chromosome in the population, and the size of each section is proportional to the chromosome's fitness. The fitter the chromosome, the higher the probability of being selected. It is true that the roulette wheel selection mechanism chooses chromosomes probabilistically, instead of deterministically. For example, although one chromosome has the highest fitness, there is no guarantee it will be selected. The only certain thing is that on the average a chromosome will be chosen with the probability proportional to its fitness. Suppose the population size is *psize*, and the fitness function for chromosome X_h is $eval(X_h)$, then the selection procedure is as follows:

Step 1: Calculate the total fitness of the population:

$$F = \sum_{h=1}^{psize} eval(X_h)$$

Step 2: Calculate the selection probability p_h for each chromosome X_h:

$$p_h = \frac{F - eval(X_h)}{F \times (psize - 1)}, \quad h = 1, 2, ..., psize$$

Step 3: Calculate the cumulative probability q_h for each chromosome X_h:

$$q_h = \sum_{j=1}^{h} p_j, \quad h = 1, 2, ..., psize$$

Step 4: Generate a random number r in the range (0, 1].
Step 5: If $q_{h-1} < r \leq q_h$, then chromosome X_h is selected.

8.3.2.2 Order Crossover

As shown in Figure 8.5, the crossover operator adopted in the GA is the classical order crossover operator. The procedure of the order crossover operation is listed as follows:

Step 1: Select a sub-string from the first parent randomly.
Step 2: Produce a proto-child by copying the sub-string into the corresponding positions in the proto-child.
Step 3: Delete those genes in the sub-string from the second parent. The resulted genes form a sequence.

Selected sub-string

Parent 1: 2.69	0.78	-0.21	0.58	-0.34	0.25	-0.67
Parent 2: 3.42	0.69	-0.31	0.49	-0.39	0.30	-0.78
Offspring 1: 3.42	0.69	-0.21	0.58	-0.34	0.30	-0.78

FIGURE 8.5 The order crossover operator.

Step 4: Place the genes into the unfilled positions of the proto-child from the left to the right according to the resulted sequence of genes in Step 3 to produce an offspring, as illustrated in Figure 8.5.

Step 5: Repeat Steps 1 to 4 to produce another offspring by exchanging the two parents.

8.3.2.3 Heuristic Mutation

A heuristic mutation [17] is designed with the neighborhood technique to produce a better offspring. A set of chromosomes transformed from a parent by exchanging some genes is regarded as the neighborhood. Only the best one in the neighborhood is used as the offspring produced by the mutation. Herein, the original heuristic mutation is modified in order to promote diversity of the population. The modification, as illustrated in Figure 8.6, is that all neighbors generated are used as the offspring. The procedure of the heuristic mutation operation is listed as follows:

Step 1: Pick up three genes in a parent at random.

Step 2: Generate neighbors for all possible permutations of the selected genes, and all neighbors generated are regarded as the offspring.

In Step 1, only three genes are selected since two genes have only one variation (one offspring) while more than three genes will generate too many offspring and it will take a very long time for computation.

8.3.2.4 Inversion Mutation

The inversion operator, as shown in Figure 8.7, selects a sub-string from a parent and flips it to form an offspring. Since the inversion operator operates with one chromosome only, it is very similar to the heuristic mutation and thus lacks interchange of the characteristics between chromosomes. Hence, the inversion operator is a mutation operation, which is used to increase the diversity of the population rather than to enhance the quality of the population.

Select 3 genes at random

Parent: 2.69	0.78	-0.21	0.58	-0.34	0.25	-0.67
Offspring 1: 2.69	0.78	-0.21	0.25	-0.34	0.58	-0.67
Offspring 2: 2.69	0.58	-0.21	0.78	-0.34	0.25	-0.67
Offspring 3: 2.69	0.58	-0.21	0.25	-0.34	0.78	-0.67
Offspring 4: 2.69	0.25	-0.21	0.78	-0.34	0.58	-0.67
Offspring 5: 2.69	0.25	-0.21	0.58	-0.34	0.78	-0.67

FIGURE 8.6 The heuristic mutation operator.

Selected sub-string

| Parent: | 2.69 | 0.78 | -0.21 | 0.58 | -0.34 | 0.25 | -0.67 |
| Offspring: | 2.69 | 0.78 | -0.34 | 0.58 | -0.21 | 0.25 | -0.67 |

FIGURE 8.7 The inversion mutation operator.

8.4 Case Study

One of the Hong Kong based companies, which has the alias named as GTL is proficient in designing and manufacturing a wide range of hand-held electronic products for consumers to acquire and to utilize information in a convenient and fast manner for education, entertainment, data storage, and communication purposes. GTL designs and manufactures a wide range of products including electronic dictionaries, personal digital assistant (PDA), translators, and electronic organizers. GTL currently employs over 3001–6000 people in China and Hong Kong. GTL founded in 1988, launched the first Instant-Dict electronic dictionary in 1989. Instead of transforming from original equipment manufacturer (OEM) or original design manufacturer (ODM) to original brand manufacturer (OBM), GTL starts OBM in the 1990s.

A major manufacturer of electronic products invests significantly in research and development activities relating to innovative product design with focus on changing customer demands. Besides transforming the customer needs to value-added features of the new product, the company utilizes the enterprise information system to support product design, procurement, production planning, and inventory management. Keeping minimum inventory level but delivering the right amount of goods to customers within the specified period is important, and precise demand forecasting is necessary for effective inventory management. Precise forecasting does not just mean that the predicted amount is equal or approximate to the actual amount. Precise forecasting also reflects that the corporation has a thorough understanding of the market trend, customer behavior, the strength and weakness of competitors, and their own corporation. GTL, which recently launches a new product, smart phone, would like to forecast the sale of the smart phone so as to formulate the strategy for gaining larger market share.

The demand of smart phones jumped 330% last year, and by 2008 sales could hit 100 million a year, according to analysts at Allied Business Intelligence in New York State [18]. The demand of smart phone is driven by personal needs and business needs. Mobility is the recent trend of living styles. In the early 1990s, mobile phones were the representative consumer products and in the late 1990s, PDA or the hand-held devices were the most popular electronic products. The hottest type of handset on the market at present is the smart phone—a device that combines the functions of a telephone and PDA [18].

8.4.1 Analytic Hierarchy Process

There are four main factors including competitors' forces, economic factors, technology development, and consumer behaviors that determine the demands of smart phone in the coming years. Each factor can further be divided into sub-factors each of which is listed in the dedicated level.

It is inevitable for manufacturing firms to face the challenges of globalization, and rapid evolution of advanced technology makes enterprises strive for excellence. The competitiveness of new entrant and existing competitors should be analyzed with competitive array so as to identify the strength and weakness of the corporation. Competitor intelligence cannot be ignored, and it is usually difficult for corporations to catch pricing policy and marketing strategy of the competitors. However, it is necessary to consider those two factors to evaluate the performance of competitors. The performance of competitors is not bound to the special features and functions of the physical product. Since many corporations may provide presales and postsales services, customer service policy is also a critical factor in determining the performance of the competitors.

Customers' habit and working style is evolving, and mobile working style leads to the increasing demand of light and slim electronic product. Smart phone also extends the entertainment features such as

Mobile TV, music, imaging applications, games, video camera, and web browser [19]. Besides personal needs, International Data Corporation (IDC) survey shows that the recent business intentions concerning mobility made companies establishing IT strategy to prepare widespread mobile device deployment in 2005 [20]. The emerging industry such as logistics industry also makes use of handheld device in the operation such as PDA with a barcode scanner for warehouse management. Both personal and corporation needs are the drive of the introduction and increasing demand of smart phone.

With regard to the factors of national economy, the following factors are to be considered: inflation rate, employment rate, and GDP growth rate. Consumers are willing to purchase the product under good economics environment and vice versa. If the unemployment and inflation rates are high, there is a great probability of decreasing the sales of high-tech electronic products. The sales of the product will be steady or decline at the mature stage and decline stage of product lifecycle.

The key elements relating to technology development are the support of network infrastructure, development of data transfer technology, and comprehensiveness of mobile solution. The smart phone has the wireless data transfer functions such as Wi-Fi and Bluetooth which require the support of network infrastructure. Besides wireless device, software is also needed to have a total mobile solution for the business environment. As a result, the provision of proficient mobile solution can also increase the sales volume of wireless device. The factors affecting the demand of smart phone are summarized in Figure 8.8.

Having constructed the hierarchical tree, logistics practitioners make use of pairwise comparison to find out the interrelation among the factors. As supply chain involves various parties including suppliers, distributors, retailers, and customers each of them can contribute to construct the hierarchy. Pairwise comparison is then made through discussion, bargaining, and persuasion. Demand forecasting is a complex issue and the value chain embraces various parties with widely varying perspectives. Having identified the overall structure of the issue shown in Figure 8.9, the group can share their experience and express the opinion about the higher order and lower order aspect of demand forecasting. Four major factors of demand forecasting include living and working style of consumer, national economics, technology development for M-commerce, and performance of competitors are identified. Group pairwise comparison shown in Figure 8.10 can be obtained by taking the geometric mean of individual judgment,

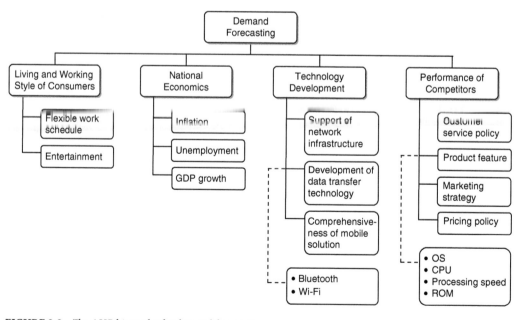

FIGURE 8.8 The AHP hierarchy for demand forecasting.

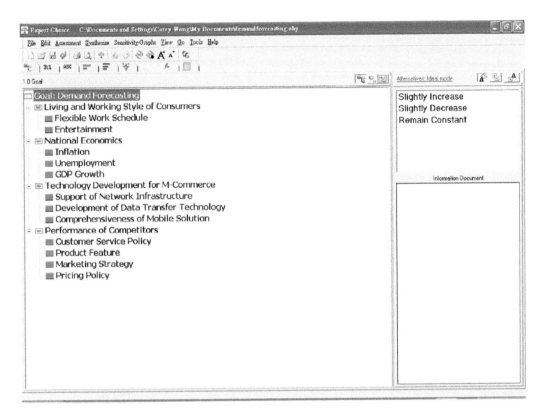

FIGURE 8.9 Hierarchy for evaluating demand forecasting.

and the final values are illustrated in Figure 8.11 which indicates that living and working style of consumer may greatly affect the sales volume of portable device. Different groups of expert may have variable preference, and it usually leads to low consistency. To tackle this problem, reviewing the factors of strong or weak importance is done and the result is found to be improved.

8.4.2 GA-Based Multiple Regression Analysis

In Section 8.4.1, several attributes affecting the demand of the smart phones are suggested. Since some of them are nonquantifiable, they cannot be considered in the GA-based multiple regression analysis. Only those that can be expressed in quantities are analyzed in the forecasting mathematical equation. The dependent variable is the demand of the smart phones, whereas the independent variables can be divided into two categories: internal and external factors. Internal factors include the advertising cost and product price. External factors include competitor's product price, GDP growth rate, inflation rate, and unemployment rate. The factors listed in Figure 8.9 are included in this quantitative approach, except the nonquantifiable factors and some without historical data.

The notation used in the forecasting mathematical equation is shown in the following. Besides, the data of both dependent and independent variables in n period of time is collected in order to formulate the mathematical equation, and most importantly, estimate the regression coefficients.

A_t = actual value of demand in period t
x_{1t} = advertising cost in period t
x_{2t} = product price in period t

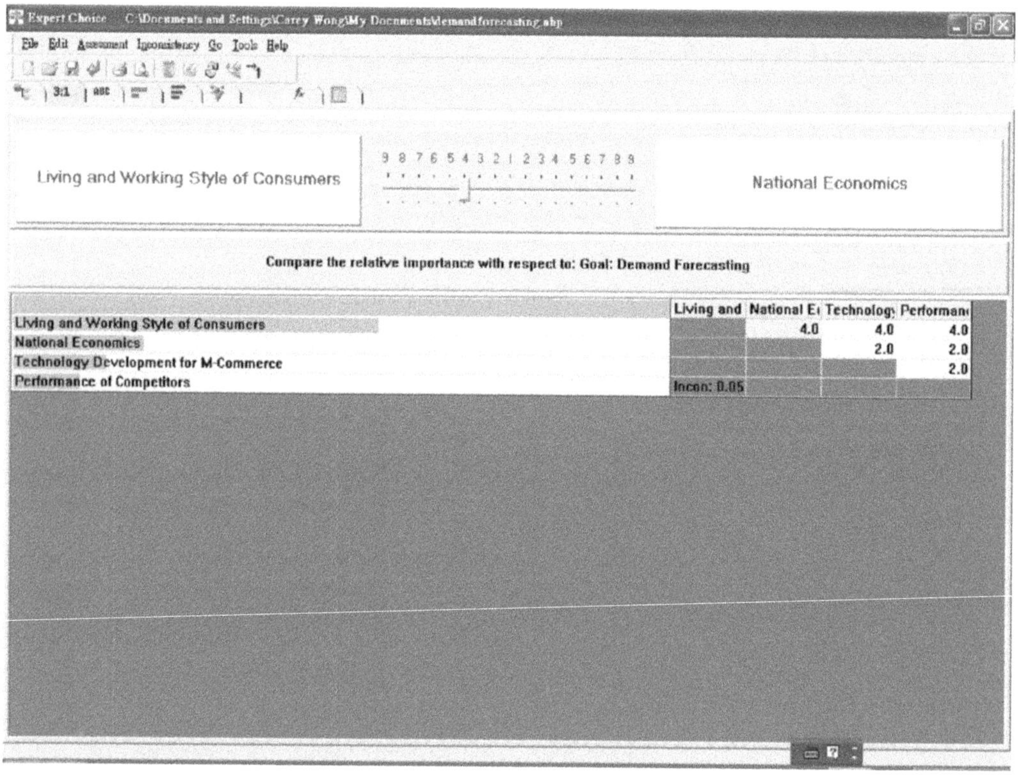

FIGURE 8.10 Matrix for comparing the relative importance of criteria.

x_{3t} = competitor's product price in period t
x_{4t} = GDP growth rate in period t
x_{5t} = inflation rate in period t
x_{6t} = unemployment rate in period t

After obtaining all historical data of the variables, the forecasted regression coefficients and constant coefficient can be generated using regression analysis software (e.g., SPSS). In the GA phase, these coefficients form the basis of initial chromosomes generation. In this case, the GA parameters are preset as population size = 25, iteration number = 100, crossover rate = 0.4, and mutation rate = 0.2. Therefore, ten chromosomes (25 × 0.4) or five pairs of chromosome are selected to perform the order crossover, whereas five chromosomes (25 × 0.2) perform the heuristic mutation and the inversion mutation. The performance of the GA is illustrated in Figure 8.12. It is found that the curve converges rapidly at the first 10 iterations, and then levels off after the 11th iteration. The final best solution obtained, that is, the MAPE is 0.3145. This result is much better than that of regression analysis (MAPE = 0.6374). The GA-based multiple regression analysis is, therefore, an accurate approach for demand forecasting.

8.5 Conclusions

This chapter presented two interrelated approaches, based on the AHP and GA, which can be deployed in a global logistics environment for demand forecasting across the firms within the supply chain. AHP approach is to collect experts' opinion, intuition, and logic in a structured manner for determining the relevant factors of demand forecasted, whereas GA-based approach is to search for the optimal coefficients

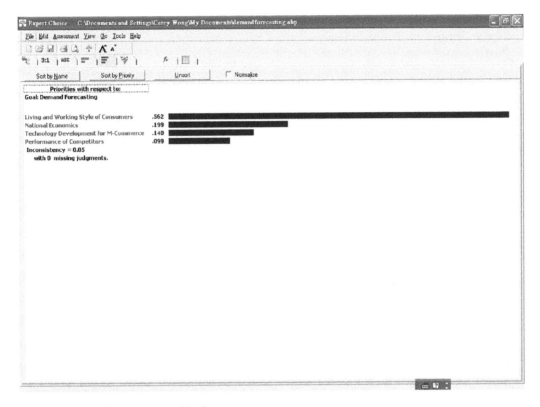

FIGURE 8.11 Relative priorities of the factors.

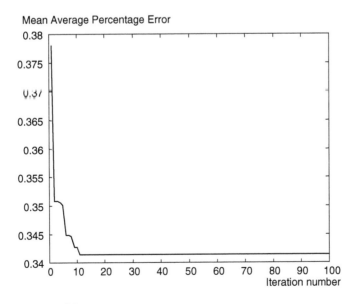

FIGURE 8.12 Performance of the GA.

of the forecasting mathematical equation so that the performance of the forecasting is enhanced or the discrepancy between the forecasted result and actual value is minimized. Using the proposed approach, the business environment can be evaluated, the factors affecting the demand of the consumer products can be assessed, and the forecasted amount of demand can be obtained. The framework of these approaches was described thoroughly with an implementation on a Hong Kong leading electronic company.

The significant contribution of this chapter is related to the effective introduction of the systematic decision-making and artificial intelligence approaches to the dispersed logistics network. These approaches, illustrated in the case study, enable the progressive inclusion of artificial intelligence features and systematic forecasting method into the demand forecasting. The solution can then form the basis of better purchasing, inventory, and transportation strategies. It is, therefore, expected that the proposed approaches can enhance the competitiveness of logistics practitioners. Further research will be focused on the refinement of the demand forecasting approaches, and reliable and "seamless" integration of the AHP and GA.

References

1. Council of Supply Chain Management Professionals. Glossary of Terms, 2006. Available at: http://cscmp.org/Downloads/Public/Resources/glossary03.pdf.
2. Stock, J.R. and Lambert, D.M., *Strategic Logistics Management*, 4th ed., McGraw-Hill, New York, 2001.
3. Heizer, J. and Render, B., *Operations Management*, 7th ed., Prentice-Hall, London, 2003.
4. Korpela, J. and Tuominen, M., Inventory forecasting with a multiple criteria decision tool, *International Journal of Production Economics*, 45, 159–168, 1996.
5. Spedding, T.A. and Chan, K.K., Forecasting demand and inventory management using Bayesian time series, *Integrated Manufacturing Systems*, 11, 331–339, 2000.
6. Kandil, M.S., El-Debeiky, S.M., and Hasanien, N.E., Overview and comparison of long-term forecasting techniques for a fast developing utility: part I, *Electric Power Systems Research*, 58, 11–17, 2001.
7. Jeong, B., Jung, H.S., and Park, N.K., A computerized casual forecasting system using genetic algorithms in supply chain management, *The Journal of Systems and Software*, 60, 223–237, 2002.
8. Snyder, R., Forecasting sales of slow and fast moving inventories, *European Journal of Operational Research*, 140, 684–699, 2002.
9. Ghobbar, A.A. and Friend, C.H., Evaluation of forecasting methods for intermittent parts demand in the field of aviation: a predictive model, *Computers and Operations Research*, 30, 2097–2114, 2003.
10. Chang, P.C., Wang, Y.W., and Tsai, C.Y., Evolving neural network for printed circuit board sales forecasting, *Expert Systems with Applications*, 29, 83–92, 2005.
11. Liang, W.Y. and Huang, C.C., Agent-based demand forecast in multi-echelon supply chain, *Decision Support Systems*, 42, 390–407, 2006.
12. Saaty, T.L., *The Analytic Hierarchy Process*, McGraw-Hill, New York, 1980.
13. Badri, M.A. and Abdulla, M.H., Awards of excellence in institutions of higher education: an AHP approach, *International Journal of Educational Management*, 18, 224–242, 2004.
14. Anderson, D.R., Sweeney, D.J., and Williams, T.A., *An Introduction to Management Science: Quantitative Approaches to Decision Making*, Thomson South-Western, Ohio, 2005.
15. Osman, I.H. and Kelly, J.P., *Meta-heuristics: Theory and Applications*, Kluwer Academic Publishers, Boston, 1996.
16. Goldberg, D.E., *Genetic Algorithms in Search, Optimization and Machine Learning*, Addison-Wesley, New York, 1989.
17. Gen, M. and Cheng, R., *Genetic Algorithms and Engineering Design*, Wiley, New York, 1997.

18. Duncan, G.R., Enjoy it while you can, *Exclusive from New Scientist Print Edition*, 1 October 2003. Available at: http://www.newscientist.com/article.ns?id=dn4224, 2003.

19. Kaiser, R.G., Hoping to dial into cell phones' future; Nokia, Finland's biggest company, betting on music, TV, games and internet; [FINAL Edition], *The Washington Post*, D.04, 2005.

20. Brown, A., IDC: IDC end-user survey shows European businesses preparing for widespread mobile device deployment in 2005, *M2 Presswire. Coventry*, 18 April 2005, 1, 2005.

9

Facilities Location and Layout Design

Benoit Montreuil
University of Laval

9.1 Introduction

Organizations and enterprises around the world differ greatly in terms of mission, scale, and scope. Yet all of them aim to deploy the best possible network of facilities worldwide for developing, producing, distributing, selling and servicing their products and offers to their targeted markets and clients. Underlying this continuous quest for optimal network deployment lies the facility location and layout design engineering that is the topic of this chapter. Each node of the network must be laid out as best as possible to achieve its mission, and similarly be located as best as possible to leverage network performance. There is a growing deliberate exploitation of the space-time continuum, which results in new facilities being implemented somewhere in the world every day while existing ones are improved upon or closed down. The intensity and pace of this flux is growing in response to fast and important market, industry and infrastructure transformations. Location and layout design is being transformed, from mostly being a cost-minimization sporadic project to being a business-enabling continuous process; a process embedded in a wider encompassing demand and supply chain design process, itself embedded in a business design process thriving for business differentiation, innovation, and prosperity. Location and layout design will always have significant impact on productivity, but it now is ever more recognized as having an impact on business drivers such as speed, leanness, agility, robustness, and personalization capabilities. The chapter grasps directly this growing complexity in its treatment of the location and layout domain, yet attempts to do so in a way that engineers will readily harness the exposed matter and make it theirs.

This chapter addresses a huge field of practice, education, and research. For example, the site www.uhd.edu/~halet, developed and maintained by Trevor Hale at the University of Houston currently provides over 3400 location-related references. Location and layout design has been a rich research domain for over 40 years, as portrayed by literature reviews such as Welgama and Gibson (1995), Meller and Gau (1996), Owen and Daskin (1998), and Benjaafar et al. (2002). It is beyond the scope of this chapter to transmit all this knowledge. It cannot replace classical books such as those by Muther (1961), Reed (1961), and Francis et al. (1992) or contemporary books by Drezner and Hamacher (2002), and Tompkins et al. (2003).

The selected goal is rather to enable the readers to leapfrog decades of learning and evolution by the academic and professional community, so that they can really understand and act upon the huge location and layout design challenges present in today's economy. The strategy used is to emphasize selected key facets of the domain in a rather pedagogical way. The objectives are on one hand to equip the reader with hands-on conceptual and methodological tooling to address realistic cases in practice and on the other hand to develop in the reader's mind a growing holistic synthesis of the domain and its evolution.

To achieve its goals and objectives, the chapter is structured as follows. Sections 9.2 through 9.6 focus on introducing the reader-design fundamentals. Aggregation and granularity are discussed in Section 9.2. It is about managing the compromise between scale, scope, and depth that is inherent in any location or layout design study given limited resources and time constraints to perform the design project. Section 9.3 is about the essential element of any location and layout design study, that is space itself, and how the designer represents it for design purposes. It exposes the key differences between discrete and continuous space representations, as well as the compromises at stake in selecting the appropriate representation in a given case. Sections 9.4 and 9.5 expose the impact of interdependencies on the design task. Section 9.4 focuses on the qualitative proximity relationships between entities to be located and laid out, as well as with existing fixed entities. Section 9.5 concentrates on the quantitative flow and traffic between these entities. Section 9.6 presents an illustrative basic layout design, exploiting the fundamentals introduced in the previous sections. The emphasis is not on how the design is generated. It is rather on the data feeding the design process, the intermediate and final forms of the generated design, and the evaluation of the design.

Sections 9.7 through 9.11 expand from the fundamentals by treating important yet more complex issues faced by engineers having to locate and lay out facilities so that the resulting design contributes as best as possible to the expected future performance of the organization or enterprise. Section 9.7 addresses how a designer can exploit the processing and spatial flexibility of the centers to be laid out and located, whenever such flexibility exists. Section 9.8 extends to describe how to deal with uncertainty when generating and evaluating designs. Section 9.9 deals with the fact that most design studies do not start from a green field, but rather from an existing design which may be costly to alter. Section 9.10 extends to dealing with the dynamic evolution of the design, which switches the output of the study from a layout or location set to a scenario-dependent time-phased set of layouts or locations. The design thus becomes more of a process than a project. Finally Section 9.11 deals with the potential offered by network and facility organization, when the engineer has freedom to define the centers, their mission, their client–supplier relationships, their processors, and so on, as part of the design generation. Overall, Sections 9.2 through 9.11 portray a rich view of what location and layout design is really about. The aim is clear. A problem well understood is a problem half solved, while attempting to solve a problem wrongly assessed is wasteful and risky in terms of consequences.

Only in Section 9.12 does the chapter directly address design methodologies. This section does not attempt to sell the latest approaches and tools generated by research and industry. It rather openly exposes the variety of methodologies used and proposed by the academic and professional communities. The presentation is structured around a three-tier evolution of proposed design methodologies, starting from the most basic and ancient to the most elaborate and emerging. This section is conceived as an eye opener on the wealth of methodological avenues available, and the compromises involved in selecting one over the other depending on the case the engineer deals with. The following Section 9.13 provides a formal mathematical modeling of location and layout design optimization. It focuses on introducing two models which give a good flavor of the mathematical complexity involved and allow to formally integrate the location and layout facets of the overall design optimization. The chapter concludes with remarks about both the chapter and the domain.

9.2 Design Aggregation and Granularity Levels

Facilities location and layout are both inherently prone to hierarchical aggregation so as to best direct design attention and harness the complexity and scale of the design space. Figure 9.1 provides an illustration of hierarchical aggregation. The entire network of facilities of an enterprise is depicted on the top portion of Figure 9.1, as currently located around the world. The company produces a core module in Scandinavia. This core module is fed to three regional product assembly plant, respectively located in the United States, Eastern Europe, and Japan. Each of these assemblers feeds a set of market-dedicated distribution centers. The middle of Figure 9.1 depicts the site of the Eastern European Assembler, located on municipal lot 62-32. The plan distinguishes seven types of zones in the site. Facility zones are segregated into three types: administration, factory, and laboratory. Transportation zones are split into two types: road zone and parking and transit zone. There is a green zone for trees, grassy areas and gardens. Finally, there is an expansion zone for further expanding activities in the future. There are two factories on the site. The lower portion of Figure 9.1 depicts the assembly factory F2, itself comprised of a number of assembly, production, and distribution centers, as well as offices, meeting rooms, laboratories, and personal care rooms.

A modular approach to represent facility networks helps navigate through various levels of a hierarchical organization. In Figure 9.1, the framework introduced by Montreuil (2006) has been used. It represents the facilities and centers through their main role in the network: assembler, distributor, fulfiller, producer, processor, transporter, as well as a number of more specific roles. A producer fabricates products, modules and parts through operations on materials. A processor performs operations on clients' products and parts. An assembler makes products and modules by assembling them from parts and modules provided by suppliers. A fulfiller fulfils and customizes client orders from products and modules. A distributor stores, prepares and ships products, modules and parts to satisfy client orders. A transporter moves, transports, and handles objects between centers according to client orders. Montreuil (2006) describes thoroughly each type of role and its design issues. Using the same terms at various levels helps the engineer comprehend more readily the nature of the network and its constituents, and leverage this knowledge into developing better designs.

Depending on the scope of design decisions to be taken, the engineer selects the appropriate level of aggregation. Yet he must always take advantage of in-depth knowledge of higher and lower levels of aggregation to leverage potential options, taking advantage of installed assets and fostering synergies.

The illustration has focused on hierarchical aggregation. In location and layout studies another type of aggregation is of foremost importance: physical aggregation. This is introduced here through a layout illustration, yet the logic is similar for multi-facility location. The layout of a facility can be represented with various degrees of physical aggregation for design purposes.

The final deliverable is to be an implemented and operational physical facility laid out according to the design team specifications. The final form of these representations is an engineering drawing and/or a 3D rendering of the facility, with detailed location of all structural elements, infrastructures, walls,

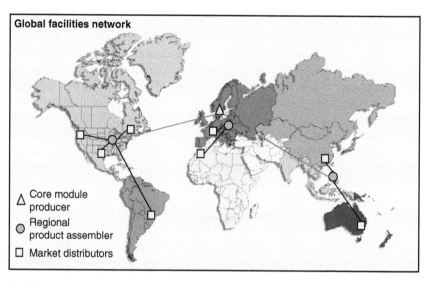

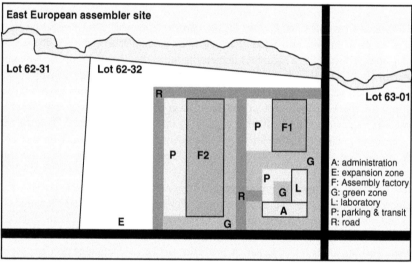

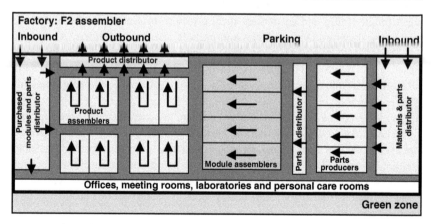

FIGURE 9.1 Hierarchical illustration of facilities network deployment, site layout, and facility layout.

machines, etc., identifying the various centers sharing the overall space. For most of the design process, such levels of details are usually not necessary and are cumbersome to manipulate.

Figure 9.2 exhibits five levels of layout representation used for design purposes. The least aggregate first level, here termed processor layout, shows the location and shape of the building, each center, each aisle

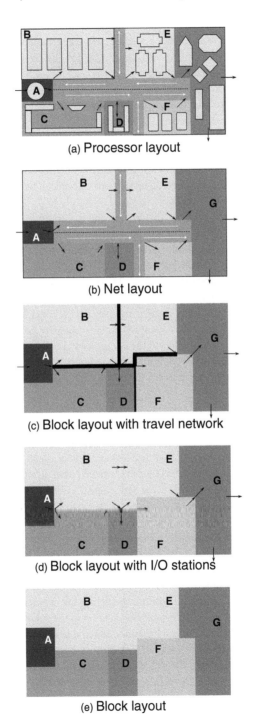

(a) Processor layout

(b) Net layout

(c) Block layout with travel network

(d) Block layout with I/O stations

(e) Block layout

FIGURE 9.2 Degrees of aggregation in layout representation for design purposes.

and each significant processor within each center (e.g., Warnecke and Dangelmaier 1982). The processor layout also locates the input and/or output stations of each center, the travel lane directions for each aisle and, when appropriate, the main material handling systems such as conveyor systems and cranes.

At the second level of aggregation lies the net layout which does not show the processors within each center (e.g., Montreuil 1991, Wu and Appleton 2002). The assumption when focusing the design process on the net layout is that prior to developing the entire layout for the facility, space estimates have been made for each center, leading to area and shape specifications, and that as long as these spatial specifications are satisfied, then the net layout embeds most of the critical design issues. The space estimation may involve designing a priori potential alternative processor layouts for each center. The transposition of the net layout to a processor layout for the overall facility is left as a detailed exercise where the layout of each center is developed given the shape and location decided through the net layout. Note that when the internal layout of the centers has influence on overall flow and physical feasibility, then basing the core of the design process on the net layout is not adequate.

At the third level of aggregation, the aisle set is not included anymore in the layout (e.g., Montreuil 1987, 1991). Instead, the space requirements for shaping each center are augmented by the amount of space expected to be used by aisles in the overall layout. For example, if by experience, roughly 15% of the overall space is occupied by aisles in layouts for the kind of facility to be designed then the space requirements of each center are increased by 15%. This percentage is iteratively adjusted as needed. The layout depicting the location and shape of the centers is now termed a block layout.

At this third level, instead of including the aisle set explicitly, the design depicts the logical travel network (Chhajed et al. 1992). This network, or combination of networks, connects the I/O stations of the centers as well as the facility entry and exit locations. There may be a network representing aisle travel, or even more specifically people travel or vehicle travel. Other networks may represent travel along an overhead conveyor or a monorail. The network is superimposed on the block layout, allowing the easy alteration of one or the other without having to always maintain integrity between them during the design process, which eases the editing process. Links of the network can be drawn proportional to their expected traffic. When transposition of a block layout with travel network into a net layout or a processor layout proves cumbersome due to the need for major adjustments, then such a level of aggregation may not be appropriate for design purposes.

At the fourth level of aggregation, the travel network is not depicted, leaving only the block layout and I/O stations (e.g., Montreuil and Ratliff 1988a). Editing such a block layout with only input/output stations depicted is easy with most current drawing packages. These stations clearly depict where flow is to enter and exit each center in the layout. Even though the I/O stations of each center can be located anywhere within the center, in practice most of the times they are located either at center periphery or at its centroid. The former is usually in concordance with prior space specifications. It is commonly used when it is known that the center is to be an assembly line, a U-shape cell, a major piece of equipment with clear input and output locations, a walled zone with access doors, etc. The latter centroid location, right in the middle of the center, is mostly used when the center is composed of a set of processors and flow can go directly to and from any of them from or to the outside of the center. It is basically equivalent to saying that one has no idea how flow is to occur in the center or that flow is to be uniformly distributed through the center.

The absence of travel network representation assumes that the design of the network and the aisle set can be straightforwardly realized afterward without distorting the essence of the network, and that flow travel can be easily approximated without explicit specification of the travel network. Normally, one of the two following assumptions justifies flow approximation. The first is that a free flow movement is representative, computed either through the rectilinear or Euclidean distance between the I/O stations between which a flow is expected to occur. Figure 9.3 illustrates these two types of free flow. Euclidean distance assumes that one can travel almost directly from one station to another while rectilinear distance assumes orthogonal staircase travel along the X and Y axes, like through a typical aisle set when one does not have to backtrack along any of the axes. The second alternative assumption is that flow travel is to occur along the center boundaries. Thus distances can be measured accordingly through the shortest path between the

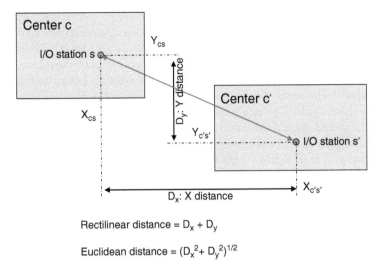

Rectilinear distance = $D_x + D_y$

Euclidean distance = $(D_x^2 + D_y^2)^{1/2}$

FIGURE 9.3   Free flow distance measured according to rectilinear or Euclidean distance.

two I/O stations of each flow, along the contour network of the facility. This network is implicitly created by inserting a node at each corner of one center and/or the facility, and inserting a link along each center or facility boundary segment between the nodes. In Figure 9.2d, a flow from the northern output station of center B to the input station of center G would be assumed to travel from the output station of B southward along the west boundary of center B, then turning eastbound and traveling along the southern boundaries of center E, and keeping straight forward to reach the input station of center G.

At the fifth level of aggregation, only the block layout is drawn in Figure 9.2e. This is the simplest representation. On the one hand it is the easiest to draw and edit. On the other hand it is the most approximate in terms of location, shape, and flow. For the last 50 years, it has been by far the most commonly taught representation in academic books and classes, often the only one (e.g., Tompkins et al. 2003), and it has been the most researched. It is equivalent to the fourth level with all the I/O stations located at the centroid of their center. The underlying assumption justifying this level of aggregation is that the relative positioning and shaping of the centers embeds most of the design value and that this positioning and shaping can be done disregarding I/O stations, travel networks, aisle sets and processors, which are minor issues and will be dealt with at later stages. While in some settings this is appropriate, in many others such an aggregation can be dangerous. It may lead to the incorrect perception that the implemented layout is optimal because its underlying block layout was evaluated optimal at the highest level of aggregation, thus limiting and biasing the creative space of designers.

It is always a worthwhile exercise, when analyzing an existing facility, to draw and study it at various levels of aggregation. Each level may reveal insights unreachable at other levels, either because they do not show the appropriate information or because it is hidden in too much detail.

9.3   Space Representation

Location and layout is about locating and shaping centers in facilities or around the world. The design effort attempts to generate expected value for the organization through spatial configuration of the centers within a facility, or of facilities in wide geographical areas. Space is thus at the nexus of location and layout design. It is therefore not surprising that representation of space has long been recognized to be an important design issue. The essential struggle is between a discrete and a continuous representation of space.

Figure 9.4 allows contrasting both types of space representation for layout design purposes (Montreuil, Brotherton and Marcotte 2002). Leftmost is the simplest and freest continuous representation of space. In the top left, the facility is depicted as a rectangle within which the centers have to be laid out. In the bottom left, an example layout is drawn. To reconstruct this layout, a designer simply has to remember the shape of each center and the coordinates of its extreme points, as well as for the facility itself. Here centers have a rectangular shape, so one needs only to remember, for example, the coordinates of their respective southwest and northeast corners. As long as the shape specifications and spatial constraints are satisfied, the designer can locate and shape centers and the facility at will.

Third from the left in Figure 9.4 is a basic example of discrete space representation. Here, the top drawing represents space as an eight-by-eight matrix of unit discrete square locations. The size requirements of each center have to be approximated so that they can be stated in terms of number of unit locations. Shape requirements express the allowed assemblies or collages of these unit blocks. For example, the blocks are usually imposed to be contiguous. The length-to-width ratio and overall shape of the block assembly are also usually constrained. The design task is to best assign center blocks to discrete locations given the specified constraints. It is common for the discrete representation to force complex shapes for the centers in order to fit in the discrete facility matrix.

At first glance it seems hard to understand why one would want to use anything but a continuous space representation as it is more representative and natural. Yet discrete space representation has a strong computational advantage, especially when a computer attempts to generate a layout using a heuristic. Manipulating continuous space and maintaining feasibility is much harder in continuous space for a computer. This is why the early layout heuristics such as CRAFT, CORELAP, and ALDEP (Armour and Buffa 1963, Lee and Moore 1967, Seehof and Evans 1967) in the 1960s have used discrete space, and why many layout software still use it and researchers still advocate it. This trend is slowly getting reversed with more powerful heuristics, optimization models, and software. Yet due to a long legacy, it is important for layout designers to master both types of representation.

Rightmost in Figure 9.4 is a more generalized nonmatrix discrete space representation. It corresponds to a facility that has a fixed overall structure characterized by a central loop aisle and centralized access on both western and eastern sides of the facility. The available space for centers becomes a set of discrete locations, each with specific dimensions. Such a kind of discrete representation is an interesting compromise, especially when space is well structured. For example, in a hospital the main aisle structure is often fixed and there are discrete rooms that cannot be easily dismantled or modified. With a discrete space representation, each room becomes a discrete space location. Even though a continuous representation can handle such cases a discrete representation can be adequate for design purposes.

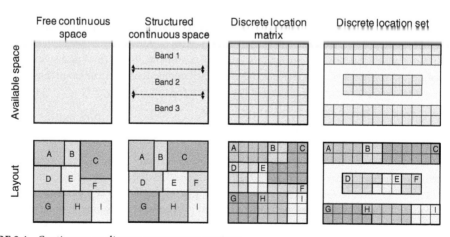

FIGURE 9.4 Continuous vs. discrete space representations.

Second from the left in Figure 9.4 is depicted a growing trend in layout software. It uses a continuous representation, yet it limits the layout possibilities through space structuring (Donaghey and Pire 1990, Tate and Smith 1995). Here the space structuring is expressed through the imposition of using three bands: a northern band, a central band, and a southern band. Within each band, centers have to be laid out side by side along the west–east axis. So the design process involves assigning centers to a band, specifying the order of centers within a band, sizing the width of each band (here its north–south length) and then shape each center within its band. The advantage of such a representation is that with a simple layout code, the entire layout can be regenerated easily, provided simple assumptions such as the sum of center areas equals the facility area. Here the code (1:A,B,C; 2:D,E,F; 3:G,H,I) enables to reconstruct the layout, provided that the facility shape is fixed. It states that the northern zone one includes centers A, B, and C ranked in this order from west to east; and similarly for the other zones. Bands are one type of space structuring. Zones and space filling curves are other well-known methods (Meller and Bozer 1996, Montreuil et al. 2002a). Besides computational advantages, an interesting feature of space structuring is that it has the potential to foster simpler layout structures.

Space representation issues also involve the decision to explicitly deal with the 3D nature of facilities or to use a limited 2D representation. Simply, one should recognize that a facility is not a rectangle, for example, but rather a cube. Then one must decide whether he treats the cube as such or reduces it to a rectangle for layout design purposes. The height of objects, centers, and facilities become important when height-related physical constraints may render some layouts infeasible and when flow of materials and people involve changes in elevation. The most obvious situation is when one is laying out an existing multi-storey facility with stairways and elevators. In most green field situations, the single-story vs. multi-story implementation is a fundamental decision. In some cases, it can be taken prior to the layout design study, in other settings it is through the layout design study that the decision is taken.

In forthcoming eras when space factories and nano factories are to be implemented, the 3D space representation will become mandatory. In space factories, the lack of gravity permits to exploit the entire volume for productive purposes, objects moving as well up and down as from left to right. In nano factories, the forces influencing movement of nano objects are such that their travel behavior becomes complex. For example, nano objects may be attracted upward by other nearby objects.

As a final edge on the discrete vs. continuous space representation choice comes the notion of space modularity. To illustrate the notion, consider a facility where space is organized as the concatenation of $10 * 10 * 10$ ft^3 cubes. Centers and aisles are assigned to groups of such cubes, charged an occupancy rate per cube. Such a modular space organization may prove advantageous in certain settings in a stochastic dynamic environment (refer to subsequent Section 9.10). In such cases, then either a zone-based continuous representation or a discrete representation can be equivalently used.

The choice between discrete and continuous space representation is also a core decision in facilities location decision making. Using a discrete representation requires to select in the early phases of the decision process, the set of potential locations to be considered. The task is then to optimize the assignment of facilities to locations. When using a continuous representation, the decision-maker limits the boundaries of the space to be considered for potential location for each facility. Then the task is to optimize the coordinates of each facility. These coordinates correspond to the longitude and latitude of the selected location, or approximate surrogates. The compromises are similar as in layout design. Making explicit the characteristics of each potential location is easier with a discrete representation, yet this representation limits drastically the set of considered locations.

9.4 Qualitative Proximity Relationships

When spatially deploying centers in a facility or locating facilities around the world, there exist relationships between them that result in wanting them near to each other or conversely far from each other. Such relationships can be between pairs of facilities or between a center and a fixed location. Each relationship exists for a set of reasons which may involve factors such as shared infrastructures, resources and personnel, organizational interactions and processes, incompatibility and interference, security and safety, as well as material and resource flow. Each case may generate specific relationships and reasons for each one.

These relations can be expressed as proximity relationships, which can be used for assessing the quality of a proposed design and for guiding the development of alternative designs. A proximity relationship is generically composed of two parts: a desired proximity and an importance level. Figure 9.5 shows a variety of proximity relationships between the 12 centers of a facility. For example, it states that it is important for centers MP and A to be near each other for flow reasons. It also states that it is very important that centers D and G be very far from each other for safety reasons. Such relationships can also be expressed with fixed entities. For example, in Figure 9.5, there are relationships expressed between a center and the outside of the facility. This is the case for center G: it is critical that it be adjacent to the periphery of the facility.

The desired proximity and the importance level can both be expressed as linguistic variables according to fuzzy set theory (Evans et al. 1987). In Figure 9.5, the importance levels used are vital, critical, very important, important, and desirable. The desired proximity alternatives are adjacent, very near, near, not far, far, and very far. Other sets of linguistic variables may be used depending on the case.

On the upper left side of Figure 9.5, the proximity relationships are graphically displayed, overlaid on the proposed net layout of the facility. Each relationship is drawn as a line between the involved entities. Importance levels are expressed through the thickness of the line. A critical relationship here is drawn as a 12-thick link while an important relationship is 3-thick. A vital relationship is 18-thick and is further highlighted by a large X embedded in the line. Gray or color tones can be used to differentiate the desired proximity, as well as dotted line patterns. Here a dotted line is used to identify a *not* distance variable such as *not far* or *not near*. In the color version of Figure 9.5, desired proximity is expressed through distinctive colors. For example, *adjacent* is black while *very far* is red. Such a graphical representation helps engineers to rapidly assess visually how the proposed design satisfies the proximity relationships. For example, in Figure 9.5, it is clearly revealed that centers E and PF do not respect the *very near* desired proximity even though it is deemed to be *critical*. Using graphical software it is easy to show first only the more important relationships, then gradually depict those of lesser importance.

Even though just stating that two centers are desired to be near each other may be sufficient in some cases, in general it is not precise enough. In fact, it does not state the points between which the distance is measured, and using which metrics. In Figure 9.5 are depicted the most familiar options within a facility. For example, inter-center distances can be measured between their nearest boundaries, their centroid, or their pertinent I/O stations. Distances can be measured using the rectilinear or Euclidean metrics, or by computing the shortest path along a travel network such as the aisle network. The choice has to be made by the engineer based on the logic sustaining the relationship. In wide area location context, distances are similarly most often either measured as the direct flight distance between the entities or through the shortest path along the transport network. This network can offer multiple air-, sea-, and land-based modes of transportation.

When evaluating a design it is possible to come up with a proximity relationship-based design score Figure 9.5 illustrates how this can be achieved. When starting to define the relationships, each importance level can be given a go/no-go status or a weight factor. In Figure 9.5, a *vital* importance results in an infeasible layout if the relationship is not fully satisfied. A critical importance level is given a weight of 64 while a desirable importance level has a weight of one. For each desired proximity variable a graph can be drawn to show the relationship satisfaction given the distance between the entities in the design. For example, in the upper right side of Figure 9.5, it is shown that the engineers have stated that a *not far* relationship is entirely satisfied within a 9-m distance and entirely unsatisfied when the distance exceeds 16 m. At a distance of 12 m it is satisfied at 50%. It is important to build consensus about the importance factors and proximity-vs.-distance satisfaction levels prior to specifying the relationships between the entities. Given a design, the distance associated with every specific relationship is computed. It results in a relationship satisfaction level. For example, it is important that centers A and C be near each other, as measured through the distance between their I/O stations assuming aisle travel. The computed distance is 12 m, which results in a satisfaction level of 10%. Since the weight associated with such an important relationship is four, the

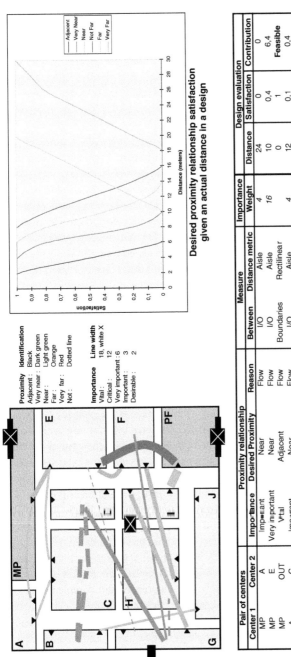

Desired proximity relationship satisfaction given an actual distance in a design

Legend (chart): Adjacent — Very Near — Near — Not Far — Far — Very Far

Proximity — Identification

Proximity	Identification
Adjacent :	Black
Very near :	Dark green
Near :	Light green
Far :	Orange
Very far :	Red
Not :	Dotted line

Importance — Line width

Importance	Line width
Vital :	18, white X
Critical :	12
Very important :	6
Important :	3
Desirable :	2

Pair of centers		Proximity relationship		Reason	Measure		Importance	Distance	Design evaluation	
Center 1	Center 2	Importance	Desired Proximity	Reason	Between	Distance metric	Weight	Distance	Satisfaction	Contribution
MP	A	Important	Near	Flow	I/O	Aisle	4	24	0	0
MP	E	Very important	Near	Flow	I/O	Aisle	16	10	0,4	6,4
MP	OUT	Vital	Adjacent	Flow	Boundaries	Rectilinear		0	1	Feasible
A	C	Important	Near	Flow	I/O	Aisle	4	12	0,1	0,4
B	C	Very important	Not far	Infrastructure	Centroid	Rectilinear	16	16	0	0
B	D	Very important	Not far	Infrastructure	Centroid	Rectilinear	16	29	0	0
B	J	Important	Near	Flow	I/O	Aisle	4	28	0	0
C	D	Very important	Near	Infrastructure	Centroid	Rectilinear	16	13	0,4	6,4
C	E	Important	Not far	Flow	I/O	Aisle	4	22	0	0
D	G	Very important	Very far	Safety	Centroid	Euclidean	16	17	0	0
D	E	Very important	Near	Flow	I/O	Aisle	16	10	0,4	6,4
E	F	Desirable	Not far	Organization	Boundaries	Aisle	1	66	0	0
E	PF	Critical	Very near	Flow+Org	I/O	Euclidean	64	24	0	4
F	G	Important	Far	Noise	Centroid	Euclidean	4	33	1	4
F	H	Very important	Not far	Process	I/O	Aisle	16	30	0	0
G	H	Important	Near	Flow	Boundaries	Aisle	4	4	1	4
G	I	Very important	Near	Flow	I/O	Aisle	16	46	0	0
G	OUT	Critical	Adjacent	Infra+Org+Flow	Boundaries	Rectilinear	64	0	1	64
H	I	Vital	Adjacent	Flow	Boundaries	Rectilinear		2	1	Feasible
I	PF	Critical	Near	Flow	I/O	Aisle	64	6	1	64
PF	OUT	Vital	Adjacent	Flow	Boundaries	Rectilinear		6	1	Feasible
					Maximum possible value :		345		Design value:	155,6
									Design score:	45,1%

FIGURE 9.5 Qualitative proximity relationships based evaluation of a layout design.

contribution of this relationship to the design score is 0.4 whereas the upper bound on its contribution is equal to its weight of four. When totaling all relationships the score contributions add up to a total of 155.6. The ideal total is equal to the sum of all weights which in this case is 345. Therefore, the design has a proximity relationship score of 45.1%. This leaves room for potential improvement.

Simplified versions of this qualitative proximity relationships representation and evaluation scheme exist. For decades, the most popular has been Muther's AEIOUX representation (Muther 1961), where the only relationships allowed are: A for absolutely important nearness, E for especially important nearness, I for important nearness, O for ordinarily important nearness, U for unimportant proximity, and X for absolutely important farness. In most computerized implementations using this representation, a weight is associated to each type of relationship and proximity is directly proportional to the distance between the centroids of the related centers. Simpler to explain and compute, such a scheme loses in terms of flexibility and precision of representation.

In general, the reliance on qualitative relationships requires rigor in assessing and documenting the specific relationships. In an often highly subjective context, the relationship set must gain credibility from all stakeholders, otherwise it will be challenged and the evaluation based on the relationship set will be discounted. This implies that the perspectives of distinct stakeholders must be reconciled. For example, one person may believe a specific relationship to be very important while another may deem it merely desirable. Some may be prone to exaggerate the importance while others may do the inverse. It is also important to realize that some relationships may be satisfied with other means than proximity. For example, two centers may be desired to be far from each other since one generates noise while the other requires a quiet environment. If noise proofing isolation is installed around the former center, then the pertinence of the proximity relationship between the two centers may disappear.

9.5   Flow and Traffic

In most operational settings, the flow of materials and resources is a key for evaluating and optimizing a layout or location decision. It is sometimes sufficient to treat it through qualitative relationships as shown in Section 9.4. However, in most cases it is far more valuable to treat flow explicitly. Flow generally defines the amount of equivalent trips to be traveled from a source to a destination per planning period. There are two basic flow issues at stake here associated with implementing a design. First is the expected flow travel or flow intensity. Second is the flow traffic. The former is generically computed by summing over all pairs of entities having flow exchanges, the product of the flow value between them and their travel distance, time or cost, depending on the setting. Flow travel has long been used as the main flow-related criterion for evaluating alternative layout and location designs (Francis et al. 1992). The goal is for the relative deployment of entities to be such that travel generated to sustain the flow is as minimal as possible. The second flow issue, flow traffic, measures the load on the travel network, through intensity of flow through each of its nodes, links, and associated aisle segments and routes. Congestion along links and at nodal intersections is aimed to be minimized by the design (Benjaafar 2002, Marcoux et al. 2005).

Table 9.1 provides the flow matrix for the case of Figure 9.5. For example, it depicts that it is expected that there will be 125 trips per period from the output station of center A to the input station of center C. The matrix also illustrates a key issue when dealing with flow: the differences and complementarities between loaded travel and empty travel. Loaded travel corresponds to trips made to transport materials, and in general resources, from their source location to their target destination location. A forklift transferring a pallet of goods from location A to location B is an example of loaded travel. Empty travel occurs when the forklift reaches location B, deposits the transferred pallet, and becomes available for transporting something else while there is currently nothing to be transported away from location B. The forklift may wait there until something is ready for transport if the expected delay is short, but in many cases it will move to another location with a load to be transported away from it, causing empty travel. A ship transporting containers from China to Canada is another example of loaded travel while the same ship traveling empty to pick up containers in Mexico is an example of empty travel.

TABLE 9.1 Illustrative Flow Estimation Matrix for the Case of Figure 9.5

From/To	MP	A	B	C	D	E	F	G	H	I	J	PF	Loaded	Empty	Total From
MP		150	50			**300**							**500**	*0*	500
A		*175*	25	125	25								175	*175*	350
B			*130*	15	10					100		5	130	*130*	260
C				*215*	40	175							215	*215*	430
D		25	10	40	*100*					25			100	*100*	200
E			45			*475*	**300**		15			**430**	**790**	*475*	1265
F							*300*		**300**				300	*300*	600
G										**350**			350	*0*	350
H							*15*	**300**	**300**				300	*315*	615
I						*170*		**500**		*500*			**500**	*670*	1170
J				35	25				20	*125*		45	125	*125*	250
PF	*500*					*145*	*335*						0	*980*	980
Loaded	**0**	**175**	**130**	**215**	**100**	**475**	**300**	**0**	**315**	**670**	**125**	**980**	**3485**		
Empty	*500*	*175*	*130*	*215*	*100*	*790*	*300*	*350*	*300*	*500*	*125*	*0*		*3485*	
Total to	**500**	**350**	**260**	**430**	**200**	**1265**	**600**	**350**	**615**	**1170**	**250**	**980**			**6970**

Entries: Trips/period.
Shaded and italics: Empty.
Bold: High relative value.
Loaded trip entries: From the output station of source center to the input station of destination center.
Empty trip entries: From the input station of source center to the output station of destination center.

In the flow matrix of Table 9.1, empty travel is written in italics. For example, it shows that 500 empty trips are to be expected from center PF to center MP. To be precise, the empty trips are from the input station of center PF to the output station of center MP, bringing transporters to enable departures from center MP. Similarly, Table 9.1 depicts that 175 trips per period are expected from the input station of center A to the output station of center A. Table 9.1 indicates a total flow of 7,320 trips per period, split equally between 3,485 loaded trips and 3485 empty trips per period. By a simple usage of bold characters, Table 9.1 highlights the most important flows for layout analysis and design.

Table 9.2 provides the distance to be traveled per trip assuming vehicle-based aisle travel in the layout of Figure 9.5. The provided distances are between the I/O stations of the centers having positive flow.

TABLE 9.2 Distance Matrix for the Layout of Figure 9.5

From/To	MP	A	B	C	D	E	F	G	H	I	J	PF
MP		24	30			10						
A		*8*	6	12	28							
B			*16*	26	42					28		56
C				*4*	16	22						
D		46	50	40	*24*					50		
E			48			*16*	10		42			24
F							*10*		30			
G								*20*		46		
H								*16*	*8*	38		
I						*12*				*14*		6
J				64	48					22	*26*	10
PF	*42*					*24*		*34*				

Entries: Trips/period.
Shaded and italics: Empty.
Loaded trip entries: Distance from the output station of source center to the input station of destination center.
Empty trip entries: Distance from the input station of source center to the output station of destination center.

The travel matrix of Table 9.3 is derived by multiplying the flow and distance for each corresponding matrix entry of Tables 9.1 and 9.2. For example, travel from the output station of center G to the input station of center I is estimated to be 16,100 m per period. The expected total travel is 151,410 m per period. So by itself the G to I flow represents roughly 11% of the total travel. Table 9.3 presents an interesting evaluation metric, which is the average travel. It simply divides the total travel by the total flow. Here it allows to state that the average travel is 22 m per trip, with 24 m per loaded trip and 20 m per empty trip. An engineer can rapidly grasp the relative intensity of travel with such a metric. Here 22 m per trip is a high value in almost every type of facility, readily indicating a strong potential for improvement. The lower portion of Table 9.3 depicts total flow, total travel, and average travel for each center. This highlights that centers E, I, and PF each have a total travel higher than 40,000 m per period, that centers MP, D, and J each have an average travel around 30 m per trip and that center G has an average travel of 40 m, making these centers the most potent sources of re-layout improvement.

Given the flows of Table 9.1 and the layout of Figure 9.5, traffic can be estimated along each aisle segment and intersection. Assuming shortest path travel, Figure 9.6 depicts traffic estimations. The aisles forming the main loop contain most of the traffic. Only one small flow travels along a minor aisle, east of centers C and H. In fact, it reveals that most of the minor aisles between centers could be deleted without forcing longer travel. The main south and east aisles get most of the traffic. The most active corners are the I-PF-J and D-E-MP intersections. Yet, the smooth distribution of traffic does not emphasize hot spots for congestions. Further analyses based on queuing theory (Kerbache and Smith 2000, Benjaafar 2002) or relying on discrete event simulations (Azadivar and Wang 2000, Huq et al. 2001, Aleisa and Lin 2005) would be required to estimate congestion effects in more depth.

Table 9.1 provides expected flows for the illustrative case. In practice, the engineer has to estimate these flows. There are basically two ways used to do so. The first is to track actual flows occurring in the actual facility during a sampling period and to extrapolate the expected flows from the sampling results, taking into account overall expected trends in demand. In technologically rich settings, precise tracking of actual flow can be achieved through the use of connective technologies such as GPS, RFID, or bar coding, using tags attached to the vehicles and/or objects being moved. In other settings, it requires people to perform trip samplings.

The second way to estimate flows is to rely on product routing and demand knowledge for estimating loaded trips and to rely on approximate analytical or simulation-based methods for estimating empty trips. Illustratively, Table 9.4 provides the planned inter-center routing and expected periodic demand for each of a set of 15 products. From these can be estimated the loaded flows of Table 9.1. For example, in Table 9.1 there is a flow of 25 loaded trips per period from center A to center D. From Table 9.4, the A to D flow is estimated through adding trips from A to D in the routings of products 5, 7, 8, and 9.

Whereas the loaded flow estimation is here rather straightforward, the estimation of empty flow requires assumptions on the behavior of vehicles when they reach an empty status and on the dispatching policy of required trips to individual vehicles. In Table 9.1, the empty flow is estimated using the two following simple assumptions. First, vehicles reaching the input station of a center are transferred in priority to the output station of that center to fulfill the needs for empty vehicles. Second, centers with exceeding incoming vehicles aim to transfer the exceeding vehicles to the nearest center having a lack of incoming loaded vehicles to fulfill its need for departing vehicles. A transportation model is used to allocate empty vehicle transfers according to center unbalances, as originally advocated by Maxwell and Muckstadt (1982). There exists a variety of alternative methods for empty travel estimation (see e.g., Ioannou 2007). It is important for the method to reflect as precisely as possible the behavior expected in the future layout implementation.

The illustrated approach for estimating flows from product routing and demand permits to highlight three fundamental issues. First, the computations divide the expected demand by the transfer lot in order to estimate the trips generated by a product routing segment. However, in practice the transfer lot is often dependent on the distance to be traveled and the type of handling system used. This illustrates the typical

TABLE 9.3 Travel Evaluation for the Layout of Figure 9.5 Based on the Flow Matrix of Table 9.1

From/To	MP	A	B	C	D	E	F	G	H	I	J	PF	Loaded	Empty	Total	Average
MP		3600	1500			3000							8100	0	8100	16
A	*1400*		150	1500	700								2350	1400	3750	11
B			*2080*	390	420					2800		280	3890	2080	5970	**23**
C					640	3850							4490	860	5350	12
D	1150		500	1600	*2400*	7600	3000		630		1250	10320	4500	2400	6900	35
E			2160			*7600*	*3000*		9000	16100			16110	7600	23710	19
F								240	*9000*	11400			**9000**	3000	12000	**20**
G										16100			**16100**		16100	**46**
H				*2240*		*2040*		240	*2400*	7000	3250	*3000*	11400	2640	14040	**23**
I									440		3250	450	3000	9040	12040	10
J										3250			4330	3250	7580	**30**
PF	21000				1200	*3480*		*11390*					0	35870	35870	37
Loaded	0	4750	4310	5730	2960	6850	3000	0	9630	27940	4050	14050	83270			**24**
Empty	21000	*1400*	*2080*	*860*	*2400*	13120	*3000*	11630	2400	7000	*3250*	0		68140		20
Total	21000	6150	6390	6590	5360	19970	6000	11630	12030	34940	7300	14050			151410	22
Average	**42**	18	**25**	15	**27**	16	10	**33**	**20**	**30**	**29**	14	**24**	20	**22**	**22**
Total flow*	1000	700	520	240	400	2530	1200	700	1230	2340	500	1960		6970		
Total travel*	29100	9900	12360	11910	12260	43680	18000	27730	26070	46980	14880	49920			151410	
Average travel*	**29**	14	**24**	**4**	**31**	17	15	**40**	21	**20**	**30**	**25**			**22**	

Entries: Trips/period, * meters/period.
Shaded and Italics: Empty.
Bold: High relative value.
Underline: Low value given high flow.

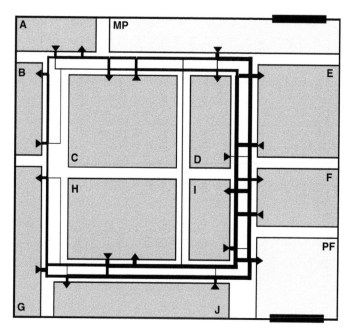

FIGURE 9.6   Expected traffic in the current layout.

chicken-and-egg phenomenon associated with layout design and material handling system design, requiring iterative design loops to converge toward realistic estimates.

Second, in the lean manufacturing paradigm, large transfer lots are perceived as an inefficiency hideout (Womack and Jones 1998), leading to the proposition that the layout be designed assuming a will to use transfer lots of one. The optimal layout assuming the stated transfer lots may well be different from the

TABLE 9.4   Product Routings and Expected Product Demand

Product	Demand	Transfer Lot	Trips/ Period	Inter-Center Routing											
1	2500	20	125	IN	MP	A	C	E							
2	560	16	35	IN	MP	B	J	C	D	C	E				
3	5250	15	350	IN	G	I	PF	OUT							
4	225	15	15	IN	MP	B	J	D	J	I					
5	120	12	10	IN	MP	A	D	A	B	C	E				
6	3000	10	300	IN	MP	C									
7	50	10	5	IN	MP	A	D	A	B	C	D	C	E		
8	30	6	5	IN	MP	A	D	A	B	D	B	PF	OUT		
9	30	6	5	IN	MP	A	D	A	B	D	B	J	D	J	PF	OUT
10	875	5	175	E	F	H	I								
11	650	5	130	E	PF	OUT									
12	25	5	5	E	B	J	D	J	I						
13	25	5	5	E	B	J	PF	OUT							
14	75	5	15	E	H										
15	625	5	125	E	F	H	I	PF	OUT						
16	175	5	35	E	B	J	PF	OUT							
17	1500	5	300	E	PF	OUT									
18	125	5	25	I	PF	OUT									

IN and OUT respectively refer to inbound from suppliers and outbound to clients.

The transfer lot expresses the number of units planned to be transported concurrently in each trip.

optimal layout assuming unitary transport. This illustrates the interaction between layout design and operating system planning, requiring a fit between their mutual assumptions.

Third, in the illustration, all trips are hypothesized equivalent. In practice, a forklift trip carrying a standard pallet is not equivalent to a forklift transporting a 10-m long full metal cylinder with a diameter of 30 cm, the latter being much more cumbersome and dangerous. Compare a forklift trip with a walking individual transporting a hammer. This is why the flow definition used the notion of equivalent trips, requiring the engineer to weigh the different types of trips so that the layout compromises adequately, taking into consideration their relative nature. Muther (1961) proposes a set of preset weights to standardize trip equivalence computations.

It is important whenever possible to transpose the travel and traffic estimations in terms of operating cost and investment estimations. This is often not a straightforward undertaking. Flow travel and traffic influence differently the operating cost and investment in a facility depending on whether handling involves trips by humans and/or vehicles or it involves items moving along a fixed system such as a conveyor. When using a conveyor to travel between two points, there is a fixed cost to implement the conveyor, then there is often negligible cost involved in actually moving specific items on the conveyor. Flow traffic along a conveyor influences investments in a staircase fashion. As traffic gets higher up to the upper bound manageable by a given technology, faster technology is required that costs more to acquire and install.

When trip-based travel is used, then flow travel increases translate more directly into cost and investment increases. First, each vehicle spends costly energy as it travels. Second, as travel requirements augment, the number of required vehicles and drivers generally augments in a discrete fashion. Third, higher traffic along aisle segments and intersections may require implementing multiple lanes, extending the space required for aisles and affecting the overall space requirements. Fourth, when using trip-based travel, the time for each trip is the sum of four parts: the pickup time, the moving time, the waiting time, and the deposit time. The pickup and deposit times are mostly fixed given the handling technology selected for each trip, and the items to be maneuvered. They range mostly from a few seconds to a minute. The only ways to reduce them are by improving the technology and its associated processes, and by avoiding making the trip. The latter can be achieved when the I/O stations are laid out adjacent to each other, or when the flow is reassigned to travel along a fixed infrastructure. The moving time depends both on the path between the entities and the speed and maneuverability of the handling technology used. The waiting time occurs when traffic becomes significant along aisles and at intersections. In small facilities, it is often the case that pickup and deposit times dominate moving and waiting times because of short distances.

In location decisions, cost estimation relative to flow travel and traffic involves making assumptions or decisions relative to the transportation mode to be used (truck, plane, boat, etc.), fleet to be owned or leased, routes to be used and contracts to be signed with transporters and logistic partners. Congestion is not along an aisle or at an intersection in a facility. It is rather along a road segment, a road intersection, at a port, at customs, etc. The geographical scope is generally wider, yet the logical issues are the same.

9.6 Illustrative Layout Design

In order to provide an example of layout design, an engineer has been mandated to spend a day trying to develop an alternative design for the case used in the previous sections. He was simply provided with the case data and given access to spreadsheet and drawing software. The case data includes the relationships of Figure 9.5, the flows of Table 9.1, and the space requirements of Table 9.5. For safety reasons, it has also been required that at least four distinct aisles provide access to the exterior of the facility.

The engineer has first developed the design skeleton of Figure 9.7. This design skeleton is simply a flow graph. The engineer has drawn nodes for each center. The node diameter is proportional to the area requirements for the center. The loaded flows have been drawn as links whose thickness is proportional to flow intensity. The engineer has placed the nodes relative to each other and the exterior so as to approximately minimize travel and to respect roughly the qualitative proximity relationships. He has also decided to split facility input between two main entrances IN1 and IN2.

TABLE 9.5   Space Requirements for the Illustrative Case

Center	Minimal Area Requirements	Length-to-Width Maximum Ratio	Fixed Shape	Fixed Relative Location of I/O Stations	Can Be Mirrored
A	18	2,0	N	N	Y
B	16	4,0	Y	Y	N
C	64	2,0	N	N	Y
D	24	3,0	N	N	Y
E	48	2,0	N	N	Y
F	30	2,0	N	N	Y
G	26	6,5	N	N	Y
H	54	2,0	N	N	Y
I	21	3,0	N	N	Y
J	39	5,0	Y	N	Y
MP	51	6,0	N	N	Y
PF	42	2,0	N	N	Y
Building		3,0	N	N	Y

Then the engineer has transformed the design skeleton into an actual layout with three self-imposed objectives: (*i*) stick as much as possible to the design skeleton relative placement, (*ii*) minimize space by avoiding unnecessary aisles, and (*iii*) keep the shape of centers and building as simple as possible. The engineer has personally decided to put priority on minimizing flow travel, qualitative proximity relationships being a second priority. While developing the design, the engineer has iteratively estimated empty travel, using two simple self-imposed rules: (*i*) give priority of destination choice to empty trips from input stations of centers with higher inbound loaded flow and (*ii*) avoid assigning more than roughly half the empty trips out of a station to any specific destination. This is a looser estimation than that made for the current design, while being defendable as a viable operating strategy to deal with empty travel. Figure 9.8 depicts the resulting alternative design preferred by the engineer.

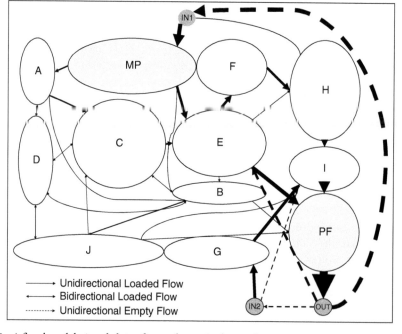

FIGURE 9.7   A flow based design skeleton for an alternative layout design.

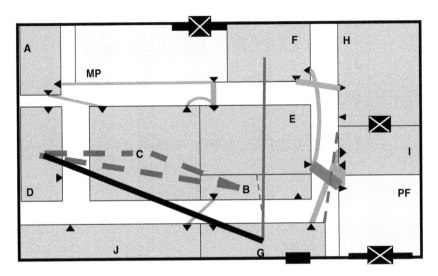

FIGURE 9.8 Alternative layout 1 with superimposed qualitative relationships.

Table 9.6 provides the flow matrix resulting from the engineer's empty travel estimation. Examination of the empty trip allocations show, for example, that the engineer has assigned empty travel out of the critical PF center to nearby centers E, G, and I, while the empty flows out of low-inbound-traffic center D have been assigned to more dispersed centers G and J, as well as to the output station of center D itself. Overall it has the same amount of empty trips. They are simply reshuffled differently given the proposed layout.

From an expected performance perspective, Table 9.7 shows that the expected total travel for the alternative design is now estimated at 84,765 m per period and the average travel is now at 12 m per trip, a 44% reduction over the current design. Table 9.8 provides its proximity relationship score of 48.9%, an 8.4% improvement over the current design. From a space perspective, the alternative layout slightly reduces the space requirements for the building. Its area shrinks from 441 to 435 square feet. This is mostly because of the reduction of unnecessary aisles in the alternative design.

9.7 Exploiting Processing and Spatial Flexibility

A key issue in location and layout design has become to exploit the flexibility offered by new technologies and means of operations. Processing flexibility allows processors to be allocated a variety of products to be treated, each with a given performance rating. Spatial flexibility occurs when management accepts to have multiple centers or processors, either identical or having intersecting capabilities, to be distributed through the facility. The combination of processing and spatial flexibility has the potential to improve significantly the design performance. Simple examples can be seen in everyday life. Switching from a single centralized toilet area or break area in a facility to multiple smaller areas spread through the facility has significant impact on people movement. A chain adding another convenience store in a city both helps it reach new customers through better convenience and reshuffles its clientele among the new and existing stores.

Exploiting flexibility makes the design process more difficult as it involves treating the flows as variables, rather than mere inputs, and dealing with capacity. The flows indeed become dependent on the relative locations and performance of entities. Thus to evaluate a design, one has to estimate how in future operations the flows will be assigned given the design and the operating policies. Given the estimated flow, one can then apply the travel and traffic scoring methods shown in Section 9.5.

For illustrative purposes, assume that in the case used in the previous sections, each center is devoted to a single process and is composed of a specific number of identical processors, as shown in processor layout 1 in Figure 9.9. For simplicity purposes, assume also that the processing times for each process are product independent as stated in Table 9.9.

TABLE 9.6 Flow Estimation for Layout Alternative

From/To	MP	A	B	C	D	E	F	G	H	I	J	PF	Loaded	Empty	Total From
MP		150	50			**300**							**500**	*0*	500
A	*70*		25	*25*	25+15						*90*		175	*175*	350
B			*15*	65	10						100	5	130	*130*	260
C	*50*	*105*	65		40+60	175		65					215	*215*	430
D		25	10	*25*				40					100	*100*	200
E	*260*		45	*25*			**300**		15		25+35	430	**790**	*475*	1265
F	*150*					150			300				300	*300*	600
G							40			350			**350**	*0*	350
H	*40*							15		300			300	*315*	615
I						300	150	60	110	180	20	**500**	**500**	*670*	1170
J			65		25			170	190			45	125	*125*	250
PF						490				320	125		**0**	*980*	980
Loaded	**500**	175	130	215	100	790	300	350	300	500	125	**500**	**3485**		
Empty	*500*	*175*	*130*	*215*	*100*	*475*	*300*	*0*	*300*	*500*	*125*	*0*		**3485**	
Total to	500	350	260	400	200	1265	600	350	615	1170	250	980			**6970**

Entries: Trips/period.
Shaded and italics: Empty.
Bold: High relative value.
Squared: Both loaded and empty travel.
Loaded trip entries: From the output station of source center to the input station of destination center.
Empty trip entries: From the input station of source center to the output station of destination center.

TABLE 9.7 Travel Evaluation for Layout Alternative

From/To	MP	A	B	C	D	E	F	G	H	I	J	PF	Loaded	Empty	Total	Average
MP		**3600**	2000			<u>1200</u>							**6800**	*0*	6800	14
A		*560*	1500	1250								2250		2800	*3080*	5880	17
B			*910*	630	320			*390*				800	110	1980	*1300*	3280	13
C		*1470*	440		440	1050								1930	*3910*	5840	14
D	*1000*	550		800	2320			*1840*				2540		2440	*4180*	6620	**33**
E	*1040*		540	*2150*	*450*		**6600**		255				**3440**	**10835**	*3190*	**14025**	11
F	*3000*						*1200*		2100	1020			2100	*4200*	6300	11
G										5600			5600		5600	16
H	*760*						1050			**3600**			3600	*3020*	6620	11
I				945	725	*1800*		*1200*	<u>*1210*</u>	*1440*		**5000**	5000	*4760*	9760	8
J			*585*					*1700*	*1520*	*1920*		*2025*	4715	*1785*	6500	**26**
PF						**3920**							0	*7540*	7540	8
Loaded	0	4150	4480	3625	2095	2250	**6600**	0	2355	10220	1450	**10575**	47800			14
Empty	*5800*	*2030*	*1495*	*2150*	*1710*	*5720*	*2250*	*5130*	*2730*	*3360*	*4140*	*0*		36965		11
Total	5800	6180	5975	5775	4255	7970	8850	5130	5085	**13580**	5590	10575			84765	12
Average	12	18	**23**	13	21	6	15	15	8	12	**22**	11	14	11	12	
Total flow*	1000	700	520	860	400	**2530**	1200	700	1230	**2340**	500	**1960**			6970	
Total travel*	12600	12060	9255	11615	10875	21995	15150	10730	11705	23340	12090	18115			84765	
Average travel*	13	17	18	14	27	9	13	15	10	10	**24**	9			12	

Entries: Trips/period, * meters/period.
Shaded and italics: Empty.
Bold: High relative value.
Underline: Low value given high flow.

TABLE 9.8 Evaluation of Layout Alternative Based on Qualitative Relationships

Pair of Centers		Proximity Relationship			Measure		Importance		Design Evalution	
Center 1	Center 2	Importance	Desired Proximity	Reason	Between	Distance Metric	Weight	Distance	Satisfaction	Contribution
MP	A	Important	Near	Flow	I/O	Aisle	4	24	0	0
MP	E	Very important	Near	Flow	I/O	Aisle	16	4	1	16
MP	In1	Vital	Adjacent	Flow	Boundaries	Rectilinear		0	1	Feasible
A	C	Important	Near	Flow	I/O	Aisle	4	10	0,45	1,8
B	C	Very important	Not far	Infrastructure	Centroid	Rectilinear	16	22	0	0
B	D	Very important	Not far	Infrastructure	Centroid	Rectilinear	16	37	0	0
B	J	Important	Near	Flow	I/O	Aisle	4	8	0,85	3,4
C	D	Very important	Not far	Infrastructure	Centroid	Rectilinear	16	15	0,1	1,6
C	E	Important	Near	Flow	I/O	Aisle	4	6	1	4
D	G	Very important	Very Far	Safety	Centroid	Euclidean	16	35	0	16
E	F	Very important	Near	Flow	I/O	Aisle	16	22	0	0
E	G	Desirable	Not far	Organization	Boundaries	Aisle	1	8	1	1
E	PF	Critical	Very near	Flow + Org	I/O	Aisle	64	8	0,2	12,8
F	G	Important	Far	Noise	Centroid	Euclidean	4	32	1	4
F	H	Very important	Far	Flow	I/O	Aisle	16	7	0,95	15,2
G	H	Important	Not far	Process	Boundaries	Aisle	4	18	0	0
G	I	Very important	Near	Flow	I/O	Aisle	16	16	0	0
G	OUT	Critical	Adjacent	Flow	Boundaries	Rectilinear	64	0	1	64
H	I	Vital	Adjacent	Infra + Org + Flow	Boundaries	Rectilinear	64	0	1	Feasible
I	PF	Critical	Near	Flow	I/O	Aisle	64	10	0,45	28,8
PF	OUT	Vital	Adjacent	Flow	Boundaries	Rectilinear	64	0	1	Feasible
						Maximum possible value:	345		Design value:	168,6
									Design score:	48,9%

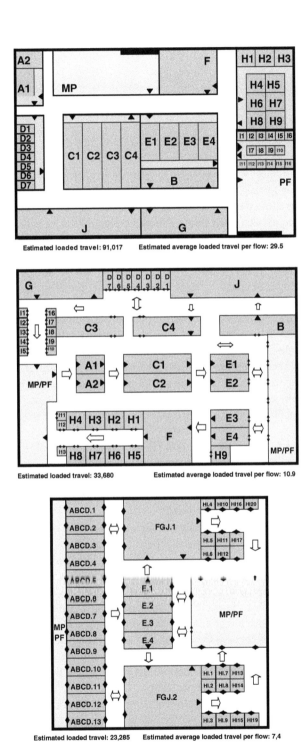

FIGURE 9.9 Processor layouts of alternatives 1, 2 with spatial flexibility and 3 with added processing flexibility.

TABLE 9.9 Elemental Process and Specialized Processor Specifications

Operation Type	A	B	C	D	E	F	G	H	I	J
Unit time (minutes)	0,6	0,5	0,9	5	0,3	0,2	0,1	5	1,5	0,7
Number of processors	2	1	4	7	4	1	1	9	13	1
Processor size	3∗2	2∗9	2∗7	1∗2	2∗5	6∗5	2∗13	2∗2	1∗1	3∗13
Expected net utilization (%)	93	68	90	88	80	31	55	91	86	96

A period is set to a 20-hour workday.
Processor efficiency is estimated at 80%.

As in Section 9.6, an engineer was again asked to generate an alternative layout exploiting this knowledge and the potential for spatially dispersing identical processors instead of grouping them in a single functional center. He was allowed to use flexible centers responsible for both inbound materials and outbound products. He generated alternative layout 2 shown in Figure 9.9. First, given the relatively small size of the case, he has decided not to create centers and has rather developed the design directly at the processor level. Second, he has indeed exploited flexibility allowed to disperse processors. He has strictly separated groups of processors of types H and I. He has contiguously laid out processors of types C and E, yet has oriented them so as to better enable efficient travel for distinct products. Third, the spatial dispersion exploited is not extreme. In fact, it is limited to a fraction of the overall design.

Assume now that there exists flexible processors capable of performing multiple processes. In fact here assume there are three flexible processor types respectively termed ABCD, FGJ, and HI capable of performing the processes embedded in their identifier. In order for the example to focus strictly on exhibiting the impact of flexibility, first the processing times are identical as in Table 9.9 and, second, the flexible processors have space requirements such that the overall space they jointly need is the same as the original specialized processors. The engineer was again required to generate an alternative layout exploiting this flexibility as well as spatial dispersion. He has designed the significantly different alternative layout 3 in the lower part of Figure 9.9.

The design scores provided under the layouts of Figure 9.9 illustrate vividly the potential of exploiting spatial dispersion and flexibility. Alternative 1 is used as a comparative basis. It has an estimated loaded travel score of 91,017. Alternative 2 exploiting spatial dispersion has an estimated loaded travel score of 33,680, slicing 63% off alternative 1's travel. Alternative 3 reduces further the estimated loaded travel score to 23,285, which slices 31% off alternative 2's travel.

The scores have been estimated by assuming that factory operating team will favor the products with high number of equivalent trips when assigning products to processors. Heuristically, the engineer has first assigned the best paths to products P3, P6, and so on, taking into consideration processor availability and processing times. For example, in alternative 2, product P3 getting out of center G is given priority for routing to processors I1 to I9 and then to the nearby MP/FP center.

When locating facilities around the world or processors within a facility, exploiting flexibility leads to what are known as location-allocation problems (Francis et al. 1992). The most well-known illustration is the case where distribution centers have to be located to serve a wide area market subdivided as a set of market zones or clients. There are a limited number of potential discrete locations considered for the distribution centers. Each distribution center is flexible, yet has a limited throughput and storage capacity which can be either a constraint or a decision variable.

The assignment of market zones to specific distribution centers is not fixed a priori. The unit cost of deserving a zone through a distribution center located at a given discrete location is precomputed for each potential combination, given the service requirements of each market zone (e.g., 24-h service). The goal is to determine the number of distribution centers to be implemented, the location and capacity of each implemented center, and the assignments of zones to centers. This can be done for single product cases and for multiple product cases. The same logic applies for flexible factories aimed to be spread around a wide market area so as to serve its production to order needs.

9.8   Dealing with Uncertainty

Explicitly recognizing that the future is uncertain is becoming ever more important in location and layout design. Such designs aim to be enablers of future performance. A design conceived assuming point estimates of demand may prove great if the future demand is in line with the forecast. However, it can prove disastrous if the forecast is off target (Montreuil 2001). Stating intervals of confidence around demand estimates may be highly beneficial to the engineer having to generate a design. For example, consider the demand estimates provided in Table 9.4. The demand for product P1 is forecast to be 2,500 per period. It makes quite a difference if the forecaster indicates that within 99% the demand is to be between 2,400 and 2,600, between 2,000 and 2,700, or between 0 and 7,500. Applied to all products, it significantly influences flow, capacity usage, and required resources to sustain desired service levels.

The case when the product mix is known, the demand for each product is known with certainty, as well as their realization processes, is fast becoming an exceptional extreme. Therefore, the engineer must gauge the level of uncertainty concerning each of these facets, and ensure that he develops a design that will be robust when faced with the uncertain future.

As proposed by Marcotte et al. (2002), Figure 9.10 depicts a graph where each dot corresponds to a design. Each design has been evaluated under a series of scenarios. The graph plots each design at the coordinates corresponding to the mean and standard deviation of its score over all scenarios. The ideal design has both lowest mean and standard deviation. However, as shown in Figure 9.10, often there is not such a single dominating design. In fact, an efficient robust frontier can normally be composed by a series of designs that are not dominated by any other design through its combination of mean and standard deviation. In the case of Figure 9.5, there are five such designs. The leftmost design along the frontier has the lowest mean and the highest standard deviation whereas the rightmost design has the highest mean and the lowest standard deviation. The mean and standard deviation for each of the five dominating designs are respectively (2,160; 315), (2,240; 235), (2,260; 215), (2,760; 180), and (3,600; 155) from left to right. The choice between the five designs becomes a risk management compromise. A more adventurous management is to opt for designs on the left while more conservative management is to opt for designs on the right. For example, if a two-sigma robustness is desired, this means that the comparison should be around the sum of the mean and two standard deviations. Here this results in looking for the minimum between (2,790; 2,710; 2,690; 3,120; 3,910). This means that the third from left design on the robustness frontier is the most two-sigma robust design. In fact, the leftmost design is the most one-half-sigma robust. The three leftmost designs are equivalent at one-sigma, and then the third from left is the most robust at two-sigma, three-sigma, and four-sigma, making a sound choice for a wide variety of risk attitudes.

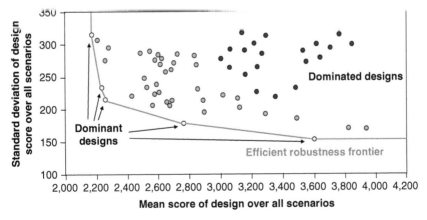

FIGURE 9.10   Efficient robustness frontier for a set of designs subject to stochastic scenarios.

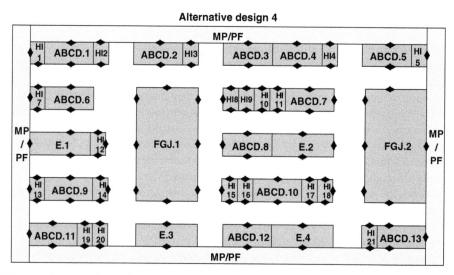

FIGURE 9.11   Alternative layout designed for high uncertainty, given the same set of processors as alternative 3.

The *a priori* acknowledgment of uncertainty should lead engineers to generate designs appropriate for the uncertainty level. As an illustration, an engineer has been requested to generate a layout using the same processors as the third alternative in Figure 9.9, but for a situation where there is complete uncertainty relative to (*i*) the product mix, (*ii*) the demand and (*iii*) the realization process for the products. He has generated a design exploiting the holographic layout concept (Montreuil and Venkatadri 1991, Marcotte et al. 1995, Montreuil and Lefrançois 1996, Lamar and Benjaafar 2005), which differs significantly from all previous alternatives. In high uncertainty contexts, the holographic layout concept suggests to strategically spread copies of the identical processors through the facility space so that from any type of processor there are nearby copies of every other type (Fig. 9.11). This distribution insures a multiplicity of short paths for a variety of product realization processes. This can be verified from Figure 9.10 by randomly picking series of processor types and attempting to find a number of alternative distinct short paths visiting a processor of each type through the facility in the randomly generated order.

9.9   Dealing with an Existing Design

The vast majority of layout and location decisions have to take into consideration the fact that there exists an implemented current design that will have to be transformed to become the selected next design. In some cases, it is an insignificant matter to relayout or to redeploy facilities. In such cases the next design can be developed without explicit consideration of the actual design.

However in most cases, reshuffling an actual implementation is not that easy. At the extreme some processors cannot be moved. They have become monuments in the facility. An example is a papermaking machine in a paper factory: once installed you do not move it. Between the two extremes of move at no cost and move at infinite cost lies an infinite spectrum of situations.

Figure 9.12 indicates graphically an interesting way to approach relayout studies when there are non-negligible moving costs. Iteratively, the engineer should generate alternative designs which take as fixed all entities having at least a specified level of moving cost. On the top portion of Figure 9.12 is displayed the current design, here assumed to be layout 2 from Figure 9.9, displaying through gray tones the expected moving cost associated with each processor. At the first level, the engineer erases only the entities that have negligible moving costs. At the second level, he erases all entities with nonimportant moving costs.

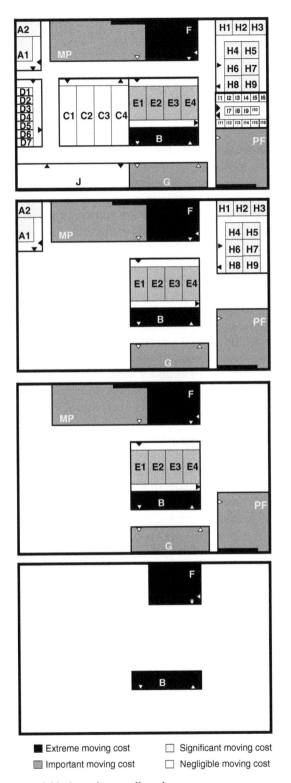

FIGURE 9.12 Design space available dependent on allowed move costs.

At the third level he erases all entities that do not have extreme moving costs. At the fourth and final level, he erases all entities.

At each level, the engineer generates a variety of alternative designs. This results in a pool of alternative designs with distinct estimated moving costs. This allows the engineer to really size the impact of moving costs. Also, when he presents them to the management, it has the potential to generate pertinent managerial discussions, beyond the current layout decision to be taken, which may open avenues in the future. Examples include aiming to implement easy-to-relocate processors, and avoiding putting monuments in the center of action of the facility.

Designs should be compared based on both the expected operating cost, but also adding to it the design transformation expenditures. Moving costs generally involve a fixed cost whenever the entity is even slightly moved. There is often a low level cost whenever the entity is moved within a nearby limited space from its current location. The cost then increases significantly when the move is outside this nearby region. The cost can be fixed as soon as there is a displacement or it can be proportional to the distance being moved. Move costs are sometimes not computable separately, entity per entity: they depend on the set of moves to be concurrently undertaken. It is interesting to assess that the cost of transforming layout 2 into layout 3 in Figure 9.9 would be astronomical given the moving cost specifications of Figure 9.12.

The second aspect relative to redesign is the timing of moves and its impact on current operations and overall implementation cost associated with transforming the current design into the prescribed design. In many settings, the space is so tight that in order to make some moves feasible, some other space has to be created a priori. This creates a cascading effect of interdependent moves which can have impact on the transformation feasibility and cost (Lilly and Driscoll 1985). Also in many settings the operations cannot be stopped for significant durations while the transformation occurs, sometimes except if stocks can be accumulated ahead of time. Some transformations may make it easy to continue operations during the moves. Others may make it very cumbersome and costly. Therefore, it is important to generate a time-phased moving plan that is proven feasible and whose cost is rigorously estimated.

9.10   Dealing with Dynamic Evolution

With the acute shrinking of product life cycles as well as the increasing pace of technological and organizational innovation, in most situations facilities should not anymore be located and laid out assuming a steady state perspective as was generally done in the past. Layout and location dynamics, explicitly considering the time-phased evolution of facilities and networks, is thus also becoming a key issue for the engineer (Rosenblatt 1986, Montreuil 2001). He has to recognize that as the current design is about to be transformed into the proposed design, this proposed design will have a finite existence. It will also have to be transformed into a subsequent design at a later time. The same will occur to this subsequent design and all subsequent others, in a repeating cycle over the entire life of the facility in layout cases or the network of facilities in location cases.

Only when relaying out or redeploying facilities involves insignificant efforts can the engineer optimize the next design strictly for the near future expectations as (*i*) there will be negligible costs in transforming the current design into the next design and (*ii*) it will later be easy to reshuffle this next design into subsequent designs as needed. In most cases, however, there are significant costs involved in dynamically altering designs. So the engineer has to explicitly deal with the dynamic evolution of his designs. This implies for him to develop a dynamic plan as illustrated in Figure 9.13, which shows a four-year layout plan for a facility. Figure 9.13 uses gray shadings to distinguish processors in terms of expected moving cost.

In this age of high market turbulence, the complexity of dealing with the dynamic nature of the design task is confounded by the fact that all demand, process, flow, and space requirements are estimates based on forecasts and that these forecasts intrinsically are known to be ever more prone to error as one looks farther into the future. For example, what will be the demand for a product family tomorrow, next month, next quarter, next year, in three years?

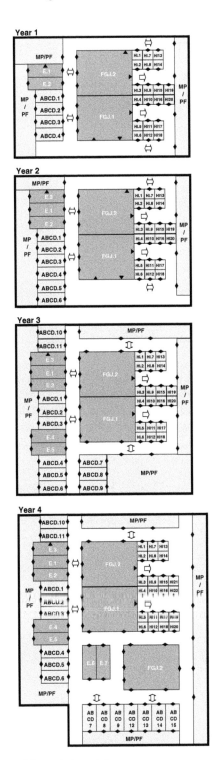

FIGURE 9.13 Myopically generated dynamic layout plan.

TABLE 9.10 Multi-Year Demand Forecasts

Product Number	Expected Daily Demand per Year			
	Y1	Y2	Y3	Y4
1	1000	2000	3000	2500
2	0	0	270	560
3	6000	5750	5500	5250
4	300	275	250	225
5	0	0	0	120
6	0	1000	2000	3000
7	0	0	0	50
8	0	0	0	30
9	0	0	0	30
10	1000	1000	950	875
11	800	750	700	650
12	25	25	25	25
13	0	0	0	25
14	0	0	0	75
15	200	300	450	625
16	0	50	100	175
17	200	500	1000	1500
18	125	110	100	125

Tables 9.10 and 9.11 illustrate this phenomenon for the products of Table 9.4. In these tables, the demand stated in Table 9.4 becomes the expected average daily demand in the fourth future year. There are also forecasts for the first three years preceding this fourth year. Table 9.10 shows that the demand for some products is forecasted to be expanding while the demand for others is forecasted to be shrinking. For each forecast of Table 9.10, Table 9.11 provides the estimated standard deviation over the forecasted mean.

TABLE 9.11 Multi-Year Uncertainty of Average Daily Demand Forecasts

Product Number	Standard Deviation of Expected Average Daily Demand			
	Y1	Y2	Y3	Y4
1	150	360	675	750
2	0	0	30	83
3	300	343	413	525
4	20	22	25	30
5	0	0	0	25
6	0	200	500	1000
7	0	0	0	12
8	5	0	0	8
9	2	0	0	3
10	50	60	71	88
11	200	225	263	325
12	5	6	8	10
13	0	0	0	5
14	0	0	0	20
15	40	72	135	250
16	0	15	38	88
17	140	420	1050	2100
18	6	6	7	12

TABLE 9.12 Multi-Year Expected Average Processor
Requirements

| Processor Type | Expected Average Processor Requirements | | | |
	Y1	Y2	Y3	Y4
ABCD	4	5	9	13
E	2	2	3	4
FGJ	2	2	2	2
HI	19	19	19	20
Total	27	28	33	39

For example, in year 1 the average daily demand for product 3 is forecasted to be 6,000 units with a standard deviation of 300 units, while in year 3 the average daily forecast is down to 5,500 units, yet with a high standard deviation of 413. Such information may come from analyzing historical forecast performance in forecasting demand for such a family respectively one year and three years ahead (Montgomery et al. 1990). This means that within two standard deviations (two-sigma) or 98% probability using normal distribution estimation, in current year zero the average daily demand for P3 is expected to be between 5,400 and 6,600 units in year 1, and between 4,674 and 6,326 units in year 3.

Using the process requirements of Table 9.4 and assuming the flexible processors introduced in the lower part of Figure 9.9, these forecasts permit to compute estimates for the average expected number of processors of each type, provided in Table 9.12. Also they allow computing robust estimations for processor requirements, such as the two-sigma robust estimates provided in Table 9.13. In year 3, for processor type ABCD, the average estimate is 9 units while the two-sigma robust estimate is 11 units. Overall the robust estimate adds up to a total of 28 processors in year 1 to a total of 47 in year 4.

Figure 9.13 provides a four-year layout plan generated in year 0 by an engineer. To help understand the compromises involved, the engineer was asked to first generate a design for year 1 based on the estimates for year 1. He had to then transform this year-1 design into a year-2 design taking into consideration the expected flows for year 2 and the cost of transforming the year-1 design into the year-2 design. He had to repeat this process for years 3 and 4. Clearly, this is a rather myopic approach because in no time was he considering the overall forecasted flows and processor requirements for the entire four-year planning horizon. Analyzing the plan, it is clear that the engineer's decision in year 1 to lay out the two FGJ processors adjacent to each other has defined a developmental pattern that has had repercussions on the designs he has produced for year 2 to year 4. Even though possible, he has not planned to move any of the processors E and FGJ once laid out in their original location, which has created a complex flow pattern in year 4, as contrasted with the elegant simplicity of the lower layout of Figure 9.9. Formally evaluating the dynamic plan requires to evaluate each design statically as described in the previous sections, and to compute the expected transformational costs from year to year. The evaluation requires the generation of demand scenarios probabilistically in line with the forecast estimates of Tables 9.10 to 9.13. Due to

TABLE 9.13 Multi-Year Two-Sigma Robust Processor
Requirements

| Processor Type | 2-Sigma Robust Processor Requirements | | | |
	Y1	Y2	Y3	Y4
ABCD	4	6	11	15
E	2	3	5	7
FGJ	2	2	2	3
HI	20	20	20	22
Total	28	31	38	47

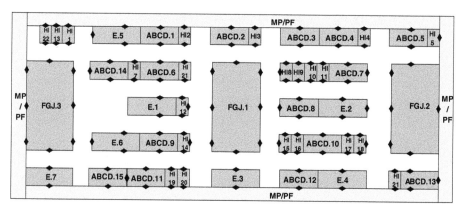

FIGURE 9.14 Illustrating the steady robust, immune-to-change, design strategy by expanding the template of Figure 9.11 to transpose it into a design for year 4 given the forecasts of Tables 9.10 to 9.13.

space constraints, the results of such an evaluation for the plan of Figure 9.13, and the generation and evaluation of alternative plans that take a more global perspective, are left as an exercise to the reader.

In practice, there are two main strategies to deal with dynamics. The first is to select processors and facilities that enable easy design transformation, and to try to dynamically alter the design so as to always be as near to optimal as possible for the forthcoming operations. Figure 9.13 can be seen as an example of this strategy. The second is to develop a design that is as robust as possible, as immune to change as possible, a design that requires minimal changes to accommodate in a satisfactory manner a wide spectrum of scenarios (Montreuil and Venkatadri 1991, Montreuil 2001, Benjaafar et al. 2002). Figure 9.14 provides an example of this strategy by simply expanding the robust design of Figure 9.11 to be able to deal with the estimated requirements for year 4. It is left as an exercise to assess how to subtract processors from Figure 9.10 to deal with the lower expected requirements for years 1 to 3.

In the above examples, a yearly periodicity has been used for illustrative purposes. In practice, the rhythm of dynamic design reassessment and transformation should be in line with the clock speed of the enterprise, in synchronization with the advent of additional knowledge about the future and the lead time required for processor and facility acquisitions and moves. Even decades ago, some companies were already reconfiguring their shop floor layouts on a monthly basis, for example, in light assembly factories dedicated to introducing new products on the market, assembling them until demand justifies mass production.

In the illustrative example of Figure 9.13, the planning horizon has been set to four years. Again, this depends on the specific enterprise situation. It can range from a few days in high flexible easy-to-alter designs to decades in rigid designs in industries with low clock speeds.

9.11 Dealing with Network and Facility Organization

Layout and location design studies often take the organization of the facility network as a given, yet organizational design has a huge impact on spatial deployment optimization. The organization of the network states for each center and/or facility its specific set of responsibilities. This bounds the type of products, processes, and clients the center is to deal with. According to Montreuil and Lefrançois (1996) the responsibility of an entity is defined by a set of combinations of markets, clients, outbound products, processes, processors, inbound products and suppliers, specified quantitatively and through time. For example, a center can be responsible for manufacturing all plastic products offered by the enterprise to the Australian market. Another center can be responsible for assembling up to 10,000 units a year of a specific product. Through the responsibility assignment process, the organizational design also defines the customer–supplier relationships among centers and facilities.

In some cases the organizational design is not complete when the layout or location design process is launched, depending on the output of this process to finalize the design. This is the case, for example, with location-allocation problems. The organizational design states, for example, that the logistic network is to comprise only distribution centers that are to be the sole source for their assigned market segments. The set of market segments is defined geographically and in terms of demand. Depending on the actual location and sizing of distribution centers, the assignment of segments to centers can be performed, completing the organizational design.

Adapted from Montreuil et al. (1998), Table 9.14 provides a responsibility based typology of centers and facilities. First, types of centers are segregated by their defining orientation. The options are product, process, project, market, and resource orientations. A product-oriented organization defines the responsibility of the center in terms of a set of products. In contrast, a process organization does not state responsibilities in terms of products; it is rather in terms of processes. The same logic holds for the three other orientations.

TABLE 9.14 Responsibility-Based Center Typology

Center Orientation	Center Type	Responsibility Set	Responsibilty in Terms of Demand Satisfaction
Product		Set of products	All or a fraction
	Product	Single product	All or a fraction
	Group	Specific group or family of products	All or a fraction
	Product fractal	Most products; generally multiple centers are replicated to meet demand	A fraction
Process		Set of processes	All or a fraction
	Function	A single function, elementary process or operations	Generally all, yet can be a fraction
	Process	A composite process composed of linked elementary processes	All or a fraction
	Holographic	A set of elementary processes, generally multiple centers are distributed to meet demand	A fraction
	Process fractal	Most processes; generally multiple centers are replicated to meet demand	A fraction
Project		Set of projects	All or a fraction
	Order or contract	A specific order, contract or, in general, project	Generally all, except for very large cases
	Repetitive project	Projects of the same that repeatedly occur through time	All or a fraction
	Program	A long-term program involving a large number of planned deliveries	Generally all
Market		Set of markets and/or clients	Generally all
	Client	A specific client	Generally all
	Client type	A set of clients sharing common characteristics and requirements	Generally all
	Market	A market or market segment, defined by geography or any other means	Generally all
Resource		Set of resources to be best dealt with	Generally all
	Inbound product	Set of inbound products needing to be processed	Generally all
	Supplier	Set of suppliers whose input has to be processed	Generally all
	Team	Set of people whose capabilites have to be best exploited and needs have to be best met	Generally as much as possible given their capacity, capabilites and preferences
	Processor	Set of processors (equipment, workstation, etc.) to be exploited as best as possible	Generally as much as possible given their capacity and capabilites

Source: Adapted from Montreuil et al. in *Material Handling Institute*, Braun-Brumfield Inc., Ann Arbor, Michigan, 1998, 353–379.

For each orientation, Table 9.14 provides a set of types of centers, stating for each its type of responsibility. Product-oriented organizations are segregated into three types: product, group, and fractal. A product center is devoted to a single product. A group center is devoted to a group or family of products. Note that product is here a generic term which encompasses materials, components, parts, assemblies as well as final products. Table 9.14 also indicates that a product or group center may be made responsible for only a fraction of the entire demand for that product or group. For example, it can be decided that there are to be two product centers mandated to manufacturing a star product. The former is to be responsible for the steady bulk of the demand while the latter is to deal with more fluctuating portion of demand above the steady quantity assigned to the former. Similarly, instead of assigning the fluctuating portion to another product center, it can assign it to a group center embracing similar situations. The possibilities are endless. A product-oriented fractal organization offers a different perspective. It aims to have a number N of highly agile centers, each capable of dealing with most products, assigning to each fractal center the responsibility of 1/N of the demand of each product. This allows operations management to dynamically assign products to centers in function of the dynamic repartition of demand among the products (Venkatadri et al. 1997, Montreuil et al. 1999). Implementing a product organization has tremendous impact on flow through the network and the constitution of each center in the network. Product centers rarely have flow of products between them, except when one provides products that are input to the other. There is more complex flow within the center as one switches from a product center to a group center and then to a fractal center. Also, when only product or group centers are used, most of the specific customer–supplier relationships are predefined. Whenever fractal centers are used, then workflow assignments become dynamic operational decisions.

Process orientations are segregated into four types: function, process, fractal, and holographic. Function, process, and fractal types are the process-oriented equivalent of the product-oriented product, group, and fractal types. For example, a process-oriented fractal center is responsible for being able to perform most elementary processes, with 1/N of the overall demand for these processes (Askin 1999). Again, adopting a process orientation has significant impact on workflow patterns. For example, function centers have minimal flow between the processors comprising them and have significant flows with other centers. Illustratively, an injection center has minimal flow between the injection moulding machines, except for the sharing of moulds, tools, and operators. In fact, when a network is comprised only of function centers, a product with P processing steps will have to travel between P distinct function centers. In such cases the relative layout of centers becomes capital in order to contain the impact on inter-center material handling/transport. Holographic organization generates a number of small centers responsible for a limited set of related processes. Most centers are replicated and strategically distributed throughout the network or facility. In fact, the robust flexible layouts of Figures 9.11 and 9.14 are exploiting a holographic organization where each processor is conceived as a small center, instead of a function organization as in the layout of Figure 9.5.

Project-oriented organizations lead to center types that are defined in terms of orders, contracts, projects, or programs. A manufacturer bids for and then wins the bid for a major contract involving a set of products and processes to be performed in given quantities according to a negotiated delivery schedule. When its managers decide to implement a facility strictly devoted to delivering this contract, the resulting facility is of the contract type. Similarly, when a factory within an automotive network is awarded a multi-year program to manufacture all engine heads of a certain type for the European market and when it devotes a center to this production, its organization now has a program center. Repetitive project centers are centers well conceived and implemented to realize specific types of projects that come up repetitively. This is common in the aeronautical industry where, for example, large centers are well equipped to perform a variety of overhauls, maintenance or assembly of airplanes depending on the flow of projects signed by the enterprise.

Resource-oriented organizations can be segregated into four types of centers. Inbound product centers and supplier centers are respectively specialized to perform operations on certain types or groups of inbound products or on all products incoming from a set of suppliers. Processor and team centers are similar conceptually, designed to exploit the capabilities and capacity of a set of processors and humans, respectively. A center grouping all the CNC machines in a factory is an example of processor center.

Table 9.14 opens a wealth of organizational design options. First, each network can be composed of any combination of centers from the various types. Second, the types provided have to be perceived as building blocks which allow the design of composite or hybrid types of centers, such as a center devoted to performing a set of processes on a group of products. Third, it can be used recursively. Higher level facilities or centers have to be organized according to a pure or hybrid type. These can be composed of a network of internal lower level centers. Each of these has to be organized, not restricted to the same type as its parent.

To illustrate the impact of network organization, for the illustrative case leading to the layouts starting in Figure 9.5, there has been the implicit assumption that the organizational design states that all the products and processes have to be performed within the same centralized facility. When this constraint is removed and further market information is provided, a network organization such as depicted in Figure 9.15 is quite possible. In the network of Figure 9.15, a global factory is proposed to manufacture products P1, P2, as well as P4 to P9. Another global factory is specialized to manufacture product P3. Three market-specific product group facilities are to be implemented. These will all make products P10 to P18. Each will be dedicated to serving a specific market: America, Europe, or Asia. Each market is to be assigned a number of regional distribution centers fed by the global P3 factory and the three P10-to-P18 factories. Now, instead of having to locate and lay out a single global facility, the design task involves locating interacting factories and distribution centers, and to organize, size and lay out each of these.

Here above the organizational emphasis has been put on the centers, stressing the importance of their specific responsibilities and their customer–supplier relationships. Figure 9.16 depicts clearly another important network organization facet: the type of organization structure of the network. Figure 9.16 provides a sample of seven types of structures resulting from organizational design of the network.

The first is termed a fixed product structure. Here the idea is that the product is brought to one location and does not move until departing the system. The processors and humans are the ones moving to, into, and away from the stationary product. The second structure type is a parallel network, where all flow is leading inbound products into one of the centers and then out of the system. The third is a flow line where each center is fed by a supplier and itself feeds a client center, this being repeated until the product gets out of the system. Centers store and/or perform operations on the product.

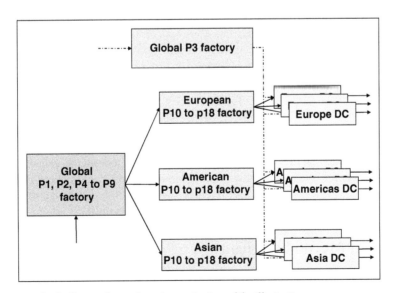

FIGURE 9.15 A multi-facility product oriented organization of the illustrative case.

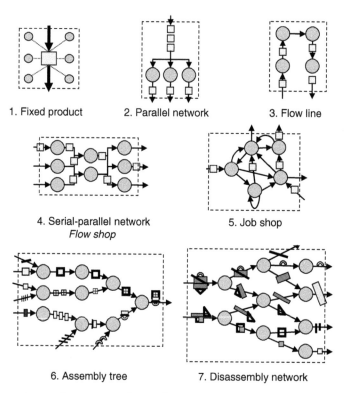

1. Fixed product   2. Parallel network   3. Flow line

4. Serial-parallel network   5. Job shop
      Flow shop

6. Assembly tree   7. Disassembly network

FIGURE 9.16 Illustrative set of organizational network structures.

The fourth structure is a serial-parallel network, typical of a flow shop. This structure combines the flow line and the parallel network. It is conceived as a series of stages. At each stage, there are parallel centers jointly responsible for delivering the stage responsibility. The fifth structure is a job shop network characterized by a profusion of inter-center flows that have no dominant serial or parallel pattern.

Whereas structures one to five can be mainly mono-echelon, the sixth and seventh example structures are multi-echelon in nature. Indeed they explicitly deal with the fact that products are needed constituents of other higher level products and organize the network around these bill-of-materials relationships among products. The sixth structure is an assembly tree. Here each center feeds a single center which later performs operations on the delivered products and/or assembles the delivered products into higher level products. The seventh structure is a disassembly network. Instead of assembling products, it disassembles them. Instead of being restricted to a directed tree, it is conceived a more flexible directed network. Here the main difference is that a center may have more than one client center, while maintaining the no back-tracking constraint of the tree structure. One can easily think of a disassembly tree structure or an assembly network structure.

Network structures have direct influence on flow patterns and therefore on layout and location decisions. In fact it can be said that the organizational combination of responsibility assignment and network structure puts the stage for layout and location studies. However, more important in a highly competitive economy is the fact that integrating the organizational, location, and layout design processes offers the potential for designing networks with higher overall performance potential.

9.12 Design Methodologies

The previous sections have focused on the essence of the location and layout design representation, stressing key facets and issues. This section focuses on methodologies used for generating the designs.

It is important to start by humbly stating that currently there is no generic automated method capable of dealing with all issues covered in the previous sections and of providing optimal, near-optimal, heuristically optimized or even provably feasible designs. It is also important to state that most issues brought forward in the previous sections are inherent parts of most location and layout design studies. Indeed facilities end up being located and laid out every day around the world, resulting into feasible yet imperfect networks which have to be adjusted to improve their feasibility and performance as their implementation and operation reveal their strengths and weaknesses, and their growing inadequacy to face evolving demands.

In this section, the emphasis is not on trying to document reported methodologies pertinent for each type of situation as defined through combinations of facets introduced in the previous sections. For example, there will be no specific treatment of stochastic dynamic layout of flexible processors in continuous space, nor of deterministic static location of unlimited capacity facilities in discrete space. The combinations are too numerous. References have already been provided through the previous sections, which propose either surveys of methods or introduce appropriate methods.

The section rather takes a macroscopic perspective applicable to most situations. It does so by mapping the evolution of the types of methodologies available to designers. Indeed, as depicted in Figure 9.17, location and layout design has evolved methodologically through the years into nine states that concurrently exist today, each with its application niches. The outer circle includes the oldest methods: manual, heuristic, and mathematical programming. The middle circle includes the more recent methods which have evolved from those in the outer circle: interactive, metaheuristic, and interactive optimization. Finally, the inner circle includes the most recent methods: assisted, holistic metaheuristic, and global optimization. The nine methodological alternatives are hereafter described.

9.12.1 Manual Design

The earliest and most enduring method is the manual method. Sheets of paper and cardboard, colored pencils, and scissors are the basic tools used. The engineer, based on his understanding of the qualitative

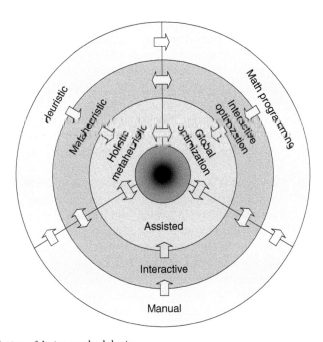

FIGURE 9.17 Evolution of design methodologies.

proximity relationships, quantitative flow estimations, cost structures and constraints, gradually draws a series of designs, from which he picks a limited set of preferred alternatives to present to the management for decision. For example, layouts are sketched on paper. They are assembled on boards using pieces of carton for each center, or using Lego-style building blocks. They are approximated using real-size flat panels on working floors through which the engineer can walk the design. Or yet they are designed by really shuffling the actual layout until a satisfying design is implemented. In practice, the evaluation of each design is often very limited and coarse, even regularly limited to a multi-criteria ranking of alternatives, where each criterion is evaluated quite subjectively or approximately.

In the manual method, computers are used quite minimally. They are exploited for reporting the preferred designs and sometimes to evaluate the final set of designs.

The manual method has the advantages of being simple and expediting. It may work well when the design complexity is low and the degrees of freedom limited. It can rapidly prove tedious and limitative as the case size and complexity increase. Yet for good or for bad, a large number of designs are still achieved this way in practice.

Starting in the 1960s, researchers have worked toward automating layout and location design. Two basic directions have been taken: simple heuristics and mathematical programming.

9.12.2   Heuristic Design

Researchers who generated heuristics for layout and location design have aimed to capture the power of the computer to generate satisfying designs by systematically searching the solution space using approximate yet rigorous methods. Kuenh and Hamburger (1963) and Nugent et al. (1968) are typical in heuristic location design while Armour and Buffa (1963) and Lee and Moore (1967) are typical in heuristic layout design. Two types of heuristics have been developed, with myriads of instances of each type and a multitude of hybrids combining both types. The two types are construction and improvement heuristics. As exemplified by ALDEP (Seehof and Evans 1967) and CORELAP (Lee and Moore 1967), a construction heuristic gradually iterates between selection and placement activities until a design is completed or infeasibility is reached. The selection activity decides on the next center to place in the design, or more generally on the order according to which centers are to be inserted in the design under construction. The placement activity locates and shapes the selected center into the partial design.

The variety of construction heuristics comes from the multiple options for selecting the next center and for placing it into the partial design. Selection can use qualitative relationships or flow, can take into consideration or not those already placed, can be deterministic or randomized, and so on. The simplest way ranks the centers in decreasing order of flow or proximity relationships intensity and then selects them for placement in that order, placing them in the best available location given its space requirements and its flow or relationships with already placed centers. When deterministic selection is used, the heuristic generates a single design. When randomized selection is used, then the heuristic generates numerous designs, scores each of them, memorizes the best N designs and then reports them at the end of the randomized sampling.

Placement is the most difficult part of construction heuristics. So as to ease the generation of feasible designs, the earlier ones relied on a discrete space representation and did not support such restrictions as having to use existing constraining buildings. In general, as the heuristic advances in its iterations, the center of the design gets occupied, leaving mostly space available at outskirts of the design, subject to ever more feasibility constraints to fit the centers in the design. In such heuristics, placement is eased when a combination of design code and filling pattern is imposed. For example, when layouts are coded as strings (ex: A-MP-F-...-PF), centers can be iteratively inserted in the string code. Then the design can be generated by systematically placing the centers according to the string code. A layout heuristic can, for example, start to place the first center in the northwest corner of an existing facility, then move left with the second and third, until it reaches the eastern boundary. Then it can move one layer southward and head back westward, zigzagging until the design is completed.

Improvement heuristics such as CRAFT (Armour and Buffa 1963) start with a given initial design. Then they iteratively generate potential local improvements to the design, estimate the improvement potential, implement the preferred improvement, evaluate the potentially improved design, set it as the incumbent design if it improves on the current best design, repeating this process until no further improvement is reachable or sufficient time has elapsed. Local improvements are typically two-way, three-way, or multiple-way exchanges. A two-way exchange of centers A and B consists of locating center A at B's current location and vice-versa. In a three-way exchange, centers A, B, and C, respectively, take on the current location of centers B, C, and A, or of centers C, A, and B. As an example, a typical heuristic based on two-way exchanges scans iteratively all pairs of centers in a predetermined order. For each pair, it estimates the design score given the exchange of center locations. If the estimation reveals a potential score improvement over the current one, two options are possible. Either the heuristic implements immediately the exchange and scores the resulting design or it keeps on testing all two-way exchanges and implements only the best potential exchange.

For computational speed reasons, an improvement heuristic evaluates the potential of an exchange rather than immediately evaluating the altered design. For example, in layout design with centers of various sizes, the impact of exchanging centers A and B may not be obvious. The simplest case is when centers A and B are of equal size and all flows and proximity relationships are assumed to be between the centers' centroids. In such a case, simply interchanging the centroid locations in the travel or proximity relationship score computations is sufficient to estimate the real impact of the interchange. In most other cases, it is not so easy. For example, exchanging centers F and J in the layout of Figure 9.4 requires altering not only the location and shape of the involved pair, but of several other nearby centers. Center J is larger than center F and it has a fixed shape with a large length-to-width ratio. While center F fits into the current location of center J, the converse is not true since center J does not fit in the current location of center F. Fitting J in the northeast region would require significant reshuffling. This is why many such heuristics, following the lead of their ancestor CRAFT (Armour and Buffa 1963), forbid interchanges involving nonadjacent different-size centers. In the illustrative case, even a simpler exchange such as centers D and I in Figure 9.6 requires to deal with the aisle segment between them, to reposition as best as possible their I/O stations, and to adjust the empty travel estimates. This is why, before realizing these tasks, a heuristic applied to this case would assume direct interchange of the centroid, I/O stations and boundaries, and would not re-estimate the empty travel, then would compute the selected design score (e.g., minimizing flow travel). Once an exchange is selected as the candidate for the current iteration, then the modifications are really made in the layout and the score more precisely computed.

Most heuristics have been implemented with a number of simplifying restrictions and assumptions. For example, in layout design, most generate a block layout instead of a more elaborate design such as a net layout. They do not support I/O stations and aisle travel. They deal only with loaded travel minimization or qualitative proximity relationship maximization assuming Muther's ABIOUX coding.

Typically a heuristic is coded in a software that allows case data entry and editing, heuristic parameter setting, and graphical solution reporting. Most such software is developed by researchers solely to support the developed heuristic. They rarely allow choosing among a variety of heuristics. Capabilities for interactive editing of produced designs are usually quite limited, the emphasis being placed on automating the design task.

9.12.3 Mathematical Programming-Based Design

Researchers have long recognized that some simple instances of location and layout design can be modeled mathematically and solved optimally in short polynomial time. A well-known example is the location of a single new facility interacting in continuous space with a set of fixed facilities so as to minimize total travel given deterministic flows (Francis et al. 1992). Another well-known location example, solvable using the classical linear assignment model (Francis et al. 1992), involves the assignment of a set of facilities to a set of discrete locations so as to minimize total travel and implementation costs, provided that at most a single facility can be assigned to any specific location and that the assignment costs can be computed a priori for each potential facility-to-location assignment, being independent of the relative assignment of facilities.

As yet another example, take dedicated warehouse layouts in which each product is assigned to a fixed set of storage locations in which no other product can be located. It is well known that, assuming deterministic demand, products can be optimally assigned to storage locations according to the cube-per-order index when all products have the same inbound and outbound behavior in terms of dock usage (Heskett 1963, Francis et al. 1992). For example, they all come in a given dock and all go out using the same other dock. The cube-per-order method (*i*) computes the expected distance travelled by a product assigned |to each storage location and then ranks the locations in nondecreasing order of expected distance, (*ii*) computes the cube-per-order index, as the ratio of product storage space requirements over product throughput, and then ranks the products in nondecreasing order of this index, and finally (*iii*) iteratively assign the first remaining location to the first nonfully assigned product, until all are assigned or no more space is available.

As a final example, given a continuous-space block layout with rectangular shaped centers, the optimal location of all I/O stations can be found in polynomial time if one aims to minimize rectilinear travel and if each station can be located anywhere within a predetermined rectangular zone (Montreuil and Ratliff 1988a).

When a design case fits exactly with a problem solvable in polynomial time, then its solution algorithm should be applied so as to get the optimal solution. Most cases do not readily fit exactly such easily solvable problems, yet if the gap is not too enormous, the case can be manually adapted to fit the problem and the optimal solution can be used as an approximate solution to the real situation, heuristically adjusted to reach satisfying feasibility. This can also be used for more complex (NP-Complete) mathematical programming problems that have been researched and for which there exist (*i*) good optimal solution algorithms exploiting techniques such as branch-and-bound, decomposition and branch-and-cut, or (*ii*) good generic heuristics capable of providing satisfying solutions.

Such an approach has led to the dominance of the quadratic assignment model in representing layout and location design problems for decades prior to the early 1990s. The model of the quadratic assignment problem (QAP) is defined as follows:

Minimize

$$\left[\sum_{\forall cl} a_{cl} A_{cl} + \sum_{\forall dc'l} c_{dc'l'} \left(A_{cl} A_{c'l'} \right) \right] \tag{9.1}$$

Subject to

$$\sum_{l \in L^c} A_{cl} - 1 \qquad \forall c \tag{9.2}$$

$$\sum_{c \in C^l} A_{cl} \leq 1 \qquad \forall l \tag{9.3}$$

Variables:

A_{cl} Binary variable equal to 1 when center c is assigned to location l, or 0 otherwise

Parameters and sets:

a_{cl} Cost of assigning center c to location l
$C_{clc'l'}$ Cost of concurrently assigning center c to location l and center c' to location l'
C^l Set of centers allowed to be located in location l
L^c Set of locations in which center c is allowed to be located

The QAP enforces the engineers to define a discrete location set, such that each center has to be assigned to a single location (constraint 9.2) and that a single center can be assigned to any location (constraint 9.3). In layout design, most of the cases researched in the scientific literature involve a M-row N-column matrix of square unit-size locations and a set of at most M*N centers with fixed square unit-size shape. The QAP problem is among the most difficult combinatorial problems. For decades, cases with at most 10 locations could be solved optimally. Even today, the largest cases optimally solved involve up to 30 locations (Anstreicher et al. 2002). Yet being such a well-known problem, the QAP has been a battling ground for researchers, leading to the availability of numerous generic heuristics and meta-heuristics applicable for location and layout cases if the engineer is capable of modeling them as a QAP (e.g., Nourelfath et al. 2007).

The early advances in the manual, heuristic, and mathematical programming based methodologies have led the way for the middle circle methodologies of Figure 9.17 described below.

9.12.4 Interactive Design

Interactive design follows directly the trail of manual design, with the difference lying in being adopted by engineers that are fluent with commercial spreadsheets such as Excel as well as with computer-aided drawing and design software (CAD) such as AutoCad, CATIA, SolidWorks and Visio, even presentation software such as PowerPoint, and with geographical information systems (GIS) such as MAPINFO or Google Earth. A spreadsheet is used for computing design scores and performing local analyses. For layout cases, the CAD software is used to draw and edit the designs, as well as to show the flows and relationships. For wide area location cases, the GIS software serves the same purpose.

Computer-aided drawing and design software has two main advantages. First, it is used for referential technical drawing of facilities in many organizations, used for keeping up to date the precise equipment, service, and utilities layout. The software and the drawings thus become freely available to the engineer for layout design purposes. Second, CAD software is often exploiting the notions of drawing object libraries and drawing layers, which speed up and ease the layout drawing effort. The main disadvantages of using CAD software are that (1) they are most often geared for precision drawing and may become cumbersome to use for design purpose, and (2) they do not understand layout design. An object is mostly a drawing object. A flow is simply a link from an object to another. The software does not embed knowledge and methods exploiting the fact that the object is a center and that the flow involves trips of products or resources between centers. The engineer must assume the sole responsibility for the representativeness of its drawn designs. The same types of advantages and disadvantages apply for GIS systems used for location purposes, adapted to a set of geographical sites rather than a set of facilities.

In the future, there will be more seamless integration of CAD and GIS software, allowing to show or edit a large-scale logistic network and to then swiftly dig into the facilities part of the network.

As generic technological capabilities increase, interactive design is enabled to achieve better representations in ever easier ways. For example, 3D drawings, renderings, and walks-throughs add significant value to an engineer involved in facilities layout. They allow dealing directly with multi-floor facilities, and in more generic terms, to exploit the cube rather than its rectangular surface. They allow a visual grasping of the facility layout which is by far superior to 2D representations. This has been well known for decades. Yet such capabilities are still very rarely used in practice because of the combination of software price, 3D drawing complexity and lack of computational power to deal with large-scale layouts. These three constraints are rapidly diminishing with new generation software. As engineers will learn to exploit them generically, they will gradually use them more for facilities layout purposes.

Interactive design is widely used in practice, second only to manual design. Both suffer from the same threat: they depend heavily on the engineer. The tools are generic and do not understand layout or location and do not have any layout and location optimization capabilities. This is why the value of both manual and interactive design depends on the engineer's mastering of the layout and location issues and on his creativity in generating great designs.

9.12.5   Metaheuristic Design

Metaheuristics have evolved from heuristics for two main reasons. The first is an attempt to get out of the local optima trap in which heuristics get stuck. This has lead to developing metaheuristics exploiting techniques such as simulated annealing (Meller and Bozer 1996, Murray and Church 1996), tabu search (Chittratanawat 1999, Abdinnour-Helm and Hadley 2000), genetic and evolutionary algorithms (Banerjee et al. 1997, Norman et al. 1998), ant colony algorithms (Montreuil et al. 2004) and swarm intelligence (Hardin and Usher 2005). The second reason is the researchers' attempt to go beyond solving the basic layout and location problems, to get away from enforcing myriads of simplifying assumptions and constraints. In location, this has led to metaheuristics for addressing complex problems (e.g., Kincaid 1992, Crainic et al. 1996, Cortinhal and Captivo 2004). In layout, researchers have attempted, for example, to integrate the automatic generation of block layouts with their travel network (e.g., Fig. 9.2c) (Norman et al. 1998). The combination of both reasons has had high stimulating impact on researchers.

Metaheuristics operate at least on two levels. The first level uses heuristics to develop a design subordinated to master decisions taken at the second level. This second level drives the overall heuristic search process, iteratively exploiting the heuristics of level 1 to scan the solution space. Complex implementations may have multiple levels, with the higher levels exploiting the lower levels in the same way as exemplified in the two-level illustration.

When trying to avoid the local optima trap, researchers have relied upon the exploitation of generic metaheuristic techniques. Genetic algorithms provide a fine example to understand how such metaheuristics are used in layout and location settings. Very shortly, genetic algorithms attempt to mimic genetic evolution leading to survival of the fittest. In layout design, members of the population are individual layouts. Used at the second level of the metaheuristic, the genetic algorithm iterates through rounds which each enact a number of immigrations, mutations, and crossovers from which is generated the next generation. At all iterations only the N best layouts are kept in to form the population of the next generation.

The key to understanding how genetic algorithms work in layout is that they exploit the notions of layout code and space structuring, both introduced in Section 9.3. Remember that the code for the three-band layout of Figure 9.4 is (1:A,B,C; 2:D,E,F; 3:G,H,I). Given this code and the knowledge that the layout is restricted to be structured into three horizontal bands, the band layout of Figure 9.4 can be reconstructed. Hence, the second level of the metaheuristics is used to search the solution space in terms of layout codes while the first level uses a heuristic or an optimization model to generate a layout from the code generated in the second level.

At the second level, the activities are simple once focused to be performed using layout codes. For example, immigration is simply achieved through the randomized generation of a new layout code. At all iterations, the genetic algorithm randomly generates a number of immigrant codes.

A mutation of the (1:A,B,C; 2:D,E,F; 3:G,H,I) code can be achieved in many ways. For example, a center can be transferred from a band to another [e.g., D in (1:A,D,B,C; 2:E,F; 3:G,H,I)], a center can be moved from its current position in the string to another position while keeping the number of centers in each band intact [e.g., D in (1:A,B,C; 2:E,F,G; 3:H,I,D)], a pair of centers can exchange positions in the code [(e.g., D and B in (1:A,D,C; 2:B,E,F; 3:G,H,I)], an entire content of two bands can be exchanged [(e.g., bands 1 and 2 in (1:D,E,F; 2:A,B,C; 3:G,H,I)]. At all iterations, the genetic algorithm randomly selects the layout codes to be mutated and the way each one is to be mutated.

A crossover involves two members of the population. As an example, consider the layout codes (1:A,B,C; 2:D,E,F; 3:G,H,I) and (1:D,H,I; 2:B,A,G; 3:E,C,F). An illustrative crossover could be formed by taking in priority the first band as in the first code, the second band as in the second band, the third band as in the first code, then assigning any unassigned center to its current ordered position in the first or second code, picking from both codes in rotating order. Here this starts the crossover-generated code with

(1:A,B,C). Second, it extends it as (1:A,B,C; 2:G). Third, it again extends it as (1:A,B,C; 2:G; 3:H,I). Fourth, it finalizes it by inserting the missing centers: (1:A,B,C; 2:D,G,F; 3:H,E,I). At each iteration, the genetic algorithm randomly selects the pairs of layout codes used for crossover purposes, and how the crossover is to be performed for each pair.

The layout code resulting from each mutation, crossover, and immigration is transferred to the level-one heuristic optimizer which generates a layout design respecting the layout code and the space structuring. This layout is scored according to the selected metric. The layout score serves for deciding which layouts are to form the next generation. The genetic algorithm keeps on searching until a time or iteration limit has been reached, or until no better layout has been generated since a specified number of iterations. The regular usage of randomization for generating layout codes, the multiplicity of ways layout codes can be generated, and the systematic screening of the score of the layout generated from each layout code augment the probability that the metaheuristic will not get stuck in local optima and thus potentially get nearer to optimality within a given solution time.

Without getting into as much detail, other metaheuristic techniques used are the following. The first and simplest to be tested has been simulated annealing, mostly used in conjunction with improvement heuristics. The second-level of the metaheuristic simply dynamically adjusts the probability that the improvement heuristic at the first level will accept to implement an exchange with negative impact on the performance of the current best design. The logic is as follows: When the heuristic finds better layouts at a good pace, the probability is kept low. When the heuristic begins to have trouble finding better layouts through local improvement, then the probability is increased, letting the improvement heuristic deteriorate temporarily its current best design so as to get away from the current local optimum region. Tabu search is another fruitful metaheuristic technique. It puts emphasis on forbidding to consider in the improvement algorithm moves that have been recently examined, speeding up the solution process by avoiding unnecessary repetitive loops examining the same potential layouts over and over.

Ant colony algorithms share with genetic algorithms the exploitation of layout code and space structuring. They differ in their second-level implementation. The underlying metaphor is to think of a resulting layout as the output of an ant looking for food. If the layout is good, then the ant leaves traces of phero-mone at milestones along the path during its return trip. Milestones depend on the metaheuristic imple-mentation: they can correspond to locating specific centers in some portion of the layout or to locating specific centers adjacent to each other. Other ants looking for food will trace a path which is influenced to some degree by the intensity of pheromone left at milestones by preceding ants, augmenting the probabil-ity that the ant will end up in hot spots for layout quality. At each iteration, the metaheuristic launches a number of ants whose job is to find a path toward a complete layout code. Then this layout code is evalu-ated by generating a layout based on this code, as is done with genetic algorithms. Dependent on the design score, various amounts of pheromone are deposited at key constructs within the design. As the metaheuristic proceeds, the aim is for the collectivity of ants to learn to avoid layout constructs which lead to bad layouts and to seek for layout constructs that are often found in great designs. In order to avoid being trapped in local optima, the amount of pheromone at each construct decays with time and the selec-tion by an ant of its next construct insertion given a partial code is made according to weighted random-ization among the possible constructs available for insertion at the current code state.

The first and second reasons driving the development and use of metaheuristics are melted in various implementations. As an example, AntZone (Montreuil et al. 2004) is a metaheuristic that is based on ant colony techniques. AntZone generates block layouts with located I/O stations with the objective of mini-mizing rectilinear travel. Its exploits space structuring by having users select among different types of band layouts: 2H-bands; 3H-bands; 3V-bands; 1V-band + 3H-bands + 1V-band; etc. For example, the second from left layout of Figure 9.4 is constructed using 3H-bands. AntZone also lets the engineer spec-ify a priori how many centers are allowed at maximum along each band and then it defines a flexible-size rectangular zone for each position along each band. A potential space structuring for the second layout from left in Figure 9.4 can be $[H_1:(Z_1,Z_2,Z_3,Z_4)/H_2:(Z_5,Z_6,Z_7,Z_8)/H_3:(Z_9,Z_{10},Z_{11},Z_{12})]$. A layout code then becomes an assignment of centers to zones. The layout code for the considered layout in Figure 9.4 is then simply (A,B,C,-,D,E,F,-,G,H,I,-). At the second level, the ant colony algorithm explores the solution space of layout codes. At the first level, a linear programming model generates the optimal block layout with located I/O stations, given a specified layout code.

Currently, most of the best-known solutions for large cases of the QAP, the block layout problem and their variants have been obtained using metaheuristics. Their advantage is their automatic capability of generating in reasonable time better designs than simpler heuristics. Their main disadvantage is their software implementation complexity, especially since most current implementations have been developed by research teams and are not widely available to practitioners.

9.12.6 Interactive Optimization-Based Design

In the early 1980s it became clear that trying to use mathematical programming for solving large realistic cases was out of reach in location and layout design involving interaction between facilities. Researchers started to look for sub-problems which could be solved optimally or near-optimally using heuristics. A design methodology emerged from this trend: termed interactive optimization-based design (Montreuil 1982). The concept is to let the engineer in the driver seat like in interactive design, while giving him access to a variety of focused optimizers supporting the various design tasks.

The earliest such methodologies used optimization to generate more advanced design skeletons than simple flow graphs and relationship graphs, from which the engineer had to interactively generate a design. The three best-known layout design skeleton-based methodologies, respectively, rely on the maximum-weighted planar adjacency graph (Foulds et al. 1985, Leung 1992), the maximum-weighted matching adjacency graph (Montreuil et al. 1987), and the cut tree (Montreuil and Ratliff 1988b).

The adjacency graph methods exploit three properties of any 2D layout. The adjacency graph property is that for any layout, one can draw an adjacency graph where each node is an entity in the layout (center, aisle segment, the outside, etc.) and each link corresponds to a pair of entities being adjacent to each other. The planar adjacency graph property states that the adjacency graph of a 2D layout is planar, meaning that it can be drawn without link crossings. Figure 9.18 illustrates these first two properties for the block layout of Figure 9.2e.

The matching adjacency graph property states that when assigning a value to each link equal to the boundary length shared by both entities defining the link, then the sum of all link values associated with a given entity is equal to the perimeter of that entity, defining the degree of the node representing the entity. For example, as shown in Figure 9.19, center A is adjacent with centers B and C and with the outside. The adjacent boundaries between A and these three entities are respectively 11.9, 10.3, and 8.8 m long. The sum of these adjacencies totals 31, which is the perimeter of center A.

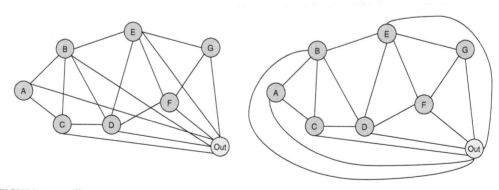

FIGURE 9.18 Illustrating the adjacency graph property and planar adjacency graph property using the block layout of Figure 9.2e.

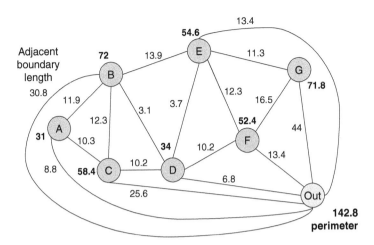

FIGURE 9.19 Illustrating the matching adjacency graph property using the block layout of Figure 9.2e.

Every layout has a planar adjacency graph. If one could find the adjacency graph of the optimal layout, then the engineer could generate the optimal layout itself. For example, given the building and center space requirements, one can use the adjacency graph of Figure 9.18 as a design skeleton from which can be drawn the layout of Figure 9.2e with much ease. Given a weight for each potential link, the weighted maximum planar graph problem (Osman et al. 2003) aims to find the planar graph whose sum of link weights is maximal. In layout design, the weight for each link corresponds either to the flow between the centers, or their qualitative proximity relationship importance expressed through the weight of their desired proximity type (e.g., adjacent: 100, very near: 50, not far: 2, very far: −50). A heuristic can be used to generate rapidly a near-optimal maximum-weighted planar graph. The engineer interactively draws the planar graph. Then he generates layouts respecting as much as possible the relative positioning of centers in the drawn graph and the adjacencies suggested by its links. This may be easy or rather difficult since not all planar graphs can be transformed in feasible layouts respecting the spatial requirements of each center and the building.

A similar approach is used when exploiting the matching adjacency graph property. The maximum-weighted b-matching problem (Edmonds 1965) can be solved optimally in polynomial time. This problem finds the graph, respecting the degree of each node while embedding links into the graph and stating a usage for each link respecting its imposed lower and upper usage bounds, which maximizes the sum over all links of the product of their usage and their value. In layout design, each node corresponds to a center; the value of each link is set as done earlier for the planar graph approach, yet here divided by the upper usage bound, the degree of each node is bounded by desired lower and upper limits imposed on the center perimeter; a positive lower bound on a link forces the centers to be adjacent to a given extent; finally, the upper bound on a link indicates the maximum allowed adjacent boundary length between two centers. For example, the maximum adjacency between a 12 × 20 rectangular center and a 15 × 30 rectangular center is at most 20 m. The b-matching algorithm finds its optimal graph which is used by the engineer as a design skeleton representing the targeted adjacency graph. The engineer interactively generates a satisfying layout by iteratively drawing and adjusting a layout respecting the matching graph as much as possible, or resolving the b-matching model with adjusted link bounds to forbid or enforce specific adjacencies.

Cut trees are another type of design skeleton used in layout design. Cut trees can be computed from a flow graph in polynomial time (Gomory and Hu 1961). Figure 9.20 depicts the cut tree for the inter-center undirected loaded flow graph extracted from Table 9.6. Montreuil and Ratliff (1988b) prove that (*i*) the cut tree is the optimal inter-center travel network when the network links are all set to a unitary-length link and the travel network is restricted to have a noncyclic tree structure and (*ii*) if the centers have to be placed in two distinct facilities with a specific pair of centers forced to be separated from each other, then

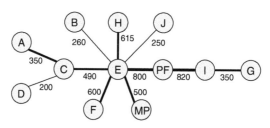

FIGURE 9.20   Inter-center cut tree based on the loaded flows of Table 9.1.

the cut tree will always indicate optimally which centers should be in each of the two facilities, assuming no restraining space constraints. For example, in Figure 9.20, if centers C and I have to be in distinct facilities, then one has simply to find the single path between C and I in the cut tree, here C-E-PF-I, then find the link with lowest value and cut it to find the optimal separation of centers. Here the lowest value link is C-E with a value of 490. Therefore, centers A, C, and D are best located in a facility and the remaining centers in the other facility. The 490 value indicates how much flow is to circulate between the facilities. In layout design, one seeks to decide what to put near each other and what to put far from each other.

As a design skeleton, the cut tree can guide an engineer into generating a layout. The cut tree can be molded at will to fit specific building constraints. The main rules are to systematically aim to locate centers so that higher value links and paths in the cut tree are as small as possible, and to avoid unnecessary link crossings. The cut tree can also be used for layout analysis, as shown in Figure 9.21 where the cut tree is overlaid on the current and alternative layouts of Figures 9.6 and 9.8, respectively. It is easy to see that the current design respects poorly the guidance of the cut tree, while the alternative layout, which has a significantly better travel score, does better even though it does not do it as best as could be.

Using design skeletons has been the first stage of interactive optimization-based design. Montreuil et al. (1993b) have later introduced a linear programming model for swiftly finding the optimal block layout, with located I/O stations, minimizing rectilinear flow travel given a set of flows and the relative positions of centers as inferred by the drawing of a design skeleton. This allows the engineer to manipulate the design skeleton, then to request a layout optimization based on the drawing of the design skeleton, and to examine a few seconds later the resulting layouts, iterating until he is satisfied with the design. Also developed were models and approaches for designing the travel network given a block layout with located I/O stations (e.g., Chhajed et al. 1992). Complementary, a linear programming

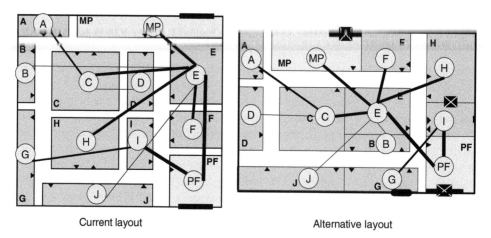

Current layout   Alternative layout

FIGURE 9.21   Cut tree over imposed on the current and alternative layouts of Figures 9.6 and 9.8.

based model was introduced for optimizing the design of a net layout given a block layout and travel network (Montreuil and Venkatadri 1988, Montreuil et al. 1993a). The model optimally shrinks cell sizes from gross to net shapes, locate aisle segments appropriately and locate I/O stations so as to minimize aisle-based travel.

The combination of these optimization models, used interactively by the engineer, allows him to generate designs that, even though they may not be globally optimal, benefit from optimized components and from the human capability to integrate them creatively. The main advantage of interactive optimization is that it enables the engineer while leaving him in the driver seat. The focused and integrated usage of optimization let him address large cases efficiently. The main disadvantage lies in the current lack of wide and open accessibility of design software capable of sustaining such rich interactive optimization.

9.12.7   Assisted Design

Introduced as one of the three methodologies in the inner circle of Figure 9.17, assisted design has evolved mainly from interactive design, and has been influenced by interactive optimization and metaheuristic design. The underlying hypothesis justifying the emergence of assisted design is that since layout and location design has high complexity, wide scope and large scale, it does not lend itself to fully automated design. Therefore, the underlying principles of assisted design are: (*i*) the engineer is to be at the core of the design process and (*ii*) it should have access to a design environment which enables him as best as possible to master the complexity, scope, and scale so as to efficiently generate high-quality creative designs. In assisted design, the focus is on (*i*) making sure the engineer is well trained into understanding the concepts, issues, and methods pertinent for his design task, (*ii*) providing him with an empowering assisted design environment, and (*iii*) training the engineer into being fluent in using the environment. As contrasted with interactive design which relies on mostly generic tools, adapted design relies on specialized knowledge-intensive software tools which have been conceived for location and layout purposes. The tools may embed generic tools, but these are seamlessly integrated and they are parameterized for layout and location purposes. In order to make clearer what assisted design is really about, below are described examples of commercial and academic assisted layout design platforms.

On the commercial side, the Plant Design and Optimization Suite from Tecknomatix, a business unit of UGS (www.ugs.com), is currently the best-known application in line with assisted design. The suite loosely couples their Plant Simulation, FactoryCAD, FactoryFlow, Factory Mockup, eM-Sequencer, and Logistics software. Of most direct interest among these is the FactoryFlow software, whose core has been developed in the late 1980s up to mid-1990s. It is introduced by the company as a graphical material handling system that enables engineers to optimize layouts based on material flow distances, frequency and costs, that allows factory layouts to be analyzed by using part routing information, material storage needs, material handling equipment specifications, and part packaging (containerization) information. Embedded in the Autocad software, www.Autodesk.com, FactoryFlow aims to assist the engineer through a series of interactive factory design features coupled with specific feasibility, material handling flow, and equipment capacity analysis tools.

On the academic side, the concept of assisted design has been investigated for a long time. In the early 1980s, Warnecke and Dangelmaier (1982) developed an early prototype. Later on, Montreuil and Banerjee (1988) and Montreuil (1990) investigated object-oriented technologies and knowledge representation toward intelligent layout design environments while Goetschalckx et al. (1990) investigated integrated engineering workstations as a platform for rapid prototyping of manufacturing facilities. The WebLayout design platform (Montreuil et al. 2002b) is perhaps the most comprehensive effort to date in making available an openassisted design environment to the community. WebLayout is conceived as a web-based platform enabling researchers, professors, and students from around the world to concurrently experiment, test, and learn the basic and latest concepts and methodologies in factory design. Figure 9.22 illustrates a design generated by a team of students using WebLayout: it includes the site, the factory building and structure, the production area layout as well as the office and service area layout. WebLayout allows

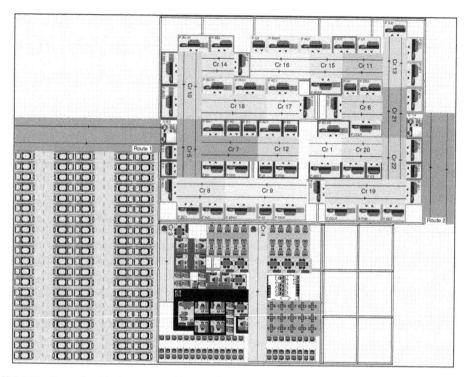

FIGURE 9.22   Example of design generated by a student team using WebLayout.

multiple levels of granularity, from multi-site block layouts to processor layouts. It supports factory organization using responsibility networks, enabling engineers to contrast for the same case factory designs based on function, process, holographic, product, group or fractal organizations, or yet any combination of the above. It accepts probabilistic demand distributions for the products and supports the analysis of their impact on production and handling resource requirements and overall expected economic performance. WebLayout is conceived from the ground up to allow various optimization, heuristic, analysis, evaluation, and validation tools to be readily integrated, so as to help engineers in various evolving ways to generate high-performance designs.

Even though they represent the current state of the art, both platforms are still primitive in terms of assisted design capabilities. There is still huge room for improvement and creativity, especially in a global, dynamic, and turbulent world where engineers have to transform their perception of location and layout design to become a process rather than a project. For example, none of the platforms currently have any significant dynamic layout capabilities, beyond letting the engineer enter an existing layout.

9.12.8   Holistic Metaheuristics

Holistic metaheuristics are significant extensions of first-generation metaheuristics. Driven by their fundamental intent to encompass a much more global design scope, they integrate a complex set of lower level focused heuristics, metaheuristic, and optimization model solvers.

A vivid example of such a holistic metaheuristic is HoloPro conceived and developed to support holographic factory design in a wide spectrum of environments (Marcotte 2005, Marcotte and Montreuil 2004, 2005). Environments range from products, processes, and demand being known deterministically to being uncertain and all the way to being basically unknown. The design task addressed by HoloPro involves the automatic generation of (*i*) set of processors to be implemented, (*ii*) set of holographic centers, each with its embedded processors, (*iii*) center and processor layout of the factory, (*iv*) expected work patterns for every center and processor, and (*v*) expected global and product-specific flow patterns. The metaheuristic is structured around a set of interacting agents, each responsible for a specific set of design tasks. Each agent relies on solving focused heuristics or optimization models. Within a few minutes, it provides sets of holographic factory designs, complete with evaluation of their relative robustness when faced with expected uncertainties.

Detailed discussion of HoloPro is beyond the scope of this chapter. What is important to grasp is the fact that researchers have begun to address more holistic location and layout design tasks, attempting to provide automated approaches for generating optimized designs, and that the current preferred means for doing so is by developing and exploiting more complex, seamlessly integrated metaheuristics. Yet this trend is barely in its infancy. It will require strong academic and industry commitment for it to grow. Indeed getting involved in researching, developing, and maintaining such holistic metaheuristics is significantly more demanding in terms of academic, technological, and financial resources than previous generation approaches addressing much more localized, aggregated, and/or simplified problems.

9.12.9 Global Optimization

Complementary to assisted design and holistics metaheuristics lies global optimization. The main drive here is to develop optimization models of more comprehensive location and layout design tasks. For a long time, this drive has been mostly associated with a desire by researchers to formally define the problems they were addressing. They knew very well that the resulting models could not be solved beyond very simple cases and that they would have to rely on approximate techniques to really solve the problem for realistic size cases. The very significant advances in performance of optimization solvers such as CPLEX (www.ilog.com) have pushed the frontier far enough that many problems considered unsolvable for realistic sizes have become amenable to solution by commercial solvers. Also, there have been advances in optimization solution techniques, such as branch-and-cut, and in heuristic solution of optimization models. All this has created a growing interest in global optimization.

In location design, the trend toward global optimization is quite evolved, especially exploiting discrete location modeling. Optimal location-allocation models have long been exploited, where facilities have to be opened or closed through a set of locations, and clients assigned to opened locations, in an attempt to minimize overall travel, opening, and closure costs. These have been extended to capacitated versions where each location has a limited capacity when opened and assigned clients use a fraction of this capacity. Revelle and Laporte (1996) propose models for capacitated facility location, including a multi-period version. Several authors have dealt with multiple stages or levels of facilities, such as factories, central distribution centers, as described by Klose (2000). Gradually this has lead to the creation of network design models, such as production–distribution networks, manufacturing and logistic networks, as exemplified by Geoffrion and Graves (1974), Geoffrion and Powers (1995), Slats et al. (1995), Cruz et al. (1999), Dogan and Goetschalckx (1999), Dasci and Verter (2001), Melkote and Daskin (2001), Martel (2005), and Paquet et al. (2007). Gradually, location decisions are considered in the midst of comprehensive supply chain design models, as illustrated by Cohen and Moon (1990), Arntzen et al. (1995), and Chopra (2003). With the globalization of the economy many such models are incorporating international issues such as differences and fluctuations in labor rates and availability, transport modes and costs, interest rates, currency rates, transfer prices, fiscal issues, country risks, as well as dealing with the geographical dispersion of markets, suppliers, and potential site locations (e.g., Goetschalckx et al. 2002, Kouvelis et al. 2004, and Martel 2005). The trend toward global optimization in location design has been fuelled by the fact that large-scale, ever more realistic and comprehensive models have been solved to optimality or near-optimality using a variety of solution techniques, such as Bender's decomposition, branch-and-cut, lagrangean relaxation, and lagrangean relax-and-cut.

In layout design, global optimization is much more embryonic, slowed by the inherent higher difficulty of solving layout models to optimality or near-optimality. Global layout optimization has started in the early 1990s. Montreuil (1991) introduced a modeling framework for integrating layout design and flow network design, leading to the modeling of net layout design. The framework scope presentation was structured in an ever increasing scope and complexity. First was introduced a mixed integer linear programming model for block layout and I/O station location which minimizes inter-station rectilinear travel. Second, the modeling was extended to take into consideration travel along a spatially fixed travel network. Third, it was again extended to allow the optional use of links along the fixed travel network. Fourth, the spatially fixed travel network constraint was relaxed to rather impose a less constraining logical travel network. Fifth, the modeling switched from designing a block layout with its travel network(s) to designing a net layout explicitly modeling the aisle system. It started doing so by imposing a set of spatially fixed interconnecting aisles. Sixth, the modeling was relaxed to replace the fixed aisle system by a spatially fixed aisle travel network which had to be transposed into an aisle system through the solution of the net layout model. Seventh, it relaxed the physically fixed aisle travel network by a logical aisle travel network. Eighth, it finally relaxed the model to allow optional aisle travel links, ending up with a net layout design modeling which only required as input a potential aisle travel network from which links could be truncated by the model, resulting in a design with less aisles while assuring minimal travel network usage and implementation cost. Heragu and Kusiak (1991) presented a continuous space layout design model that offers an alternative to the first model in the framework. To this day, several variants of the first model are now being solved optimally or near optimally for small cases and some medium-size cases (Montreuil et al. 2002a, Sherali et al. 2003, Anjos and Vannelli 2006). The more encompassing models are yet subject to investigation by the research community to enable the solution of realistic size cases. The modeling framework has been used as a problem formalization template by the layout research community over the years. Researchers such as Barbosa-Póvoa et al. (2001), Marcotte (2005), and Ioannou (2007) are now embarking on new global optimization modeling avenues.

9.13 Integrated Location and Layout Design Optimization Modeling

Location and layout are tightly interlaced and complementary. This section introduces two design optimization models which should formalize the relationship between layout and location. The models are part of the global optimization trend described earlier. They both deal explicitly with dynamics and uncertainty. The first is a dynamic probabilistic discrete location model whereas the second is a dynamic probabilistic discrete location and continuous layout model. The exposition of these models aims to counterbalance the design issue orientation of Sections 9.2 to 9.11 and the solution methodology orientation of Section 9.2 by taking a formal mathematical modeling orientation. In no way should these two models be perceived as the models. They must rather be understood as two examples of a vast continuum of potential models to formalize the design issues described in Sections 9.2 through 9.11.

9.13.1 Dynamic Probabilistic Discrete Location Model

This model optimizes the dynamic assignment of a set of centers (or facilities) to a set of discrete locations. The model supports a number of predefined future scenarios, each with a number of successive periods covering the planning horizon. The occurrence of a future is probabilistic. For each period in each future, each center has specific space requirements and pairwise unitary flow travel (or proximity relationship) costs are defined. Each center has also a fixed cost for being assigned in a specific location during a period of a future. Each location can be dynamically made available, expanded or contracted through time, with associated costs. Centers can be moved from a period to another, incurring a moving cost. The model recognizes that decisions relative to the first period are the only rigid ones, as all others will be revisable later on based on further information, as future scenarios will either become past, present, or nearer future scenarios. Below are first exposed the objective function and the constraints, followed by definitions for variables, parameters, and sets. Then the model is described in detail.

Minimize

$$\sum_f P_f \left[\sum_{\forall d, t>1} a_{dtf} A_{dtf} + \sum_{\forall dc'lt'} c_{dc'l'tf}(A_{dtf} A_{c'l'tf}) + \sum_{\forall dl't} m_{dl'tf} A_{d,t-1,f} A_{d'tf} \right]$$

$$+ \sum_l (e_l^+ E_l^+ + e_l^- E_l^-) + \sum_d l_{d} L_d + \sum_f P_f \left[\sum_{\forall l,t>1,f} (s_{ltf} S_{ltf} + s_{ltf}^+ S_{ltf}^+ + s_{ltf}^- S_{ltf}^-) \right] \qquad (9.4)$$

Subject to

$$\sum_{l \in L^c} A_{dtf} = 1 \qquad \forall c,t,f \qquad (9.5)$$

$$A_{d1f} = L_d \qquad \forall c,l,f \qquad (9.6)$$

$$\sum_{c \in C^l} r_{ctf} A_{dtf} \le S_{ltf} \le s_l^m \qquad \forall l,t,f \qquad (9.7)$$

$$S_{ltf} = S_{l,t-1,f} + S_{ltf}^+ - S_{ltf}^- \qquad \forall l,t > 1,f \qquad (9.8)$$

$$S_{l1f} = E_l = s_l^0 + E_l^+ - E_l^- \qquad \forall l,f \qquad (9.9)$$

Where the variables are

A_{dtf} Binary variable equal to 1 when center c is assigned to location l in period t of future f, or 0 otherwise.

E_l, E_l^+, E_l^- Continuous non-negative variables deciding the space availability, expansion, and contraction of location l in period 1.

L_{cl} Binary variable equal to 1 when center c is assigned in location l in period 1, or 0 otherwise.

$S_{ltf}, S_{ltf}^+, S_{ltf}^-$ Continuous non-negative variable deciding space availability, expansion, and contraction of location l in period t of future f.

While the parameters and sets are

a_{dtf} Cost of assigning center c to location l in period t of future f.

$c_{dc'l'tf}$ Cost of concurrently assigning center c to location l and center c' to location l' in period t of future f.

C^l Set of centers allowed to be located in location l.

e_l, e_l^+, e_l^- Unit space availability, expansion, and contraction costs for location l in period 1.

l_{cl} Cost of assigning center c to location l in period 1.

L^c Set of locations in which center c is allowed to be located.

$m_{dl'tf}$ Cost of moving center c from location l to location l' in period t of future f.

P_f Probability of occurrence of future f.

r_{ctf} Space requirements for center c in period t of future f.

S_l^0 Initial space availability at location l.

$S_{ltf}, S_{ltf}^+, S_{ltf}^-$ Unit space availability, expansion, and contraction costs for location l in period t of future f.

The objective function 9.4 minimizes the overall actualized marginal cost, here described along two lines. The first line includes the sum of three cost components over all probable futures, weighted by the probability of occurrence of future each future f. The first sums over all allowed combinations, the cost of assigning a center c to a location l in period $t > 1$, independent of where other centers are located in this period t and of where center c was located the previous period. The second sums, over all allowed combinations, the cost associated with concurrently locating center c in location l and center c' in location l' in period t. This is generically the cost associated with relations, interactions, and flows between centers. The third sums over all allowed combinations, the cost of moving center c from its location l in period $t - 1$ to location l' in period t. This is generically the dynamic center relocation cost.

The second line of the objective function includes three cost components. The first two add up all immediate transition costs from the actual state to the proposed state in period one. First is the cost associated with the space of each location as proposed for the first period. It includes the cost of making this space available and the cost of either expanding or contracting the location from its actual state. Second is the cost of implementing each center in its proposed location. These two components do not explicitly refer to specific futures as they are common to all futures since they are a direct result of the location decisions and will be incurred in all futures. The third component is similarly the space availability, expansion, and contraction cost for all locations in all later periods of all futures.

Constraint set 9.5 makes sure that a center c is located in a single location l in each period t of each future f. Constraint set 9.6 attaches the location decisions made for time period 1 over all probable futures. So in the first period, each center c is assigned to the same location l in all futures. These are the decisions that have to be taken now, that will definitely lead to implementation. These decisions cannot be altered afterward. In all later periods, the location decisions are allowed to vary from one future to another. They define a probabilistic plan that will be alterable subsequently, in light of further information availability, until they are associated to the first period in the revised model and become the hard location decision leading to immediate implementation.

Constraint set 9.7 insures that the space availability constraint of each location l is respected at each period t of each future f, constraining that the sum of the required spaces of each center assigned to a location l does not exceed its space availability at that time. This availability is bounded for each location l to a specified maximum. The space availability of a location l can vary from one period to the next. For each future, constraint set 9.8 keeps an account of planned expansions and contractions of each location at all periods except the first. As constraint set 9.6 does for the location assignments of period one, constraint set 9.9 deals with the incumbent expansion or contraction of each location in the forthcoming first period, common to all futures for each location.

When the space requirement and availability parameters are restrained to one, there is a single time period and a single future, and no location expansion or contraction is allowed, then this model reduces to the well known QAP.

9.13.2 Dynamic Probabilistic Discrete Location and Continuous Layout Model

This model generalizes the above model by allowing to treat each discrete location as a facility within which its assigned centers have to be laid out. The model thus explicitly deals with center shaping and location within facilities, and with avoidance of spatial interference between centers. Centers are restricted to rectangular shapes. They are allowed to be moved between facilities and within facility from a period to the next in a future. Below are first exposed the objective function and the constraints, followed by definitions for variables, parameters, and sets. Then the model is described in detail.

Minimize

$$(4) + \sum_f p_f \left(\sum_{csc's't} d_{csc's'tf} D_{csc's'tf} + \sum_{cst} m_{cs} M_{cstf} \right) \tag{9.10}$$

Subject to (9.5) to (9.9) and

$$\left(X_{ctf}^{u} - X_{ctf}^{l}\right)\left(Y_{ctf}^{u} - Y_{ctf}^{l}\right) = r_{ctf} \qquad \forall c,t,f \tag{9.11}$$

$$\left(X_{ltf}^{u} - X_{ltf}^{l}\right)\left(Y_{ltf}^{u} - Y_{ltf}^{l}\right) = S_{ltf} \qquad \forall l,t,f \tag{9.12}$$

$$\frac{\left(Y_{etf}^{u} - Y_{etf}^{l}\right)}{f_{etf}} \le \left(X_{etf}^{u} - X_{etf}^{l}\right) \le f_{etf}\left(Y_{etf}^{u} - Y_{etf}^{l}\right) \qquad \forall e \in (C \cup L), t, f \tag{9.13}$$

$$X_{ltf}^{l} - m\left(1 - A_{ctf}\right) \le X_{ctf}^{l} \le X_{ctf}^{u} \le X_{ltf}^{u} + m\left(1 - A_{ctf}\right) \qquad \forall c,l,t,f \tag{9.14}$$

$$Y_{ltf}^{l} - m\left(1 - A_{ctf}\right) \le Y_{ctf}^{l} \le Y_{ctf}^{u} \le Y_{ltf}^{u} + m\left(1 - A_{ctf}\right) \qquad \forall l,t,f \tag{9.15}$$

$$x_{l}^{l} \le X_{ltf}^{l} \le X_{ltf}^{u} \le x_{l}^{u} \qquad \forall l,t,f \tag{9.16}$$

$$y_{l}^{l} \le Y_{ltf}^{l} \le Y_{ltf}^{u} \le y_{l}^{u} \qquad \forall l,t,f \tag{9.17}$$

$$X_{ctf}^{l} \le X_{cstf}^{s} \le X_{ctf}^{u} \qquad \forall c,s,t,f \tag{9.18}$$

$$Y_{ctf}^{l} \le Y_{cstf}^{s} \le Y_{ctf}^{u} \qquad \forall c,s,t,f \tag{9.19}$$

$$\left(X_{c'tf}^{l} - X_{ctf}^{u}\right) \ge m\left(P_{cc'tf}^{x} - 1\right) \qquad \forall c,c't,f \tag{9.20}$$

$$\left(Y_{c'tf}^{l} - Y_{ctf}^{u}\right) \ge m\left(P_{cc'tf}^{y} - 1\right) \qquad \forall c,c't,f \tag{9.21}$$

$$P_{cc'tf}^{x} + P_{c'ctf}^{x} + P_{cc'tf}^{y} + P_{c'ctf}^{y} \ge A_{ctf} + A_{c'ltf} - 1 \qquad \forall c < c'l,t,f \tag{9.22}$$

$$X_{e1f}^{l} = X_{e}^{l} \qquad \forall e \in (C \cup L), f \tag{9.23}$$

$$Y_{e1f}^{l} = Y_{e}^{l} \qquad \forall e \in (C \cup L), f \tag{9.24}$$

$$X_{e1f}^{u} = X_{e}^{u} \qquad \forall e \in (C \cup L), f \tag{9.25}$$

$$Y_{e1f}^{u} - Y_{e}^{u} \qquad \forall e \in (C \cup L), f \tag{9.26}$$

$$X_{cstf}^{s} - X_{c's'tf}^{s} = D_{csc's'tf}^{x+} - D_{csc's'tf}^{x-} \qquad \forall c,s,t,f \tag{9.27}$$

$$Y_{cstf}^{s} - Y_{c's'tf}^{s} = D_{csc's'tf}^{y+} - D_{csc's'tf}^{y-} \qquad \forall c,s,t,f \tag{9.28}$$

$$D_{csc's'tf} \ge D_{csc's'tf}^{x+} + D_{csc's'tf}^{x-} + D_{csc's'tf}^{y+} + D_{csc's'tf}^{y-} - m\left(2 - A_{ctf} - A_{c'ltf}\right) \qquad \forall c,s,l,t,f \tag{9.29}$$

$$0{,}5\left(\left(X_{ctf}^{l} + X_{ctf}^{u}\right) - \left(X_{c,t-1,f}^{l} + X_{c,t-1,f}^{u}\right)\right) = M_{ctf}^{x+} - M_{ctf}^{x-} \qquad \forall c,t,f \tag{9.30}$$

$$0{,}5\left(\left(Y_{ctf}^{l} + Y_{ctf}^{u}\right) - \left(Y_{c,t-1,f}^{l} + Y_{c,t-1,f}^{u}\right)\right) = M_{ctf}^{y+} - M_{ctf}^{y-} \qquad \forall c,t,f \tag{9.31}$$

$$M_{ctf} \ge M_{ctf}^{x+} + M_{ctf}^{x-} + M_{ctf}^{y+} + M_{ctf}^{y-} - m\left(2 - A_{ctf} - A_{cl,t-1,f}\right) \qquad \forall c,l,t,f \tag{9.32}$$

Where new variables are

X_e^l, X_e^u, Y_e^l, Y_e^u	Continuous variables for the coordinates of the lower and upper boundaries of the sides of entity e along the X and Y axes in period 1 for all futures, where an entity is either a center or a location.
X_{etf}^l, X_{etf}^u, Y_{etf}^l, Y_{etf}^u	Continuous variables for the coordinates of the lower and upper boundaries of the sides of entity e along the X and Y axes in period t of future f, where an entity is either a center or a location.
X_{cstf}^s, X_{cstf}^s	Continuous variables for the X and Y coordinates of I/O station s of center c in period t of future f.
$D_{csc's'tf}$	Continuous non-negative variable for the rectilinear distance between station s of center c and station s' of center c' in period t of future f.
$D_{csc's'tf}^{x+}$, $D_{csc's'tf}^{x-}$, $D_{csc's'tf}^{y+}$, $D_{csc's'tf}^{y-}$	Continuous non-negative variables for the positive and negative components along the X and Y axes of the rectilinear distance between station s of center c and station s' of center c' in period t of future f.
M_{ctf}^{x+}, M_{ctf}^{x-}, M_{ctf}^{y+}, M_{ctf}^{y-}	Continuous non-negative variables for the positive and negative components along the X and Y axes of the rectilinear move of center c in period t of future f from its coordinates in the previous period of future f.
M_{ctf}	Continuous non-negative variables for the rectilinear move of center c in period t of future f from its coordinates in the previous period of the same future, whenever center c is assigned to the same location in periods t and $t-1$.
$P_{cc'tf}^x$, $P_{cc'tf}^y$	Binary variables stating whether or not center c is to position lower than center c' along axes X and Y whenever both centers are assigned to the same location in period t of future f.

While new parameters are

$d_{csc'tf}$	Unitary positive interaction cost associated to the rectilinear distance between station s of center c and station s' of center c' whenever both centers are assigned to the same location in period t of future f.
f_{etf}	Maximum allowed ratio between the longest and shortest sides of rectangular entity e, which is either a location or a center, this ratio can be distinct for each period of each future except for the first period when it has to be the same for all futures.
m_{ctf}	Unitary positive move cost associated with the rectilinear displacement of center c in period t of future f from its coordinates in the previous period of the same future, whenever center c is assigned to the same location in periods t and $t-1$.
m	A very large number.
x_l^l, x_l^u, y_l^l, y_l^u	Lower and upper limits for location l along the X and Y axes.

The objective function 9.10 minimizes the sum of objective function 9.4 and the overall expected actualized interaction and move costs. These costs result from the summation over all futures, weighted by their probability of occurrence, of their future-specific costs. When laid out in the same location (site, building, etc.), pairs of centers having significant interactions (flows, relationships) incur a cost when their involved I/O stations are positioned a positive distance from each other. For example, if there is flow from the output station of center A to the input station of center B, then a unitary cost is specified for this pair. Then the interaction cost associated with the pair is the product of their unitary interaction cost and their rectilinear distance. The move cost for a center is computed over all periods of a future, multiplying the rectilinear displacement of its centroid from a period to the next by the unitary move cost specified for this center.

The constraint set includes previously defined constraints 9.5 to 9.9. The new constraints 9.11 to 9.32 are associated with the actual layout of centers assigned to the same location, where they have to share space without interfering with each other while satisfying their shape requirements. Each center and location is restricted to have a rectangular shape, and to be orthogonally laid out relative to each other. Each is defined through the positioning of its lower X and Y axis corner and its upper X and Y axis corner.

Constraints 9.11 and 9.12, respectively, enforce that each center and location respect its specified area requirements. These quadratic constraints can be linearized using a set of linear approximation variables and constraints (e.g., Sherali et al. 2003). Constraints 9.13 impose maximal form ratio between the longest and smallest sides of each center and location.

Constraints 9.14 and 9.15 ensure that whenever a center is assigned to a location, then it is to be laid out within the rectangular area of the location. Constraints 9.16 and 9.17 guarantee that each location is itself located within its maximal allowed coordinates. For example, a building cannot be extended beyond its site boundaries. Similarly, constraints 9.18 and 9.19 impose that each I/O station of a center be positioned within the center's rectangular area.

Constraints 9.20 to 9.22 insure no physical overlap between centers assigned to the same location in a specific period of a future. They do so by imposing that for any two such centers, the former is either lower or upper along the X axis, or lower or upper along the Y axis.

Similarly, to constraint 9.6, constraints 9.23 to 9.26 recognize that the first layout decisions are imposed to all futures, to be immediately implemented while all other layout decisions can be subsequently altered depending on future information.

Constraints 9.27 to 9.29 compute the rectilinear distance between any two I/O stations of centers having positive interactions. The first two constraints linearize the computation of the rectilinear distance by adding its positive and negative components along the X and Y axes respectively, while the latter adds up all these components to get the overall rectilinear distance. Constraints 9.30 to 9.32 similarly compute the rectilinear displacement of the centroid of each center from its previous position to its current position. Constraints 9.27 to 9.32 assume positive unitary interaction and move costs. When negative unitary costs are involved, such as when one wants two centers to be far from each other, then the constraints have to be altered using binary variables to adequately compute the rectilinear distances and displacements.

When all centers are a priori assigned to the same location and the layout is to be fixed over the entire planning horizon, then the model simplifies to the static continuous block layout model introduced by Montreuil (1991).

9.14 Conclusion

From the offset, the chapter has warned the reader that location and layout design complexity would be addressed straight in the face, in a hand-on-usual way, with the objectives of providing the reader with a holistic vision and equipping him or her to be able to deliver designs that address the real issues at hand. This has been a demanding task as most of the sections end up presenting material rarely or never yet presented in such a way, often starting with levels of elevation normally achieved only in research papers or in the conclusive remarks of textbooks. As much as possible in such a chapter, practical examples have been provided. Several of these examples are highly elaborate to guarantee that the reader can transpose the material for usage in realistic cases. The overall bet is that the reader will be capable of mastering the essence of the material, and achieve levels of design performance much higher than with a more traditional approach.

Even though the chapter is quite long, it has been subject to critical editing choice among the huge number of potential topics. Perhaps the most difficult has been the continuous struggle between presenting more location or layout examples and material, aiming to strike for the right balance. It should be clear that this chapter could easily be rewritten without ever mentioning the world layout, or similarly the word location. Yet both domains constitute a continuum where location is present both at the macro level and micro level surrounding layout. Each center must be located in the network, assigned to an existing or new facility. The union over all interacting centers defines the highly strategic and global location design challenge. At each site location, the facilities must be laid out so as to best deploy their assigned centers, hence defining a layout design challenge at each site. Then given that the main location and layout designs have been set, there appears at the micro level the need for locating a variety of resources through the network and the sites. There is much in common between the two interlaced domains. Yet there are also differences which are most evidenced when presenting an example. It should be clear that layout examples have taken a dominant position in the chapter, in an effort to use in many contexts the same basic case. This is surely because of the author's background. Hopefully the overall balance does not penalize too strongly the location facets.

One of the purposeful omissions in the core of the chapter has been a section on the global comparative evaluation of design alternatives. The justification is that its application is much wider in scope than location and layout design. However, as this chapter reaches its closure, it becomes important to address it briefly. Whenever possible all nondominated design alternatives should be evaluated financially. Their expected return on investment should be computed, as well as their economic value added, taking into consideration all impacts on potential revenues, costs, and investments, as well as the inherent identified risks involved. Furthermore, all the nonfinancial criteria should be analyzed, weighted in terms of their relative importance, and each nondominated alternative design should be evaluated relative to each criterion. Then typical multicriteria decision-making techniques should be used to merge the financial and nonfinancial evaluations to end up with relative rankings of alternatives, as well as sensitivity analyses, so as to best feed the decision-makers (Gal et al. 1999). A wide variety of criteria has been listed through the chapter, yet many more can be found on the reference material. Overall, criteria fall in two categories: performance criteria and capability criteria. All criteria should be in line with the strategic intent of the enterprise. Also all key stakeholders should be taken into consideration when setting the set of criteria. For example, employees will motivate safety, quality-of-life and visibility criteria. Clients will motivate lead time and flexibility capability criteria. Suppliers may motivate vehicle access criteria. Headquarters will motivate financial performance and may motivate agility and personalization capabilities. The regional community may motivate environment criteria. Such lists of stakeholders and associated criteria are highly case dependent and should be carefully investigated.

It has been said that location and layout design has become a mature domain subject to limited room for significant innovation and impact. The chapter has hopefully contributed to challenge this somber assessment and prove that the domain is highly pertinent and challenging, and that there is a lot of room for professional, academic, and technological research and innovation. Overall the two main keys appear to take a performance and capability development perspective in line with the strategic intents of the organization, and to think of location and layout design has being a continuous process rather than a punctual project, always aiming to proactively adjust to relentless dynamics and turbulence in the organization and in the environment.

Acknowledgments

This work has been supported by the Canada Research Chair in Enterprise Engineering, the NSERC/Bell/Cisco Business Design Research Chair, and the Canadian NSERC Discovery grant Demand and Supply Chain Design for Personalized Manufacturing. In addition, the author would like to thank Edith Brotherton and Caroline Cloutier, both research professionals in the CIRRELT research center at Laval University, for their help in the validation and proofing of this chapter.

References

Abdinnour-Helm S. and S.W. Hadley (2000). _Tabu search based heuristics for multi-floor facility layout_, International Journal of Production Research, v38, no2, 365–383.

Aleisa, E.E. and L. Lin (2005). _For effective facilities planning: layout optimization then simulation, or vice versa?_, Proc. of 2005 Winter Smulation Conference, 1381–1385.

Anjos M.F. and A. Vannelli (2006). *A new mathematical-programming framework for facility-layout design*, INFORMS Journal on Computing, v18, no1, 111–118.

Anstreicher, K.M., N.W. Brixius, J.P. Goux and J. Linderoth (2002). *Solving large quadratic assignment problems on computational grids*, Journal Mathematical Programming, v91, no3, 563–588.

Armour, G.C. and E.S. Buffa (1963). *A heuristic algorithm and simulation approach to the relative location of facilities*, Management Science, v9, no2, 249–309.

Arntzen, B.C., G.G. Brown, T.P. Harrison and L.L. Trafton (1995). *Global supply chain management at Digital Equipment Corporation*, Interfaces, v25, no1, 69–93.

Askin, R.G. (1999). *An empirical evaluation of holonic and fractal layouts*, International Journal of Production Research, v37, no5, 961–978.

Azadivar, F. and J. Wang (2000). *Facility layout optimization using simulation and genetic algorithms*, International Journal of Production Research, v38, no17, 4369–4383.

Banerjee P., Y. Zhou and B. Montreuil (1997). *Genetically assisted optimization of cell layout and material flow path skeleton*, IIE Transactions, v29, no4, 277–292.

Barbosa-Póvoa A.P., R. Mateus and A.Q. Novais (2001). *Optimal two-dimensional layout of industrial facilities*, International Journal of Production Research, v39, no12, 2567–2593.

Benjaafar, S. (2002). *Modeling and analysis of congestion in the design of facility layouts*, Management Science, v48, 679–704.

Benjaafar, S., S.S. Heragu and S. Irani (2002). *Next generation factory layouts: research challenges and recent progress*, Interfaces v32, no6, 58–76.

Chhajed, D., B. Montreuil and T.J. Lowe (1992). *Flow network design for manufacturing systems layout*, European Journal of Operational Research, v57, 145–161.

Chittratanawat, S. (1999). *An integrated approach for facility layout, P/D location and material handling system design*, International Journal of Production Research, v37, no3, 683–706.

Chopra, S. (2003). *Designing the distribution network in a supply chain*, Transportation Research Part E: Logistics and Transportation Review, v39, no2, 123–140.

Cohen, M.A. and S. Moon (1990). *Impact of production scale economies, manufacturing complexity, and transportation costs on supply chain facility networks*, Journal of Manufacturing and Operation Management, v3, 269–292.

Cortinhal M.J. and M.E. Captivo (2004). *Genetic algorithms for the single source capacitated location problem*, Metaheuristics: Computer Decision-Making, Kluwer, 187–216.

Crainic T.G., M. Toulouse and M. Gendreau (1996). *Parallel asynchronous tabu search for multicommodity location-allocation with balancing requirements*, Annals of Operations Research, v63, 277–299.

Cruz, F.R.B., J. MacGregor Smith and G.R. Mateus (1999). *Algorithms for a multi-level network optimization problem*. European Journal of Operational Research, v118, no1, 164–180.

Dasci, A. and V. Verter (2001). *A continuous model for production-distribution system design*, European Journal of Operational Research, v129, no2, 287–298.

Drezner, Z. and H.W. Hamacher, eds (2002). Facility Location, Applications and Theory, Springer-Verlag, Berlin, Germany.

Dogan, K. and M. Goetschalckx (1999). *A primal decomposition method for the integrated design of multi-period production-distribution systems*, IIE Transactions, v31, no11, 1027–1036.

Donaghey C.E. and V.F. Pire (1990). Solving the Facility Layout Problem with BLOCPLAN, Industrial Engineering Department, University of Houston, TX, 1990.

Edmonds, J. (1965). *Paths, trees and flowers*, Canadian Journal of Mathematics, v17, 449–467.

Evans, G.W., M.R. Wilhelm and W. Karwowski (1987). *A layout design heuristic employing the theory of fuzzy sets*, International Journal of Production Research, v25, no10, 1431–1450.

Foulds, L.R., P. Giffons and J.Giffin (1985). *Facilities layout adjacency determination: an experimental comparison of three graph theoretic heuristics*, Operations Research, v33, 1091–1106.

Francis R.L., L.F. McGinnis and J.A. White (1992). Facility Layout and Location: An Analytical Approach, 2nd edition, Prentice-Hall, Englewood Cliffs, NJ, USA.

Gal, T., T.J. Stewart and T. Hanne, eds (1999). Multicriteria Decision Making: Advances in MCDM Models, Algorithms, Theory and Applications, Kluwer Academic Publishers, Boston, USA.

Geoffrion, A.M. and G.W. Graves (1974). *Multicommodity Distribution System Design by Benders Decomposition*, Management Science, v20, no5, 822–844.

Geoffrion, A.M. and R.F. Powers (1995). *Twenty years of strategic distribution system design: an evolutionary perspective,* Interfaces, v25, no5, 105–127.

Goetschalckx, M., L.F. McGinnis and K.R. Anderson (1990). *Toward rapid prototyping of manufacturing facilities,* Proceedings of the IEEE First International Workshop on Rapid System Prototyping, Research Triangle Park, NC, USA.

Goetschalckx, M., C.J. Vidal and K. Dogan (2002). *Modeling and design of global logistics systems: a review of integrated strategic and tactical models and design algorithms,* European Journal of Operational Research, v143, no1, 1–18.

Gomory R.E. and T.C. Hu (1961). *Multi-terminal network flows,* Journal SIAM on Computing, v9, no4, 551–570.

Hardin, C.T. and J.S. Usher (2005). *Facility layout using swarm intelligence,* Proceedings of 2005 IEEE Swarm Intelligence Symposium, 424–427.

Heragu S. and A. Kusiak (1991). *Efficient models for the facility layout problem,* European Journal of Operational Research, v53, 1–13.

Heskett, J.L. (1963). *Cube-per-Order Index–A Key to Warehouse Stock Location,* Transportation and Distribution Review, v3, 27–31.

Huq, F., D.A. Hensler and Z.M. Mohamed (2001). *A simulation analysis of factors influencing the flow time and through-put performance of functional and cellular layouts,* Journal of Integrated Manufacturing Systems, v12, no4, 285–295.

Ioannou, G. (2007). *An integrated model and a synthetic solution approach for concurrent layout and material handling system design,* Computers and Industrial Engineering, v52, no4, 459–485.

Kerbache, L. and L.M. Smith (2000). *Multi-objective routing within large scale facilities using open finite queueing networks,* European Journal of Operational Research, v121, 105–123.

Kincaid, R.K. (1992). *Good solutions to discrete noxious location problems via metaheuristics,* Annals of Operations Research, v40, no1, 265–281.

Klose, A. (2000). *A lagrangean relax-and-cut approach for the two-stage capacitated facility location problem,* European Journal of Operational Research, v126, no 2, 408–421.

Kouvelis, P., M.J. Rosenblatt and C.L. Munson (2004). *A mathematical programming model for global plant location problems: analysis and insights,* IIE Transactions, v36, no2, 127–144.

Kuenh A. and M.J. Hamburger (1963). *A heuristic program for locating warehouses,* Management Science, v9, 643–666.

Lamar, M. and S. Benjaafar (2005). *Design of dynamic distributed layouts,* IIE Transactions, v37, 303–318.

Lee, R.C. and J.M. Moore (1967). *CORELAP—computerized relationship layout planning,* Journal of Industrial Engineering, v18, 195–200.

Leung J. (1992). *A new graph-theoretic heuristic for facility layout,* Management Science, v38, no4, 594–605.

Lilly, M.T. and J. Driscoll (1985). *Simulating facility changes in manufacturing plants,* Proceedings of the 1st International Conference on Simulation in Manufacturing, Standford Upon Avon, UK.

Marcotte, S. (2005). *Optimisation de la conception d'aménagements holographiques* Holographic Layout Design Optimization, Ph.D. Thesis, Administration Sciences, Laval University, Québec, Canada.

Marcotte, S. and B. Montreuil (2004). *Investigating the impact of heuristic options in a metaheuristic for agile holographic factory layout design,* Proceedings of Industrial Engineering Research Conference 2004, Houston, Texas, 2004/05/15–19.

Marcotte, S. and B. Montreuil (2005). *Factory design robustness: an empirical study,* Proceedings of Industrial Engineering Research Conference 2005, Atlanta, Georgia, 2005/05/14–18.

Marcotte, S., B. Montreuil and P. Lefrançois (1995). *Design of holographic layout of agile flow shops,* Proceedings of International Industrial Engineering Conference, Montréal, Canada.

Marcotte S., B. Montreuil C. and Olivier (2002). Factory design fitness for dealing with uncertainty, Proceedings of MIM 2002: 5th International Conference on Managing Innovations in Manufacturing, Milwaukee, Wisconsin, 2002/09/9-11.

Marcoux, N., A. Langevin and D. Riopel (2005). *Models and methods for facilities layout design from an applicability to real-world perspective*, Logistics Systems: Design and Optimization, edited by A. Langevin and D. Riopel, Springer, USA, 123–170.

Martel, A. (2005). *The design of production-distribution networks: a mathematical programming approach*, Supply Chain Optimization, Geunes, J. et Pardalos, P. (eds), Springer, 265–306.

Maxwell, W.L. and J.A. Muckstadt (1982). *Design of automated guided vehicle system*, IIE Transactions, v1, no2, 114–124.

Melkote, S. and M.S. Daskin (2001). *Capacitated facility location/network design problems*, European Journal of Operational Research, v129, no3, 481–495.

Meller, R.D. and Y.A. Bozer (1996a). *A new simulated annealing algorithm for the facility layout problem*, International Journal of Production Research, v34, 1675–1692.

Meller R.D. and Y.K. Gau (1996b). *The facility layout problem: recent and emerging trends and perspectives*, Journal of Manufacturing Systems, v15, 351–366.

Montgomery, D.C., L.A. Johnson and J.S. Gardiner (1990). Forecasting and Time Series Analysis, 2nd edition, McGraw-Hill, NY, USA.

Montreuil, B. (1982). Interactive Optimization Based Facilities Layout, Ph.D. thesis, ISYE School, Georgia Institute of Technology, Atlanta, GA, USA.

Montreuil, B. (1987). *Integrated design of cell layout, input/output station configuration, and flow network of manufacturing systems*, Intelligent and Integrated Manufacturing Analysis and Synthesis, ed. by C.R. Liu, A. Requicha and S. Chandrasekar, ASME PED, v25, 315–326.

Montreuil, B. (1990). *Representation of domain knowledge in intelligent systems layout design environments*, Computer Aided Design, v22, no2, 90/3, 97–108.

Montreuil, B. (1991). *A modelling framework for integrating layout design and flow network design*, Progress in Material Handling and Logistics, v2, ed. by J.A. White and I.W. Pence, Springer-Verlag, 95–116.

Montreuil, B. (2001). *Design of agile factory networks for fast-growing companies*, Progress in Material Handling: 2001, ed. by R. J. Graves et al., Material Handling Institute, Braun-Brumfield Inc., Ann Arbor, Michigan, U.S.A.

Montreuil, B. (2006). Facilities network design: a recursive modular protomodel based approach, Progress in Material Handling Research: 2006, ed. by R. Meller et al., Material Handling Industry of America (MHIA), Charlotte, North Carolina, U.S.A., 287–315.

Montreuil, B. and P. Banerjee (1988). *Object knowledge environment for manufacturing systems layout design*, International Journal of Intelligent Systems, v3, 399–410.

Montrouil B., E. Brotherton, N. Ouazzani and M. Nourelfath (2004). *AntZone layout metaheuristic: coupling zone-based layout optimization, ant colony system and domain knowledge*, Progress in Material Handling Research: 2004, Material Handling Industry of America (MHIA), Charlotte, North-Carolina, U.S.A. p. 301–331.

Montreuil, B. and H.D. Ratliff (1988a). *Optimizing the location of input/output stations within facilities layout*, Engineering Costs and Production Economics, v14, 177–187.

Montreuil, B. and H.D. Ratliff (1988b). *Utilizing cut trees as design skeletons for facility layout*, IIE Transactions, v21, no2, 88/06, 136–143.

Montreuil B. and U. Venkatadri (1988). *From gross to net layouts: an efficient design model*, Document de travail FSA 88-56, Faculté des sciences de l'administration, Université Laval, Québec, Canada.

Montreuil, B. and U. Venkatadri (1991). *Scattered layout of intelligent job shops operating in a volatile environment*, Proceedings of International Conference on Computer Integrated Manufacturing, Singapore.

Montreuil, B., E. Brotherton and S. Marcotte (2002a). *Zone based layout optimization*, Proceedings of Industrial Engineering Research Conference, Orlando, Florida, 2002/05/19–22.

Montreuil, B., N. Ouazzani and S. Marcotte (2002b). *WebLayout: an open web-based platform for factory design research and training*, Progress in Material Handling : 2002, ed. by R. J. Graves et al., Material Handling Institute, Braun-Brumfield Inc.

Montreuil, B. and P. Lefrançois (1996). *Organizing factories as responsibility networks*, Progress in Material handling: 1996, ed. by R. J. Graves et al., Material Handling Institute, Braun-Brumfield Inc., Ann Arbor, Michigan, U.S.A., 375–411.

Montreuil, B., H.D. Ratliff and M. Goetschalckx (1987). *Matching based interactive facility layout*, IIE Transactions, v19, no3, 271–279.

Montreuil B. Y. Thibault and M. Paquet (1998). *Dynamic network factory planning and design*, Progress in Material handling: 1999, ed. by R. J. Graves et al., Material Handling Institute, Braun-Brumfield Inc., Ann Arbor, Michigan, U.S.A., 353–379.

Montreuil, B. U. Venkatadri and E. Blanchet (1993a), *Generating a net layout from a block layout with superimposed flow networks*, Document de travail GRGL 93-54, Faculté des sciences de l'administration, Université Laval, Québec, Canada.

Montreuil, B., U. Venkatadri and H.D. Ratliff (1993b). *Generating a layout from a design skeleton*, IIE Transactions, v25, no1, 93/1, 3–15.

Montreuil, B., U. Venkatadri and R.L. Rardin (1999). *The fractal layout organization for job shop environments*, International Journal of Production Research, v37, no3, 501–521.

Murray A. T. and R.L. Church (1996). *Applying simulated annealing to location-planning models*, Journal of Heuristics, v2, no1, 31–53.

Muther, R. (1961). Systematic Layout Planning, Industrial Education Institute, Boston, USA.

Norman B.A., A.E. Smith and R.A. Arapoglu (1998). *Integrated facility design using an evolutionary approach with a subordinate network algorithm*, Lecture Notes in Computer Sciences, v1498, Springer-Verlag, London, UK, 937–946.

Nourelfath M., N. Nahas and B. Montreuil (2007). *Coupling ant colony optimization and the extended great deluge algorithm for the discrete facility layout problem*, Engineering Optimization, forthcoming.

Nugent C.E., T.E. Vollmann and J. Ruml (1968). *An experimental comparison of techniques for the assignment of facilities to location.* Operations Research, v16, 150–173.

Osman I.H., B. Al-Ayoubi and M. Barake (2003). *A greedy random adaptive search procedure for the weighted maximal planar graph problem*, Computers and Industrial Engineering, v45, no4, 635–651.

Owen, S.H. and M.S. Daskin (1998). *Strategic facility location: a review*, European Journal of Operational Research, v111, no3, 423–447.

Paquet, M., A. Martel and B. Montreuil (2007). *A Manufacturing Network Design Model Based on Processor and Worker Capabilities.* International Journal of Production Research, 27 p., forthcoming.

Reed, R. (1961). Plant Layout: Factors, Principles and Techniques, R.D. Irwin Inc., Homewood, IL, USA.

Revelle, C.S. and G. Laporte (1996). *The Plant Location Problem : New Models and Research Prospects*, Operations Research, v44, no 6, 864–874.

Rosenblatt M.J. (1986). *The dynamics of plant layout*, Management Science, v32, no1, 76–86.

Seehof, J.M. and W.O. Evans (1967). *Automated layout design program*, Journal of Industrial Engineering, v18, 690–695.

Sherali H.D., B.M.P. Fraticelli and R.D. Meller (2003). *Enhanced Model Formulations for Optimal Facility Layout*, Operations Research, v51, no4, 629–644.

Slats, P.A., B. Bhola, J.J.M. Evers and G. Dijkhuizen (1995). *Logistic Chain Modelling*, European Journal of Operational Research, v87, no1, 1–20.

Tate D.M. and A.E. Smith (1995). *Unequal-area Facility Layout by Genetic Search.* IIE Transactions, v27, 465–472.

Tompkins, J.A., White, J.A., Bozer, Y.A. and Tanchoco, J.M.A. (2003). Facilities Planning, 3rd edition, John Wiley and Sons, Inc., USA.

Venkatadri U., R. L. Rardin and B. Montreuil (1997). *A Design Methodology for Fractal Layout Organization*, IIE Transactions (Special Issue of Design and Manufacturing on Agile Manufacturing), v29, no10, 911–924.

Warnecke H.J. and W. Dangelmaier (1982). *Progress in Computer Aided Plant Layout*, Technical report, Institute of Manufacturing Engineering and Automation – Fraunhofer, Germany.

Welgama P.S. and P.R. Gibson (1995). *Computer-aided facility layout – A status report*, International Journal of Advanced Manufacturing Technology, v10, no 1, 66–77.

Womack, J.P. and D.T. Jones (1998). Lean Thinking, Free Press, USA.

Wu Y. and E. Appleton (2002). *The optimization of block layout and aisle structure by a genetic algorithm*, Computers and Industrial Engineering, v41, no4, 371–387.

10

Inventory Control Theory: Deterministic and Stochastic Models

Lap Mui Ann Chan
Virginia Polytechnic Institute and State University

Mustafa Karakul
York University

10.1 Introduction

We encounter an inventory problem whenever physical goods are stocked for anticipated demand. Inventory is often necessary when there is uncertainty in demand. However, even when demand is known for certain, inventory is built up to satisfy large demands when production is time-consuming. Stocking can also be used as a strategy to take advantage of the economies of scale since suppliers often offer discount to encourage large orders and administrative costs can be saved by combining orders. Another critical reason for keeping high inventory is the loss of customer goodwill when shortages occur.

On the other hand, inventory ties down capital and incurs storage costs and property taxes. Appropriate cost functions are included in inventory models to capture the trade-off between overstocking and shortage. An optimization of the total profit or cost function generates a best ordering policy that specifies the quantities and times of replenishments.

In this chapter, we consider the problem of keeping inventory for different situations. Deterministic models with known demand and then stochastic models that involve uncertainty in demand are discussed in two separate sections.

10.2 Deterministic Models

Inventory models can be classified into two categories according to the review policy. In continuous review models, inventory is tracked continuously and replenishment is possible at any moment. The second category is periodic review models, in which inventory is checked at prespecified regular epochs, such as the end of a week, and replenishment can be done only at these check points. For continuous review models, we start our discussion with a single facility for a single product and continue to more complicated multi-facility and multi-product systems. On the other hand, for periodic review models, we focus on deriving an optimal inventory policy for stocking a single product for a single facility only.

10.2.1   Economic Order Quantity Model

A classical continuous review inventory problem is the economic order quantity (EOQ) model. This basic model considers a single product that has a known continuous demand of d units per unit time. The cost of replenishment is a fixed set up cost k plus a per unit variable cost of c per unit ordered. The cost of holding each unit of the product is h per unit time. Replenishments are instantaneous. The objective is to find a replenishment policy that satisfies the demand without delay so as to minimize the average replenishment and holding cost over the infinite horizon.

There exists an optimal replenishment policy that has a couple of nice properties. If replenishment is made when there is a positive inventory of the product, we can adjust the order quantities to make sure that replenishment occurs only when inventory is down to zero. Specifically, for an inventory policy that orders at time t_r for $r = 0, 1, 2, \ldots$ with $t_0 = 0$, the adjusted order quantity is $d(t_{r+1} - t_r)$ at t_r. Note that for a feasible policy, the inventory level after replenishment at t_r is no less than $d(t_{r+1} - t_r)$.

Thus, this adjustment does not affect the replenishment cost but reduces the inventory holding cost as illustrated in Figure 10.1a. Hence, to find an optimal replenishment policy, we only need to consider policies in which replenishments are made only when inventory is down to zero. This is called the zero-inventory-ordering (ZIO) property.

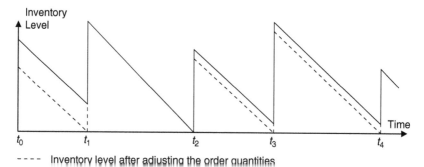

---- Inventory level after adjusting the order quantities

(a) Adjusting a Feasible Policy to a Zero-Inventory-Ordering Policy

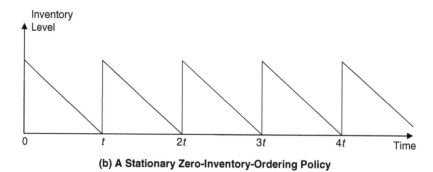

(b) A Stationary Zero-Inventory-Ordering Policy

FIGURE 10.1   A feasible policy and a stationary zero-inventory-ordering policy.

For the second property of an optimal replenishment policy, consider a fixed order quantity of Q in a ZIO policy. The replenishment cost of this order is $k + cQ$, the time till the next replenishment is Q/d and the average inventory until the next replenishment is $Q/2$. Together, we have the average cost of this replenishment:

$$AC(Q) = (k + cQ)/(Q/d) + hQ/2 = kd/Q + hQ/2 + cd \qquad (10.1)$$

Suppose replenishments of different quantities are made in a ZIO policy, then there must be one among all these replenishments that is associated with the smallest average cost. Thus, a ZIO replenishment policy in which every order is for a quantity that is the same as this smallest average cost replenishment has a smaller average cost than the one with different order quantities. Specifically, the average cost of the inventory policy shown in Figure 10.1b is no more than that of the policies shown in Figure 10.1a when $t = t_{k+1} - t_k$ with $AC(d[t_{k+1} - t_k]) \leq AC(d[t_{r+1} - t_r])$ for $r = 1, 2, 3, \ldots$. Hence, to find an optimal policy, we only need to consider ZIO policies that always order the same quantity. This is called the stationary property.

To find an optimal stationary ZIO policy, it remains to obtain the best order quantity that minimizes the average cost provided in Equation 10.1. Since the first derivative of $AC(Q)$ with respect to Q is $AC'(Q) = -kd/Q^2 + h/2 = 0$ when $Q = (2kd/h)^{1/2}$ and the second derivative of $AC(Q)$ with respect to Q is $A''(Q) = 2kd/Q^3 > 0$, $AC(Q)$ is a convex function that attains its minimum possible value of $(2kdh)^{1/2} + cd$ at $Q^* = (2kd/h)^{1/2}$. In summary, an optimal policy is to order $(2kd/h)^{1/2}$ unit of the product, when inventory is down to zero, every $(2kd/h)^{1/2}/d = [2k/(dh)]^{1/2}$ units of time with an average cost of $(2kdh)^{1/2} + cd$ per unit time. Note that in the existence of a constant replenishment lead time l, orders are placed a lead time ahead to make sure that they arrive when the inventory level is down to zero in a ZIO policy. Hence, constant lead time has no effect on the optimal order quantity Q^* or the optimal reorder interval Q^*/d.

Various efforts have been made by researchers to extend the EOQ model from the stocking of a single product for a single facility to more complicated systems. However, optimality results are elusive. Roundy (1985) introduces the class of near optimal power-of-two policies, which are stationary ZIO policies with reorder intervals that are power-of-two multiples of each other. We discuss the derivation of near optimal power-of-two policies for a two facility in-series and then a multi-product assembly system in the following.

10.2.2 Economic Order Quantity Model for a Series of Two Facilities

Consider a retailer who faces the demand of a product that occurs at a constant rate of d per unit time. The retailer obtains the product from a warehouse at a cost of k_1 per order and holds inventory at a cost of H_1 per unit product per unit time. The warehouse in turn obtains the product from a supplier at a cost of k_0 per order and holds inventory at a cost of H_0 per unit product per unit time. Orders are satisfied instantaneously for both the warehouse and the retailer. The objective is to obtain an ordering policy that satisfies the demands at the warehouse and retailer without delay so as to minimize the long-run average ordering and holding cost for both the warehouse and retailer over the infinite horizon.

The retailer faces an EOQ problem. However, since the warehouse receives discrete orders from the retailer, he does not face an EOQ problem. On the other hand, if the warehouse does not consider the retailer as a separate facility and considers the product held at the retailer as part of its own inventory, then the warehouse is facing an EOQ problem with fixed order cost k_0, per unit holding cost rate of $h_0 = H_0$ and demand that occurs at a constant rate of d. As the warehouse accounts for a per unit holding cost rate of h_0 for the inventory at the retailer, the retailer has to pay a holding cost of only $h_1 = H_1 - h_0$ per unit product per unit time for its inventory. $h_j, j = 0, 1$ are referred to as echelon holding costs. The EOQ problem faced by the retailer is modified to one with fixed order cost k_1, per unit holding cost rate of h_1 and demand that occurs at a constant rate of d. From the analysis of the EOQ model, the optimal

ZIO policy have reorder intervals of $T_j^* = [2k_j/(dh_j)]^{1/2}$ with an average cost of $(2k_jdh_j)^{1/2}$, $j = 0,1$ for these two EOQ problems.

If $T_0^* = T_1^*$, then the warehouse and retailer can synchronize with each other by ordering dT_0^* units of the product simultaneously when inventory is down to zero, every T_0^* units of time. Note that in ordering simultaneously, the product is delivered to the retailer through the warehouse, but is never stored there. In implementing an optimal policy for each one of the two EOQ models, system-wide average cost is minimized.

Since inventory is kept at the retailer but not at the warehouse, the average cost of $(2k_0dh_0)^{1/2} + (2k_1dh_1)^{1/2}$ for this optimal policy should be a function of H_1 but not of H_0. To rewrite the average cost in terms of H_1 only, note that since

$$[2k_0/(dh_0)]^{1/2} = T_0^* = T_1^* = [2k_1/(dh_1)]^{1/2},$$

$$T_0^* = \{2(k_0 + k_1)/[d(h_0+h_1)]\}^{1/2} = [2(k_0 + k_1)/(dH_1)]^{1/2} \text{ and the optimal average cost}$$

$$(2k_0dh_0)^{1/2} + (2k_1dh_1)^{1/2} = dT_0^*(h_0 + h_1) = dT_0^*H_1 = [2(k_0 + k_1)dH_1]^{1/2}. \tag{10.2}$$

We use this observation to help determine an optimal policy for the case when $T_0^* < T_1^*$.

If $T_0^* < T_1^*$, then the warehouse and retailer cannot synchronize with each other to implement the optimal ZIO policies for the two EOQ models simultaneously. The problem is $T_0^* = [2k_0/(dh_0)]^{1/2}$, a decreasing function of $h_0 = H_0$, is too small. In other words, the per unit holding cost rate h_0 at the warehouse is too large. Consider a duplicate system with the same H_1 but a smaller holding cost rate $h_0' = H_0'$ at the warehouse so that $[2k_0/(dh_0)]^{1/2} < \{2(k_0 + k_1)/[d(h_0 + h_1)]\}^{1/2} = [2(k_0 + k_1)/(dH_1)]^{1/2} = [2k_0/(dh_0')]^{1/2}$. As discussed earlier, an optimal policy for this duplicate system is for the warehouse and the retailer to order simultaneously every $[2(k_0 + k_1)/(dH_1)]^{1/2}$ units of time when inventory at the retailer is down to zero for an average cost of $[2(k_0 + k_1)dH_1]^{1/2}$. Since H_1 is the same for both systems, in following the same policy for the original system, inventory is kept only at the retailer and the average cost is:

$$\{(k_0 + k_1) + h_1d[2(k_0 + k_1)/(dH_1)]/2\}/[2(k_0 + k_1)/(dH_1)]^{1/2} = [2(k_0 + k_1)dH_1]^{1/2}.$$

Note that it is the same as Equation 10.2. Since the costs for the original system are no less than the duplicate system and the optimal policy for the duplicate system results in the same average cost for the original system, it is an optimal policy for the original system as well.

Note that in keeping H_1 constant and reducing the holding cost H_0 by δ, h_0 is reduced by δ while h_1 is increased by δ. In other words, reducing the holding cost by δ at the warehouse is equivalent to redistributing δ units of the echelon holding cost at the warehouse to the retailer. Furthermore, if the warehouse and the retailer have the same reorder interval, then inventory is kept only at the retailer and the average cost is not affected by this redistribution of the echelon holding cost from the warehouse to the retailer. These observations are used in the discussion of the multi-product systems.

If $T_0^* > T_1^*$ and the retailer orders every T_1^* units of time, then the warehouse can synchronize with the retailer and place an order every T_0^* units of time only if $T_0^* = rT_1^*$ for some positive integer r. In that case, optimality is achieved by placing every order from the warehouse simultaneously with an order from the retailer, since optimal ZIO policies are implemented for the two EOQ models. On the other hand, in case T_0^* is not an integer multiple of T_1^*, Roundy (1985) suggests a heuristic from the class of power-of-two policies, which satisfy the ZIO and stationary property with the reorder interval for the warehouse equals to a power-of-two multiple of the reorder interval for the retailer. In particular, let $2^mT_1^* \le T_0^* < 2^{m+1}T_1^*$ for some non-negative integer m. If $T_0^*/(2^mT_1^*) \le 2^{m+1}T_1^*/T_0^*$, then the warehouse places an order every $T_0 = 2^mT_1^*$ units of time. Otherwise, the warehouse places an order every $T_0 = 2^{m+1}T_1^*$ units of time. In either case, every order from the warehouse is placed simultaneously with one from the retailer to make sure that ZIO policies are implemented for the two EOQ models. However, optimality is achieved for only one of the two EOQ models to attain an average cost of $(2k_1dh_1)^{1/2}$. For the other EOQ model, the reorder interval is T_0 and the corresponding average cost is $k_0/T_0 + h_0dT_0/2$. For the effectiveness of this power-of-two policy, note that if $T_0^*/(2^mT_1^*) \leq 2^{m+1}T_1^*/T_0^*$, then $T_0^*/T_0 = T_0^*/(2^mT_1^*) \leq 2^{m+1}T_1^*/T_0^* = 2T_0/T_0^*$ and hence $1 \leq T_0^*/(2^mT_1^*) = T_0^*/T_0 \leq 2^{1/2}$. Otherwise, $T_0/T_0^* = 2^{m+1}T_1^*/T_0^* \geq T_0^*/(2^mT_1^*) = 2T_0^*/T_0$ and hence $1 \leq 2^{m+1}T_1^*/T_0^* = T_0/T_0^* \leq 2^{1/2}$. Together, we have $2^{-1/2} \leq T_0/T_0^* \leq 2^{1/2}$. Since $(k_0/T_0 + h_0dT_0/2)/(2k_0dh_0)^{1/2} = (T_0^*/T_0 + T_0/T_0^*)/2$ is a convex function of T_0/T_0^* that attains its minimum value at $T_0/T_0^* = 1$, we have

$$(k_0/T_0 + h_0dT_0/2)/(2k_0dh_0)^{1/2} \leq (2^{-1/2} + 2^{1/2})/2 \sim 1/0.94.$$

This implies that the average cost of an optimal policy is at least 94% of that of the power-of-two policy. In other words, this power-of-two policy is 94% optimal.

By adjusting the reorder interval for the warehouse only, an optimal policy is used for one EOQ model, while 94% optimality is achieved for the other one. Roundy (1985) suggests another power-of-two policy that is obtained by adjusting the reorder intervals for both the warehouse and retailer in order to minimize the total cost of optimality for the two EOQ models. This more complicated power-of-two policy is 98% optimal.

The results for power-of-two policies can be extended to systems with facilities that form an acyclic network. We illustrate this by considering a multi-product assembly system.

10.2.3 Economic Order Quantity Model for a Multi-Product Assembly System

Consider a manufacturer of n products. Demand of each product occurs at a constant rate. By scaling, we can assume without loss of generality that the demand rate of each product is 2 units per unit time. Each product i is manufactured by a number of assemblies of parts specified by an assembly directed network $T_i = (N_i, A_i)$. N_i represented the set of parts involved in the production of product i. We will refer to product i also as a part. Hence, i is in N_i. Node i has no successor, while each one of the other nodes in N_i has exactly one immediate successor in T_i. Each part j in N_i is produced by assembling the parts in the set P_j^i of its immediate predecessors in T_i. Figure 10.2 illustrates the production assembly networks for products 1 and 2. The holding cost rate of each part is linear. For each part j required for the production of part i, let H_j^i be the holding cost of part j per unit production of product i per unit time. For example, if the demand rate of product 2 is 6 lb per day, the holding cost of part 4 is \$2/lb per day, and ¼ lb of part 4 is required per pound production of product 2; then in using each day as a time unit, product 2 is measured in units of 3 lbs, and part 4 for the production of product 2 is measured in units of $(3)(¼) = ¾$ lbs with $H_4^2 = \$2(¾) = \$3/2$. Independent of the amount of part j to produce, each assembly is instantaneous and incurs a setup cost of k_{ij}. The objective is to obtain a production policy that satisfies the demands without delay so as to minimize the long-run average setup and holding cost over the infinite horizon.

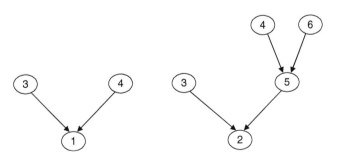

FIGURE 10.2 The assembly networks for products 1 and 2.

Similar to the earlier discussion of the two facilities in-series system, to obtain a 94% optimal EOQ production policy for this model, we first transform the problem into EOQ models. Then, the assembly policies are synchronized by redistributing the echelon holding costs and adjusting the inter-setup intervals to power-of-two multiples of each other.

10.2.3.1 Constructing the EOQ Models Network

To transform the problem into EOQ models, each node j in N_i considers the part j designated to the production of product i still in the system, either as part j or assembled inside other parts already, as its inventory. As an illustration, the echelon inventory of part 4 for the production of part 2 includes the quantity of part 4 that is designated for the production of part 2, the quantity of part 4 inside part 5 that is designated for the production of part 2, and the quantity of part 4 inside part 2 that is still in the system. Since the holding cost of each part l in P_j^i required for the production of part j is already accounted for by its predecessors in T_i, the echelon holding cost for node j in N_i is

$$h_j^i = H_j^i - \sum_{l \in P_j^i} H_l^i .$$

Thus, each node j in N_i corresponds to an EOQ model with a setup cost of k_j, a holding cost of h_j^i per unit product per unit time, and a demand rate of 2 units per unit time for the product.

However, a part j might be required by different products for production. To avoid multiple counts of setup cost for an assembly, each part j that is required for the production of multiple products is identified with an EOQ model with a setup cost of k_j, no holding cost, and a demand rate of 2 units per unit time for the product. At the same time, for each N_i that includes j, the setup cost is removed from EOQ model corresponding to node j in N_i. That is, node j in N_i corresponds to an EOQ model with zero setup cost, a holding cost of h_j^i per unit product per unit time, and a demand rate of 2 units per unit time for the product.

These EOQ models are presented in the EOQ models network for the system. The EOQ models network for the system is a directed network $G^E = (N^E, A^E)$ with

$$N^E = U_i\{i_j : j \in N_i\} U \{j : j \in N_i \text{ for at least 2 different } i\} \text{ and}$$

$$A^E = U_i\{(i_j, i_l): (j, l) \in A_i\} U \{(i_j, j): j \in N_i \text{ and } j \in N^E\}.$$

Associated with each node x in N^E is an order pair $(k^E(x), h^E(x))$ that represents the EOQ model, with a setup cost of $k^E(x)$, a holding cost of $h^E(x)$ per unit product per unit time and a demand rate of 2 units per unit time for the product, associated with node x. In particular,

$$k^E(x) = k_j \text{ if } x = j \in N_i \quad \text{for some } i, \quad \text{or} \quad x = i_j \text{ for some } j \in N_i \text{ and } j \notin N^E;$$

otherwise,

$$k^E(x) = 0$$

and

$$h^E(x) = h_j^i \quad \text{if } x = i_j \in N_i \text{ for some } i;$$

otherwise,

$$h^E(x) = 0.$$

To illustrate this with an example, the EOQ models network for the two products with the assembly networks in Figure 10.2 is shown in Figure 10.3. In Figure 10.3, $x:k^E(x),h^E(x)$ is shown inside each node x.

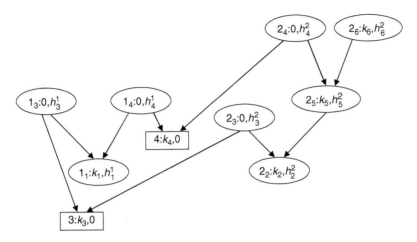

FIGURE 10.3 The EOQ models network for products 1 and 2.

Redistributing the Echelon Holding Costs

The optimal reorder interval for an EOQ model with a setup cost of $k^E(x)$, a holding cost of $h^E(x)$ per unit product per unit time and a demand rate of 2 units per unit time for the product is $[k^E(x)/h^E(x)]^{1/2} \cdot [k^E(x)/h^E(x)]^{1/2}$ is infinite if $h^E(x) = 0$. For any (x, y) in A^E, the optimal assembly policies corresponding to nodes x and y cannot be synchronized if $[k^E(x)/h^E(x)]^{1/2} < [k^E(y)/h^E(y)]^{1/2}$. As discussed earlier for the two facilities in-series system, the problem that $h^E(x)$ is too large can be rectified by redistributing some of the echelon holding cost $h^E(x)$ from node x to node y. Note that while redistributing some of the echelon holding cost $h^E(i_j)$ from node i_j to node i_l for some product i and $(j,l) \in A_i$ is equivalent to reducing the holding cost of part j, that is, designated for the production of product i, redistributing some the echelon holding cost $h^E(i_j)$ from node i_j to node j for some product i and $j \in N_i$ is equivalent to not changing that part of the holding cost of part j that is designated for the production of product i.

For any subset N of N^E and the corresponding subnetwork $G = (N, A)$ of G^E with

$$A = \{(x,y): x,y \in N \text{ and } (x,y) \in A^E\}.$$

It is optimal to assemble the parts corresponding to the nodes in N simultaneously, if the echelon holding costs can be redistributed from predecessors to successors in G until the resulting echelon holding costs $h(x)$ satisfies $[k^E(x)/h(x)]^{1/2}$ is a constant for all the nodes in G with $h(x) = 0$ in case $k^E(x) = 0$. That is, for each $x \in N$ with $k^E(x) > 0$,

$$h(x)/k^E(x) = \sum_{x \in N} h(x)/\sum_{x \in N} k^E(x) = \sum_{x \in N} h^E(x)/\sum_{x \in N} k^E(x).$$

Such an even redistribution of the echelon holding costs is possible if and only if we can flow the amount of excess echelon holding cost, $h^E(x) - h(x)$, from the source nodes x with $h^E(x) > h(x)$ to cover the lack of echelon holding cost, $h(y) - h^E(y)$, at the sink nodes y with $h^E(y) > h(y)$ through the network G. In particular, the maximum flow network is $G^F = (N^F, A^F)$ with

$$N^F = N \cup \{s,t\},$$

$$A^F = A \cup \{(s,x): h(x) > h^E(x)\} \cup \{(y,t): h(y) < h^E(y)\}.$$

In addition, the capacity $c(x, y)$ associated with each arc in A^F is infinite if $(x, y) \in A$. Each $(s, x) \in A^F$ has a capacity $c(s, x) = h^E(x) - h(x)$, while each $(y, t) \in A^F$ has a capacity $c(y, t) = h(x) - h^E(x)$. The objective is to maximize the flow from node s to t through the network G^F, where there is a capacity $c(x, y)$ on the flow to send along arc (x, y). The maximum s-t flow problem is a typical application of linear programming (LP). In solving the LP for the maximum s-t flow problem, either the optimal objective flow value $= \Sigma_{(s,x) \in A^F} c(s, x)$, then an even redistribution of the echelon holding cost is possible. Otherwise, the dual minimum s-t cut (X, X') with $s \in X$ and $t \in X'$ partitions N into two sets $N^1 = X \backslash \{s\}$ and $N^2 = X' \backslash \{t\}$. Since excess echelon holding costs, that cannot flow to cover the lack of echelon holding cost at the nodes in N^2, are still available at the nodes in N^1; nodes in N^2 are predecessor of the nodes in N^1. That is, predecessors do not have enough while successors have too much echelon holding costs. In other words, there is no problem of a predecessor having a smaller optimal inter-setup interval than its successor between the nodes in N^2 and N^1, and redistribution of echelon holding cost can be considered separately for the two sets of nodes.

Start with $N = N^E$. Solve the maximum flow problem for the network subnetwork G, if an even redistribution of echelon holding cost is possible for G, then set $h^F(x) = h(x)$ for each $x \in N$. Otherwise, partition N into two sets N^1 and N^2 according to the optimal dual minimum cut and repeat the process for $N = N^1$ and $N = N^2$ until $h^F(x)$ is determined for each $x \in N^E$. As indicated by the earlier discussion, these redistribution of echelon holding costs results in $h^F(x)$, $x \in N^E$ that satisfy $h^F(x) = 0$ in case $k^E(x) = 0$, and $[k^E(x)/h^F(x)]^{1/2} \geq [k^E(y)/h^F(y)]^{1/2}$ for each $(x, y) \in A^E$ with $k^E(x) > 0$ and $k^E(y) > 0$.

Adjusting the Inter-setup Intervals of the Assemblies

Since $[k^E(x)/h^F(x)]^{1/2} \geq [k^E(y)/h^F(y)]^{1/2}$ for each $(x, y) \in A^E$ with $k^E(x) > 0$ and $k^E(y) > 0$, $\min\{[k^E(x)/h^F(x)]^{1/2}: k^E(x) > 0$ and $x \in N^E\} = [k^E(z_z)/h^F(z_z)]^{1/2}$ for some product $z = 1, 2, \ldots, n$. Assemble product z every $T_z = T_z^* = [k^E(z_z)/h^F(z_z)]^{1/2}$ units of time.

For any $x = i_j \in N^E$ with $k^E(i_j) > 0$ for some part j and product i, or $x = j \in N^E$ for some part j, let $T_j^* = [k^E(x)/h^F(x)]^{1/2}$ and $2^{m(j)}T_z^* \leq T^* < 2^{m(j)+1}T_z^*$ for some positive integer $m(j)$. If $T_j^*/(2^{m(j)}T_z^*) \leq 2^{m(j)+1}T_z^*/T_j^*$, then part j is assembled every $T_j = 2^{m(j)}T_z^*$ units of time. Otherwise, the part j is assembled every $T_j = 2^{m(j)+1}T_z^*$ units of time. For any $i_j \in N^E$ with $k^E(i_j) > 0$ for some part j and product i, let $T_j^i = T_j$.

For any $x = i_j \in N^E$ with $k^E(i_j) = 0$ for some part j and product i, assembly inter-setup time is set backward for successors first then predecessors up the assembly network for product i. Let l be the unique immediate successor of j in the assembly network for product i, then part j designated for the production of product i is assembled every $T_j^i = \max\{T_j, T_l^i\}$ units of time.

The assemblies are synchronized by assembling $2T_j^i$ units of part j designated for the production of product i simultaneously. Then, $2T_j^i$ units of part j designated for the production of product i is assembled every T_j^i units of time. Since $[k^E(x)/h^F(x)]^{1/2} \geq [k^E(y)/h^F(y)]^{1/2}$ for each $(x, y) \in A^E$ with $k^E(x) > 0$ and $k^E(y) > 0$ implies that $T_j^i \geq T_l^i$ any product i and $(j, l) \in A_i$, inventory is down to zero at every assembly of part j designated for the production of product i.

Since echelon holding cost is redistribution from a node x to a node y in N^E only when they have the same corresponding assembly inter-setup time, accounting for the redistributed part of the holding cost at the assembly corresponding to node x or that at node y makes no difference to the average cost of the assembly policy. Hence, the average setup and holding cost of the power-of-two policy is $\Sigma\{k^E(x)/T_j + h^E(x)T_j: x = i_j \in N^E$ with $k^E(i_j) > 0$ for some part j and product i, or $x = j \in N^E$ for some part $j\}$.

For any $x = i_j \in N^E$ with $k^E(i_j) > 0$ for some part j and product i, or $x = j \in N^E$ for some part j, since $2^{-1/2} \leq T_j/T_j^* \leq 2^{1/2}$ by the choice of T_j. $\Sigma\{k^E(x)/T_j + h^E(x)T_j: x = i_j \in N^E$ with $k^E(i_j) > 0$ for some part j and product i, or $x = j \in N^E$ for some part $j\} \leq [(2^{-1/2} + 2^{1/2})/2]\Sigma\{2[k^E(x)h^E(x)]^{1/2}: x = i_j \in N^E$ with $k^E(i_j) > 0$ for some part j and product i, or $x = j \in N^E$ for some part $j\}$.

In other words, it is a 94% optimal policy. A 98% optimal power-of-two policy can be obtained using a more complicated adjustment of the assembly inter-setup intervals.

10.2.3.2   Multi-Period Inventory Model

A general periodic review inventory model considers the problem of satisfying the demand of a single product without delay for T periods of time. Replenishment can be made at the beginning of each period and used to satisfy demand in that and later periods. Holding cost of a period is charged against inventory left at the end of the period. For each period $t = 1, 2, \ldots, T$, the demand is d_t, the cost of ordering Q_t units is $C_t(Q_t)$, and the cost of holding I_t units of inventory is $H_t(I_t)$. It is assumed that d_t is a nonnegative integer whereas $C_t(Q_t)$ and $H_t(I_t)$ are nondeceasing functions for $t = 1, 2, \ldots, T$, as is often true in practice. The objective is to find a replenishment policy that satisfies the demand without delay so as to minimize the total replenishment and holding cost over the T periods.

Typically, multi-period inventory problem is formulated as a dynamic program.

The optimal value function: Let $F_t(I_{t-1})$ be the minimum cost of satisfying the demand from period t to T starting with an inventory of I_{t-1} at the beginning of period t.

The boundary condition: Since the replenishment and holding costs are nondecreasing, holding inventory at the end of period T will not lower the cost of a replenishment policy. Hence, we only need to consider replenishment policy that does not hold inventory at the end of period T to find an optimal one and set $F_{T+1}(0) = 0$.

The recursive formula: Since we only consider policies that end with no inventory at period T, the starting inventory, I_{t-1}, at the beginning of period t is no more than t to T. The total demand for periods starting with an inventory I_{t-1} at the beginning of period t, the decision is on how much to order. A quantity of at least $d_t - I_{t-1}$ must be ordered to satisfy the demand at period t without delay. Furthermore, since we only consider policies that end with no inventory at period T, at most the total demand from period t to T minus I_{t-1} units of the product will be ordered in period t. In ordering Q_t units of the product, the replenishment cost at period t is $C_t(Q_t)$, whereas the inventory at the end of period t is $I_t = I_{t-1} + Q_t - d_t$. Hence, the holding cost at period t is $H_t(I_{t-1} + Q_t - d_t)$, while the minimum cost for periods $t+1$ to T is $F_{t+1}(I_{t-1} + Q_t - d_t)$. $F_t(I_{t-1})$ is obtained by selecting the order quantity, Q_t, that minimizes the total cost at period t, $C_t(Q_t) + H_t(I_{t-1} + Q_t - d_t)$, and the remaining periods $t+1$ to T, $F_{t+1}(I_{t-1} + Q - d_t)$. That is, for $I_{t-1} = 0, 1, \ldots, \Sigma_{t \le i \le T} d_i$,

$$F_t(I_{t-1}) = \text{Min}\{C_t(Q_t) + H_t(I_{t-1} + Q_t - d_t) + F_{t+1}(I_{t-1} + Q_t - d_t):$$

$$\text{Max}\{0, d_t - I_{t-1}\} \le Q_t \le \sum\nolimits_{t \le i \le T} d_i - I_{t-1}\} \tag{10.3}$$

An optimal policy: To obtain an optimal replenishment policy, we start with setting $F_{T+1}(0) = 0$. Using the recursive formula, we calculate backwards the function $F_t(I_{t-1})$ and store the corresponding optimal order quantity choice $Q_t(I_{t-1})$ for $t = T, T-1, \ldots, 2$ and $I_{t-1} = 0, 1, \ldots, \Sigma_{t < i < T} d_i$. For an initial inventory level of I_0, we can then find $F_1(I_0)$ and the corresponding optimal order quantity $Q_1^* = Q_1(I_0)$ for period 1 using Equation 10.3. In ordering Q_1^* units of the product in period 1, the inventory at the end of period 1 is

$$I_1^* = I_0 + Q_1^* - d_0$$

Hence, the optimal order quantity at period 2 is $Q_2^* = Q_1(I_1^*)$. We then continue for $t = 3, \ldots, T$ in using

$$I_{t-1}^* = I_{t-2}^* + Q_{t-1}^* - d_{t-1}$$

and

$$Q_t^* = Q_t(I_{t-1}^*)$$

to obtain an optimal replenishment policy Q_t^*, $t = 1, 2, 3, \ldots, T$.

10.2.3.3 Multi-Period Inventory Model with Concave Costs

Economies of scale often exist for large quantities. Incremental discount is a popular model that reflects this phenomenon. An incremental discount cost model $C(Q)$ is associated with $B + 1$ quantities $0 = Q_0 < Q_1 < \cdots < Q_B$. The cost for the jth unit of product is c_b if $Q_{b-1} \leq j < Q_b$ for some $b = 1, 2, 3, \ldots, B$ and c_{b+1} if $Q_B \leq j$ with $c_1 > c_2 > \cdots > c_{b+1}$. An incremental discount model with $B = 3$ is illustrated in Figure 10.4. Since the incremental discount model has a nonincreasing marginal cost, it is a concave function. Concave cost functions are very popular and have many nice properties that a lot of research has been focused on.

A nice property of concave function is that the linear combination of a set of concave functions results in a concave function. Another nice property is that it induces consolidation. To illustrate this, consider buying a product from two different sources that offer concave cost models. The cost of buying Q_j units of the product from source j is $C_j(Q_j)$ for $j = 1, 2$. Suppose a nonnegative quantity Q_j of product is bought from source j for $j = 1, 2$. Since $C_j(Q_j)$, $j = 1, 2$ have nonincreasing marginal values, $C_1(Q_1) + C_2(Q_2) > C_1(Q_1 + Q_2)$ if $C_1(Q_1) - C_1(Q_1 - 1) < C_2(Q_2) - C_2(Q_2 - 1)$, otherwise $C_1(Q_1) + C_2(Q_2) \geq C_2(Q_1 + Q_2)$. Hence,

$$C_1(Q_1) + C_2(Q_2) \geq \min\{C_1(Q_1 + Q_2), C_2(Q_1 + Q_2)\}.$$

In other words, multiple sourcing does not result in lower cost than single sourcing, and we only need to consider single sourcing policy to obtain a minimum cost policy.

For a multi-period inventory model with concave functions C_t and H_t for $1, 2, \ldots, T$, the first property implies that the cost of having the product available at period t from an ordering in each period $j = 1, 2, \ldots, t$ is a concave function. In addition, the second property implies that the product available at a period can be consolidated to come from a single order. In other words, there exists an optimal replenishment policy that satisfies the ZIO property.

Thus, to obtain an optimal replenishment policy for the multi-period inventory model with concave costs, we can use the following dynamic program which determines an optimal ZIO replenishment policy.

The optimal value function: Since an order will be made only when there is no inventory at the beginning of a period, we only need to identify the periods with positive orders to fully determine a ZIO policy. Let F_t be the minimum cost of satisfying the demand from period t to T starting with no inventory at the beginning of period t.

The boundary condition: Since the replenishment and holding costs are nondecreasing, holding inventory at the end of period T will not lower the cost of a replenishment policy. Hence, we only need to consider the replenishment policies that do not hold inventory at the end of period T to find an optimal one and set $F_{T+1} = 0$.

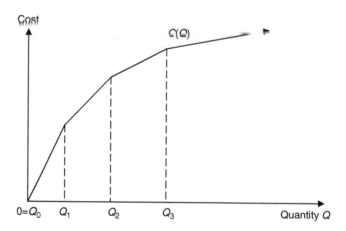

FIGURE 10.4   An incremental discount model.

The recursive formula: The next replenishment can be at any one of the future periods $t + 1$ to T. If the next replenishment is made at period i, then the order placed at period t is for a quantity that equals to the total demand from period t to $i-1$.

$$F_t = \text{Min}_{t < i \leq T+1} \left\{ C_t \left(\sum_{t \leq j \leq i-1} d_j \right) + \sum_{t \leq i \leq i-2} H_l \left(\sum_{l+1 \leq j \leq i-1} d_j \right) + F_i \right\} \qquad (10.4)$$

An optimal policy: To obtain an optimal ZIO policy, we start with setting $F_{T+1} = 0$. Using the recursive formula (10.4), we calculate backwards the function F_t and store the corresponding optimal next order period $P(t)$ for $t = T, T - 1, \ldots, 1$. To obtain the optimal order periods through the function P, start with $t_1^* = 1$ and for $j = 1, 2, 3, \ldots$, set $t_{j+1}^* = P(t_j^*)$ until $t_{j+1}^* = T + 1$. Let $t_{p+1}^* = T + 1$. Then the optimal policy is to place an order at period t_i^* for a quantity equal to the total demand from period t_i^* to $t_{i+1}^* - 1$.

10.3 Stochastic Models

In Section 10.2 we considered inventory models which assume that demand is known. Without uncertainty in demand, these models focus on balancing the trade-offs between setup and inventory holding costs. However, in a lot of real life situation, demand is forecasted with quite a lot of uncertainty as reflected by the main principles of forecasting: (*i*) forecasts are always wrong and (*ii*) forecasts weaken as the length forecast horizon increases. When demand is uncertain, besides the trade-offs among setup and inventory holding costs, one has to consider the costs related to possible shortages. In this section, we consider several inventory control models that incorporate demand uncertainty. Instead of assuming that demand is known, it is assumed that demand is a random variable with known probability distribution.

We start with a discussion of the classical newsvendor problem (a.k.a. newsboy problem), the simplest yet, possibly, the most celebrated and powerful of all the single-period stochastic inventory control models in Section 10.3.1. Then, we extend this model in several ways. Section 10.3.2 discusses the scenarios where price is also a decision variable. In Section 10.3.3, we focus our attention on the multiple-period stochastic inventory control of a single product.

10.3.1 Newsvendor Problem

Consider any retailer who needs to make a single procurement decision for a perishable product that is sold over a single period during which demand is uncertain. There are several examples of such businesses. A newsvendor sells newspapers in a day and weekly magazines over a week. A retailer sells summer clothing over a summer season, or T-shirts and hats for the Super Bowl football event. A manufacturer who designs, produces and sells winter fashion items such as old jackets, coats, etc. over a winter season. The main characteristics of such businesses are: first, the products are perishable. That is, at the end of the selling period the excess inventory is not of any use in the current market; a day-old newspaper cannot be sold as newspaper anymore but can be disposed as recycle paper or possibly sold to rural areas where paper is not delivered daily; summer clothing is not generally for sale in winter unless the excess stock is shipped over to the other parts of the world; T-shirts and hats for 2006 Super Bowl are not in demand after the event. Similarly, winter fashion items are not generally sold after the season is over, they are either shipped off to discount stores or cleared through sales. Second, the procurement lead time is assumed to be too long to make secondary procurements. Hence, there is only one procurement opportunity before the sales season and the retailer has to commit himself to a certain procurement quantity well in advance.

Based on realized demand from past sales, current economic conditions, and expert judgment, randomness in the demand, D, is assumed to follow a known product-specific demand distribution $F(\cdot)$. Our discussion in this section will assume continuous distributions unless otherwise stated. Products are procured at a per unit cost of c, sold at a per unit price of r. Due to the randomness in demand there could be excess inventory or demand at the end of the sales season. Excess inventory is assumed to be returned to the supplier or salvaged at a per unit price of v, which is less than c, and excess demand is assumed to be lost causing not only a loss of the possible profit , $r - c$, but also a possible shortage cost of s dollars per unit that represents the loss of goodwill. Note that, $r > c > v$, otherwise, the problem can trivially be solved by either ordering as much as necessary if $v > c$, or not ordering at all if $c > r$.

Since demand is random, the procurement decisions are very much dependant on the risk averseness of the retailer. In this chapter, we only consider the risk-neutral decision-makers. Hence, our risk-neutral retailer needs to determine a procurement quantity Q such that the single period expected total inventory ordering, holding, and shortage cost is minimized, or, equivalently, the single-period expected profit is maximized. That is, the retailer needs to solve

$$\max_{Q \geq 0} \Pi(Q)'$$

where the expected profit $\Pi(Q)$ can be expressed as

$$\Pi(Q) = rE[\min(Q,D)] - cQ + vE[\max(0,Q-D)] - sE[\max(0,D-Q)],$$

The operator $E(\cdot)$ denotes the expectation. Each expectation, respectively, represents the expected sales, excess inventory, and excess demand for any given Q. This model is known as the newsvendor model (or, more commonly, newsboy problem). Note that, for any procurement quantity Q and any realization d of the random demand D,

$$\min(Q, d) = d - \max(0,d-Q) \text{ and } Q = d + \max(0,Q-d) - \max(0,d-Q).$$

Hence,

$$E[\min(Q,D)] = E[D] - E[\max(0,D-Q)]$$

and

$$Q = E[D] + E[\max(0,Q-D)] - E[\max(0,D-Q)].$$

Substituting these identities in the given equation, the expected profit function can be rewritten as

$$\Pi(Q) = E[D](r-c) - (c-v)E[\max(0,Q-D)] - (r+s-c)E[\max(0,D-Q)]. \qquad (10.5)$$

Interpretation of this function is interesting by itself. The first term is the riskless profit for the equivalent certainty problem that experiences a known demand of $E[D]$. The second term represents the total expected holding cost, which is the per unit holding (overage) cost of $c_o = c - v$ charged against every unit of excess inventory $E[\max(0, Q - D)]$. And finally, the third term is the total expected shortage (underage) cost, which is the per unit shortage cost of $c_u = r + s - c$ (where $r - c$ is the lost sales profit) charged against each unit of the excess demand $E[\max(0, D - Q)]$. In the literature (see Silver and Peterson 1985) total expected cost

$$L(Q) = (c-v)E[\max(0,Q-D)] + (r+s-c)E[\max(0,D-Q)] \qquad (10.6)$$

is known as the single-period *loss function*. Since riskless profit $E[D](r - c)$, which would occur in the absence of uncertainty, is independent of Q, maximizing $\Pi(Q)$ is equivalent to minimizing $L(Q)$. Before finding the optimal procurement policy, let us write $L(Q)$ explicitly as

$$L(Q) = (c-v)\int_{x=0}^{Q} (Q-x)dF(x) + (r+s-c)\int_{x=Q}^{\infty} (x-Q)dF(x).$$

Taking the derivative of $L(Q)$ with respect to Q and applying Leibnitz' Rule, the first-order optimality condition can be written as:

$$(c - v)\Pr(D \le Q) - (r + s - c)\Pr(D \ge Q) = 0.$$

This condition suggests that the optimal procurement quantity S is such that the marginal cost of overage, which is the probability of a shortage multiplied by the unit overage cost $(c - v)$, is equal to the marginal cost of underage, which is the probability of a shortage multiplied by the unit cost of a shortage $(r + s - c)$.

Solving this equation for Q, the optimal procurement quantity S is found from the fractile formula

$$F(S) = \frac{c_u}{c_u + c_o} \text{ that is } F(S) = \Pr(D \le S) = \frac{r + s - c}{r + s - v}.$$

The assumption $r > c > v$ implies that the right-hand-side of the formula is greater than 0 and less than 1; $F(\cdot)$ is continuous nondecreasing function, and hence a finite positive S always exists. Furthermore, the second derivative of $L(Q)$, $(c - v)f(Q) + (r + s - c)f(Q) \ge 0$ for all $Q \ge 0$, implies the convexity of $L(Q)$. In addition, $L(Q)$ has a negative slope at $Q = 0$, $-(r + s - c)$, and a positive slope, $c - v$, as Q tends to ∞, implying that $L(Q)$ has a finite minimizer S over $(0, \infty)$.

Sometimes customers order in bulk. In such cases, number of customers might be low and their demand structure might not assume a continuous distribution. Also, some products such as planes, trains, and so on cannot be ordered in fractions. An airline can order an integral number of jumbo jets, but it does not quite make sense to order 0.11 planes! Hence, the assumption of a continuous demand distribution might not make sense for all cases. Luckily, for the newsvendor model, this is not a problem. If demand distribution F is actually discrete, the above analysis follows similarly with a small adjustment. The expectation terms in the loss function has to be explicitly represented by summations rather than the integrals. That is, let F be a discrete distribution with probability density function (pdf)

$$f(d_j) = q_j, \ j = 1, 2, ... N, \text{ and } \sum_{j=1}^{N} q_j = 1.$$

Without loss of generality, one can assume that $d_1 = 0$ and $d_1 < d_2 < \cdots < d_N < \infty$. Then, the loss function for $Q \in [d_j, d_{j+1}]$ for any $j = 1, 2, \ldots, N$ is

$$L(Q) = (c - v)\sum_{i=1}^{J}(Q - d_i)q_i + (r + s - c)\sum_{i=j+1}^{N}(d_i - Q)q_i,$$

which is a piece-wise linear convex function of Q. Analyzing the first derivative of $L(Q)$ this property can be easily observed:

$$L'(Q) = (c - v)\sum_{i=1}^{j} q_i - (r + s - c)\sum_{i=j+1}^{N} q_i$$
$$= (c - v)\Pr(D \le Q) - (r + s - c)\Pr(D > Q)$$

which is constant for all $Q \in [d_j, d_{j+1}]$, meaning that $L(Q)$ is linear over this range. For $j = 1$, that is, for all $Q < d_1$, the derivative is a negative constant, $-(r + s - c) < 0$. Hence, $L(Q)$ is a decreasing function at

$Q = 0$. As j increases, $L'(Q)$ is nondecreasing (increasing if all $q_j > 0$) because $\Pr(D \le Q)$, which multiplies the positive quantity $(c - v)$, increases or stays the same and $\Pr(D > Q)$, which multiplies the positive quantity $(r + s - c)$, decreases or stays the same. Hence, $L(Q)$ has nondecreasing first derivative, and thus is a convex function. Since $L(Q)$ is decreasing at $Q = 0$ and increasing at $Q = d_N$, a minimizer of this function exists.

Finally, realize that when the demand distribution is discrete, the optimal quantity is equal to a possible demand point d_j. Furthermore, this demand point is easily found by finding the smallest index such that $L'(Q) > 0$. Note that, as j increases, $L'(Q)$ increases from a negative value $-(r + s - c)$ to a positive value $(c - v)$. Hence, the optimal procurement quantity $S = d_z$ where z is the smallest j such that

$$(c - v) \sum_{i=1}^{j} q_i - (r + s - c) \sum_{i=j+1}^{N} q_i > 0$$

There are several tacit assumptions in the earlier analysis: first, there is no initial inventory; second, there is no fixed ordering cost; third, the excess demand is lost; fourth, price is exogenous; fifth, salvage value is guaranteed to be achieved. The first three of these assumptions can easily be dealt with by making some observations in the earlier analysis, but we will discuss the other two assumptions in more detail in the coming subsections.

Let us assume that before the retailer places an order, which costs her a setup cost of k dollars per order (paper work, labor etc.), she realizes that there are I units of the product in her warehouse. If the retailer would like to increase the inventory level to Q, the expected cost of procuring $(Q - I)$ units is $k - cI + L(Q)$, which is still minimized by S if we actually decide to procure any units at all. Setup cost k is only incurred if we decide to procure any item at all, and hence if we do not procure any units on top of I, k is not incurred. Under what conditions the retailer should decide to procure on top of the initial inventory I? There are two cases: (*i*) If $I > S$, no units should be procured, (*ii*) If $I < S$, then the retailer needs to compare the cost of procuring the extra $S-I$ units, that is $k - cI + L(S)$, with the cost of not procuring any extra units at all, that is $-cI + L(I)$. If $k + L(S) < L(I)$, $S-I$ units should be procured, otherwise none should be procured.

If we let s to be a value such that $k + L(S) = L(s)$, the earlier discussion suggests that the optimal procurement policy is an (s, S) policy. That is, procure $S-I$ if the initial inventory I is less than or equal to s, otherwise do not procure. Quantity S is known as the order-up-to level, and s is known as the reorder point. Note that, if $k = 0$, $s = S$, this kind of a procurement policy is known as the base-stock policy. That is, if the initial inventory level I is less than S, procure $S-I$, otherwise do not procure at all.

Let us now consider the case where the excess demand is not lost, but backordered, and the shortage cost not only reflects the loss of goodwill but also the emergency shipment costs. In this case, the single-period loss function is

$$L(Q) = (c - v) \int_{x=0}^{Q} (Q - x)dF(x) + (s - c) \int_{x=Q}^{\infty} (x - Q)dF(x),$$

which is almost identical to the lost sales case except that the shortage cost, $s - c$, does not include the lost revenue anymore. Hence, the optimal procurement quantity is found from

$$F(S) = \frac{s - c}{s - v}.$$

Example 1

A hot dog-stand at Toronto SkyDome, home of Blue Jays baseball club, sells hot dogs for $3.50 each on game days. Considering the labor, gas, rent, and material, each hot dog costs the vendor $2.00 each. During any game day, based on past sales history, the daily demand at SkyDome is found to be normally distributed with mean 40 and standard deviation 10. If there are any hot dogs left at the end of the day, they can be sold at the entertainment district for $1.50 each. If the vendor sells out at SkyDome, she closes shop and calls it a day (lost sales).

(a) If the vendor buys the hot dogs daily, how many should she buy to maximize her profit?
The optimal procurement level S satisfies

$$F(S) = \frac{r + s - c}{r + s - v}$$

where $r = 3.50$, $c = 2.00$, $s = 0$, $v = 1.50$, and $F(\cdot)$ is normally distributed. That is, S satisfies $P(D \leq S) = 1.5/2.0 = 0.75$. Standardizing the normal distribution, we have $P(Z < (S - 40)/10) = 0.75$. From the normal table or MS Excel $z = 0.675$ and $S = 40 + 10(0.675) = 46.75$. Rounding up, the vendor should procure 47 hot dogs with an expected profit of $53.64.

(b) If she buys 55 hot dogs on a given day, what is the probability that she will meet all day's demand at SkyDome?
She needs to determine the probability that demand is going to be less than or equal to 55. This is easily done by calculating $Pr(D \leq 55) = Pr(Z \leq (55 - 40)/10) = Pr(Z \leq 1.5) = 0.9332$. Hence, she has 93.32% chance that she will satisfy all the demand at SkyDome and have an expected profit of $51.92.

(c) If we assume that the vendor can purchase hot dogs from the next hot dog stand for $2.50 each in case she sells out her own stock (backorder case), how many hot dogs should she buy?
In the backorder case, the critical fractile is found as $(s - c)/(s - v)$, where $s = 2.50$. Hence, $Pr(D \leq S) = (2.5 - 2)/(2.5 - 1.5) = 0.5$. Standardizing the normal distribution $P(Z < (S - 40)/10) = 0.50$. From the normal table or MS Excel $z = 0.0$ and $S = 40 + 10(0.0) = 40$. The vendor should procure 40 hot dogs with an expected profit of $65.98.

10.3.2 Joint Pricing and Inventory Control in a Newsvendor Setting

In this section, we assume that the retailer has the capability of setting the price as well as the procurement quantity of the product. For now, we consider the lost sales case with zero setup cost and initial inventory. This can very well be the case for many innovative companies who introduce the product first to the market and have some patent rights to charge the price they would like. Even though they might charge any price they wish, companies still need to consider the effect of the price on demand. Companies need to jointly determine the optimum price and procurement quantity with respect to the demand-price relationship that they assume in order to maximize their expected profit.

In the operations management and economics literature, demand is often modeled in an additive or a multiplicative fashion and the randomness in demand is assumed to be price independent. Specifically, demand is defined as $D(r, \varepsilon) = m(r) + \varepsilon$ in the additive case and $D(r, \varepsilon) = m(r)\varepsilon$ in the multiplicative case, where $m(r)$ is a decreasing function that captures the price–demand relationship and ε is a random variable defined over $[0, \Delta]$ with mean μ. Note that, if ε is a random variable defined over $[A, B]$, it can easily be converted to another random variable defined over $[0, B - A]$. We will consider $m(r) = a - br$ $(a > 0, b > 0)$ in the additive case and $m(r) = ar^b$ $(a > 0, b > 1)$ in the multiplicative case. Both representations of $m(r)$ are popular in the economics literature, with the former representing a linear demand curve and the second representing an iso-elastic demand curve. Due to several reasons, there might be bounds on the price charged, that is, $r_L \leq r \leq r_U$. Note that, any realization of the demand needs to be non-negative, there might be profit margin requirements from upper management, and finally, competitive or government forces might not allow you to charge any price you would like. Hence, the retailer needs to solve the non-linear program

$$\max \; \Pi(Q,r)$$
$$st \quad r_U \geq r \geq r_L$$
$$Q \geq 0.$$

The expected single-period profit very much depends on the demand–price relationship. Each demand–price relationship scenario needs independent treatment in the lost sales case. However, a unified approach is possible in the backorder case.

10.3.2.1 Lost Sales Models

Additive Demand–Price Relationship

In the joint pricing and procurement problem, minimizing the single-period loss function $L(Q)$ is not equivalent to maximizing the single-period expected profit. Hence, the retailer needs to maximize her profit which is identical to the newsvendor profit in (10.5) except that demand D is replaced by $D(r, \varepsilon)$ which is equal to $a - br + \varepsilon$.

$$\Pi(Q,r) = E[D(r,\varepsilon)](r-c) - (c-v)\int_{x=0}^{Q-m(r)}(Q-x-m(r))dF(x) - (r+s-c)\int_{x=Q-m(r)}^{\Delta}(x+m(r)-Q)dF(x)$$

As opposed to the exogenous price case, this expected profit function is not necessarily concave for all possible values of the parameters. However, it is shown by Karakul (2007) if demand distribution satisfies a weak condition, it is still a well-behaved function and it has a unique stationary point in the feasible region which is also the unique local maximum. That is, it is a unimodal function. To see this, we first introduce a change of variable $u = Q - m(r)$ which is interpreted as a safety stock factor representing the type 1 service level, that is, the probability of not stocking out. For given u, the service level is $F(u)$, but the procurement quantity Q does not have this one-to-one correspondence with the service level: for given Q, the service level is $F(Q - m(r))$ and is dependent on the price. Carrying out this change of variable, the expected profit in terms of u and r is:

$$\Pi(u,r) = E[D(r,\varepsilon)](r-c) - (c-v)\int_{x=0}^{u}(u-x)dF(x) - (r+s-c)\int_{x=u}^{\Delta}(x-u)dF(x),$$

where expectations are taken over the random variable ε. Now consider the first-order conditions of this function with respect to r and u.

$$\frac{\partial \Pi(u,r)}{\partial r} = -2br + a + bc + \mu - \int_{x=u}^{\Delta}(x-u)dF(x) = 0 \tag{10.7}$$

$$\frac{\partial \Pi(u,r)}{\partial u} = (r+s-v)(1-F(u)) - (c-v) = 0. \tag{10.8}$$

From Equation 10.7, optimal price r as a function of u is found as:

$$r(u) = \frac{a + bc + \mu - \int_{x=u}^{\Delta}(x-u)dF(x)}{2b}.$$

Substituting this in (10.8) and assuming that the demand distribution $F(\cdot)$ has a hazard rate $z(\cdot) = f(\cdot)/(1-F(\cdot))$ such that $2z(x)^2 + dz(x)/dx > 0$ for all $x \in (0, \Delta)$,[*] Karakul (2007) shows that there is a unique solution that satisfies the first-order conditions and it corresponds to a local maximum.

Define $\Pi_u = d\Pi(u, r(u))/du$ and consider its first and second derivatives

$$d\Pi_u / du = \frac{-f(u)}{2b}\left[2b(r(u)+s-v) - \frac{(1-F(u))}{z(u)} \right],$$

$$d^2\Pi_u / du^2 = \frac{df(u)/d(u)}{2b}\left[2b(r(u)+s-v) - \frac{(1-F(u))}{z(u)} \right] - \frac{f(u)(1-F(u))}{2bz(u)^2}\left[2z(u)^2 + dz(u)/du \right].$$

Note that, any stationary point of Π_u (not any stationary point of Π) needs to satisfy the first-order condition $d\Pi_u/du = 0$, and hence

$$d^2\Pi_u / du^2 \big|_{d\Pi_u / du=0} = - \frac{f(u)(1 - F(u))}{2bz(u)^2}\left[2z(u)^2 + dz(u)/du \right] < 0$$

if $2z(u)^2 + dz(u)/du > 0$ for all $u \in (0, \Delta)$. This suggests that all stationary points of Π_u (the total derivative of Π) are local maxima, which means that it actually has a unique stationary point and it is a local maximum. This implies that Π_u can vanish at most twice over $[0, \Delta]$ and consequently, Π might have two stationary points with the larger one being the local maximum over this range. However, $\Pi_u(0) = r(0) + s - c > 0$ and hence Π_u equals zero at most once in $(0, \Delta)$ proving the unimodality of $\Pi(u, r)$.

The optimal stocking factor and price (u^*, r^*) can be found by first solving the nonlinear equation,

$$\frac{\partial\Pi(u,r(u))}{\partial u} = (r(u) + s - v)(1 - F(u)) - (c - v) = 0$$

with respect to u to obtain u^* and then, substituting u^* in $r(u)$ to obtain r^*. The optimal procurement quantity is calculated as $S = a - br^* + u^*$.

Example 2

Consider the hot dog-stand example. Assuming that excess demand is lost and there is not any competition, the vendor would like to determine the best price and procurement level. Luckily, the vendor has an operations research background and she was able to figure out that the demand is a linear function of the price: $100 - 10r + \varepsilon$ where ε is a random variable with a normal distribution 40 and standard deviation 10. What is her best price and procurement quantity?

Remember that $c = 2, s = 0, v = 1.5$. Solving

$$\frac{\partial\Pi(u,r(u))}{\partial u} = (r(u) + 2 - 1.5)(1 - F(u)) - (2 - 1.5) = 0$$

---

[*] Note that, all log-concave distribution functions, that is, the distribution functions whose logarithm are concave, satisfy this condition (see An 1995 for a discussion of log-concave distributions which include Normal, gamma, Erlang and many other well-known distributions).

for u, we find $u^* = 54.25$. Note that F represents the normal distribution and it is necessary to use a package like Maple or Matlab to solve this nonlinear equation. Optimal price is $r(54.25) = \$7.98$ and optimal order quantity is $S = 100 - 10 * 7.98 + 54.25 = 74.45$. Closest integer value is 74, and hence the vendor should order 74 hot dogs and charge them $7.98 each for a total profit of $350.62.

The rounding of the order quantity is not necessarily up or down always. Since in this case, a continuous distribution is used to approximate a discrete one, the integer number that is closest to S is more likely to bring the highest profit. Note that, the hot dogs would be quite expensive if there is not competition and the demand–price relationship is given by $100 - 10r$. (What would the price be if demand–price relationship was $100 - 15r$?)

Multiplicative Demand–Price Relationship

In case the demand and price have a multiplicative relationship, the change of variable is somewhat different. We define $u = Q/m(r)$. Substituting $D(r,\varepsilon) = m(r)\varepsilon$ for D and $u = Q/m(r)$ in the objective function of the newsvendor problem in Equation 10.5, the single-period expected profit function is

$$\Pi(u,r) = E[D(r,\varepsilon)](r - c) - m(r)\{(c - v)\int_{x=0}^{u}(u - x)dF(x) + (r + s - c)\int_{x=u}^{\Delta}(x - u)dF(x)\}$$

$$= m(r)\left\{E[\varepsilon](r - c) - (c - v)\int_{x=0}^{u}(u - x)dF(x) - (r + s - c)\int_{x=u}^{\Delta}(x - u)dF(x)\right\}.$$

As in the additive case, this expected profit function is not necessarily concave or convex for all parameter values, but is unimodal for the demand distributions considered earlier. From the first-order condition that $\partial\Pi(u, r)/\partial r = 0$, the optimal price r as a function of u can be obtained as:

$$r(u) = \frac{bc}{(b-1)} + \frac{b\left[(c - v)\int_{x=0}^{u}(u - x)dF(x) + s\int_{x=u}^{\Delta}(x - u)dF(x)\right]}{(b - 1)\left[\mu - \int_{x=u}^{\Delta}(x - u)dF(x)\right]},$$

substituting this into

$$\frac{\partial\Pi(u,r)}{\partial u} = m(r)[(r + s - v)(1 - F(u)) - (c - v)] = 0,$$

assuming that the demand distribution $F(\cdot)$ has a hazard rate $z(\cdot) = f(\cdot)/(1-F(\cdot))$ such that $2z(x)^2 + dz(x)/dx > 0$ for all $x \in (0, \Delta)$ and following similar ideas as in the proof for the additive case, one can show that $d\Pi(u, r(u))/du$ is increasing at $u = 0$, decreasing at $u = \Delta$, and is itself a unimodal function over $[0, \Delta]$ for $b > 2$. This proves that there is a unique solution that satisfies the first-order conditions and it corresponds to a local maximum (see Petruzzi and Dada 1999 for a proof). Hence, the optimal stocking factor and price (u^*, r^*) can be found by first solving the nonlinear equation:

$$\frac{d\Pi(u,r(u))}{du} = m(r(u))[(r(u) + s - c)(1 - F(u)) - (c - v)F(u)] = 0$$

with respect to u to obtain u^* and then, substituting u^* in $r(u)$ to obtain r^* and in $u = Q/m(r(u))$ to obtain the optimal procurement quantity $S = u^*ar^{*-b}$.

10.3.2.2 Backorder Models

The analysis of the joint pricing and procurement problem of a single product with random demand follows a different approach when it is assumed that the excess demand is backlogged rather than lost.

As in the discussion of the backorder case in the newsvendor problem, per unit shortage cost is now represented by s and it does not consider the loss of profit $(r - c)$. Note that, s does not only represent the loss of goodwill but also the cost of fulfilling the unmet demand with an emergency order and $s > c$ is a reasonable assumption. Furthermore, by defining $h = h^+ - v$ as the per unit adjusted holding cost (which can be a negative value because it is defined as the real holding cost h^+ minus the salvage v) and realizing that the expected sales is equal to the expected demand, the single-period profit is:

$$\Pi(Q,r) = E[rD(r,\varepsilon)] - cQ - E[h\max(0,Q - D(r,\varepsilon)) + s\max(0,D(r,\varepsilon) - Q)].$$

For some specific demand–price relationships, further analysis is possible. Let the demand function satisfy $D(r, \varepsilon) = \alpha m(r) + \beta$, where $\varepsilon = (\alpha, \beta)$, α is a non-negative random variable with $E[\alpha] = 1$ and β is a random variable with $E[\beta] = 0$. By scaling and shifting, the assumptions $E[\alpha] = 1$ and $E[\beta] = 0$ can be made without loss of generality. Furthermore, assume that $m(r)$ is continuous and strictly decreasing, and the expected revenue $R(d) = dm^{-1}(d)$ is a concave function of the expected demand d. Note that, $D(r, \varepsilon) = a - br + \beta$ $(a > 0, b > 0)$ and $D(r, \varepsilon) = \alpha ar^{-b}$ $(a > 0, b > 1)$ are special cases that satisfy these conditions.

Since there is a one-to-one correspondence between the selling price r and the expected demand d, the single-period expected profit function can be equivalently expressed as:

$$\Pi(Q,d) = R(d) - cQ - E[h\max(0,Q - \alpha d - \beta) + s\max(0,\alpha d + \beta - Q)]$$

Observing that $h\max(0, y) + s\max(0, -y)$ is a convex function of y, one can see that $h\max(0, Q - \alpha d - \beta) + s\max(0, \alpha d + \beta - Q)$ is a convex function of (Q, d) for any realization of α, β (see Bazaraa et al. 1993, page 80). Furthermore, taking expectation over α and β preserves convexity and hence, $H(Q) = E[h\max(0, Q - \alpha d - \beta) + s\max(0, \alpha d + \beta - Q)]$ is convex in (Q, d). This proves that $\Pi(Q, d)$ is a concave function and the optimal expected demand, d^*, and procurement quantity, Q^*, can be obtained from the first-order conditions. Optimal price is determined as $r^* = m^{-1}(d^*)$. In the existence of initial inventory, it is shown by Simchi-Levi et al. (2005) that the optimal procurement quantity is determined by a base-stock policy. That is, if the initial inventory I is less than the optimal procurement level S, then we replenish our stock to bring the inventory level up to S, otherwise we do not order. The optimal price is determined as a nonincreasing function of the initial inventory level.

There are several extensions to the given single-period joint pricing and inventory control problems with stochastic demand. Karakul and Chan (2004) and Karakul (2007) consider the case in which the excess inventory is not salvaged for certain, but they are sold at a known discounted price to a group of clients who exhibit a discrete demand distribution for this excess stock. This case is known as the newsvendor problem with pricing and clearance markets. Cachon and Kok (2007) analyze the importance of estimating the salvage price correctly. Karakul and Chan (2007) consider the product introduction problem of a company which already has a similar but inferior product in the market. Authors consider a single-period model that maximizes the expected profit from the optimal procurement of these two products and the optimal pricing of the new product. A detailed review of the inventory control of substitutable products that include the seats in flights, hotel rooms, technologically improved new products, fashion good, etc. can be found there.

10.3.3 Multiple Period Models

In this section we extend the newsvendor model such that the retailer needs to make procurement decisions for a specific product over the next N periods. At the beginning of each period t, for example, day, week, month, the inventory amount of the product is counted and noted as I_t. Then, an order of size Q_t may be placed or not depending on the quantity on hand. Initially, we assume that the orders are filled instantly, that is, the lead time is zero. A discussion of the nonzero lead times will be provided at the end of this section.

Although the analysis can be carried out for time-varying demand distributions, for the sake of simplicity, we assume that demand at each period D is independent and identically distributed following the continuous distribution $F(\cdot)$ defined over a bounded non-negative region $(0, \Delta)$. We focus on backorder models in this section.

Although the costs involved in this model are very similar to the newsvendor model, they might have a different interpretation. Initially, let us assume that setup cost k is zero. There is a non-negative holding cost h for each unit of the excess inventory at the end of each period, this can be thought of as the capital, insurance, handling cost per unit carried in inventory. For each unit of demand that is not met at the end of the period, the retailer incurs s dollars of backorder penalty cost.

Since price is exogenous retailer needs to determine the optimal procurement quantities Q_t for $t = 0, 1, 2 \ldots, N-1$ that minimize the total expected cost

$$TC(\vec{Q}) = \sum_{t=0}^{N-1} \left\{ cQ_t + hE\left[\max(0, I_t + Q_t - D)\right] + sE\left[\max(0, D - I_t - Q_t)\right] \right\}.$$

Most natural and appropriate technique to solve this problem is Dynamic Programming (DP). The appropriate DP algorithm has the following cost-to-go function:

$$J_t(I_t) = \min_{Q_t \geq 0} \left\{ cQ_t + H(I_t + Q_t) + E\left[J_{t+1}(I_t + Q_t - D)\right] \right\}, \tag{10.9}$$

where

$$H(y) = hE\left[\max(0, y - D)\right] + sE\left[\max(0, D - y)\right].$$

The cost-to-go function represents the minimum expected cost from periods $t, t+1, \ldots, N-1$ for an initial inventory of I_t at the beginning of period t and optimal procurement quantities Q_j $j = t, t+1, \ldots, N-1$. Note that, the inventory at the beginning of period $t+1$ is found as $I_{t+1} = I_t + Q_t - d$, where d is a realization of the demand variable D. Assuming that any excess inventory at the end of period N is worth nothing, the DP algorithm has the boundary condition:

$$J_N(I_N) = 0.$$

A change of variables is useful in analyzing (10.9). We introduce the variable $y_t = I_t + Q_t$ that represents the inventory level immediately after the order in period t is placed. With this change of variable, right-hand-side of Equation 10.9 can be rewritten as:

$$\min_{y_t \geq y_t} \left\{ cy_t + H(y_t) + E[J_{t+1}(y_t - D)] \right\} - cI_t$$

The function H is easily seen to be convex because, for each realization of D, $\max(0, y - D)$ and $\max(0, D - y)$ are convex in y and taking expectation over D preserves convexity. If we can prove that J_{t+1} is convex, the function in the curly brackets, call it $G_t(y_t)$, is convex as well. Then the only result that remains to be shown is $\lim_{|y| \to \infty} G_t(y) = \infty$ which proves the existence of an unconstrained minimum S_t. If these properties are proven, which we will shortly, then a base-stock policy is optimal. That is, if S_t is the unconstrained minimum of $G_t(y_t)$ with respect to y_t, then, considering the constraint $y_t = I_t$, a minimizing y_t equals S_t if $I_t \leq S_t$ and equals I_t otherwise. Using the reverse transformation $Q_t = y_t - I_t$, the minimum of the DP Equation 10.9 is attained at $Q_t = S_t - I_t$ if $I_t \leq S_t$, and at $Q_t = 0$ otherwise. Hence, an optimal policy is determined by a sequence of scalars $\{S_0, S_1, \ldots, S_{N-1}\}$ and has the form

$$Q_t^*(I_t) = \begin{cases} S_t - I_t, & \text{if } I_t < S_t \\ 0, & \text{if } I_t \geq S_t \end{cases} \tag{10.10}$$

where each S_t, $t = 0, 1, \ldots, N - 1$ solves

$$G_t(y) = cy + H(y) + E[J_{t+1}(y - D)].$$

The earlier-discussed convexity and existence proofs are done inductively. We have $J_N = 0$, so it is convex. Since $s > c$ and the derivative of $H(y)$ tends to $-s$ as $y \to -\infty$, $G_{N-1}(y) = cy + H(y)$ has a negative derivative as $y \to -\infty$ and a positive derivative as $y \to \infty$. Therefore, $\lim_{|y| \to \infty} G_{N-1}(y) = \infty$ and the optimal policy for period $N - 1$ is given as:

$$Q_{N-1}^*(I_{N-1}) = \begin{cases} S_{N-1} - I_{N-1}, & \text{if } I_{N-1} < S_{N-1} \\ 0, & \text{if } I_{N-1} \geq S_{N-1} \end{cases},$$

where S_{N-1} minimizes $G_{N-1}(y)$. From the DP Equation 10.9 we have

$$J_{N-1}(I_{N-1}) = \begin{cases} c(S_{N-1} - I_{N-1}) + H(S_{N-1}), & \text{if } I_{N-1} < S_{N-1} \\ H(I_{N-1}), & \text{if } I_{N-1} \geq S_{N-1} \end{cases},$$

which is a convex function because: first, both $H(I_{N-1})$ and $c(S_{N-1} - I_{N-1}) + H(S_{N-1})$ are convex; second, it is continuous; and finally, at $I_{N-1} = S_{N-1}$ its left and right derivatives are both equal to $-c$. For $I_{N-1} < S_{N-1}$, J_{N-1} is a linear function with slope $-c$ and, as I_{N-1} approaches S_{N-1} from right-hand-side, its derivative is $-c$ because S_{N-1} minimizes the convex function $cy + H(y)$ whose derivative $c + H'(y)$ vanishes at $y = S_{N-1}$ (see Fig. 10.5).

Note that if the initial inventory at the beginning of period $N - 1$ is greater than the unconstrained minimizer S_{N-1}, we do not order anymore, and hence do not incur any extra procurement cost, but rather face the possible holding or shortage cost H. On the contrary, if the initial inventory is less than the unconstrained minimizer S_{N-1}, then we procure enough to increase the on-hand inventory level to S_{N-1}. Hence, we not only incur the procurement cost $c(S_{N-1} - x_{N-1})$ but also the possible holding or shortage cost $H(S_{N-1})$.

Hence, given the convexity of J_N, we proved that J_{N-1} is convex and $\lim_{|y| \to \infty} J_{N-1}(y) = \infty$. This argument can be repeated to show that if J_{t+1} is convex for $t = N - 2$, $N - 3$, $\ldots$, 0, $\lim_{|y| \to \infty} J_{t+1}(y) = \infty$ and $\lim_{|y| \to \infty} G_t(y) = \infty$, then

$$J_t(I_t) = \begin{cases} c(S_t - I_t) + H(S_t) + E[J_{t+1}(S_t - D)], & \text{if } I_t < S_t \\ H(I_t) + E[J_{t+1}(I_t - D)], & \text{if } I_t \geq S_t, \end{cases}$$

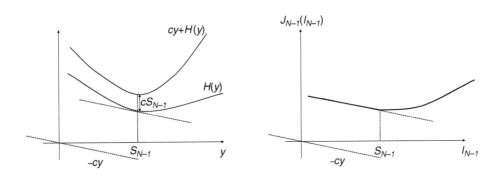

FIGURE 10.5 Structure of the cost-to-go function.

where S_t minimizes $cy + H(y) + E[J_{t+1}(y - D)]$. Furthermore, J_t is convex, $\lim_{|y| \to \infty} J_t(y) = \infty$ and $\lim_{|y| \to \infty} G_{t-1}(y) = \infty$. This completes the proof that J_t is convex for all $t = 0, 1, \ldots, N = 1$ and a base-stock policy is optimal.

Analysis is more complicated in the existence of a positive setup cost k.

10.3.3.1 Positive Setup Cost

If there is a setup cost for any non-negative procurement quantity Q_t, then the procurement cost is:

$$C(Q) = \begin{cases} k + cQ, & \text{if } Q > 0 \\ 0, & \text{if } Q = 0. \end{cases}$$

The DP algorithm has the following cost-to-go function:

$$J_t(I_t) = \min_{Q_t \geq 0} \left\{ C(Q_t) + H(I_t + Q_t) + E\left[J_{t+1}(I_t + Q_t - D) \right] \right\}, \tag{10.11}$$

with the boundary condition $J_N(I_N) = 0$.

Considering the functions $G_t(y) = cy + H(y) + E[J_{t+1}(y - D)]$ and the piecewise linear procurement cost function $C(Q)$,

$$J_t(I_t) = \min\{G_t(I_t), \min_{Q_t > 0}[k + G_t(I_t + Q_t)]\} - cI_t,$$

or by the change of variable $y_t = I_t + Q_t$,

$$J_t(I_t) = \min\{G_t(I_t), \min_{y_t > I_t}[k + G_t(y_t)]\} - cI_t.$$

If G_t can be shown to be convex for all $t = 0, 1, \ldots, N - 1$, then it can be easily seen that an (s, S) policy will be optimal. That is,

$$Q_t^*(I_t) = \begin{cases} S_t - I_t, & \text{if } I_t < s_t \\ 0, & \text{if } I_t \geq s_t \end{cases}$$

would be optimal, where S_t minimizes $G_t(y)$ and s_t is the smallest y value such that $G_t(y) = k + G_t(S_t)$. Unfortunately, if $k > 0$, it is not necessarily true that G_t is convex. However, it can be shown that G_t is still a well-behaved function that satisfies the property

$$k + G_t(z + y) \geq G_t(y) + z\left(\frac{G_t(y) - G_t(y - b)}{b} \right), \qquad \text{for all } z \geq 0, b > 0, y.$$

Since the proof is mathematically involved, we skip the proof and refer the interested readers to Bertsekas (2000). Functions that satisfy the stated property are known as K-convex functions. There are several properties of K-convex functions, which we provide in the next lemma without its proof [for proofs, see Bertsekas (2000), pp. 159–160], that help us show that (s, S) policy is still optimal in the existence of a non-negative fixed ordering cost.

Lemma 1: Properties of K-convex functions

(a) *A real-valued convex function g is also 0-convex and hence also K-convex for all $K >= 0$.*

(b) *If $g_1(y)$ and $g_2(y)$ are K-convex and L-convex ($K \geq 0$, $L \geq 0$), respectively, then $ag_1(y) + bg_2(y)$ is $(aK + bL)$-convex for all $a > 0$ and $b > 0$.*

(c) *If $g(y)$ is K-convex and w is a random variable, then $E\{g(y - w)\}$ is also K-convex, provided $E\{|g(y - w)|\} < \infty$ for all y.*

(d) *If g is a continuous K-convex function and g(y) − > ∞, then there exists scalars s and S with s < S such that*

   i.   *g(S) ≤ g(y), for all y.*
   ii. *g(S) + K = g(s) < g(y), for all y < s.*
   iii. *g(y) is a decreasing function on (−∞,s).*
   iv. *g(y) ≤ g(z) + K for all y, z with s ≤ y ≤ z.*

Part (a) is a technical result showing the relationship between convex and K-convex functions. Part (b) extends a result that holds for the convex functions to K-convex ones, that is, affine combination of K-convex functions is still K-convex (with a different K). Part (c) states that the expectation operator preserves K-convexity. Finally, part (d) gives the results that are necessary to see that an (s, S) policy is optimal if J_t for all $t = 0, 1, \ldots, N-1$ are K-convex.

Following similar lines of the proof in the zero setup cost case and using the K-convexity properties, optimality of the (s, S) policy can be shown inductively. Since $J_N = 0$, it is convex. As in the pervious case, $G_{N-1}(y) = cy + H(y)$ is convex [hence K-convex from Lemma 1(a)] and $\lim_{|y| \to \infty} G_{N-1}(y) = \infty$. Since we have

$$J_{N-1}(I_{N-1}) = \min\left\{ G_{N-1}(I_{N-1}), \min_{y \geq x_{N-1}} \left[k + G_t(y) \right] \right\} - cI_{N-1},$$

it can be seen that

$$J_{N-1}(I_{N-1}) = \begin{cases} k + G_{N-1}(S_{N-1}) - cI_{N-1}, & \text{if } I_{N-1} < s_{N-1} \\ G_{N-1}(I_{N-1}) - cI_{N-1}, & \text{if } I_{N-1} \geq s_{N-1}, \end{cases}$$

where S_{N-1} minimizes $G_{N-1}(y)$ and s_{N-1} is the smallest value of y such that $G_{N-1}(y) = k + G_{N-1}(S_{N-1})$. Note that, for $k > 0$, $s_{N-1} < S_{N-1}$. Furthermore, the derivative of G_{N-1} at s_{N-1} is negative and hence the left derivative of J_{N-1} at s_{N-1}, $-c$, is greater than the right derivative, $-c + G'_{N-1}(s_{N-1})$, which implies that J_{N-1} is not convex (but it is continuous, see Fig. 10.6). However, based on the K-convexity of G_{N-1}, it can be shown that J_{N-1} is also K-convex. Using part (c) of the lemma, G_{N-2} is a K-convex function whose limit is infinity as $|y|$ approaches infinity. Repeating the earlier-stated arguments, J_{N-2} is K-convex. Continuing in this manner one can show that for all t, G_t is a K-convex and continuous function which approaches ∞ as $|y|$ approaches ∞. Hence, by using part (d) of the lemma, an (s, S) policy is optimal.

So far we assume that demands are independent and identically distributed, cost parameters c, h, s are time-invariant, excess demand is backordered, total expected holding and shortage cost is convex, there is no capacity constraints, time horizon is finite, decision-maker is risk-neutral, and price is exogenous.

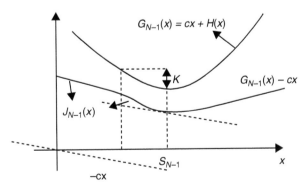

FIGURE 10.6 Structure of the cost-to-go function with positive setup cost.

All these assumptions can be relaxed, and it can actually be proven that an (s, S) type policy is still optimal. But due to the similarity of the proofs and conciseness concerns, we leave them as an extra reading to the reader. For models with time-invariant parameters, capacity constraints, and exogenous pricing assumptions see Simchi-Levi et al. (2005); lost sales and correlated demand see Bertsekas (2000); quasi-convex loss functions see Veinott (1966) or a slightly simplified version of it in Simchi-Levi et al. (2005); infinite horizon see Zheng (1991); risk-averse decision making see Agrawal and Seshadri (2000) and Chen et al. (2004).

10.4   Case Study

AMS is a growing fashion house. It started as a small family business in selling novelty T-shirts a couple of decades ago. Nowadays, it is a recognized forerunner in the global casual apparel industry. Its products are divided into two categories: novel and basic. Novel products are designed to put in the market for one season only, while basic products are offered for at least two seasons. Unlike the basic products that might have inventory left over from previous years, all the excess novel product are salvaged at the end of their selling season. Furthermore, all the novel products are produced before their selling season.

Five new novelty T-Shirts are designed for the next season. The cost of each T-shirt is $3. Traditionally, cost is 30% of the selling price and the quantity of production is the average of the modes of the forecast. According to a $10 selling price, the expert forecasts of their independent demand are shown in Table 10.1.

Steven James is a product manager just hired to work under the Director of Novel Products and asked to report on the sales plans of these T-shirts. As a top graduate from an Industrial Engineering department who has a keen interest in inventory and pricing models, Steven is very enthusiastic in his new job and is confident that he will contribute significantly to AMS. After checking the current sales plans for the new novelty T-shirts, he wants to improve the current production plan and also try to convince his boss that a better pricing scheme should be implemented. In order to achieve these objectives, he needs to answer the following questions in his report.

- What is the expected profit for the current sales plan?
- What is the optimal production plan for the current pricing scheme? What is the corresponding expected profit?
- What is the potential increase in expected profit in deploying a different pricing scheme?

Discussion with the Sales Department reveals that excess novelty T-shirts has a salvage value of $.5. Furthermore, for a selling price from $5 to $15, each independent demand can be approximated by an additive model with a 10% drop in the $10 low-demand estimate per dollar increase in selling price. That is, demand $= a - br + \varepsilon$ for $5 \leq r \leq 15$ with a, b and ε as shown in Table 10.2.

10.4.1   Exercises

1. Suppose that the demand for a product is 20 units per month and the items are withdrawn at a constant rate. The setup cost each time a production run is undertaken to replenish inventory is $10.

TABLE 10.1   Forecasts of Low, Medium, and High Demand for the Novelty T-Shirts

T-Shirts	Demand (Probability)		
Swirl	10,000 (.2)	40,000 (.5)	80,000 (.3)
Strip	5000 (.25)	10,000 (.25)	50,000 (.5)
Sea	4000 (.1)	7000 (.5)	15,000 (.4)
Stone	3000 (.3)	9000 (.4)	20,000 (.3)
Star	8000 (.4)	10,000 (.4)	12,000 (.2)

TABLE 10.2 Parameters for the Additive Demand-Price Models

T-Shirts	a	b		ε (Probability)	
Swirl	20,000	1000	0 (.2)	30,000 (.5)	70,000 (.3)
Strip	10,000	500	0 (.25)	5000 (.25)	45,000 (.5)
Sea	8000	400	0 (.1)	3000 (.5)	11,000 (.4)
Stone	6000	300	0 (.3)	6000 (.4)	17,000 (.3)
Star	16,000	800	0 (.4)	2000 (.4)	4000 (.2)

TABLE 10.3 Requirement and Production Information

Month	Requirement	Setup Cost ($)	Production Cost ($)
1	2	500	800
2	4	700	900
3	3	400	900

The production cost is $1 per item, and inventory holding cost is $0.20 per item per month. Assuming shortages are not allowed, determine the optimal production quantity in a production run. What are the corresponding time between consecutive production runs and average cost per month?

2. Consider a situation in which a particular product is produced and placed in in-process inventory until it is needed in a subsequent production process. The number of units required in each of the next three months, the setup cost, and the production cost that would be incurred in each month are shown in Table 10.3. There is no inventory of the product, but 1 unit of inventory is needed at the end of the three months. The holding cost is $200 per unit for each extra month the product is stored. Use dynamic programming to determine how many units should be produced in each month to minimize the total cost.

3. In Example 2, if the demand–price relationship was $(100r^{-3})\varepsilon$ what would the optimal price and procurement level be?

4. Consider the hot dog-stand example (Example 1 in Section 10.3.1). Now suppose that we would like to determine the optimal procurement policy over the next week (assume four games a week and we are only concerned about the game days). Each order costs the vendor $10.00 for gas and parking. Assume that any hot dog left at the end of the day is stored for next game day and are not sold at the entertainment district. Each excess hot dog costs us $0.50 for handling and proper refrigeration. Also, let us assume that there are other vendors next door. In case of a shortage, extra hot dogs can be purchased from the neighboring hot dog vendors for $2.50 each and hence, no demand is lost. Find the optimal procurement policy for the vendor over the next four sales periods.

5. Following the outline given in Section 10.3.2, prove that J_t for $t = 0, 1, \ldots, N - 1$ is K convex when order setup cost k is positive.

References

Agrawal, V. and S. Seshadri (2000). Impact of uncertainty and risk aversion on price and order quantity in the newsvendor problem. *M&SOM*, 2(4):410–423.

An, M. (1995). Log-concave probability distributions: theory and statistical testing. Technical Report NC 27708-0097, Department of Economics, Duke University.

Bazaraa, M.S., H.D. Sherali, and C.M. Shetty (1993). *Nonlinear Programming: Theory and Algorithms*. 2nd Edition, Wiley, New York, NY.

Bertsekas, D. (2000). *Dynamic Programming and Optimal Control*, 2nd Edition, Athena Scientific, Belmont, MA.

Cachon, G.P. and A.G. Kok (2007). Implementation of the newsvendor model with clearance pricing: How to (and how not to) estimate a salvage value, M & SOM, 9(3):276–290.

Chen, X., M. Sim, D. Simchi-Levi, and P. Sun (2004). Risk aversion in inventory management. Working paper, Massachusetts Institute of Technology.

Karakul, M. (2007). Joint pricing and procurement of fashion products in the existence of a clearance market. *International Journal of Production Economics*, forthcoming.

Karakul, M. and L.M.A. Chan (2004). Newsvendor problem of a monopolist with clearance markets. *Proceedings of YA/EM 2004 on CD.*

Karakul, M. and L.M.A. Chan (2007). Analytical and managerial implications of integrating product substitutability in the joint pricing and procurement problem. *European Journal of Operational Research*, doi: 10.1016/j.ejor.2007.06.026.

Petruzzi, N. and Dada, M. (1999). Pricing and the newsvendor problem: a review with extensions. *Operations Research*, 47:183–194.

Roundy, R. (1985). 98%-effective integer-ratio lot-sizing for one-warehouse multi-retailer systems. *Management Science*, 31:1416–1430.

Silver, E.A. and R. Peterson (1985). *Decision Systems for Inventory Management and Production Planning.* John Wiley, New York.

Simchi-Levi, D., X. Chen, and J. Bramel (2005). *Logic of Logistics: Theory, Algorithms, and Applications for Logistics and Supply Chain Management.* 2nd Edition, Springer Verlag, New York, NY.

Veinott, A. (1966). On the optimality of (*s*, *S*) inventory policies: new condition and a new proof. *SIAM Journal of Applied Mathematics*, 14:1067–1083.

Zheng, Y.S. (1991). A simple proof for the optimality of (*s*, *S*) policies for infinite horizon inventory problems. *Journal of Applied Probability*, 28:802–810.

11

Material Handling System

Sunderesh S. Heragu
University of Louisville

11.1 Introduction

Material handling systems are hardware systems that move material through various stages of processing, manufacture, assembly, and distribution within a facility [1]. Material movement occurs everywhere in a factory or warehouse—before, during, and after processing. The cost of material movement is estimated to be anywhere from 5% to 90% of overall factory cost with an average around 25% [2]. Material movement typically does not add value in the manufacturing process. However, this step is necessary to make a product.

The increasing demand for high product variety and short response times in today's manufacturing industry emphasizes the importance of highly flexible and efficient material handling systems. The operation of the material handling system is determined by product routings, factory layout, and material flow control strategies. Most existing textbooks cover just parts of these aspects. In this chapter, we try to introduce the material handling system from an integrated system point of view and include most factors related to the material handling system. In Section 11.2, 10 principles of the material handling system are discussed. They provide some general guidelines while selecting equipment, designing layout, in standardizing, managing, and controlling the material as well as the handling system. Section 11.3

discusses the material handling equipment topic. The multiple types of equipment and how to select these equipments are discussed in this section. Section 11.4 discusses the material handling equipment selection problem. An analytical model for the material handling selection is presented in Section 11.5. Warehousing and its functions are presented in Sections 11.6 and 11.7. Case studies illustrating applications in material handling and warehousing are presented in Sections 11.8 and 11.9.

11.2 Ten Principles of Material Handling

The 10 principles of material handling developed by the Material Handling Industry of America are: planning, standardization, work, ergonomic, unit load, space utilization, system, automation, environmental, and life cycle. A multimedia education CD explaining various aspects of the 10 principles is available upon request (see [3]).

11.2.1 Planning

A material handling plan is a prescribed course of action that specifies the material, moves, and the method of handling in advance of implementation. Five key aspects need to be considered in developing a sound materials handling plan.

1. The communication between designers and users is very important in developing the plans for operations and equipments. For large-scale material handling projects, a team including all stakeholders is required.
2. The materials handling plan should incorporate the organization's long-term goals and short-term requirements.
3. The plan must be based on existing methods and problems, subject to current physical and economic constraints, and meet organizational requirements and goals.
4. The plan should build in flexibility so that sudden changes in the process can be assimilated.

11.2.2 Standardization

Standardization is a way of achieving uniformity in the material handling methods, equipment, controls and software without sacrificing needed flexibility, modularity, and throughput. Standardization of materials handling methods and equipment reduces variety and customization. This is a benefit so long as overall performance objectives can be achieved. The key aspects of achieving standardization are as follows:

1. The planner needs to select methods and equipment that can perform a variety of tasks under a variety of operating conditions and anticipate changing future requirements. Therefore, the methods and equipment can be standardized at the same time ensuring flexibility. For example, the conveyor system in Figure 11.1 can carry different sizes of parcels.
2. Standardization can be applied widely in material handling methods such as the sizes of containers and other characteristics as well as operating procedures and equipment.
3. Standardization, flexibility, and modularity need to complement each other, providing compatibility.

11.2.3 Work

Material handling work is equal to the product of material handling flow (volume, weight, or count per unit of time) and distance moved. It should be minimized without sacrificing productivity or the level of service required of the operation. The work can be optimized from three aspects.

FIGURE 11.1 Conveyor system. (Courtesy of Vanderlande Industries, The Netherlands. With permission.)

1. Combine, shorten, or eliminate unnecessary moves to reduce work. For example in dual command storage and retrieval cycles, two commands, storage or retrieval, are executed in one trip so it has less work than single storage and retrieval cycles.
2. Consider each pick-up and set-down or placing material in and out of storage, as distinct moves and components of distance moved.
3. Material handling work can be simplified and reduced by efficient layouts and methods (Fig. 11.2). Gravitational force is used to reduce material handling work.

FIGURE 11.2 Gravity roller conveyor. (Courtesy of Sunderesh S. Heragu, 10 Principles of Materials Handling, CD. With permission.)

11.2.4   Ergonomics

Ergonomics is the science that seeks to adapt work and working conditions to suit the abilities of the worker. It is important to design safe and effective material handling operations by recognizing human capabilities and limitations.

1. Select equipment that eliminates repetitive and strenuous manual labor and that user can operate effectively. Equipment specially designed for material handling is usually more expensive than standard equipment. But using standard equipment will result in fatigue, hurt particular part of the worker's body and result in error and low-operating efficiency. Therefore it may be necessary to select specialized equipment to minimize long-term costs and injury.
2. In material handling systems, ergonomic workplace design and layout modification, it is important to pay more attention to the human physical characteristics. For example, in Figure 11.3 the work place design on the left does not provide toe place for the worker requiring him or her to bend forward. Maintaining this posture will produce fatigue and injury. The modified work place with toe space is more comfortable for the worker because his or her body is in an erect position (see right side in Fig. 11.3).
3. The ergonomics principle embraces both physical and mental tasks. For example, when a printed label or message must be read quickly and easily, the plain and simple type font should be chosen preferentially. Less familiar designs and complex font may result in errors, especially when read in haste. Aesthetic fonts are poor choices. Obviously, extremes like Old English should never be used. In one word, keep it simple.
4. Safety is the priority in workplace and equipment design.

11.2.5   Unit Load

A unit load is one that can be stored or moved as a single entity at one time, regardless of the number of individual items that make up the load. When unit load is used in material flow, the following key aspects deserve attention:

1. Less effort and work are required to collect and move a unit load than to move many items one at a time. But this does not mean bigger unit load size is always better. As the unit load size increases,

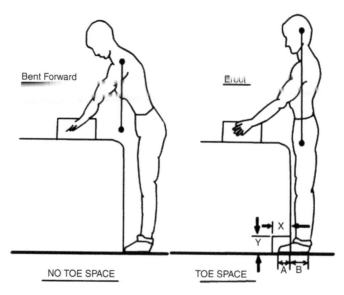

FIGURE 11.3   Modified work place. (From DeLaura, D. and Kons, D., *Advances in Industrial Ergonomics and Safety II*, Taylor & Francis, 1990. With permission.)

the total transportation cost decreases. This decrease is offset by the increase in the inventory cost. Figure 11.4 shows the relationship between the two.

2. Load size and composition may change as material and product move through various stages of manufacturing and the resulting distribution channels.

3. Large unit loads of raw material are common before manufacturing and also after manufacturing when they comprise finished goods.

4. During manufacturing, smaller unit loads, sometimes just one item, yield less in process inventory and shorter item throughput times. From Little's law [4], when a system has reached steady state, the average number of parts in the system is equal to the product of the average time per part in the system and its arrival rate.

5. Smaller unit loads are consistent with manufacturing strategies that embrace operational objectives such as flexibility, continuous flow, and just-in-time delivery.

11.2.6 Space Utilization

A good material handling system should try to improve the effectiveness and efficiency of all the available space. There are three key points for this principle.

1. In work areas, eliminate cluttered, unorganized spaces, and blocked aisles. For example, blocked aisle will add more material flow work. In Figure 11.5, the product on the floor will force the forklift to pick the product on the shelf using a longer material flow path, while the storage in Figure 11.6 will result in inefficient use of vertical storage space waste (called honeycombing loss).

2. In storage areas, the objective of maximizing storage density must be balanced against accessibility and selectivity. If items are going to be in the warehouse for a long time, storage density is an important consideration. If items enter and leave the warehouse frequently, their accessibility and selectivity are important. If the storage density is too high to access or select the stored product, high storage density may not be beneficial.

3. Cube per Order Index (COI) storage policy is often used in a warehouse. COI is a storage policy in which each item is allocated warehouse space based on the ratio of its storage space requirements (its cube) to the number of storage/retrieval transactions for that item. Items are listed in a nondecreasing order of their COI ratios. The first item in the list is allocated to the required number of storage spaces that are closest to the input/output (I/O) point; the second item is allocated to the required number of storage spaces that are next closest to the I/O point, and so on. Figure 11.7

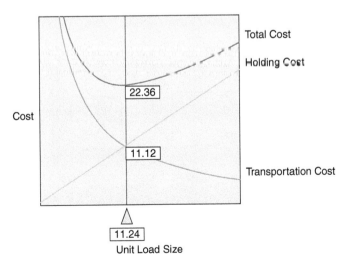

FIGURE 11.4 Trade-off between unit load and inventory costs. (Courtesy of Sunderesh S. Heragu, 10 Principles of Materials Handling, CD. With permission.)

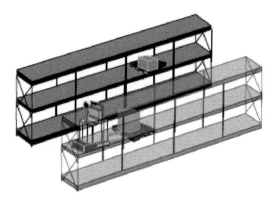

FIGURE 11.5   Retrieving material in blocked aisles. (Courtesy of Sunderesh S. Heragu, 10 Principles of Materials Handling, CD. With permission.)

shows an interactive "playspace" in the "10 Principles of Materials Handling" CD that allows a learner to understand the fundamental concepts of the COI policy.

11.2.7   System

A system is a collection of interdependent entities that interact with each other. The main components of the supply chain are suppliers, manufacturers, distributions, and customers. The activities to support materials handling both within and outside a facility need to be integrated into a unified material handling system. The key aspects of the system principle are:

1. At all stages of production and distribution, minimize inventory levels as much as possible.
2. Even though high inventory allows a company to provide a higher customer service level, it can also conceal the production problems which, from a long-term point of view, will hurt the company's operations. These problems can eventually result in low production efficiency and high product cost.
3. Information flow and physical material flow should be integrated and treated as concurrent activities. The information flow typically follows material flow.

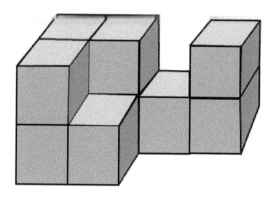

FIGURE 11.6   Honeycombing loss. (Courtesy of Sunderesh S. Heragu, 10 Principles of Materials Handling, CD. With permission.)

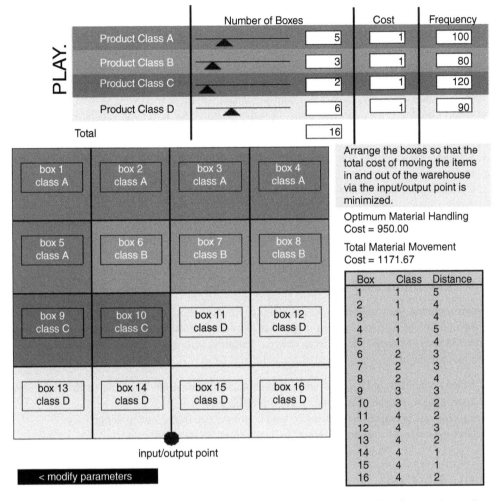

FIGURE 11.7 Example of COI policy. (Courtesy of Sunderesh S. Heragu, 10 Principles of Materials Handling, D. With permission.)

4. Materials must be easily identified in order to control their movement throughout the supply chain. For example, bar coding is the traditional method used for product identification. Radio frequency identification (RFID) uses radio waves to automatically identify people or objects as they move through the supply chain. Due to two unique product identification mandates, one from the private sector (Wal-Mart) and another from the public sector (Department of Defense), RFID has become very popular in recent years. The big difference between the two automatic data capture technologies is that bar codes is a line-of-sight technology. In other words, a scanner has to "see" the bar code to read it, which means people usually have to orient the bar code towards a scanner for it to be read. RFID tags can be read as long as they are within the range of a reader even if there is no line of sight. Bar codes have other shortcomings as well. If a label is ripped, soiled, or falls off, there is no way to scan the item. Also standard bar codes identify only the manufacturer and product, not the unique item. The bar code on one milk carton is the same as every other, making it impossible to identify which one might pass its expiration date first. RFID can identify items individually.

5. Meet customer requirements regarding quantity, quality, and on-time delivery and fill orders accurately.

11.2.8   Automation

Automation in material flow system means using electro-mechanical devices, electronics and computer-based systems with the result of linking multiple operations to operate and control production and service activities. These automated devices and systems are usually controlled by programmed instructions. Automation enables equipment or systems to run with little or no operator intervention. It improves safety, operational efficiency, consistency, and predictability, while increasing system responsiveness. Automation also decreases operating costs. In order to make the automation serve the material flow system properly, the following key aspects should be considered.

1. Simplify pre-existing processes and methods before installing mechanized or automated systems.
2. Consider computerized material handling systems where appropriate for effective integration of material flow and information management.
3. In order to automate handling, items must have features that accommodate mechanization.
4. Treat all interface issues in the situation as critical to successful automation.

11.2.9   Environmental

The environmental principle in materials handling involves designing material handling methods, selecting and operating equipment in a way that preserves natural resources and minimizes adverse effects on the environment coming from material handling activities. The following three key aspects need to be considered.

1. Design container, pallets, and other products used in materials handling so they are reusable and/or biodegradable. For example, use recyclable pallets.
2. By-products of materials handling, should be considered in the system design.
3. Give special handling considerations to hazardous materials handling.

11.2.10   Life Cycle

Life cycle costs include all cash flows that occur between the time the first dollar is spent on the material handling equipment or method until its disposal or replacement. Its key aspects are:

1. Life cycle costs in material handling system include: capital investment; installation, setup, and equipment programming; training, system testing, and acceptance; operating, maintenance, and repair; and recycle, resale, and disposal.
2. Plan for preventive, predictive, and periodic maintenance of equipment. Include the estimated cost of maintenance and spare parts in the economic analysis. There are three types of equipment failures that occur over the equipment's useful life—early failures when the product is being debugged, constant failures associated with the normal use of equipment, and increasing failure rate during the wear-out stage, when products fail due to aging and fatigue. A sound maintenance program will postpone the wear-out period and extend the useful life of equipment. Maintenance cost should be considered in the life cycle.
3. Prepare a long-range plan for equipment replacement.
4. In addition to measurable cost, other factors of a strategic or competitive nature should be quantified when possible.

The 10 principles are vital to material handling system design and operation. Most are qualitative in nature and require the industrial engineer to employ these principles when designing, analyzing, and operating material handling systems.

11.3 Material Handling Equipment

In this section we list the various equipments that actually transfer materials between different stages of processing. In manufacturing companies, various material handling devices (MHDs) are used and together they constitute a material handling system (MHS). If we regard materials as the blood of a manufacturing company, then MHSs are the vessels that transport blood to the necessary parts of the body. The major function of MHS is to transport parts and materials; this type of activity does not add any value to products and can be regarded as a sort of "necessary waste." However, in some cases, MHSs perform value added activities. MHS is an important subsystem of the entire manufacturing system; it interacts with the other subsystems. Thus, when we try to design or run a MHS, we should look at it from a system perspective. If we isolate MHS from other subsystems, we might get an optimal solution for MHS itself, but one that is suboptimal for the entire system.

In the following sections, we will first introduce seven basic types of MHDs. We then discuss how to choose the "right" equipment and how to operate equipment in the "right" way.

11.3.1 Types of Equipment

Several different types of MHDs are available for manufacturing companies to choose. These companies need to consider a number of factors including size, volume of loads, shape, weight, cost, and speed. As mentioned in the introduction, we need to consider the entire system when we try to make our choices. Of course, in order to make good decision, we need to have an overview of different MHDs. There are seven basic types of MHDs [1]: conveyors, palletizers, trucks, robots, automated guided vehicles, hoists cranes and jibs, and warehouse material-handling devices. We will introduce these types one by one briefly.

11.3.1.1 Conveyors

Conveyors are fixed path MHDs. They are only used when the volume of material to be transported is large and relatively uniform in size and shape. Depending upon the application, many types of conveyors are possible, including: accumulation conveyor; belt conveyor, bucket conveyor, can conveyor, chain conveyor, chute conveyor, gravity conveyor, power and free conveyor, pneumatic or vacuum conveyor, roller conveyor, screw conveyor, slat conveyor, tow line conveyor, trolley conveyor, and wheel conveyor. Pictures of a few conveyors are shown in Figure 11.8. The above list is not complete. Readers can refer to www.mhia.org for additional information on conveyors (and other types of MHDs).

FIGURE 11.8 Various conveyors types and their applications in material movement and sortation (a–d). (Courtesy of FKI Logistex, Dematic Corporation. With permission.)

(b)

(c)

(d)

FIGURE 11.8 (continued)

11.3.1.2 Palletizers

Palletizers are used to palletize items coming out of a production or assembly line so that unit loads can be formed directly on a pallet. Palletizers are typically automated, high speed MHDs with a user-friendly interface so that operators can easily control them. Another type of equipment that is related to a pallet is pallet lifting device. This MHD is used to lift and/or tilt pallets and raise or lower heavy cases to desired heights so that operators can pick directly from the pallets. A palletizer is shown in Figure 11.9.

11.3.1.3 Trucks

Trucks are particularly useful when the material moved varies frequently in size, shape, and weight, when the volume of the parts/material moved is low and when the number of trips required for each part is relatively few. There are many different types of trucks in the market with different weight, cost, functionality, and other features. A sample is shown in Figure 11.10.

11.3.1.4 Robots

Robots are programmable devices that mimic the behavior of human beings. With the development of artificial intelligence technology, robots can do a number of tasks not suitable for human operators. However, robots are relatively expensive. But they can perform complex or repititive tasks automatically. They can work in environments that are unsafe or uncomfortable to the human operator, work under extreme circumstance including very high or low temperature, and handle hazardous material.

11.3.1.5 Automated Guided Vehicles

Automated guided vehicles (AGVs) have been very popular since they were introduced about 30 years ago and will continue to be an important MHD in the future. AGVs can be regarded as a type of specially designed robots. Their paths can be controlled in a number of different ways. They can be fully automated or semiautomated. They can also be embedded into other MHDs. A sample of AGVs and their applications is illustrated in Figure 11.11.

FIGURE 11.9 Palletizer. (Courtesy of FKI Logistex, Dematic Corporation. With permission.)

(a)
(b)

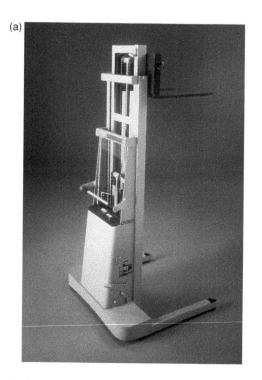

FIGURE 11.10   Order-picking trucks. (Courtesy of Crown Corporation. With permission.)

11.3.1.6   Hoists, Cranes, and Jibs

These MHDs use the overhead space. The movement of material in the overhead space will not affect production process and worker in a factory. Typically, these MHDs are expensive and time consuming to install. They are preferred when the parts to be moved are bulky and require more space for transportation (Fig. 11.12).

11.3.1.7   Warehouse Material-Handling Devices

Warehouse material-handling devices are also referred to as storage and retrieval systems. If they are highly automated, they are referred to as automated storage and retrieval systems (AS/RSs). The primary function of warehouse material-handling devices are to store and retrieve materials as well as transport them between the pick/deposit (P/D) stations and the storage locations of the materials. Some AS/RSs are shown in Figure 11.13.

(a)

FIGURE 11.11   Application of AGVs (a and b).

(b)

FIGURE 11.11 (continued)

11.4 How to Choose the "Right" Equipment

Apple[5] has suggested the use of the "material handling equation" in arriving at a material handling solution. As shown in Figure 11.14, it involves seeking thorough answers to six major questions—why (select material handling equipment), what (is the material to be moved), where and when (is the move

(a)

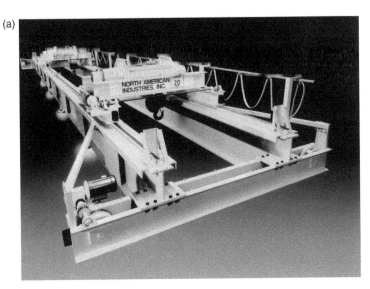

FIGURE 11.12 Gantry Crane and Hoist (a and b). (Courtesy of North American Industries and Wallace Products Corporation. With permission.)

FIGURE 11.12   (continued)

to be made), how (will the move be made), and who (will make the move). It should be emphasized that all the six questions are extremely important and should be answered satisfactorily. Otherwise, we may end up with an inferior material handling solution. In fact, it has been suggested that analysts come up with poor solutions because they jump from the what to the how question [5].

The material handling equation can be specified as: *Material + Move = Method* as shown in Figure 11.14. Very often, when the *material* and *move* aspects are analyzed thoroughly, it automatically uncovers the appropriate material handling *method*. For example, analysis of the type and characteristics

FIGURE 11.13   Automated storage and retrieval systems (AS/RS). (Courtesy of Jervis B. Webb Company. With permission.)

FIGURE 11.14 Material handling "equation." (From J.M. Apple, *Plant Layout and Modeling*, Wiley, New York, 1977. With permission.)

of *material* may reveal that the material is a large unit load on wooden pallets. Further analysis of the logistics, characteristics and type of *move* may indicate that 20 feet load/unload lift is required, distance traveled is 100 feet, and some maneuvering is required while transporting the unit load. This suggests that a fork lift truck would be a suitable material handling device. Even further analysis of the method may tell us more about the specific features of the fork lift truck. For example, narrow aisle fork lift truck, with a floor load capacity of 200 pounds, and so on.

11.5 Analytical Model for Material Handling Equipment Selection

Several analytic approaches have been proposed to select the required number and type of MHDs and to assign them to material-handling moves so that different objectives are achieved optimally. These models fall into three catalogs: deterministic approach, probabilistic approach, and knowledge-based approach. A deterministic model is presented below.

The objective of the model for simultaneously selecting the required number and type of MHDs and assigning them to material-handling moves is to minimize the operating and annualized investment costs of the MHDs. A material-handling move or simply a move is the physical move that a MHD has to execute in order to transport a load between a pair of machines. The number of moves depends upon not only the volume and transfer batch size of each part type manufactured, but also the number of machines it visits. All candidate MHD types that can perform the moves are evaluated and an optimal selection and assignment is determined by this model. If necessary, we can modify the objective function of the model to incorporate equipment idle time in conjunction with capital and operating costs. Before presenting the model, we define its variables and parameters.

i part type index, $i = 1, 2, ..., p$

j machine type index, $j = 1, 2, ..., m$

l MHD type index, $l = 1, 2, ..., n$

L_i set of MHDs that can be used to transport part type i

H length of planning period

D_i number of units of part type i required to be produced

K_{ij} set of machines *to* which part type i can be sent from machine j for the next processing step

M_{ij} set of machines *from* which part type i can be sent to machine j for the next processing step

A_i set of machine types required for the first operation on part type i

B_i set of machine types required for the last operation on part type i

V_l purchase cost of MHD H_l

T_{ijkl} time required to move one unit of part type i from machine type j to k using MHD l

C_{ijkl} unit transportation cost to move part type i from machine j to k using MHD l

X_{ijkl} number of units of part type i to be transported from machine j to k using MHD l

Y_l number of units of MHD type l selected

Model

$$\text{Minimize} \quad \sum_{l=1}^{r} V_l Y_l + \sum_{i=1}^{p} \sum_{j=1}^{m} \sum_{k \in K_{ij}} \sum_{l \in L_i} C_{ijkl} X_{ijkl} \tag{11.1}$$

$$\text{Subject to} \quad \sum_{j \in A_i} \sum_{k \in K_{ij}} \sum_{l \in L_i} X_{ijkl} = D_i \tag{11.2}$$

$$\sum_{k \in M_{ij}} \sum_{l \in L_i} X_{ijkl} - \sum_{k \in K_{ij}} \sum_{l \in L_i} X_{ijkl} = 0 \qquad i = 1, 2, ..., p; \; j : j \notin A_i \cup B_i \tag{11.3}$$

$$\sum_{j \in B_i} \sum_{k \in M_{ij}} \sum_{l \in L_i} X_{ijkl} = D_i \quad i = 1, 2, ..., p \tag{11.4}$$

$$\sum_{i=1}^{p} \sum_{j=1}^{m} \sum_{l \in K_i} T_{ijkl} X_{ijkl} \leq HY_l \quad l = 1, 2, ..., n \tag{11.5}$$

$$X_{ijkl} \geq 0 \quad i = 1, 2, ..., p; \; j = 1, 2, ..., m; \; k = 1, 2, ..., n; \; l = 1, 2, ..., n \tag{11.6}$$

$$Y_l \geq 0 \text{ and integer} \quad l = 1, 2, ..., n \tag{11.7}$$

The objective function of the above model minimizes not only the operating costs (measured as a function of the move transportation costs), but also the MHD purchase costs. When only one part type is being considered, the above model is a fixed charge network flow problem in which the number of nodes depends upon the number of machines and MHDs capable of processing and transporting the part type. In the network flow context, Constraint 11.2 ensures that the "flow" generated at the first (supply) node is equal to the number of parts to be processed. In other words, the number of units of each part type leaving their respective machines after the first operation should be equal to the number of units of that part type to be produced. These part types are "absorbed" at the last (demand) node as enforced by Constraint 11.4. In other words, the number of units of each part type coming to their respective machines for the last operation should be equal to the number of units of that part type produced. Constraint 11.3 is a material balance expression and ensures that for each intermediate transhipment node corresponding to the machines required for the in between operations, that is, other than first and last, all the units received are passed on to node(s) at the next stage. Thus, all the parts received at each intermediate machine are sent to the appropriate machine(s) for the next processing step. Constraint 11.5 imposes that the MHD capacity not be exceeded. Because Constraint 11.6 is an integer constraint, the required number of each type of MHD necessary for transporting material between machines will be selected. It can be shown that X_{ijkl} variables will automatically be integers in the optimal solution. Hence, no additional integer restrictions for these variables are necessary. When there are load limits on the MHDs, these can be enforced by introducing capacities on the appropriate arcs. This means that the corresponding X_{ijkl} variables will have an upper bound as well.

11.6 Warehousing

Many manufacturing and distribution companies maintain large warehouses to store in-process inventories or components received from an external supplier. Businesses that lease storage space to other companies for temporary storage of material also own and maintain a warehouse. In the former case, it has been argued that warehousing is a time consuming and nonvalue adding activity. Because additional paperwork and time are required to store items in storage spaces and retrieve them later when needed, the JIT manufacturing philosophy suggests that one should do away with any kind of temporary storage and maintain a pull strategy in which items are produced only as and when they are required; that is, it should be produced at a certain stage of manufacturing, only if it is required at the next stage. Moreover, the quantity produced should directly correspond to the amount demanded at the next stage of manufacturing. JIT philosophy requires that the same approach be taken towards components received from suppliers. The supplier is considered as another (previous) stage in manufacturing. However, in practice, because of a variety of reasons including the need to maintain sufficient inventory of items because of the unreliability of suppliers, and to improve customer service and respond to their needs quickly, it is not possible or at least, not desirable to completely do away with temporary storage.

Consider the following situation in Nike, a company that makes athletic wear. Nike has recently built a large distribution warehouse in Belgium because one of their main business objectives is to serve 75%

of their customers within 24 hours. Without appropriate warehousing facilities, it is impossible for Nike to achieve this objective because many of their manufacturing plants and suppliers are overseas—in the Far East! Members Club stores such as Sam's Club, Costco, and B.J.'s Warehouse Club have found a niche in the consumer retailing business in the past decade. These stores provide memberships to businesses and their employees or friends and allow only members to shop in their stores. They generally sell merchandise in bulk and directly out of their warehouse eliminating the need to build and maintain costly retail stores. While this significantly reduces overhead costs for the warehouse, for the consumer, it typically costs to less to shop in such stores than in traditional malls because s/he buys in bulk. The primary function in such warehouse stores is not warehousing but retailing!

The above two examples amply demonstrate the need for establishing warehouses to satisfactorily service end customers despite the lack of value added services in many of them. This chapter is devoted to warehouse and storage design and planning.

11.7   Warehouse Functions

As seen in the previous section, there are several reasons for building and operating warehouses. In many cases, the need to provide better service to customers and be responsive to their needs appears to be the primary reason. While it may seem that the only function of a warehouse is warehousing, that is, temporary storage of goods, in reality, many other functions are performed. Some of the more important ones are listed and briefly discussed below [6].

Temporary storage of goods: To achieve economies of scale in production, transportation and handling of goods, it is often necessary to store goods in warehouses and release them to customers as and when the demand occurs.

Put together customer orders: Warehouses, for example, the Nike distribution center in Laakdal, Belgium, receives shipments in bulk from overseas and using an automated or manual sortation system puts together individual customer orders and ships them directly to the stores.

Serve as a customer service facility: Because warehouses ship goods to customers and therefore are in direct contact with them, a warehouse can serve as a customer service facility and handle replacement of damaged or faulty goods, conduct market surveys and even provide after sales service. For example, many Japanese electronic goods manufacturers let warehouses handle repair and after sales service in North America.

Protect goods: Because warehouses are typically equipped with sophisticated security and safety systems, it is logical to store manufactured goods in warehouses to protect against theft, fire, floods, and weather elements.

Segregate hazardous or contaminated materials: Safety codes may not allow storage of hazardous materials near the manufacturing plant. Because no manufacturing takes place in a warehouse, this may be an ideal place to segregate and store hazardous and contaminated materials.

Perform value added services: Many warehouses routinely perform several value added services such as packaging goods, preparing customer orders according to specific customer requirements, inspecting arriving materials or products, testing products not only to make sure they function properly but also to comply with federal or local laws, and even assemble products. Clearly, inspection and testing do not add value to the product. However, we have included them here because they may be a necessary function because of company policy or federal regulations.

Inventory: Because it is difficult to forecast product demand accurately, in many businesses, it may be extremely important to carry inventory and safety stocks to allow them to meet unexpected surges in demand. In such businesses, not being able to satisfy a demand when it occurs may lead to a loss in revenues or worse yet, may severely impact customer loyalty towards the company. Also, companies that produce seasonal products, for example, lawn-mowers and snow-throwers, may have excess inventory left over at the end of the season and have to store the unsold items in a warehouse.

A typical warehouse consists of two main elements:

Storage medium
Material handling system

Of course, there is a building that encloses the storage medium, goods and the storage/retrieval (S/R) system. Because the main purpose of the building is to protect its contents from theft and weather elements, it is made of strong, light weight material. Warehouses come in different shapes, sizes, and heights depending upon a number of factors including the kind of goods stored inside, volume, type of S/R systems used. The Nike warehouse in Laakdal, Belgium covers a total area of 1 million square feet. Its high-bay storage is almost 100 feet in height, occupies roughly half of the total warehouse space and is served by a total of 26 man-aboard stacker cranes. On the other hand, a "members club" store may have a total warehouse space of 200,000 square feet with a building height of 35 feet.

11.8 Material-Handling System Case Study

The European Combined Terminals (ECT) in Rotterdam, the Netherlands is the largest container terminal in the world. Goods to and from Europe are transported to the outside world primarily via two types of containers—large and small. The newer docks have has been built on reclaimed land in the North Sea (Fig. 11.15a). Trucks arriving from Belgium, Germany, France, the Netherlands, and other countries wait their turn in a designated spot for their load, that is, container, to be picked up by a straddle carrier (Fig. 11.15b). The straddle carrier holds the load under the operator and moves it (Fig. 11.15c and 11.5d) to a temporary hold area from where it is loaded on to ships (Fig. 11.15e). Containers are usually held for two days in this area. When they are ready to be loaded on to ships, mobile overhead gantry cranes that move on tracks and have special container holding attachments, lift the containers from above and take them to another location where AGVs are waiting to receive the load (Fig 11.15d and 11.5f). A fleet of AGVs then transports the containers to tower cranes (Fig. 11.15g). The tower cranes are positioned very close to the loading area of the ships. Moreover, one of their arms can be tilted upward at a 90° angle to allow for tall ships to pass under them (Fig. 11.15f). Using overhead cranes, the containers are picked up from the AGV and transported one by one to the ship deck (Fig. 11.15g). While the figures illustrate how ships are loaded, unloading is done in a similar manner—only the steps are reversed. Effective use of AGVs, cranes and trucks allows ECT to load or unload a ship in about one day

11.9 Automated Storage and Retrieval Systems Case Study

Phoenix Pharmaceuticals, a German pharmaceutical company established in 1994 has a 150,000 square feet warehouse in Herne, Germany. This warehouse, which has an annual turnover of $400 million, receives pharmaceutical supplies from 19 plants all across Germany and distributes them to area drug stores. Phoenix has a 30% market share and is a leader in the pharmaceutical business. Due to competitive and other business reasons, the company must fill each order from drug stores and ship it in less than 30 minutes. There are roughly 87,000 items stored in the warehouse of which 61% is pharmaceutical and the remainder is cosmetic supplies. The number of picks range anywhere from 150 to 10,000 in any given month. If Phoenix did not have warehouses located at strategic locations, it will obviously not be able to respond to its customers, that is, fill and ship orders, accurately and adequately. Not only it is very costly for the company to ship the pharmaceutical supplies from the plant to each drug store directly, but also not possible to do so because of the distances.

(a)

(b)

(c)

FIGURE 11.15   MHSs in action (a–g). (From Sunderesh S. Heragu, *Facilities Design*, iUniverse, Lincoln, NE. With permission.)

(d)

(e)

(f)

FIGURE 11.15 (continued)

(g)

FIGURE 11.15   (continued)

Order picking in Phoenix is done using three levels of automation.

1. Manual order picking using flow racks
2. Semiautomated order picking using an automatic dispensing system
3. Full automation using robotic orderpicker

Incoming customer orders are printed on high-speed printers and the orders are attached manually to totes and sent via conveyors (Fig. 11.16a) to manual order picking areas. Here, operators pick items specified in an order from flow racks, fill the container and send it to shipping areas from where it is sent to the customer (i.e., drug stores). Order picking in Phoenix is done manually for bulky items that are not suitable for the AS/RS.

Semiautomated order picking is used for small items (e.g., a box containing a few dozen aspirin tablets, nasal spray medicine, etc.), which are stacked up on the outside of automatic vertical dispensers (Fig. 11.16b) in their respective columns. The dispenser has several columns—one for each brand of medicine picked. The dispensers are inclined over a conveyor forming an A shape and a computer controlled mechanism kicks items specified in an order from their respective columns on to the moving conveyor belt (Fig. 11.16c). The items then proceed to the end of the conveyor line, where they are dropped into a waiting tote. Each tote corresponds to a specific order. The tote (similar to those shown in Fig. 11.16a) are at a lower level than the conveyor line. Hence, there is no need for manual handling of the picked items. A light signal (Fig. 11.16b) tells the operators when items need to be replenished—typically when the item has reached or gone below its safety stock level. The automatic dispensing mechanism is very effective for picking a large variety of items for which the picking frequency is medium. The automation level with the dispenser mechanism is medium. It is relatively inexpensive and the order picking is done at a much faster rate than manual order picking. The degree of accuracy is also very high. However, it usually can be used only for handling relatively small items.

The third level of order picking in Phoenix is done via expensive robotic orderpicker. Phoenix has two sets of robots—one for storage and another for retrieval. The retrieval robots (Fig. 11.16d) pick items from narrow aisles whose width is just a little over that of the robot (Fig. 11.16e). Equipped with computers (Fig. 11.16f) and optical scanners, the robot retrieves items specified in an order and puts them into one of several compartments (see the circular compartmentalized drum in the middle of Fig. 11.16f). Each compartment corresponds to a customer order. The required items are picked from their respective locations, loaded on to the compartments and taken to a conveyor line (see the right side of Fig. 11.16g) where they are dropped into waiting totes. Each compartment in the circular drum has a metal flap at the bottom that automatically opens and allows all the items in an order to be dropped into its specified tote.

FIGURE 11.16 Order picking in Phoenix Pharmaceuticals warehouse, Germany (a–i). (From Sunderesh S. Heragu, *Facilities Design*, illniverse, Lincoln, NE. With permission.)

(e)

(f)

(g)

FIGURE 11.16   (continued)

(h)

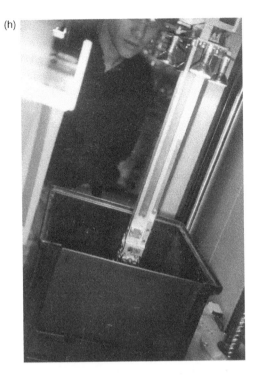

(i)

FIGURE 11.16 (continued)

The storage robots (Fig. 11.16g) have a deck that can hold large bins. Items to be stored in racks are put into these bins which are then loaded on the robot deck one at a time. A robot arm plunges into the bin and picks items using vacuum suction cups (again one at a time—see Fig. 11.16h). The items are then put into their respective storage bins using robot arms equipped with optical scanners (Fig. 11.16i). The bins are then transported and stored by the robot.

11.10   Summary

Material handling system is a complex system that provides a vital link between successive workstations. In this chapter, we tried to introduce it from the 10 principles point of view, different types of available material handling equipment, a model for selection MHDs, warehousing, its functions and presented two case studies to illustrate MHSs in action.

References

1. S. S. Heragu, (2006), *Facilities Design*, 2nd Edition, iUniverse, Lincoln, NE.
2. J. A. Tompkins, J. A. White, Y. A. Bozer, and E. H. Frazeller, (2003), *Facilities Planning*, 3rd Edition, Wiley, New York.
3. S. S. Heragu, "10 Principles of Materials Handling," Educational CD available upon request.
4. J. D. C. Little (1961), A proof for the queuing formula $L = \lambda W$, *Operations Research*, Vol. 9, pp. 383–385.
5. J. M. Apple, (1977), *Plant Layout and Material Handling*, 3rd Edition, Wiley, New York.
6. R. A. Kulwiec, (1980), *Advanced Material Handling*, The Material Handling Institute, Charlotte, NC.
7. DeLaura, D. and Kons, D., (1990), *Advances in Industrial Ergonomics and Safety II*, Taylor & Francis.

12

Warehousing

Gunter P. Sharp
Georgia Institute of Technology

12.1 Introduction

This chapter presents a description of a small, fictitious warehouse that distributes office supplies and some office furniture to small retail stores and individual mail-order customers. The facility was purchased from another company, and it is larger than required for the immediate operation. The operation, currently housed in an older facility, will move in a few months. The owners foresee substantial growth in their high-quality product lines, so the extra space will accommodate the growth for the next few years. The description of the warehouse is of the planned operation after moving into the facility.

The purpose of this chapter is to introduce the reader to the operations of warehouses. Basic functions are described, typical equipment types are illustrated, and operations within departments are presented in some detail so that the reader can understand the relationships among products, orders, order lines, storage space, and labor requirements. Storage assignment and retrieval strategies are briefly discussed. Evaluation of the planned operation includes turnover, performance, and cost analyses. Additional information can be found in other chapters of this volume and in the reference material.

12.1.1   Role of the Warehouse in the Supply Chain

Warehouses can serve different roles within the larger organization. For example, a stock room serving a manufacturing facility must provide a fast response time. The major activities would be piece (item) picking, carton picking, and preparation of assembly kits (kitting). A mail-order retailer usually must provide a great variety of products in small quantities at low cost to many customers. A factory warehouse usually handles a limited number of products in large quantities. A large, discount chain warehouse typically "pushes" some products out to its retailers based on marketing campaigns, with other products being "pulled" by the store managers. Shipments are often full and half truckloads. The warehouse described here is a small, chain warehouse that carries a limited product line for distribution to its retailers and independent customers.

The purpose of the warehouse is to provide the utility of time and place to its customers, both retail and individual. Manufacturers of office supplies and furniture are usually not willing to supply products in the quantities requested by small retailers and individual customers. Production schedules often result in long runs and large lot sizes. Thus, manufacturers usually are not able to meet the delivery dates of small retailers and individuals. The warehouse bridges the gap and enables both parties, manufacturer and customer, to operate within their own spheres.

12.1.2   Product and Order Descriptions

12.1.2.1   Product Descriptions

The products handled include paper products, pens, staplers, small storage units, other desktop products, low-priced media like CD and DVD blanks, book and electronic titles, and office furniture. High-value electronic products are delivered directly from other distributors and not handled by the warehouse. One would say that the warehouse handles relatively low-value products from the viewpoint of manufacturing cost.

Products are sold by the warehouse as pieces, cartons, and on pallets. Figure 12.1 shows the relationships among these load types. Individuals usually request pieces; retailers may also request pieces of slow movers, products that are not in high demand. Retailers usually request fast movers, products that are in high demand, in carton quantities. Bulky products like large desktop storage units may be in high enough demand so that they are sold by the warehouse in pallets. Furniture units are also sold on pallets for ease of movement in the warehouse and in the delivery trucks. Table 12.1 shows the number of products to be stored and the number of storage locations needed. The latter issue is discussed in Section 12.3.

The typical dimensions of a piece is $10 \times 25 \times 3.5$ cm, with a typical volume of 0.875 liters. A carton has typical dimensions of $33 \times 43 \times 30$ cm, with a typical volume of 42.6 liters. Thus, a typical carton contains 48.7 pieces. The typical dimension of a pallet is 80 × 170 × 140 cm, with the last dimension being

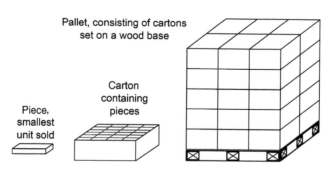

Pallet, consisting of cartons
set on a wood base

Carton
containing
pieces

Piece,
smallest
unit sold

FIGURE 12.1   Load types.

TABLE 12.1 Product Storage Requirements Summary

	Piece Pick, Slow Movers	Piece Pick, Fast Movers	Carton Pick	Pallet Pick	Total
Number of products	1000	500	500	140	2140
Number of pick locations	1000	500	540	208	2248
Number of total locations	1050	550	1620	1560	4780

the height. The pallet base is about 10 cm high, so the typical product volume is 1.25 m³, corresponding to 29.3 cartons. The pallet base allows for pickup by forklift truck from any of the four sides. Table 12.2 summarizes these values. Different products, of course, have different dimensions and relationships. The conversion factors can vary depending on whether the product is sold mainly in piece, carton, or pallet quantities. We will not introduce further complexity here and use the values given here for determining storage and labor requirements.

12.1.2.2 Order Descriptions

There are two types of orders processed at the warehouse. Large orders are placed by the retailers who belong to the same corporation; these are delivered by less-than-truckload (LTL) carrier. Small orders are placed by individuals, and these are delivered by package courier service like United States Postal Service (USPS), United Parcel Service (UPS), and Federal Express (FedEX). Large orders contain more products and the quantity per product is greater than for small orders, as shown in Table 12.3.

12.2 Functional Departments and Flows

An overall view of the functions that represent the distribution center is shown in Figure 12.2, the function flow map of the operations in the facility. This diagram shows the logical flow of products all the way from receiving through storage and retrieval to shipping. Solid arrows represent main flows, and dashed arrows show minor and occasional flows. We maintain a distinction between functional departments and physical areas. A functional department, although it may be affected by a physical area boundary, is not restricted by the ordinary physical boundaries that might appear on a layout plan.

12.2.1 Receiving and Stowing

Products enter the facility at the receiving and/or shipping dock after being unloaded from trucks with the use of forklift trucks. Figure 12.3 shows a typical vehicle that can be used to load or unload trucks and store pallets in storage racks up to about 4 m high (load support height). Products are inspected using vehicle-mounted and hand-held barcode scanners that contain integrated radio-frequency (RF) communication devices (see Fig. 12.4). If the product does not match an incoming purchase order, or if inspection and/or quarantine are needed, the product is moved to the inspect or quarantine area. This

TABLE 12.2 Product Dimensions and Conversion Factors

Unit	Width, cm	Length, cm	Height, cm	Volume, Liters	Units in Next Larger Unit
Piece	10	25	3.5	0.875	48.7
Carton	33	43	30	42.6	29.3
Pallet	80	120	130	1250	—

Note: About 10 cm needs to be added to pallet height for the base.

TABLE 12.3 Order Characteristics

	Order Size	From Piece Pick, Slow Movers	From Piece Pick, Fast Movers	From Carton Pick	From Pallet Pick	Total
Lines/order	Small	2	6	2	0.1	10.1
	Large	3	9	30	1	43
Quantity/line	Small	2	2	1.5	1	6.5
	Large	6	6	2.5	1	15.5

happens infrequently, and most products are either moved to the inbound staging area or staged in the dock area.

From inbound staging, products are moved to storage locations and stowed. Pallets are moved by forklift truck to pallet reserve storage areas. Exceptions may occur if a corresponding product pick location in either the carton pick area or the piece pick area is empty. In that situation, the pick area is replenished first, using one or more cartons from the incoming pallet, and the remainder is sent to a pallet storage area. Products that are received in carton quantities are moved by either pallet jack or cart to a piece pick area. Table 12.4 shows the daily quantities of receipts, number of trips, and labor hours needed.

Products in the pallet reserve storage area (see Fig. 12.5) are assigned locations using a shared storage concept, with the more active products located closer to the receiving or shipping dock. The storage area is divided into three areas, (A, B, and C), corresponding to (fast, medium, slow). An incoming lot of pallets of identical product is classified as (A, B, or C), on the basis of adjusted turnover of the lot:

$$\text{Adjusted turnover} = \text{pallet sales per period/number of pallets in lot} \tag{12.1}$$

The incoming lot is assigned to the first available space in its area (A, B, or C). This method of storage assignment is called class-based storage. It combines the advantages of shared and dedicated

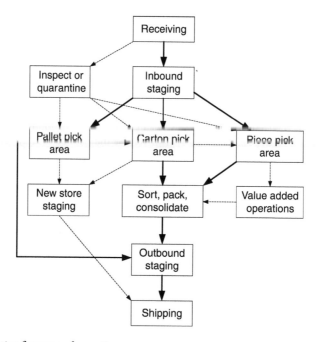

FIGURE 12.2 Function flow map of operations.

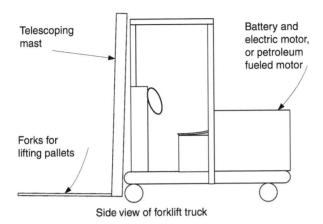

Side view of forklift truck

FIGURE 12.3 Typical forklift truck.

storage: lower overall space needs due to sharing and faster cycle times because A products are in better locations (Tompkins et al. 1996).

The pallet rack area (see Fig. 12.6) is a hybrid area, with picking (retrieval) by carton from the lower level, and the upper levels used for full pallet reserve storage. The reserve pallets for a product are stored, to the extent possible, in the same aisle and near the lower level location where the product is picked. The lower level positions are dedicated: each product is assigned a fixed location, with a few fast movers being assigned two locations. The upper level positions are shared. The classification of products is based on the number of access trips per period.

The piece pick area is divided into fast and slow movers: carton flow rack for fast movers and bin shelving for slow movers. The products are given dedicated assignments, using one of the two indices. More details on these methods of storage assignment are in Goetschalckx and Ratliff (1990), Sharp (2001), and Bartholdi and Hackman (2006).

$$Cube\ per\ order\ index = \frac{Access\ trips\ per\ period}{Maximum\ storage\ space\ needed} \tag{12.2}$$

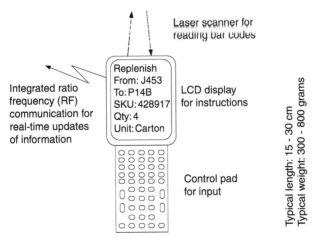

FIGURE 12.4 Handheld barcode scanner with integrated RF device.

TABLE 12.4   Receiving Operations, Daily Summary

Storage Area	Receive Units	Equivalent Pick Units	cu.m.	Method	Capacity per Trip	Number of Trips	Time per Trip	Labor Hours
Piece pick	40.4 cartons	1968 pieces	1.72	Cart, batch if possible	5 cartons	8.08	10 min.	1.35
Carton pick	50.4 pallets	1476 cartons	62.9	Forklift and pallet jack	1 pallet	50.4	4 min.	3.36
Pallet pick	22 pallets	22 pallets	27.5	Forklift	1 pallet	22	3 min.	1.10
							Total	5.81

Note: Approximately the same labor is needed for loading outbound LTL carriers.

$$Viscosity\ index = \frac{Retrieval\ visits\ per\ period}{(Cubic\ volume\ of\ product\ retrieved\ per\ period) \wedge 0.5} \tag{12.3}$$

12.2.2   Piece Pick Operations

In the carton flow (see Fig. 12.7) and bin shelving (see Fig. 12.8) areas, order pickers move along the product locations and select items in response to customer orders. The carton flow area is for relatively fast moving and bulkier products, according to one of the methods given above, and the bin shelving area is for slower moving and smaller products. Most of the products in the carton flow area are not stored anywhere else in the warehouse. The products are received as cartons and brought to the replenishment (back) sides of the flow racks, inserted, and selectively picked from the front end. Some fast moving products may be picked as pieces in this area and as cartons in the pallet rack area. For these products, there is a replenishment movement from the pallet rack area to the carton flow rack, instead of from the receiving dock. Nearly all of the products in the bin shelving area are received as cartons and moved directly from receiving to the storage area.

The main purpose of the piece pick operation is to enable the transformation of carton quantities of product into piece quantities. Some pieces are picked by order using a cart, and these move to packing and consolidation. Others are picked using batch picking, where requests from several small orders are combined into one pick list to minimize total travel during the picking process. If the items are not kept separate on the cart, they must first undergo sorting before going to packing and then consolidation.

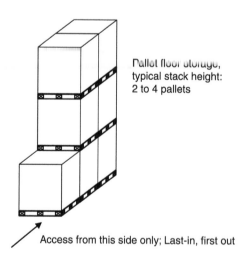

Pallet floor storage, typical stack height: 2 to 4 pallets

Access from this side only; Last-in, first out

FIGURE 12.5   Pallet reserve storage area, floor stacking.

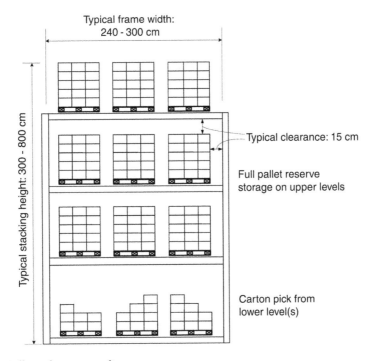

FIGURE 12.6 Pallet rack storage medium.

12.2.3 Carton Pick Operations

The lower level of the pallet rack area is used for selective retrieval (picking) of cartons in response to customer orders. The purpose of the carton pick operation is to enable the transformation of pallet quantities into carton quantities. Some cartons are picked by order using a pallet jack, and these move directly to consolidation and outbound staging. Other cartons are picked using batch picking and these

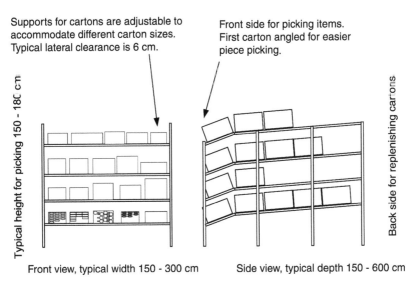

FIGURE 12.7 Carton flow storage medium.

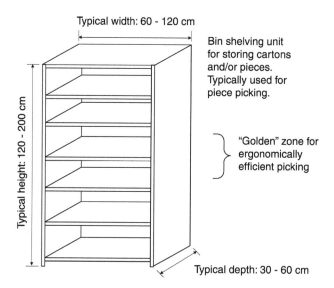

FIGURE 12.8   Bin shelving storage medium.

must first go to sorting before going to consolidation. If the total volume of activity is small, sorting and consolidation can be combined.

When the pick location at the lower level of the pallet rack becomes empty, a replenishment operation moves a full pallet from an upper level to the lower one and removes the empty pallet base. These operations are anticipated based on the orders to be filled during the next time window.

12.2.4   Pallet Pick Operations

Full pallet picking is done primarily in the floor storage area and occasionally in the pallet rack area. These pallets move directly to outbound staging. A forklift truck has the capacity to transport one pallet at a time. Travel within the pallet floor storage area follows the rectilinear distance metric (Francis et al. 1992).

12.2.5   Sorting, Packing, Staging, Shipping Operations

Pieces and cartons that are picked using batch picking must first be sorted by order before further processing. The method of batch picking, described in the following, is designed to facilitate this process without requiring extensive conveyor equipment. In addition, all pieces must be packed into overpack cartons, and these are then consolidated with regular (single product) cartons by order. Some cartons and overpacks move to outbound staging for package courier services like USPS, UPS, and FedEx. Others move to outbound staging for LTL carrier service. The package courier services load their vehicles manually, and the LTL carriers are loaded by warehouse personnel using either forklift trucks or pallet jacks.

12.2.6   Support Operations, Rewarehousing, Returns Processing

At irregular times, the warehouse staff must perform additional functions that are not part of the normal process. Whenever a new store is being prepared for opening, a large quantity of product, for the full product line, must be picked and staged. There is a separate area set aside for this staging.

Occasionally, some products need to be repackaged and/or labeled for retail stores. This value-added processing is performed between picking and packing. Returned merchandise must be inspected, possibly repackaged, and then returned to storage locations. The volume is not significant, and it is handled in the value-added area. Periodically, product locations must be changed to reflect changing demand. This rewarehousing is performed during slack periods so as not to require additional labor.

In addition, the warehouse contains an office for management and sales personnel, toilets for both staff and truck drivers, and a break room with space for vending machines and dining. There is a battery charging room for the electric batteries used by forklifts and pallet jacks, and a small maintenance room.

12.3 Storage Department Descriptions and Operations

This section presents details on the individual storage departments and their operations. Here we determine the storage space requirements, and we describe the pick methods and obtain labor requirements.

12.3.1 Bin Shelving

The bin shelving area contains 1000 slow moving products that are picked as pieces. They are housed in shelving units that are 40 cm deep, 180 cm high, and 100 cm wide, for a cubic volume of 0.72 m³. Using a cubic space utilization factor of 0.6 to allow for clearances and mismatches of carton dimensions with the shelves, each shelving unit can accommodate on average $0.72 \times 0.6/0.0426 = 10.14$ cartons. If each product requires at most one carton, then we need $1000/10.14 = 98.6$ or 99 shelving units. Rounding this to 100 units implies a pick line $100/2 = 50$ m. One way to implement this is to establish two pick aisles, each 25 m long, as shown in Figure 12.9. In the final layout, the system is expanded to a length of 30 m. In addition, space is provided for two future aisles. Although all the products stored here are considered slow movers, with some exceptions for products with small total required inventory measured in cubic volume, the principle of activity-based storage is extended further to identify the faster moving products (among the slow movers). These are placed in the ergonomically desirable golden zone (see Fig. 12.8).

The small number of requests per order for slow moving products (see Table 12.3) makes it appropriate to use a sort-while-pick (SWP) method for retrieval. An order picker uses a cart with multiple compartments (see Fig. 12.10) to pick items for several orders on one trip past the shelves. The compartments prevent items for different orders being mixed. Later, when the cart is moved to sorting, consolidation, and packing, there is actually little sorting work to do, but mainly consolidation and packing.

12.3.1.1 Time Windows

The warehouse operates one shift per day, with two time windows: an A.M. (morning) and a P.M. (afternoon) window. This reflects a balance between having a short response time at the warehouse and some fixed truck departure times, especially the LTL carriers. Most orders that are received before 6:00 A.M. are processed during the morning window; those that cannot and those that arrive during the morning are processed in the afternoon window. Table 12.5 shows how the 60 orders per day are split between A.M. and P.M., and between large and small orders.

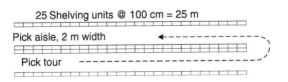

FIGURE 12.9 Bin shelving area layout.

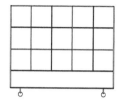

Sort-while-pick (SWP)
cart for piece picking items for
more than one order at a time.
Approximate dimensions: 150 cm
long, 60 cm wide, 120 cm high.
Cart has 15 compartments on
each side, for a total of 30.

FIGURE 12.10   Sort-while-pick (SWP) cart.

12.3.1.2   Operations Analysis

The small orders during the a.m. window represent 20 orders, 40 order lines (lines), and 80 total pieces. These are picked on one U-shaped tour, using the SWP method, with a cart. At a rate of 30 lines per hour, this translates into 1.3 labor hours. In a similar manner, the large orders during the a.m. window are picked on one U-shaped tour. The cart is similar to that used for small orders, but it has fewer and larger compartments. In the p.m. window, the process is repeated, with the result that four pick tours per day, using SWP, are made in the slow moving, bin shelving area. These results are summarized in Table 12.6. The employees who work in this area move to the sorting, consolidation, and packing area and continue with the same orders, to the extent possible. This allows for easier tracking of quality problems, such as errors in selecting the wrong item, the wrong quantity, or errors in consolidation.

12.3.2   Carton Flow Rack

The carton flow area contains 500 fast moving products, housed in carton flow rack frames that are 250 cm deep, 180 cm high, and 200 cm wide. Each frame is 4 levels high, and on average 5 lanes wide, thus containing 20 lanes. The staggering of the levels means that each lane is less than 250 cm deep, but closer to 220 cm, and thus able to accommodate 5 cartons. For the 500 products, 540 lanes are needed since some products need more than one lane. Thus, 540/20 = 27 frames are needed. This is rounded up to 30 frames that are arranged in a single aisle 30 m long, as shown in Figure 12.11. Any future expansion would be in the pallet floor storage area, where another 30 m long aisle could be placed. The adjusted turnover principle of the golden zone is also applied here.

The retrieval process for small orders is similar to that in the bin-shelving area: SWP using a cart with multiple compartments. For the large orders, there is enough volume and the length of the pick line (30 m) is short enough so that a single-order-pick (SOP) method with a cart can be used. The results are shown in Table 12.7. The employees who work in this area also move to the sorting, consolidation, and packing area and continue with the same orders, to the extent possible.

In many warehouses, there is the design question of which products to assign to an area naturally suited for piece picking and how much space to allocate to each product. If the product is also stored in a carton pick area, there is always the possibility of retrieving pieces from that area, with some loss of efficiency. If the replenishment of the piece pick area is from carton picking, then this is an example of the forward-reserve problem (Bartholdi and Hackman, 2006). The essence of this problem is how to maximize the gains from improved picker efficiency in the forward area, like bin shelving or carton flow rack, with the number of replenishment trips from the reserve area, like carton pick. Three questions

TABLE 12.5   Window Characteristics, Number of Orders

Window	Small Orders	Large Orders	Total for Window
AM, W1	20	8	28
PM, W2	22	10	32
Total for day	42	18	60

TABLE 12.6 Piece Pick Operations, Slow Movers, Bin Shelving, Daily Summary

Order Size	Window	Number of Orders	Lines per Order	Total Lines	Lines per Hour	Labor Hours	Pick Method	Qty. per Line	Total Pieces	Number of Trips
Small	W1	20	2	40	30	1.3	SWP	2	80	1
	W2	22	2	44	30	1.5	SWP	2	88	1
Large	W1	8	3	24	30	0.8	SWP	6	144	1
	W2	10	3	30	30	1.0	SWP	6	180	1
Total		60		138		4.6			492	4

can be posed for such a problem: (*i*) Which products should be assigned to the forward area? (*ii*) How much space should be assigned to each product in the forward area? and (*iii*) How large should the forward area be?

12.3.3 Pallet Rack

The pallet rack area physically has the appearance of one storage area. In fact, it consists of two functional areas, a carton pick area at the first level and a pallet reserve storage area directly above. This is a common arrangement. The requirement is for 540 pick locations for 500 products; some products move faster and require two pick locations to avoid replenishment delays. The second, third, and fourth levels of the pallet rack provide $3 \times 540 = 1620$ pallet reserve positions. This number exceeds the 1040 required; this is a consequence of the hybrid configuration where one floor-level position means three positions in the upper levels.

The structure of a pallet rack frame is shown in Figure 12.6. The frame width accommodates three pallets on each level, and the frames are connected back-to-back for stability. Because of the relatively low activity, the products are classified by adjusted turnover, and the assignment of classes is shown in Figure 12.12, which shows a typical layout of the pallet rack area. Each aisle contains 20 frames and contains $20 \times 3 = 60$ pick locations. Within an activity class, the assignment is by product number. The number of frames needed is $540/3 = 180$, corresponding to 9 aisles. The actual layout has 10 aisles. Specifying such a large area inevitably means that some adjustment and fitting must be done so that the aisles don't contain building columns, there is sufficient space for main circulation aisles, and so forth.

The retrieval process for small orders is similar to that in the carton flow area: SWP using a pallet jack to select items for five orders at a time, resulting in four trips during the A.M. window and 5 in the P.M. window. For large orders, the SOP method is used, resulting in 16 trips (two per order) in the A.M. and 20 (two per order) in the P.M. window, as shown in Table 12.8. The assignment of products into classes (A, B, and C) by adjusted turnover has some benefit here, but not as much as would be expected. When batch picking is used for small orders, the number of stops on a pick list increases, and this

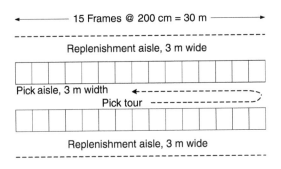

FIGURE 12.11 Carton flow area layout.

TABLE 12.7   Piece Pick Operations, Fast Movers, Carton Flow Rack, Daily Summary

Order Size	Window	Number of Orders	Lines per Order	Total Lines	Lines per Hour	Labor Hours	Pick Method	Qty. per Line	Total Pieces	Number of Trips
Small	W1	20	6	120	30	4.0	SWP	2	240	2
	W2	22	6	132	30	4.4	SWP	2	264	2
Large	W1	8	9	72	30	2.4	SOP	6	432	8
	W2	10	9	90	30	3.0	SOP	6	540	10
Total		60		414		13.8			1476	22

reduces the benefit from activity-based storage assignment. (It is a coincidence that the daily sum of 1476 cartons is the same as the daily sum of 1476 pieces in the carton flow rack.)

Replenishment activity occurs when a low-level pick location is empty or will become empty during the next time window. The warehouse management system (WMS) triggers a replenishment move from an upper level to the low level, and the removal of the empty pallet base. Table 12.9 reflects this activity.

12.3.4   Pallet Floor Storage

The pallet floor storage area is for products that move in pallet quantities and that can be stacked in pallets; these products do not need to be stored in pallet racks, although that is always an option. There is a requirement for storing a maximum of 140 products and 1560 total pallets, with an average stacking height of 2.5. Using lanes that are 3 pallets deep (see Fig. 12.13), each lane holds 7.5 pallets. Thus, 208 lanes are needed. The actual area assigned has considerably more space, to allow for future activity increase. As is the situation for pallet rack, the large area means that adjustment and fitting must be made to avoid structural columns, allow for main circulation aisles, and so forth. The storage assignment in this area is similar to that in the carton pick area, that is, class based by adjusted turnover. Within a class, storage assignment is based on the first available location of an empty lane: when a new lot is received, it is stored in the first available location in its activity class. The retrieval activity in this area is straightforward: the lane containing the oldest product is identified, and the first accessible pallet is removed and taken to the outbound staging area. The activity in this area is included in Table 12.9.

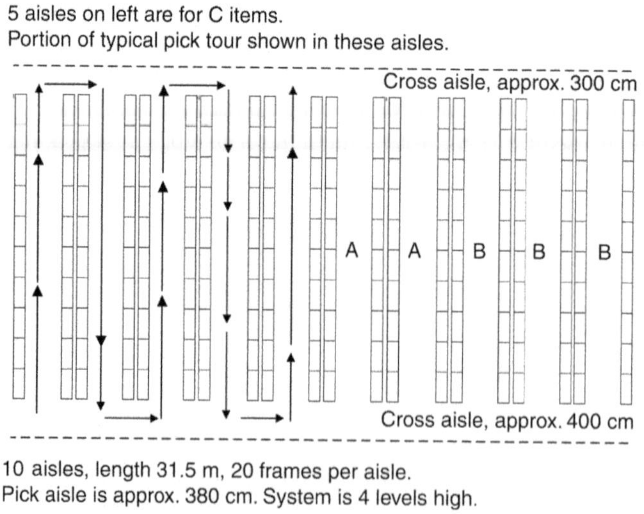

5 aisles on left are for C items.
Portion of typical pick tour shown in these aisles.

Cross aisle, approx. 300 cm

A   A   B   B   B

Cross aisle, approx. 400 cm

10 aisles, length 31.5 m, 20 frames per aisle.
Pick aisle is approx. 380 cm. System is 4 levels high.

FIGURE 12.12   Pallet rack area layout.

TABLE 12.8 Carton Pick Operations, Lower Level of Pallet Rack, Daily Summary

Order Size	Window	Number of Orders	Lines per Order	Total Lines	Lines per Hour	Labor Hours	Pick Method	Qty. per Line	Total Cartons	Number of Trips
Small	W1	20	2	40	15	2.7	SWP	1.5	60	4
	W2	22	2	44	15	2.9	SWP	1.5	66	5
Large	W1	8	30	240	15	16.0	SOP	3	600	16
	W2	10	30	300	15	20.0	SOP	2.5	750	20
Total		60		624		41.6			1476	45

12.4 Sorting, Packing, Consolidation, and Staging Descriptions

The description given is for the A.M. time window; the P.M. window is similar but has slightly higher volume. To gain efficiency in the retrieval process, the SWP method is used extensively. This means that the items placed into the SWP carts must then be sorted, consolidated, and packed by order. The sorting of pieces is really more like consolidation: the items in the carts are not mixed since each compartment holds items for only one order or part of an order. There are only 4 carts and 4 pallets that undergo this process (see Table 12.10):

    1 cart from bin shelving for small orders
    2 carts from carton flow for small orders
    4 pallets from carton pick for small orders
    1 cart from bin shelving for large orders

The three carts for small orders are staged before the pack stations, and the items for the different orders are removed and packed into overpack cartons. These overpack cartons are then consolidated with regular (single product) cartons from the four pallets. Since most small orders are shipped by package courier, the cartons (overpack and full product) for those orders then move to the staging area for package courier. In Figure 12.14 this flow is to the right.

The one cart for the large orders is staged before a pack station, and the items are removed and packed into overpack cartons. These overpack cartons then move to the left. Also to the left are:

    8 carts for individual, large orders (SOP)
    16 pallets for individual, large orders (SOP, 2 per order)

Again, the sorting of pieces in the carts is more like consolidation, since the items for different orders are not mixed in the same vehicle. The items in the carts are packed into overpack cartons, and then all three flows are consolidated onto pallets and staged for the LTL carriers.

    Overpack cartons for items from bin shelving, from the 1 cart
    Overpack cartons from carton flow, from the 8 carts
    Full cartons from carton pick, from the 16 pallets

It should be mentioned here that the nature of the work in this area depends on the way that items are picked. If the SOP and SWP methods are used, then the work is mainly packing and consolidation.

TABLE 12.9 Pallet Handling, Internal, Daily Summary

Operation Type	Equivalent Pick Units	Units Handled	Pallets per Hour	Labor Hours
Pick customer orders	22 pallets	22	15	1.5
Replenish carton pick slots	1476 cartons	50.4	20	2.5
Total		72.4		4.0

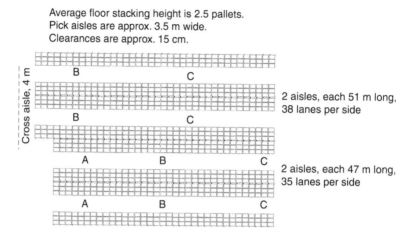

Average floor stacking height is 2.5 pallets.
Pick aisles are approx. 3.5 m wide.
Clearances are approx. 15 cm.

FIGURE 12.13   Pallet floor stacking area layout.

On the other hand, if batch picking is used, where an order picker selects items for more than one order into a container or onto a conveyor, then items must be sorted, either manually or mechanically. The choice of which method(s) to use is not always obvious. In many situations, there is more than one cost-effective solution, whereas in others a detailed comparison of alternatives is needed.

12.5   Warehouse Management

The operation of the warehouse requires careful and constant management. The scanning of received products is just one example of the functions performed by the WMS. It is beyond the scope of this chapter to present details of a typical WMS. However, some main features should be mentioned here. The tracking of flows throughout the warehouse is one of the basic functions of a WMS. This can be done manually, but most facilities today use barcode scanners, and many use barcode scanners integrated with radio-frequency transmitters (RFID) to allow for real-time updates of the underlying database. A typical WMS enables the functions listed below. These requirements are not inclusive, but only indicate the types of functions desired. Further details are in (Sharp, 2001).

The WMS should enable scheduling of personnel, including regular full-time employees and temporary and part-time employees. Tracking of employee productivity is useful for training and workload balancing. Workload scheduling should be linked to forecast information, and the conversion of product volumes should be automatically translated to labor hours by function and employee productivity.

In the receiving function, the WMS should have on-line verification of expected receipts; it should flag out-of-stock conditions, process partial receipts, and quarantine products requiring inspection. It should generate labels for pallets and cartons with data on SKU (unique product type), description, date received, lot or purchase order number, expiration code(s), and location code(s). It should assign storage location recognizing physical characteristics of product, physical characteristics of location, environmental restrictions, and stock rotation. It should also have the ability to send products directly to out-bound vehicles (cross-docking). The ability to schedule trucks and assign them to docks is also useful.

Control of storage and inventory, one of the most important functions of a WMS, includes confirmation of stow (storage) action, updating of inventory upon stow, stock reservation capability, and provision for cycle counting. The WMS should support more than one location per SKU and more than one SKU per location. Report generation should include stock activity reports (fast, medium, slow, dead), empty location reports, and anticipated replenishment of forward pick areas.

TABLE 12.10 Sort, Pack, Consolidate, Stage, Daily Summary

Incoming Vehicles	Order Size	Window	Pick Area, Method	Optn. Type	Number of Orders	Lines per Order	Total Lines	Qty. per Line	Total Qty.	Rate	Apply to	Labor Hours
1 cart	Small	W1	Bin shelving, SWP	Sort, pack	20	2	40	2	80	30	Lines	1.3
2 carts	Small	W1	Carton flow, SWP	Sort, pack	20	6	120	2	240	30	Lines	4.0
4 pallets	Small	W1	Carton pick, SWP	Sort, cons.	20	2	40	1.5	60	40	Cartons	1.5
1 cart	Large	W1	Bin shelving, SWP	Sort, pack	8	3	24	6	144	30	Lines	0.8
8 carts	Large	W1	Carton flow, SOF	Cons. pack	8	9	72	6	432	30	Lines	2.4
16 pallets	Large	W1	Carton pick, SOF	Cons.	8	30	240	2.5	600	40	Cartons	15.0
			AM Window subtotal		28		536		1556			25.0
1 cart	Small	W2	Bin shelving, SWF	Sort, pack	22	2	44	2	88	30	Lines	1.5
2 carts	Small	W2	Carton flow, SWF	Sort, pack	22	6	132	2	264	30	Lines	4.4
5 pallets	Small	W2	Carton pack, SWF	Sort, cons.	22	2	44	1.5	66	40	Cartons	1.7
1 cart	Large	W2	Bin shelving, SWP	Sort, pack	10	3	30	6	180	30	Lines	1.0
10 carts	Large	W2	Carton flow, SCP	Cons. pack	10	9	90	6	540	30	Lines	3.0
20 pallets	Large	W2	Carton pick, SCP	Cons.	10	30	300	2.5	750	40	Cartons	18.8
			PM Window subtotal		32		640		1888			30.3

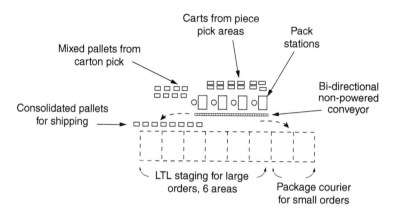

FIGURE 12.14   Sorting, packing, consolidation, and staging area layout.

An important function of the WMS is order processing. The WMS should support on-line verification of item availability, on-line verification of customer credit status, and inventory reservation at time of order entry. It should validate quantity restrictions, suggest the next quantity multiple, support quantity price breaks, and allow for flexibility in pricing by customer and order type. It should record priority and shipping methods, generate invoices, have flexibility for partial and split shipments, and have flexibility for shipping charges (customer pays or warehouse pays).

Order picking usually involves the largest labor component in a warehouse and offers the greatest opportunity for savings. Because of the potential complexity of order picking, this area is one of the most crucial aspects of a WMS. At a minimum, the WMS should support SOP, SWP, and batch picking, with flexibility for changing from one mode to another. Batch picking may require grouping of orders based on criteria like shipping deadline, truck route, and storage locations. Orders might be picked in waves corresponding to time windows. Consolidated pick documents need to be generated, considering route optimization, container capacities, and workload balancing among pickers. Often, labels need to be generated and packing instructions issued. Last, truck loading instructions need to be generated.

Hardware requirements and compatibility present further questions, such as processor type (PC, main frame), operating system (Windows, Unix, Linux), network compatibility, support for RF terminals, support for pick-to-light displays, support for voice prompt and voice recognition systems, and support for RF tags. Summarizing, the selection and implementation of a WMS is a major decision that requires time, money, and expert advice.

12.6   Facility Layout and Flows

12.6.1   Translation of Abstract Flow Diagram to Layout

Warehouse layout planning differs from traditional factory layout planning in several respects. First, one or two large storage departments usually account for more than half the total space. Second, the locations of the receiving and shipping docks are often dictated by the surrounding roads and site topography. These first two factors mean that often there are only a few ways the layout can be arranged. Third, except for pallets, the actual cost of moving product from one department to another is relatively small compared to the cost of processing within departments. This means that it is not so important where these departments, especially those for piece picking, are located. Fourth, unlike some manufacturing equipment, many storage media can be configured in a variety of ways without greatly affecting

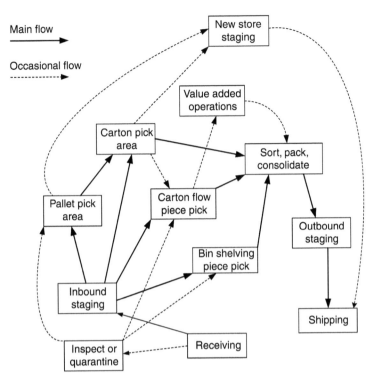

FIGURE 12.15   Abstract flow diagram for facility layout.

the equipment cost or operating efficiency. These last two factors give the designer more flexibility in determining the final layout without having to be too concerned about efficiency of flow between departments.

Using the descriptions of product flows given earlier, an abstract flow diagram is constructed, as shown in Figure 12.15. Solid lines indicate regular, daily flow, while dashed lines show occasional flow. It is possible to construct this abstract diagram so that no product flows cross. This suggests that a layout can be constructed with the same characteristic, that is, with no product flows crossing. The resulting area layout is shown in Figure 12.16. By using major circulation aisles, it is possible to keep product flows from crossing. In addition to department boundaries, some outlines of equipment units are shown, as well as major aisles. More detail is shown in the earlier illustrations for the individual departments. Inevitably, some departments were enlarged so that boundaries would follow column lines or major aisles.

12.6.2   General Building Description

The overall building is a rectangle of dimensions 80×100 m, with a column grid on 20×20 m spacing. The receiving and shipping docks are combined, with a total of 8 dock doors. Receiving is on the left side, LTL shipping in the center and right, and package courier shipping on the extreme right. The lower left section is devoid of storage media: most of the area is for pallet floor storage, with some small sections for inbound staging, inspect and/or quarantine, and new store staging. The pallet rack area (carton pick on lower level, reserve pallet storage on upper levels) occupies most of the upper part of the layout. Along the right side is an area for office, toilets, and break room; these areas are not shown in detail. However, it is preferable for the toilets and the break room to be near the dock so visiting truck drivers

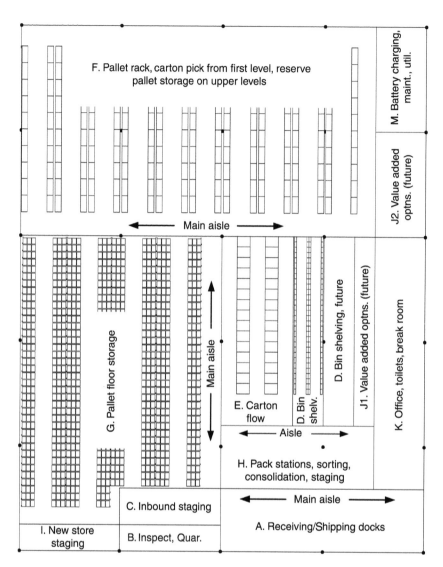

FIGURE 12.16 Facility area layout.

can easily access them. The forklift battery charging room and maintenance area is in the upper right, mainly because this reduces the length of expensive electric conductor from the nearest utility pole. It also keeps the room out of the way of product flows. The gap between the office and the battery charging room is designated for future value-added activities. It could also be used for expansion of the office.

The area to the left of the office is where most of the action is in this warehouse. The piece pick areas are vertically aligned so that the output from those areas flows down to the pack stations, and then to the shipping dock. The flow layout of the pack stations is described earlier and shown in Figure 12.14. There is space allocated for a doubling of the bin-shelving area. The narrow strip to the left of the office is only 4 m wide. It could possibly be used for value-added operations in the future, but at least part of it needs to be a personnel aisle. Table 12.11 presents a summary of department areas. The major circulation aisles are not separately specified but included in the large departments they serve. Table 12.12 is a summary of storage capacity by department.

TABLE 12.11 Department Area Summary

Dept. ID	Department	Dimensions, m	Area, sq m
A	Receiving/shipping	40 × 12	480
B	Inspect, quarantine	20 × 5	100
C	Inbound staging	20 × 7	140
D1	Bin shelving, piece pick, slow movers	36 × 6	216
D2	Bin shelving, future	36 × 6	216
E	Carton flow rack, fast movers	36 × 14	504
F	Pallet rack	70 × 40	2800
G	Pallet floor storage (rectangle includes B, C, I)	40 × 60	2060
H	Pack stations, sorting, consolidation, staging	30 × 12	360
I	New store staging	20 × 5	100
J1	Value added operations, future	36 × 4	144
J2	Value added operations, future	20 × 10	200
K	Offices, toilets, break room	48 × 10	480
M	Battery charging, maintenance, utilities	20 × 10	200
		Total	8000

Department areas include circulation space.

12.6.3 Flows and Circulation

The general flow of product within the building is clockwise, starting at the receiving dock. There are no product flow crossings except within the pack stations, where it is unavoidable. At the back of the dock is a horizontal circulation aisle (shown by a two-headed arrow). There is a vertical circulation aisle from the dock to the pallet rack area, and horizontal aisle along the lower edge of the pallet rack. Between the piece pick areas and the pack stations there is a horizontal aisle.

12.7 Performance and Cost Analyses

Evaluating warehouse operations is done from three perspectives: inventory turnover, productivity, and cost. The first perspective reveals opportunities for improving the purchasing function in the organization. Warehouses that have high turnover usually have higher productivity and lower unit costs. Productivity is usually based on labor hours required to process orders and order lines. Cost follows from capital assets and labor productivity.

These types of analyses have many potential pitfalls. For example, the inventory turnover at a warehouse may depend on purchasing decisions made at the corporate level, and this is often beyond the

TABLE 12.12 Storage Capacity Summary

Department	Storage Media	Width, cm	Depth, cm	Height, cm	Number of Media Units	Unit Stored	Capacity, Units Stored	Unit Picked	Capacity, Units Picked
Bin shelving, slow piece pick	Shelving unit	100	40	180	120	Carton	1217	Piece	59,258
Carton flow, fast piece pick	Carton flow frame	200	220	180	30	Carton	3000	Piece	146,100
Carton pick, including reserve storage above	Pallet rack frame	315	140	465	200	Pallet	2400	Carton	70,320
Pallet floor storage	Floor storage lane	135	85	420	292	Pallet	2190	Pallet	2190

Note: Pallet floor storage has an average height of 2.5 pallets.

control of the warehouse manager. In a multi-level distribution system, the lowest level usually stocks only the faster moving products while the regional and national levels stock slower moving items. Data for two large facilities that had 10% or more of the products not sold during a 12-month period was verified. In most situations, those products would be considered "dead" and candidates for removal. However, in both situations the mission of the warehouse was to stock spare parts for expensive industrial equipment that had a useful life of 20 years. From that perspective, the slow overall turnover was unavoidable. In another facility, a global warehouse for a large manufacturer of construction and earth-moving equipment with sales and support services around the world, 10% of the orders were rush orders. These were for products that were not stocked at local, regional, or national distributors. Clearly, the high fraction of rush orders leads to higher overall costs per order. From the perspective of global logistics, however, the overall approach seems sensible.

Several benchmarking studies have been made of warehouse operations (Schefczyk, 1993; Hackman et al. 2001; Chen, 2004; Frazelle, 2006). Some of these studies include extremely wide ranges of parameters. For example, in one study, the lines shipped per product per year ranged from 1000 to 900,000; the number of products ranged from 250 to 225,000; the inventory turns per year ranged from 2 to 60. This diversity poses challenges in interpreting any comparisons.

12.7.1 Turnover Analysis

We will perform the turnover analysis by estimating the average inventory for each product set to be half the design capacity. This is an approximate method to get some quick results. A more detailed method would require actual data on inventory. The operation is scheduled to move from an old facility into the one being described. Any inventory data from the old facility reflects constraints on purchasing decisions, and thus is not directly usable. Similarly, capacity that exceeds the design requirements can lead to purchasing decisions that take advantage of special discounts; such action can "fill" the available capacity. Another way to estimate average inventory would be to establish the safety stock or reorder point for each product and use that information with a mathematical inventory model (Nahmias, 2005).

The bin-shelving area for slow piece picking requires 100 shelving units, each of which holds on average 10.14 cartons. The resulting 1014 cartons correspond to a maximum inventory of $1014 \times 48.7 = 49382$ pieces, and average of 24691. With 250 operating days per year, the daily sales of 492 pieces results in an average time in storage of $24,691/492 = 50$ days. This corresponds to 5.0 inventory turns per year. The carton flow area for fast piece picking requires 27 frames. Since each frame holds 20 lanes $\times$ 5 cartons, the maximum inventory is 2700 cartons, and the average is 1350 cartons, or $1350 \times 48.7 = 65,745$ pieces. Daily sales of 1476 pieces results in an average storage time of 44.5 days, and 5.6 turns per year. The pallet rack area for carton pick needs 1620 locations. The average inventory in cartons is $1620 \times 29.3 \times 0.5 = 23,733$. Daily sales of 1476 cartons result in an average storage time of 16 days and 16 turns per year. The pallet rack area needs a maximum storage capacity of 1560 pallets, or average of 780, so the daily sales of 22 results in an average storage time of 35 days and 7 turns per year. These results are summarized in Table 12.13.

12.7.2 Productivity Analysis

The performance analysis is done at the department level for pick operations and at the warehouse and facility levels for the entire operation. Detailed performance analysis could be done for each individual operation, such as unloading trucks, stowing products, packing orders, and so forth. Since the operation will move into the facility in a few months, only data for the planned operation are available. Thus, detailed performance is reflected in the productivity rates used in the tables. These include pallet trips at 15–20 per hour, carton stow at 30 per hour, piece line retrieval at 30 per hour, carton line retrieval at 15 per hour, line pack at 30 per hour, and carton sort and consolidate at 40 per hour.

At the department level, we obtain productivity per order, per line, and per piece, carton, or pallet. These results are shown in Table 12.14. For example, in the bin-shelving area, each labor hour corresponds to 13.0 orders, 30.0 lines, and 107 pieces. At the warehouse level, we reflect all direct labor, including that

TABLE 12.13 Turnover Analysis

Department	Units	Average Inventory	Daily Sales	Ave. Time in Storage, Days	Turns per Year
Bin shelving, slow piece pick	Piece	24,692	492	50.2	5.0
Carton flow, fast piece pick	Piece	65,745	1476	44.5	5.6
Carton pick, lower level of pallet rack	Carton	23,733	1476	16	16
Pallet floor storage	Pallet	780	22	35	7

used for unloading and loading trucks (11.6 h), stowing (included in 11.6 h), replenishing (4.0 h), and sorting and packing (55.3 h). Another factor that must be considered is that in planning the operation, the labor hours represent effective hours. A warehouse employee typically works 6.5 effective hours on an 8-h shift. The rest of the time is spent preparing to receive instructions, meetings, breaks, and idle time due to the irregular schedule of activities. Further, employees are paid for holidays. Thus, the value 80 direct labor hours for bin shelving, carton flow, carton pick, and pallet floor stack, reflects this ratio of 8 paid hours for 6.5 effective hours applied to the 64.0 h, rounded to an integer number of 10 people. Including the warehouse indirect labor increases this number to 176 h, reflecting an additional 12 people. Productivity at the warehouse (total labor) level is 0.3 orders per hour and 6.8 lines per hour.

At the facility level, we also reflect management labor of 13 people, which consists of supervisory, maintenance, and sales staff (see Table 12.15). It is not unusual for the administrative labor to be more than the direct labor for a small operation like this one. These values can then be used for benchmarking the operation with other facilities.

12.7.3 Cost Analysis

The natural extension of productivity analysis is to cost analysis. Table 12.15 shows the investment costs for building, equipment, their annual maintenance costs, and the translation into annual costs, with and without the time value of money (TMV) of 15% per year. These costs reflect only the storage requirements for the immediate future in the pallet rack, carton flow, and bin-shelving areas, based on the design requirements. Labor costs for the facility are as follows:

Order pickers	10 @ $45,000	$450,000
Other WH labor	12 @ $38,000	$456,000
Administrative	13 @ $61,000	$793,000
	Total	$1,699,000

In addition, there are $200,000 annual costs for utilities and other administrative expenses. Considering the fixed investment costs, the annual costs of labor and equipment maintenance, utilities, administrative, and the time value of money, the total cost per order line is $8.35, and $167 per order.

TABLE 12.14 Performance Analysis, Daily Average

Department	Hours	Orders	Orders per Hour	Lines	Lines per Hour	Unit Type	Units	Units per Hour
Bin shelving	4.6	60	13.0	138	30.0	Piece	492	107.0
Carton flow	13.8	60	4.3	414	30.0	Piece	1476	107.0
Carton pick	41.6	60	1.4	624	15.0	Carton	1476	35.5
Pallet floor stack	4.0	22	5.5	22	5.5	Pallet	22	5.5
Warehouse, direct	80	60	0.8	1198	15.0			
Warehouse, total	176	60	0.3	1198	6.8			
Facility, total	280	60	0.2	1198	4.3			

TABLE 12.15   Cost Data

Item	Qty.	Unit Price	Initial Investment	Life-Time	Annual Cost, no TMV	Maint., Annual	Total Annual Cost, no TMV	Total Annual Cost, w. TMV 15%
Building, sq.m.	8000	350	2,800,000	40	70,000	56,000	126,000	477,574
Pallet rack	1620	60	97,200	20	4860	972	5832	16,501
Carton flow rack	30	5000	150,000	20	7500	1500	9000	25,464
Bin shelving	100	500	50,000	20	2500	500	3000	8488
Fork lift truck	3	30,000	90,000	5	18,000	18,000	36,000	44,848
Pallet jack	3	2500	7500	5	1500	1125	2625	3362
Pick cart	10	1000	10,000	5	2000	1000	3000	3983
Pallet base, extra	1000	20	20,000	5	4000	3000	7000	8966
Pack stations	4	2000	8000	10	800	1200	2000	2794
Other, misc.	1	20,000	20000	5	4000	3000	7000	8966
Totals			3,252,700		115,160	86,297	201,457	600,948

These costs are on the high side compared to other facilities, but they reflect the relatively low volume of operations, with anticipated growth, and the nature of the high-quality product line. Further, they include all costs of the facility operation, whereas many benchmark figures report only direct labor in the warehouse.

12.8   Summary

Warehouse operations are much more complex than when they appear at first glance. Profiling (partitioning) of products and orders leads to a potential multitude of warehouses inside the warehouse. The ingenuity of manufacturers to develop new technology, along with rapid advances in data processing (WMS) and mobile communications (RFID) present an ever-changing set of alternatives for storing products and retrieving items for customer orders (Kulwiec, 1982). This chapter is an attempt to present an introduction to warehousing using a case example with sufficient detail to illustrate the main concepts.

The variety of storage and retrieval technologies makes the equipment selection process difficult for the designer. At the same time, the variety of storage assignment and retrieval methods presents a challenge to both the facility designer and operator. In most circumstances, it is not possible within the limits of time and budget to investigate all possible alternatives. Instead, a guided selection process for functional departments and retrieval processes is recommended (McGinnis et al. 2005).

Acknowledgments

This research was supported by the National Science Foundation under grant EEC-9872701, the W. M. Keck Foundation, the Ford Motor Company, and The Logistics Institute at the Georgia Institute of Technology. Discussions with my colleagues Leon McGinnis and Marc Goetschalckx have provided valuable advice in the approach to the warehouse design problem.

References

Bartholdi, J.B. and S. Hackman, (2006), *Warehousing and Distribution Science*, text in progress, Atlanta, Georgia.

Chen, W.C., (2004), Available at: http://www2.isye.gatech.edu/ideas/

Francis, R.L., L.F. McGinnis, Jr., and J.A. White, (1992), *Facility Layout and Location: An Analytical Approach,* 2nd Ed., Prentice Hall, Englewood Cliffs, NJ.

Frazelle, E.H., (2006), Available at: http://www.logisticsvillage.com/MediaCenter/Documents/Presentations/*2005WarehouseBenchmarkingRpt.pdf*

Goetschalckx, M. and H.D. Ratliff, (1990), Shared Storage Policies Based on the Duration Stay of Unit Loads, *Management Science,* Vol. 36–9: 53–62.

Hackman, S.T., E.H. Frazelle, P.M. Griffin, S.O. Griffin, and D.A. Vlatsa, (2001). Benchmarking Warehousing and Distribution Operations: An Input-Output Approach, *Journal of Productivity Analysis,* Vol.16: 79–100.

Kulwiec, R.A., (1982), *Material Handling Handbook,* John Wiley & Sons, New York.

McGinnis, L.F., D. Bodner, M. Goetschalckx, T. Govindaraj, and G. Sharp, (2005), Toward a Comprehensive Descriptive Model for Warehouses, in Meller, R., et al. (Eds.), *Progress in Material Handling Research: 2004,* MHI, Charlotte, pp. 265–284.

Schefczyk, M., (1993), Operational Performance of Airlines: An Extension of Traditional Measurement Paradigms, *Strategic Management Journal,* Vol.14 (4): 301.

Sharp, G.P., (2001), Warehouse Management, in Salvendy, G., (Ed.), *Handbook of Industrial Engineering,* 3rd Ed., John Wiley & Sons, New York, pp. 2083–2109.

Steven, N., (2005), *Production & Operations Analysis,* 5th Ed., McGraw-Hill/Irwin, New York.

Tompkins, J.A., J.A. White, Y.A. Bozer, E.H. Frazelle, J.M.A. Tanchoco, and J. Trevino, (1996), *Facilities Planning,* John Wiley & Sons, New York.

13

Distribution System Design

Marc Goetschalckx
Georgia Institute of Technology

Distribution system design is the strategic design of the logistics infrastructure and logistics strategies to deliver products from one or more sources to the customers. Because of the long-term impact of the distribution system, the interrelated design decisions, and the different objectives of the various stakeholders, designing a distribution system is a highly complex and data intensive engineering design effort. A large variety of mathematical programming models has been developed to provide decision support to the design engineer. The results of the models and tools have to be very carefully validated. The uncertainty of the forecasted data has to be explicitly incorporated through sensitivity and risk analysis. The final configuration is often based on the balance between many different factors and many alternative configurations may exist. However, modeling-based design is the only available method to generate high-quality distribution system configurations with quantifiable performance measures.

13.1 Introduction

In today's rapidly changing world, corporations face the continuing challenge to constantly evaluate and configure their production and distribution systems and strategies to provide the desired customer service at the lowest possible cost. Distribution system design focuses on the strategic design of the logistics infrastructure and logistics strategies to deliver the products from one or more sources to its customers at the required customer service level. Typically, it is assumed that the products, the sources of the products (manufacturing plants, vendors, and import ports), the destinations of the products (customers), and the required service levels are not part of the design decisions but constitute constraints or parameters for the system. Distribution system design focuses on the following five interrelated decisions:

1. Determining the appropriate number of distribution centers
2. Determining the location of each distribution center
3. Determine the customer allocation to each distribution center
4. Determine the product allocation to each distribution center
5. Determine the throughput and storage capacity of each distribution center

A schematic illustration of the questions in distribution system design is shown in Figure 13.1. Decisions on delivery by direct shipping and transportation mode selection are part of the overall distribution system design.

The objective of the distribution system design is to minimize the time-discounted total system cost over the planning horizon subject to service-level requirements. The total system cost includes facility costs, inventory costs, and transportation costs. It should be noted that the detailed inventory and transportation planning decisions are made at the tactical or even operational level, but that aggregate values for the corresponding costs and capacity parameters are used in the strategic design. The facility costs include labor, facility leasing or ownership, material handling and storage equipment, and taxes.

It is clear from the description that designing a distribution system involves making numerous trade-offs. Let us assume that transportation from the manufacturing facilities to the distribution center occurs in relatively larger quantities at a relatively lower cost and that delivery from the distribution center to the customer occurs in smaller quantities at a higher cost rate. Increasing the number of distribution centers typically has the following consequences:

- Customer service levels improve because the average transportation time to the customers is smaller.
- Outbound transportation costs decrease because the local delivery area for each distribution center is smaller.
- Inbound transportation costs increase because the economies of scale of the transportation to the distribution centers are reduced.

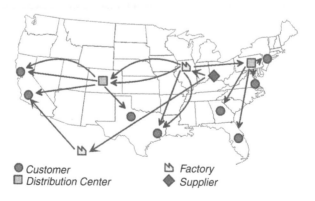

FIGURE 13.1   Distribution system schematic.

- Inventory costs increase because there are more inventory stocking locations and there is less opportunity for risk pooling so that the required safety stocks increase.
- Facility costs increase because of the overhead associated with each facility and increased handling costs because the economies of scale of handling inside the distribution centers are reduced.

13.2 Engineering Design Principles for Distribution System Design

13.2.1 Heterogeneous Data for Distribution System Design

Engineering design of any system is based on data and models for the particular system area and the design of distribution systems is no different. However, because of the large number and variety of participants in the system, the long planning horizon, and the large variety of possible distribution systems and strategies, the data for distribution system design is highly diverse and highly uncertain. This is in contrast with the more focused data and models in other engineering disciplines, such as for the design of a bridge in civil engineering, a pump in mechanical engineering, or an integrated circuit in electrical engineering.

To make the proper trade-offs, a large amount of data from a variety of sources is required. This includes:

1. Data on the customer demand for products for all the time periods in the planning horizon.
2. Product characteristics such as monetary value and physical dimensions.
3. Geographical location data for all the product sources such as manufacturing facilities and import ports, for the distribution center candidate locations, and for the customers.
4. Transportation cost rates by transportation mode and by origin and destination point.
5. Fixed facility operating costs associated with the distribution centers. Different costs can be caused by different land and construction costs in function of location or by different equipment costs in function of technology and size of the center.
6. Variable facility operating costs associated with labor and material handling costs inside the distribution centers.
7. Order processing and information technology costs associated with each distribution center.
8. Capacity constraints on the throughput and storage of various possible sizes of the distribution centers.
9. Required service levels by customer and product combination. This may include maximum delivery time to the customer form the distribution center or minimum acceptable fill rate in the distribution center.

These data have to be extracted from a variety of sources. A basic list of data sources is given next in order of decreasing data specificity and accuracy. The most relevant and accurate data are based on the in-house databases of historical transactions. Prime examples of such databases are customer sales orders, customer data, facility data, and freight bills. While relevance, in-house availability, and accuracy are the main advantages of this type of date, data volume, historical time frame, and availability for the current system only are the main disadvantages. The detailed information in these databases can be overwhelming and it has to be aggregated in order for it to be used in a strategic design model. Fundamentally, the data provides highly detailed information on what the corporation did in the past. This type of data is most suited for the restructuring of an existing distribution system. A second source of data is contained in corporate documents such as the annual report and the corporate strategic plan. These documents contain aggregate data such as the cost of capital to the corporation or the corporate service level goals.

The previous two types of data sources are specific to the logistics organization or corporation itself. The next sources of data are reports and databases on the general business area. They include logistics performance ratios for "best-of-class" corporations, databases of the aggregate industry, and detailed forecast reports for the industry. Corporations typically have access to such data through membership in trade organizations. Their membership may oblige them in turn to report their activities to the association. Another source of this type of data is provided by specialized consulting or trade organizations that produce either reports for sale or publish annual rankings and reviews [see e.g., Trunick (2006)]. This type of data allows the corporation to compare or benchmark their own logistics operations and costs against their competitors and provides them with aggregate data on opportunities for business expansion. This date is very useful in a distribution design project for a new product, new customer group, or new geographical area. However, this data may be very expensive to acquire.

The last class of data sources is provided by governmental organizations and provides data on the overall status and characteristics of the economy and the population. In the United States, the Department of Commerce and the Census collect large amounts of data and provide statistical summaries free of charge or for a modest price.

Validation and reconciliation of data may expose significant incompatibilities and inconsistencies between various stakeholder organizations in the corporation. The process of assembling a single data set on which the design will be based and on which all stakeholder groups agree, is time-consuming and expensive. It is not unusual that 60% to 80% of the design project cost and duration is spent on collecting, validating, and aggregating the data. A single point of authority and funding is required to bring the data collection phase to a successful completion.

It is crucial for the success of the designed distribution system to realize during the design that the distribution system will be constructed and implemented in the near future, while it is intended to operate and serve for an extended period into the far future. Virtually all the data used in the design project are based on forecasts of economic, commercial, industrial, and population parameters. The error ratios of these forecasts may easily be thousands of percentages. Corporations and design engineers are under enormous pressure to design a system that will operate with minimum cost. However, the resulting designs often are lacking in flexibility and robustness. Selecting the trade-off between efficiency and robustness is typically done by the senior management of the corporation. However, providing the decision-makers with performance metrics for the various designs is the task of the engineering design group. Failing to incorporate the inherent uncertainty of the data in those evaluations may expose the design engineers to liability.

13.2.2   Engineering Design Principles

Three well-established principles are essential for the successful completion of a design project for a distribution system: (*i*) data synthesis and validation, (*ii*) successive model refinement, and (*iii*) sensitivity and risk analysis reporting. The essence of the first principle is captured by the popular acronym GIGO, which stands for "garbage in, garbage out." The distribution system design based on faulty data will not satisfy the design requirements regardless of the sophistication and validity of the design model. The essence of the second principle is captured by the popular acronym KISS, which stands for "keep it simple and stupid." The distribution system design generated by an integrated and comprehensive model is nearly impossible to validate, unless simpler models can be used. For example, in prior work, a model with more than 1.2 million variables was executed repeatedly to select one preferred configuration of the distribution system [see Santoso et al. (2005)]. A completely separate program was created to check the feasibility and cost of the generated designs. To my knowledge, a similar catchy acronym does not exist for the third principle of analysis that explicitly incorporates data uncertainty.

These three principles are further explored in detail in the following sections. Distribution network design also is discussed in several books on logistics and supply chains such as Simchi-Levy et al. (2003), Shapiro (2001), Ballou (2004), Wood et al. (1999), Stadtler and Kilger (2003), and Robeson and Copacino (1994).

13.3   Data Analysis and Synthesis

The data analysis and synthesis phase of the design project will be highly dependent on the individual project. The overall goal of this phase is to create a data set that contains valid and agreed-upon data for all the major components in the design project. For many objects in the data set there may be only a single data value, for example, the longitude and latitude coordinates of city. For other objects, the data can only be described by statistical distributions and their characteristics in function of possible scenarios. For example, the demand of a particular customer area for a particular product may be stored as its statistical distribution type, mean, and standard deviation for the worst-case, best-guess, and best-case scenarios.

13.3.1   Logistics Data Components

All the data for a distribution system design project is typically stored in a single database. Using a database allows the use of database validation tools and consistency checks. The data is organized in function of objects and their characteristics. Similar objects are collected in classes. The most important objects in a distribution system design project and some of their characteristics are described in the following.

13.3.1.1   Time Periods

Planning and design of logistics systems occurs at the strategic, tactical, and operational level. The different planning levels are distinguished by their duration. The various time period(s) are fundamental components in the logistics planning. If only a single time period exists, the planning or model is said to be static. If there exist multiple periods the model is said to be dynamic. For a strategic planning project such as the design of a distribution system often there are five periods of one year, corresponding to a five-year strategic plan. For a tactical planning model the periods are often months, quarters, or semesters. If the system is highly seasonal, the strategic design may be based on five cycles of seasons.

13.3.1.2   Geographical Locations

Logistics components exist at a particular location in a geographical or spatial area. Typically, the geographical areas become larger in correspondence to longer planning periods. For a strategic model, the areas may be countries or states in the United States. If there exist only a single country the system is said to be domestic, if more than one country exists the system is said to be global.

The combination of a country and a yearly period is used very often to capture the financial performance of a logistics system. The combination typically has financial characteristics such as budget limitations, taxation, depreciation, total system cost, and net cash flow.

13.3.1.3   Products

The material being managed, stored, transformed, or transported is called a product. An equivalent term is commodity. It should be noted that the term material is here applied very loosely and applies to discrete, fluid, and gaseous materials, livestock, and even extends to people. If only a single material is defined, the system is said to be single commodity. If multiple materials are defined the model is said to be multi-commodity.

It is very important to determine the type of material being considered. A first-level classification is into people, livestock, and products. The products are then further classified as commodity, standard, or specialty. Different types of products will have different service level requirements, which in turn dictate the overall structure of the distribution system. A product is said to be a commodity if there are no distinguishable characteristics between quantities of the same product manufactured by different producers. Examples of commodities are low-fat milk, gasoline, office paper, and poly-ethylene. Consumers acquire products solely on the basis of price and logistics factors such as availability and convenience.

A product is said to be a standard product if there exist comparable and competing products from different manufacturers. However, the products of different producers may have differences in functionality and quality. Examples are cars, personal computers, and forklift trucks. Consumers make acquisitions based on trade-offs between functionality, value, price, and logistics factors. A product is said to be a specialty or custom product if it is produced to the exact and unique specifications of the customer. Examples are machines, printing presses, and conveyor networks. The product is described by a technical specification and the supplier is selected by reputation, price, and logistics factors.

If one or more products are transformed into or extracted from another product, the products are said to have a bill of materials. The presence of bill of materials makes the design problem significantly more complicated. While value adding operations such as labeling are common in distribution centers, bill of materials are more common in supply chain design projects that also include the configuration of the manufacturing system.

13.3.1.4 Facilities

The locations in the logistics network where material can enter, leave, or be transformed are called facilities and are typically represented by the nodes of the logistics network. Suppliers are the source of materials and customers are the sink for materials. The internal operation of suppliers or customers is not considered to be relevant to the planning problem. The other facilities are called transformation facilities.

13.3.1.5 Customers

The customer facilities in the network have the fundamental characteristic that they are the final sink for materials. What happens to the material after it reaches the customer is not considered relevant to the planning problem. The customer facilities can be different from the end customers that use the product, such as the single distribution center for the product in a country, the dealer, or the retailer.

For every combination of products, periods, and customers there may exist a customer demand. A demand has a pattern, be it constant, with linear trend, or seasonal. Service level requirements are a complicating characteristic of customers in logistics planning. Prominent service level requirements are single sourcing, minimum fill rate, maximum lead time, and maximum distance to the serving distribution center. The single sourcing service requirement requires that all goods of single product group or manufacturer are delivered in a single shipment to the customer. Single sourcing makes it easier to check the accuracy of the delivery versus the customer order and it reduces the number of carriers at the customer facility where loading and unloading space often is at a premium. As a consequence single sourcing is a very common requirement. A customer may have a required fill rate, which is the minimum acceptable fraction of goods in the customer order that are delivered from on-hand inventory at the immediate distribution center for this customer. Finally, there may be a limit on the lead time between order and delivery based on competitive pressures. This limits in turn the maximum distance between a customer and the distribution center that services this customer. But this maximum distance depends on the selected transportation mode.

13.3.1.6 Suppliers

The supplier facilities in the network have the fundamental characteristic that they are the original source of the materials. What happens to the material before it reaches the supplier and inside the supplier facility is not considered relevant to the planning problem. The supplier facilities can be different from the raw material suppliers that produce the product, such as the single distribution center for the product in a country. For every combination of supplier facility, product, and time period there may exist an available supply.

Quantity discounts are one of the complicating characteristics of suppliers in logistics planning. A supplier may sell a product at lower price if this product is purchased in larger quantities during the corresponding period. This leads to concave (nonlinear) cost curves in function of the quantity purchased.

13.3.1.7 Transformation Facilities

The transformation facilities in the network have the fundamental characteristic that they have incoming and outgoing material flow and that there exists conservation of flow over space (transportation) and time (inventory) in the facility. Major examples of transformation facilities are manufacturing and distribution facilities, where the latter are also denoted as warehouses.

For every combination of transformation facility, time period, and product there may exist incoming flow, outgoing flow, inventory, consumption of component flow, and creation of assembly flow. All of these are collectively known as the production and inventory flows. A facility may have individual limits on each of these flows.

Transformation facilities have two types of subcomponents: machines and resources. Machines represent major transformation equipment such as bottling lines, assembly lines, process lines, and are more common in supply chain design projects. A resource is multi-product capacity limitation. Typical examples of resources are machine hours, labor hours, and material handling hours.

13.3.1.8 Transportation Channels

Transportation channels, or channels for short, are transportation resources that connect the various facilities in the logistics system. Examples are over-the-road trucks operating in either full truck load (TL) or less-than-truck-load (LTL) mode, ocean-going and inland ships, and railroad trains.

For every combination of transportation channel, time period, and product there may exist a transported flow. A channel may have individual limits on each of these flows. A major characteristic of a channel is its conservation of flow, that is, the amount of flow by period and by product entering the channel at the origin facility equals the amount of flow exiting the channel at the destination facility. A second conservation of flow relates channel flows to facility throughput flow and storage. The sum of all incoming flow plus the inventory from the previous period equals the sum of all outgoing flow plus the inventory to the next period. The channels represent material flow in space, while the inventory arcs represent material flow in time. Note that such period-to-period inventory is extremely rare in strategic logistics systems and should only be included for highly seasonal systems that use seasons as strategic time periods.

A channel has two types of subcomponents: carriers and resources. A carrier is an individual moving container in the channel. The move from origin to destination facility has a fixed cost, regardless of the capacity utilization of the carrier, that is, the cost is by carriers and not by the quantities of material moved on the carrier. Examples are a truck, intermodal container, or ship. A carrier may have individual capacities for each individual product or multi-product weight or volume capacities. A resource is multi-product capacity limitation. Examples of resources are cubic feet (meters) for volume, tons for weight, or pallets. Truck transportation may be modeled as a carrier if a small number of trucks are moved and cost is per truck movement, or it may be modeled as a resource if the cost is per product quantity and a large or fractional number of trucks are allowed.

There exist several complicating characteristics for modeling transportation channels. The first one is the presence of economies of scale for transportation costs. The second one is the requirement that an integer number of carriers have to be used. Since typically a very large number of channels exist, this creates a large number of integer variables. Less common is the third complicating factor, which requires a minimum number of carriers or a minimum amount of flow if the channel is to be used.

All of the logistics components described so far have characteristics. For example, most of the facilities, channels, machines and their combinations with products and periods have a cost characteristic. Sales have a revenue characteristic. The financial quantities achieved in a particular country and during a particular period are another example of characteristics.

13.3.1.9 Scenarios

So far, all the logistics components described were physical entities in the logistics system. A scenario is a component used in the characterization of uncertainty.

Many of the parameters used in the planning of logistics systems are not known with certainty but rather have a probability distribution and are said to be stochastic. For example, demand for a particular product, during a period by a particular customer may be approximated by a normal distribution with certain mean and standard deviation. In a typical logistics planning problem there may be thousands of stochastic parameters. The combination of a single realization or sample of each stochastic parameter with all the deterministic parameters is called a scenario. Each scenario has a major characteristic, which is its probability of occurring. However, this probability may not be known or even not be computable.

A very large amount of data for the logistics objects defined earlier and their attributes has to be generated, collected, and validated. To reduce the data acquisition and management burden, to reduce the forecast errors, and to provide better insight, the logistics objects have to be aggregated. Customers are aggregated by customer class and then by geographical proximity. Products are aggregated by physical characteristics and demand patterns. A very small number of transportation modes are considered, such as TL and LTL. Often sufficient accuracy can be achieved with a few hundred customers and a few tens of products.

13.4   Distribution System Design Models

Once the data have been collected, validated, and aggregated, the next task is to determine high-quality configurations for the distribution system. Because of the large size of the data set and the heterogeneous nature of the requirements and objectives, an objective engineering design has to use design models.

A meta-model is an explicit model of the components and rules required to build specific models within a domain of interest. A logistics planning meta-model can be considered as model template for the domain of activity planning for logistics systems. The planning models for the design of distribution centers belong to the logistics planning meta-models. They have the following general structure:

Decide on	1) Transportation activities, resources, and infrastructure
	2) Inventory levels, resources, and infrastructure
	3) Transformation activities, resources, and infrastructure
	4) Information technology systems
Objective	1) Minimize the risk-adjusted total system cost over the planning horizon
Subject to	1) Capacity constraints such as demand, infrastructure, budget, implementation time
	2) Service level constraints such as fraction of demand satisfied, fill rate, cycle times, and response times
	3) Conservation of flow constraints in space, over time, and including bill of materials
	4) Additional extraneous constraints, which are often mandated by corporate policy
	5) Equations for the calculation of intermediate variables such as the safety inventory, achieved fill rate, and other performance measures

The two major types of design decisions are (*i*) the status of a particular facility and relationships or allocations during a specific planning period and (*ii*) the product flows and storage quantities (inventory) in the distribution system during a planning period. For example, if the binary variable y_{klt} equals one, this may indicate that a facility of type l is established and functioning at location or site k during time period t. Similarly, the binary variable w_{pklts} indicates if product p is assigned or allocated to a distribution center at site k and of type l during period t in scenario s or not. The continuous variable x_{ijmpts} indicates the product flow of product p from facility i to facility j using transportation channel m during time period t in scenario s.

Distribution systems are typically designed to minimize the time discounted total system cost over the planning horizon, denoted by *NPVTSC*. Often, the system is designed with an expected value objective and then evaluated with respect to more complicated objectives through simulation. Let cdf_t denote the capital discount factor for a period t and $\mathbf{E}[\cdot]$ denote the expectation operator. Then the objective of the strategic distribution system design is $\min\{E[NPVTSC]\}$.

If the capital discount factor remains constant over the planning horizon, the expression for the *NPVTSC* simplifies to

$$NPVTSC = \sum_{t=1}^{T} TSC_t \cdot (1 + cdf)^{-t} = \sum_{t} \left(\sum_{c \in C} \frac{TSC_{ct}}{er_{ct}} \right) \cdot (1 + cdf)^{-t} \tag{13.1}$$

TSC_{ct} is the total system cost for a country in the currency of the country during a particular time period (year) and it is based on all the facilities in operation or being established in that country during that time period. er_{ct} is the exchange rate for the currency of country c expressed in the currency of the home country. If only one country is involved in the distribution system, the first expression can be used to calculate the *NPVTSC*.

The strategic design of distribution and supply chains is based on the application of a sequence of models with increasing realism and complexity. The results of using the previous model in the sequence are used to validate the current model. The first two models are the K-median and Location-Allocation (LA) models. The next model is called the Warehouse Location Problem (WLP). A more comprehensive variant of the WLP was first published by Geoffrion and Graves (1974). Finally, a number of comprehensive models for single country and global logistics have been developed. See, for example, Dogan and Goetschalckx (1999), Vidal and Goetschalckx (2001), and Santoso et al. (2005).

13.4.1 K-Median Model

The K-median model is used to determine the number and location of distribution centers and the customer allocations with respect to a set of customers in order to minimize the total system cost.

13.4.1.1 K-Median Formulation

$$Min \sum_{i=1}^{N} \sum_{j=1}^{N} c_{ij} x_{ij} \tag{13.2}$$

$$s.t. \sum_{j=1}^{N} x_{ij} = 1 \quad i = 1 \dots N \tag{13.3}$$

$$\sum_{j=1}^{N} y_j \le K \quad j = 1 \dots N \tag{13.4}$$

$$x_{ij} \le y_j \quad i, j = 1 \dots N \tag{13.5}$$

$$x_{ij} \ge 0, y_j \in \{0,1\} \quad i, j = 1 \dots N \tag{13.6}$$

Where
y_j 1 if a distribution center is established at customer location j, zero otherwise.
x_{ij} 1 if customer i is serviced from the distribution center at location j.
c_{ij} Cost to service customer i completely from the center at location j.
K Maximum number of distribution centers to be established.

The objective 13.2 is to minimize the sum of the costs to service the customers (median problem). There are two types of decisions, the first one selects which distribution centers will be established, and the second one assigns customer to the centers. Constraint 13.3 ensures that each customer has to be served from a center. Constraint 13.4 ensures that the number of distribution centers is no larger than the maximum number (K). Constraint 13.5 allows customer i only to be served from location j if the center at location j is established. The distribution centers are assumed to have no capacity restrictions. It is also assumed that the set of customers covers all of the distribution area. The status of a distribution center is a binary variable since a center cannot be fractionally open and thus this problem has to be solved with a mixed-integer programming solver. It should be noted that the customer assignment variable x is modeled as a nonzero continuous variable without upper bound, but the optimal solution of this uncapacitated problem will yield automatically zero and one values for the assignments barring alternative optimal solutions. If a fractional optimal solution is generated, the assignment variables can also be declared as binary variables with the lowest branching priority in the mixed integer programming solver. The K-median problem has been studied extensively [see e.g., Francis et al. (1992)], and can be solved reasonably efficiently for realistic problem sizes. Observe that an upper bound on the number of established centers is required; otherwise the optimal solution would be to open a center at every customer location. Often determining this upper bound is part of the design project. This can be achieved by running the model for a series of acceptable upper bounds and to compare the resulting configurations. The formulation has the advantage that the assignment costs c are completely under the designer's control. They can be proportional to the customer demand size, transportation distance between center and customer, the product of the two, or any problem-specific value. The formulation has the disadvantage that no site-specific costs can be incorporated. The model is highly aggregate and usually only a single time period is used.

To yield reasonable configurations, the formulation assumes that a customer exists in every section of the design area, so that a center could be established there. If no such coverage of the design area exists, then the following LA formulation is more appropriate.

13.4.2 Location-Allocation Model

The LA model considers manufacturing facilities (plants), customers, and distribution centers (depots). It determines the location of the distribution centers and the allocation of customers to distribution centers based on transportation costs only. The distribution centers can be capacitated and flows between the distribution centers are allowed.

The algorithm starts with an initial solution in which the initial location of the distribution centers is specified. This initial location can be random, specified by the user, or the result of another algorithm. Based on this initial location, the network flow algorithm computes the transportation distances d and then assigns each customer to a distribution center with sufficient capacity by solving the following network flow problem

13.4.2.1 LA Formulation (Allocation Phase)

$$\min \sum_{i=1}^{M} \sum_{j=1}^{N} c_{ij} d_{ij} w_{ij} + \sum_{j=1}^{N} \sum_{k=1}^{L} c_{jk} d_{jk} v_{jk} \tag{13.7}$$

$$s.t. \ \sum_{j=1}^{N} v_{jk} = dem_k \qquad k = 1 \dots L \tag{13.8}$$

$$\sum_{j=1}^{N} w_{ij} \leq cap_i \qquad i = 1 \dots M \tag{13.9}$$

$$\sum_{k=1}^{L} v_{jk} \leq cap_j \qquad j = 1 \dots N \qquad (13.10)$$

$$\sum_{i=1}^{M} w_{ij} - \sum_{k=1}^{N} v_{jk} = 0 \qquad j = 1 \dots N \qquad (13.11)$$

$$w_{ij} \geq 0, v_{jk} \geq 0 \qquad (13.12)$$

Where

w_{ij}, v_{jk} The product flows from plant i to distribution center j and from distribution center j to customer k, respectively.

c_{ij}, c_{jk} The transportation costs per unit flow and per unit distance from plant i to distribution center j and from distribution center j to customer k, respectively.

d_{ij}, d_{jk} The inter-facility transportation distances from plant i to distribution center j and from distribution center j to customer k, respectively.

cap_i, cap_j Throughput capacity of plant i and distribution center j, respectively.

dem_k Demand of customer k.

Constraint 13.8 ensures that each customer receives its full demand. Constraints 13.9 and 13.10 ensure that the capacity of the plants and distribution centers is observed. Constraint 13.11 ensures that the total inflow into a distribution center is equal to the total outflow, that is, that conservation of flow is maintained. This network flow formulation can be very efficiently solved by a linear programming solver for all realistic problem sizes. The result of the allocation phase is the assignment of customers to distribution centers as given by the flow variables.

After all the customers have been allocated to a distribution center with available capacity, a second sub-algorithm locates the distribution centers so that the sum of the weighted distances between each source and sink facility is minimized for the given flows. This problem is formulated as a continuous, multiple facility weighted Euclidean minisum location problem.

13.4.2.2 LA Formulation (Location Phase)

$$\text{Min } f(x, y) = \sum_{i=1}^{M} \sum_{j=1}^{N} c_{ij} w_{ij} \sqrt{(x_j - a_i)^2 + (y_j - b_i)^2}$$

$$+ \sum_{j=1}^{N} \sum_{k=1}^{L} c_{jk} v_{jk} \sqrt{(x_j - a_k)^2 + (y_j - b_k)^2} \qquad (13.13)$$

Where

$(a_i, b_i), (a_k, b_k)$ the (known) Cartesian location coordinates of customers i and plants k.

(x_j, y_j) the location coordinate variables of distribution center j.

$d_{ij} = \sqrt{(x_j - a_i)^2 + (y_j - b_i)^2}$ Euclidean distance norm.

The solution to this unconstrained continuous optimization problem can be found by setting the partial derivatives equal to zero and solving the resulting equations iteratively [see Francis et al. (1992)] for further details]. However, an approximate solution can be obtained by computing the center of gravity solution.

$$x_j = \frac{\sum_{i=1}^{M} c_{ij} w_{ij} a_i + \sum_{k=1}^{L} c_{jk} v_{jk} a_k}{\sum_{i=1}^{M} c_{ij} w_{ij} + \sum_{k=1}^{L} c_{jk} v_{jk}} \tag{13.14}$$

$$y_j = \frac{\sum_{i=1}^{M} c_{ij} w_{ij} b_i + \sum_{k=1}^{L} c_{jk} v_{jk} b_k}{\sum_{i=1}^{M} c_{ij} w_{ij} + \sum_{k=1}^{L} c_{jk} v_{jk}} \tag{13.15}$$

The solution provided by 13.14 and 13.15 is optimal with respect to the squared Euclidean distance norm, and it provides sufficient accuracy at this level of a strategic design project for the Euclidean distance norm. It should be noted that the iterative solution algorithm based on the partial differential equations is usually started with this center of gravity solution as starting point. The location phase provides new locations for the distribution centers. Based on these new locations the distances between the various facilities can be updated.

The algorithm iteratively cycles through its allocation and location phase until the network flows remain the same between subsequent iterations. The obtained solution is dependent on the initial starting locations for the distribution centers, so several different starting configurations should be used and the best final solution retained.

This model is again highly aggregate and usually only a single time period is modeled. The model has the advantage that it can locate distribution centers in locations where no customers are present. Capacities of the plants and distribution centers can be incorporated. The model assumes that distribution centers can be located anywhere within the boundaries of the feasible domain, which may not be feasible because of geographical infeasible regions such as oceans, lakes, and mountain ranges. The model has the disadvantage that no site-dependent costs can be incorporated. The solutions are only approximate and indicate a general area for the location of the distribution centers. This model is called a site-generating model since it creates the solution locations.

13.4.3 Warehouse Location Problem

In the WLP model the distribution centers can only be established in a finite number of given locations. The model is called a site-selection model since it selects center locations from a list of candidate locations. Because the candidate locations are known in advance, site-dependent costs can now be included in the model. The number of warehouses to establish is based on the cost trade-off between fixed facility costs and variable transportation costs. Establishing an additional distribution center yields higher fixed facility costs and lower variable transportation costs.

13.4.3.1 Warehouse Location Problem Formulation

$$\min \ z = \sum_{j=1}^{N} \left(f_j y_j + \sum_{i=1}^{M} c_{ij} x_{ij} \right) \tag{13.16}$$

$$s.t. \ \sum_{j=1}^{N} x_{ij} = 1 \qquad i = 1 \dots M \tag{13.17}$$

$$\sum_{i=1}^{M} x_{ij} - My_j \le 0 \qquad i = 1 \dots M, \ j = 1 \dots N \tag{13.18}$$

$$y_j \in \{0,1\}, x_{ij} \ge 0 \qquad j = 1 \dots N, \ i = 1 \dots M \tag{13.19}$$

Where in addition to the definitions for the K-median problem the following parameter is defined:

f_j fixed cost for establishing a distribution center at candidate location j.

The objective 13.16 is to minimize the sum of the costs of the facilities and the costs to service the customers. There are two types of decisions, the first one selects which distribution centers will be established, and the second one assigns customer to the centers. Constraint 13.17 ensures that each customer has to be served from a center. Constraint 13.18 allows customer i only to be served from location j if the center at location j is established. The distribution centers are assumed to have no capacity restrictions.

An alternative formulation for the WLP replaces Constraint 13.18 with a larger number of the following constraints, where each constraint has fewer variables.

$$x_{ij} - y_j \le 0 \qquad j = 1 \dots N, i = 1 \dots M \tag{13.20}$$

Historically, this has yielded faster solution times, but contemporary mixed-integer programming solvers recognize the structure of constraints of type 13.18 and have optimized their solution algorithms so that the differences in solution times have become negligible.

This formulation has the advantage that site-dependent costs can be incorporated. But the formulation only makes the trade-off between facility costs and the transportation costs. The throughput capacities are not incorporated.

Based on a currently existing configuration or a baseline design configuration, it is possible to evaluate the relative savings of establishing a new distribution center based on its site-relative cost $\rho_j (\mathbf{U})$.

$$\rho_j(\mathbf{U}) = f_j + \sum_{i=1}^{M} \min\{0, c_{ij} - u_i\} \tag{13.21}$$

Where
u_i Current cost for servicing customer i.
$\rho_j (\mathbf{U})$ Site-relative cost for opening warehouse j based on the current customer service cost u_i.

Note that both u_i and c_{ij} are the cost for servicing the total demand of a customer. Candidate sites with a large negative cost, which is equivalent to large positive savings, are highly desirable sites for establishing a distribution center. Candidate sites with a large positive cost are undesirable for a new distribution center. The current cost for servicing customer i is the sum of its transportation cost and its allocated share of the fixed cost of the center that currently services it. A common cost allocation is to make the cost shares proportional to the annual demand of the customers serviced by the center. The site-relative cost provides an efficient mechanism to rank potential candidate locations, without having to resolve the base WLP. Further information can be found in Francis et al. (1992).

13.4.4 Geoffrion and Graves Distribution System Design Model

The K-median and the WLP models ignore the capacity restrictions of distribution centers. All of the previous models considered only a single product and this ignore the single-sourcing customer service constraints. Geoffrion and Graves (1974) developed a model that incorporated both capacity

Arc-based   Path-based

FIGURE 13.2   Illustration of arc and path-based transportation flows.

and single-sourcing constraints. One of its fundamental characteristics was that the flow was modeled along a complete path from supplier, through the distribution center, and to the customer by a single flow variable. Formulations of that type are called path-based. If a flow variable exists for each transportation move, then the formulations are said to be arc-based. The difference between path-based and arc-based formulations is illustrated in Figure 13.2. Path-based formulations have many more variables than arc-based formulation for the equivalent system. On the other hand, arc-based formulations have to include the conservation of flow equations for each commodity and each intermediate node of the logistics network.

13.4.5   Geoffrion and Graves Formulation

$$\min \sum_{ijkp} c_{ijkp} x_{ijkp} + \sum_{j} \left(f_j z_j + h_j \sum_{kp} dem_{kp} y_{jk} \right) \tag{13.22}$$

$$s.t. \sum_{jk} x_{ijkp} \le cap_{ip} \qquad \forall ip \tag{13.23}$$

$$\sum_{i} x_{ijkp} = dem_{kp} y_{jk} \qquad \forall jkp \tag{13.24}$$

$$\sum_{j} y_{jk} = 1 \qquad \forall k \tag{13.25}$$

$$TL_j z_j \le \sum_{pk} dem_{kp} y_{jk} \le TU_j z_j \qquad \forall j \tag{13.26}$$

$$x_{ijkp} \ge 0, \; y_{jk} \in \{0,1\}, \; z_j \in \{0,1\} \tag{13.27}$$

Where the following notation is used:

c_{ijkp}	Unit transportation cost of servicing customer k from supplier i through depot j for product p.
f_j	Fixed cost for establishing a distribution center at candidate location j.
h_j	Unit handling cost for distribution center at candidate location j.
cap_{ip}	Supply availability (capacity) of product p at supplier i.
dem_{kp}	Demand for product p by customer k.
TL_j, TU_j	Lower and upper bounds on the flow throughput of distribution center at candidate location j.
z_j	Status variable for distribution center at candidate location j, equal to 1 if it is established, zero otherwise.
y_{jk}	Assignment variable of customer k to distribution center at candidate location j, equal to 1 if the customer is single-sourced from the center, zero otherwise.
x_{ijkp}	Amount of flow shipped by supplier i through distribution center j to customer k of product p.

The objective 13.22 minimizes the sum of the transportation cost, fixed facility costs, and distribution center handling costs. Constraint 13.23 ensures sufficient product availability at the suppliers. Constraint 13.24 ensures that the customer demand is met for each product and ensures conservation of flow for each product at the distribution centers. Constraint 13.25 forces every customer to be assigned to a distribution center. Constraint 13.26 ensures that the flow through the distribution centers does not exceed the throughput capacity and that, if a distribution center is established, it handles a minimum amount of flow.

The above formulation captured many of the real-world constraints and objectives of distribution system design. The formulation can be solved with an efficient but complex solution algorithm based on Benders' decomposition that requires significant experience in mathematical programming and computer programming. It allows the solution of real-world problem instances with limited computational resources. At the current time, very sophisticated commercially available mixed-integer programming solvers and powerful computer processors have made use of the Benders' decomposition algorithm unnecessary except for all of the largest problem instances. Using a path-based or arc-based formulation for distribution systems design has become largely a matter of designer preference.

The Benders' decomposition solution algorithm is still used when the designer wants to incorporate data uncertainty explicitly in the model through the use of scenarios. Instead of having a single demand value per customer and product, a number of demand scenarios are included in the model. Common choices for scenarios are best-guess (the most likely scenario), best-case, worst-case, and so on. In the formulation stated earlier a scenario is represented by an additional subscript s for all parameters and variables except the facility status variables z. The objective for the scenario-based model becomes

$$\min \sum_s p_s \left[\sum_{ijkp} c_{ijkps} x_{ijkps} + \sum_j \left(f_{js} z_j + h_{js} \sum_{kp} dem_{kps} y_{jks} \right) \right] \tag{13.28}$$

with

p_s Probability of scenario s.

It is often very difficult to determine the scenario probabilities accurately. The values may be based on imprecise managerial judgment. From the modeling point of view, the scenario probabilities have to satisfy the following constraint.

$$\sum_s p_s = 1 \tag{13.29}$$

13.5   Sensitivity and Risk Analysis

In addition to the scenarios discussed earlier, the distribution system configuration should be further evaluated to measure its response to small variations in the parameter values. The data sets for this evaluation are created by random sampling from the probability distribution for each of the data parameters. The material flows are then determined for the given distribution system configuration and a particular sampled data set by a minimum cost network flow optimization. The formulation for the network flow problem is identical to the Geoffrion and Graves model, but with the status and assignment variables fixed by the given configuration.

Based on the sensitivity analysis discussed, a particular configuration of the distribution system has a certain expected value and standard deviation of the *NPVTSC*. A classical risk analysis graph can be plotted where each candidate configuration is placed according to two dimensions: one axis representing the expected value and the other axis the variability or risk measure. Often the corporation does not

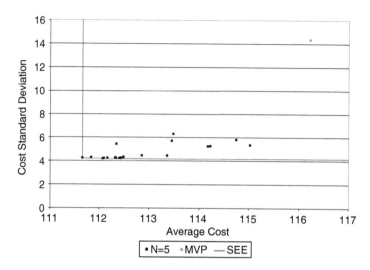

FIGURE 13.3   Risk analysis graph for distribution system design.

know explicitly its risk preferences and is interested in identifying several alternative high-quality distribution system configurations with various risk performances. The efficiency frontier is the collection of distribution system configurations that are not Pareto-dominated by any other configuration. For any efficient or non Pareto-dominated configuration, no configuration exists that has simultaneously a smaller expected value and a smaller variability value. For a given set or sample of distribution system configurations that are located in the risk analysis graph, the sample efficiency envelope (SEE) of those configurations can be determined by connecting efficient configurations. This SEE is an approximation of the efficiency frontier. The risk analysis graph for an industrial case with the standard deviation chosen as risk measure and including the SEE is shown in Figure 13.3 [see Santoso et al. (2005)]. Note that the best distribution system configuration for the most likely value of the parameters (MVP scenario) is indicated by the square. The performance of the MVP in this example illustrates the often-observed fact that the best (optimal) distribution system configuration for the best-guess value of the parameters may have a performance far away from the efficiency frontier. The risk analysis graph is a very powerful communications tool with corporate executives since it displays in a concise manner the expected yield and risk of several possible candidates. It is the function of the design engineer to perform all the calculations, optimizations, and simulations that are then synthesized into this graph. The preferred distribution system configuration is then selected by senior management from the configurations close to the SEE.

13.6   Distribution Design Case

The following distribution design case is based on a real-world design project; however, the company name and some of the details have been obscured or changed to protect confidentiality. MedSup, a subsidiary of a larger corporation, delivers medical supplies to primary care providers in the continental United States, which include general and specialized physician offices, small surgery centers, and specialty clinics. Hospitals and large surgery centers as well as home care and long-term care facilities are not part of the customer base since they are served by other subsidiaries. MedSup has a current distribution system with 13 distribution centers, four of which are exclusively used for the primary care customers, while the others are shared with the other subsidiaries that deliver to the other customer classes. Competitive pressures have established next-day delivery as the required customer service standard. The system has to handle a large number of relatively small customer orders with a very short turn-around time. MedSup anticipates an increased demand from primary care providers for their products based on the aging of the general population with its corresponding increase in health care requirements. MedSup also expects that the current demographic relocations will become even more pronounced in the future. The objective was to design a distribution system that maintains the customer service standard at the lowest possible cost for the current and future customer base. Specific questions to be answered by the design project are the number and locations of new distribution centers and the identification of any current distribution centers to be closed.

The design utilized two primary types of data. The first data is the current and future population distribution in the continental United States. This data was obtained from Microsoft MapPoint, which contains U.S. Census population data from 1998, 2000, and 2002. The second data set is the geographical distribution of primary care practices in the United States, where the practices are categorized by medical specialty. In addition, the current configuration of the distribution system is provided.

MedSup decided to focus in the first phase on the general configuration of the distribution system, since no site-specific data for distribution center establishment and operation were immediately available. To support this high-level view, it was decided to aggregate customers by 3-digit zip codes (ZIP3). There are 878 ZIP3 zones in the continental United States. The K-median model was used as objective to the sum of the weighted distances. The K-median model was used rather than the LA model because the ZIP3 zones sufficiently covered the continental United States and because of the reduced programming requirements for the solution algorithm.

The distances were computed with the great circle distance norm between central locations in each ZIP3 zone. The great circle distance norm was used because the location data was available as longitude and latitude coordinates. The great circle distance norm computes the distance along a great circle on the surface of the earth between two points with latitude and longitude coordinates (lat_i, lon_i) and (lat_j, lon_j) with the following formula, where R denotes the world radius. The earth radius is approximately 6366.2 km or 3955.8 miles.

$$d_{ij}^{GC} = R \cdot \arccos(\cos(lat_i)\cos(lat_j)\cos(lon_i - lon_j) + \sin(lat_i)\sin(lat_j)) \qquad (13.30)$$

The exact computation method for the weights for the K-median formulation is case specific and different formulas should be used during the sensitivity analysis. The exact weight formula for this case is proprietary. The weight is proportional to the population, the number of primary care practices, plus an additional weight for specific types of practices in the ZIP3. Parallel to the modeling effort, the marketing and operations organizations in MedSup were interviewed to identify possible locations for new centers. The K-median model was first solved without and then later with the current distribution configuration as constraints. The model was solved with a commercial mixed-integer programming solver and required about 30 min of computation time per run. The model contained 878 binary variables, 770,004 continuous variables and terms in the objective function, and 771,763 constraints. The maximum number of distribution centers (K) varied from 12 through 16. When K was systematically increased, the majority of distribution centers remained in the same ZIP3 zones and the splits of customer zones appeared logically to MedSup. The system configurations were compared with the system configurations determined by the marketing and operations departments. The configurations were nearly identical, if center locations that were in different ZIP3 zones in the two configurations were considered identical if they were located in the same metropolitan area. Finally, candidate locations were ranked by how many times they appeared in the model solution, by preference of management, by population growth, and by practice count. Sensitivity analysis was performed on the relative weights of those factors. Three metropolitan areas (Houston, TX; Chicago, IL; and Oakland, CA) ranked consistently first through third, but no single location was preferred for all values of the weight factors. The objective function value decreased from 8.3% to 4.5% if a distribution center in those three locations was established in addition to the 13 currently existing centers. The next phase of the design project will require the collection of detailed site-specific cost and capacity data for those three locations and a more comprehensive model such as Geoffrion and Graves.

13.7   Conclusions

This design project illustrated again the following observations about strategic distribution system design. First, without modeling-based decision support, the configuration of a distribution system is essentially reduced to intuition or guesswork. Second, the concept of a single "optimal" distribution system configuration generated by deterministic optimization is an illusion. Third, through careful modeling-based sensitivity analysis a limited number of high-quality candidate configurations can be identified and submitted for final selection.

Several major factors such as cycle and safety inventory and taxation have not been discussed so far. More comprehensive models that incorporate these factors have been developed, but such models must be used with extreme care and typically have a steep learning curve. Their use can be only recommended if the models will be used repeatedly.

Three phases are essential for the successful completion of a distribution system design project. During the first phase, data from a variety source is collected, validated, aggregated, and synthesized. This activity is time- and resource-consuming, but it provides the foundation on which the rest of the design project is based. In the second phase, a series of design models is formulated and solved. The models become increasingly comprehensive, require more sophisticated and computationally expensive algorithms, become more difficult to validate, and the results become more difficult to interpret. Validation and interpretation of the current model must be completed before the next level model can be used. In the third phase, sensitivity analysis is used extensively. The models are solved with a large variety of data values and the results are statistically analyzed. In the end, a limited number of high-quality configurations is identified and presented to the upper management for final selection.

Clearly, a strategic distribution design project is time- and resource-intensive activity. But a properly executed project can reduce the distribution costs by 5% to 10%.

References

Ballou, R. H., (2004). *Business Logistics Management*, 5th Edition, Pearson Education, Upper Saddle River, New Jersey.

Dogan, K. and M. Goetschalckx, (1999). A Primal Decomposition Method for the Integrated Design of Multi-Period Production-Distribution Systems, *IIE Transactions*, Vol. 31, No. 11, pp. 1027–1036.

Francis, R. L., L. F. McGinnis, and J. A. White, (1992). *Facility Layout and Location: An Analytical Approach*, 2nd Edition, Prentice-Hall, Englewood Cliffs, New Jersey .

Geoffrion, A. M. and G. W. Graves, (1974). Multicommodity Distribution System Design by Benders Decomposition, *Management Science*, Vol. 20, No. 5, pp. 822–844.

Robeson, J. and W. Copacino, (Eds.), (1994). *The Logistics Handbook*. Free Press, New York, New York.

Santoso, T., S. Ahmed, M. Goetschalckx, and A. Shapiro, (2005). A Stochastic Programming Approach to Designing Strategic Supply Chains Under Uncertainty, *European Journal of Operational Research*, Vol. 167, No. 1, pp. 96–115.

Shapiro, J. F., (2001). *Modeling the Supply Chain*, Duxbury Press, Pacific Grove, California.

Simchi-Levi, D., P. Kaminsky, and E. Simchi-Levi, (2003). *Designing and Managing the Supply Chain: Concepts, Strategies, and Case Studies*, 2nd Edition, McGraw-Hill, New York, New York.

Stadtler, H. and C. Kilger, (2003). *Supply Chain Management and Advanced Planning*, 3rd Edition, Springer, Heidelberg, Germany.

Trunick, P. A., (2006). Don't Oversimplify Site Selection, *LogisticsToday*, March 2006, pp. 28–30.

Vidal C. and M. Goetschalckx, (2001). A Global Supply Chain Model with Transfer Pricing and Transportation Cost Allocation, *European Journal of Operational Research*, Vol. 129, No. 1, pp. 134–158.

Wood, D. F., D. L. Wardlow, P. R. Murphy, J. Johnson, and J. C. Johnson, (1999). *Contemporary Logistics*, Seventh Edition, Prentice-Hall, Englewood Cliffs, New Jersey.

14

Transportation Systems Overview

Joseph Geunes
University of Florida

Kevin Taaffe
Clemson University

Transportation systems form a vital backbone of economic activity, enabling the movement of people and goods required for providing goods and services. Effective creation and management of transportation systems can provide a substantial competitive advantage for a firm in the private sector, and can drastically influence a nation's productivity and global competitiveness from a public-sector perspective. This chapter provides a foundation for understanding critical factors in efficient transportation system development, as well as the complexities that lead to challenging decision problems in transportation service delivery.

14.1 Introduction and Motivation

Transportation systems, broadly defined, encompass the collective infrastructure, equipment, and processes utilized in the movement of people and goods among different geographic locations. The relative economic importance of transportation systems is evidenced by the fact that between 1990 and 2001, the cost of transportation equipment, service, and infrastructure ranged between 10.2% and 10.9% of the United States Gross Domestic Product (GDP), with transportation's contribution to the GDP totaling more than $1 trillion per year since 1999 (in 2005 dollar value).* Passenger transportation expenditures in 1999 exceeded $936 billion, while freight expenditures topped $560 billion.† Transportation, therefore, accounts for a significant portion of the U.S. economy, and the same holds for the majority of industrially developed nations. This investment in transportation is a substantial factor in enabling the United States to lead the world in real GDP per capita.‡ In addition to the impacts of transportation systems on productivity, these systems also contribute to the quality of life of consumers in the form of leisure travel (tourism-related goods and services recently topped $1 trillion annually in the United States§).

The focus of logistics engineering in this domain is on identifying the most efficient methods for establishing and utilizing transportation infrastructure and equipment. The chapters in the following section of this handbook discuss methods for a variety of transportation planning decision contexts and problems. The intent of this chapter is to provide an overarching foundation for the scope of relevant issues in the study of transportation systems and to characterize the range of decision types in this field.

Within the transportation context, it is important to distinguish between the roles and functions of carriers and shippers. A carrier performs the transportation function and must therefore concern itself with issues such as managing and operating a transportation fleet and associated support equipment and facilities. A shipper, on the other hand, has a need to move a good from place to place, but does not perform the transportation function (except in cases where the shipper and carrier are the same, that is, a shipper maintains and manages an internal fleet of vehicles for goods transport). The shipper is therefore concerned with the cost, quality, responsiveness, and reliability of the transportation service (which is provided by a carrier or a set of carriers). This distinction will play an important role in characterizing the relevant issues an organization faces with respect to transportation systems.

The organization of this chapter is as follows: Section 14.2 begins by characterizing the important differences in transportation systems that cater to transporting people versus those that focus on moving freight. There we identify the factors that differentiate the challenges faced in designing and operating these distinct types of transportation systems. Transporters face the challenge of determining the most effective mode for moving a good, which we discuss in Section 14.3. Section 14.4 then considers the importance of transportation infrastructure in enabling productivity and competitiveness in a global economy. For transporters of both people and goods, forecasts of transportation demands drive transportation investment, as well as the ultimate utilization of the resulting transportation equipment and infrastructure. These factors in turn directly impact the return on transportation investment as well as the efficiency (and congestion, or associated loss of efficiency) of the transportation system. We consider the complexities involved in forecasting transportation demands in Section 14.5. Section 14.6 presents a case example highlighting the importance of transportation systems planning in practice, and Section 14.7 provides concluding remarks.

14.2   Moving People versus Moving Goods

When considering the movement of people (as opposed to goods), the distinction between carriers and shippers does not play an important role. In this context, we focus on transportation carriers, who

---

* *Source:* U.S. Department of Transportation, Bureau of Transportation Statistics: www.bts.gov.

† *Source:* Eno Transportation Foundation, Inc., Transportation in America, 2001 (Washington, DC, 2000).

‡ Data as of 2004. *Source:* U.S. Department of Labor Bureau of Labor Statistics Office of Productivity and Technology: http://www.bls.gov/fls/

§ *Source:* U.S. Bureau of Economic Analysis: www.bea.gov.

typically offer one mode of transport (e.g., an airline, rail company, bus company, etc.), and their concerns lie in providing an efficient (and profitable) means of moving these individuals within their system network. An exception to this would be regional mass transit systems, which can offer several modes for people to travel within an urban area. Hensher and Button (2000) and Hall (2002) characterize key issues in modeling transportation systems. While these references also offer an introduction into freight transportation [a subject which is covered in greater detail by, e.g., Friesz (2000) and Crainic (2002)], they provide a much more thorough treatment of passenger transportation (or movement of people).

In contrast, when considering the movement of goods or freight, carriers and shippers have differing and unique roles. Carriers and shippers engage in cooperative partnerships (to varying degrees, depending on the context), much like retailers and suppliers in a supply chain setting. Each must remain competitive within its own line of business, yet they often depend on each other to achieve their desired levels of performance. An exception to this would be those companies who own their transport fleet for moving their goods. Most freight transportation modeling in the operations literature adopts the viewpoint of the carrier, where the focus is on determining an appropriately designed system that can provide transport for a wide range of consumers or shippers.

Transportation problems have been studied for many decades. Applications in the airline industry, for example, have led to the introduction of a number of operations research techniques for solving various types of transportation problems, including schedule generation and fleet assignment [e.g., Lohatepanont and Barnhart (2004)], crew scheduling [e.g., Hoffman and Padberg (1993)], and yield management problems [e.g., McGill and van Ryzin (1999)]. The majority of this work focuses on systems that transport people or passengers. Barnhart and Talluri (1997) details an excellent introduction to this field. Barnhart et al. (2003) also provide a recent survey of operations research applications in the airline industry. Until recently, less attention has been paid to the cargo and freight side of the airline industry. As one would imagine, many similar issues exist, especially since the same fleet is used for transporting both passengers and cargo. Still, modeling air cargo decisions introduces new and different objectives and constraining factors. For freight operations that operate independent of any systems that move people, there are even more clear distinctions in the associated transport systems. We address these broader issues and design challenges in this section.

14.2.1 Differences and Similarities in Systems

The most clear distinction between transportation systems that move people versus freight would most likely be evident in regional mass transit systems. Here, a transportation system offers regularly scheduled operations with many intersecting routes, allowing people to easily connect to other routes in reaching their final destinations. The system provides a daily capacity (based on scheduled routes using assigned vehicles), and individuals typically do not purchase advance tickets that reserve them any specific portion of this capacity. However, they are free to ride on any part of the system at any time that they choose, provided that capacity is available. The closest analog in the movement of goods would likely be the transportation networks of parcel and package delivery firms. These firms have the flexibility to determine the routing of the items in their networks (except for the origin and destination points), while the items themselves (people) determine the routes they take in regional mass transit systems.

Regional mass transit systems often include a mix of modes, such as rail, light rail, elevated/underground trains, and buses, and the design of these systems must take into consideration the needs of passengers. Capacity is typically measured in terms of the number of passengers that can be accommodated (e.g., the number of passengers that can be moved between an origin–destination pair per unit time). In contrast, systems that move goods may have different temperature and space usage requirements, and can utilize space more efficiently in the movement of inanimate and durable objects. Moreover, capacity in this context is often measured in terms of the volume (or weight) of freight that can be moved between locations per unit time.

When traveling outside of urban areas without private vehicles, various scheduled transportation systems typically carry passengers, nearly all of which require reservations of space on the mode of travel. For long-distance travel, available mode choices include roadway (bus or car), rail, sea, or air; however, the predominant means of travel across long distances is either rail or air. The choice of mode is not only influenced by cost and convenience, but also by the transportation infrastructure within the particular country or region, as well as the prevailing culture within that region.

Freight transportation modes are slightly more diverse, involving roadway (truck), rail, inland waterways, pipelines, sea, or air. Road and rail transportation represent a significant percentages of total freight moved, and the use of road transportation has steadily increased in recent years [see UNECE (2001) and Eurostat (2002)]. In the European Union (EU) countries, for example, 77% of freight was moved by roadway and 15% was moved by rail in 1999.

While ocean transport for moving people typically only applies to leisure travel, it is extremely important for shipping materials from heavy industries (where ocean transport is the only viable alternative) and for shipping low-cost items over long distances. Barge transportation on inland waterways provides a similar service as an alternative to road or rail for cross-country transports. For each mode of freight travel, the carrier typically has a volume capacity, and depending on the size and weight of the products to be shipped, each system's capacity can vary. Thus, in addition to its origin and destination, a particular good may also dictate the mode of travel based on its size, weight, and value.

14.2.2 Differences and Similarities in Performance Measures

Clearly, moving people and goods involve differing performance measures and objectives. Passengers typically would like to spend as little time as possible in a transportation system, although they recognize the trade-offs between cost and convenience. For example, the cost of a cross-country bus ticket in the United States is much lower than that of a flight, although the former may require days while the latter can be completed in less than half a day. The individual must therefore consider the overall utility gained from a bus trip versus a flight when determining how to go cross country. The transportation carrier's performance when transporting people is often a function of individuals' perceptions of the overall value of a form of transportation. A number of elements determine this overall value including safety, monetary cost, time, and value-added services.

When it comes to freight, on the other hand, the items being moved do not experience the trip, and the key performance measures involve cost, trip duration, and reliability. Unlike people, goods in transport accrue "pipeline" inventory holding costs that are typically proportional to the duration of time in the transportation system (which might be roughly analogous to the value of time for a person in transit; in either case, an investment opportunity cost is incurred). The shipper must therefore consider the trade-off between transportation and inventory costs when making transportation mode decisions. While transportation modes with long lead times are often less costly, they lead to higher pipeline inventory costs. Moreover, for mass merchandise with uncertain consumer demand, longer transportation lead times imply greater inventory safety stock costs to buffer against uncertain lead-time demand. An additional complicating factor affecting inventory cost, which is discussed in greater detail in Section 14.3, is the reliability of a given transportation mode. Less reliable modes (where reliability might be measured, e.g., by on-time performance or by the standard deviation from the average delivery lead-time value) naturally lead to increased buffer stocks to provide insurance in cases where deliveries are late.

A somewhat unique performance factor within freight shipping is the notion of "empty-balancing." Due to trade imbalances between countries and geographic regions within a particular country, vehicles sometimes need to travel empty in order to rebalance the system. This need exists at a much smaller level in passenger transport systems, in the form of "dead-heading" crew or vehicles to realign the system. For example, while there may be imbalances in mass transit travel between the morning (into the city) and evening (out of the city) rush hours, people generally return to their point of origin at some point during their journey. The same is true for passengers using air or rail transportation for work or leisure. Freight transporters often seek out shippers who can utilize excess capacity in return trips, while passenger transporters may utilize pricing to increasing utilization on under-utilized trips.

14.2.3 Shared Systems

The most common form of a shared transportation system for people and goods would be commercial aviation. While the system network is designed to provide passengers a means of reaching their destinations in an acceptable travel time, the airlines can also provide cargo capacity on these same flights for those products that have a time-sensitive component. As previously mentioned, the airline industry has been developing mathematical programming solutions for passenger travel and cargo for the past few decades. However, cargo research has gained more interest in recent years as the airlines attempt to identify new revenue streams.

To a lesser degree, there is some shared travel by rail and sea. In particular, cruise ships can provide some point-to-point freight capacity as these ships travel between their ports of call.

14.2.4 System Design Challenges

Given the differences and similarities in how people and goods prefer to travel, several challenges arise when designing a transportation system. Adopting the designs for a passenger transport system will not apply in many cases for freight transportation. For logistics companies deciding on what type of system to provide, the choice will often depend on whether it wants to offer high weight capacity, express deliveries, custom routing, or door-to-door services. Each of these may drive a different set of customers, so the logistics provider must have a comprehensive understanding of the needs and preferences of the customer base that it wishes to serve.

For example, overnight shipping providers (or carriers) have similar objectives to those in passenger travel. While the intermediate destinations are not necessarily important, the freight must reach its destination by specific time-sensitive deadlines. Other carriers may focus on providing shipping without time-sensitive freight, and such carriers are primarily concerned with meeting promised delivery dates. These characteristics of the customer needs and expectations can make a substantial difference in the requirements of the fleet capacity.

From fleet capacity and route structure to empty-balancing and multi-mode solutions, there are many issues that face any potential freight carrier, and many of these solutions will be unique to the freight industry. The models that would be designed to provide such tactical design solutions will also be reliant on quality freight forecast data. We will address the issues of modes, infrastructure, and demand forecasting in the remaining sections.

14.3 Transportation Modes

As discussed in Section 14.2.1, there are many modes of transport available for freight: road, rail, maritime, air, and pipelines. We briefly discuss the characteristics of each mode, as well as the situations in which one of these modes would be considered the preferred method for transporting freight. Then, we motivate the need for multi-mode infrastructure solutions in successful logistics engineering.

14.3.1 Mode Characteristics

Let us briefly examine the modes of travel available for people and freight, and discuss the characteristics of each of these modes. As stated earlier, available options for transporting people include roadway, rail, sea, or air. Within each mode, varying levels of flexibility exist. Roadway and air offer the highest level of flexibility in terms of schedule options. Private automobiles can provide virtually any door-to-door service they desire. Buses still offer flexibility based on the number of stops included in their route structure. Since highway networks are remarkably well connected and developed, many destinations can be reached. However, the most rapidly growing mode has been air travel. Travelers can reach a growing number of destinations by air, which dramatically reduces trip times in many instances. When using a travel mode such as air, however, it is likely that the passenger will require a multi-mode solution to reach his or her final destination. This could include light rail, bus, mass transit, rental cars, or taxis.

In Section 14.2.1, we also noted the available modes of travel when shipping freight, which include roadway, rail, inland waterways, pipelines, sea, or air. Road and rail transportation represent significant percentages of total freight moved, and the use of road transportation has steadily increased in recent years [see UNECE (2001) and Eurostat (2002)]. As we noted previously, as of 1999, for example, 77% of freight was moved by roadway and 15% was moved by rail in the EU countries.

Road and rail transportation require capital-intensive projects to expand existing networks. Road networks offer high flexibility, and are primarily used for light industries, which require frequent, timely deliveries. Rail networks are not quite as flexible, yet the ability to containerize goods has allowed this industry to connect to sea or maritime transportation. Maritime and rail transportation are typically associated with heavy industries, and due to the volume of goods shipped by sea, this is another reason why connecting these modes is advantageous. As an example, excluding Mexico and Canada, over 95% of U.S. foreign trade tonnage is shipped by sea, and 14% of U.S. inter-city freight is transported by water [U.S. House Subcommittee (2001)]. Compared to other modes of transportation, shipment by waterways is generally less expensive, safer, and less polluting. Still, there can be substantial costs associated with port terminal operations, mostly due to port charges for shipping/receiving and inventory costs.

Air transportation can offer a method for transporting freight with either a time-sensitive nature or high value associated with it. Due to the high cost of this mode and the relatively limited capacity per vehicle (when compared to rail or water transport options), it is still used in low volumes compared to other shipping options, although it has the highest reliability among transportation mode choices.

14.3.2 Mode Selection

Transportation mode decisions for personal transport are a function of individual preferences and resources, that is, what the individual is able to afford, how the individual values his or her time, and the degree of utility the individual derives from the travel itself. We therefore focus our discussion on transportation mode decisions for goods in this section.

A highly stylized and simplified analysis of the mode decision for point-to-point delivery of a single good would proceed as follows. Suppose we manage a stock of goods that require periodic replenishment from a supplier, and we must meet demand that occurs at a constant rate of λ units per year. We have M possible modes of transportation from which to choose, and we pay for items at the time they are shipped, in addition to the shipping cost (here a mode might imply any multi-mode transportation solution). Selecting mode m implies that Q_m units will be delivered at equally spaced time intervals of Q_m/λ (equivalently, on average we receive λ/Q_m deliveries per year). The delivery lead time of mode m is L_m time units, which we assume for simplicity is a constant. The per shipment cost of mode m is f_m (independent of the quantity delivered), and we are also charged c_m per unit in transportation cost. Thus our average annual transportation cost for mode m is given by

$$\left(\frac{f_m}{Q_m} + c_m\right)\lambda. \tag{14.1}$$

Because we receive a shipment of size Q_m every Q_m/λ time units, and because it is optimal to receive these shipments precisely when our inventory on-hand hits zero, on average we will carry $Q_m/2$ units of cycle stock in inventory. If H denotes the cost to hold a unit of inventory on-hand for one year, then our average annual cycle stock cost for holding inventory locally when using mode m is given by

$$\frac{HQ_m}{2}. \tag{14.2}$$

Every unit of demand we meet in a year requires transportation from our supplier and spends L_m time units in the pipeline. If H_{pl} denotes the holding cost per unit of item in transit (or in the pipeline) per year, then the average annual pipeline inventory cost for mode m is given by

$$H_{pl}L_m\lambda. \tag{14.3}$$

If, for example, Q_m denotes the *economic order quantity** (EOQ) associated with mode m, that is, $Q_m = EOQ_m = \sqrt{2f_m\lambda/H}$, then the average annual cost per unit associated with mode m (which equals total cost divided by annual demand, λ) can be written as

$$c_m + H_{pl}L_m + \sqrt{\frac{2f_mH}{\lambda}}. \tag{14.4}$$

Equation 14.4 illustrates a basic trade-off in modes choices, as we would like to select the mode m from among the M choices that minimizes (14.4). In particular, those modes with short lead times (L_m) typically have high shipping costs, as reflected in the fixed (f_m) and/or variable (c_m) shipping costs. Longer lead-time mode choices, on the other hand, increase the pipeline holding cost term, while reducing the shipping cost terms.

To introduce the effects of uncertain demand without obscuring the analysis too greatly, we suppose that a positive safety stock level is required at the stocking point, and that the safety stock level is proportional to the standard deviation (uncertainty) of demand during the replenishment lead time {this is not an uncommon approach to setting safety stock levels in practice where, for example, we set some minimum level on the probability of not stocking out in any replenishment cycle; the associated probability is sometimes referred to as a cycle service level [see, e.g., Chopra and Meindl (2004)]}. In this setting, λ denotes the average annual demand, and safety stock is set equal to $k\sigma_L$, where k is a prescribed safety factor corresponding to the desired cycle service level, and σ_L is the standard deviation of demand during the replenishment lead time. If σ is the standard deviation of annual demand (and this annual demand is composed of a contiguous and statistically independent demand intervals) then we can write the standard deviation of demand during lead time as $\sqrt{L_m}\sigma$, and then the average annual safety stock cost equals $Hk\sqrt{L_m}\sigma$. Defining $cv \equiv \sigma/\lambda$ as the coefficient of variation, our average cost per unit becomes

$$c_m + H_{pl}L_m + \sqrt{\frac{2f_mH}{\lambda}} + Hk\sqrt{L_m}cv. \tag{14.5}$$

Equation 14.5 illustrates that long lead-time values increase not only pipeline holding costs, but also increase the required safety stock holding cost for meeting a prescribed service level at the stocking point. This effect is compounded by products with high coefficient of variation values (or equivalently, products with a high degree of demand uncertainty).

The stylized model we used to illustrate important trade-offs in selecting a transportation mode employs a number of simplifying assumptions, including that of a constant lead time. With less reliable

---

* This analysis assumes that the EOQ is feasible, or less than any capacity limit associated with the mode. Similar insights apply under capacity limits, although our intent here is to highlight the trade-offs associated with costs and lead times, and how these drive mode choices.

transportation modes, where the lead time itself is unpredictable, we tend to see an increased uncertainty of lead-time demand, which increases the impact on safety stock cost incurred in meeting a desired service level.

Our analysis also considered transportation of a single product that can be shipped in batches equal to the economic order quantity, whereas practical contexts often call for multiple products sharing shipping costs and capacity limits. Additional practical factors include risk of damage, trade tariffs and duties in international transport, and nonstationary product demands. While in principle the analysis can be generalized to account for such assumptions, the basic trade-offs between transportation and inventory costs when making mode decisions remain essentially the same, and include transportation costs, inventory costs due to economies of scale in shipping (cycle stock), and inventory costs due to uncertainty in demand and less-than-perfect reliability (safety stock).

14.3.3 Multi-Mode Transportation

In seeking an end-to-end transportation solution, the most cost-effective option often involves a mix of different transportation modes. Economies of scale in transportation often lead to highly utilized transportation equipment and links between major metropolitan areas, although the metropolitan areas themselves are often not the origin and/or destination points in the end-to-end solution sought. Transportation to and from regions surrounding major metropolitan areas is then accomplished by regional transporters who focus on the economics of regional transportation. Therefore, different organizations focus on efficiency within a different piece of the multi-mode puzzle, which permits finding cost-effective door-to-door solutions.

When considering the transport systems available for people, we focus on two areas: mass transit solutions for commuting, and air travel for business or leisure. Mass transit systems typically will include one or more of the following modes of travel: bus, subway, light rail, and commuter rail. In larger cities, these systems are designed in such a way that the commuter has the ability to easily connect between one system and another (e.g., a commuter from a suburb can take a commuter rail to the city or business district and transfer to either a bus or subway to reach a particular destination). While a commuter rail can only serve a small set of station locations in a region, bus and subway systems still provide commuters access to the majority of locations in a region. For air travel, passengers commute to an airport via personal vehicles, bus/rail, or other ground transportation options.

Freight transportation also involves logical multi-mode options. As previously mentioned in Section 14.3.1, freight can be moved between rail and barges/ships through port terminals that can handle containerized goods. Again, this allows heavy goods that travel by sea to reach various land-based locations by rail. Similarly, there is a logical connection between air cargo and trucking. For heavy freight, these connections often occur as hand-offs between different firms who specialize in managing and operating a single transportation mode. Coordination among these different carriers is often achieved by logistics service providers such as TNT Logistics, who often typically do not own transportation equipment, but serve to ensure that producers and distributors can achieve economical door-to-door deliveries. For small packages, this multi-mode service is most often seen with express overnight carriers such as FedEx and UPS, each of which owns a fleet of ground and air vehicles and provides door-to-door transportation solutions.

14.4 Importance of Transportation Infrastructure

An effective regional transportation infrastructure allows businesses in that region to compete in a global economy and allows consumers of the region to access goods from the rest of the world. Given the existence of free trade zones and additional markets being opened for the first time, it is extremely important for a region to determine the degree of transportation infrastructure to provide in order to connect to various parts of the world. Without adequate infrastructure, carriers cannot provide the type of service customers demand in a global economy, which can leave regional suppliers at a severe competitive disadvantage. How different locations are connected can vary greatly depending on the regional demographics and industry, as well as the economic goals of the carrier(s) providing service in the region. Because governments, consumers, and regional firms have a stake in the overall public transportation infrastructure (e.g., public roadways), it is natural to have conflicting economic and service objectives. While this public infrastructure enables commerce, it also has environmental as well as quality of life impacts for a region. Thus, the collective interests of the stakeholders in a region as well as the economic and social tradeoffs must be weighed when making public infrastructure investment decisions. For these reasons, it is extremely important to accurately assess both public and private needs across the system. Numerous political and social factors affect public transportation infrastructure decisions, which partially determine the transportation capabilities of private transportation firms. Because of this, this section focuses on the infrastructure decisions over which a private firm has control, examining important considerations when designing private infrastructure in a transportation network. The case study in Section 14.6 provides an interesting illustration of the potential for conflicting objectives and diverse interests involved in public infrastructure investment decisions.

14.4.1 Scope of Transportation Solutions Provided

The nature of the transportation solutions offered by a firm affects its need for infrastructure investment. When a firm has a product to ship from destination A to destination B, it may be presented with several options for choosing the method of shipment. Regardless of the method, the firm needs to complete the entire transaction. Some carriers may actually provide the entire freight shipping service, depending on the two locations of A and B and the type of business that the carrier intends to provide. Point-to-point shipping is defined as moving goods between any two "major" locations. Often, these locations will be warehousing or cross-docking facilities that serve many local destinations. Certain providers will focus on providing transportation between these point-to-point trips, and their system infrastructure will reflect this. That is, their infrastructure investments will focus on equipment and facilities that provide high economies of scale in shipping and terminal operations.

Other carriers focus on providing door-to-door shipping service. Such carriers are responsible for picking up the product at the shipping location of a customer (which does not need to be a centrally located warehouse with consolidated goods), and delivering it to that customer's destination of choice. Providing door-to-door service can certainly drive a different business model for freight carriers. Multi-mode express freight carriers such as FedEx and UPS tend to build extremely large system networks, with separate fleets (trucks and aircraft) to expedite the delivery of product and meet the service requirements of customers. The infrastructure provided by each of these companies allows their customers to ship products door-to-door around the world in one to two days. Other door-to-door carriers may not have the same massive infrastructure in place, so they may need to wait for enough demand to materialize before consolidating on a vehicle for transport. Such carriers have a different business model, usually moving less time-sensitive materials. However, they still must meet customer service goals, which will include predetermined delivery due dates. Note that this door-to-door service need not necessarily be provided by only one carrier. In fact, several carriers may be involved in the shipment of a good from destination A to destination B, although the shipper interfaces with a single carrier. Such carriers manage the coordination among a number of firms involved in the actual transport. The infrastructure investment for such firms might involve regional transportation networks and equipment, interfacing with the networks of long distance carriers who manage networks that interconnect major metropolitan areas.

14.4.2 Consolidation versus Operational Frequency

As mentioned in the previous section, carriers must consider how frequently to provide service between any two points. This delivery frequency has important implications for investment in equipment and facilities. One approach is to consolidate demands from several points until this accumulated demand reaches the capacity of the transporting vehicle. This "on-demand" approach is desirable for carriers because it ensures low unit transportation costs and high capacity utilization. The investment in vehicle capacity is therefore lower than would be required when shipping partially loaded vehicles at prescheduled times. This approach may, however, be quite undesirable for customers with time-sensitive delivery requirements or with high-value goods that have high associated inventory holding costs.

In this "on-demand" context, the carrier may have the flexibility of delaying individual customer delivery requests until the carrier can generate sufficient revenue to warrant the entire shipment. The unscheduled nature of such shipments can also cause problems depending on the mode of travel. For example, the freight and passenger rail industry often share the same service network (i.e., the same track). As the number of scheduled operations that use a common infrastructure increases, it becomes increasingly critical to know when to expect these "on-demand" shipments. System capacity simply may not be able to accommodate them at the point in time when the vehicle capacity is reached and the "on-demand" shipment is ready for transport.

Another option is for the carrier to provide scheduled operations that match its customer shipping requirements. In other words, the carrier can schedule a particular delivery once a day, once a week, and so on. Uncertainty in shipping requirements will often result in under-utilized capacity under this approach, which will require a higher capacity investment (as compared to "on-demand" shipping). At the same time, the carrier will have a predictable schedule of operations, and hence anticipated shipping arrival dates will be more accurate. In either of these two cases ("on-demand" or scheduled frequency), their ability to achieve efficiency in transportation is driven by the accuracy of freight demand forecasts, and how these forecasts are used in making capacity investments. The issue of forecasting is addressed in Section 14.5.

14.4.3 Domestic and International Infrastructure

Domestic infrastructure in developed regions of the world is typically quite effective in enabling transportation to virtually anywhere in the region. Customers have the ability to move goods between almost any two points that they desire, and numerous carriers offer services to do this. The level of customer service provided by these carriers must meet the standards expected for the particular country, based on the cultural and political nature of that country. When connecting domestic infrastructure with international infrastructure, a host of new issues arise.

From a carrier point-of-view, not all carriers are equipped to expand their businesses internationally. If their primary mode of transport is roadway and rail, they may be limited to providing additional long-haul services to land-based destinations (i.e., it may be difficult to venture into maritime or air travel). A carrier's success may be driven by the unique environment of its domestic operation. Serving new markets may require a change in the carrier's economic and customer service objectives.

Carriers who are willing to make such changes must also now deal with the additional logistical issues with moving goods across borders. In general, it can be much harder to provide accurate delivery dates when the goods must be cleared through customs. Each country has its own rules for how this process works, and likewise, the modeling requirements for determining appropriate transport system requirements can be case-specific. Nonetheless, there is an enormous market for companies that provide international logistics solutions; in particular, freight forwarders, who specialize in moving goods between countries, can serve as valuable partners for firms seeking global expansion.

14.4.4 Global Infrastructure Example: FedEx

The FedEx Corporation provides an excellent example of a global logistics network.* FedEx coordinates deliveries throughout a global network broken down into five regions:

- Asia-Pacific
- Canada

---

* *Source:* FedFx Corporate Web Site: www.fedex.com.

- Europe, Middle East, and Africa
- Latin America-Caribbean
- United States

Their operation began by providing service to 25 cities in the United States via a single hub in Memphis using a fleet of 14 planes in 1973. Packages were flown to and from Memphis from each of the connecting cities on a daily basis, and couriers transported packages on the ground within a 25-mile radius of each connecting airport. In 1977, FedEx was successful in lobbying the U.S. Congress to allow private cargo airlines to purchase larger planes, which led to the purchase of seven Boeing 727 aircraft that year, and paved the way for the unprecedented growth in air cargo that followed.

In 1981, FedEx began international service to Canada in cooperation with Cansica, a Canadian licensee. By the end of the 1980s they had purchased three Canadian air cargo firms and were providing full service to Canada. Their 1984 purchase of Gelco expanded operations into Europe and Asia. In 1987, FedEx acquired Island Airlines, which provided air cargo service to the Caribbean. FedEx subsequently acquired Tiger International (owner of the air cargo firm Flying Tigers) in 1989, becoming the world's largest cargo airline. Flying Tigers' existing business throughout the world led to a substantial global expansion for FedEx, who was now the largest air cargo carrier in South America, and was providing service to Europe and Asia.

Today, Fedex serves over 220 countries and employs roughly 260,000 employees and contractors throughout the world. Their express package delivery network (FedEx Express) reaches 375 airports using 10 air express hubs and 677 aircraft, with approximately 43,000 motor vehicles. FedEx Ground uses ground transportation for package delivery in the United States, Canada, and Puerto Rico. The ground fleet contains 18,000 motor vehicles connecting 29 ground hubs and 500 pickup/delivery terminals. For larger packages, FedEx Freight provides trucking services using more than 10,000 tractors throughout the United States, Canada, Mexico, South America, Europe, and Asia, with 321 service centers. Its collective global network and companies handle approximately 6 million deliveries per day throughout the world.

Clearly, FedEx provides an example of an organization that has successfully confronted the complexities discussed throughout this chapter. As a door-to-door delivery service provider who serves countries all over the globe, they must achieve economies of scale between metropolitan areas using a combination of air and truck travel modes, while effectively scheduling time-sensitive local deliveries in any region using smaller motor vehicles. Their massive global infrastructure permits reaching almost anywhere in the developed world in a very short time and serves as a vital component of the supply chains of firms in many countries.

14.5 Difficulties in Forecasting Freight Demand

Transportation infrastructure planning and development is primarily driven by forecasted demand for transportation. Governments have faced this challenge for many years in developing public transportation infrastructure, and this has driven a continuous stream of transportation pattern studies as well as research on methods for predicting transportation demands. A greater amount of public information is available concerning methods for predicting passenger travel than freight demand, as passenger travel studies are often sponsored by governments, while competitive firms do not necessarily make their freight demand studies available to the public. As a result, and because well-developed large-scale freight transportation networks do not have an extremely long history, research on freight demand modeling remains at a relatively early stage as a discipline.

Because transportation of freight forms the backbone of a large percentage of economic activity, forecasting freight demand can be as complex as forecasting the performance of an economy. For example, if we could accurately forecast freight flows between two countries, we could then likely provide an accurate estimate of the trade balance between the two countries. It is no surprise, therefore, that advanced methods for predicting freight demand between countries utilize well-developed methods from international economics [see Haralambides and Veenstra (1998)] and time-series forecasting. Zlatoper and Austrian (1989) provide an excellent characterization of econometric models for transportation forecasting.

Winston (1983) characterizes predictive freight flow models as either aggregate or disaggregate models. The aggregate models consider a geographic region and attempt to predict the percentage of total flow that utilizes a given mode in the region. Such aggregate models found in the literature typically employ log-linear regression, with a mode's relative market share dependent on relative price and other independent quantitative and qualitative factors, including population demographics and economic indices [see Regan and Garrido (2001) for a more detailed discussion of these methods, in addition to an excellent survey of freight demand modeling research]. Disaggregate models focus on the individual decision-maker's choice of mode, and incorporate individual decision factors in forecasting methods. Clearly, the disaggregate models require substantially more data and understanding of individual utility factors. These models characterize a probability distribution of the utility value an individual derives from a given mode, assuming some deterministic utility factor (e.g., that depends on the corresponding good and industry), and a stochastic error term that accounts for variations among different consumer preferences. The probability distributions of mode utilities are then used to characterize the probability one mode will be preferred to another by a typical shipper (or passenger).

As the foregoing discussion indicates, freight transport demand modeling requires an understanding of economic, behavioral, and demographic factors, as well as advanced statistical forecasting methods. The number of factors affecting transportation demand and the complexity of the interrelationships among these factors can make accurate freight modeling as difficult as predicting the stock market. This section highlighted the complexities inherent in freight demand modeling and provided some basic characterizations of effective approaches. [For more details on freight demand modeling, see Regan and Garrido (2001).]

14.6   Case Study: Dutch Railway Infrastructure Decisions

The Dutch railway system in the mid-1990s provides an excellent example of the conflicting objectives and trade-offs inherent in transportation systems decisions.* Highly congested roadways in the 1990s, which had negative implications for the economy and the environment, forced the Dutch government to explore ways to improve their transportation system. Increasing the use of rail for both passengers and freight was an attractive alternative in terms of reducing the environmental impact of transportation, but the rail system capacity was hardly able to handle the impact. Moreover, long delays (due to insufficient capacity) and relatively high prices for rail travel discouraged both passengers and freight shippers from choosing this option.

Control of railway infrastructure development and funding rested with the government, and prior to 1997 the monopolist firm Nederlandse Spoorwegen (NS) operated the railway with a goal of maximizing profit (subject to certain restrictions on infrastructure and service areas). Therefore, the government's infrastructure investment decisions (and government requirements to keep certain lines open whether or not they were profitable) constrained NS's profitability. In 1997, the EU also required privatizing rail operations and allowing competition in this industry. This created added complexity for the government, which now had to assign infrastructure to multiple competing firms.

To address their transportation infrastructure shortcomings, the Dutch government allocated about $9 billion to improve rail infrastructure between 1985 and 2010. In addition to reducing roadway congestion, the government had several additional priorities, including:

- Stimulating regional economies by providing rail connections to large metropolitan areas.
- Increasing the amount of freight carried via rail.
- Reducing the need for short flights to other countries in Europe.

---

* A detailed discussion of this case context can be found in Hooghiemstra et al. (1999).

Private rail operating firms, on the other hand, are interested in profit maximization, and therefore would like to see faster travel times and reduced delays on their networks, which could be achieved to a significant degree through equipment and infrastructure (both rail-line and energy) improvements. The Dutch government relied on an independent organization called Railned to provide recommendations on how best to invest the funds allocated to rail infrastructure improvement. A team of government, Railned, and private operating firm representatives assembled to tackle the problem of maximizing the return on infrastructure investment. The team developed three sets of project portfolio options (that they called cocktails), each emphasizing a different investment focus. These three focus areas were (*i*) metropolitan area, (*ii*) main port, and (*iii*) regional (non-metropolitan) development. To evaluate the investment options, they drew on a sequence of decision support models.

Given an investment option (which consists of a set of potential projects), the first step required estimating the demand for rail service if a given set of projects was implemented. For this they used the Dutch National Mobility Model (DNMM), which uses econometric regression to determine the demand for travel using a mode, given the service level provided by the mode, population sizes, and various economic factors associated with regions served by the mode. After estimating travel demands, these demands provide input to an optimization model that determines trip frequencies and equipment required to meet demands on the rail network links at minimum operating cost. Inherent in this optimization model is a utility value for each transportation alternative that is used to estimate the percentage of total passengers who will choose a given route. Although our discussion here greatly simplifies the description of the forecasting and optimization models employed (which include a large-scale integer-linear programming model), it is a quite complex system and requires a heuristic solution in order to obtain a good feasible solution in reasonable computing time.

Because of the number of qualitative factors affecting the attractiveness of a solution (from a government and social perspective), further evaluation (beyond profitability) was required subsequent to implementing the forecasting and optimization models. The team of analysts performed a cost-benefit analysis for each investment alternative by assigning a monetary value to each of the important qualitative factors. This analysis allowed the team to quantify the value and utility of each of the possible investment alternatives, and to recommend a course of action to the government. In the end, the government selected the second-ranked alternative (involving metropolitan area development), because of the apparent perception on the part of the government that the profitability of private firms received too high a weight in the cost-benefit analysis, while the qualitative impacts of metropolitan area congestion received too low a weight. The operations research-based analysis served an extremely valuable purpose in this context, providing the team with a methodologically based tool for evaluating a set of extremely complex decision alternatives. Moreover, the system continues to pay dividends through repeated analysis of additional transportation infrastructure investment options.

14.7 Concluding Remarks

Our goal in this chapter was to provide a general framework for understanding the issues and tradeoffs inherent in transportation systems decisions. This chapter lays the groundwork for studying the detailed problem classes addressed in the following section on transportation management. We have taken a necessarily broad overview in this discussion, highlighting many of the qualitative factors that lead to conflicting objectives, and make transportation systems decisions a complex field of study. Because transportation systems affect the economic performance of a region, as well as the daily lives of nearly all people, our discussion focused on the systems and infrastructure and how they affect the movement of goods and passengers.

The FedEx global network discussed in Section 14.4 provides a nice snapshot of the progress that has been made in developing transportation systems in the last 30 years. The ability to ship a package anywhere in the United States within 24 h, a luxury many now take for granted, is quite remarkable, particularly in light of the fact that this reach extends far beyond the U.S. borders. This progress could not have been accomplished without the collective infrastructure investment made by developed and developing countries during this time frame. A possibly equally important factor in accelerating this development has been the advancement of enabling information and communication technologies, a topic that is discussed in greater detail in Part IV of this book. The willingness of the people and governments of countries around the world to invest in both transportation and information infrastructure has certainly led to a tighter economic integration among countries, and has opened new markets. The continued development of new markets in Asia, Eastern Europe, South America, and Africa will likely lead to changes in transportation systems over the next 30 years that are as interesting as those of the past three decades.

References

Barnhart, C., P. Belobaba, and A.R. Odoni. Applications of operations research in the air transport industry. *Transportation Science*, 37(4):368–391, 2003.

Barnhart, C. and K. Talluri. Airline operations research. In C. ReVelle and A.E. McGarity, editors, *Design and Operation of Civil and Environmental Engineering Systems*, pages 435–469. Wiley, 1997.

Chopra, S. and P. Meindl. *Supply Chain Management: Strategy, Planning, and Operations*. Prentice-Hall, Upper Saddle River, New Jersey, 2nd edition, 2004.

Crainic, T.G. Long-haul freight transportation. In R.W. Hall, editor, *Handbook of Transportation Science*. Kluwer Academic Publishers, Boston, Massachusetts, 2nd edition, 2002.

Eurostat. *Transport and environment: statistics for the transport and environment reporting mechanism (TERM) for the European Union*, January 2002.

Friesz, T.L. Strategic freight network planning models. In D.A. Hensher and K.J. Button, editors, *Handbook of Transport Modelling*, chapter 32, pages 181–195. Pergamon, 2000.

Hall, R.W. editor. *Handbook of Transportation Science*. Kluwer Academic Publishers, Boston, Massachusetts, 2nd edition, 2002.

Haralambides, H. and A. Veenstra. Multivariate autoregressive models in commodity trades. In *8th World Conference on Transportation Research*, Antwerp, Belgium, 1998.

Hensher, D.A. and K.J. Button, editors. *Handbook of Transport Modelling*. Pergamon, New York, New York, 1st edition, 2000.

Hoffman, K.L. and M. Padberg. Solving airline crew scheduling problems by branch-and-cut. *Management Science*, 39(6):657–682, 1993.

Hooghiemstra, J.S., L.G. Kroon, M.A. Odijk, M. Salomon, and P.J. Zwaneveld. Decision support systems support the search for win–win solutions in railway network design. *Interfaces*, 29(2):15–32, March–April 1999.

Lohatepanont, M. and C. Barnhart. Airline schedule planning. Integrated models and algorithms for schedule design and fleet assignment. *Transportation Science*, 38(1):19–32, 2004.

McGill, J.I. and G.J. van Ryzin. Revenue management: Research overview and prospects. *Transportation Science*, 33:233–256, 1999.

Regan, A.C. and R.A. Garrido. Modeling freight demand and shipper behaviour: State of the art, future directions. In D. Hensher, editor, *Travel Behaviour Research: The Leading Edge*, pages 185–216. Pergamon Press, Oxford, 2001.

United Nations Economic Commission for Europe (UNECE). *Annual bulletin of transport statistics for Europe and North America*, July 2001.

U.S. House Subcommittee on Coast Guard and Maritime Transportation. *Joint Hearing on Port and Maritime Transportation Congestion*, May 2001.

Winston, C. The demand for freight transportation: Models and applications. *Transportation Research Part A*, 17(6):419–427, 1983.

Zlatoper, T. and Z. Austrian. Freight transportation demand: A survey of recent econometric studies. *Transportation*, 16:27–46, 1989.

III

Topics in Transportation Management

15

Real-Time Dispatching for Truckload Motor Carriers

Warren B. Powell
Princeton University

15.1 Introduction

Truckload trucking may be the simplest operational problem in freight transportation. Shippers use truckload motor carriers to move large quantities of freight which require hiring an entire truck to move a load of goods from one location to another. Similar to taxi operations for passengers, a shipper will call a carrier with information about a load of freight that needs to be moved from one city to another. If the carrier agrees to move the load, he sends out a driver with a tractor (and possibly a trailer) who then picks up a trailer loaded with freight. When the driver delivers the freight at the origin, it is now the responsibility of the company to figure out what to do with the driver next. Although there are many one-man trucking companies, our focus is on operations that manage fleets that may range from several dozen to over 10,000 drivers.

At the heart of real-time operations in truckload trucking is a disarmingly simple problem: given a set of drivers and loads, which driver should be assigned to which load? We could address the problem from the perspective of a particular driver (what is the best load?) or a particular load (what is the best driver?), but the real problem requires juggling the needs of multiple drivers and loads. A company with 100 drivers, faced with the decision of which driver to assign to each of 100 loads, can choose from among $100 \times 99 \times 98 \times \cdots \times 2 \times 1 \approx 10^{158}$ possible solutions.

In the 1970s, as part of a consulting project with a young trucking company called Schneider National (today, the largest truckload carrier in the United States), Dr. Richard Murphy, then a faculty member at the University of Cincinnati, recognized that the load-matching problem was a special type of linear programming problem known as a pure network. These are best visualized using the network in Figure 15.1. In this representation, drivers are represented as nodes on the left, with a flow of 1 unit entering each driver node. Links join driver nodes to load nodes, with an additional "load link" from each load node to a "supersink" from which all the flow leaves the network. An upper bound on each load link prevents more than one driver from covering each load. The mathematical structure of the problem guarantees that we would never assign a fraction of a driver to a load.

At that time, this observation meant that it was possible to solve very large problems with exceptionally powerful algorithms that were extremely fast (even on computers of that era). Just as important, the model provided for a surprisingly high level of detail in how the costs were calculated. The cost $c_{r\ell}$ of assigning driver r to load ℓ could include the cost of driving empty from the current location of the driver to where the load had to be picked up, along with a variety of other factors. For example, we could add artificial penalties for assigning drivers who would pick up the load after its pickup appointment, or if the driver would arrive so early that he would have to sit for several hours. We could put a bonus (negative cost) for desirable assignments such as putting sleeper teams (pairs of drivers who swap driving so that the truck does not sit idle while a driver sleeps) on long loads (where the team gets higher utilization). In real applications, the list of such issues can be quite long, and yet this simple model can handle a broad range of these operational goals and constraints.

Linear programming models and the associated algorithms looked like the perfect match of a new technology with an industrial application. They offered to overcome what appeared to be a major limitation of human dispatchers—the ability to consider all the drivers and loads at the same time when making a decision. Humans tend to break problems down into small pieces. What is the best driver to move a particular load? What load should I assign this available driver to? Juggling the assignment of multiple drivers and loads at the same time is beyond the problem solving skills of most people.

News of this model spread quickly, and suddenly other carriers wanted their own optimization models for solving the load-matching problems. A small cottage industry of consulting firms popped up in the late 1980s and 1990s to sell this technology. The promise of the technology closely matched the

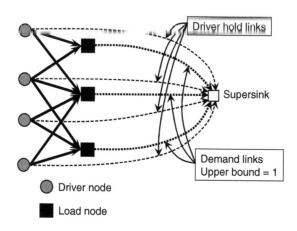

FIGURE 15.1 Network model for driver assignment problem.

early promises of robots building cars in the 1980s. What we learned is that the technology is promising, but the problem has proven to be far more difficult than anyone realized.

15.2 Basic Load-Matching Model

The network model shown in Figure 15.1 is easily modeled as an assignment problem involving the assignment of driver r (we think of drivers as the resources we are managing) to load l. Such models would define c_{rl} to be the cost of assigning driver r to load l, and we would define a decision variable x_{rl} where $x_{rl} = 1$ means we have decided to assign driver r to load l. In this section, we adopt a somewhat different notation that will prove to be much more general, allowing us to easily represent other issues that are not captured by this basic model. We describe a driver using a vector of attributes we denote by a, such as:

$$a = \begin{pmatrix} a_1 \\ a_2 \\ a_3 \\ a_4 \\ a_5 \\ a_6 \\ a_7 \\ \vdots \\ a_n \end{pmatrix} = \begin{pmatrix} \text{Location} \\ \text{ETA} \\ \text{Home domicile} \\ \text{Equipment} \\ \text{Team?} \\ \text{Days from home} \\ \text{DOT hours} \\ \vdots \\ a_n \end{pmatrix}$$

Here, location might represent his exact current location, his last reported location, or the location to which he is headed (it would be his current location if the driver is sitting still). If the driver is enroute, "estimated time of arrival " (ETA) represents when he is expected to arrive at his destination. Locations can be represented at a number of different levels of aggregation. Equipment might capture the type of trailer (and even tractor) he is pulling. "Team?" is an indicator variable that tells us whether it is a single driver or a pair of drivers who trade off between driving and sleeping. The attribute a_7, "DOT hours" is actually a vector that tells us how long the driver has been driving today, how long he has been on-duty, and how many hours he was on-duty for each of the last seven days. These attributes are used to enforce federal Department of Transportation rules on how much a driver can work in a given day.

Similarly, we let $b = (b_1, b_2, ..., b_m)$ be a vector of attributes describing a load. Attributes might include origin, destination, pickup and delivery time windows, equipment characteristics, shipper (or shipper priority), and any other information needed to describe the load.

We need to represent how many drivers and loads we have of each type. We let A be the set of all possible attribute vectors for drivers, B be the set of all possible load attributes. We then define:

$$R_{ta}^D = \text{Number of drivers of type } a \text{ at time } t.$$
$$R_t^D = \left(R_{ta}^D \right)_{a \in A} = \text{Driver resource vector.}$$
$$R_{tb}^L = \text{The number of loads of type } b \text{ at time } t.$$
$$R_t^L = \left(R_{tb}^L \right)_{b \in B} = \text{Load resource vector.}$$
$$R_t = \left(R_t^D, R_t^L \right) = \text{System resource vector.}$$

Throughout our discussion, R_t describes the state of all our drivers and loads at time t. We need to emphasize that in a software implementation, we would never explicitly store the entire vector R_t^D or R_t^L.

Instead, it makes more sense to define a set R_t^D where $r \in R_t^D$ is a particular driver with attribute vector a_r. However, the notation we have adopted will prove more convenient as we progress.

Rather than assigning a particular driver to a particular load, we adopt the convention that we are acting on a driver (or resource) with attribute a using a decision of type d chosen from a set of possible decision types, given by the set D. There are different classes of decisions, which we define using

D^L = Decisions to move a type of load. Each element of D^L corresponds to an element in the set of load attributes **B**.

D^M = Decisions to move empty to another location (perhaps in anticipation of loads that might become available in the future).

D^H = Decision to "go home" and go off duty for a period of time.

d^ϕ = The decision to "do nothing" (sit and wait).

$D = D^L \cup D^M \cup D^H \cup d^\phi$.

This notation is useful since it allows us to easily add new decision classes (e.g., repair or clean a trailer, maintain a tractor) without fundamentally changing our model. The set D is all of our types of decisions. We then define

x_{tad} = Number of times we cat on a driver of type a with a decision of type d at time t.

$x_t = (x_{tad})_{a \in A, d \in D}$ = The decision vector.

As a rule, x_{tad} will be 0 or 1, but our notation allows us to aggregate drivers. $x_{tad} = 1$ when $d \in D^L$ means that we have assigned a driver to a load of type b_d. We also note that $x_{tad} = 1$ does not mean that we are moving a driver at time t. It only means we are making the decision at time t. We may be preassigning a driver due to arrive later in the afternoon to a load, but we are making the decision in the morning.

The effect of a decision is captured using the modify function, which we write as follows:

$$M(a,d) = (a', c, \tau).$$

The modify function is a set of rules that specifies that if we act on a driver with attribute vector a with a decision of type d, we produce a modified driver with attribute vector a' and generate a contribution (cost if we are minimizing) c, where the time required to complete the action is given by τ. The completion time τ is also captured by one of the attributes of a' (the "estimated time of arrival" field). It is also useful to define

$a^M(a,d)$ = The terminal attribute function, which is the attribute a' produced by decision d.

$\delta_{a'} = \begin{cases} 1 & \text{If } a^M(a,d) = a' \\ 0 & \text{Otherwise} \end{cases}$

c_{ad} = Contribution generated by acting on a driver of type a with a decision of type d.

The attribute transition function $a^M(a,d)$ gives the attribute a' produced by decision d.

We can now state our basic problem (depicted in Fig. 15.1) as the following mathematical model:

$$\max_x \sum_{a \in A} \sum_{d \in D} c_{ad}\, x_{tad}. \tag{15.1}$$

This is solved subject to the following constraints

$$\sum_{d \in D} x_{tad} = R_{ta}^D \tag{15.2}$$

$$\sum_{a \in A} x_{tad} \le R^L_{tbd} \qquad d \in \mathrm{D}^L \tag{15.3}$$

$$x_{tad} \le 0 \qquad \text{(and integer)} \tag{15.4}$$

Equation 15.1 is our objective function, which we have written in the form of maximizing total contribution. Equation 15.2 is conservation of flow for drivers. Note that because "doing nothing" is an explicit decision, this must hold with equality. Equation 15.3 is conservation of flow for loads (we cannot move loads we do not have). Equation 15.4 states that flows must be non-negative integers.

The optimization problem in 15.1–15.4 is powerful in part because it provides for a very high level of detail in the representation of drivers and loads, but it also has another extremely useful property. In practice, it is very common for issues not captured by the computer to prevent the assignment of a particular driver to a particular load. In commercial implementations, it is standard practice to produce a ranked list of options by using dual variables. Let v^L_a be the dual variable for each driver constraint 15.2 and let v^D_b be the dual variable for each load constraint 15.3. We can now compute the reduced cost of each decision using

$$\bar{c}_{tad} = c_{tad} + (v^L_{b_d} - v^D_a)$$

where b_d is the attribute of the load corresponding to decision $d \in D^L$. If $x_{tad} > 0$, then $\bar{c}_{tad} = 0$. If $x_{tad} = 0$, then $\bar{c}_{tad} \le 0$, which means that contributions are reduced if we increase flow on these links. It is not unusual for $\bar{c}_{tad}$ to be zero (or very close to zero) for decisions that the model does not recommend, indicating that these are very close to being optimal. Errors in costs or missing data can easily change these recommendations, so we normally provide dispatchers with a ranked list of recommendations. In fact, it is common practice to choose a value, say $10, where we would say that if the dispatcher chooses a decision where $\bar{c}_{tad} \le 10, they we would say that the dispatcher "agrees with" the model.

15.3 Variations and Extensions

The initial appeal of the load-matching problem is quickly tempered by the realities of actual operations. In this section, we briefly review a number of operational issues that our basic model does not consider. We divide our discussion between more complex operational problems in Section 15.3.1, followed by a discussion of the challenge of working with forecasted demand.

15.3.1 More Complex Operational Problems

This section briefly describes some operational issues that arise in real applications that cannot be handled by our basic load-matching model.

15.3.1.1 Short-Haul Loads

The load-matching problem was first solved in the context of a long-haul truckload carrier. In this setting, loads typically require a day or more to complete. Since we cannot accurately predict what will happen a day from now, it makes sense to assign a driver to at most one load, and then wait until he is at least close to completing the load before assigning him to another load. Many trucking companies pull a significant amount of short-haul movements, and even long-haul carriers have to execute a number of short movements. Since short movements can be completed quickly, we have the ability to plan a sequence of movements for a single driver through several loads. Figure 15.2 illustrates a problem with four drivers and six loads, three of which are quite short. If we use a load-matching model, where a driver can be assigned to at most one load, we obtain the solution in Figure 15.2a. If we plan a tour using

(a)

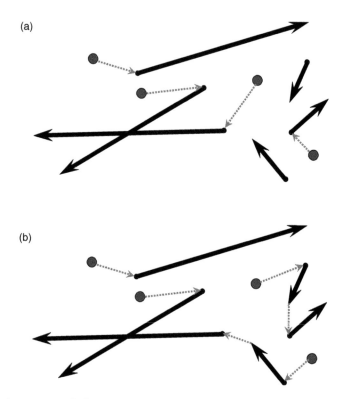

(b)

FIGURE 15.2   (a) Assigning each driver to one load, (b) assigning drivers to full tours.

all the loads we know about, we might obtain the solution in Figure 15.2b. The difference is significant, and remains a source of complaint about commercial load-matching systems.

15.3.1.2   Managing Trailers

It is common to assume that we are assigning "drivers" to "loads," but exactly what do we mean by a driver? We generally assume that a driver also has a tractor, but what about a trailer? A truckload carrier might have twice as many trailers as drivers. It is common, for example, for a driver to drop a loaded trailer in a shipper's lot, and then drive just his tractor to another shipper where he picks up a trailer that has just been loaded. At a later time, the first trailer will be unloaded, and either added to the shippers pool of trailers, or someone will have to pull the trailer out empty.

In the language of resource management, modeling just the drivers and loads represents a two-layer problem. If we explicitly model trailer inventories, we have a three-layer problem. Computationally, this can be much more difficult. It is fairly easy to model trailers in a simple way. For example, if a driver does not have a trailer, but the load requires that the driver bring a trailer to pick up the load, then the cost of assigning a driver to a load requires that we consider routing the driver through a trailer pool. But a real model of trailers would also make recommendations to move trailers from one pool to the next in order to manage trailer inventories.

15.3.2   Looking into the Future

A second complication arises when decisions now have to consider what might happen in the future. There are three issues that we need to consider that require us to look into the future:

1. Should we commit to move this load? We often have the ability to decline loads (when they are first offered), and we have to think about whether we will have too many drivers at the destination of the load to use them productively.

2. Is this the right type of driver? Teams like long loads. Regional drivers like short loads that keep the driver close to home. A driver pulling a refrigerated trailer would like to move to locations where this is needed. Not every region has the same mix of freight.

3. Will we be able to get the driver home? A long-haul carrier might keep a driver away from home for several weeks. The carrier might like to get a driver home every week or two (less often for some drivers). What is the likelihood that we can get a driver to his home if we assign him to a particular load?

These issues require that we think about loads that we do not yet know about. In other words, we have to forecast future demands, a problem we address in Section 15.4. But it is not enough to forecast demands; we also have to think about how we might manage the drivers in the future. This problem is addressed in Section 15.5.

15.4 Forecasting Demand

Looking into the future requires that we estimate customer requests before they become known. This section provides a brief overview of issues that arise when forecasting for a truckload operation.

15.4.1 Elementary Forecasting

Let $\hat{D}_{tij}$ be a random variable representing the number of loads that will need to be moved from location i to location j on day t. $\hat{D}_{tij}$ is a random variable that we might assume takes the form

$$\hat{D}_{tij} = \mu_{tij} + \varepsilon_{tij} \tag{15.5}$$

where μ_{tij} is the mean of the random variable and ε_{tij} is a random error around the mean which we assume has zero mean and some variance. We will never be able to guess ε_{tij}, but we would like to try to estimate μ_{tij}. There is a vast array of forecasting techniques, but the simplest and most widely used is exponential smoothing. Let $\bar{\mu}_{t-1,ij}$ be our estimate of the mean after day $t - 1$, and let $\hat{D}_{tij}$ be our observation of the actual demand on day t. Using exponential smoothing, we would update the mean using

$$\bar{\mu}_{t,ij} = (1 - \alpha)\bar{\mu}_{t-1,ij} + \alpha\hat{D}_{tij} \tag{15.6}$$

where α, a parameter between 0 and 1, is variously known as the smoothing parameter, learning rate or stepsize.

If only forecasting were this easy. One challenge is that it is very common to have forecasts (for the flow from a particular origin to destination) to be a fraction less than one. Carriers will often forecast the total loads out of an origin, but this means they have no idea where the loads are going. This problem arises because of the common misconception that to forecast demand means to estimate μ_{tij}, which is known as a point forecast. More modern tools forecast the actual distribution of demand (e.g., the probability that there will be 5 loads from i to j).

The following two sections address two important issues: (*i*) the challenge of forecasting daily demand and (*ii*) methods for handling advance bookings.

15.4.2 Challenge of Forecasting Daily Demand

In an operational model, it is necessary to forecast demand on a particular day. This introduces not only the problem of handling day-of-week effects, but also the more challenging problem of holidays and other "special days"—end of month, end of quarter, the Monday after Thanksgiving, the fifth of July when the fourth of July is on a Thursday, and so on.

It is popular in industry to use techniques such as averaging the last four Mondays to forecast the next Monday. This might capture day of week effects, but requires that we go back a month, which means we might be missing out on seasonal effects such as those that typically arise around the Christmas season. It also ignores holidays and special days. An alternative that we have found effective is to use a model of the form (dropping the indices i and j):

$$\bar{\mu}_t = b_t \theta_t^{dow} \theta_t^{wom} \theta_t^{sd}$$
(15.7)

where b is the baseline, θ_t^{dow} is a day-of-week adjustment factor, θ_t^{wom} is a week-of-month adjustment factor, and θ_t^{sd} is a factor for special days. For example, $\theta_t^{dow} = 1.07$ if the day-of-week effect for day t is 7% higher than normal. θ_t^{sd} is particularly challenging to estimate, since we might observe a particular "special day" only once a year [see Godfrey and Powell (2000) for methods to update this model].

15.4.3   Handling Pre-Booked Loads

A particular challenge in forecasting demand in truckload trucking is the fact that customers will book orders in advance. We refer to a demand process where there is a gap between when we know about the demand, and when we can act on it, as a lagged information process. This is modeled using the notation:

$\hat{D}_{tt'}$ = The number of new demands that first become known between $t-1$ and t that need to be served at time t'.

$f_{tt'}$ = Point forecast of $\hat{D}_{tt'}$, made before time t.

$F_{tt'}$ = Forecast of the total demand for time t' using the information available at time t.

Here, time t refers to both a day as well as time of day. If today is Tuesday, our forecast of loads on Thursday depends on whether it is 10:00 A.M. on Tuesday or 4:00 P.M. Because of the need to have a truly continuous time forecast, we have to view the time index t as being continuous. However, when we prepare a forecast of the total loads to be served on Thursday, time t' needs to be viewed as an entire day. We have generally found it best to first forecast assuming time t represents an entire day, and then forecast an hour-of-day distribution. Thus, if we forecast that 10 loads will be called in on Tuesday to be served on Thursday, we can use a separate hour-of-day distribution to determine how many of these loads would be known by 11:30 A.M. on Tuesday.

It is important to phase in known demands with your forecast, so that at a time t, we take advantage of what we know with what we do not yet know. The biggest mistake is to forecast demand, and then update the forecast by subtracting what is already known. To illustrate, we might forecast that we will pick up 40 loads in a region. Assume we already have 27 booked loads. It is tempting to update the forecast so that we assume that we have 27 known loads and 40 – 27 = 13 forecasted loads, giving us a total forecast of 27 + 13 = 40. We hope the error in this process is apparent.

There are two methods for phasing in known demands, and we generally have to use both. The first is primarily suited for phasing in demands from numerous small customers, while the second is particularly important when phasing in demands from a small number of large customers.

15.4.3.1   Forecasting Small Customers

When we forecast demand from numerous small customers, we assume they behave like a model known as a Poisson process. This model assumes that the number of calls made, say, before noon on Tuesday, is completely independent of the number of calls made afternoon on Tuesday (or the number of calls yet to be made on Wednesday). For this discussion, assume that we are forecasting the total demand on *day* t' in the future, but that we are forecasting demand at hour t, since as a rule, we need to update forecasts continuously during a day.

$f_{tt'}$ is our estimate of the total number of phone calls that will arrive between $t - 1$ and t. At time t, our forecast of the total number of loads that have to be served at time t' is given by

$$F_{tt'} = \sum_{t'' \leq t} D_{t'',t} + \sum_{t'' > t} f_{t'',t} \qquad (15.8)$$

Thus, our forecast of t' combines what we know as of time t, plus a forecast of what is not yet known. Note that this only produces a point forecast. We can also produce estimates of the variance of the forecast.

15.4.3.2 Forecasting Big Customers

The method described in the previous section does not work for large customers who tend to make a single phone call at some point during the day. They may make their phone call early in the day, but other days the phone call may come in later. The way that Equation 15.8 merges known and forecasted demands does not work for this type of process. Instead, it is better to have a separate forecast for each of these big accounts. Since we know when the account has made its orders known, we can simply use the forecasted before the orders are entered, and use the actual after the orders are entered.

15.5 Capacity Forecasting

Now that we have a basic method for forecasting demand, we next have to forecast capacity, which is the movement of trucks in the future. This is important if we want to know, for example, if we have too many trucks in a region compared to the demand. It is also needed if we want to estimate our ability to get drivers home.

Capacity forecasting is subtle. Older models would project capacity into the future by solving a large space–time network such as that depicted in Figure 15.3, where nodes represent points in space and time, solid links represent the movement of loads (known or forecasted), and dashed lines representing either holding in a location, or moving empty. A point in space is typically one "region," where the continental United States might be divided into 100 regions.

The advantage of space-time networks is that they are easy to understand and communicate, and can be solved as linear programs using commercially available solvers. The problem is that they can do an extremely poor job of modeling the real system, seriously biasing the forecasts. One problem is that they completely ignore the uncertainty in demand forecasts, where the number of forecasted loads from one

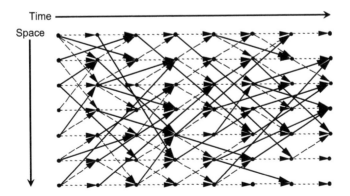

FIGURE 15.3 Illustrative space-time network.

location to another is likely to be a fraction such as 0.17. The second and perhaps more serious error is that the network models the flow of trucks, not drivers, and is unable to capture limits on how much a driver can move, or any other issues such as getting a driver home. We have found that such models can dramatically overestimate the capacity of a fleet, disguising capacity problems in the future.

In the sections which follow, we briefly outline a strategy that is able to forecast capacity into the future, without any loss of detail in how drivers are represented. Section 15.5.1 begins by describing a simple, myopic simulator. Section 15.5.2 shows how we can take this basic simulator and produce a solution that looks into the future.

15.5.1   Simulating a Myopic Policy

The easiest way to forecast capacity in the future is to simply simulate the dispatch process, which we can do by applying the model we first introduced in Section 15.2 iteratively into the future. To describe this more formally (which we need for our discussion in Section 15.5.2), let $X_t^\pi (R_t)$ be a function that represents solving the load-matching optimization problem given by Equations 15.1 through 15.4 at time t, producing a vector x_t that satisfies the constraints 15.2 through 15.4. The problem is solved given the resource state vector R_t that tells us the status of all drivers and loads at time t. The superscript π is an index in a set Π so that we can represent the fact that this is not one function, but a family of functions from which we can choose (often referred to as policies). The model in Section 15.2 is one of these models which we represent using $\pi = M$, where M denotes a myopic policy (i.e., a rule for making decisions that ignores the future).

In a real-time system, we would solve Equations 15.1 through 15.4 where $t = 0$. In this case, $R_0 = (R_0^D, R_0^L)$ tells us the status of all the drivers and loads that we know about right now. We again emphasize that this does not mean all drivers and loads that can be assigned right now. The driver's ETA attribute, and the pickup window of a load, may specify that a driver may not arrive for two days, and the load has to be picked up a week from now. Solving $X_0^\pi (R_0)$ returns a decision vector x_0. We can now combine this information with our modify function. If $x_{0ad} = 1$ (which means that $R_{0a} > 0$), then we now have a driver with attribute $a' = a^M (a, d)$. We write the effect of these decisions on our resource state vector using

$$R_{0a'}^{D,x} = \sum_{a \in A} \sum_{d \in D} x_{0ad} \delta_{a'}(a, d) \tag{15.9}$$

We refer to $R_0^{D,x} = (R_{0a}^{D,x})_{a \in A}$ as the post-decision resource vector for drivers. Equation 15.9 describes how our decisions impact drivers. We also have to model the effect of decisions on loads, which is quite simple. If we assign a driver to a load, the load leaves the system; otherwise it remains in the system, a process that is written as

$$R_{0bd}^{L,x} = R_{0bd}^L - \sum_{a \in A} x_{0ad} \quad d \in D^L \tag{15.10}$$

Now we are going to make the transition from $t = 0$ to $t = 1$ which might represent, for example, a point in time 4 hours later. During this time, we might have a set of phone calls. Earlier, we represented new demands using $\hat{D}_t$. Here, we slightly revise this notation so that our model can easily handle phone calls that provide updates to demands (new orders, changes in orders) as well as updates to drivers (drivers being added to the system, drivers leaving the system, and updates to existing drivers, such as delays in arrival times). These updates are represented as the following random variables:

$\hat{R}_{ta}^{D}$ = Change in the number of drivers with attribute a due to exogenous information that arrived between $t - 1$ and t.

$\hat{R}_{ta}^{L}$ = Change in the number of loads with attribute b due to exogenous information that arrived between $t - 1$ and t.

$\hat{R}_{t} = \left(\hat{R}_{ta}^{D}, \hat{R}_{ta}^{L} \right)$ = Exogenous information arriving between $t - 1$ and t.

We note that $\hat{R}_{t}$ is a function that depends on the drivers and loads already in the system, as well as exogenous information that arrives to tell us how the system is changing. With this notation, we can describe how our system evolves forward in time:

$$R_{t+1,a}^{D} = R_{ta}^{D,x} + R_{t+1,a}$$ (15.11)

$$R_{t+1,a}^{L} = R_{ta}^{L,x} + R_{t+1,a}$$ (15.12)

Equations 15.9 through 15.11, combined with 15.11 through 15.12, tell us how the resource state vector R_{t} evolves from time t to time $t + 1$, given a decision x_{t} and the exogenous information $\hat{R}_{t+1}$.

We know how to find x_{t} (by solving the Equation 15.1 through 15.4). How do we actually obtain $\hat{R}_{t+1}$? We do this by forecasting future updates to the system, and then randomly sampling from this forecast. In a basic model, we might ignore any random events happening to drivers, and simulate only the random arrival of new customer orders. If we have forecasted, say, 0.20 orders will arrive to move from Dallas to New York this Thursday, then we would generate a random integer whose mean is 0.20. For example, we might generate a random variable between 0 and 1; if the random variable is less than 0.20, then we would set the random demand to 1, and otherwise set it to 0. To allow for means greater than 1, we might treat the forecast as the mean of a Poisson random variable and sample from this. A number of popular simulation textbooks describe this process.

Since we have to sample randomly, it is generally a good idea to perform repeated sample realizations and average any statistic that is desired from the model. Let $\hat{R}_{t}^{n}$ be the nth sample of the random information that arrived between $t - 1$ and t. We refer to the sequence $\left(\hat{R}_{1}^{n}, \hat{R}_{2}^{n}, ..., \hat{R}_{t}^{n}, ..., \hat{R}_{T}^{n} \right)$ as a sample path, which is to say a single set of realizations over all the time periods we are interested in.

We now have a process for simulating our system as far into the future as we would like. The only weakness is that our decision function $X_{t}^{\pi}(R_{t})$ always ignores the impact of decisions now on the future. The following section addresses this problem.

15.5.2 Approximate Dynamic Programming Solution

There is a simple way to make our myopic decision function much more sophisticated. Instead of solving the objective function given by Equation 15.1, assume instead that we solve the problem

$$\max \sum_{a \in A} \sum_{d \in D} c_{ad} x_{tad} + \bar{V}_{t}(R_{t}^{x}(R_{t}, x_{t}))$$ (15.13)

where $\bar{V}_{t}\left(R_{t}^{x}(R_{t}, x_{t}) \right)$ approximates the value of being in resource state $R_{t}^{x}(R_{t}, x_{t})$ which depends, of course, on R_{t} and x_{t}. For this class of problems, a reasonable approximation is a linear function, given by

$$\bar{V}_{t}(R_{t}^{x}(R_{t}, x_{t})) = \sum_{a'} R_{ta'}^{D,x} \bar{v}_{ta'}$$ (15.14)

where $\bar{v}_{ta'}$ can be thought of as the marginal value of drivers with attribute a'. Combining 15.9 and 15.14, and using the definition of our terminal attribute function $a^M(a,d)$, allows us to write

$$\bar{V}_t(R_t^x(R_t, x_t)) = \sum_{a' \in A}\left(\sum_{a \in A}\sum_{d \in D} x_{tad}\delta_{a'}(a,d)\bar{v}_{ta'}\right)$$

$$= \sum_{a \in A}\sum_{d \in D} x_{tad}\bar{v}_{t,a^M(a,d)}$$

(15.15)

Combining 15.13, 15.14, and 15.15, with a bit of algebra, produces

$$\max \sum_{a \in A}\sum_{d \in D} c\left(c_{ad} + \bar{v}_{t,a^M(a,d)}\right)x_{tad}$$

(15.16)

which we solve subject to the flow conservation constraints 15.2 through 15.4. The only question we have not answered is: how do we get the values $\bar{v}_{ta'}$?

Fortunately, this is the easy part. When we solve problem 15.1 through 15.4, or problem 15.16 subject to 15.2 through 15.4 using any commercial linear programming package, we also obtain a dual variable that for the flow conservation constraint on drivers Equation 15.2, that tells us the marginal value of a driver with attribute a. We have to keep in mind that we are solving these problems iteratively, where at iteration n we use the sample realization $\hat{R}_t^n$. Let $\bar{v}_{t-1,a}^{n-1}$ be our estimates of the marginal values after iteration $n-1$. Let $\hat{V}_{ta}^n$ be the dual variable we obtained during the nth iteration of our simulation. We then apply exponential smoothing to obtain our value function approximation, using

$$\bar{v}_{t-1,a}^n = (1-\alpha)\bar{v}_{t-1,a}^{n-1} + \alpha\hat{v}_{ta}^n$$

(15.17)

We note that $\hat{v}_{ta}^n$ is used to update $\bar{v}_{t-1,a}^{n-1}$. For further background on this subtle bit of modeling, see the discussion in Powell et al. (2007).

Recall that the attribute a can be quite complicated. Although we do not write it explicitly, while $\hat{v}_{ta}^n$ will depend on the full attribute vector, our value function approximation $\bar{v}_t^n,a$ depends only on a subset of attributes such as location, the domicile of the driver and perhaps his equipment type.

15.5.3 Getting Drivers Home

One of the real challenges of load-matching models is getting drivers home. While this can be quite difficult, our dynamic programming approximation of the previous section already accomplishes this for us, as long as we retain driver domicile as one of the attributes of the value function. In addition, it is also necessary to include logic in the load-matching problem that recognizes that a driver may be close to his home, or is assigned to a load that allows a driver to pick up a load, move to his home, spend a day or two and still deliver on time. The model must include rewards for getting drivers home, or penalties for keeping drivers away from home.

15.6 Demand Management

The focus of real-time dispatch systems is typically on what we should do with a driver, but it is usually the case that we can have a much bigger impact on the company by controlling which loads are accepted. This is particularly true since many orders are booked several days in advance.

Carriers typically accept or reject loads based on issues such as, (*i*) Is this a major account? (*ii*) Is the rate being offered (often expressed in units of dollars per mile) above a minimum? (*iii*) Do I have enough capacity (relative to demand) out of the region where the load originates? and (*iv*) If I accept the load, will this create a situation where I have too many drivers at the destination of the load?

The capacity forecasting logic described in Section 15.5 can produce estimates of the number of drivers (and loads) out of an origin or into a destination several days into the future. In addition, it can even account for the fact that while we may be sending too many drivers into New Jersey, we have a shortage of drivers in nearby eastern Pennsylvania, which means that we may have more capacity in a region than would be forecasted if we simply add up the number of loads terminating in a region.

15.7 Implementation

The use of computers to not only store information about drivers and loads, but also to recommend how drivers should be managed, seems like it should be a major application of operations research. A number of issues have resulted in extremely slow adoption. One problem is that current commercial packages do not handle problems such as trailers, routing drivers through a sequence of several loads, and the uncertainty of forecasts in the future.

15.7.1 Computer Integration

A real-time dispatch system requires having up-to-date information about drivers and loads, information that carriers enter into the computer. There are a number of commercial management information systems designed specifically for truckload motor carriers, and as of this writing, none include an automated driver assignment module. The reason for this is simple. There are thousands of trucking companies with less than 50 drivers, and this is the market when a company starts to use computers. It is only when a company gets a fleet of at least 200 drivers that an automated dispatch system starts to make sense.

Since the 1980s, a small cottage industry emerged to install real-time load-matching systems for truck-load carriers (two of these companies, Princeton Transportation Consulting Group and Transport Dynamics, were founded by students under the supervision of this author). Without question, the Achilles heel of these systems was the interface between the optimization model and the dispatch system. To be successful, this interface has to be seamless, with rapid transmission of information, something that has been achieved only in highly customized applications (and very high cost).

15.7.2 Problem of Data

Not surprisingly, automated dispatch systems depend on quality data. But data "errors" can be quite subtle. Consider the problem of assigning five drivers to five loads as depicted in Figure 15.4. Take a minute to solve the problem in your mind before turning the page.

The mathematically guaranteed, optimal solution provided by the computer is given in Figure 15.5. This may surprise the reader, but it is because the reader has not been provided all the information. The problem is first noticed by the company (which cannot look at the entire solution) when driver B calls in and the dispatcher sees that he has been assigned to load 1 instead of load 2, which seems closer. The problem, actually, is with the data associated with load 3. When this load was first called in, the traffic manager asked "Is it possible to pick the load up before noon? We get busy in the afternoon." Dispatch systems allow the specification of pickup time windows, but these are hard constraints; failure to pick the load up within the window is viewed as a service failure which can lead to the loss of the contract. Only driver A in this group is arriving early enough to meet this constraint. As we can see, however, the request to pick it up before noon was only a preference.

15.7.3 Measuring Compliance

Dispatch systems are successful only if people actually do what the systems tell them to do. Not surprisingly, it is common to measure compliance, which is a statistic that describes the number of driver assignments

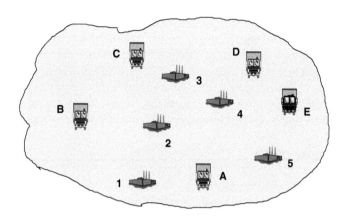

FIGURE 15.4 A driver assignment problem.

where the decision made by the dispatcher agrees with the model. Unsuccessful implementations (where the company runs the model but no-one uses the recommendations) tend to observe around 30–35% compliance. Carefully calibrated models with strong management support might average 80–85% compliance (with some users over 90%). However, users can manipulate these numbers if they feel they are being judged by them.

The issue of user noncompliance has received relatively little attention from the academic community since it is an issue that only arises in a field implementation. Powell et al. (2000b) studied the effect of user noncompliance. After solving the myopic optimization model defined by Equations 15.1 through 15.4, the value of assigning a driver of type a to a load of type b is given by a formula based on the reduced cost:

$$\bar{c}_{ab_d} = c_{ab_d} + \theta \left(v_a^D - v_{b_d}^L \right)$$

where c_{ab_d} is the direct contribution of assigning a driver of type a to a load of type b_d, v_a^D is the dual variable for the driver node of type a (the dual for Equation 15.2), v_{bs}^L is the dual for the load node for a load of type b, and θ is a scaling factor. Our rule is to assign a driver to the load with attribute b_d with the highest value of $\bar{c}_{ab_d}$. We can model user compliance by randomly deciding whether the recommendation is "acceptable" to the dispatcher; if this assigned is (randomly) judged to be unacceptable, we go to the second-ranked load, and so on. If we are modeling perfect user compliance, and if $\theta = 1$, then we are

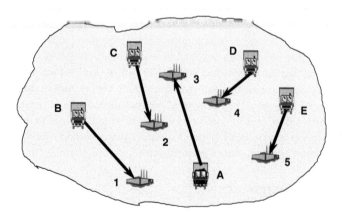

FIGURE 15.5 The computer generated solution.

implementing basically the same solution recommended by solving the global optimization model in Equations 15.1 through 15.4. If we use $\theta = 0$, then we are ignoring the dual variables, and implementing a greedy solution where we do the best for each driver. Intermediate policies are obtained using $0 < \theta < 1$. Figure 15.6 shows total profits as a function of the level of user compliance for $\theta = 0$, 0.75, and 1.00. Note that at an 80% compliance rate (considered quite good in actual applications), the difference between the globally optimal solution and the greedy solution is not large, but there is a noticeable improvement if we use $\theta = 0.75$.

The issue of user compliance is a serious one. We have to recognize that the computer simply does not have all the information needed to make perfect decisions. We have often found that senior management is attracted to models since it provides them some level of control over the decisions made on the dispatch floor. One vice president used the term "dispatcher savant" to describe talented people in operations who otherwise could not be controlled. Responding to the challenges of changing operating philosophies on the dispatch floor, another senior manager remarked "Sometimes, when you can't change the people, you have to change the people." The problem that managers face is identifying the best dispatchers. Each manager works with a different region of the country, or different groups of drivers, making direct comparisons impossible. Too much emphasis on user compliance produces behavior where dispatchers "game the system," manipulating the process to produce the best score. Despite these qualifications, automated systems can add real value in the following ways:

- While it is possible to put too much emphasis on "matching the computer," it is generally the case that the best dispatchers have the highest compliance, but it is very important that the model be of high quality, capturing most operational situations. The early commercial models, despite their tremendous promise, did possess serious limitations.
- Models provide a useful benchmark. Comparing user performance (empty miles, on-time service, getting drivers home) against model performance for the same region and/or the same group of drivers, provides a benchmark that adapts to the unique situations faced by each dispatcher.

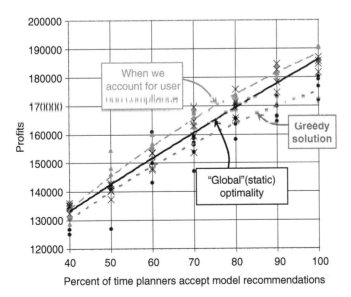

FIGURE 15.6 Value of global optimization in the presence of user noncompliance. (From Powell, W.B., et al., *Transportation Service*, 34, 1, 2000b.)

- Filling in for vacations and departures—Models can be of significant value during times such as when a dispatcher leaves or takes a vacation. Experienced dispatchers can often outperform a model, but the model can provide a genuine safety net for new people.

15.8   Case Study—Burlington Motor Carriers

I knew the Burlington Motor Carriers (BMC) from when they had four or five employees, until they finally closed their doors. The company was started in the mid-1980s by the Burlington Northern Railroad by Dr. Michael Lawrence, a Ph.D. economist with a dream of merging a series of truckload carriers to gain the economies of larger networks. Dr. Lawrence hired Michael Crowe from Schneider National, then (and now) the nation's largest truckload motor carrier, and one with a long history of innovation. Mike had a Master's in Operations Research, and had acquired during his years at Schneider a vision of how operations research could be used to help run a truckload operation.

The story of BMC unfolded in two acts, which we refer to as "Round I" and "Round II."

15.8.1   Operations Research Models—Round I

The first attempt to implement operations research models at BMC occurred while the company was first being formed, combining an entirely new information system with the purchase of four or five companies with established operations. Mike Crowe's vision of what models to use and how to use them demonstrated a deep understanding of the right way to use models within truckload trucking. Rather than focus on using models to assign drivers to loads (a technology pioneered by Richard Murphy at Schneider National), the vision was to focus more on demand management and capacity management. You can have a much greater impact by managing demand than managing capacity. This was aligned with some of our recent research at Princeton (see Powell, 1987, 1996), which focused more on looking into the future (while capturing uncertainty) and less on modeling individual drivers. The ideas were embodied in a software package we dubbed LOADMAP, which was also implemented at the same time at North American Van Lines [Powell et al. (1988)].

The system that Mike Crowe had designed was brilliant, and at least 10 years ahead of its time. Although the strategy emphasized network-wide profits rather than micro-level decisions, it was still important to know basic data such as where the driver was located, where he was headed to, and when he was likely to arrive. Today, many long-haul carriers use satellite systems to provide two-way communication with drivers, but this did not come until the late-1980s. The computers and communication technologies we take for granted today were just being invented in 1985. I realized the project was headed to failure when I met Mike after a conference call with field operations, and listened to his frustration getting basic data such as when a driver might arrive at the destination.

15.8.2   Real-Time Dispatch System—Round II

In 1994, under new management, BMC again attempted a project to perform real-time dispatch. Before describing the details of the project, some background is needed.

15.8.2.1   Bit of History

Our first efforts at fleet management focused on more aggregate level capacity measurements—how many drivers were in a region, how many loads were booked out of a region, and how many loads were booked into a region. Our model would make recommendations such as "move two drivers loaded from region A to region B." These instructions were met with complete mystery by the dispatchers, who would immediately respond "which two drivers?" For them, every assignment was unique. This driver needed to get home to Dallas, another driver was supposed to be available at 2:00 P.M. but was notoriously unreliable, a third driver needed to go on rest for the remainder of the day because he had hit the limit on the number of hours he could drive. Regions were good for planning, but it made a big difference if a driver was on the northern boundary of eastern Pennsylvania or was south of Philadelphia (all in the same region). A driver might be sitting in a yard staring at a trailer in the process of being loaded that had not yet been entered into the computer system. It may be that aggregate, network-level flows drives the economics of a carrier, but the devil is in the details, and our first effort completely ignored this.

By the early 1990s, BMC had moved to its third headquarters, a low-rent office building in a cornfield north of Indianapolis. With a much lower cost structure, the company began to thrive, growing to a respectable $300 million annual revenue and a fleet of about 500 trucks. Under new management, I was contacted again, but this time to implement a real-time dispatch system. Knowing many of the uncertainties, I offered to take on the task as a research project through Princeton University, which introduced its own special issues. While negotiating the contract, the Princeton University grants office first insisted that BMC was welcome to use the results of our work in their research, but had to pay royalties if they wanted to actually use the system. After getting over the vision of a small trucking company writing journal publications, I had to explain that the project was field research. I wanted to observe the process of implementation, take measurements, and publish the results. This is exactly what happened, and the results were published in the prestigious journal *Operations Research* [Powell et al. (2002)].

15.8.2.2 Dispatch System

Through the 1980s and 1990s, one of our most significant achievements was the development of a model which combined the real-time assignment of individual, rather than aggregated, drivers and loads which also looked into the future and captured the uncertainty of future demands. In a separate breakthrough (at the time), we also found a way to route drivers through a sequence of two or more loads, rather than restricting each driver to being assigned to at most one load [Powell et al. (2000a)]. The challenge was solving these problems in real-time (updates could not take more than a second or two), using available computers and algorithms. Just as important—it was not enough just to tell dispatchers what load a driver should pick up, it was still necessary to provide a ranked list of options, a feature from the basic load-matching model that was critical to field implementation.

The project proceeded smoothly. BMC's capable vice president of information technology, Mr. Robert Lamere, handled all aspects of the interface between the corporate information system and the model. However, due to the nature of the technology, the resulting solution was hardly pretty. As with many truckload carriers, BMC used a very popular computer developed by IBM called an AS/400, a fantastic machine for processing data but which would not run languages such as C or C++. The preferred implementation platform for models at the time was Unix-based machines, in our case a Silicon Graphics workstation. Bob worked out a solution where data would be passed from the AS/400 to a PC which then talked to the Unix workstation. Ugly, but it worked. The system went into production about a year after the project started, and we began the painful process of gaining user acceptance.

15.8.2.3 Oh, but could you help us with…

In the middle of the project, I received a call from senior management asking if I could take a look at their network profitability and pricing policies. I had developed a model for this purpose that had been applied to several other carriers, and I offered to apply it to their network. Normally the model would produce a profit-and-loss statement fairly close to actuals, but after several weeks of fiddling, I had to call to tell them that the model did not seem to be working. It was producing results where the operating ratio (total costs over total revenue) was 110 (i.e., costs were running 10% higher than revenues)! I was politely informed "that was about right" !!!

I learned in that phone call that the reason the company had funded the dispatch project was a belief that the reason their costs were 10% higher than revenues was problems with the dispatchers. By this time, we knew that dispatch systems could reduce operating costs by 1% or 2%. They do this by reducing the empty miles traveled, but these are typically only 10% of total miles traveled. A 1% or 2% reduction in total costs is a big number if your profit margin is only 3% or 4%. But they will never drop costs by 10%. The phrase "Houston, we have a problem" was appropriate here.

We had to put the dispatch project on the shelf for about six months to focus on their profitability problem. Truckload trucking is famous for lines such as "we lose money on every load but make it up in volume." Approximately 95% of the total costs of a truckload operation are what economists would call short-term variable costs (i.e., directly related to miles traveled and the size of the fleet). BMC did not have an operational problem. They had a pricing and marketing problem. They were carrying the wrong loads at the wrong price.

Using my network planning model, BMC shrank the company by over 25%, reducing their operating ratio from 110 to below 100 (which is to say, profitable). This is extremely difficult in most companies, but is surprisingly easy to do in the truckload industry. It is not that hard to shrink a company to profitability as long as assets such as fleets are also reduced. Of course, corporate overhead has to be reasonable. With the big problem solved, we returned to the dispatch system.

15.8.2.4   Implementing the Dispatch System

The remainder of the project progressed smoothly. System compliance was tracked daily, and dispatchers were given bonuses for higher levels of compliance. Overall compliance exceeded 80% after six months of use (Fig. 15.7). The vice president of operations particularly enjoyed the sudden control he was given over the behavior of the dispatchers. A major frustration in managing a room full of dispatchers is that they are notoriously independent. It is important for a company to balance on-time service, operating costs (measured primarily through the miles a driver moves empty to pick up a load), equipment productivity (often miles per driver per week), and keep drivers happy (which translates to putting them on long loads and getting them home on time). Not surprisingly, these goals are often competing, and management may decide to shift emphasis from one goal to another as business conditions warrant. The model, however, allowed management to easily raise or lower penalties and bonuses to emphasize different management objectives, and Burlington management took advantage of this feature.

This control, however, proved somewhat illusory. One week we found that system compliance had dropped noticeably. We found that some of the parameters controlling the importance the model puts on these soft issues had been adjusted somewhat dramatically. We reset the parameters (we had direct access to the Silicon Graphics workstation where the model was run) and called the vice president of

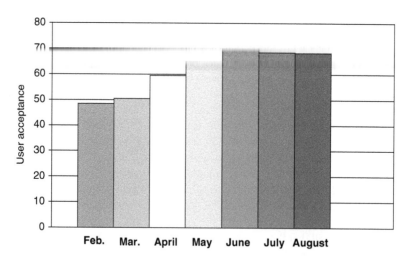

FIGURE 15.7   Evolution of system usage over first six months.

operations to explain that optimization models are a little like artificial hearts. You can use these machines to increase or decrease the speed of a patient's heart, but it is also a nice way to give the patient a stroke. Models work the same way. Just because you adjust the parameter does not mean that all the dispatchers adjust equally quickly.

15.8.2.5 Measuring the Impact

We carefully documented the impact of the model by collecting an extremely valuable dataset. Since the model ran in real-time, we not only received updates to drivers and loads, we were also given, in real-time, actual assignments of drivers to loads. We stored every transaction during a day for approximately 50 days out of a six-month period. This also allowed us to make changes to the model, and then resimulate an entire day using actual transactions. With this system, we could make certain measurements from actual decisions and compare them to what the model recommended at that point in time. We measured empty miles, on-time performance and our ability to get drivers home on time. Figure 15.8 reports reduction in empty miles and improvement in getting drivers home for each day that we captured. We found that the model consistently produced improved performance.

15.8.2.6 Prologue

At first, this seemed like a completely successful project in every measure. In addition, one of my graduate studies, Derek Gittoes, had just started a new consulting firm called Transport Dynamics to implement and maintain these systems. Although the project was quite successful, they did not want to maintain an ongoing research relationship, and a university was not the right organization to provide maintenance and support. By this time, the company was convinced that it was not possible to run a profitable truckload carrier without models to identify profitable accounts and to help the dispatchers. But facing continuing cost pressures, they decided to move forward with the load profitability and driver dispatch systems without maintenance.

A few years later (post year 2000) I received a call saying that the system had not survived the post-2000 transition. Our model was specifically designed to be Y2K compliant (i.e., we could handle dates in two-digit or the workstation format, such as 00, 01, 02), but we suspected a problem either in the PC interface between the AS/400 and the workstation. But when I asked what their level of user compliance was, they reported that it had dropped to around 35%, a level that I understood to mean that they were no longer using the dispatch system. Knowing that they did not have a budget for maintenance, I recommended that they simply shut off the dispatch system, which they did.

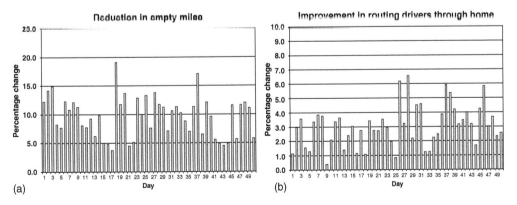

FIGURE 15.8 Day-by-day comparison of model recommendations against actual performance for (a) reducing empty miles and (b) getting drivers home.

Three months later, I received another phone call from the president asking how we could get the system running again. The problem, he explained, was that while the experienced dispatchers seem to do fine, they "got killed" each time one of them went on vacation or left the company. This was the first time I had solid evidence that a real-time dispatch system offered the kind of value that a manager could clearly and unambiguously recognize.

With a renewed commitment to the use of models, the company looked at bids from Transport Dynamics and a competitor. They had just hired a new manager from a competitor, and not surprisingly decided to go with the competitor. BMC closed its doors and liquidated two years later. Sigh.

References

Godfrey, G. and W.B. Powell, Adaptive estimation of daily demands with complex calendar effects, *Transportation Research*, Vol. 34, No. 6, pp. 451–469 (2000).

Powell, W.B. An operational planning model for the dynamic vehicle allocation problem with uncertain demands, *Transportation Research*, Vol. 21B, No. 3, pp. 217–232 (1987).

Powell, W.B., A stochastic formulation of the dynamic assignment problem, with an application to truckload motor carriers, *Transportation Science*. Vol. 30, No. 3, pp. 195–219 (1996).

Powell, W.B., W. Snow, and R. K.-M. Cheung, Adaptive labeling algorithms for the dynamic assignment problem, *Transportation Science*, Vol. 34, No. 1, pp. 67–85 (2000a).

Powell, W.B., M.T. Towns, and A. Marar, On the value of globally optimal solutions for dynamic routing and scheduling problems, *Transportation Science*, Vol. 34, No. 1, pp. 50–66 (2000b).

Powell, W.B., A. Marar, J. Gelfand, and S. Bowers, Implementing operational planning models: a case application from the motor carrier industry, *Operations Research*, Vol. 50, No. 4, pp. 571–581 (2002).

Powell, W.B., B. Bouzaiene-Ayari, and H.P. Simao, Dynamic Models for Freight Transportation," Handbooks in Operations Research and Management Science: Transportation (G. Laporte and C. Barnhart, eds.), Elsevier, Amsterdam (2007).

Powell, W.B., Y. Sheffi, K. Nickerson, K. Butterbaugh, and S. Atherton, Maximizing profits for North American van lines' *Truckload Division: A New Framework for Pricing and Operations, Interfaces*, Vol. 18, No. 1, pp. 21–41 (1988).

16

Classic Transportation Problems

K. Bulbul
Gunduz Ulusoy
A. Sen
Sabanci University

In this chapter, a brief introduction into classical transportation problems underlying logistics engineering is presented. A broad range of the major classical transportation problems has been included in this chapter although the list may still not be complete. These problems can rarely be employed in their generic form to solve practical problems of logistics engineering, but still they constitute building blocks for modeling real-life problems. For each type of problem, a verbal definition of the problem in its generic form is given followed by some extensions of the generic form and by possible application areas in practice. For some of the problems, a mathematical formulation is also provided. These problems have been investigated for some time now particularly by the Operations Research community and thus a rich body of solution procedures exists. Due to space limitations, only one or a few representative solution approaches have been included here for each problem type. The chapter concludes with a case describing the distribution operations at a central warehouse in a major Turkish retail chain.

16.1   Shortest Path Problem

The shortest path problem (SPP) is one of the fundamental network flow problems. A large number of problems from diverse areas such as routing in telecommunication networks to DNA sequencing can be formulated as a SPP or as a variant [1]. Furthermore, the shortest path formulation can also serve as a sub-model in larger, more complex models, such as, for example, in determining an optimal or near-optimal integer solution to the set covering formulation of the capacitated vehicle routing problem [2]. It has a simple structure and is relatively easy to solve. SPP will be defined here over directed graphs with no negative length cycle(s). Efficient algorithms exist for this version of SPP. The problem of determining the shortest path on a graph with negative length cycle(s), on the other hand, is an $\mathcal{NP}$-complete problem. The basic SPP can then be defined as follows: Given a directed graph $G = (V, A)$, where V is the set of vertices, and A is the set of arcs (i, j) each with a length c_{ij}, find the shortest directed path from vertex s to vertex t, $s, t \in V$. Note that this definition assumes the existence of a directed path between any two nodes i and j of the graph. If that is not the case, then an arc (i, j) with a relatively high c_{ij} value is added to the graph. The arc length c_{ij} can represent other relevant measures such as cost or duration depending on the nature of the SPP investigated.

A linear programming (LP) problem formulation for SPP is given in the following, which represents SPP as a minimum cost network flow problem in which one unit of flow is sent from the source s to the sink t. Thus, all vertices except s and t are transshipment vertices with one unit of flow entering and leaving. The variable x_{ij} denotes the flow on arc (i, j), and the cost of sending one unit flow on arc (i, j) is given by c_{ij}.

$$\min \sum_{(i,j) \in A} c_{ij} x_{ij} \tag{16.1}$$

$$\sum_{(s,j) \in A} x_{sj} = 1 \tag{16.2}$$

$$-\sum_{(i,j) \in A} x_{ij} + \sum_{(j,i) \in A} x_{ji} = 0 \qquad \forall j \in V \setminus \{s, t\} \tag{16.3}$$

$$-\sum_{(j,i) \in A} x_{jt} = -1 \tag{16.4}$$

$$x_{ij} \geq 0 \qquad \forall (i, j) \in A \tag{16.5}$$

Note that the variables x_{ij} are continuous in the formulation given. However, all variables in all extreme point optimal solutions of this linear program are either 0 or 1 because the constraint matrix is totally unimodular. Therefore, we can state that x_{ij} is 1 if the shortest path from s to t includes arc (i, j), and 0 otherwise. An optimal solution to this linear program may be obtained by standard LP solvers. However, many more effective algorithms are available for different kinds of SPP. Some of these polynomial time algorithms will be introduced in the remainder of this section.

16.1.1   Single Source SPP

Three algorithms are presented in this subsection, which determine the shortest paths from one source vertex to all other vertices on a directed graph. The first two are label setting algorithms and the last one is a label correcting algorithm [1]. They are based on an important property of shortest paths on a directed graph with no negative length cycles: Given a shortest path P from a vertex s to another vertex k, then the path from s to any other vertex h on P is the shortest path from s to h. This property implies the existence of a shortest path tree in the form of an out-tree emanating from the source vertex and reaching all other vertices of the graph. In the following, $d(j)$ represents the path length from the source vertex to vertex j. Alternatively, $d(j)$ is also referred to as the distance label for vertex j. In order to construct the shortest paths once the algorithm terminates, predecessor indices are maintained and updated along with the distance labels.

16.1.1.1 SPP on Acyclic Graphs

The vertices of an acyclic graph can be ordered such that if $(i, j) \in A$, then $i < j$. Such an ordering is referred to as topological ordering. The shortest paths from the source vertex 1 to all other vertices can be found by examining all nodes one by one in topological ordering because the shortest path to vertex $i + 1$ can only go through the vertices $1, ..., i$. Thus, in iteration i, we scan all arcs (i, j) emanating from i, and update the path length $d(j)$ from vertex 1 to vertex j by $\min\{d(j), d(i) + c_{ij}\}$. Upon completion of iteration i, $d(j)$ denotes the shortest path length from 1 to j that only goes through the vertices $1, ..., i$. The algorithm terminates after $|V|$ iterations when all arcs are examined and is indeed the most efficient possible technique for obtaining the shortest paths from one vertex to all other vertices in acyclic graphs. This algorithm remains valid even if one or several arc lengths are negative.

16.1.1.2 Dijkstra's Algorithm

Dijkstra's algorithm finds the shortest paths from one vertex to all other vertices of the graph with non-negative arc lengths [3], and it allows for directed cycles in the graph. The algorithm assigns one of the two types of labels to each node: the permanent distance label and the temporary distance label. A permanent (temporary) distance label for a vertex i represents the length (the upper bound on the length) of the shortest path from the source vertex s to i. Initially, all vertices i except the source s are assigned the temporary distance label $d(i) = \infty$, and s is assigned the permanent distance label $d(s) = 0$.

In each iteration of the algorithm, the label of the vertex i with the minimum temporary distance label is made permanent. Then, as in the algorithm described earlier for acyclic graphs, all arcs (i, j) emanating from i are scanned, and for all vertices j with a temporary distance label and so that $(i, j) \in A$, the temporary distance labels are updated as $\min\{d(j), d(i) + c_{ij}\}$. The algorithm terminates when all vertices receive permanent distance labels.

For directed graphs with non-negative arc lengths, Dijkstra's algorithm is the most efficient algorithm for finding the shortest path from a given vertex to all other vertices of the graph. Several implementation suggestions for improvements on running time performance have been proposed for Dijkstra's algorithm (see, e.g., Dial [4], Fredman and Tarjan [5], Ahuja et al. [6]). Among these, the Fibonacci heap implementation suggested by Fredman and Tarjan results in the best available strongly polynomial time running time.

Note that in both algorithms stated earlier the shortest path from an origin vertex to a destination vertex can be obtained by terminating the algorithm once the destination vertex has been reached. This is a basic feature of the label setting algorithms.

16.1.1.3 Ford-Bellman-Moore Algorithm

The Ford-Bellman-Moore algorithm allows for arbitrary arc lengths and is a label correcting algorithm with a refined version stated by Pape [7,8]. The graph can have negative arc lengths given that no negative length cycle exists. A dynamic list of vertices called the queue is maintained, and initially, the only element of the queue is the source vertex s. Furthermore, $d(s) = 0$, and $d(j) = \infty$ for all other vertices j. At each iteration, the vertex at the head of the queue, say vertex i, is selected. For each arc $(i, j) \in A$, if the condition $d(j) > d(i) + c_{ij}$ is satisfied, then $d(j)$ is updated as $d(j) = d(i) + c_{ij}$. If j is not an element of the queue, then j is added to the queue. Note that a vertex j can enter and leave the queue several times throughout the implementation of the algorithm. Thus, no distance label becomes permanent until the algorithm terminates, that is, the queue is empty. At termination, $d(i) + c_{ij} - d(j) = 0$ for all $j \in V$ and $(i, j) \in A$. In other words, no reduction in path length $d(j)$ is possible anymore by arriving at vertex j using any other arc than the current one.

16.1.1.4   Some Variants of the Single Source SPP

A minor modification of Dijkstra's algorithm is applied to generate the shortest paths from every vertex of a graph to a single vertex t, called the sink. This time the distance label $d(i)$ is associated with the path length from vertex i to sink t. Assume that the distance label for vertex j is declared as permanent. Then, each incoming arc (i, j) is examined for possible update of temporary labels $d(i)$ to $\min\{d(i), c_{ij} + d(j)\}$. The algorithm terminates when all distance labels become permanent.

In the *time-constrained* SPP, arc traversal times t_{ij} are defined in addition to the arc lengths c_{ij}. The problem is then to determine the shortest paths from a source s to all other vertices with the additional constraint that the duration of no-directed path exceeds an upper bound T.

16.1.2   All-Pairs SPP

A trivial solution to this problem would be the repeated application of the Dijkstra's algorithm stated earlier. But a better, more efficient algorithm developed for that purpose is the Floyd–Warshall algorithm [9] which belongs to the class of label correcting algorithms. The algorithm allows for negative arc lengths as long as no negative length cycle exists. It is built on the following necessary and sufficient optimality conditions for the all-pairs SPP: $d[i, j] \leq d[i, k] + d[k, j]$ $\forall$ $i, j, k \in V$ in an optimal solution, where $d[i, j]$ represents the length of some directed path from vertex i to vertex j.

The Floyd–Warshall algorithm operates on a distance matrix of size $|V| \times |V|$, where the element (i, j) of this matrix is $d[i, j]$. Initially, $d[i, i] = 0$ for all $i \in V$, $d[i, j] = c_{ij}$ for all $(i, j) \in A$ and $d[i, j] = \infty$ otherwise. In its initial form, a finite entry (i, j) of the distance matrix represents the shortest path between vertices i and j consisting of a single arc. The distances in the distance matrix are further improved by making use of the optimality conditions by checking all possibilities for new alternate paths between nodes i and j. This is accomplished through a loop over the vertices of the graph, where for each vertex k of the loop, each element $[i, j]$ of the distance matrix is updated by $d[i, j] = \min\{d[i, j], d[i, k] + d[k, j]\}$. Thus, at iteration k of the algorithm, we check whether inserting vertex k into the path between vertices i and j decreases the path length $d[i, j]$. The algorithm terminates once the loop is fully executed over all vertices. The shortest paths may be constructed by maintaining predecessor indices and updating them along with the distance labels.

16.1.3   Finding the K-Shortest Paths

In certain applications, knowing the shortest path may not be enough. Information on the second, third, ..., Kth shortest path between two vertices might be useful. For example, distributing the traffic over K alternate routes may require such information. Knowing the next shortest paths would also be useful when developing contingency plans for the case where the shortest path might not be available for some reason. Algorithms have been developed for finding the K-shortest paths from a vertex to all other vertices and from every vertex to every other vertex on a graph.

16.1.3.1   Single Source K-SPP

The K-shortest paths algorithm to be introduced here is the double sweep algorithm [10]. The underlying idea of this algorithm is similar to that of the Floyd–Warshall algorithm. The algorithm operates on a matrix D^m, and each element d_{ij}^m in D^m is a $(1 \times K)$ vector in which each component d_{ijh}^m represents the hth shortest path from vertex i to vertex j, and m denotes the iteration number. Initially, a feasible D^0 would be obtained by setting $d_{ii}^0 = 0$ for all $i \in V$, by entering the lengths of the existing arcs as the first components of the corresponding K-shortest paths vectors, and by setting all other components of the vectors equal to ∞.

The algorithm employs generalized addition and minimization operations in order to handle the operations on the K-dimensional vectors. At each iteration of the double sweep algorithm, the already existing K-shortest paths from the source vertex 1 to each vertex j are tested whether shorter paths can replace them. This is accomplished by replacing the already existing K-shortest paths with shorter paths which may be obtained by combining paths from vertex 1 to h and from vertex h to j. Considering all vertices $h \in V, h \neq j$, completes a loop. Executing this loop for all vertices $j \in V$ constitutes an iteration. The double sweep algorithm terminates once two consecutive iterations produce identical results.

16.1.3.2 All-Pairs K-SPP

For finding the K-shortest paths from each vertex of a graph to all other vertices of the graph, the generalized version of the Floyd–Warshall algorithm is employed [11]. The algorithm remains basically the same, except that the addition and minimization operations are replaced by their generalized versions employed in the double sweep algorithm as well.

16.2 Minimum Spanning Tree Problem

Many infrastructure design problems involve connecting a set of spatially distributed points by installing links so that the total cost of constructing a connected network is minimum. Building transportation networks, for example, highways and railroads, is a prime example for such problems and plays an important role in strategic logistics system design and planning. For instance, Goodaire and Parmenter [12] discuss an example in which a county administration faces the problem of paving a subset of the county roads so that any two towns in the county are connected by paved roads. In graph theory, this problem is known as the minimum spanning tree (MST) problem. In a spanning tree, all nodes of a network are connected by unique paths between every pair of nodes. A graph with five nodes and a corresponding spanning tree is illustrated in Figure 16.1. In the MST problem, we search for a spanning tree which minimizes the sum of the arc costs in the tree. The arc costs may be related to link construction costs, travel times, etc.

The MST problem arises in direct applications of network design as discussed earlier; however, MST does also play an important role as a subproblem in the algorithms designed for other important combinatorial optimization problems. For instance, Christofides' and MST-based heuristics for the traveling salesman problem (TSP) discussed in "Constructive Heuristics for TSP," a subsection of Section 16.5.3.2, require an MST in order to construct good TSP tours.

16.2.1 Solution Methods for MST

The MST problem is considered to be one of the cornerstones of combinatorial optimization and was pioneered by Boruvka as early as in 1926, and during 1950s the problem was attacked by many

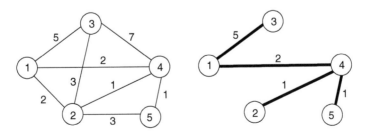

FIGURE 16.1 Example of a network and a corresponding spanning tree.

researchers [13]. This problem can be solved effectively even for very large instances. Kruskal's algorithm [14] and Prim's algorithm [15] are the two best-known algorithms for this problem, and we discuss them briefly. In addition, there are also hybrid algorithms that borrow ideas from these two basic algorithms, for example, Sollin's algorithm [16]. For more recent and faster algorithms, the reader is referred to Chazelle et al. [17], Chazelle [18], Pettie and Ramachandran [19], Karger et al. [20] and the references in these papers.

Both Kruskal's algorithm and Prim's algorithm are greedy in nature. In other words, at each iteration an arc with minimum cost is selected from a list of candidate arcs and added to the solution. The two algorithms differ in how they build the candidate list. In Kruskal's algorithm, the arcs are initially sorted in nondecreasing order of their costs. Then, arcs are added to the solution one by one from this list, and we skip an arc in the list if it forms a cycle with the arcs already present in the solution. (Note that a tree does not contain a cycle by definition.) In Prim's algorithm, we start with a single node i and add the arc (i, j) with lowest cost incident to i to the solution. In the next iteration, we pick an arc with lowest cost incident to either i or j and add it to the solution. We continue until all nodes in the network are spanned.

Although these algorithms are very simple in nature, many elaborate implementations with advanced data structures exist that improve on the running times of the basic algorithms stated earlier [1].

16.3 Transportation Problem

The transportation problem is one of the first problems that was studied in detail in the operations research literature. It has direct applications; however, even more importantly, it appears as a subproblem that needs to be solved frequently in more elaborate logistics design and operation problems. We describe one such application in Section 16.3.2.2.

The transportation problem was introduced by Hitchcock [21] in 1941 and is concerned with satisfying the demands of m demand nodes for a single commodity by shipments from n supply nodes. The demand of demand node $j = 1, ..., m$, is denoted by d_j while s_i represents the total supply of supply node $i = 1, ..., n$. The unit transportation cost and the shipment quantity between supply node i and demand node j are denoted by c_{ij} and x_{ij}, respectively. The objective is to minimize the total transportation costs so that the demands are satisfied while no supply node ships more than its capacity [22,23]. The corresponding network representation is given in Figure 16.2. Note that in different contexts, supply nodes may be referred to as sources, suppliers, production centers, factories, origins, etc. Similarly, demand

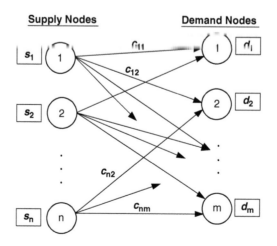

FIGURE 16.2 Generic transportation problem.

nodes may be called as customers, sinks, warehouses, etc., depending on the specific application. For calculating the cost of shipping between supply and demand nodes, Simchi-Levi et al. [24] define transportation rates. The transportation problem can be stated as a linear program:

$$\min \sum_{i=1}^{n} \sum_{j=1}^{m} c_{ij} x_{ij} \tag{16.6}$$

$$\sum_{j=1}^{m} x_{ij} = s_i \quad i = 1, 2, ..., n \tag{16.7}$$

$$\sum_{i=1}^{n} x_{ij} = d_j \quad j = 1, 2, ..., m \tag{16.8}$$

$$x_{ij} \geq 0 \quad i = 1, 2, ..., n \quad j = 1, 2, ..., m \tag{16.9}$$

In the model above, the problem is uncapacitated because there are no upper bounds on the variables x_{ij}. If the demand and supply constraints are stated as equalities as in 16.7 and 16.8, we have a balanced transportation problem, and the total demand $\sum_{j=1}^{m} d_j$ must equal the total supply $\sum_{i=1}^{n} s_i$ in order for the problem to have a feasible solution. If $\sum_{i=1}^{n} s_i \neq \sum_{j=1}^{m} d_j$, and one wishes to use the model stated earlier, then a dummy supply or demand node, as appropriate, and the corresponding arcs are added to the network. For instance, if $\sum_{i=1}^{n} s_i < \sum_{j=1}^{m} d_j$, then the original problem is infeasible. To restore feasibility, the network is augmented with a dummy supply node $n + 1$ with $s_{n+1} = \sum_{j=1}^{m} d_j - \sum_{i=1}^{n} s_i$, and one arc with high cost to each demand node from node $n + 1$. Thus, we attempt to satisfy as much of the demand as possible with minimum cost [25].

The constraint matrix of the linear program model given earlier is totally unimodular, that is, any extreme point optimal solution of the linear program given earlier is integral for all objective functions as long as all supply and demand values are integers. Hence, even if integrality of the variables x_{ij} is required, these constraints can safely be ignored [1,26,27].

16.3.1 Variants of the Transportation Problem

16.3.1.1 Assignment Problem

The special case of the transportation problem with unit supply and demand quantities is referred to as the assignment problem. For instance, if n jobs are to be performed on m machines, then the problem of assigning these jobs to these machines is an assignment problem where the cost of assigning job i to machine j is c_{ij}. The Hungarian method [28] solves this problem efficiently.

16.3.1.2 Transshipment Problem

In the transportation problem, the network is bipartite, that is, no arcs are present between two supply or two demand nodes. However, such arcs are allowed in the transshipment problem in addition to transshipment nodes that are neither source nor demand points. For a transshipment node i, $d_i = s_i = 0$, and the total inflow must equal the total outflow. In other words, when in an application shipment of goods is allowed or desired between two source and/or two demand nodes, or if some nodes are only used as transfer locations to final destinations, then we have to use a transshipment model. Note that a transshipment problem can be converted into a corresponding transportation problem in which only the demand and supply nodes are present. For obtaining the arc costs in this transportation problem, one needs to find the shortest paths between all pairs of demand and supply nodes in the original transshipment network in which the arc lengths are the unit shipment costs (for details, see William [29]).

16.3.2 Applications of the Transportation Problem

16.3.2.1 Warehouse Layout

Consider a warehouse with n items to be stored and m docks for loading and unloading. Let c_{ij} be the cost of moving one unit of item i from dock j to its corresponding storage area. Then the warehouse layout problem is to assign items to docks in order to minimize the total cost of carrying goods inside the warehouse. The carrying cost can be related to forklift usage [1].

16.3.2.2 Storage Space Allocation

Consider a scenario in which storage space in a container terminal is determined and allocated to keep transportation and handling costs as low as possible and to satisfy the throughput requirements of the shipment plan. This problem is modeled on a rolling horizon basis in two stages which requires solving a transportation problem at each iteration [30].

This application accounts for all aspects of storage and material handling at a container terminal, such as allocation of storage space and operation of cranes and transporters in the terminal. In the first stage, the authors solve for the total number of containers to be placed on storage blocks in order to balance the workload. Then, in the second stage the number of containers for each vessel in each block is determined in order to minimize the total container movement where a transportation problem is solved. This application is illustrated in Figure 16.3.

16.3.3 Solution Methods for the Transportation Problem

Transportation problems were among the first linear programs that were explicitly stated, studied, solved, and used in the industry [31]. Note that the transportation problem is a special case of the minimum cost network flow problem for which there are a number of effective algorithms, for example, the out-of-kilter and auction algorithms [1,26,27,32]. In addition, the LP model stated in Section 16.3 can be solved by readily available LP solvers. So, there are many ways of solving the transportation problem using more general techniques; however, there are also very effective algorithms developed specifically for solving the transportation problem and its variants.

16.3.3.1 Classical Methods

One essential solution technique for solving transportation problems is the transportation simplex algorithm. It incorporates two main steps as many other optimization techniques. First, an initial feasible solution is obtained, and then we move to a better solution at each iteration until no further improvement in the objective function is possible.

Different techniques exist for obtaining a starting feasible solution in the transportation simplex algorithm, such as the Northwest (NW) corner [33], minimum cost, and Vogel's techniques [34]. In the NW corner technique, the cost of shipping goods is not considered while the minimum cost technique takes into account the absolute values of the unit shipping costs. On the other hand, Vogel's technique is based on the relative magnitudes of the unit shipping costs rather than their absolute values.

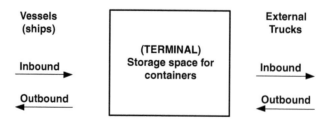

FIGURE 16.3 Operation of a container terminal.

Once an initial feasible solution is available, one constructs a feasible solution with a lower total cost at each iteration of the transportation simplex algorithm. The stepping stone technique [33] is the best-known method to obtain an improved feasible solution iteratively and is based on the duality theory of linear programming. Arshom and Khan [35] present an alternative to the stepping stone technique.

16.3.3.2 Exterior Point Methods

Classical methods start with a primal feasible solution and look for dual feasibility while the exterior point simplex algorithms (EPSA) [36] start with a dual feasible solution and may violate this condition at intermediate steps until it is restored again at termination. Papamanthou et al. [36] present an EPSA and demonstrate that such techniques may prove more effective with appropriate initializations. Their results suggest that their method is on average 4.5 times faster than the classical primal simplex methods.

16.4 Arc Routing Problems

Arc routing problems (ARPs) deal with satisfying the demand for service in a least-cost manner with or without constraints over the edges and/or arcs of a graph $G = (V, E, A)$, where V, E, A represent the vertex set, the undirected edge set, and the directed edge (arc) set, respectively. Examples of practical problems in this context can be listed as mail delivery, snow removal, and school busing, among others. Excellent surveys of ARPs and their real-life applications are provided by Assad and Golden [37], Eiselt et al. [38,39], and in the book edited by Dror [40]. Each edge $(i, j) \in E$ has a traversal cost $c_{ij} \geq 0$ (in general, an additive non-negative real-cost function). Beyond actual traversal cost these costs can represent different measures such as edge traversal duration or length. Unless otherwise stated, the cost matrix is symmetric and it satisfies the triangle inequality: $c_{ik} + c_{kj} \leq c_{ij}$ for all vertices $i, j, k \in V$. In general, the edge traversal cost differs depending on whether service is provided or not when traversing an edge. Edge traversal without providing service is called deadheading. In logistics problems, there can be one or more vehicles involved and the vehicle(s) may be capacitated or not. The vehicles in the fleet may be identical (homogeneous fleet) or the fleet might consist of more than one type of vehicle (heterogeneous fleet).

In the following, ARPs will be studied under three different main headings: the Chinese postman problem (CPP), the rural postman problem (RPP), and the capacitated arc routing problem (CARP).

16.4.1 Chinese Postman Problem (CPP)

In the CPP, the aim is to determine the least-cost tour obtained by starting from a given vertex on a graph and returning to the same vertex after traversing all the edges of the graph at least once. The problem was first proposed by the Chinese mathematician Meigu Guan in 1962 [41]: "A mailman has to cover his assigned segment before returning to the post office. The problem is to find the shortest walking distance for the mailman." The CPP defined on a graph $G = (V, E)$ is called the Undirected CPP (UCPP). Similarly, the CPP defined on a graph $G = (V, A)$ is called the Directed CPP (DCPP). Graph G then defines a road network with one-way streets. When due to physical and/or traffic conditions, the road network consists of both one-way and two-way roads, that is, $G = (V, E, A)$, the problem becomes the mixed Chinese postman problem (MCPP).

Obviously, if it is possible to construct a tour such that all the edges are traversed only once, then this tour is the least-cost tour. A tour that starts in one of the vertices of a connected graph and returns to the starting vertex after traversing each edge once and only once is called the Eulerian cycle and the connected graph is said to be unicursal or Eulerian. When solving CPP on a graph G, one would, in general, not expect to deal with an Eulerian graph G_E. If indeed this is not the case, then one needs to transform the existing graph G into G_E by replicating some of the edges of G resulting in the least cost augmentation. Hence, it is important to know the necessary and sufficient conditions for a connected graph to be Eulerian.

These conditions were first proven by Euler (1736) for an undirected graph. The necessary and sufficient conditions for undirected, directed, and mixed graphs are stated as follows [42]:

The necessary and sufficient condition for an undirected and connected graph G to be Eulerian is that each node of G must have an even degree, that is, an even number of incident edges.

If G is directed and strongly connected, G is said to be Eulerian, if and only if the number of arcs entering and leaving each node are equal. A graph is strongly connected, if there is a directed path between any two nodes of the graph.

If G is a mixed and connected graph, G is said to be Eulerian, if and only if it is even and balanced. A graph is even, if every vertex is incident to an even number of directed and undirected edges. A graph is balanced, if for every subset V' of vertices, the difference between the number of directed edges from V' to the remaining set of vertices V'' and the number of directed edges from V'' to V' is less than or equal to the number of undirected edges joining V' and V''.

16.4.1.1   Undirected Chinese Postman Problem UCPP

The least cost augmentation so as to transform the existing graph G into an Eulerian graph G_E can be achieved in polynomial time by solving a matching problem [43]. For obtaining the Eulerian cycle, several algorithms have been suggested (Fleischner [44]). The following algorithm is one proposed by Edmonds and Johnson [43].

Step 1. Trace a simple tour that may not contain all vertices. If all edges have been included in the tour, stop.
Step 2. Consider any vertex v on the tour incident to an edge not on the tour. Form a second tour starting at v not overlapping with the first one.
Step 3. Merge the two tours into a single tour. If all edges have been traversed, stop. Otherwise go to Step 2.

16.4.1.2   Directed Chinese Postman Problem DCPP

The least cost augmentation for transforming the existing graph G into an Eulerian graph G_E can be achieved in polynomial time for DCPP as well. Recall that for a solution to exist, G_E has to be strongly connected. To obtain the least cost augmentation, a transportation problem is solved [43,45,46]. Let the nodes for which the number of incoming arcs exceeds the number of outgoing arcs be the source nodes with a supply equal to this difference. Similarly, let the nodes for which the number of outgoing arcs exceeds the number of incoming arcs be the demand nodes with a demand equal to this difference. The problem is balanced in the sense that total supply is equal to total demand. The objective is to minimize the total cost of arcs augmented to obtain an Eulerian graph G_E. Once G_E is obtained, then the following algorithm reported in Edmonds and Johnson [43] with a reference to van Aardenne Ehrenfest and de Bruijn [47] is applied.

Step 1. Construct a spanning arborescence rooted at any vertex v_r.
Step 2. Label all the arcs as follows: order and label the arcs originating from v_r in an arbitrary fashion; order and label the arcs of any other vertex consecutively in an arbitrary fashion, as long as the last arc is the arc used in the arborescence.
Step 3. Obtain an Euler tour by first following the lowest labeled arc emanating from an arbitrary vertex; whenever a vertex is entered, it is left through the arc not yet traversed having the lowest label. The procedure ends with an Euler circuit when all arcs have been covered.

16.4.1.3   Mixed Chinese Postman Problem MCPP

In MCPP, a major line of research to obtain the least cost augmentation while assigning a direction to every edge has been to formulate and solve the problem as an integer linear programming problem. An example would be the branch and cut approach proposed by Grötschel and Win [48]. The van Aardenne-Ehrenfest and de Bruijn algorithm described earlier can then be applied to determine the Euler tour on the augmented directed and symmetric graph. Another optimal solution procedure is proposed by Nobert and Picard [49]. Since MCPP is $\mathcal{NP}$-hard [50], heuristic solution procedures have also been proposed (Frederickson [51], Christofides et al. [52], Pearn and Liu [53], and Raghavachari and Veeresamy [54]).

16.4.2 Rural Postman Problem (RPP)

Rural postman problem is introduced by Orloff [45]. As opposed to CPP, in RPP only a subset of the edges is to be traversed. In a real-life context, RPP has more potential applications since it allows for the case where some of the edges do not request service.

The undirected version of RPP (URPP) can be stated as follows: Given a connected graph $G = (V, E)$ with non-negative edge traversal costs and a set of edges $E_R \subseteq E$ requesting service, determine the least cost tour starting and ending at a given vertex and traversing edges in E_R at least once. The directed version of RPP (DRPP) is defined over the graph $G = (V, A)$ with the set of arcs requesting service being $A_R \subseteq A$. Note that if $E_R = E$ ($A_R = A$), then the problem reduces to CPP (DCPP). Both the undirected and the directed versions of RPP are shown to be $\mathcal{NP}$-hard [55]. A comprehensive survey of RPP is given by Eiselt et al. [39]. Note that given the edges (arcs) requesting service constitute a connected (strongly connected) graph G^*, then a UCPP (DCPP) can be solved on G^*.

Solution procedures for both undirected and directed cases are developed making use of 1-matching and transportation problems, respectively [45]. More recently, URPP has been solved to optimality by Corberan and Sanchis [56], and DRPP by Christofides et al. [57]. Heuristic procedures have been proposed by Pearn and Wu [58] and Hertz et al. [59].

The mixed RPP (MRPP) is defined on a mixed graph $G = (V, E, A)$, where there are two sets of edges $E_R \subseteq E$ and arcs $A_R \subseteq A$ with a demand for service. In cases where the edges and arcs to be traversed include all the edges and arcs of the graph, then the problem becomes identical with MCPP. Hence, MRPP is $\mathcal{NP}$-hard. A constructive type heuristic and a tabu search approach employing both short- and long-term memory are developed by Corberan et al. [60].

16.4.3 Variants of the ARP

Capacitated CPP (CCPP)

In CPP, a single vehicle with infinite capacity is assumed. This assumption, of course, is far from representing reality. Christofides [61] formulates CCPP by imposing a finite capacity W on the vehicles employed in a UCPP. The heuristic solution procedure suggested develops feasible cycles without exceeding W and removes the edges of the cycles generated from the graph. When feasible cycles cannot be generated anymore, then the graph is augmented to create an Euler tour, and the procedure continues until all arcs are served.

A related problem is the m–CPP, where there are m–postmen to traverse the edges. m–CPP is of interest both from theoretical as well as practical point of view. This problem is formulated for the undirected case by Frederickson et al. [62] with the objective of minimizing the length of the longest route. They show that this problem is $\mathcal{NP}$-hard and propose a heuristic.

Windy Postman Problem (WPP)

Windy postman problem was first introduced by Minieka [63]. In this problem, the cost of traversing an edge depends on the direction of traversal, that is, the cost matrix is not symmetric. It has been shown that WPP is $\mathcal{NP}$-hard [64,65]. If the graph on which WPP is defined is Eulerian, then the problem can be solved in polynomial time [66]. An integer LP formulation is given by Grötschel and Win [48]. They propose a cutting plane algorithm producing good results. Several heuristics are also developed among others by Pearn and Li [67].

Hierarchical CPP (HCPP)

In HCPP, the edges of a graph G are partitioned into clusters or classes, and a precedence relation specifies the order in which the clusters are to be traversed starting and ending at a given vertex on G. In HCPP, a linear precedence relation means that all the edges of a cluster E_i have to be traversed before those of the cluster E_{i+1} it precedes. Although the HCPP is $\mathcal{NP}$-hard in general, if the precedence relation is linear, and each cluster E_i is connected, then a polynomial time solution is possible [68]. An improved polynomial time algorithm is presented by Ghiani and Improta [69]. An example of HCPP would be snow plowing with the streets classified into main and secondary streets [70]. A weaker form of precedence relation is treated by Alfa and Liu [71], where the traversal of the edges of a cluster E_i can start before that of any edge of the cluster E_{i+1} and finishes before the end of the traversal of E_{i+1}.

Hierarchical RPP

This version of RPP has been defined by Dror and Langevin [72] as an RPP in which each connected component of arcs to be serviced has to be completely serviced before servicing another component. Dror and Langevin suggest an exact solution procedure based on a polynomial transformation of the clustered RPP into a generalized TSP and then solving it using the exact procedure suggested by Noon and Bean [73].

Maximum Benefit CPP (MBCPP)

Maximum benefit CPP has been introduced by Malandraki and Daskin [74] for the directed graphs and by Pearn and Wang [75] for the undirected graphs. In MBCPP, the requirement that each edge must be traversed at least once is relaxed. Each edge is associated with a service cost for which service is provided while traversal, a deadhead cost for the traversal with no service, and a set of benefits. Each time an edge is traversed, a benefit is generated and a cost is incurred. The benefits for an edge are assumed to be non-increasing in the number of traversals of that edge. The objective is to find a postman tour starting from a depot and traversing a selected set of edges and returning to the same depot so as to maximize the total net benefit accrued.

Time-Constrained CPP

In this formulation of CPP, a time window may be associated with an edge, in which the first traversal of that edge has to occur [76].

U-Turns and Turn Penalties

In practice, there can be turn penalties associated with making left turns and U-turns on the road network. The DRPP with turn penalties is treated by Benavent and Soler [77]. The case of MRPP with turn penalties is formulated by Corberan et al. [78]. They show that the problem is $\mathcal{NP}$-hard and propose a polynomial time transformation of the problem into an asymmetric TSP, which then can be solved either by heuristic or exact solution procedures available.

16.4.4 Capacitated Arc Routing Problem

Capacitated arc routing problem is essentially an extension of RPP to the case where the vehicles are restricted to have a finite capacity [79]. In CARP, given an undirected connected graph $G = (V, E)$ and a homogeneous fleet of vehicles each with capacity W, the objective is to determine the minimum cost tour such that all edges requiring service $E_R \subseteq E$ are traversed at least once. Vehicles start and end at the depot, and the total demand serviced by a vehicle does not exceed its capacity. CARP is an $\mathcal{NP}$-hard problem [79]. An exact solution procedure is provided by Belenguer and Benavent [80]. A large number of heuristic solution procedures have been developed, such as the simple constructive algorithms by Golden et al. [81] and Golden and Wong [79], a route-first, cluster-second algorithm by Ulusoy [82], a tabu search algorithm by Hertz et al. [59], and a guided local search algorithm by Beullens et al. [83].

The case with the fleet size and its mix of vehicles with different capacities being decision variables is investigated by Ulusoy [82]. The problem is extended to the case where there can be upper bounds on the number of vehicles with given capacities using a branch and bound method.

A bi-objective formulation of CARP is presented by Lacomme et al. [84], where the objectives are the minimization of the total cost of the routes and the cost of the longest route. They employ an adapted version of the nondominated sorted genetic algorithm (NSGA II).

16.5 Traveling Salesman Problem

The TSP is one of the most celebrated problems in operations research. In this problem, a salesman who lives in city 0 needs to visit n cities for business purposes and then come back to his home town. Each of the n cities must be visited exactly once, and the objective is to minimize the total cost of traveling. Since the cities may be visited in any order, there are $n!$ possible tours, and it becomes almost impossible to enumerate all tours to determine the least cost tour when n is large. In fact, TSP is hard both theoretically and computationally, and it belongs to the set of $\mathcal{NP}$-complete problems [85,86]. For a more detailed discussion, the reader is referred to Lawler et al. [86], Gutin and Punen [85], and Schrijver [87].

Historically, TSP is related to many famous problems in discrete mathematics. For instance, in the knight's tour problem introduced by Euler in 1759, a knight must visit all of the squares on a chessboard exactly once. Another example is the Icosian Game by the Irish mathematician Hamilton. In this game, a solution requires finding a tour that goes through 20 points. Identifying such a tour is known as the Hamiltonian circuit problem and is closely related to TSP [86]. Menger's discussion of a variant of TSP known as the Messenger problem in 1932 is considered to be the first published mathematical treatment of TSP [85,88]. Schrijver presents a compact historical perspective on TSP starting with Hamilton and Kirkman in the nineteenth century and continuing up to 1960s. The reader is referred to http://www.tsp.gatech.edu for a list of milestones in solving TSP instances from 1950s till the present. It is worthwhile to note here that while Dantzig et al. [89] presented the optimal solution to a 49 city instance in their seminal paper, recently Applegate et al. [90] solved a TSP instance with 24,978 cities to optimality.

All variants of TSP are classified into two main categories depending on whether the cost c_{ij} of traveling from city i to city j is equal to the cost of traveling in the reverse direction. If $c_{ij} = c_{ji}$ for all pairs of cities i and j, then the problem is called symmetric, and asymmetric otherwise. Generally, different traffic regulations such as one-way streets may give rise to asymmetric costs. Also, if uncertainty in problem parameters is incorporated explicitly in the problem statement, then we have a stochastic TSP. The binary programming formulation for the deterministic version of symmetric TSP is given in the following:

$$\min \frac{1}{2} \sum_{j=0}^{n} \sum_{k \in J(j)} c_k x_k \tag{16.10}$$

$$\sum_{k \in J(j)} x_k = 2 \qquad j = 0, 1, \ldots, n \tag{16.11}$$

$$\sum_{k \in E(S)} x_k \leq |S| - 1 \qquad \forall S \subset \{0, 1, 2, \ldots, n\}, S \neq 0 \tag{16.12}$$

$$x_k \in \{0, 1\} \qquad \forall k \tag{16.13}$$

In the formulation 16.10–16.13, the cost of an edge k between a pair of cities is c_k, and x_k takes the value of 1 if edge k is included in the tour, and 0 otherwise. The set of edges that are incident to city j are indicated by $J(j)$, and constraints 16.11 subscribe that a tour enters and leaves a city exactly once. Constraints 16.12 are referred to as clique packing or subtour elimination constraints and were introduced by Dantzig et al. [89]. A subtour which consists of the cities in a proper subset S of all n cities is constructed if $|S|$ edges are included in a solution so that these edges originate and terminate in S. Constraints 16.12 prevent such subtours in which $E(S)$ denotes the set of edges with both end points in S. We note that the number of subtour elimination constraints is exponential in n. In the formulation for the asymmetric cost structure, x_{ij} is set to 1 if the edge from city i to city j is included in the tour, and the subtour elimination constraints 16.12 are replaced by

$$\sum_{i \in S} \sum_{j \in S} x_{ij} \leq |S| - 1 \qquad \forall S \subset \{0,1,2,\ldots,n\}, S \neq \emptyset \qquad (16.14)$$

In addition, constraints 16.11 need to be expressed by appropriate constraints involving the variables x_{ij}.

16.5.1 Variants of TSP

Below, we briefly describe some of the more common variants of TSP. For others such as the period TSP, the delivery man problem, the minimum latency problem, the black and white TSP, the angle TSP, etc., the reader is referred to Gutin and Punnen [85].

In the maximum TSP, the tour cost is to be maximized, and this problem can easily be transformed into the minimum cost TSP discussed earlier [85]. This problem is also referred to as the taxicab ripoff problem.

In the bottleneck TSP, a tour is constructed so that the largest edge cost in the tour is minimized [85].

In the TSP with multiple visits, the salesman is allowed to visit a city more than once, but all cities must be visited at least once [85].

In the messenger problem [88], we look for a Hamiltonian path with minimum cost between cities i and j. In other words, we look for a minimum cost path that originates at city i and terminates at j. The messenger does not return to i in this problem.

In the clustered TSP, the cities are partitioned into k clusters, and the cities in each cluster must be visited consecutively on a tour [91,92].

In the time dependent TSP, time is discretized into periods, and the cost of traveling from city i to city j may change over time. The salesman may only travel from one city to the next in a single period, and the objective is to minimize the total travel cost as usual [93,94].

The generalized TSP is similar to the clustered TSP except that exactly one city from each cluster must be visited [95].

In the m-salesmen problem, all salesmen must visit at least one city, and all cities must be visited exactly once. The objective is to partition the cities into m groups so that each partition is visited by one salesman, and the total distance traveled by all salesmen is minimized [85].

16.5.2 Applications of TSP

Traveling salesman problem has stirred so much interest not only because it is a theoretically interesting and challenging problem in itself, but also because it has applications in very diverse set of areas. Some examples include vehicle routing, machine scheduling, cellular manufacturing, arc routing, clustering, computer wiring, card board and wall paper manufacturing, and frequency assignment problems [85, 86,87]. In addition, many logistics design and operation problems yield TSP as a subproblem. For instance, Diaz et al. [96] propose a hierarchical framework for the milk collection and distribution operations of a large dairy firm in Spain. Once the customers are assigned to area representatives, the bottom level in the modeling hierarchy schedules drivers for the milk producers where a TSP with time windows is solved for each driver. The time window constraints are required due to the perishable nature of milk products. Clearly, the vehicle routing problem (VRP) is the most significant domain in logistics in which TSP appears as a subproblem, and we devote considerable attention to VRP in Section 16.6. Also, some ARP may be transformed into related TSPs as discussed in Section 16.4.2.

16.5.3 Solution Methods for TSP

Traveling salesman problem is an $\mathcal{NP}$-complete problem except for some special cases [85], and generally only complete enumeration of all possible tours guarantees an optimal solution. Unfortunately, even for a small 25-city problem and with the capability of calculating one billion tours per second, it would approximately take 20 million years to evaluate all possible tours. Hence, many heuristics for TSP have been designed in addition to optimal algorithms. For complexity issues regarding TSP, the reader is referred to Korte and Vygen [97], Ausselio et al. [98], Gutin and Punnen [85], Schrijver [87], and Lawler et al. [86]. A comprehensive treatment of the solution approaches for TSP is beyond the scope of this text, and we only discuss some of the more well-known approaches briefly and provide references otherwise.

16.5.3.1 Optimal Algorithms

The branch-and-bound (B&B) and branch-and-cut techniques developed for TSP originated in the seminal work of Dantzig et al. [89]. An overview of this line of research is presented by Schrijver [87]. Note that the number of subtour elimination constraints 16.12 in the formulation 16.10–16.13 for the symmetric TSP is exponential in the number of cities. Therefore, these constraints are typically excluded from consideration in a B&B approach initially. Then, once integer solutions are obtained for the relaxation, cuts for violated subtour elimination constraints are identified and added as necessary.

Other bound improvement techniques for TSP include 1-trees and Lagrangean relaxation, 2-factor constraints and clique tree inequalities. Held and Karp [99,100] describe a method based on a 1-tree relaxation and subgradient optimization. The 2-factor constraints and the clique tree inequalities [87] yield valid facet defining inequalities for the TSP polytope and improve the bounds in the search tree. Gutin and Punnen [85] present a thorough list of bounding approaches based on polyhedral studies for TSP. Comb, star, path, clique tree, bipartition and ladder inequalities are among the valid inequalities developed for TSP.

16.5.3.2 Heuristics for TSP

Constructive Heuristics for TSP

Some of the well-known constructive heuristics for TSP are explained in the following. For others, such as nearest insertion, furthest insertion, double minimum spanning tree, strip and space filling curve heuristics, the reader is referred to Lawler et al. [86].

The nearest neighbor heuristic [98] is a greedy algorithm which starts at city 0 and proceeds by traveling to the closest unvisited city from the current city at each iteration. At the nth iteration, a Hamiltonian path is obtained, and the tour is completed by adding an edge from the nth city visited to the initial city 0. Note that this final edge may be very long.

Starting with a minimum spanning tree, the minimum spanning tree-based heuristic [86] produces a tour in which some cities may be visited more than once. Then, a simple shortcutting strategy removes additional visits to a city, and a traveling salesman tour is obtained. In fact, the length of such a tour is no more than twice the optimal tour length if the distance matrix satisfies the triangle inequality. The nearest merger, nearest addition, nearest insertion, cheapest insertion algorithms are in essence similar algorithms [86].

Christofides' heuristic [101] starts with a minimum spanning tree like the minimum spanning tree-based heuristic discussed earlier. The tree is then converted into an Eulerian graph by solving a minimum matching problem, and the obtained Eulerian tour is transformed into a traveling salesman tour by removing redundant visits to cities that are visited several times. If the triangle inequality holds for the distance matrix, this heuristic provides a solution with an objective value no larger than 1.5 times the optimal objective value. This is currently the best-known worst-case performance bound for TSP [24].

Improvement Heuristics for TSP

Starting from one or several initial tours typically obtained by constructive heuristics, improvement heuristics for TSP generate shorter tours by a series of moves defined by some local neighborhood. Most common improvement heuristics are k-opt methods. Given an initial tour, a k-opt algorithm replaces a set of k edges in the tour by another set of k edges while improving the objective function value. This is called a k-move. Increasing k improves solution quality [98], but also increases solution time. Often, $k = 3$ is a good choice, and 2.5-opt [102] and Or-opt [103] algorithms aim at reducing the complexity of the 2-opt algorithm. Gutin and Punnen [85] provide more on k-opt procedures and their variations.

The Lin-Kernighan algorithm [104] is a very famous improvement heuristic for TSP, and it is still considered to be one of the best heuristics for this problem. In this algorithm, consecutive 2-opt moves are employed to generate k-opt moves where k is determined dynamically throughout the iterations.

Recent applications of local search heuristics to TSP use more sophisticated neighborhood structures [85]. Some of these modern approaches include variable neighborhood search, sequential fan, filter and fan, (chain and iterated) Lin–Kernighan and ejection chain methods. For these relatively new methods, the reader is referred to Gutin and Punnen [85].

Metaheuristics for TSP

Simulated annealing is one of the first heuristics applied to TSP [86]. Tabu search methods [85], genetic algorithms [105], evolutionary algorithms using scatter search and path relinking [85] and ant colony optimization [106] techniques have also been developed for TSP.

16.6   Vehicle Routing Problem

The VRP introduced by Dantzig and Ramser [107] in 1959 is a very important and hard combinatorial optimization problem which finds applications in many logistics systems [108]. For instance, consider the problem of serving a number of customers from a warehouse (depot) by dispatching a limited number of identical vehicles, for example, trucks with limited capacity. Assume that the vehicles are at the warehouse initially, and routes connect the warehouse to the customers, and customers to customers. A simple objective for this scenario is to find the optimal routes that minimize the total distance traveled by the vehicles. In this problem, a route must start and finish at the depot, and a customer is visited by exactly one vehicle. The total demand of customers serviced by one vehicle cannot exceed the vehicle's capacity. This problem with a single depot and identical vehicles of limited capacity is considered as the most basic VRP and referred to as the single depot capacitated vehicle routing problem (CVRP) [24]. In fact, VRP is an extension of the well-known TSP (see Section 16.5) which may be considered as a VRP with a single vehicle with enough capacity to serve all customers. (See Fig. 16.4 for an illustration of CVRP.)

For CVRP, let $V = \{0, \ldots, n\}$ be the set of customers, where customer 0 corresponds to the depot, m be the number of vehicles, Q be the capacity of a vehicle, d_i be the demand of customer i, and c_{ij} be the cost of traveling from customer i to customer j. The parameter c_{ij} is commonly related to the time of travel between customers i and j, and if $c_{ij} = c_{ji} \: \forall i, j \in V$, the problem is said to be symmetric, and asymmetric otherwise. A tour R is an ordered sequence of customers starting and terminating at the depot so that no customer (other than the depot) appears more than once in the sequence. Then, any route R_r is a sequence of the form $(n_{r,0}, n_{r,1}, n_{r,2}, \ldots, n_{r,i-1}, n_{r,i}, n_{r,i+1})$ where $n_{r,k}$ is the kth customer to be visited on

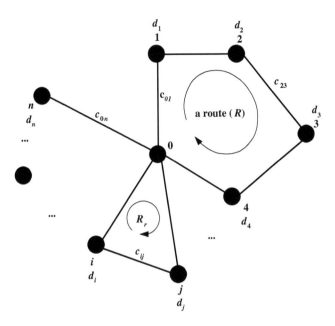

FIGURE 16.4 Vehicle routing problem.

route R_r, and $n_{r,0}$, $n_{r,i+1} = 0$. The cost of route R_r is calculated as $C(R_r) = \sum_{k=1}^{k=i} c_{k, k+1}$. The mathematical programming formulation for the symmetric version of CVRP is given in the following [109] in which x_{ij} denotes the number of times the link (i, j) is included in any route.

$$\min \sum_{r=1}^{m} C(R_r) = \sum_{i \in V \setminus \{n\}} \sum_{j>i} c_{ij} x_{ij} \tag{16.15}$$

$$\sum_{h<i} x_{hi} + \sum_{j>i} x_{ij} = 2 \quad i \in V \setminus \{0\} \tag{16.16}$$

$$\sum_{j \in V \setminus \{0\}} x_{0j} = 2m \tag{16.17}$$

$$\sum_{i \in S} \sum_{\substack{h<i \\ h \notin S}} x_{hi} + \sum_{i \in S} \sum_{\substack{j>i \\ j \notin S}} x_{ij} \geq 2r(S) \quad \forall S \subseteq V \setminus \{0\}, |S| \geq 1 \tag{16.18}$$

$$x_{ij} \in \{0,1\} \quad \forall i,j \in V \setminus \{0\}, i < j \tag{16.19}$$

$$x_{0j} \in \{0,1,2\} \quad \forall j \in V \setminus \{0\} \tag{16.20}$$

In the symmetric VRP, the routes have no orientation, that is, a link $(i, j) \in R_r$ may be traversed in either direction. Therefore, the variables x_{ij} are only defined for $i < j$ in the formulation given earlier. The objective 16.15 minimizes the total traveling cost of all vehicles. Any customer is visited exactly once as enforced by the degree constraints 16.16. However, each route starts and terminates at the depot, and the degree of the depot is $2m$ as specified by 16.17. The parameter $r(S)$ on the right hand side of constraints

16.18 denotes the minimum number of vehicles required to satisfy the demand of the customers in the set S, and is given by the optimal solution of a related bin packing problem (BPP) in which the size of item $i \in S$ is d_i, and the bin capacity is Q. Thus, the so-called capacity-cut constraints 16.18 enforce both the connectivity of the routes and the vehicle capacities. BPP is an $\mathcal{NP}$-complete problem in itself; however, effective solution methods are available for this problem [110]. Except when a route visits a single customer, a link may appear in any route only once. Therefore, only links that originate from the depot may take a value of 2 as stated by the constraints 16.19 and 16.20.

The model given in 16.15–16.20 is a commonly used two-indexed formulation of the problem that does not associate a specific vehicle with any route. Therefore, in some extensions of CVRP a third index is included in order to keep track of the specific vehicle traversing the link (i, j) [109].

Today, the management of almost any transportation or distribution system incorporates solving some variant of VRP. Examples include delivery of consumer products from warehouses to retail stores, collection of products such as milk from big farms and the delivery of industrial gases. See Ball et al. [111] for further examples. In addition to their direct applications in transportation science, VRP models are also useful in other types of problems, for example, in designing and operating material handling systems in automated production facilities, in multi-facility production systems, and in circuit board manufacturing [112].

16.6.1   Variants and Applications of VRP

In practice, the basic CVRP discussed earlier may not account for some of the essential aspects of real-life problems. For instance, if the vehicles have different capacities, then we have a heterogeneous VRP while if the demand of a customer may be satisfied by more than one visit, we refer to this problem as VRP with split deliveries.

The basic CVRP allows only one route per vehicle and constrains the vehicle capacity while some extensions allow several routes for one vehicle. In addition, other restrictions on the routes exist in some circumstances. For instance, the vehicles servicing the customers may be asked to return to the depot by a given deadline, or the distance traveled by a vehicle may not exceed a limit. These types of VRPs have minimax objectives [113,114,115].

In urban areas, travel times generally depend on the time of the day due to accidents, heavy traffic during rush hour, etc. If these issues are significant enough to be incorporated in the formulation, then we have a VRP with time-dependent travel times [116].

In many cases, a company outsources its distribution to a third party. For instance, this may be desirable due to highly volatile demand. In such problems, we typically have less-than-truckload (LTL) deliveries [117].

Several important variants of VRP are discussed in more detail in the following. For a more in-depth discussion about the complications and the different types of constraints that may arise in VRPs, the reader is referred to Ball et al. [111].

16.6.1.1   VRP with Time Windows

Typically, grocery stores in crowded urban areas do not allow deliveries outside a time interval in the early morning hours (see the case in Section 16.7). This is an example of VRP with time windows (VRPTW) in which customer i may only be visited during a time window $[a_i, b_i]$ [24,108,118]. In some applications, if a vehicle arrives at customer i earlier than a_i, then it must wait until time a_i, while in some situations a delivery later than b_i is acceptable with an associated penalty [108].

16.6.1.2   VRP with Pickup and Delivery

Sometimes, the vehicles do not only deliver goods to the customers, but they also collect items, for example, defective products, from them [119]. If the customers are partitioned into two mutually exclusive subsets so that linehaul customers receive deliveries while backhaul customers send back products, and in addition backhaul customers are only served after the linehaul customers, then this problem is known as VRP with backhauls (VRPB) [109].

If every customer i has a demand d_i from some other node O_i, and every customer has a supply p_i to be delivered at node D_i, then this problem is known as VRP with pickup and delivery (VRPPD). If the demands have a common destination or origin, such as the depot, then this problem is known as VRP with simultaneous pickup and delivery (VRPSPD). A formal description of VRPSPD is given by Bianchessi and Righini [120], and Nagy and Salhi [119] include a detailed survey on VRPPD and associated solution techniques. A related problem is discussed by Gronalt et al. [121] who consider a variant of VRPPD with time-windows and restricted route lengths where the objective is to minimize the movements of empty vehicles.

The reader is referred to Figure 1.1 [109] for a graphical representation of the relationships between the fundamental VRP variants.

16.6.1.3 Inventory Routing Problem

Rapid progress and wide availability of high-tech hardware and associated software at reasonable costs enables the use of vendor-managed inventory systems in logistics [110]. In these systems, suppliers monitor their customers' inventory levels and send replenishments as necessary. Such applications give rise to an extension of VRP known as the inventory routing problem (IRP). IRP is defined as the continuous delivery of one product to many customers from a single warehouse using a given set of vehicles over a planning horizon. If the planning horizon is only one day, the problem reduces to VRP. In this setting, the main goal of the supplier is to keep each customer's inventory above a certain level by sending shipments at appropriate times. Then, the objective function of IRP is to minimize average distribution costs over the planning horizon while all customers are prevented from facing stockouts. The decisions involve the timing and amount of the deliveries to the customers in addition to the routes to be used [109,110,111]. It is important to note that while customers place orders in VRP, the warehouse (depot) tracks the inventory levels at the customers in IRP and sends replenishments as necessary.

The IRP literature often assumes that the customers consume at constant rates which is hardly the case. If the rates change over time, then we have a dynamic IRP (DIRP) while stochastic rates give rise to the stochastic IRP (SIRP). In SIRP, stockout penalties need to be included explicitly in the objective function because it is not possible to eliminate stockouts completely in this case.

Inventory routing problem is hard to formulate and almost impossible to solve optimally for a long planning horizon. So far, planning horizons up to a few days have been considered due to the complexity of the problem. As mentioned earlier, a planning horizon of one day corresponds to VRP, and mathematical programs for IRPs involving longer planning horizons are solved using decomposition techniques [109].

Applications of IRP are common in supermarket chains, in automotive parts distribution, as well as in the gas and beverage industries [109,110,111,122].

16.6.1.4 VRP with Profits

Many extensions of VRP discussed so far require that all customers are served in the planning horizon. However, in some variants of the problem, a value is assigned to each potential service, and some customers may be rejected if rendering service to them decreases the overall profit. This extension of VRP is known as VRP with profits. A detailed survey and a discussion of applications and solution methodologies for this problem are presented recently by Feillet et al. [123].

16.6.1.5 Dynamic Vehicle Routing Problem

All variants of VRP considered so far, except for IRP, ignore the dynamic nature of the problems faced in logistics planning and execution. In other words, these models take a snapshot of the system, and provide solutions based on the values of the problem parameters at present time. However, in many real applications, these parameters change over time. For instance, traffic conditions may change or new customer requests may arrive in real time. The problems in this setting which consider real-time decision making are classified as dynamic vehicle routing and dispatching problems (DVRP) which belong to the broader family of dynamic transportation problems [110]. DVRPs are commonly classified as local or wide area problems depending on their area of coverage. A second factor that affects the nature of a DVRP is whether a vehicle is dispatched to a single customer or whether shipping is done on an LTL basis.

If a warehouse sends one vehicle to each store in a wide area, then routing is trivial while it is an integral part of the problem if LTL shipments are considered. In wide area problems with full truck load shipping, the main issue is where to station a vehicle after service completion in anticipation of future orders. This is similar to the facility location problems faced by emergency services in local DVRPs as discussed in the following.

The planning horizon for many DVRPs is a single day. For instance, consider any courier service where the dispatching manager receives requests from customers in real time during the course of the day. In this setting, it is essential for the dispatcher to be able to either decline an incoming pickup or delivery request because there is no capacity available to respond to the new customer or to modify the current vehicle routes as required. This problem may be modeled as a local DVRP. Local DVRPs are typically classified based on their delivery and pickup routines and vehicle capacities. If, for example, the dispatching center receives a pickup and (probably different) delivery location for each order, then this problem is a many-to-many DVRP. If pickup occurs only once, and delivery is made to multiple customers, we have a one-to-many problem, whereas in a many-to-one problem pickups from several locations are delivered to a single location. While the vehicle capacity is typically a major concern in vehicle routing, there are examples of DVRP in which the vehicle capacity may be disregarded for all practical purposes, for example, in express mail delivery systems. An important type of the local DVRP is the dispatch of emergency vehicles, for example, fire engines or ambulances. In these problems, routing is not important because each vehicle attends to one customer location only. The reader is referred to Crainic and Laporte [110] for a detailed classification of local DVRPs.

Other important applications of DVRP include dial-a-ride, repair, courier, and express mail services. Dial-a-ride problems are local area, many-to-many and capacitated DVRPs with time window constraints. The defining characteristic of repair service systems is the service time at the customer site. In these problems, one can think of the pickup and delivery as occurring at the same location. Note that in most other vehicle routing problems delivery and pickup locations are different.

16.6.1.6 Fleet Size and Mix Problem

One important facet of vehicle routing is customer demand. Depending on the magnitude and mix of demand, a company may need different types of vehicles with certain specifications. If the demand pattern changes over time, a tactical decision involves determining the type and number of vehicles to buy (vehicle mix problem). In addition, a third-party carrier may be employed which again requires determining the number of vehicles to request from the service provider. Campbell and Hardin [124] investigate the problem of minimizing the number of vehicles for periodic deliveries, while Chu [117] addresses the problem of vehicle mix. Note that the fleet size and mix problem considers a medium- to long-range planning horizon in contrast to the other types of VRPs discussed before which typically have planning horizons of at most up to a few days.

16.6.2 Solution Methods for VRP

The extensive literature on VRPs considers many different solution techniques. These approaches generally fall into one of the following three categories [109,125]: exact algorithms, classical heuristics and metaheuristics. Each of these classes are discussed in the following three subsections.

16.6.2.1 Exact Algorithms

The basic capacitated vehicle routing problem (CVRP) discussed at the beginning of Section 16.6 is a combination of two $\mathcal{NP}$-complete problems: the BPP and the TSP. Thus, CVRP itself is an $\mathcal{NP}$-complete combinatorial optimization problem [126]. Although Cordeau et al. [125] describe VRPs as still far from being consistently solved using exact algorithms for problem instances having more than 50 customers, this size increased recently due to improved lower bounding techniques and higher computational power. For instance, Tarantilis et al. [112] solve problems with up to 75 customers. The tightness of the lower bound employed is probably the most crucial factor for the effectiveness of any B&B algorithm, and several approaches exist for obtaining lower bounds in VRPs [109,111]. The applicability and effectiveness of a lower bounding approach depends on the type of VRP at hand and the properties of the problem parameters. A lower bound that is effective under a symmetric cost structure may not yield good results if the costs are asymmetric. Some of the more well-known relaxations are based on dropping certain constraints from the mathematical formulation 16.15 through 16.20 stated previously. For instance, dropping the capacity cut constraints 16.18 yields an assignment relaxation for the asymmetric CVRP while a matching relaxation is obtained in the symmetric case [109].

In many B&B methods for VRP, a Lagrangean relaxation technique with subgradient optimization is employed. For example, Fisher [127] develops a K-tree approach for CVRP which is an extension of the one-tree approach for TSP while Fisher et al. [118] introduce two techniques based on Lagrangean relaxation for VRPTW. Researchers have also experimented with different branching schemes in B&B algorithms, for example, arc-based or customer-based branching [110].

In addition, Toth and Vigo [109] present several branch-and-cut algorithms and approaches based on set covering formulations for solving VRPs optimally.

16.6.2.2 Classical Heuristics

Vehicle routing problems are both theoretically and computationally challenging which prompted the development of many heuristics since the early days of research in this area. The classical VRP heuristics fall into one of three categories: constructive, two-phase, and improvement heuristics. Constructive heuristics build a solution iteratively by checking how the objective changes at each step, and they terminate when a complete solution is obtained. The most famous constructive heuristic is the sweep algorithm of Clarke and Wright [128]. Improvement heuristics require an initial solution which is typically obtained by a constructive heuristic and search the neighborhood of the current solution for better solutions until some termination criterion is met. Two phase heuristics consider the bin packing and traveling salesman aspects of VRP separately and in sequence. Route first-cluster second and cluster first-route second heuristics are in this category. These heuristics can also be regarded as constructive heuristics as they do not implement an improvement step.

The savings algorithm of Clarke and Wright [128] is probably the best known heuristic for VRP [109] because it is simple to understand and quick to implement. The basic idea in this heuristic is to combine two existing routes $R_1 = (0, i, ...)$ and $R_2 = (0, ..., j, 0)$ into a single route $R = (0, ..., i, ...)$ if this is feasible with respect to the vehicle capacity and the potential savings $c_{0i} + c_{j0} - c_{ji}$ is positive. Several improvements and modifications of the savings algorithm are reported in the literature [24,109,121,129,130]. Toth and Vigo [109] report parallel and serial implementations of the savings algorithm and benchmark these implementations using standard test instances. The savings algorithm is also very popular for obtaining an initial solution for improvement heuristics as well as for metaheuristics [131].

In route first-cluster second heuristics, customers are ordered according to a traveling salesman tour ignoring their associated demands, and then customers are divided into clusters based on their demands and vehicle capacities. For instance, the sweep algorithm of Gillett and Miller [132] and the optimal partitioning heuristic of Beasley [133] are in this category.

In cluster first-route second heuristics, customers are first assigned to regions depending on some criterion, and then routing is performed separately for each region. The cluster first-route second heuristics include the generalized assignment heuristic of Fisher and Jaikumar [134], the location based heuristic of Bramael and Simchi-Levi [135] and the truncated branch-and-bound algorithm of Christofides et al. [136].

Improvement heuristics require an initial solution which is typically provided by a constructive heuristic, and they improve the objective function by exploring some neighborhood of the current solution. For VRP, the neighborhoods are typically generated by a sequence of vertex and/or arc exchanges in and/or between routes identified so far. These modifications may be performed on a single route at a time or on several routes simultaneously. For instance, the l-opt heuristic of Lin [137] removes l edges from a single route and then investigates whether combining the parts of the tour in a different order decreases the objective function. For multi-route improvements, operations such as string cross, string exchange, string relocation, and string mix may be used (for details, see Toth and Vigo [109]).

16.6.2.3　Metaheuristics

Metaheuristics can be considered as more elaborate improvement techniques [24]. Metaheuristics focus on the rules for traversing solution neighborhoods, memory structures and recombination of solutions. Although alternate classifications for metaheuristics may be considered [112], we divide metaheuristics into three categories based on the search procedures employed: local search metaheuristics, population search metaheuristics, and hybrid algorithms. Simulated annealing (SA), deterministic annealing (DA), and tabu search (TS) fall into the first category while genetic algorithms (GA) and ant systems (AS) fall into the second.

The main strength of metaheuristics is that they explore the solution space thoroughly. While doing so, they may allow the deterioration of the objective function or even infeasible solutions at intermediate steps. Some metaheuristics incorporate classical heuristics at their initialization phase [24,109]. For instance, the savings algorithm is employed by Prins [131] for the initialization of a GA, by Homberger and Gehring [138] and by Golden et al. [115] for the initialization of a TS algorithm. Similarly, the sweep algorithm by Gillett and Miller [132] is used by Baker and Ayechew [139] for obtaining an initial population in their GA and by Montané and Galvão [140] for the initialization of their TS. For recent surveys about metaheuristics in VRP, please refer to Li et al. [130] and Nagy and Salhi [119]. Some of the well-known metaheuristics for VRPs are briefly discussed in the following text.

Simulated annealing allows moves to inferior solutions with some probability in order to escape local minima. The most successful SA heuristic reported for VRP is by Osman [141]. For other implementations and details of these algorithms such as neighborhood structures, stopping criteria, cooling schemes, etc., the reader is referred to Toth and Vigo [109] and Tarantilis et al. [112]. Although Osman's SA algorithm is used for benchmarking purposes in the literature, no new SA algorithms are reported recently. This may be partly due to the established success of tabu search and hybrid algorithms in recent years [120,125,131,138,139,140,142]. Deterministic annealing is similar to SA except that the rules which specify whether a nonimproving solution is accepted or not are not probabilistic in nature. Examples of DA for solving VRPs are presented by Tarantilis and Kiranoudis [143,144].

Tabu search is a local search metaheuristic, and most of the promising results for VRPs are obtained by TS implementations [109,125,131]. In TS, some properties of the current solution are declared as tabu for a number of iterations in order to prevent false convergence and to avoid being trapped at an inferior local minimum. This property can be considered as short-term memory [112]. TS also incorporates many other tools such as diversification and intensification strategies. Gendreau et al. [145] and Toth and Vigo [146] provide examples of tabu search applied to VRPs and discussions of the important aspects of a tabu search algorithm such as memory usage, neighborhood definition (creation) tools, diversification strategy, tabu approach, and identification of nonpromising solutions. Some successful TS implementations to VRPs in the literature are due to Osman [141], Taillard [147], Gendreau et al. [145], Golden et al. [148], Xu and Kelly [149], Rego and Roucairol [150], Rego [151], Toth and Vigo [146], Barbarosoğlu and Özgür [152], Rego [153], Cordeau et al. [125] and Ho and Haugland [142]. Tabu search has also been used in hybrid algorithms, for example, by Bouthillier and Crainic [154] and Homberger and Gehring [138].

Genetic algorithms are population-based neighborhood search strategies. In GA, members of an existing set (population) of solutions (chromosomes) are recombined (coupled) in order to produce new solutions (offsprings). Random modifications (mutations) to solutions are performed for diversification purposes, and inferior chromosomes are replaced by offsprings. Generally, the initial population of chromosomes is generated randomly. The application of GA to VRPs is limited. Baker and Ayechew [139] employ a pure GA to the basic CVRP and compare their results with TS and SA on benchmark instances. Haghani and Jung [116] report benchmark results for their earlier GA [155] using randomly generated test problems of dynamic VRP with time-dependent travel times with up to 70 customers where uncertainty in travel times is also incorporated. Usually, GA is combined with other techniques in order to obtain hybrid metaheuristics. Prins [131] proposes a very successful hybrid GA and reports comparisons with other successful metaheuristics on some classical benchmark problems improving the best-known solutions for very large-scale instances of Golden et al. [115].

16.6.3   VRP Software

There are many software products available for vehicle routing operations. Comprehensive surveys about VRP software appear in *OR/MS Today* with the recent one being by Hall [156]. This survey reports that the minimum service level acceptable by customers increases as the capabilities of distribution software grow. Furthermore, VRP software is getting cheaper over time as the technology develops. These surveys point out that VRP software helps allocating drivers and vehicles to appropriate duties in addition to programming pickup and delivery times and locations. Hall lists 21 software products and compares them on the basis of available platforms (Windows, Linux, etc.), solution techniques and capabilities, features (maps, compatibility with geographic information systems), and application areas. Most of these products are designed for end users, for example, dispatchers. It appears that one of the most important challenges facing VRP software vendors in the future is dynamic (real time) vehicle routing and detailed planning.

16.7   VRP Application in a Retail Chain in Turkey

In this section, we describe the distribution operations at a central warehouse in a major Turkish retail chain. Our objective here is not to provide the details of a full-scale solution and implementation, but rather to frame the problem with respect to the existing literature and identify some interesting features of this application that may motivate further research.

16.7.1   Company Background

Migros Turk is a major retail chain store company established in 1954 and operating 191 Migros stores and 310 Sok stores in Turkey. It also owns and operates 60 stores under the name Ramstore in Russia, Bulgaria, Azerbaijan, Kazakhstan, and Macedonia. Migros stores are classified according to their sizes and the range of products they carry as M, MM, MMM, and Hypermarkets. Currently there are 79 M type, 79 MM type, 33 MMM type and 3 Hypermarkets located in 41 cities in 7 geographical regions in Turkey. Sok chain stores are discount stores designed to stock all household needs of consumers. Sok stores are situated in 24 cities in 5 geographical regions.

16.7.2   Warehouse and Supply of the Stores

The stores in Turkey are supplied from five regional warehouses. Each warehouse supplies a set of Migros stores and each store is supplied by only one warehouse. Not all deliveries to the stores are made from the warehouses. Deliveries of all frozen products (at −18°C), and daily dairy products (daily milk, yogurt, dairy desserts, etc.), poultry, fresh meat, daily magazines, and newspapers are made into the stores directly by the suppliers.

The warehouse to be investigated here is the one in Sekerpinar, Istanbul and it is the largest of all warehouses operated by Migros Turk for supplying the Migros and Sok stores in the Marmara region. Marmara region is the most densely populated region in Turkey and includes Istanbul province with a population of roughly 13 million. Two major industrial cities, Bursa and Izmit producing among themselves more than 40% of industrial output in Turkey are located in this region. Sekerpinar warehouse is situated close to the eastern border of Istanbul province. It serves 273 stores in the Marmara region out of which 211 are spread throughout Istanbul province. It is about 20 to 50 miles away from the stores in Istanbul. The rest of the stores are mostly located in Edirne, Bursa, Izmit, and Sakarya provinces.

Sekerpinar warehouse has 62,000 m² of covered space. Initially, supply of Migros stores and Sok stores were run as separate operations under the same roof. Later, these operations were consolidated saving 16,000 m² of covered space and removed excessive inventories due to stock keeping unit (SKU) duplication. The space saved as a result of this consolidation has been rented to a major tire company for use as a warehouse. Thus, currently, 30,000 m² of this covered space is employed for core logistics operations. Sekerpinar warehouse carries approximately 9000 SKUs and, on the average, keeps 8–10 days of inventory measured on a revenue basis and stored in 550,000 standard units. Fresh vegetables/fruits are stored separately in the warehouse and are kept close to the loading docks in order to minimize their handling. The non-perishable goods (all goods other than the fresh vegetables/fruits) are located in the warehouse according to nine different order picking groups with each group having one or two aisles with racks on both sides dedicated to it. Both fresh vegetables/fruits and non-perishable products may be loaded on the same truck. Approximately 500 tons of fresh vegetables/fruits are shipped from Sekerpinar warehouse daily. On the average, 10,000 trucks are loaded/unloaded monthly at Sekerpinar warehouse. It has been observed that in general the demand variability of stores is relatively low measured on volume basis expressed in number of roll cages.

16.7.3 Fleet

Around 95% of the stores do not have docks, which make it impossible to deliver goods on pallets. Therefore, the goods are delivered on roll cages. The trucks are equipped with lifts that can load and unload roll cages. The fleet may be considered as homogeneous, where 90% of the fleet is comprised of 10-wheel trucks with a storage cabin kept at +4°C. The capacity of a 10-wheel truck is 21 roll cages. Small stores are served by 16 m³ trucks, which hold 12 roll cages. Special trucks need to be dispatched to a small set of stores due to physical limitations such as low garage entrance, etc.

The roll cages have several advantages: Since the roll cages used are collapsible, they can be folded to occupy less space when empty. They are reusable and sturdy. Storage space in the stores is limited, and material handling is manual (no automatic rack system). Therefore, deliveries are usually brought directly onto the shop floor from the truck on roll cages. Since the delivery and backhauling operations are generally handled together, the items on the truck have to be rearranged during each delivery. This task is simplified by the use of roll cages.

A subcontractor operates the fleet. The subcontractor is paid a constant fee per truck dispatched. Currently, the average fill rate of the trucks over all operations in Turkey is 98%. In the near past, the subcontractor expenses have been consistently less than 1% of the revenue generated, which is below the world standard of 1% for the retail sector. This value is in the range of 1–1.5% for other retailers in Turkey.

16.7.4 Operational Constraints

Deliveries to the stores are made daily except for stores with very small volumes. Daily replenishment to most stores is required mainly because of limited in-store storage capacity. Replenishment orders for the next day are accepted until 10.00 P.M. on the previous day.

A small number of stores require special trucks due to physical limitations, and some stores request small trucks.

Some stores are not accessible for delivery trucks, and the goods have to be carried into the store on roll cages from a parking lot nearby.

The stores have time window constraints. During morning rush hours from 7.00 A.M. to 10.00 A.M., trucks are not allowed in the city except for the main arteries. On a given route, a store may be given priority. For instance, the delivery to stores at or near junction nodes with heavy traffic must be done as early as possible before the morning rush hour, and such a store must be the first stop on a route. Stores in residential neighborhoods cannot receive deliveries before 10.00 A.M. In addition to traffic bans for trucks during certain hours, lack of alternate routes beside the main arteries also makes it difficult to satisfy the delivery constraints.

A truck route contains at most four stores. No more than one high priority store is allowed on a truck route. Note that demands may be split as the demand of a store may exceed the truck capacity.

Generally, returns are collected on the same visit to a store after the delivery is completed. However, if the time-window constraints cannot be satisfied, then the truck pays a second visit to the store around noon in order to collect the returned items. This is a rather unusual aspect of this problem, and we will revisit this issue at the end of this section.

16.7.5 Vehicle Loading and Routing at the Warehouse

In this section, we briefly explain the procedure employed for vehicle loading and routing at Sekerpinar warehouse. In accordance with the contractual agreement with the subcontractor stated earlier, the objective is formulated as minimizing the number of trucks dispatched subject to the constraints stated earlier. In line with their corporate social responsibility rules, Migros Turk has voluntarily imposed the secondary objective of minimizing the total mileage traveled by the trucks.

16.7.5.1 Picking and Loading at the Warehouse

Order data from stores are transferred to the warehouse over the company network automatically at midnight. Radio frequency handheld devices are used for picking. The worker starts from one end of the aisle and fills the roll cage in the same order as the items are listed on the handheld screen visiting the shelves in the storage blocks on both sides of the aisle. Once the items on the pick list are all picked, the roll cage is delivered to the assigned dock. The loading dock has space for 27 trucks to be loaded/unloaded simultaneously. It takes approximately 30 min to load a truck. The first wave of trucks leaves the warehouse at 5.00 A.M. in the morning.

Stores have dedicated docks. The stores are ordered in nondecreasing order of their distances to the warehouse, and then assigned to the docks in this order. If there are n docks, the $(n + 1)$ store is assigned to the first dock, $(n + 2)$ store is assigned to the second dock, etc. The rationale is that the trucks destined to the first store will be dispatched before it is time to load the truck for the $(n + 1)$ store.

This procedure for the dock assignment corresponds to a clustering of stores according to their distances from the warehouse and among each other. Some stores may change from one cluster to neighboring clusters on different days.

Once the products for a number of stores at adjacent dock locations are organized onto roll cages, these roll cages are loaded onto the trucks. In order to maximize the space usage in the trucks, the deliveries to large stores are combined with deliveries to smaller stores.

16.7.5.2 Routing

Since the number of stores that a truck visits is small, and some stores have priorities, the routing problem for each truck is relatively easy to solve once the store assignment to each truck is completed by the clustering procedure. On the average, 2.6 trucks per store are dispatched. This procedure fits into the cluster first—route second framework.

16.7.6 Modeling and Distinctive Features

This application can be modeled as a VRP with Pickup and Delivery with Time Windows (VRPPDTW) [109] in which all deliveries originate from the depot and all pickups are destined for the depot. However, two additional aspects need to be taken into account. First, as mentioned in Section 16.7.4, a store may have to be visited during a second time window, if there is no time to complete the pickup at the time of delivery. To the best of our knowledge, this extension of VRPPDTW has not been considered in the literature before. Second, the loading capacity of the depot plays a significant role in vehicle dispatching. The heuristic approach employed by Migros Turk tackles this issue by a cluster first-route second approach as discussed in Section 16.7.5. However, it is far from clear that this is the best way of handling this aspect of the problem, and an integrated approach which combines the material handling at the depot with vehicle dispatching could be investigated.

References

1. R.K. Ahuja, T.L. Magnanti, and J.B. Orlin. *Network Flows*. Prentice Hall, Englewood Cuffs, New Jersey, 1993.
2. K.L. Hoffman and M. Padberg. Solving airline crew scheduling problems by branch-and-cut. *Management Science*, 39:657–682, 1993.
3. E. Dijkstra. A note on two problems in connection with graphs. *Numerische Mathematik*, 1:269–271, 1959.
4. R. Dial. Algorithm 360: Shortest path forest with topological ordering. *Communications of ACM*, 12:632–633, 1969.
5. M.L. Fredman and R.E. Tarjan. Fibonacci heaps and their uses in improved network optimization algorithms. *Journal of ACM*, 34:596–615, 1987.
6. R.K. Ahuja, K. Mehlhorn, J.B. Orlin, and R.E. Tarjan. Faster algorithms for the shortest path problem. *Journal of ACM*, 37(2):213–223, 1990.
7. U. Pape. Implementation and effciency of Moore-algorithms for the shortest route problem. *Mathematical Programming*, 7:212–222, 1974.
8. U. Pape. Algorithm 562: Shortest path lengths. *ACM Transactions on Mathematical Software*, 6(3):450–455, 1980.
9. R.W. Floyd. Algorithm 97: Shortest path. *Communications of ACM*, 5:345, 1962.
10. D.R. Shier. Iterative methods for determining the K shortest path in a network. *Networks*, 6:205–229, 1976.
11. E. Minieka. *Optimization Algorithms for Networks and Graphs*. Marcel Dekker, Inc., New York, 1978.
12. E.G. Goodaire and M.M. Parmenter. *Discrete Mathematics with Graph Theory*. Prentice Hall, Englewood Cuffs, New Jersey, 2002.
13. J. Nesetril. A few remarks on the history of MST problem. *Archivum Mathematicum*, 33:15–22, 1997.
14. J.B. Kruskal. On the shortest spanning tree of graph and the traveling salesman problem. *Proceedings of the American Mathematical Society*, 7:48–50, 1956.
15. R.C. Prim. Shortest connection networks and some generalizations. *Bell System Technical Journal*, 36:1389–1401, 1957.
16. M. Sollin. Le trace de canalisation. In C. Berge and A. Ghouila-Houri, editors, *Programming, Games, and Transportation Networks*. Wiley, New York, 1965.
17. B. Chazelle, R. Rubinfeld, and L. Trevisan. Approximating the minimum spanning tree weight in sublinear time. *SIAM Journal on Computing*, 34(6):1370–1379, 2005.
18. B. Chazelle. A minimum spanning tree algorithm with inverse–Ackermann type complexity. *Journal of ACM*, 47:1028–1047, 2000.

19. S. Pettie and V. Ramachandran. An optimal minimum spanning tree algorithm. *Automata, Languages and Programming*, pages 46–60. Number 1853 in Lecture Notes in Computer Science. Springer–Verlag, Berlin, 2000.

20. D.R. Karger, P.N. Klein, and R.E. Tarjan. A randomized linear-time algorithm to find MST. *Journal of ACM*, 42:321–328, 1995.

21. F.L. Hitchcock. The distribution of a product from several sources to numerous localities. *Journal of Mathematics and Physics*, 20:224–230, 1941.

22. T.C. Koopmans. Optimum utilization of the transportation system. *Econometrica*, 17, 3/4: 136–146, 1949.

23. G.B. Dantzig. Programming of interdependent activities: II mathematical model. *Econometrica*, 17(3/4):200–211, 1949.

24. D. Simchi-Levi, X. Chen, and J. Bramael. *The Logic of Logistics: Theory, Algorithms, and Applications for Logistics and Supply Chain Management*. Springer Verlag, Berlin, 2005.

25. W.L. Winston and M. Venkataramanan. *Introduction to Mathematical Programming*. Thomson, Brooks/Cole, Pacific Grove, California, 4th edition, 2003.

26. Don T. Phillips and Alberto Garcia-Diaz. *Fundamentals of Network Analysis*. Prentice Hall, Englewood Cliffs, New Jersey, 1981.

27. S.B. Bazaraa, J.J. Jarvis, and H.D. Sherali. *Linear Programming and Network Flows*, 2nd edition Wiley, New York, 1990.

28. H.W. Kuhn. The Hungarian method for the assignment problem. *Naval Research Logistics Quarterly*, 2:83–97, 1955.

29. H.P. Williams. *Model Building in Mathematical Programming*. Wiley, New York, 1999.

30. C. Zhang, J. Liu, Y. Wan, G.K. Murty, and R.C. Linn. Storage space allocation in container terminals. *Transportation Research Part B*, 37:883–903, 2003.

31. Leonid N. Vaserstein. *Introduction to Linear Programming*. Prentice Hall, Englewood Cliffs, New Jersey, 2003.

32. Dimitri P. Bertsimas. *Network Optimization, Continuous and Discrete Models*. Athena Scientific, Nashua, New Hampshire, 1998.

33. A. Charnes and W.W. Cooper. The stepping stone method of explaining linear programming calculations in transportation problems. *Management Science*, 1(1):49–69, 1954.

34. W.L. Winston and M. Venkataramanan. *Operations Research—Applications and Algorithms*. Thomson, Brooks/Cole, Pacific Grove, California, 2004.

35. H. Arshom and A.B. Khan. A simplex type algorithm for general transportation problem: an alternative stepping-stone. *Journal of Operational Research Society*, 40:581–590, 1989.

36. C. Papamanthou, K. Paparrizos, and N. Samaras. Computational experience with exterior point algorithms for the transportation problem. *Applied Mathematics and Computation*, 150.459 175, 2004.

37. A. Assad and B. Golden. Arc routing methods and applications. In M. Ball, T. Magnanti, C. Monma, and G. Nemhauser, editors, *Handbooks in Operations Research and Management Science: Network Routing*, volume 8, pages 375–483. Elsevier, Amsterdam, 1995.

38. H.A. Eiselt, M. Gendreau, and G. Laporte. Arc routing problem, part I: The Chinese postman problem. *Operations Research*, 43(2):231–242, 1995.

39. H.A. Eiselt, M. Gendreau, and G. Laporte. Arc routing problem, part II: The rural postman problem. *Operations Research*, 43(3):399–414, 1995.

40. M. Dror, editor. *Arc Routing: Theory, Solutions and Applications*. Kluwer Academic Publishers, Boston, 2000.

41. M. Guan. Graphic programming using odd or even points. *Chinese Mathematics*, 1:273–277, 1962.

42. L.R. Ford and D.R. Fulkerson. *Flows in Networks*. Princeton University Press, Princeton, New Jersey, 1962.

43. J. Edmonds and E.L. Johnson. Matching, Euler tours and the Chinese postman problem. *Mathematical Programming*, 5:88–124, 1973.

44. H. Fleischner. Eulerian graphs and related topics (part 1, volume 2). *Annals of Discrete Mathematics*, 50(3):399–414, 1991.

45. C.S. Orloff. A fundamental problem in vehicle routing. *Networks*, 4:35–64, 1974.

46. E.L. Beltrami and L.D. Bodin. Networks and vehicle routing for municipal waste collection. *Networks*, 4:65–94, 1974.

47. T. van Aardenne-Ehrenfest and N.G. de Bruijn. Circuits and trees in oriented linear graphs. *Simon Stevin*, 28:203–217, 1951.

48. M. Grötschel and Z. Win. A cutting plane algorithm for the windy postman problem. *Mathematical Programming*, 55:339–358, 1992.

49. Y. Nobert and J.-C. Picard. An optimal algorithm for the mixed Chinese postman problem. *Networks*, 27:95–108, 1996.

50. C.H. Papadimitriou. On the complexity of edge traversing. *Journal of ACM*, 23:544–554, 1976.

51. G.N. Frederickson. Approximation algorithms for some postman problems. *Journal of ACM*, 26:538–554, 1979.

52. N. Christofides, E. Benavent, V. Campos, A. Corberan, and E. Mota. An optimal method for the mixed postman problem. In P. Thoft-Christensen, editor, *Systems Modeling and Optimization*, volume 59 of *Lecture Notes in Control and Information Sciences*, pages 641–649. Springer Verlag, Berlin, 1984.

53. W.L. Pearn and C.M. Liu. Algorithms for the Chinese postman problem on mixed networks. *Computers and Operations Research*, 22:479–489, 1995.

54. B. Raghavachari and J.J. Veeresamy. A 3/2 approximation algorithm for the mixed postman problem. *SIAM Journal on Discrete Mathematics*, 12:425–433, 1999.

55. J.K. Lenstra and A.H.G. Rinnooy Kan. On general routing problems. *Networks*, 6:273–280, 1976.

56. A. Corberan and J.M. Sanchis. A polyhedral approach to the rural postman problem. *European Journal of Operational Research*, 79:95–114, 1994.

57. N. Christofides, V. Campos, A. Corberan, and E. Mota. An algorithm for the rural postman problem on a directed graph. *Mathematical Programming Studies*, 26:155–166, 1986.

58. W.L. Pearn and T.C. Wu. Algorithms for the rural postman problem. *Computers and Operations Research*, 22:819–828, 1995.

59. A. Hertz, G. Laporte, and M. Mittaz. A tabu search heuristic for the capacitated arc routing problem. *Operations Research*, 48:129–135, 2000.

60. A. Corberan, R. Marti, and A. Romero. Heuristics for the mixed rural postman problem. *Computers and Operations Research*, 27:183–203, 2000.

61. N. Christofides, The optimum traversal of a graph. *OMEGA*, 1:719 727, 1973.

62. G.N. Frederickson, M.S. Hecht, and C.E. Kim. Approximation algorithms for some routing problems. *SIAM Journal on Computing*, 1978.

63. E. Minieka. The Chinese postman problem for mixed networks. *Management Science*, 25: 643–648, 1979.

64. P. Brucker. The Chinese postman problem for mixed graphs. In H. Noltemeier, editor, *Graph Theoretic Concepts in Computer Science*, pages 354–366, Springer Verlag, Berlin, 1981.

65. M. Guan. On the windy postman problem. *Discrete Applied Mathematics*, 9:41–46, 1984.

66. Z. Win. On the windy postman problem on Eulerian graphs. *Mathematical Programming*, 44: 97–112, 1989.

67. W.L. Pearn and M.L. Li. Algorithms for the windy postman problem. *Computers and Operations Research*, 21:641–651, 1994.

68. M. Dror, H. Stern, and P. Trudeau. Postman tour on a graph with precedence relation on arcs. *Networks*, 17:283–294, 1987.

69. G. Ghiani and G. Improta. An algorithm for the hierarchical Chinese postman problem. *Operations Research Letters*, 26:27–32, 2000.

70. P.F. Lemieux and L. Campagna. The snow plowing problem solved by a graph theory algorithm. *Civil Engineering Systems*, 1:337–341, 1984.

71. A.S. Alfa and D.Q. Liu. Postman routing problem in a hierarchical network. *Engineering Optimization*, 14:127–138, 1988.

72. M. Dror and A. Langevin. A generalized traveling salesman problem approach to the directed clustered rural postman problem. *Transportation Science*, 31:187–192, 1997.

73. C.E. Noon and J.C. Bean. A Lagrangean based approach to the asymmetric generalized traveling salesman problem. *Operations Research*, 39:623–632, 1991.

74. C. Malandraki and M.S. Daskin. The maximum benefit Chinese postman problem and the maximum benefit traveling salesman problem. *European Journal of Operational Research*, 65:218–234, 1993.

75. W.L. Pearn and K.H. Wang. On the maximum benefit Chinese postman problem. *Omega*, 31: 269–273, 2003.

76. H.-F. Wang and Y.-P. Wen. Time-constrained Chinese postman problem. *Computers and Mathematics with Applications*, 44:375–387, 2002.

77. E. Benavent and D. Soler. The directed rural postman problem with turn penalties. *Transportation Science*, 33:408–418, 1999.

78. A. Corberan, R. Marti, E. Martinez, and D. Soler. The rural postman problem on mixed graphs with turn penalties. *Computers and Operations Research*, 29:887–903, 2002.

79. B.L. Golden and R.T. Wong. Capacitated arc routing problems. *Networks*, 11:305–315, 1981.

80. J.M. Belenguer and E. Benavent. A cutting plane algorithm for the capacitated arc routing problem. *Computers and Operations Research*, 30:705–728, 2003.

81. B.L. Golden, J.S. DeArmon, and E.K. Baker. Computational experiments with algorithms for a class of arc routing problems. *Computers and Operations Research*, 10:47–59, 1983.

82. G. Ulusoy. The fleet size and mix problem for the capacitated arc routing problem. *European Journal of Operational Research*, 22:329–337, 1985.

83. P. Beullens, M. Muyldermans, D. Cattrysse, and D. van Outheusden. A guided local search heuristic for the capacitated arc routing problem. *European Journal of Operational Research*, 147:629–643, 2003.

84. P. Lacomme, C. Prins, and M. Sevaux. A genetic algorithm for a bi-objective capacitated arc routing problem. *Computers and Operations Research*, 33(12):3473–3493, 2006.

85. G. Gutin and A.P. Punnen, editors. *The Traveling Salesman Problem and Its Variations*. Combinatorial Optimization. Kluwer Academic Publishers, Dordrecht, 2002.

86. E.L. Lawler, J.K. Lenstra, A.H.G.R. Kan, and D.B. Shmoys, editors. *The Traveling Salesman Problem: A Guided Tour of Combinatorial Optimization*. Wiley, Chichester, 1985.

87. A. Schrijver. *Combinatorial Optimization: Polyhedra and Efficiency*. Springer Verlag, Berlin, 2003.

88. K. Menger. Das botenproblem. *Ergebnisse Eines Mathematischen Kolloquiums*, 2:11–12, 1932.

89. G.B. Dantzig, D.R. Fulkerson, and S.M. Johnson. Solution of a large scale traveling salesman problem. *Operations Research*, 2:393–410, 1954.

90. D. Applegate, R. Bixby, V. Chvatal, W. Cook, and K. Helsgaun. Optimal tour of Sweden., www.tsp.gatech.edu/sweden/index.html, 2004.

91. K. Jongens and T. Volgenant. The symmetric clustered traveling salesman problem. *European Journal of Operational Research*, 19:68–75, 1985.

92. G. Laporte and U. Palekar. Some applications of the clustered travelling salesman problem. *Journal of the Operational Research Society*, 53:972–976, 2002.

93. L. Gouveia and S. Voss. A classification of formulations for the (time-dependent) traveling salesman problem. *European Journal of Operational Research*, 83:69–82, 1995.

94. R.J. Vander Wiel and N.V. Sahinidis. An exact solution approach for the time-dependent traveling-salesman problem. *Naval Research Logistics*, 43(6):797–820, 1996.

95. M. Fischetti, J.J. Salazar, and P. Toth. A branch-and-cut algorithm for the symmetric generalized traveling salesman problem. *Operations Research*, 45:378–394, 1997.

96. B.D. Diaz, M. Gonzalez, and E. Garcia. A hierarchical approach to managing dairy products. *Interfaces*, 28(2):21–31, 1998.

97. B. Korte and J. Vygen. *Combinatorial Optimization, Theory and Algorithms*. Springer Verlag, 2nd edition, 2002.

98. G. Ausiello, P. Crescenzi, G. Gambosi, A. Marchetti-Spaccamela, and M. Prabasi. *Complexity and Approximations*, Springer, 1999.

99. M. Held and R.M. Karp. The TSP and minimum spanning trees. *Operations Research*, 18: 1138–1162, 1970.

100. M. Held and R.M. Karp. The TSP and minimum spanning trees: part II. *Mathematical Programming*, 1:6–25, 1971.

101. N. Christofides. Worst-case analysis of a new heuristic for the traveling salesman problem. Technical Report CS-9313, Carnegie Mellon University, Pittsburgh, Pennsylvania, 1976.

102. J.L. Bentley. Fast algorithms for geometric traveling salesman problems. *ORSA Journal on Computing*, 4:387–411, 1992.

103. I. Or. *Traveling Salesman-Type Combinatorial Problems and their Relation to the Logistics of Regional Blood Banking*. PhD thesis, Department of Industrial Engineering and Management Sciences, Northwestern University, Evanston, Illinois, 1976.

104. S. Lin and B.W. Kernighan. An effective heuristic algorithm for the traveling salesman problem. *Operations Research*, 21:972–989, 1973.

105. Z. Michalewicz. *Genetic Algorithms + Data Structures = Evolution Programs*. Springer Verlag, Berlin, 3rd edition, 1996.

106. M. Dorigo and T. Stützle. *Ant Colony Optimization*. MIT, Cambridge, Massachusetts, 2004.

107. G.B. Danztig and J.H. Ramser. The truck dispatching problem. *Management Science*, 6(1):80–91, 1959.

108. D. Mester and O Bräaysy. Active guided evolution strategies for large scale VRPTW. *Computers and Operations Research*, 32:1593–1614, 2005.

109. P. Toth and D. Vigo, editors. *The Vehicle Routing Problem*. SIAM, Philadelphia, Pennsylvania, 2001.

110. T.G. Crainic and G. Laporte. *Fleet Management and Logistics*. Kluwer Academic Publishers, Boston, 1998.

111. M.O. Ball, T.L. Magnanti, C.L. Monma, and G.L. Nemhauser, editors. *Network Routing*, volume 8 of *Handbooks in Operations Research and Management Science*. Elsevier, Amsterdam, 1995.

112. C.D. Tarantilis, G. Ioannou, and G. Prastacos. Advanced vehicle routing algorithms for complex operations management problems. *Journal of Food Engineering*, 70:455–471, 2005.

113. E.M. Arkin, R. Hassin, and A Levin. Approximations for minimum and min-max vehicle routing problems. *Journal of Algorithms*, 59(1):1–18, 2006.

114. C. Bazgan, R. Hassin, and J. Monnota. Approximation algorithms for some vehicle routing problems. *Discrete Applied Mathematics*, 146:27–42, 2005.

115. B.L Golden, E.A. Wasil, J.P. Kelly, and I.M. Chao. The impact of metaheuristics on solving the VRP: algorithms, problem sets, and computational results. In T.G. Crainic and G. Laporte, editors, *Fleet Management and Logistics*, pages 33–56, Kluwer Academic Press, Boston, 1998.

116. A. Haghani and S. Jung. A dynamic vehicle routing problem with time-dependent travel times. *Computers and Operations Research*, 32:2959–2986, 2005.

117. C-W Chu. A heuristic algorithm for the truckload and less-than-truckload problem. *European Journal of Operational Research*, 165:657–667, 2005.

118. M.L. Fisher, K.O. Jörnsten, and O.B.G. Madsen. Vehicle routing with time windows: two optimization algorithms. *Operations Research*, 45(3):488–492, 1997.

119. G. Nagy and S. Salhi. Heuristic algorithms for single and multiple depot vehicle routing problems with pickups and deliveries. *European Journal of Operational Research*, 162:126–141, 2005.

120. N. Bianchessi and G. Righini. Heuristic algorithms for the vehicle routing problem with simultaneous pick-up and delivery. *Computers and Operations Research*, 2005.

121. M. Gronalt, R.F. Hartl, and M. Reimann. New savings based algorithms for time constrained pickup and delivery of full truckloads. *European Journal of Operational Research*, 151:520–535, 2003.

122. V. Gaur and M.L. Fisher. A periodic inventory routing problem at a supermarket chain. *Operations Research*, 52:813–822, 2004.

123. D. Feillet, P. Dejax, and M. Gendreau. Traveling salesman problems with profits. *Transportation Science*, 39(2):188–205, May 2005.

124. A.N. Campbell and J.R. Hardin. Vehicle minimization for periodic deliveries. *European Journal of Operational Research*, 165:668–684, 2005.

125. J-F. Cordeau, M. Gendraeu, G. Laporte, J-Y. Potvin, and F. Semet. A guide to vehicle routing heuristics. *Journal of Operational Research Society*, 53:512–522, 2002.

126. J. Lenstra and A. Rinnooy Kan. Complexity of vehicle routing and scheduling problems. *Networks*, 11:221–227, 1981.

127. M.L. Fisher. Optimal solution of vehicle routing problems using minimum K-trees. *Operations Research*, 42:626–642, 1994.

128. G. Clarke and J.V. Wright. Scheduling of vehicles from a central depot to a number of delivery points. *Operations Research*, 12:568–581, 1964.

129. D.O. Casco, B.L. Golden, and E.A. Wasil. Vehicle routing with backhauls: models, algorithms, and case studies. In B.L. Golden and A.A. Assad, editors, *Vehicle Routing: Methods and Studies*, pages 127–147, North Holland, Amsterdam, 1988.

130. F.Y. Li, B. Golden, and E. Wasil. Very large–scale vehicle routing: new test problems, algorithms, and results. *Computers and Operations Research*, 32:1165–1179, 2005.

131. C. Prins. A simple and effective evolutionary algorithm for the vehicle routing problem. *Computers and Operations Research*, 31:1985–2002, 2004.

132. B.E. Gillett and L.R. Miller. A heuristics algorithm for the vehicle dispatch problem. *Operations Research*, 22:340–349, 1974.

133. J.E. Beasley. Route first–cluster second methods for vehicle routing. *Omega*, 11:403–408, 1983.

134. M.L. Fisher and R. Jaikumar. A generalized assignment heuristic for the vehicle routing problem. *Networks*, 11:109–124, 1981.

135. J. Bramael and D. Simchi-Levi. A location based heuristic for general routing problems. *Operations Research*, 43:649–660, 1995.

136. N. Christofides, A. Mingozzi, and P. Toth. The vehicle routing problem. In N. Christofides, A. Mingozzi, P. Toth, and C. Sandi, editors, *Combinatorial Optimization*, pages 313–338. Wiley, New York, 1979.

137. S. Lin. Computer solutions of the TSP. *Bell System Technical Journal*, 44:2245–2269, 1965.

138. J. Homberger and H. Gehring. A two–phase hybrid metaheuristic for the vehicle routing problem with time windows. *European Journal of Operational Research*, 162:220–238, 2005.

139. B.M. Baker and M.A. Ayechew. A genetic algorithm for the vehicle routing problem. *Computers and Operations Research*, 30:787–800, 2003.

140. F.A.T. Montané and R.D. Galvão. A tabu search algorithm for the vehicle routing problem with simultaneous pick-up and delivery service. *Computers and Operations Research*, 33(3):595–619, March 2006.

141. I.H. Osman. Metastrategy, simulated annealing and tabu search algorithms for the vehicle routing problem. *Annals of Operations Research*, 41:421–451, 1993.

142. S.C. Ho and D. Haugland. A tabu search heuristic for the vehicle routing problem with time windows and split deliveries. *Computers and Operations Research*, 31:1947–1964, 2004.

143. C.D. Tarantilis and C.T. Kiranoudis. An efficient metaheuristic algorithm for routing product collecting vehicles of dehydration plants. I. Algorithm development. *Drying Technology*, 19:965–985, 2002.

144. C.D. Tarantilis and C.T. Kiranoudis. An efficient metaheuristic algorithm for routing product collecting vehicles of dehydration plants. II. Algorithm performance and case studies. *Drying Technology*, 19:987–1004, 2002.

145. M. Gendreau, A. Hertz, and G. Laporte. A tabu search heuristic for the vehicle routing problem. *Management Science*, 40:1276–1290, 1994.

146. P. Toth and D. Vigo. The granular tabu search (and its application to the VRP). *Technical Report OR/98/9, DEIS, Universita di Bologna,* Italy, 1998.

147. E.D. Taillard. Parallel iterative search methods for vehicle routing problems. *Networks*, 23: 661–673, 1993.

148. B.L Golden, G. Laporte, and E.D. Taillard. An adaptive memory heuristic for a class of vehicle routing problems with minmax objective. *Computers and Operations Research*, 24(5):445–452, 1997.

149. J. Xu and J.P. Kelly. A new network now based tabu search heuristic for the VRP. *Transportation Science*, 30:379–393, 1996.

150. C. Rego and C. Roucairol. Parallel tabu search algorithm using ejection chains for the VRP. In I.H. Osman and J.K. Kell, editors, *Metaheuristics: Theory and Applications*, pages 661–675, Kluwer Academic Publishers, Boston, 1996.

151. C. Rego. A subpath ejection method for the VRP. *Management Science*, 44:1447–1459, 1998.

152. G. Barbarosoğlu and D. Özgür. A tabu search algorithm for the vehicle routing problem. *Computers and Operations Research*, 26:255–270, 1999.

153. C. Rego. Node-ejection chains for the VRP sequential and parallel algorithms. *Parallel Computing*, 27:201–222, 2001.

154. A.L. Bouthillier and T.G. Crainic. A cooperative parallel metaheuristic for the vehicle routing problem with time windows. *Computers and Operations Research*, 32:1685–1708, 2005.

155. S. Jung and A. Haghani. A genetic algorithm for the time dependent VRP. *Journal of Transportation Research Board*, 1771:161–171, 2001.

156. R.W. Hall. On the road to recovery: vehicle routing software survey. *OR/MS Today*, June 2004.

17

Pricing and Rating

Ryan E. Maner
Tarek T. Taha
Gary L. Whicker
J.B. Hunt Transport

17.1 Types of Truck Transportation

Truck transportation represents a large percentage of national supply chain expenditures. According to U.S. Census Bureau figures, truck transportation firm revenues for 2003 totaled over $171 billion. For 2005, these transportation expenditures represent 3.36% of sales for a representative company involved in shipping freight, according to Establish, Inc. and Herbert W. Davis and Company survey information.

The most common truck transportation modes can be classified into one of three types: parcel package shipments, less-than-truckload (LTL) quantity shipments, and full truckload (TL) quantity shipments. Parcel carriers convey individual packages. LTL carriers move larger items, often in pallet quantities, but not enough to fill the typical truck trailer. In both operations, a form of hub-and-spoke operation is utilized. Individual shipments are gathered in a local market, sorted and consolidated by common destination, transferred via a line haul move to the destination, where they are de-consolidated and re-sorted for local delivery. Truckload transportation differs in that it operates under a point-to-point system where the full load is moved directly to the final destination without any additional handling of the freight.

In the full load, or truckload, segment, further variations emerge. Transit can be made using a single driver who makes regular rest stops or a tandem team of two drivers who never stop driving. Additionally, full truckload shipments can be performed using a combination of rail and truck service. This is commonly referred to as intermodal truckload service. Shippers can also arrange to have multiple pickups and/or deliveries combined into one full truckload. No sorting is done, only multiple stops to pick up or deliver freight.

Each of the three truck service types are unique and require different assets and processes. Most truck transportation providers, or carriers, offer only one of the three service types. Some carriers do provide multiple services; generally through separate divisions. Truckload and LTL modes are the focus of the remaining discussion.

17.2   Common Pricing Components in Truckload Rating

This difference in service types translates into unique and distinctive pricing structures for each form of truck transportation. Most transportation pricing is based on a combination of activity-based costing principles, coupled with competitive market forces and existing freight flows. For truckload service, pricing is generally presented as a combination of line-haul or tariff rate, accessorial charges, and fuel surcharges.

Tariff rates cover the service of moving a shipment from origin to destination, and can be considered as the base rate. Move to location, load, transport, and deliver activity costs are typically included in the tariff rate. These rates are quoted in either a flat dollar amount or a rate per mile where some agreed-upon mileage standard is used to determine the mileage basis for the shipment. While other transportation services are priced and rated on a per weight basis or a combination of weight and distance pricing structures, truckload pricing assumes the full trailer is consumed to haul the shipment, therefore pricing is quoted on a per mile or per shipment basis. Activity-based costs along with competitive market factors are used to determine the quoted tariff rate.

In addition, accessorial charges are quoted to cover incidental services carriers may provide in addition to the base service of conveying the shipment. These charges include things like spending extra time loading or unloading, having the driver perform some part of the unloading or loading task, or making an additional stop to either pick up or deliver a portion of the shipment. As with the tariff rate, activity-based costing principles are used to determine the rate per occurrence for the various accessorial items. Accessorial rates typically add an additional 2% to 4% of the base tariff rate to the cost of truckload transportation.

The final truckload rating component is a fuel surcharge. Most truckload tariff pricing is set on a base fuel price. A fuel surcharge is then developed to either collect more or charge less to cover fluctuating fuel prices over the life of the rate quoted. Government fuel price indices are typically used to calculate a fuel surcharge (or rebate) based on the difference between the base fuel rate at the time the tariff rates were established and the current fuel price. A combination of assumed fuel consumption per mile, the miles traveled, and the fuel price difference are used to calculate the fuel surcharge amount for a given shipment. Fuel surcharges vary as the price of diesel fuel swings. In constrained energy markets, fuel surcharges can exceed 30% of the base tariff rate.

The combination of these three pricing and rating components: base or tariff rate, accessorial charges, and fuel surcharges, determines the overall rate charged for each shipment serviced

17.3   Truckload Costing and Pricing Factors

Pricing and rating transportation services is a more unique and diverse activity than pricing most products and many services. For example, an automobile manufacturer builds a standard car for any given make and model. Due to modern manufacturing processes, the costs and therefore the quoted rates for the car plus options are predetermined and highly controlled. There is a base price associated with the car, plus additional charges for any desired option.

In contrast, truckload transportation is a customized service nearly every time. Moving two different shipments the same distance may have radically different associated costs, and therefore quoted rates, depending on the shipment characteristics. Some major characteristics include:

1. Distance transported, commonly referred to as the length of the haul (LOH)
2. Specific points of origin and destination

3. Expected loading and unloading activities
4. Consistency and seasonality of shipment volumes
5. Commodity characteristics
6. Equipment type requirements
7. Cargo claims exposure
8. Fuel cost basis

The first characteristic determines a majority of the cost to transport. Driver wages, fuel, and maintenance expense are all directly related to the length of haul; the further transported, the greater the costs. Origin and destination points are also critical pricing criteria. Balance is a must. Carriers look for shipments that allow them to freely move equipment into and out of markets. When imbalances in freight flows exist, the carrier must factor in repositioning costs to move the equipment to a market with offsetting flows. These costs are usually determined by the miles moved and the opportunity cost of the time moving from one area to another.

Loading and unloading activities are another key cost driver. Typically, pickups and deliveries are made in one of two ways: live or drop. "Live" activities involve the driver and tractor in the loading or unloading process. "Drop" activities decouple the trailing equipment from the driver and tractor. An equipment pool is maintained so that the dock activity occurs without the driver and tractor. Once the particular loading or unloading event is completed, the trailing equipment is staged for the driver to simply drop one piece of trailing equipment and hook to another. In industry parlance, this is a "drop and hook." Activity costs for the live activities include driver and tractor time, while costs for drop events include extra trailing equipment at the shipping or receiving location along with equipment repositioning costs to maintain the equipment pool levels. For live activities, the duration of the dock activity is the principal cost-driver. Additionally, the amount of physical involvement required of the driver to load or unload the product is a further determinant of cost.

The consistency and seasonality of shipment volumes are also factors that drive transportation costs. Consistent (not highly variable) freight volumes are less costly because carriers can plan equipment needs with a high degree of certainty, reducing the spoilage costs associated with holding equipment and not using it when demand drops. In the reverse case, when demand surges, the carrier has extra cost to reposition additional equipment from other markets to cover the spike in demand. Seasonality is a similar cost-driver when volumes ebb and flow over the course of the year.

The type of commodity hauled also determines cost. For instance, certain hazardous materials require special driver certifications and possibly routes to be traveled increasing costs. Cargo claim exposure is a factor of both the commodity (high-dollar electronics versus inexpensive wood pallets) value and the particular requirements by shippers regarding cargo liability amounts. The higher the value of the cargo and the higher the cargo liability requested by the shipper, the more the cost to transport the shipment.

Equipment type also drives costs. Truckload carriage can be defined by three primary trailing equipment types: dry vans, flatbeds, and temperature-controlled trailers. The typical dry van and flatbed trailers are equivalent in cost, although the loading of a flatbed requires additional driver involvement to secure and protect the shipment. Temperature-controlled trailing equipment is significantly more costly to purchase (greater than a factor of two times), operate, and maintain than the dry van or flatbed trailers.

A final major cost driver for any given truckload shipment tariff or base rate is the assumed fuel cost basis. Fuel surcharges are driven by comparing the fuel cost index against a base fuel rate. Rates can be constructed with a low base rate, meaning fuel surcharges will apply immediately or with a current market rate, meaning fuel surcharges will not apply until the fuel cost index moves significantly. A low base rate can dramatically lower the tariff rate. A corresponding increase to the fuel surcharge component of the rate will apply, making the total transportation cost theoretically neutral regardless of choosing a high or low fuel base rate.

17.4   Truckload Rate Construction

As mentioned earlier, truckload services are highly unique and customizable. To automate and increase accurate invoicing and payment, complex rate management systems have been developed. A general hierarchy of (*i*) contract, (*ii*) section, (*iii*) item, (*iv*) lane, and (*v*) rate has become the dominant structure for storing and accessing truckload rates. In addition to the rate structure schema, two additional features are needed to fully define and apply truckload rates systematically. The first is a means of defining applicable geometry and the second is a method to calculate mileage (distance) between any two origin and destination pairs. Each of the three major building blocks to construct truckload rates will be discussed in the following sections in more detail.

17.4.1   Rate Structure

The contract identifies the particular commercial agreement, usually between a carrier and a specific customer. The section segment is a further breakdown which usually denotes the different business segments serviced under one master commercial agreement. The item segment refers to a group of similar rates, usually based on geographic definitions of origin and destination points. The lane level information refers to the origin and destination pairs for which rates are established. The rate or charge itself and specifications on how the rate is applied makes up the final truckload rate information category. Figure 17.1 illustrates the conceptual structure most truckload rates are built upon.

17.4.2   Geographic Definition

Geographic definitions are critical to truckload rating systems. Shippers and carriers have developed methods of defining the boundaries of various geographic areas using postal service codes. Typically, the largest geographic area definition is a state or province, while the smallest level of geography for truckload purposes is a low-level postal code. For the United States, as an example, the USPS five-digit zip code is commonly the lowest level of geographic specificity used, while a state is likely to be the highest level. In between, states can be broken into regions. For example, California may be split into southern and northern regions. Another common intermediate definition is the three-digit USPS zip code. Any combination of these and other geographic definitions can be used to define specific lanes, or origin and destination pairs. Table 17.1 illustrates the geographic definition combinations commonly used to define lanes in truckload transportation settings.

The principle of rate precedence follows from this illustration. The specific origin and destination for any shipment is known at the five- or nine-digit zip code level. Given these zip code pairs, most truckload rating processes search for the most specific geographic combination for which pricing is established (i.e., published) first. In Table 17.1, the most specific definition is the five-to-five digit zip code combination. Summing the digits for each possible combination in Table 17.1 gives a relative precedence level. In the case of the five-digit to five-digit lane (also referred to as a point-to-point lane), the sum is 10. The process is to search the database for lanes and corresponding rates that match the specifics of the particular shipment. If no published lanes and corresponding rates are found matching the five-digit zip code pair for the shipment, the next most specific lane definition is searched. In this case, the five-digit to three-digit and three-digit to five-digit combinations are checked next. This process repeats until the most geographically specific published rate is found that matches the shipment's lane.

Most commercial agreements follow a Pareto-like principle when negotiating and establishing rates. In all likelihood, 20% of the lanes will represent 80% of the shipments. It is sensible to establish specific point-to-point rates for this volume. The remaining 80% of the lanes with only 20% of the shipment volume can be defined using a broader approach which includes several levels of geographic territory definition. An example of this concept is a raw material supplier's distribution center (origin) going into a manufacturing plant (destination). These points are known, fixed, and have steady flows. It is beneficial to analyze this lane at a point-to-point level. In cases where various supply points may be used at any given time, a broader

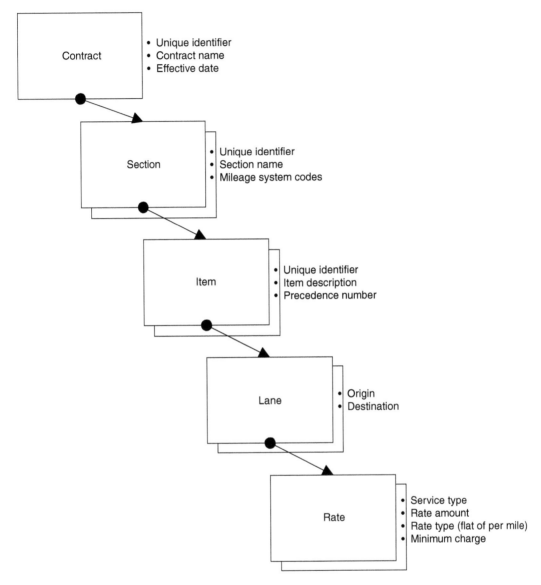

FIGURE 17.1 Typical truck load rate definition structure.

geographic definition for the origin makes sense. Assume that the raw material is a wood product coming from many possible supply points located in northern Wisconsin versus one distribution center. In this case, a region of state geographic definition can be used to cover all possible combinations without the need to analyze, negotiate, and administer every single point-to-point combination.

17.4.3 Mileage Basis

Rates are quoted and applied in one of two methods, either a flat charge to move from point A to point B or by a rate per mile to move from point A to point B. Logically, the flat charge is simply the precalculated charge determine by applying rate per mile to the transit miles:

$$\text{Rate per mile} \times \text{Miles shipped} = \text{Flat charge}$$

TABLE 17.1 Geographic Definition Combinations Commonly Used to Define Lanes in Truckload Transportation Settings

From\To	State	Region of State	3-Digit Zip Code	City/State	5-Digit Zip Code
State	1 + 1	1 + 2	1 + 3	1 + 4	1 + 5
Region of State	2 + 1	2 + 2	2 + 3	2 + 4	2 + 5
3-Digit Zip Code	3 + 1	3 + 2	3 + 3	3 + 4	3 + 5
City/State	4 + 1	4 + 2	4 + 3	4 + 4	4 + 5
5-Digit Zip Code	5 + 1	5 + 2	5 + 3	5 + 4	5 + 5

To harmonize rating processes, both carrier and shipping customer must have an agreed-upon standard to determine the transit mileage between any two points. Several commercial software packages provide this standardized approach to calculating mileage. Rand McNally & Company and ALK Technologies, Inc. are two of the leading providers of standardized and automated mileage calculation tools. Once a mileage system is chosen, other mileage basis factors to choose and agree upon include which route type to apply and which version of the particular software tool will be used to calculate mileage for rating purposes. Typically, either shortest or practical routes are available. In shortest routes, the mileage is the least, meaning the route for which the mileage is based upon is the most direct path from origin to destination. A practical route, in contrast, is the one that will generally follow the most accessible and well-traveled roads, foregoing any "short-cuts" that are impractical like a windy, narrow mountain road versus the longer, more practical route on the interstate around the mountain. As routes are updated with changes in the physical road network, the software companies release new versions of their products. Both shipper and carrier must agree upon which version the rates are based upon.

17.5 Key Rating Data Elements

The contract segment of basic truckload rate data, also referred to as publication, is a collection of rates that apply to a single customer or business entity. This section contains a unique identifier for the contract, customer name, effective date denoting when the rates are applicable, expiration date stating when the rates no longer are valid, and other customer-specific information relating to the commercial agreement.

The section segment of the rates provides a mechanism for separating rates for business or operating units within a particular contract. Basic data kept at the section level includes a unique section identifier code, a section name describing the business covered by this particular section, and mileage system basic data like which system version and route type applies to the rates included in the section.

The item segment is a grouping of related lanes, usually based on geographic specificity of the lanes. Basic data elements contained in the item segment include a unique identifier code, a description of the lanes usually containing some indication of the geography type of the lanes contained within the item, and a revision number representing the number of times that changes have been made to the item since it was originally created. Additionally, the rate precedence sequence may be kept at this level. This number represents the precedence of the item within the section.

Also, a minimum charge which applies to any lane contained within this item may be stored at this level. For example, suppose the geographic definition is a very broad state-to-state definition and that rates are stored and applied in a dollar per mile fashion and this rate is $2.00 per mile. For some moves where the origin and destination points are on either side of the state border, the mileage might be quite short. If the mileage is 50 miles for a shipment, the charge would be $100 which is below the minimum cost to pick up, transport, and deliver the load. Many truckload rates are subject to a minimum charge to ensure adequate compensation on shorter length of haul shipments like this.

The lane segment contains the particular origin and destination pair information for which rates have been negotiated. The first data element at the lane level is a geographic description of the origin point of the lane. The point may be a city or state, five-digit zip code, three-digit zip code, state, or any combination of these types of points. The second data element is the same information for the destination point.

The last rate structure segment contains information related to the rate: the specific rate amount or charge, whether the rate is in a flat charge or rate per mile format, and effective and expiration dates indicating the duration of the period for which the rates are applicable. Also, a code indicating the mode of service for which the rate applies is stored at the rate level as well. As example, modes of service indicators may identify single versus team-driver service or highway-only versus intermodal service levels.

17.6 LTL Rating

In the LTL environment, numerous factors influence the decision of what rate to charge. Building density by combining multiple shippers' business is the objective for LTL carrier. When reviewing potential business, an LTL carrier will balance the desire to generate the most revenue with the goal of reducing expense and maximizing the amount of freight on each trailer. Typically, the carrier will require a representative sample of shipments from the customer based on the potential business. Using the shipment sample, analysis will be performed to project the activity-based costing, the impact to the freight network, and the estimated revenue. At the highest level, the rating process includes the lane in which the freight travels, the type and amount of freight tendered, and any special services required. The rating of LTL freight can be a complex process. To negotiate a fair price, the paying party must be mindful of several key components.

17.6.1 Freight Characteristics

The basic information required is the weight, piece count, origin, and destination of the freight. The lane information provides the number of miles required to haul the freight in addition to which terminals will handle the freight. Since an LTL shipment is moved via a "hub and spoke" network, freight may need to move over several terminals to reach its final destination. In an LTL network, a shipment is picked up at the origin and then moved to a distribution center. At the distribution center, multiple origins are consolidated together to make loads to either another distribution center or to the final destination. As the shipment gets handled more often, the touch or product handling labor expense will increase. Also, some terminals may have more efficient operations causing variations in local costs. Each zip-to-zip combination has its own rate and cost structure.

The zip-level rate is also reflective of the freight balance in that lane, that is, how many trailers go into the market versus the number coming out. To keep from collecting at a terminal, equipment sometimes must be moved out whether fully loaded or not. Drivers also must be moved once rested to avoid paying additional wages. A market is identified as imbalanced if empty trailers must be moved in to or out of the market to compensate the lack of freight in that direction. For imbalanced markets, the costs will be higher on the dominant direction which is called headhaul or fronthaul. Carriers will provide discounted rates in the weak or backhaul direction because empty trailer space already exists. Moreover, highly congested areas such as Manhattan, NY will have inflated rates to compensate for higher labor expense involved with pick up and delivery.

17.6.2 Freight Mix/Density

The density of the freight factors into the rates because it determines whether or not the shipment is weight-dominant or cube-dominant. LTL carriers consolidate freight from multiple customers and are very sensitive to trailer utilization. The trailer dimensions along with federal regulations limit the total cube and weight loaded on each trailer. Freight can complement other freight, or "comingle," such that the trailer is completely utilized, but most freight will "cube out" a trailer prior to reaching the maximum weight. In general, as the density gets lower the rates will be higher. The carrier can estimate the amount of space necessary to move the freight so that trailer utilization (or load factor as it is commonly referred to) will impact the price. If the density characteristics deteriorate, carriers may approach the customer with a rate increase to compensate for the additional trailer space occupied by their freight.

17.6.3 Class

The freight mix can typically be accounted for by looking at the class or commodity. The National Motor Freight Classification (NMFC) code groups items by value and freight characteristics. There is a freight class for every NMFC item. Dense, less-valuable freight has a low class and a lower freight charge. Less dense and more valuable commodities will have a higher class and higher rate.

Shippers can negotiate reductions in rates by using a provision called Freight All Kinds (FAK). This mechanism allows a shipper to hedge the risk of varying rates based on NMFC item and class by having everything rated at the same class. For example, an FAK 70 guarantees that all shipments will be rated as Class 70 regardless of the actual class. An FAK 70 would work in favor of the shipper if the majority of the freight is class 70 and above.

17.6.4 Discount/Waived Charges

The larger LTL carriers will publish new rate bases every year. The rates take into account the aforementioned attributes: lane, weight, class, etc. Rather than a customer paying the full rate, a discount is applied and is used as a means of negotiation among the carriers. The discount may be off of the carrier's current rate base, a past rate base, another carrier's rate base, or a third party rate scale produced by an independent company. Besides the base rate and discount, the customer's tariff will show provisions for special services called accessorial charges. Some examples of common LTL accessorials include residential delivery/pick up, inside delivery/pick up, ground delivery, storage, change in billing, and assured service. Accessorials are not discounted or waived unless agreed upon by both parties. For example, customers may want to have the ground delivery (GRD) fee waived or discounted if the majority of their shipments require lift-gate service for delivery. Otherwise, they would get charged for the full special service amount on each shipment that contained a GRD. The carrier will offer a waived GRD in exchange for a lower discount.

Two charges that are applicable to nearly every customer are fuel surcharge (FSC) and absolute minimum charge (AMC). To compensate the carrier for the cost of fuel, a surcharge is added to the freight bill and is calculated as a percentage of the freight charges. As diesel prices change, the FSC is adjusted accordingly. The AMC is the minimum amount that will be charged to the customer. This compensates the carrier for certain fixed costs in case they receive a small shipment which produces a low rate. Both the FSC and AMC are negotiable with the carrier and can be viewed as another way to discount the base rate.

Today's LTL customer needs to be aware that their transportation charges are based on more than a simple discount. To evaluate the impact of all of the rating components, carriers use an internal effective discount. Customers should also consult a similar method in determining their "true discount." The effective discount computes a discount number based on the current year's base rates, taking into account all concessions involving the discount, class, and special services. For example, if the customer has a 60% discount on a past rate base, waived FSC, and a $50 AMC, this may equate to an effective discount of 80% off of the present full rate. In other words, the same charges would be paid with an 80% discount on today's rate base and a full FSC and AMC. With the knowledge of what impacts LTL rates, a customer will know how to negotiate towards a complete rate.

17.7   Industry Application

Transportation buyers procure transportation services through prenegotiated agreements for known, steady business, as well as occasional and sporadic open-market, "day of use" purchases. Once rates are established, business is conducted between shipper and carrier. A typical process for most truckload transportation service procurement and execution actions is to:

1. Identify shipping lanes and volumes
2. Generate and administer a request for proposals from various carriers
3. Analyze and select primary and secondary carriers by lane
4. Implement agreements and rates
5. Execute shipments
6. Arrange "day of use" transportation services as needed (exceptions)
7. Audit freight charges and pay carrier

Many transportation users repeat the procurement process outlined in steps one through four every one to three years. As shipping patterns and needs change and carrier capabilities evolve, a periodic checking and re-setting of transportation services is sound business practice.

Depending on the size and complexity of a particular shipper's network, software tools can greatly facilitate this process. Companies like Manhattan Associates and CombineNet, Inc. offer optimization-based decision support tools to help gather, administer, and analyze transportation service procurement events. Once the analysis is complete, a routing guide is generally the final output of steps one through four. A routing guide, in its simplest form, is a listing of carriers, rates, and committed capacity by lane in a ranked order of preference.

This routing guide is then used to direct which carriers should service which shipments. In practice, this guide can be a paper document or a component of a transportation management system which automates the transportation execution process. With the rapid evolution of technology via the web as well as commercial purchased software packages, most transportation services consumers use some form of automated transportation management systems.

These systems tender shipment offers to carriers based on the routing guide, and facilitate the freight payment process by comparing the rates negotiated and on file with the actual carrier invoice. When a shipment moves in a lane with a carrier for which no prenegotiated agreement is in place, up-front agreement on the rate, the accessorial charges that may apply, and any fuel surcharge provisions is paramount.

The complexity and volume of transportation service is best managed with information systems. Many providers of commercial transportation management systems are available for a wide range of needs. Most such systems will administer multiple modes of transportation: truckload, LTL, parcel, air, and ocean. Codifying the agreement prior to execution by both parties is critical to sound transportation management, and transportation management systems bring a structure and discipline to ensure this work is done up-front. These transportation management systems generally provide a convenient database structure to store the necessary rating information data elements, making a complex process quite manageable.

17.8   Truckload Procurement Case Study

As a result of a recent merger, a leading consumer packaged goods manufacturing company approached truckload carriers in the Fall of 2005 with a bid package requesting pricing on new lanes. Because of the merged operations, the bidding company identified new transportation service needs consisting of approximately 100 lanes (origin and destination pairs) representing 85,000 annual truckload shipments. A large shipper before the merger, with an annual transportation spend exceeding one-half billion dollars, the bid was a key step in ensuring the new operations and transportation services meet the same service, quality, and cost targets as the existing operation. The bid was sent to the existing carrier base of both companies. Over 100 carriers were invited to submit pricing proposals. J.B. Hunt Transport, Inc. was one of the invited carriers. From initial letting of the bid to final negotiation and award, the process to procure over eighty-five million dollars of truckload services took just over three months.

17.8.1 Bid Preparation Process

The bidding company managed all bid information through a service offering from a company called CombineNet. CombineNet's Truckload Manager helps companies streamline their truckload bid process, giving insight into how to reduce costs and improve delivery performance. The bidding company provides specific requirements, constraints and preferences on a lane-by-lane basis to the participating carriers. The Truckload Manager web interface presents carriers with a transportation contract, facility profiles, accessorial charge requirements and lane information with historical volumes. After the proposals are received, the tool allows the bidding company to perform "what if" analysis to determine the best overall transportation solution for the company.

The quality of the bid data is directly proportional to the quality of the proposals received. In this case, the bidding company had complete and accurate information regarding the business being bid due to the analysis process used to merge the operating entities. Origin and destination points, average volumes, day of week and week of year seasonality impacts, operating characteristics regarding loading and unload procedures, site profiles (company shipping location specifics) as well as forward-looking estimates of shipment volumes were provided to all bidders. In addition, the bidding company identified the service performance characteristics for all incumbent carriers to aid in the quantitative analysis of each proposal's overall value when considering service, quality, and capacity against cost.

17.8.2 Pre-Bid Meeting

Once the bid process and data were finalized, the bidding company arranged a pre-bid meeting. Often, these meetings are conducted via web-based meetings. On-site meetings with all participants are also a popular method of communicating to the bidders the guidelines and expectations for submitting acceptable proposals. Carriers typically have a bid or pricing analyst who will take responsibility for preparing and submitting the bid response to the bidding company. This bid owner is the person who must understand what the bidding company expects in detail, and usually represents the carrier in the pre-bid meeting.

In this case, a web seminar was held to communicate to all bidding carriers. In less than two hours, the bid process and timelines were communicated to and reviewed with all participating carriers. The corporate transportation procurement staff for the bidding company led the meeting, with support from CombineNet staff members. A period of three weeks was allowed to prepare proposals. All proposals were asked to offer the best price first, as no further price negotiations (bidding rounds) were planned.

17.8.3 Bid Preparation and Submission

From the bidding carrier perspective, the goal of this process is to submit pricing on those lanes where the bidder can provide the required service at a reasonable profit. The J.B. Hunt Transport bid analyst utilizes an in-house software tool to review lane information in the bid file and attach unique lane identifiers that indicate whether a lane is best served by Truck (highway) or Intermodal (rail) service. A large amount of data manipulation and analysis is often necessary to translate the data from the bidding company format to the standard internal carrier format. No two bidders are exactly alike, nor are any two carriers. A common problem area is geographic descriptors—city names, zip codes, etc. are often abbreviated, misspelled, or inconsistent and must be corrected before pricing work can begin. This conversion and translation process is a necessary evil, but the negative impacts are minimized with well-prepared bids. With this particular bid, the data formats were clean and simple and the translation process went smoothly.

Next, a request-for-pricing (RFP) number is assigned and the bid package is then sent to the Truck and Intermodal service pricing departments to establish capacity and price commitments on their respective lanes. The majority of the proposal response time is spent understanding the requested services, how they complement the existing carrier network, and the likely impact on carrier profitability for a given proposed price. An ABC-type analysis is often used, identifying A lanes that are attractive and will be priced more aggressively, C lanes that do not fit the network or are costly to serve and will be priced higher, and B lanes that are in the middle; neither attractive nor unattractive. This analysis depends on both the freight characteristics and the carrier-specific situation. A wide range of pricing on the same freight is to be expected from multiple carriers.

Once this step is complete, the bid analyst readies the bid file for upload by converting it to the specified format. Along with a cover letter summarizing the bid information, the proposed pricing file is uploaded to the CombineNet website. For this particular bid, a hard due date and time were set by the bidding company, and the bid was uploaded on the final day with an hour to spare.

17.8.4   Bid Proposal Analysis

With all proposals in hand, the bidding company used CombineNet to analyze the proposals from all carriers; determining which carrier proposals provide the best combination of service, quality, and capacity (value) for the price quoted. The analysis process for this bid lasted approximately one month. The bidding company used the CombineNet tools to analyze various "what if" scenarios to ensure the decision reached met their desired performance objectives at the lowest cost.

17.8.5   Award

J.B. Hunt Transport was awarded specific lanes of business for a period of one year. Capacity and rate commitments were reviewed and mutually approved. An implementation team was formed, and preparations were made to finalize the service and pricing terms, establish equipment pools at the manufacturing sites where the freight originated, and to implement the necessary electronic data interchange protocols to support automated processing for load tendering, shipment status, and freight invoicing and payment.

17.8.6   Result

The bid process spanned better than three months in total. During that time, the bidding company successfully secured value-adding transportation services for their new shipping needs; services they expect will create a competitive advantage in terms of the service level to their customers at a market-competitive cost.

18

Management of Unbalanced Freight Networks

G. Don Taylor, Jr.
*Virginia Polytechnic Institute and
State University*

18.1 Introduction to Unbalanced Freight Networks

Freight networks are inherently unbalanced. Some areas have a large manufacturing base relative to their population and are thus net freight sources. Others have a large population of consumers but relatively little manufacturing or distribution infrastructure and are thus net freight sinks. Price can be used, to a certain extent, to drive shipper behavior and to overcome the added costs associated with imbalance, but this is not true in all cases and the problems are worse in some industries than in others.

Nowhere is the freight imbalance problem more significant than in the truckload trucking industry in North America. Because of the challenges posed by this difficult environment, this industry group is selected to provide a context for the remarks within this chapter. Although some of the imbalance issues discussed herein are specific to the truckload trucking industry, others will have much broader implications to freight transit in general. For a general overview of freight imbalance and empty repositioning in the rail industry, the reader is referred to Sherali and Suharko (1998). In maritime transport, imbalance issues are partially addressed by Cheung and Chen (1998) and by Crainic et al. (1993).

Although logistics experts agree that freight imbalance is a huge issue, and governmental agencies publish at least aggregate information regarding its significance, it is difficult to find published imbalance information at the correct level of resolution to be especially effective for a specific industry group or company. Hall (1999) offers one of the few exceptions to this rule by publishing an overview of the existing historical literature on stochasticity and imbalance in freight networks. Hall also discusses the difference between temporal imbalance (short-term imbalance that must be defeated by repositioning moves) and lane restriction imbalance (rules that may force earlier or more frequent returns). Of greater importance, at least to the less-than-truckload (LTL) industry, is the quantification of imbalance that is provided. Hall states that 90% of all LTL terminals have inbound to outbound ratios between 0.5 and 2.0 while 50% have ratios between 0.67 and 1.5. Hall also advocates the aggregation of terminals into groups to reduce the effects of imbalance. Terminal groups have a tighter range of inbound and/or outbound ratios with 95% of terminal groups having a ratio between 0.67 and 1.5 and with 50% having ratios between 0.9 and 1.2. Airlines also utilize "grouping" in the form of industry alliances to develop transportation networks that are more complete and better balanced (see Abeyratne, 2000).

In most imbalanced freight networks, one of the greatest challenges is that of returning empty equipment from poor (backhaul) markets to good (headhaul) markets. Clearly, this repositioning must occur before additional revenue can be obtained and the freight company will not be paid for repositioning moves. The problem is exacerbated in the truckload trucking industry due to the fact that drivers as well as equipments must be returned profitably to headhaul markets and eventually back to their domicile. The difficulty associated with returning drivers to their domiciles contributes to one of the greatest problems in the industry, that of retaining drivers. The truckload trucking industry is plagued with driver turnover rates that are routinely above 100% per year (see Min and Emam, 2003).

Clearly, it is desirable to use variable pricing to make trips into backhaul markets more economically feasible, but this cannot be done to the extent desirable. Rietveld and Roson (2002) provide an excellent analysis of using price to drive behavior and maximize revenue, but in a monopolistic public transport setting. Airlines also utilize price to partially control demand and to maximize revenue. As competition increases, the ability to use variable pricing to achieve desired results decreases. The truckload trucking industry is very far toward the other end of the spectrum. In the United States alone, almost 700,000 active interstate truck and bus companies exist, with almost eight million large trucks on the road (U.S. DOT, 2006). With this environment of high competition, it is very difficult indeed to shape customer behavior with price or to use pricing as an especially effective means of paying for driver and equipment repositioning. This high level of cost competition also causes carriers to focus heavily on the minimization of empty repositioning moves in driver dispatching, which leads to excessively long driver tour lengths, and further exacerbates the driver retention problem.

Thus, although freight imbalance is inherent in almost all producer and/or consumer networks, it negatively affects the truckload trucking and related industries. It adds to cost, complicates the dispatching task, and makes fair pricing difficult. In the truckload trucking industry it is a particularly daunting problem, adding to driver tour lengths, affecting driver pay, and contributing heavily to a very serious driver retention problem.

Furthermore, using Hall's definitions of imbalance, the imbalance problem is both lane restriction based and temporal (Hall, 1999). In truckload trucking, loads must be picked up and delivered at very precise locations and cost constraints do not permit large empty repositioning movements. Therefore, the imbalance is lane based over the longer term at a high level of resolution with literally thousands of delivery lanes for national or continental carriers, with each lane having a more or less known average imbalance. The real complication is due to the temporal imbalance. The lane volume on any given day

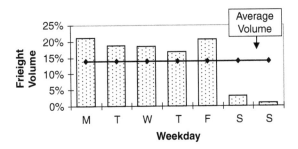

FIGURE 18.1 Daily freight volume distribution.

is highly stochastic for most lanes, so in addition to dealing with the high-level aggregate planning issues, trucking companies must also overcome the daily, weekly, monthly, and seasonal imbalances. As an example of temporal imbalance, consider the percentage of demands that are called in on each day of the week as reported in Powell (1996). Figure 18.1 graphically illustrates the data reported in Powell (1996) and makes it clear that temporal imbalance can exist in even well-established lanes. As reported in Fite et al. (2002), it is difficult to forecast freight demand using almost any leading or lagging economic index even on a strategic level. It is almost impossible to do so on a tactical or operational basis.

18.2 What Is Freight Density and How Do We Find It?

As pointed out in Taylor (2002a), the availability of truckload freight in North America appears to be very random in terms of both location and timing. Major shippers tend to exploit the highly competitive environment by treating freight transportation as a commodity purchase. Shippers often wait until the last minute to procure a carrier for their loads, making it difficult to preplan, and there is little price differentiation among the carriers. Closer examination reveals that there may be exploitable trends in this seemingly random freight. These exploitable traits build upon the fact that there are dense freight nodes (origins and destinations), dense "pass through" or transshipment points, and dense freight lanes. Although actual freight density is a function of manufacturing and population demographics that cannot be changed, it is possible to partition existing freight into administrative or operational groupings that enable freight carriers to more fully exploit freight trends and better utilize existing or planned infrastructure. Concurrently, understanding the density characteristics of existing freight enables the development of marketing and pricing strategies that help to create new markets and shape customer behavior.

One of the great underlying principles of freight transit is that economies of scale exist when freight density can be found. The recent tremendous growth of the third-party logistics (3PL) industry demonstrates this fact well, because one of the primary reasons for 3PL success has been the ability to combine the freight of multiple customers to find density. Density with respect to economies of scale means that appropriate modal choices can be made to reduce transportation charges. For example, parcel shipments can be consolidated, in a dense freight network, to permit conversion to LTL shipments. These shipments, in turn, can be consolidated into full truckloads. In many cases, truckloads can be further consolidated for long-haul intermodal transit via rail.

Another advantage of dense freight regions is that they can help to eliminate empty repositioning costs. If a load is delivered into dense freight network, it is much more likely that a suitable "next" load can be found to help in eliminating the empty repositioning costs associated with the next dispatch or to help in returning a driver to his or her domicile.

The meaning of freight density and the reasons for the availability of freight density differ by industry. Toh and Higgins (1985) argue, for example, that the hub and spoke systems employed by airlines following government deregulation in the United States often enabled a particular carrier to dominate a regional market, thus creating density as a function of regional marketing and negotiated airport gate availability. To a lesser extent, the LTL trucking industry can also partially dominate a region based on its decisions regarding infrastructure location. The truckload trucking industry offers one of the more interesting sets of challenges relative to freight density. This is partially because the industry requires very little regional infrastructure in comparison with the LTL or airline industries. There is no need for freight consolidation and consequently no need to have physical resources in place to assist in regional domination, especially for single-mode or short-haul freight. For these reasons, it is harder to exploit freight density in the truckload trucking industry in comparison with other logistics networks.

An individual truckload trucking customer's freight may be quite predictable, but other customers may have very erratic demand for freight transportation. Even so, loads can often be found that have both origins and destinations within a small radius of one another. The net effect of this phenomenon is the presence of dense freight "lanes." These lanes, if dense enough, become the basic building blocks of regular tours that ensure driver domicile returns and therefore enhance driver retention.

It is not always possible to find dense lanes, but it is almost always possible to find dense "hubs," where a hub is defined as a region in which many loads originate, conclude, or both. These hubs, as a minimum, help us to estimate the expected regional driver waiting time and deadhead (empty repositioning) charges for a next dispatch, thus enabling appropriate activity-based costing and pricing. Effective marketing strategies are also supported by the determination of dense freight hubs, as are revenue management (RM) strategies that either encourage or discourage freight into a particular region via recommended pricing.

The issue of freight balance is a very important consideration on either a hub or a lane basis. Some lanes may have a somewhat equal distribution of freight in each direction on the lane while others may have vastly different directional volumes. Similarly, some hubs have much more outbound than inbound freight (headhaul markets) whereas others have much more inbound freight than outbound freight (backhaul markets). Prices tend to be low into headhaul markets because the probability of quickly obtaining a high paying outbound load is very good. Prices are generally high into backhaul markets where a driver is likely to have a long outbound dwell time or a lengthy deadhead repositioning move.

If we are able to find and exploit dense freight regions, all parties benefit. Carriers can achieve higher utilization and can better manage the inherent freight imbalance that exists in almost any freight network. Drivers can use the density to return more frequently to their domicile, thus aiding immeasurably in job satisfaction and driver retention. Customers benefit from improved dispatching methods that lead to lower freight shipment costs and better overall service. Furthermore, as customers grow in understanding the operational efficiencies to be gained from exploiting density, they tend to become more of a partner with their carrier and they begin to shape their shipment behavior to better fit the operational paradigms that density permits.

18.2.1 Locating Areas of Dense Freight Activity: *Hub_Finder* Software System

The author has developed a suite of software tools to find and exploit various types of density. The easiest form of density to find, understand, and exploit is that of hub-based or node-based density. This is simply finding dense freight hubs or nodes that have high volume in terms of freight origins or destinations. The author has developed a software tool called *Hub_Finder* to assist in finding these dense hubs. *Hub_Finder* is a multi-criteria, agglomerative, hierarchical clustering method that finds hub centroids to maximize the "coverage" of freight origins and destinations within specified radii of freight centroids. These hubs then become building blocks for systems that exploit freight density.

Inputs to *Hub_Finder* include load information files and user input parameters. Load information can be based on recent historical data if the past is deemed a reasonable estimator of future demand or it can be based on some other form of long- or short-term demand forecasting system. A typical input record includes a location identifier (such as a postal zip code), information about load quantity inbound to the location, information regarding the quantity of loads outbound from the location, and positional information regarding the centroid of the location (likely in latitude and longitude). *Hub_Finder* then combines the various freight locations into hubs.

The selection of what type of location identifiers to be used determines the resolution of *Hub_Finder* solutions. The use of large identifiers (say a city or state) would likely not provide adequate freight density information, while the use of very small identifiers (perhaps each anticipated load) would likely lead to computational inefficiencies. The default input is that of five-digit postal zip codes.

User input parameters include the minimum and maximum allowable hub radii (in miles), the minimum number of load occurrences (the sum of loads inbound to and outbound from the hub) to produce a viable hub, a distance increment that defines how many possible hub radii are to be considered between the minimum and maximum sizes, and an indication of whether or not the user desires to use freight balance considerations in assigning freight origins and destinations to hubs.

The algorithm driving the software begins by retrieving the highest total volume freight centroid. This location becomes a seed location for a hub centroid. All other freight centroids within some user-specified minimum radius of this seed location are merged into the existing hub and a new weighted freight centroid location is calculated. If the user specifies that freight imbalance is not to be considered, the process is completed for all remaining freight centroids using a greedy algorithm (highest volume locations first). In the event that hub imbalance is to be considered, the hub radius is temporarily increased in size according to a user-specified incremental size increase. All of the freight within that increased radius is evaluated to determine whether or not inclusion would make freight imbalance better at the hub. If the answer is yes, the freight is included in the hub and the freight centroid and imbalance information for the hub is recalculated. If the answer is no, the freight is only included in the hub if the hub freight volume falls below a user-specified minimum hub volume. This process is repeated in increments until either the maximum allowed physical hub size is reached or until a point is reached in which further expansions do not contribute to the minimization of freight imbalance.

Figure 18.2 depicts graphically a situation in which a hub with a 50-mile radius is increased in size to a 60-mile radius based upon resulting improvements to imbalance. The same figure indicates that it would not be wise to increase the hub size to 70 miles.

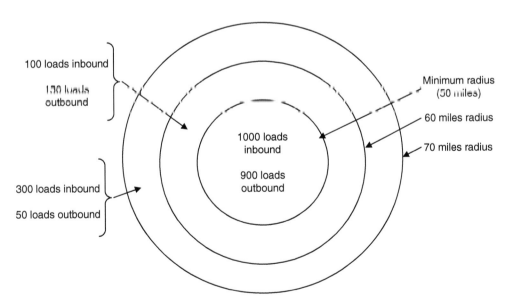

FIGURE 18.2 Freight imbalance considerations in hub size determination.

Hub_Finder system outputs include a listing of hubs including each hub number, the zip codes included in the hub, the latitude and longitude of the freight location weighted hub centroid, the hub radius in miles, the total hub volume (inbound plus outbound) and the hub imbalance (inbound minus outbound). The determination of how many hubs to maintain is a subjective task and is dependent upon the uses for the hubs. The basic trade-off is between freight "coverage" and solution tractability for various hub uses. Because freight needs tend to shift over time, the hubs should be re-established periodically to ensure that they accurately reflect current freight availability.

Hubs can provide a foundation for many types of management decisions. The *Hub_Finder* system has been used to provide hubs in support of pricing, RM decision making, and alternative dispatching methodologies. *Hub_Finder* hubs have also had uses that were initially unanticipated. For example, the hubs can easily provide a basis for marketing decisions by identifying and reinforcing knowledge regarding the extent of imbalance in a particular area. *Hub_Finder* has also been used as the first half of a "cluster-first, route-second" vehicle routing problem (VRP) heuristic for dispatching trucks in a city pick up and delivery environment.

18.2.2 Locating Dense Freight Lanes: *Lane_Finder* Software System

Another type of density is in the form of freight lanes or network arcs. This density can be found using the *Lane_Finder* software tool. The *Lane_Finder* software system is similar in function to the *Hub_Finder* system except that the focus is on finding dense freight lanes (network arcs) instead of freight hubs (network nodes). For freight loads to be aggregated into a single dense lane configuration, they must share both a common origin and a common destination. The user specifies a minimum radius for aggregation and a maximum radius for aggregation at each potential lane endpoint. Once a targeted lane volume is reached and minimum service area sizes are reached at both lane endpoints, only improvements to lane imbalance will cause the service area size to be increased and additional lane volume to be added.

Lane_Finder inputs include load information files and user input parameters. A typical load record includes an origin location identifier (zip code or hub), the origin latitude and longitude, a destination identifier (zip code or hub), the destination latitude and longitude, the total "out" volume (origin to destination volume over the planning period) and the total "in" volume (destination to origin volume over the planning period). User input parameters include the minimum and maximum service area sizes at lane endpoints and the target minimum load volume.

Beginning with the highest volume lane, all other data records are compared to determine how many of them have both common endpoints with that lane. All lanes that have origins and destinations within the minimum specified radius of the large lanes are combined with the larger lane and new inbound and outbound volumes and freight centroids are calculated. This process is repeated for each remaining lane, from largest to smallest, until all lanes have been considered for aggregation. The next step is to determine whether or not these aggregated lanes should be further combined into higher volume lanes with service area radii up to the user-specified maximum radius on both ends of the lane. This further aggregation takes place only if it would help the lane to reach desired minimum lane flow volume or if it would lead to improvements in lane balance.

Figure 18.3 graphically depicts one possible outcome of *Lane_Finder* activities. In this case, two lanes are aggregated with a larger "seed" lane based on their geographical proximity to the endpoints of the seed lane. Note that this aggregation takes place even though the resulting lane imbalance becomes worse as a result. This is because the lanes have endpoints within the specified minimum radius (20 miles in this case). This aggregated lane then becomes a candidate for expansion to a maximum radius (35 miles in this case). Whether or not this additional aggregation would take place depends upon whether or not the lane is already larger than the specified target value and whether or not the imbalance is improved as a result.

Lane_Finder output includes a listing of each aggregate lane complete with information about lane volume, lane imbalance, and the aggregate weighted average location of freight endpoints specified in

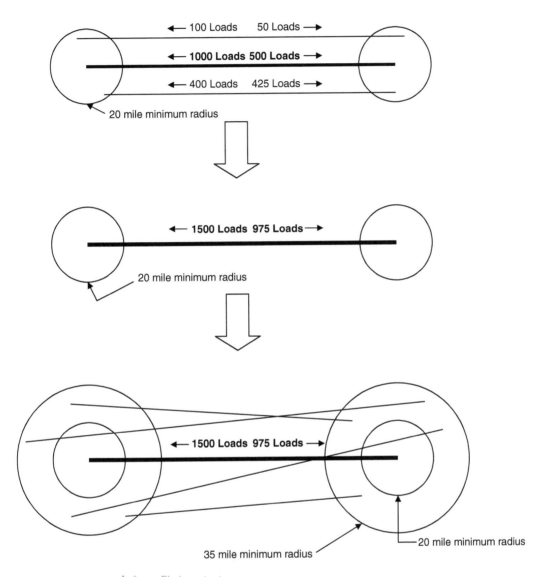

FIGURE 18.3 Example *Lane_Finder* output.

terms of latitude and longitude. These freight lanes can be used in support of several important activities. Well-balanced lanes can be used directly as a means of driver tour regularization in which a subset of drivers would operate solely on the specified lane with a high probability of regular return freight. Even those lanes that are not well balanced can be used as building blocks in regular tour development tools as specified in Section 18.4. The lanes can also be used as a means of simplifying backhaul availability data files in support of pricing and new customer bid development activities.

18.2.3 Locating Areas of Dense Intermodal Activity: *Ramp_Finder* Software System

The *Ramp_Finder* software system is designed to assist in determining "near optimal" freight aggregation points for situations in which small local freight unit loads are combined into larger unit loads for efficient long-haul transport. Examples of this type of transport include LTL city pickups being aggregated at a breakbulk terminal for truckload shipment or the aggregation of truckload shipments into a long-haul rail shipment for intermodal delivery.

In this case, the objective is to determine "near optimal" locations for rail ramps (the point of transfer between truck and rail) given a specific freight data set. Obviously, a carrier may have several intermodal ramp options in an urban setting and very few in rural settings. In either case, the ramp locations tend to be more or less fixed. Therefore, *Ramp_Finder* is a tool for assisting in ramp selection strategies in cases where choices exist. Also, trucking companies can work in coordination with rail companies to develop new ramp locations to service regions that may have dense intermodal freight activity. When used to locate LTL breakbulk terminals or manufacturing distribution centers, the output of *Ramp_Finder* can support location decisions more directly.

Ramp_Finder data input files are similar to those used in *Lane_Finder*. Each record includes an origin zip code identifier, the origin latitude and longitude, a destination zip code identifier, the destination latitude and longitude, and the volume of freight (in unit loads) from origin to destination. *Ramp_Finder* must interpret inputs in terms of both distance and direction in an effort to locate ramps that eliminate out-of-route circuity associated with freight movements. User input parameters include the specification of a maximum service area radius for building aggregate freight lanes, the specification of a similar maximum service area radius for establishing ramps, and the specification of a maximum "sweep angle" for ramp consolidation. The sweep angle is used to determine the freight lanes originating or ending at a common dense intermodal freight region that can be serviced by a single ramp based on commonality in direction.

The *Ramp_Finder* program begins very similarly to *Lane_Finder* by aggregating all freight lanes that share common endpoints. Next, each aggregated lane is given a "direction" value (in degrees) to indicate its slope. For example, an eastbound lane is designated as 0° while return freight on the same lane is designated as 180°. Next, *Ramp_Finder* locates any "existing" ramps that are within a specified ramp service area and that are directionally appropriate for the lane as specified by the sweep angle. This process is illustrated in Figure 18.4. If the lane can be serviced by an existing ramp, the ramp statistics are updated according to the added lane volume. If no existing ramp exists on either end of the lane, a new "desired" ramp is established. This process is repeated until all freight lanes are serviced by some ramp.

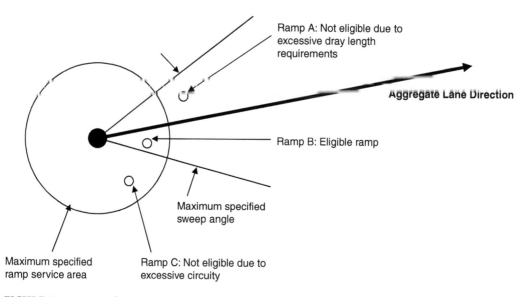

FIGURE 18.4   Ramp selection example.

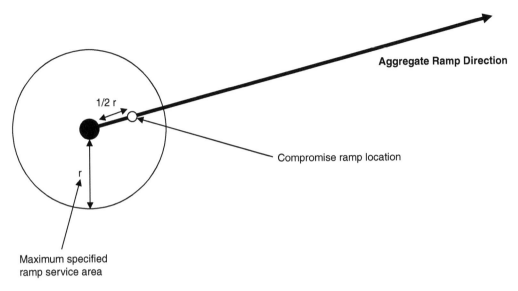

FIGURE 18.5   Establishing the ramp location.

Each ramp maintains statistics that represent the aggregate freight centroid and the aggregate freight direction. In the last step, the actual ramp locations are determined by locating the ramps in a way that represents a compromise between the desire to minimize circuity and the desire to minimize dray movements as indicated in Figure 18.5.

Ramp_Finder output includes a ramp information and lane information. The lane information is primarily a listing that indicates the inbound and outbound ramp locations that service each lane. The ramp location information includes information regarding location in terms of latitude and longitude and information regarding ramp volume and imbalance. Additional information is available in Taylor et al. (2002b).

18.2.4   Locating Areas of Dense Pass-Thru Activity: *Domicile_Finder* Software System

The *Domicile_Finder* software system is designed to locate regions with dense inbound, outbound or pass-thru freight activity to help determine the location of driver domiciles. Other possible uses include the determination of geographical locations for LTL breakbulks, manufacturing distribution centers, or freight exchange points for "drop and swap" load dispatch strategies.

Inputs to *Domicile_Finder* include freight data, seeded domicile locations, proximity definitions, and procedural parameters. The supporting data are freight lane records. Each record includes an origin location (zip code, hub, or latitude/longitude coordinate), a destination location, and the volume of freight (in loads) traveling from the origin to the destination in the course of the selected planning horizon. Potential domicile locations can be input by the user or can be defaulted. If candidate domicile locations are entered, the software seeks to determine which of the predefined candidate locations (i.e., cities, infrastructure locations, highway intersections, intermodal ramps, etc.) are best suited for driver domiciles. When existing domicile locations are not entered, the software uses a 1° by 1° latitude/longitude grid to establish default locations. The software ultimately "assigns" freight to the candidate domiciles. The domiciles with the most "assigned" freight then become the locations specified as the "best" locations to domicile drivers. The proximity definitions include the specification of the maximum allowable distance between domicile locations and freight lane endpoints and the maximum circuity permitted to define pass-thru nodes. User input procedural parameters specify which procedures are to be used in making domicile decisions. These inputs include the following:

- Ownership Specification: If ownership is "off," freight will be assigned to all candidate domicile locations that are within a specified distance from either the origin or the destination of the freight. Also, the freight will be assigned to all intermediate candidate locations that can be passed through without excessive circuity. If ownership is "on," freight can be assigned to only one candidate location.
- Load Weighting: Two methods of load weighting are permitted. The first partitions the total weight among inbound, outbound, and pass-thru locations. For example, a weighting of (0.35 in, 0.50 thru, 0.15 out) would indicate that low pass-thru circuity is the most important consideration with 50% of the total weight. Proximity to the in-bound freight destination is weighted 35%, and proximity of the location to the out-bound origin holds a 15% weighting. The freight would be assigned to the location with the smallest weighted distance value. In the second method, the user assigns priorities of 0, 1, 2, or 3 for variables associated with inbound, outbound, and pass-thru priorities. For example, if the values for inbound, pass-thru, and outbound priorities are 1, 2, and 3, respectively, inbound freight takes priority over pass-thru and outbound freight in determining freight "ownership." If there is at least one candidate domicile within a minimum distance of the inbound destination, the freight is assigned to the shortest distance location. The second choice is a pass-thru node and the third choice is an out-bound location. If values of 0, 1, and 0 are entered for inbound, pass-thru, and out-bound freight, respectively, only pass-thru domicile locations would be considered in the analysis.
- Capacity Limits: When this feature is activated, a maximum number of loads (miles) can be assigned to any given domicile. When this number is exceeded, no additional loads can be assigned to that domicile.
- Imbalance Limits: When this feature is activated, freight that increases the absolute value of imbalance at a domicile cannot be added to that domicile.

The *Domicile_Finder* software system examines each lane, from highest to lowest volume, in order, and finds all domicile locations (according to the user input) that are within minimum radius or circuity requirements. This process is repeated for all freight lanes until all lanes are either assigned to a domicile or marked as lanes that cannot meet the criteria of any domicile.

Output includes the total loads and miles examined in the total data set, the number and percentage of loads and miles that are ultimately assigned to a particular domicile (i.e., those that meet all assignment criteria), and a breakdown of miles and loads assigned to each hub in terms of those that are assigned as outbound, inbound, or pass-thru miles. Relative to driver needs, the output includes an estimate of the number of drivers required to handle the miles assigned to each domicile and an estimate of the number of drivers required to handle the remaining loads not assigned to any domicile. Also, the load imbalance (loads in minus loads out) is reported for each domicile. Additional information regarding *Domicile_Finder* output is presented in the case study following this chapter.

18.2.5 Importance of Finding Freight Density

This section has stressed the importance of finding freight density based on the premise that economies of scale exist when freight density can be found. Each of the types of density discussed in this section (hub-based, lane-based, intermodal, and pass-through) can be exploited to produce more efficient and more "regular" or repeatable driving routes in a traditional dispatching environment. Furthermore, the software tools introduced in this section can be used to support new dispatching methodologies that more fully exploit the possibilities brought about by having knowledge of overall dense freight patterns. These methodologies are discussed in the following sections of this chapter.

18.3 Exploiting Freight Density

This section focuses on several ideas to assist in exploiting freight density; how to find additional back-haul freight, how to find and exploit freight density via alternative dispatching strategies, how to develop RM strategies to assist with freight management in the presence of imbalance, and how to exploit existing pricing structures. Each of these strategies provides an opportunity to partially defeat the difficulties brought about by imbalance. The following discussion includes some new ideas but it is built largely around information and techniques from previously published articles by the author, most notably from Taylor (2003).

18.3.1 Finding Backhaul Freight

Perhaps the best way to defeat imbalance is to make it go away. This cannot be achieved across the entire freight transportation network because freight is inherently imbalanced. It is also not likely that perfect balance can be achieved on an operational level. Even so, individual companies can improve their network balance through effective marketing and pricing. It is certainly possible to target marketing activities toward improving backhaul opportunities. If a company has profitable inbound freight in a particular "lane" or to a particular "node," backhaul freight should be sought from that destination region. There are numerous examples in which successful trucking companies have built good two-way or three-way traffic lanes. This targeted marketing requires no sophisticated techniques. It simply requires a motivated and informed sales staff to contact shippers in the area that are known to have freight that either returns to a desired location or that can help in repositioning equipment to a better marketing area where greater choice in freight selection is possible.

The process of targeted backhaul marketing is greatly enhanced by the development of a "backhaul database." Such a database includes historical records of past load data. This data includes fields to indicate which customers in a specific region have a history of moving loads in a desired freight lane and also includes contact information for freight solicitation. It is particularly helpful when the carrier seeks freight in a very specific lane to assist in immediate driver domicile returns or in completing a tour in a dedicated contract service environment when "expected" freight does not materialize.

The use of dedicated fleets is a growing trend for carriers as many companies are making the decision to abandon private fleets but are not yet comfortable with the idea of reliance on the random over-the-road (OTR) network with multiple carriers. A third party dedicated fleet can often achieve better performance than a private fleet, largely because they are better prepared to handle freight imbalance based on their access to the freight of other customers. In cases where it is possible to match inbound and outbound movements, discounted prices can be offered to customers on both ends of a freight lane while the carrier concurrently makes a higher profit with lower risk.

Another recent innovation supporting the ability of carriers to find backhaul freight is that of electronic marketplaces where shippers can post freight movement needs and carriers can bid on the work. Interesting discussions of these innovations appear in Lin et al. (2002) and in Song and Regan (2001). These marketplaces serve a purpose similar to the backhaul databases discussed earlier. They are likely much more thorough, but they are less likely to provide a competitive advantage for an individual carrier.

18.3.2 Alternative Dispatching Strategies to Assist in Imbalance Management

Historical truckload trucking dispatching strategies have been heavily skewed toward minimizing carrier cost as opposed to improving driver satisfaction or even customer service. This means that the strategies have attempted to minimize empty repositioning miles as opposed to a more balanced approach that would minimize driver tour length. As the driver market tightened in the late 1980s and early 1990s, this traditional means of dispatch was replaced with dispatching tools that include more driver considerations and more insight into the future consequences of current dispatching decisions. In an industry plagued by more than 100% annual driver turnover, it makes financial sense to do so. In fact, Ronen (1997) points out that distance minimizing approaches are 35% more expensive than corresponding total cost minimizing dispatching approaches. Even so, these total cost strategies likely do not go far enough in terms of managing imbalance.

Clearly, alternative forms of dispatch are needed in order to find ways to return drivers home more frequently while maintaining the viability and profitability of the carrier. One way of doing this is through alternative dispatching methods that partition drivers into two sets: "regular route" drivers and remaining random OTR drivers. The dispatching methods must improve the quality of life and reduce tour lengths for the regular route drivers without significantly harming the quality of life for the remaining drivers.

Over the past few years, the author has been involved with several methods to make this driver partition by dispatching drivers in creative new ways. Broadly speaking, these dispatching methods fall into two categories: regular route methods and random route dispatching. The regular route methods will be discussed in Section 18.4. The random route methods discussed in Section 18.3 include hub and spoke dispatching (Taha and Taylor, 1994; Taha et al., 1996), zone dispatching (Taylor et al., 1999; Taylor et al., 2001), regional dispatching (Taylor et al., 2006a), "pipeline" dispatching (similar to intermodal transportation with rail, but using trucks for all dray and line haul movements, as in Taylor et al., 2007), "popcorn" dispatching (a form of random OTR dispatch with limited driver destinations permitted), and weekend yard stacking, among others.

The hub and spoke dispatching method discussed in Taha and Taylor (1994) and Taha et al. (1996) mimic LTL delivery networks in a truckload trucking environment. These methods provide better driving jobs for many drivers in the full delivery network but they do not necessarily solve the dispatching problem for the carrier. Excessive delay time is encountered for many of the long-haul loads due to the multiple dispatches required for each load. Even so, it appears that a limited number of exchanges could lead to improvement. This motivates zone dispatching and some of the regular route alternatives discussed in Section 18.4.

The zone dispatching methods discussed in Taylor et al. (1999) and Taylor et al. (2001) provide much more dramatic improvements with respect to defeating imbalance. In that work, drivers are assigned to geographical zones that they do not leave. Freight crossing zone boundaries is dropped at a "swap" yard for pick up by a driver in the next zone. Drivers are domiciled at the swap yards and are ensured of having frequent domicile returns. The research addresses several ways to define zone boundaries but the most successful method is one in which integer programming is used to build zone boundaries within relatively wide constraints with an objective of minimizing imbalance between the zones. When empty travel is required for a domicile return, the moves are small for individual drivers because of the geographical restriction on their driving area. An example of zone use on a national scale in the United States is depicted graphically in Figure 18.6.

The regional dispatching methods discussed in Taylor et al. (2006a) make use of a driver partitioning system that utilizes a series of regional fleets for some subset of drivers and related freight volume. In this dispatch system, regional drivers would deliver loads that originate and deliver wholly within their region and OTR drivers would continue to carry the longer hauls that do not fit wholly within any region. The idea is that regional drivers would have better retention rates due to their frequent domicile returns and familiarity with their driving conditions. The loads left for the partitioned OTR driver fleet would tend to be long-haul loads, thus adding to the ability of OTR drivers to have consistent earnings while on the road. Taylor et al. (2006a) find that:

- Regional fleet service areas should be on the order of 300 miles in radius.
- Fleets should be domiciled in 10 to 14 locations in the continental United States.

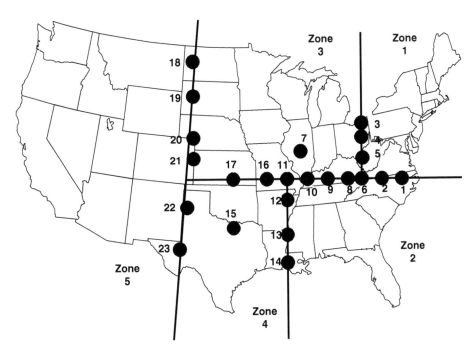

FIGURE 18.6   Zone dispatching.

- Regional fleets work well in both backhaul and headhaul markets.
- Regional fleets help the remaining OTR fleet more than they hurt it, both in terms of miles/driver/day and loaded miles per dispatch.
- Regional drivers will likely not be able to drive sufficient miles to make the switch from OTR attractive without significant changes in wage structures.
- Fleet centroid locations are best determined via driver domicile needs and pass-thru freight availability (*Domicile_Finder*) as opposed to endpoint freight density (*Hub_Finder*) or geographical/market concerns.
- Regional fleets require only minor changes to existing dispatching methods.

The best regional dispatching locations in the continental United States for the case environment in the truckload trucking industry are listed in Taylor et al. (2006a) and are shown in Table 18.1.

Pipeline dispatching involves the use of delivery "pipelines," where pipelines are defined as delivery lanes with dense flow volumes. In this case, drivers and loads are partitioned into two sets: those that utilize pipelines, and the remaining set of random OTR drivers who are dispatched under traditional methods. Loads with sufficient delivery slack time that can travel down these pipeline lanes without encountering excess circuity (out-of-route miles) are dispatched as pipeline loads. In this case, the load would be picked up by a local or regional driver for a "dray" move to the start of the pipeline. From there, the load would be taken to the other end of the pipeline by a "linehaul" driver for ultimate delivery by another local or regional "drayman." Operationally, the method is similar to intermodal transportation with rail, only the rail segment is replaced with a pipeline driver. The advantage of this type of dispatch is that some subset of drivers can be partitioned as either draymen or linehaul drivers. Linehaul drivers have very frequent domicile returns. The local or regional drivers performing dray moves likewise get home more frequently, particularly if they are domiciled near the endpoint of the pipeline that they support. These local or regional drivers may or may not be dispatched independently from the remaining OTR fleet. Taylor et al. (2007) indicate that delivery pipelines can also be operated with minimal effects on the remaining fleet. The downside is that each pipeline load requires three dispatches per load.

TABLE 18.1 Best Truckload Regional Locations in the Continental United States

Regional Location	Centroid Latitude	Centroid Longitude
Cincinnati, Ohio	39.10	−84.30
New York, New York	40.80	−74.00
Atlanta, Georgia	33.80	−84.20
Memphis, Tennessee	35.10	−90.00
Chicago, Illinois	41.80	−87.70
Pittsburgh, Pennsylvania	40.30	−80.00
Charlotte, North Carolina	35.20	−80.90
Dallas, Texas	31.80	−97.00
Kansas City, Missouri	39.10	−94.40
Washington, District of Columbia	38.90	−77.10
Louisville, Kentucky	38.10	−85.90
Columbus, Ohio	39.90	−83.00
Indianapolis, Indiana	39.80	−86.20
Charleston, WV	38.40	−81.50

Since pipeline delivery is so similar to intermodal delivery with truck and rail, it should be noted that intermodal dispatching also offers a simple way to handle imbalance with lower costs, especially when rail is used for loads inbound to backhaul markets. In this case, it is less expensive to reposition a container via rail than a trailer or container via truck. This strategy can only be employed in situations in which product packaging, lane length of haul, and temporal delivery requirements permit it.

Popcorn dispatching methods permit short-term temporal imbalance as mentioned in Hall (1999), but do not permit long-term imbalance. Popcorn dispatching represents a direct compromise between traditional "random" dispatching and the regular route methods discussed in Section 18.4. The idea is that if drivers can be partitioned such that some subset operates solely within a very limited network of permissible high-volume destinations, those drivers will return more frequently to their domicile, even if permitted to "bounce" randomly (like popcorn in a pan) between destinations without explicitly being forced to their domicile. This is especially true if domiciles are located in network locations with dense pass-through or return freight. Several variations of the popcorn dispatching method exist, ranging from fully random to using heuristics that force the driver home after a certain number of miles or dispatches. In the strictest form, drivers can venture only one move from their domicile before returning.

A final form of alternative dispatch recently examined by the author (see Humphrey et al., 2007) takes advantage of the weekly temporal imbalance previously discussed in coordination with Figure 18.1. The idea is that weekly temporal imbalance can be reduced if excess Friday freight can be serviced on Saturday or Sunday. In this case, drivers arriving inbound to a location on Friday may be asked to pick up two loads. The first would be drayed to a local "drop yard" and left for pick-up on Saturday or Sunday by another driver. The second would be a "normal" dispatch for the Friday inbound driver. An inbound driver arriving on Saturday or Sunday would then have an outbound load waiting at the drop yard. Humphrey et al. (2007) indicate that while this strategy is relatively neutral for drivers, it can be profitable for carriers.

18.3.3 Using RM to Combat Imbalance

The state-of-the-art in RM strategies for truckload trucking lags well behind that of RM strategies in other industry segments, but this chapter would be incomplete without at least a cursory look at this important and relevant topic. RM has been used to great advantage to partially defeat imbalance problems in other industries such as airlines, LTL carriers, hotels, etc. Similar strategies are lacking in truckload trucking because of differences in how capacity is sold and how customer relationships have evolved as a result. Customers tend to wait until the last minute and then request freight movements that are seemingly random. This is a far more difficult set of circumstances for RM than encountered in industries with fixed schedules and limited, known destinations. Furthermore, when a truckload movement is sold, the entire transportation entity is sold all at once. If one thinks of truckload capacity in an aggregate sense (i.e., total available trucks within a region), however, it is possible to apply some RM techniques.

RM strategies seek to solve the imbalance problem in truckload trucking by eliminating or ameliorating the effects of imbalanced freight inbound to a backhaul market. On the one hand, RM models can seek to maximize profit subject to a perfectly balanced freight network. In this case, carriers simply refuse freight that is not profitable from a network balance viewpoint. On the other hand, RM models might accept a temporary imbalance in return for a very favorable freight rate for certain loads.

At the heart of RM strategies is the ability to fix pricing. Because profit margins are so low in the industry, only a narrow band of price changes is possible. Even so, these subtle changes can go far in terms of shaping customer behavior and adding discipline to carrier load acceptance policies. The carrier must learn that there is some freight that is simply not wanted at the market rate. As Barker et al. (1981) point out, management science techniques are showing us how easy it is to solicit potentially unprofitable freight. The freight might look great in terms of being profitable as a single move, but each move repositions aggregate system capacity in ways that may be very unprofitable globally. As more carriers begin to honestly evaluate their freight needs from a holistic view of their entire freight network, customers must learn that they must increasingly be prepared to pay a rate that more realistically covers carrier costs.

Procedurally, a RM model to support the truckload trucking industry may involve something as simple as a linear programming model to maximize revenue over a tactical or strategic planning horizon while being constrained by network balance and minimum/maximum lane volumes at various assumed prices. The output would then include recommendations regarding how much freight is desired on each lane. Shadow pricing information would tell us how much of a price increase would be required to make poor lanes (from a network balance and profitability viewpoint) desirable or how much of a price reduction could be provided as an incentive to customers to obtain more freight on good lanes. At an operational level, the model might permit short-term network imbalance by using penalty functions in the objective function to drive the capacity network back to a "stable" position.

18.3.4 Exploiting Imbalance-Based Pricing Structures

Jordan and Burns (1984) are perhaps the first to make a convincing argument that backhauling should become an important factor in determining terminal location and in selecting suppliers. Taylor et al. (2006b) quantify the effects of taking their advice by providing methods to locate distribution centers within backhaul markets instead of the traditional approach of building them in locations that minimize delivery time and distance. The idea is that backhaul markets offer low outbound freight pricing. The results show that clever selection of locations on problems of national or continental scale can lead to tremendous annual savings with little negative effect to delivery time and distance. The authors acknowledge that mass movement of distribution centers to backhaul locations would alter pricing, but also argue that such movement would ultimately lead to improved network balance.

18.4 Dedicated Fleets and "Regularized" Tour Development

The previous section focused on exploiting freight density and included alternative dispatching methods that build upon existing random dispatching methods. In this section, dispatching alternatives that make use of "regular route" dispatching are examined.

18.4.1 Benefits of "Regularization"

Route "regularization" can be defined as the establishment of high density driving tours, starting and ending at the same point (preferably the driver domicile), that are traveled repeatedly and perhaps exclusively by a single subset of drivers. Many benefits exist associated with route regularization. Perhaps, most importantly is that regular route drivers return to their domiciles more frequently and this leads to greater driver retention. Evidence that domicile returns are important to driver retention is readily available. In exit interviews with drivers who are quitting their jobs, the reasons most frequently cited for their decision include pay, the quality of life while on tour, and domicile returns. Drivers who return home frequently tend to stay with carriers longer and will do so even with less pay.

In addition to more frequent domicile returns, regular driving routes also tend to improve operational safety. Because regular driving routes are repeated frequently by the same drivers, they tend to become familiar with their surroundings. Thus, they can more competently plan rest stops, better position themselves in appropriate traffic lanes at intersections, and are more familiar with roadway obstacles such as low clearances or weight restricted bridges. They are also likely to know when and where to expect traffic delays and are more knowledgeable about the effects of construction projects. Consequently, the driver is safer and more comfortable on his or her job.

The establishment of regular and disciplined tours can also lead to improved customer service, particularly when the customer becomes a partner in the endeavor by striving to have loads available on-time and according to schedule to support regularized routes for drivers that become well-known representatives of the carrier.

The type of density-based regularization described in this section is typical of the strategies employed by major carriers in dedicated contract services. In this case, a customer makes a long-term contract with a carrier for a fleet of trucks and drivers dedicated solely to their freight (with perhaps some additional freight from other customers to reduce empty backhaul costs). The use of dedicated contract services can benefit both the shipper and the carrier. Shippers benefit by having a fleet of trucks and drivers that operate as a company-owned private fleet, but with the professional driving fleet, experienced management, backhaul density, and information systems that a carrier can provide. Carriers benefit from the increased ability to preplan activities, plan routes, and perhaps reduce empty miles.

18.4.2   Selecting Freight for Regular Route Development

Obviously, not all freight will fit nicely into regular driving patterns. No matter how carefully we plan, it is impossible to make all random freight fit into regular tours. As with the random dispatch alternatives discussed previously, freight must be selected for inclusion in regular route development that does not create dispatching difficulties for the remaining OTR freight.

It is best to begin the search for candidate freight for regularization in dense freight regions. The tools introduced previously assist in this search. It is important that regular tours have freight endpoints in dense freight regions or "hubs." In the case of dedicated contract services, some level of freight availability at hubs is guaranteed by the customer. Other possible hubs include large cities or terminal cities that have evolved within a particular carrier as a location of marketing emphasis or regional domination. The *Hub_Finder* system is particularly helpful in finding dense freight regions.

The *Lane_Finder* tool and other tools from the previous section are also useful. Recall that the output of the *Lane_Finder* system is dense freight lanes in which all loads included have common endpoints on both ends (hubs). When strung together, these lanes can become the basic building blocks for regular driving tours. Even so, it is not possible to utilize all of the freight selected for lanes by *Lane_Finder* in the development of regular tours. If we desire to keep drivers loaded on all legs of a tour, the tour is limited by the smallest volume arc on the proposed tour. In seeking regular routes, it is important to find the right balance between regularization and efficiency. If we do not regularize enough freight, we miss an important opportunity to improve the quality of driver life and thereby retain drivers. If we regularize too much, we may introduce marginally efficient routes while potentially removing "good" freight in high density regions that would be suitable for OTR domicile returns or other dispatching efficiencies.

One other issue related to freight selection for route regularization is that of determining where we want to aggressively seek freight growth. Strategic decisions in this regard were discussed previously in the RM discussion, but some issues specific to dedicated fleet management are also relevant. It is now common for large shippers to share anticipated freight volume information with carriers. Carriers can then make bids for some or all of this freight. This freight should complement existing freight in terms of imbalance, but also in terms of temporal variance. Relative to imbalance, carriers can make very low bids for freight that would enable them to reposition loaded trucks into better markets on lanes in which they currently operate in an empty status. This adds profitability to the carrier and also provides a reduced rate to the shipper. Relative to variance, especially in a bid for dedicated contract services, it is desirable to maintain a steady workload to support the fleet of drivers. Therefore, if a number of lanes are available for bid, the carrier should attempt to select those lanes that tend to produce a regular driving schedule for a statically sized fleet.

A technique has been developed to assist in selecting freight lanes to minimize the daily or periodic variance associated with the selection. It is called the variance optimizing technique (VOT), and is based on integer programming. For persons interested in mathematical formulations, the following is a formulation of VOT:

Minimize:

$$\sum_{\forall i} P_i + \sum_{\forall i} N_i \tag{18.1}$$

Subject to:

$$\sum_{\forall j} V_{ij} X_j - Pi - Ni \quad A \pm \epsilon \quad \forall_i \tag{18.2}$$

$$\sum_{\forall j} \left(\sum_{\forall i} V_{ij} \right) X_j = I(A) \tag{18.3}$$

$$A \geq A_{min} \tag{18.4}$$

$$P_i \geq 0, \text{ and } N_i \leq 0, \quad \forall i \tag{18.5}$$

$$X_j = 0, 1 \text{ and integer} \quad \forall j \tag{18.6}$$

Where:

A = The overall average freight per time period based on the lanes selected by the model.
A_{min} = The minimum acceptable value for A as specified by the user.
ϵ = The maximum acceptable deviation from A permitted during any given time period.
I = The number of time periods in the study.
N_i = The negative deviation of the freight selected for time period i in comparison to the average freight across all periods (if any).
P_i = The positive deviation of the freight selected for time period i in comparison to the average freight across all periods (if any).
V_{ij} = The volume of loads typically available during time period i (likely day of week) on lane j.
X_j = A binary decision variable that is 1 if lane j is selected by the carrier, else 0.

The objective function (Equation 18.1) minimizes the total deviation from the mean freight volume across all time periods involved in the study, perhaps days of the week. The Equations presented in (18.2) ensure that the freight lanes selected do not result in freight volumes that deviate from the overall mean by more than a user-specified amount ε during any time period. Equation 18.3 calculates the overall average freight volume per time period based on the lanes selected. Equation 18.4 ensures that the freight selected exceeds some overall minimum in each time period. Finally, Equations 18.5 and 18.6 specify that P_i values are positive, N_i values are negative, and that X_j values (the primary decision variables) are binary integers.

18.4.3   Hub Efficiency Analysis Tool

Once freight is selected for regularization, or perhaps before, it is necessary to undertake the important job of determining how drivers will be dispatched and how their tours will be operated. The author's version of a Hub Efficiency Analysis Tool (HEAT) software system is one means of making this determination. HEAT attempts to build regular tours, originating and ending at one point, preferably the driver domicile.

Inputs to the HEAT system include the location of hubs (freight endpoints, perhaps from *Hub_Finder* output), and the availability of dense freight lanes (from *Lane_Finder* output). The desired outcome is the development of one or more regular tours of five types, dubbed CL2, CL3, J21, J31, and J22, where CL indicates "closed loop" and J indicates "jump." In a CL2 tour, the driver travels to a distant location and is immediately dispatched to his or her city of origin to create a fully loaded "out and back" tour. CL3 tours are triangular with an additional dispatch between the original and final movements. Jump tours utilize some loaded legs and some unloaded legs. In a J21 tour, for example, the driver is dispatched loaded to some distant location where he or she then makes a brief empty move to another nearby region. From there, the driver is dispatched home in a loaded state. Even in closed loop tours, the driver must undertake empty repositioning moves, but these local moves are relatively insignificant in length in comparison with the loaded moves. In jump tours, the empty portion of the tour is more significant. J31 tours utilized three loaded legs with one regional empty repositioning move and J22 tours utilize two loaded moves with two regional empty repositioning moves.

Software options in the HEAT system permit the user to prioritize the types of tours desired through specification of the order in which tours are built. For example, a user desiring more CL2 or CL3 tours would specify that these tours should be developed first. In this way, these tour types have access to all available freight during the tour building process. Remaining tour types would have access to only that freight that is available after all possible CL2 and CL3 tours are found. Within a tour type, the software utilizes a "greedy" heuristic to first find those tours that have the highest value of volume times miles. Alternatively, the user can specify that no tour type should have priority over another. In this case, the software uses the same greedy heuristic to find high volume and/or long mileage tours regardless of their type.

To further illustrate the HEAT system, consider the example presented graphically in Figure 18.7. This example is typical of a realistic trucking setting in which few of the market areas (nodes) are balanced. Node 2, for example, has many more loads in than out while Node 3 has many more loads out than in. The lanes also exhibit a great deal of imbalance. This indicates that much of the freight cannot be used for the highly desirable CL2 tours. Even so, lanes with one-way freight can often be used for CL3 or "jump" tours. Note that the example has four candidate lanes for regional jump moves, 1–7, 2–8, 3–6, and 4–5.

In the example problem, the HEAT program is run under the assumption that the desired tour order is CL2, CL3, J21, J31, and then J22. Figure 18.8 provides the HEAT output for the example problem. Default output includes summary information regarding the total miles that are included in regular tours as a percentage of total miles available in the data set plus summary information for each tour type. In this case, the bulk of miles regularized are in CL2 tours. This is partially because of the specified search order, but partially because CL2 tours are simple to build. The tour summary information includes the tour routing, the volume information (the number of times this tour is to be driven during the planning period) and the tour mileage (per trip).

It is important to note that HEAT is a tactical planning tool, not an operational dispatching tool. In practice, the recommended tours often perform very well, but still require significant management to achieve suitable results. Tours that seem to be strong candidates for regularization are sometimes not well suited to operational effectiveness. Consider, for example, the operational effectiveness of CL3 tours

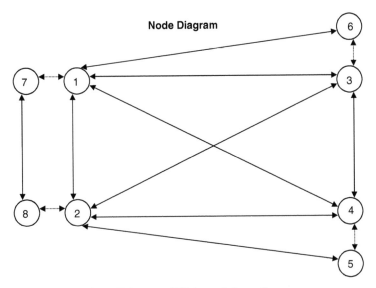

FIGURE 18.7 HEAT example.

Lane Volume and Distance Information

Start Node	End Node	Lane Volume	Lane Distance		Start Node	End Node	Lane Volume	Lane Distance
1	2	4210	239		3	6	0	50
1	3	2493	779		4	1	1646	843
1	4	1052	843		4	2	1001	788
1	6	1000	779		4	3	342	184
1	7	0	90		4	5	0	60
2	1	395	239		5	2	0	788
2	3	653	796		5	4	0	60
2	4	70	788		6	1	0	779
2	5	900	788		6	3	0	50
2	8	0	80		7	1	0	90
3	1	2029	779		7	8	0	239
3	2	290	796		8	2	0	80
3	4	4828	184		8	7	500	239

when compared to J21 or J22 tours. Obviously, we would prefer to move loaded on all tour legs as in a CL3 tour. In practice, however, J21 and even J22 tours often perform better than CL3 tours because it is easier to de-couple arriving and departing loads from a temporal perspective.

18.4.4 Optimal Seeking Tour Development Tools

It is possible to approach the tour design problem using optimal seeking approaches such as integer programming. This approach provides optimal tours but is computationally intensive and difficult to formulate and solve. For those readers interested, consider the following formulation from Taylor and Whicker (2002c):

Maximize:

$$\sum_j \sum_k \sum_\ell \sum_{m \neq \ell} X_{jk\ell m}(Z_{\ell m}) - \sum_j \sum_k \sum_\ell \sum_{m \neq \ell} Y_{jk\ell m}(Z_{\ell m}) \tag{18.7}$$

Summary Information
- *Total Miles Available:*   11,051,200.00
- *Total Miles Regularized:*   8,678,556.00
- *Percent of Miles Regularized:*   78.53

CL2 Summary
- *Total Miles Used in CL2 Tours*   5,821,520.00
- *Percent of Miles Used in CL2 Tours*   52.68
- *Tour Summaries:*

1-3-1	2,029 Loads	1,558 Miles
4-1-4	1,052 Loads	1,686 Miles
2-3-2	290 Loads	1,592 Miles
1-2-1	395 Loads	478 Miles
3-4-3	342 Loads	368 Miles
4-2-4	70 Loads	1,576 Miles

CL3 Summary
- *Total Miles Used in CL3 Tours*   1,479,768.00
- *Percent of Miles Used in CL3 Tours*   13.39
- *Tour Summaries:*

1-3-4-1	464 Loads	1,806 Miles
3-4-2-3	363 Loads	1,768 Miles

J21 Summary
- *Total Miles Used in J21 Tours*   895,168.00
- *Percent of Miles Used in J21 Tours*   8.10
- *Tour Summaries:*

4-2-5-4	568 Loads	1,576 Miles

J31 Summary
- *Total Miles Used in J31 Tours*   243,100.00
- *Percent of Miles Used in J31 Tours*   2.20
- *Tour Summaries:*

4-1-2-5-4	130 Loads	1,870 Miles

J22 Summary
- *Total Miles Used in J22 Tours*   239,000.00
- *Percent of Miles Used in J22 Tours*   2.16
- *Tour Summaries:*

1-2-8-7-1	500 Loads	478 Miles

FIGURE 18.8 HEAT system output for example problem.

Subject to:

$$\sum_{j}\sum_{k} X_{jk\ell m} \leq A_{\ell m} \qquad \forall \ell, m \neq \ell \tag{18.8}$$

$$\sum_{\ell \neq j}\sum_{m \neq \ell} X_{jk\ell m} + \sum_{\ell \neq j}\sum_{m \neq \ell} Y_{jk\ell m} = 0 \qquad \forall j, k = 1 \tag{18.9}$$

$$\sum_{\ell \neq m} X_{jk\ell m} + \sum_{\ell \neq m} Y_{jk\ell m} - \sum_{\ell \neq m} X_{j(k+1)m\ell} - \sum_{\ell \neq m} Y_{j(k+1)m\ell} = 0 \quad \forall j, k < K, m \neq j \tag{18.10}$$

$$\sum_{\ell \neq m}\sum_{m \neq j} X_{jk\ell m} + \sum_{\ell \neq m}\sum_{m \neq j} Y_{jk\ell m} = 0 \qquad \forall \, j, k = K \tag{18.11}$$

$$X_{jk}\ell_m = \text{Integer} \quad \forall\, j, k, \ell, m \tag{18.12}$$

$$Y_{jk}\ell_m = \text{Integer} \quad \forall\, j, k, \ell, m \tag{18.13}$$

Where:

$X_{jk}\ell_m$ = The number of times during the planning horizon that some driver domiciled at city j makes their kth move from city ℓ to city m in a loaded status.

$Y_{jk}\ell_m$ = The number of times during the planning horizon that some driver domiciled at city j makes their kth move from city ℓ to city m in an unloaded status.

$Z\ell_m$ = Miles (or revenue minus costs) from city ℓ to city m.

$A\ell_m$ = Maximum allowable moves from city ℓ to city m.

The objective function in Equation 18.7 maximizes the loaded miles minus empty miles, which is directly proportional to profit (carriers are normally not paid for empty repositioning moves). Actually $Z\ell_m$ values can take on profit (revenue minus cost) values for a slightly different objective function that would penalize empty moves more heavily. In this chapter, we assume that $Z\ell_m$ holds mileage values for city-to-city pairs. The first constraint (Equation 18.8) is an expression that restricts network flow to known or assumed lane capacity based on the total number of shipments available during the time period under consideration. In other words, the carrier cannot move freight that does not exist but can use empty repositioning moves once freight on a particular lane is exhausted. Equation 18.9 ensures that all drivers begin their tours at their domicile by requiring that the sum of all empty or loaded moves for the first dispatch is zero when the dispatch is not from the driver domicile. Equation 18.10 ensures that all transshipment nodes (excluding the domicile) in each driver tour maintain a balance of capacity. Each driver that enters a node that is not his or her domicile must leave that node on the next dispatch. Drivers reaching their domicile prior to the kth dispatch are not required to leave on the next dispatch. The next constraint (Equation 18.11) ensures that each driver must return to his or her domicile prior to the end of the planning period. Actually, the constraint requires that the sum of moves during the last dispatch is zero at every node except the driver domicile. To ensure that drivers return to their domicile according to carrier goals, tour length is controlled by specification of the number of allowed dispatches per tour via specification of the upper bound, K, on the driver subscript, k, representing the dispatch number. Finally, Equations 18.12 and 18.13 specify that $X_{jk}\ell_m$ and $Y_{jk}\ell_m$ are positive integers.

18.5 Summary

This chapter discusses the idea that freight imbalance is a fact of life that is inherent in delivery networks. It cannot be defeated but its negative effects can be somewhat mitigated through creative thinking and effective management strategies. This chapter also stresses the importance of finding freight density in all of its various forms; dense freight activity locations, dense freight lanes, dense intermodal activity areas, and dense pass-through activity. Tools are discussed to assist with finding each of these types of freight density.

Using the dense freight activity information, this chapter discusses various ways that density can be exploited to support backhaul solicitation, RM, and pricing. Also discussed is the opportunity to exploit density in support of two forms of dispatch; those that make use of driver partitions that continue to use traditional random dispatching tools and those that make use of regular driving tours suitable for dedicated contract services or regularized driving jobs. In some cases, such as zone and pipeline dispatching, the dispatching methods themselves assist in establishing artificial freight density at transshipment points.

Hopefully, the reader will find that in aggregate this chapter addresses issues that are not well supported in the literature. The net effect of using all of the tools presented in this chapter is that carriers can improve marketing capabilities, raise the bar in terms of strategic planning ability, and ultimately achieve greater operational profitability. Even though the chapter builds upon the author's expertise in truckload trucking, the application of the techniques employed herein should be equally applicable to other types of delivery networks.

18.6 Case Study

18.6.1 Finding Driver Domiciles

This case study utilizes the *Domicile_Finder* software system presented within the chapter to locate the best places in North America for driver domiciles. The software provides the opportunity to make this determination based on several alternative solution approaches in terms of freight "ownership," node capacity, node imbalance, and load type weighting. In this case study, the goal is to provide a "wide open" answer that is based primarily on the volume of pass-thru freight in generic regions. The input parameters that were used in the case study include the following:

- Allowable out-of-route miles (circuitous miles) are fixed at 50 miles.
- No "ownership" of freight is permitted. Therefore, each load can contribute to the pass-thru volume of multiple nodes.
- Node capacity is assumed to be infinite. There is no limit to the number of loads that can be assigned to pass-thru status.
- Node imbalance is not considered. All pass-thru freight is assigned to each suitable node regardless of the freight imbalance at the node.
- No initial terminal locations are specified. This means that the default 1° by 1° lat/long grid will be used as domicile seed locations.

For more information about these parameters, please refer to the more detailed description of them within the chapter.

The data to support the case study has been provided by J.B. Hunt Transport, Inc. (JBHT). The data consists of more than 19,000 individual records, each representing the hub-to-hub volume (in full truckloads) for a one-year period. The approximately 140 hubs used by JBHT have been established using the *Hub_Finder* software system as described in the chapter. Each data record contains five fields, the origin latitude and longitude, the destination latitude and longitude, and the volume from the origin to the destination. The data set excludes intermodal freight that would travel primarily on rail at most pass-thru locations.

Although it can be argued that the solution obtained in this case study is highly data dependent, the results are expected to be fairly characteristic of what might be found in general because of the size of the JBHT data set. As will be indicated below, the highest density pass-thru node has more than 40,000 loads passing within 50 miles in a period of one year.

Figure 18.9 graphically depicts the results of running *Domicile_Finder* to find generic pass-thru freight density for the JBHT data set. As the figure indicates, the most frequently passed pass-thru location in North America is at 39°N latitude and −84°W longitude, near Dayton and Cincinnati, Ohio. This region is surrounded by a relatively large geographical area in which 30,000 to 40,000 loads per year pass by each node in the general 1° by 1° lat/long grid. This larger region includes most of Ohio, Indiana, and Kentucky, but also extends into Illinois, Tennessee, West Virginia, and western Virginia. This area includes the cities of Indianapolis, IN, Louisville, KY, Nashville, TN, and Columbus, OH.

Two distinct regions have between 20,000 and 30,000 pass-thru truckloads per year. The first is a large region surrounding the previously discussed dense freight regions. This extended region includes much of the mid-west United States, the northern parts of the southeast region, and the western sections of many of the Atlantic states. The second region is in southern California, centered around Los Angeles.

Finally, Figure 18.9 shows two locations that offer 10,000 to 20,000 pass-thru loads per year. The first is a vast area encompassing most of the eastern United States and even a small portion of southern Ontario

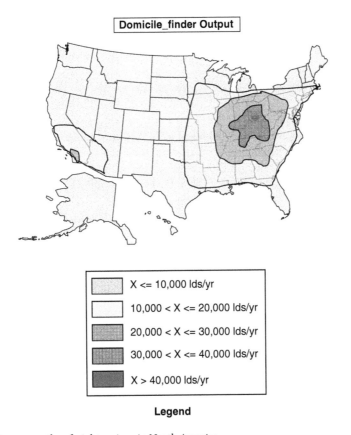

Legend

FIGURE 18.9   Dense pass-thru freight regions in North America.

in Canada. The second region includes most of southern California, approximately one-third of Arizona including Phoenix and Tucson and a small part of southern Nevada including the city of Las Vegas.

The results indicate those regions most likely to support driver domiciles for efficient dispatching. By hiring or locating drivers in major cities or highway intersections in dense pass-thru regions, dispatchers are assured of having greater opportunity to return drivers efficiently to their domiciles at the end of their tours.

References

Abeyratne, R.I.R. (2000), Strategic Alliances of Airlines and Their Consequences, *Journal of Air Transportation World Wide*, Vol. 5, No. 2, pp. 55–71.

Barker, H.H., Sharon, E.M., and Sen, D.K. (1981), From Freight Flow and Cost Patterns to Greater Profitability and Better Service for a Motor Carrier, *Interfaces*, Vol. 11, pp. 4–20.

Cheung, R.K. and Chen, C.Y. (1998), Two-Stage Stochastic Network Model and Solution Methods for the Dynamic Empty Container Allocation Problem, *Transportation Science*, Vol. 32, pp. 132–162.

Crainic, T.G., Gendreau, M., and Dejax, P. (1993), Dynamic and Stochastic Models for the Allocation of Empty Containers, *Operations Research*, Vol. 41, pp. 102–106.

Fite, J.T., Taylor, G.D., Usher, J.S., English, J.R., and Roberts, J.N. (2002), Forecasting Freight Demand Using Economic Indices, *International Journal of Physical Distribution and Logistics Management*, Vol. 32, No. 4, pp. 299–308.

Hall, R.W. (1999), Stochastic Freight Flow Patterns: Implications for Fleet Optimization, *Transportation Research: Part A*, Vol. 33, pp. 449–465.

Humphrey, A.S., Taylor, G.D., Usher, J.S., and Whicker, G.L. (2007), Evaluating the Efficiency of Truckload Operations with Weekend Freight Leveling, in review for publication in the *International Journal of Physical Distribution and Logistics Management*.

Jordan, W.C. and Burns, L.D. (1984), Truck Backhauling on Two Terminal Networks, *Transportation Research: Part B*, Vol. 18, pp. 487–503.

Lin, I.I., H.S. Mahmassani, P. Jaillet, and C.M. Walton (2002), Electronic Marketplaces for Transportation Services: Shipper Considerations, *Transportation Research Record*, Vol. 1790, pp. 1–9.

Min, H. and Emam, A. (2003), Developing the Profiles of Truck Drivers for Their Successful Recruitment and Retention: A Data Mining Approach, *International Journal of Physical Distribution and Logistics Management*, Vol. 33, No. 2, pp. 149–162.

Powell, W.B. (1996), A Stochastic Formulation of the Dynamic Assignment Problem With An Application to Truckload Motor Carriers, *Transportation Science*, Vol. 30, No. 3, pp. 195–219.

Rietveld, P. and Roson, R. (2002), Direction Dependent Prices in Public Transport: A Good Idea? The Back Haul Pricing Problem for a Monopolistic Public Transport Firm, *Transportation*, Vol. 29, No. 4, pp. 397–417.

Ronen, D. (1997), Alternate Mode Dispatching. *Journal of the Operational Research Society*, Vol. 48, pp. 973–977.

Sherali, H.D. and Suharko, A.B. (1998), Tactical Decision Support System for Empty Railcar Management, *Transportation Science*, Vol. 32, pp. 306–329.

Song, J. and A.C. Regan (2001), Transition or Transformation? Emerging Freight Transportation Intermediaries, *Transportation Research Record*, Vol. 1763, pp. 1–5.

Taha, T.T. and Taylor, G.D. (1994), An Integrated Modeling Framework for the Evaluation of Hub and Spoke Networks in Truckload Trucking, *The Logistics and Transportation Review*, Vol. 30, No. 2, pp. 141–166.

Taha, T.T., Taylor, G.D., and Taha, H.A. (1996), A Simulation-Based Software System for Evaluating Hub-and-Spoke Transportation Networks, *Simulation Practice and Theory*, Vol. 3, pp. 327–346.

Taylor, G.D. (2002a), Improved Trucking Operations Through Creative Exploitation of Freight Information, *Proceedings of the 7th International Logistics Conference*, Prague, Czech Republic, April 9–10, 2002, pp. 12–16.

Taylor, G.D. (2003), Managing Freight Imbalance in Truckload Trucking, *Proceedings of the 5th International Industrial Engineering Conference* (on CD-ROM), File:/papers/441.pdf, Quebec City, Canada, Oct. 26–29, 10 p.

Taylor, G.D. and Whicker, G.L. (2002c), Optimization and Heuristic Methods Supporting Distributed Manufacturing, *International Journal of Production Planning and Control*, Vol. 13, No. 6, pp. 317–328.

Taylor G.D., DuCote, W.G., and Whicker, G.L. (2006a), Regional Fleet Design in Truckload Trucking, *Transportation Research: Part E.*, Vol. 42, No. 3, pp. 167–190.

Taylor, G.D., Onggowijaya, S.N., and Whicker, G.L. (2006b), The Tradeoff Between Delivery Cost and Time/Distance Minimization in Distribution Network Design, in *Progress in Material Handling Research: 2006*, R.D. Meller, M.K. Ogle, B.A. Peters, G.D. Taylor, and J.S. Usher, Eds. Material Handling Industry, Charlotte, NC, pp. 475–492.

Taylor, G.D., Whicker, G.L., and Usher, J.S. (2001), Multi-Zone Dispatching in Truckload Trucking, *Transportation Research: Part E*, Vol. 37, pp. 375–390.

Taylor, G.D., Whicker, G.L., and DuCote, W.G. (2007), Design and Analysis of Delivery 'Pipelines' in Truckload Trucking, Accepted for publication and to appear in *Transportation Research: Part E*.

Taylor, G.D., Broadstreet, F., Meinert, T.S., and Usher, J.S. (2002b), An Analysis of Intermodal Ramp Selection Methods, *Transportation Research: Part E*, Vol. 38, pp. 117–134.

Taylor, G.D., Meinert, T.S., Killian, R.C., and Whicker, G.L. (1999), Development and Analysis of Efficient Delivery Lanes and Zones in Truckload Trucking, *Transportation Research: Part E*, Vol. 35, pp. 191–205.

Toh, R.S. and Higgins, R.G. (1985), The Impact of Hub and Spoke Network Centralization and Route Monopoly on Domestic Airline Profitability, *Transportation Journal*, Vol. 14, No. 4, pp. 16–27.

United States Department of Transportation (2006), Available at: http://www.fmcsa.dot.gov/facts-research/facts-figures/analysis-statistics/cmvfacts.htm, Accessed November 21, 2006.

19

Revenue Management and Capacity Planning

Douglas R. Bish
Ebru K. Bish
Virginia Polytechnic Institute and State University

Bacel Maddah
American University of Beirut

19.1 Introduction

Broadly defined, revenue management (RM)* is the process of maximizing revenue from a fixed amount of perishable inventory using "market segmentation" and "demand management" techniques. While RM is not new (in fact it is as old as commerce, e.g., haggling in a market can be considered a form of RM), the theory and practice of RM have seen significant scientific and practical advances in the last few decades, starting with the Airline Deregulation Act in 1978, which opened the door for RM in the airline industry. It is not surprising that airlines adopted RM, as most of the market characteristics conducive to RM are present. RM is considered an essential function of any airline due to the highly uncertain and competitive marketplace.

Consequently, airlines have some of the most sophisticated RM implementations around. As we discuss RM in more detail, we will illustrate concepts using examples (mainly) from the airline industry. We do this because of the importance of the airlines to the development of RM, and because air travel is common enough that most people have experienced airline RM (perhaps unknowingly). In addition, we have some industry background in airline RM. Despite this focus on the airline industry, we note that RM has expanded to many different industries, starting with industries that share similar characteristics with the airline industry, such as hotels and car rental agencies (Boyd and Bilegan 2003),

---

* "Yield management" is another common terminology for RM. For details on terminology and the scope of RM, we refer the interested reader to Talluri and van Ryzin (2005) and Weatherford and Bodily (1992).

TABLE 19.1 Who Uses RM?

Airlines	All
Hotels	Hyatt, Mariott, Hilton, Sheraton, Forte, Disney
Vacation	Club Med, Princess Cruises, Norwegian
Car rental	National, Hertz, Avis, Europcar
Washington Opera	
Freight	Sea-Land, Yellow Freight, Cons, Freightways
Television	CBS, ABC, NBC, TVNZ, Ads Aus7
UPS, SNCF	
Retail	Retek, Khimetrics
Real estate	Archtone
Natural gas	
Texas children's hospital	

Source: Bell P. Revenue Management. Presentation at Vision 2020 Conference, Ahmedabad, January 2005. Available online at http://www.ivey.uwo.ca/faculty/Peter_Bell/

and then to many other industries, including retailing and manufacturing industries; see Table 19.1 for a sample of industries that have implemented RM-based approaches.

RM applications have been highly successful, with benefits in billions of dollars in some cases. For example, American Airlines reported an estimated $1.4 billion from applying RM techniques over three years in the early 1990s (Smith et al. 1992), and has later reported an estimated annual benefit of $1 billion from implementing RM (Cook 1998). Boyd 1998 estimates an increase in revenue in the order of 2–8% due to implementing RM in an airline. In the rental car business, National was able to escape liquidation and generate $56 million incremental revenue due to RM (Geraghty and Johnson 1997). Moreover, Hertz indicated that the implementation of an RM system yielded an increase in revenue in the order of 1–5% (Carrol and Grimes 1995). Other successful examples of RM implementation are abundant.

In the following, we first present the terminology that will be used throughout this chapter, and then discuss market segmentation and the other market characteristics often associated with RM.

19.1.1   Terminology

Since we will illustrate RM concepts using airline examples, we first present some basic (airline) RM terminology that will be used throughout the chapter.

Fare-class (class): Each market segment is represented by a fare-class. We will index fare-classes such that a lower index refers to a higher valued customer segment, that is, fare-class 1 has the highest ticket price or fare of any class.

Itinerary: The set of specific flights a traveler uses to fly between his/her origin and destination.

Product: A combination of an itinerary and a fare-class.

Booking limit: The maximum number of tickets (seats) that can be sold to each fare-class for a particular flight.

Overbooking limit: The total number of tickets (seats) that can be sold for a particular flight; this limit is typically larger than the aircraft's capacity in anticipation of travelers canceling their reservations or not showing up for their flights.

19.1.2   Market Characteristics Conducive to RM

Market segmentation is an essential part of RM and is hence discussed in some detail. Market segmentation depends on a heterogeneous customer base with diverse consumer preferences. The goal of market segmentation is to take what might seem an identical product or service, and somehow differentiate it from the consumers' perspective. A good example is a coach seat on any flight. Despite the fact that the service the customer receives (i.e., flying with a coach seat) is nearly identical, there can be a great disparity on the price paid for a seat on a flight. This is because airlines try to segment the market into "business" and "leisure" customers, based on certain likely characteristics of each segment. Leisure customers usually book earlier, are more flexible concerning travel times, are more likely to stay at their destination over a weekend, are more certain of the trip and thus do not require refundable tickets, and are more price sensitive than business customers. Airlines therefore design their fare structures and booking rules (e.g., advanced purchase requirements, refundability, Saturday night stay) to segment the customer base, and thus charge business customers a premium (as they are usually less price sensitive), that is, this is why a fully refundable ticket, bought six days before departure, for a trip without a Saturday night stay, is more expensive than a nonrefundable ticket, bought a month before departure, with a Saturday night stay.

It is interesting to consider these segmentation rules. A fully refundable ticket is obviously a more expensive product for the airline to offer than a nonrefundable ticket, as the airline faces the risk of an empty (and unpaid) seat if the customer decides to cancel the trip in the last minute. Likewise, it is only sensible for the airline to save a seat for a business passenger booking six days before departure if they pay more than the leisure passengers, as the airline risks not selling the seat. (This decision of how many products to reserve for the higher valued classes, often termed "capacity control," is where the fixed amount of perishable inventory comes into play.) In contrast, the Saturday night stay requirement is solely for segmentation purposes; it does not impact the airline in any other way. As can be seen, these "fences" (restrictions) are constructed so as to prevent customers of a high-valued class from "leaking" from their segment and buying at lower prices (although this remains possible).

Here, we will discuss, in more detail, the market characteristics that tend to favor the use of RM: (*i*) *Perishable inventory*: The products perish after a certain date. For example, an airline seat has no value after the flight departs; it cannot be "stored" for use later. A night stay at a hotel must be used on the given night or, otherwise, the revenue opportunity from that room on that night will be gone. Other examples include seats for a sporting event, space on any means of transportation, electricity and other utilities, etc. [Weatherford and Bodily (1992)]. Obviously, the concept of perishable inventory applies to service industries. What is not so obvious is that it may also apply to the manufacturing industry. Products themselves (e.g., cars, computers) perish after a certain date (last year's computer might be nearly worthless now). Manufacturers producing customized products based on orders (i.e., on a "make-to-order" basis) do not typically carry finished-good inventories; hence, their production capacity is perishable. The concept of perishable inventory also applies to retailers selling, for example, fashion items, seasonal items, or perishable grocery items. (*ii*) *Fixed (limited) inventory*: Obviously, RM is relevant only if capacity is scarce with respect to demand. For example, an airline having airplanes large enough, to the extent that demand never exceeds capacity, need not worry about protecting seats for business travelers. (*iii*) *Low marginal costs*: When accommodating an additional customer costs very little compared to the fixed cost of establishing the product, it becomes very important to sell to the highest possible number of customers (while, of course, satisfying the fixed capacity limit). For example, selling one more flight seat on a flight or one more room in a hotel will cost very little compared to other overhead costs. (*iv*) *Demand uncertainty*: The limited ability to predict the future demand complicates demand management decisions (e.g., determining the appropriate "booking limit" for each fare-class). Most RM applications rely on probabilistic demand models that attempt to maximize the expected profit.

Individual RM implementations also depend on other market characteristics such as the consumers' buying behavior, seasonality in demand, substitution/complementarity between the different products sold in the market, sales channels available to the firm, marketing and sales policies, relation of supply with respect to demand, and competition. As one might imagine, RM systems have become highly sophisticated, driven by intense competition, and enabled by scientific advances in the related disciplines as well as advances in information technology, which makes it possible to store, retrieve, and analyze vast amounts of data, and to implement complex algorithmic approaches to demand management decisions.

19.1.3 Overview

In this chapter, we present a representative cross-section of RM models. Our objective is to give the reader a basic understanding of how RM effectively utilizes operations research (OR) techniques and methodology, while introducing the reader to the fundamentals of RM methodology. Specifically, we focus on three areas of RM, which we believe are the most related to OR: pricing, capacity control, and overbooking. Our presentation of pricing in Section 19.2 is cursory, and is mainly included to emphasize the benefits of price differentiation between customer segments. Our intention here is to illustrate how RM exploits a segmented market to maximize returns under fixed capacity by charging each customer "the right price," which matches the customer's willingness to pay. In Section 19.3, we discuss capacity control in some detail, as this is an area that has received the most attention in RM. In Section 19.4, we discuss the benefits, necessity, and practice of "overbooking." Finally, in Section 19.5, we share our thoughts on the current and future challenges for RM.

19.2 Pricing

Consider an airline selling seats on a single flight to n fare-classes. Let C denote the capacity of the aircraft (i.e., the total number of seats available) and p_i denote the ticket price for fare-class i, $i = 1, \ldots, n$, with $p_1 > p_2 > \cdots > p_n$. Assume that each fare-class is characterized by a *deterministic* demand function, $d_i(p_i)$, $i = 1, \cdots, n$. That is, if the price is set at p_i, then the demand for fare-class i is $d_i(p_i)$. (Observe that the assumption that the demand functions are deterministic does not generally hold in practice, and is mainly made to simplify the problem and gain some insights.) With the ability to segment the market, and the existence of a fixed capacity and low marginal costs, as discussed in Section 19.1, the revenue management pricing problem with market segmentation under deterministic demand functions can be expressed as follows (see, e.g., Bell 2004):

$$\max_{p_1, p_2, \ldots, p_n} \Pi^S = \sum_{i=1}^{n} p_i d_i(p_i)$$

$$\text{subject to } \sum_{i=1}^{n} d_i(p_i) \leq C. \tag{19.1}$$

The objective function in Model 19.1, denoted by Π^S, is the revenue generated from all fare-classes. The demand function, $d_i(p_i)$, is usually a decreasing function of p_i. Therefore, setting p_i too low will produce a high demand, but might not maximize the revenue. On the other hand, setting p_i too high would reduce demand, resulting in little or no revenue. The constraint in Model 19.1 reflects the fact that different fare-classes are competing for the limited capacity. This indicates that the airline is using prices to manage demand in order to match it with supply.

Suppose now that the airline is not willing or is unable to segment passengers into different fare-classes. In this case, the firm charges the same price, p, for all customers. The firm's pricing problem without market segmentation then reduces to the following problem:

$$\max_{p} \Pi^{NS} = \sum_{i=1}^{n} p d_i(p)$$

$$\text{subject to } \sum_{i=1}^{n} d_i(p) \leq C. \tag{19.2}$$

It is easy to show that $\Pi^S \geq \Pi^{NS}$, that is, customer segmentation increases the airline's revenue. Ignoring customer segmentation, as is done in Model 19.2, results in missed revenue opportunities. We illustrate this point further with an example.

Example 1

Suppose that Fly High Airlines (FHA) can segment the market for a particular flight into two distinct fare-classes (e.g., business versus leisure), with demand curves given by $d_1(p_1) = 100 - 2p_1$ and $d_2(p_2) = 200 - 10p_2$, as illustrated in Figure 19.1. Suppose also that the aircraft assigned to this flight has a capacity of $C = 150$ seats. Then, the pricing problem can be solved using Model 19.1, which has the following form:

$$\max_{p_1, p_2} \ p_1(100 - 2p_1) + p_2(200 - 10p_2)$$
$$\text{subject to } (100 - 2p_1) + (200 - 10p_2) \leq 150.$$

Solving for the optimal prices under customer segmentation, we obtain $p_1^* = \$25$ and $p_2^* = \$10$. The corresponding demands (i.e., number of seats sold to classes 1 and 2, respectively) are $d_1(p_1^*) = 50$ and $d_2(p_2^*) = 100$, with an optimal revenue (with segmentation) of $2,250. On the other hand, if the airline charges the same price for both fare-classes, then from Model 19.2, the optimal price will be the solution to:

$$\max_{p} \ p(100 - 2p) + p(200 - 10p)$$
$$\text{subject to } (100 - 2p) + (200 - 10p) \leq 150.$$

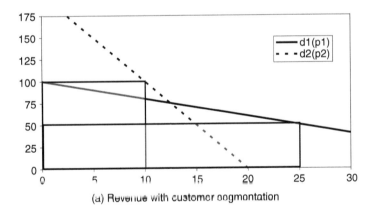

(a) Revenue with customer segmentation

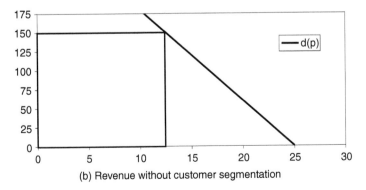

(b) Revenue without customer segmentation

FIGURE 19.1 Illustration of customer segmentation.

In this case the optimal price with no customer segmentation is $p_1^* = \$12.5$, with corresponding sales quantities $d_1(p^*) = 75$ and $d_2(p^*) = 75$. The optimal revenue without customer segmentation is $1875. Thus, customer segmentation increases revenue from $1875 to $2250, by 20%. Figure 19.1 illustrates this revenue increase graphically, where the areas of the rectangles represent revenue, and the function, $d(p) \equiv d_1(p)+d_2(p)$, in Figure 19.1b represents the total market demand without customer segmentation.

The assumptions in Models 19.1 and 19.2 seldom apply in real life. First, demand is generally uncertain, time-dependent, and depends on the price(s) of all similar products sold by the firm as well as on other factors (such as competition, weather, special events) in a complex way. Firms engaged in RM use demand models that are far more sophisticated than these linear models. They often gather historical demand data and utilize sophisticated forecasting models to estimate the "form" (i.e., distribution and parameters) of the uncertain future demand.

Second, the ability to segment customers and determine the price-dependent demand function for each segment is not a straightforward task. As discussed in Section 19.1, firms need to design their fare structures (i.e., construct fences) that prevent the high-valued customers (such as those of segment 1 in Example 1) from buying the products at prices set for the less-valued customers (such as those of segment 2 in Example 1, since $p_2^* < p_1^*$). Recall that in the airline industry, this is done by requiring the low-fare customers to book in advance, have a Saturday night stay, and pay high penalties in the events of cancellation or no-show. Nonetheless, leakage between the different segments remains possible, and further complicates the demand management problem.

As a result of the complexities in the demand and the business environment discussed earlier, most firms applying the RM methodology make their pricing and capacity decisions separately. In the remainder of this chapter, we will assume that prices have been determined, and study the problems of capacity control and overbooking.

19.3   Capacity Control

In this section, we assume that the airline has determined the price for each fare-class, and is now attempting to maximize revenue by controlling the availability of its seats (which are perishable and limited in number). This involves determining whether or not to sell tickets for a certain fare-class at a given point in time (under the assumption of advanced purchase), or equivalently, determining how much inventory to reserve for each segment. As an example, consider an airline that offers two fare-classes on a given flight, with class 1 fare higher than class 2 fare as discussed earlier. Then the airline should never reject a class 1 customer as long as there is capacity available. Then the question that naturally arises is when to accept a class 2 customer. As discussed, selling a seat to a class 2 customer runs the risk of not having a seat available for a **higher paying** class 1 customer in the future (i.e., a business customer might be "spilled"). On the other hand, rejecting a class 2 customer could result in the plane flying with empty "spoiled" seats. The capacity control decision revolves around this trade-off.

In Section 19.3.1, we discuss the single-resource (i.e., single flight) multi-class problem. Although many RM problems in reality involve networks, and hence require multiple resources (e.g., consider a customer who needs to take multiple flights between her origin and destination), it is not uncommon to solve such problems as a series of single-resource problems due to simplicity and flexibility. This, of course, translates into assuming that all resources are independent. In addition, these single-resource models provide basic insights into the aforementioned trade-off. Then, in Section 19.3.2 we study the multi-resource multi-class problem, also known as the "network revenue management," "network capacity control," or "origin/destination control" problem in the RM literature.

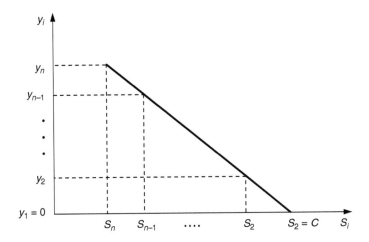

FIGURE 19.2 Relationship between booking limits, S_i, and protection levels, y_i, $i = 1, \ldots, n$.

19.3.1 Single-Resource Problem

Consider that the airline offers n fare-classes with $p_1 > p_2 > \cdots > p_n$, and assume that class n demand is realized first (i.e., class n customers buy tickets first), followed by class $n - 1$, then class $n - 2$, etc., until class 1 demand is realized. This assumption is fairly realistic given the way the fare-classes are designed (see Section 19.2). Denote the demands for the different classes by independent random variables X_i, $i = 1, \ldots, n$. In addition, assume that there are no cancellations or overbooking. The problem is to determine how many ticket requests to accept for each fare-class.

The optimal solution to this problem can be obtained by dynamic programming (see, e.g., Brumelle and McGill 1993). In particular, the structure of the optimal solution involves n booking limits, S_n, $S_{n-1}, \ldots, S_1$, with $S_1 = C$, such that the airline accepts up to S_i customers from class i, $i = 1, \ldots, n$, depending on the number of seats left after satisfying the demands for classes n, $n - 1$, $\ldots$, $i + 1$ (each up to its own booking limit, of course). Observe that the booking limits result in a "nested" protection structure, $y_1 \leq y_2 \leq \ldots \leq y_n$, with y_i, $i = 1, \ldots, n$, denoting the number of seats protected from (i.e., unavailable to) classes i to n. Then $y_i = C - S_i$, $i = 1, \ldots, n$, see Figure 19.2.

There are certain drawbacks of the dynamic programming approach, however. As the number of fare-classes gets large (which is usually the case for most major airlines), so does the size of the resulting dynamic program, hence the computational times required to obtain the optimal booking limits. Consequently, in the following, we first present a special case with only two fare classes (for this case, the optimal solution can be easily determined using the properties of the expected profit function), and then present a heuristic procedure for the general case having more than two classes.

19.3.1.1 Single-Resource Two-Class Problem

We first consider a special case of the single-resource problem with two fare-classes only ($n = 2$). The following model was first suggested by Littlewood (1972), and is one of the earliest models for capacity control in RM. Recall that X_1 and X_2, respectively, denote the demand for classes 1 and 2. X_1 and X_2 are both assumed to be non-negative, independent, and continuous* random variables, with respective probability density functions $f_{X1}(\cdot)$ and $f_{X2}(\cdot)$. As stated earlier, the form of the optimal policy is to sell

---

* This assumption is made to simplify the analysis; similar results can be obtained for the case where X_1 and X_2 are discrete random variables.

S_2 seats to class 2 customers (under the assumption that the airline can sell as many class 2 tickets as it wants) and then "close" class 2 and accept only class 1 demand (up to capacity). Therefore, the airline's expected profit for a given S_2 is

$$E[\Pi(S_2)] = p_2 S_2 + p_1 E[\min(X_1, C - S_2)]$$

$$= p_2 S_2 + p_1 \left(\int_0^{C-S_2} x_1 f_{X_1}(x_1)dx_1 + (C - S_2) \int_{C-S_2}^{\infty} f X1(x_1)dx_1 \right). \qquad (19.3)$$

The second term in the right-hand-side of Equation 19.3 follows because the number of class 1 seats sold equals to $C - S_2$ if X_1 exceeds $C - S_2$, and equals to X_1, otherwise. Upon simplification, Equation 19.3 reduces to

$$E[\Pi(S_2)] = p_2 S_2 + p_1 \left(E[X_1] + \int_{C-S_2}^{\infty} (C - S_2 - x_1) f_{X_1}(x_1)dx_1 \right). \qquad (19.4)$$

Recall that the problem is to determine the optimal booking limit, S_2^*, that maximizes the airline's expected profit. It can be easily verified that function $E[\Pi(S_2)]$ is strictly concave in S_2. Hence, the optimal solution is unique, and the first-order optimality condition, given by $\left. \frac{\partial E[\Pi(S_2)]}{\partial S_2} \right|_{S=S_2^*} = 0$, is necessary and sufficient to determine the optimal solution, S_2^*. Setting $\left. \frac{\partial E[\Pi(S_2)]}{\partial S_2} \right|_{S=S_2^*} = 0$ in Equation 19.4 implies that $p_2 - p_1 \left(\int_{C-S_2^*}^{\infty} f_{X_1}(x_1)dx_1 \right) = 0$, or equivalently,

$$F_{X_1}(C - S_2^*) = \frac{p_1 - p_2}{p_1}, \qquad (19.5)$$

where $F_{X1}(\cdot)$ is the cumulative density function (CDF) of X_1. Finally, rewriting Equation 19.5 as

$$p_2 = p_1 \bar{F}_{X_1}(C - S_2^*), \qquad (19.6)$$

where $\bar{F}_{X_1}(x_1) = 1 - F_{X_1}(x_1) = P(X_1 > x_1)$, allows for another interesting interpretation of (19.5). The interpretation, which is due to Belobaba (1989), is as follows: Accept a class 2 request as long as its price is greater than or equal to the expected marginal seat revenue (EMSR) of class 1, given by $EMSR_1(C - S_2) = p_1 \bar{F}_{Y_1}(C - S_2)$. Note that (19.6) implies that $p_2 > EMSR_1(C - S_2)$ for $S_2 < S_2^*$ (see Fig. 19.3 for a graphic illustration). (This interpretation is the basis for the heuristic for the single-resource multi-class problem discussed in Section 19.3.1.2.)

Remark 1

This problem is equivalent to a well-known inventory problem, the "newsvendor problem," in which a newsvendor sells a daily newspaper. At the start of each day, the newsvendor must decide on the number of newspapers to purchase from the publisher at a price of c per paper. Then during the day, she observes the random demand, D, which is modeled as a continuous* random variable with CDF $F_D(\cdot)$, and sells papers at a price of r per paper. At the end of each day, the newsvendor can salvage any unsold newspapers for a price of v per paper. The parameters are such that $r > c > v$ (otherwise, the problem becomes either trivial or ill-defined). Then it can be shown that the newsvendor's optimal order quantity, y^*, satisfies

---

* It is easy to extend the results to the case where the demand, D, follows a discrete distribution.

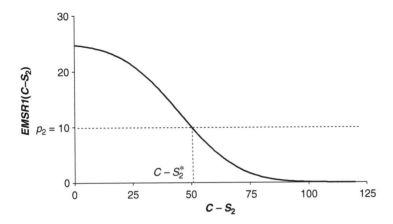

FIGURE 19.3 Determining the optimal booking limit in Example 2.

$$F_D(y^*) = \frac{r-c}{r-v} = \frac{c_u}{c_o + c_u},$$

where c_u can be interpreted as the "underage cost," that is, the cost incurred per unit of unsatisfied demand, and c_o can be interpreted as the "overage cost," that is, the cost incurred per unit of positive inventory remaining at the end of the period, with $c_u = r - c$ and $c_o = c - v$ (explain why). Then setting the "overage" cost, c_o, to p_2, and the "underage" cost, c_u, to $p_1 - p_2$ establishes the equivalence between the single-resource two-class problem discussed earlier and the newsvendor problem.

We conclude this section with an example on the evaluation of S_2^*.

Example 2

Consider again FHA, which has an aircraft with capacity $C = 150$ and offers two fare-classes at prices $p_1^* = \$25$ and $p_2^* = \$10$, as determined in Example 1. However, FHA has now better information about the demand, and has postulated that demand for class 1, X_1, can be modeled as a Normal random variable with mean $\mu_1 = 45$ and standard deviation $\sigma_1 = 20$. FHA must now decide on S_2^*, the optimal booking limit for class 2.

It follows from 19.5 that

$$S_2^* = C - (\mu_1 + Z\sigma_1), \tag{19.7}$$

where $Z = \Phi^{-1}((p_1 - p_2)/p_1)$ and $\Phi^{-1}(\cdot)$ is the inverse of the standard Normal CDF. Therefore, $S_2^* \approx 150 - (45 + 0.253 \times 20) = 99.93$. Since in reality the airline will be restricted to discrete units, the airline should accept the first 99 requests [since $EMSR_1(C - 100) > EMSR_1(C - 99.93) = p_2$, see Fig. 19.3] from class 2 customers and reject the rest. In other words, the airline should protect $C - S_2^* = 51$ units for class 1 customers; see Figure 19.3 for a graphical illustration of this solution.

19.3.1.2 Single-Resource Multi-Class Problem

We now revisit the single-resource problem with n fare-classes. As mentioned earlier, the structure of the optimal solution involves n booking limits, $S_n, S_{n-1}, \ldots, S_1$, with $S_1 = C$, which can be determined exactly by dynamic programming. However, as the number of fare-classes gets large, the computational times required to obtain the optimal booking limits with dynamic programming increase significantly. Therefore, in the following we will present an efficient heuristic, termed EMSR-B (see Belobaba 1992), which is widely used in practice. This heuristic generalizes the EMSR rule proposed in Equation 19.6. In particular, the booking limit for class $i = 2, ..., n$ (with $S_1 = C$) is given by

$$p_i = \bar{p}_{i-1}\bar{F}_{W_{i-1}}(C - S_i^*),$$

(19.8)

where $\bar{p}_{i-1}$ is the "average" fare for classes $1, ..., i-1$, and W_{i-1} is the sum of the demands of classes $1, ..., i-1$, that is,

$$\bar{p}_{i-1} = \frac{\sum\limits_{j=1}^{i-1} p_j E[X_j]}{\sum\limits_{j=1}^{i-1} E[X_j]}, \ W_{i-1} = \sum\limits_{j=1}^{i-1} X_j, \text{ and } \bar{F}w_{i-1}(w) = P(W_{i-1} > w).$$

[Compare (19.8) with (19.6).] It is commonly assumed that X_i, $i = 1, ..., n$, is a normal random variable with mean μ_i and standard deviation σ_i. In this case, W_{i-1} is also normal with mean $\sum_{j=1}^{i-1} \mu_j$ and standard deviation $\sqrt{\sum_{j=1}^{i-1} \sigma j^2}$. Then, S_i^* can be evaluated easily using a formula similar to (19.7) in Example 2.

19.3.2   Multi-Resource (Network) Problem

Not surprisingly, when several products share two or more resources, the RM capacity control problem becomes more complicated. For example, if an airline offers trips from city A to city B and from city B to city C, then "connecting" passengers going from A to C through B are a possibility, with B acting as a "hub" (examples of this problem in other industries include multi-night stays at hotels or multi-day car rentals). This complicates the capacity control problem. We must now consider the question of how to value a connecting passenger. A connecting passenger might have a relatively high fare; so is accepting a connecting passenger on the flight from A to B using the rules presented in Section 19.3.1 based on this relatively high fare a good decision? What factors should be considered? Now imagine an airline with multiple hubs and thousands of flights a day. Clearly this is a difficult problem, which does not lend itself to an optimal solution with any reasonable assumptions. As such, in the following, we will present two commonly used heuristic approaches for this problem: "bid price control" and "displacement-adjusted virtual nesting." In fact, both heuristics have many variations in practice. Here we only describe basic versions of each.

19.3.2.1   Bid Price Control Heuristic

The first approach models this capacity control problem on the origin/destination (OD) network (with multiple, dependent resources and multiple fare-classes) as a network flow maximization problem using expected demands from each product offered by the airline (see, e.g., Boyd and Bilegan 2003):

$$\max_{x_i, i \in I} \sum_{i \in I} p_i x_i$$

$$\text{subject to } \sum_{i \in I(l)} x_i \leq C_l, \ l \in L, (\lambda_l)$$

(19.9)

$$x_i \leq E[X_i], \ i \in I,$$

$$x_i \geq 0, \ i \in I,$$

where I is the set of products offered, L is the set of flight legs (resources) in the network, I(l) is the set of products utilizing leg l, $l \in L$, C_l is the available capacity of leg l, $l \in L$, E[X_i] is the expected demand for product i, $i \in I$, and p_i is the price of product i, $i \in I$. The decision variables, x_i, represent the number of reservations accepted for product i, $i \in I$.

Instead of utilizing the primal formulation given in Model 19.9 (and the corresponding decision variables, x_i, $i \in I$), a common approach is to solve the corresponding dual problem and obtain λ_l, the dual variable corresponding to the capacity constraint on leg l, $l \in L$. The dual variable λ_l, $l \in L$, represents the "displacement cost" of accepting a passenger on leg l, or equivalently, the minimum acceptable "bid price" for leg l. Then, product i is made available for sale if its fare is greater than or equal to the sum of the bid prices for the legs it utilizes. That is, the "bid price control" policy is to accept a request for product i if $p_i \geq \sum_{l \in L(i)} \lambda_l$, where $L(i)$ is the set of flight legs in product i (see, e.g., Boyd and Bilegan 2003; Talluri and van Ryzin 2005). In practice, the expected demand estimates, $E[X_i]$, and the available capacities, C_l, are frequently updated as the departure time approaches and more demand information is obtained, and new values of λ_l are obtained (and hence, a new control policy leading possibly to closing some low-price fares is developed).

19.3.2.2 Displacement-Adjusted Virtual Nesting Heuristic

Displacement-adjusted virtual nesting (DAVN) is another commonly used heuristic for network capacity control [see, e.g., Talluri and van Ryzin (2005)]. The idea is to determine a displacement cost for each leg [using, for example, the formulation in Model (19.9)], and then to decide on the capacity control of each leg separately utilizing the single resource methods discussed in Section 19.3.1. In particular, given displacement costs, λ_l, $l \in L$, we first calculate a "displacement-adjusted revenue," p_{il}, for each product $i \in I$ and each leg $l \in L$, $p_{il} = p_i - \sum_{k \in L} (i) \setminus \{l\} \lambda_k$, which approximates the net revenue for accepting product i on leg l. Given the large number of products that use a given leg, a common approach is to cluster products into "virtual buckets" based on their displacement-adjusted revenues (this is also known as "virtual nesting"). Each virtual bucket is then treated as a separate product and booking limits are obtained for each bucket on each leg.

Under virtual nesting, a request for a product will be rejected if it falls in a bucket that has received reservations that exceed its booking limit on any of the leg that the product utilizes. Consider again the airline example given earlier, and suppose that the product with origin A, destination C, and class 2 has been assigned to bucket 7 on leg AB and to bucket 4 on leg BC. Then, a request for this itinerary will be rejected if either bucket 7 on leg AB or bucket 4 on leg BC has exceeded its booking limit.

Example 3

Fly High Airlines operates between three cities on the East Coast, Boston (BOS), New York (JFK), and Washington DC (IAD) (see Fig. 19.4). FHA utilizes IAD as a hub and flies one round-trip daily between IAD and each of the other two cities; see Table 19.2 for the flight information and capacities of the aircraft assigned to the flights. FHA offers tickets in two fare-classes, 1 and 2. As a result, it offers 12 products (i.e., itineraries for six OD pairs, each offered in two fare-classes; see the first column in Table 19.3, where each product is denoted by indices ij, with i denoting the OD pair, and j denoting the fare-class).

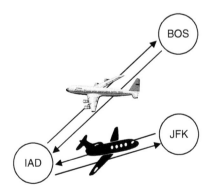

FIGURE 19.4 FHA's flight network.

TABLE 19.2 FHA's Flight Schedule and Capacity Assignment

Flight	Origin	Destination	Departs	Arrives	Capacity*
200	JFK	IAD	8:00 A.M.	9:30 A.M.	70
250	IAD	JFK	11:00 A.M.	12:30 P.M.	70
300	BOS	IAD	8:00 A.M.	9:30 A.M.	50
350	IAD	BOS	11:00 A.M.	12:30 P.M.	50

*This capacity might have been adjusted up from the actual physical plane capacity to account for no-shows and cancellation. See Section 19.4 on overbooking.

It is February 1 and FHA is determining its capacity control policy for flights on February 15. As of February 1, no bookings have been received for flights on February 15. The demand forecast for each product is broken into two periods (i.e., weeks), with period 1 preceding period 2. The fares and the period demand forecasts for each of FHA's 12 products are given in Table 19.3, where $N(\mu, \sigma)$ denotes a Normal random variable with mean μ and standard deviation σ. FHA uses a bid price control policy as described in Model 19.9. Thus FHA's network capacity control problem on February 1 can be formulated as follows:

$$
\begin{aligned}
\max \quad & 203x_{11} + 63x_{12} + 303x_{21} + 93x_{22} + 204x_{31} + 44x_{32} + 304x_{41} \\
& + 94x_{42} + 203x_{51} + 53x_{52} + 204x_{61} + 64x_{62}
\end{aligned}
$$

$$
\begin{aligned}
\text{subject to} \quad & x_{11} + x_{12} + x_{21} + x_{22} \le 70 && (\lambda_{JFK-IAD}) \\
& x_{31} + x_{32} + x_{41} + x_{42} \le 50 && (\lambda_{BOS-IAD}) \\
& x_{41} + x_{42} + x_{61} + x_{62} \le 70 && (\lambda_{IAD-JFK}) \\
& x_{21} + x_{22} + x_{51} + x_{52} \le 50 && (\lambda_{IAD-BOS}) \\
& x_{11} \le 13; \quad x_{12} \le 39 \\
& x_{21} \le 11; \quad x_{22} \le 28 \\
& x_{31} \le 14; \quad x_{32} \le 40 \\
& x_{41} \le 12; \quad x_{42} \le 31 \\
& x_{51} \le 12; \quad x_{52} \le 39 \\
& x_{61} \le 11; \quad x_{62} \le 38 \\
& x_{ij} \ge 0, \ i = 1, \dots, 6, \ j = 1, 2
\end{aligned}
$$

(19.10)

TABLE 19.3 FHA's Demand Forecasts and Fares

Product	Orig.	Dest.	Demand—Class 1		Demand—Class 2		Fares	
			Period 1	Period 2	Period 1	Period 2	Class 1	Class 2
11, 12	JFK	IAD	N(4, 1)	N(9, 3)	N(30, 7)	N(9, 3)	$203	$63
21, 22	JFK	BOS	N(3, 1)	N(8, 2)	N(20, 4)	N(8, 2)	$303	$93
31, 32	BOS	IAD	N(4, 1)	N(10, 3)	N(30, 7)	N(10, 3)	$204	$44
41, 42	BOS	JFK	N(3, 1)	N(9, 3)	N(22, 4)	N(9, 3)	$304	$94
51, 52	IAD	BOS	N(3, 1)	N(9, 3)	N(30, 7)	N(9, 3)	$203	$53
61, 62	IAD	JFK	N(3, 1)	N(8, 2)	N(30, 7)	N(8, 2)	$204	$64

TABLE 19.4 FHA's Capacity Control Policy

Product	Orig.	Dest.	Displacement Cost	Fares		Accept Reservation?	
				1	2	1	2
11, 12	JFK	IAD	$\lambda_{JFK-IAD} = \$40$	$203	$63	Yes	Yes
21, 22	JFK	BOS	$\lambda_{JFK-IAD} + \lambda_{IAD-BOS} = \$93$	$303	$93	Yes	Yes
31, 32	BOS	IAD	$\lambda_{BOS-IAD} = \$30$	$204	$44	Yes	Yes
41, 42	BOS	JFK	$\lambda_{BOS-IAD} + \lambda_{IAD-JKF} = \$94$	$304	$94	Yes	Yes
51, 52	IAD	BOS	$\lambda_{IAD-BOS} = \$53$	$203	$53	Yes	Yes
61, 62	IAD	JFK	$\lambda_{IAD-JKF} = \$64$	$204	$64	Yes	Yes

Solving Model 19.10 (which can be done using an optimization software such as AMPL, see http://www.ampl.com) gives:

$$\lambda_{JFK-IAD} = \$40$$
$$\lambda_{BOS-IAD} = \$30$$
$$\lambda_{IAD-JFK} = \$64$$
$$\lambda_{IAD-BOS} = \$53$$

Consequently, under the bid price control policy, FHA would accept reservations for a product whose fare is greater than or equal to the sum of the bid prices for the flights it utilizes. Then, the bid price control policy for FHA on February 1 is to accept reservations for all products for February 15, see Table 19.4.

19.4 Overbooking

Airline RM systems are based on advance reservations for a future travel itinerary. In many cases, customers have the right to cancel their reservation with little or no penalty. In other instances, customers may simply not show up for a flight (e.g., due to the vagaries of airline RM, one-way tickets are often not discounted, and thus are more expensive than round-trip tickets. This makes it cheaper for passengers to buy the round-trip ticket and not show up for the return flight. Can you guess why one-way tickets would be more expensive?). This can be a significant source of lost revenue. In fact, recent studies in the airline and rental car industries (Smith et al. 1992; Geraghty and Johnson 1997) report that on average only 50% of all reservations "survive" (i.e., the customer actually uses the product). To avoid this revenue loss, airlines commonly allow reservations to exceed capacity in anticipation that some reservations will not survive. This business practice is known as "overbooking." Obviously, the drawback of overbooking is that it can lead to more products sold than capacity, hence some customers being denied service. Therefore, it is important to set an "overbooking limit" appropriately in order to utilize most of the available capacity while honoring the reservations of most of the customers. RM focuses on setting the overbooking limits so as to maximize revenue while considering such "service level" constraints.

Historically, overbooking has its roots in the airline industry. However, it dates back to the 1960s and 1970s, prior to the deregulation of airlines and the subsequent development of modern RM [see Rothstein (1984) for an extensive historical exposure]. In the 1960s and 1970s, airlines used to engage in overbooking in a discrete manner without informing the customers of its consequences. A law suite won by Ralph Nader in 1976 changed this practice. The airlines became obliged to inform customers about overbooking (which they still do on the back of each ticket). Airlines also started developing innovative ways to make service denials more acceptable to customers. Motivated by research in economics [e.g., Simon (1968)], some airlines currently manage overbooking as an auction. They offer compensation (such as a travel voucher of some monetary value, to be used for future travel) to get volunteers for service denials on flights for which more travelers than seats show up at the time of departure.

In the remainder of this section we present two simple, static (i.e., they ignore the dynamics of cancellations and new reservations over time) models for determining the overbooking limit to introduce the reader to some overbooking concepts. As always, in practice overbooking models are more sophisticated and are usually integrated with the other models discussed here.

19.4.1 Distribution of Shows (Survivals)

Suppose it is estimated that a passenger will "show up" for the flight with probability q independently of the other passengers, that is, q is the probability that a reservation will "survive." Suppose also that y customers have reservations at a given time. Then, out of the y reservations, the probability that z reservations survive is

$$P(Z(y) = z) = \binom{y}{z} q^z (1-q)^{y-z}, \tag{19.11}$$

where $Z(y)$ is the random variable representing the number of surviving reservations. This is known as the "binomial model" because $Z(y)$ follows a binomial distribution. This model is attributed to Thompson (1961). Although this model is based on several simplifying assumptions (e.g., it is static, it ignores people traveling in groups, who need to cancel their reservations together), it is desirable due to its simplicity. In the following, we present two methods to determine the overbooking limit: based on service level and expected profit.

19.4.1.1 Overbooking Limit Based on Service Level

Using the binomial model, we can determine the corresponding service level. Suppose that the overbooking limit is L ($L > C$). There are two commonly used service levels (see, for instance, Talluri and van Ryzin 2005).

1. Type 1 service level: the probability that at least one customer will be denied service, that is

$$s_1(L) = P(Z(L) > C) = \sum_{k=C+1}^{L} \binom{L}{K} q^k (1-q)^{L-k} . \tag{19.12}$$

2. Type 2 service level: the fraction of customers who are denied service, that is

$$s_2(L) = \frac{E[Z(L) - C)^+]}{E[Z(L)]} = \frac{\sum_{k=C+1}^{L} (k-C)\binom{L}{k} q^k (1-q)^{L-k}}{Lq}, \tag{19.13}$$

where $x^+ = \max(0, x)$.

A firm will set a desired service level (e.g., the probability that at least one customer is denied service is less than 1%, the percentage of customers who are denied service is less than 2%). The corresponding overbooking limit can then be calculated by solving for L in (19.12) or (19.13).

19.4.1.2 Overbooking Limit Based on Expected Profit

Suppose that each customer denied service incurs a cost G. For example, in the airlines, G is the cost of a full refund and an additional reward ticket. Let p be the price of the product. Then, the airline's expected profit given that the airline sells L tickets, where L is the overbooking limit, is given by

$$pL - G \sum_{k=C+1}^{L} (k-C) \binom{L}{k} q^k (1-q)^{L-k}. \tag{19.14}$$

Then, it can be shown that the optimal booking limit, L^*, is the largest value of L that satisfies*

$$qP(Z(L-1) \geq C)G \leq p, \tag{19.15}$$

or equivalently,

$$qG \sum_{k=C}^{L-1} \binom{L-1}{k} q^k (1-q)^{L-1-k} \leq p \tag{19.16}$$

The left-hand side of (19.15) reflects the fact that in order for the Lth reserving customer to be denied service, (i) there should be enough survivals from the first $(L-1)$ reservations to utilize all the capacity, and (ii) the Lth customer should survive (with probability q).

Remark 2
When dealing with several customer segments that will show up for service with different probabilities, a common approach is to approximate the survival probability, q, by a weighted average of the survival probabilities of the segments [Talluri and van Ryzin (2005)].

Example 4
Fly High Airlines actually flies a 43-seat Embraer RJ145 between BOS and IAD. FHA estimates that the survival probability for this leg is 0.86. The overbooking limit of 50 used in Example 3 was obtained utilizing a simple heuristic, which used the ratio of the actual capacity to the survival probability (i.e., 43/0.86 = 50). FHA now wants to use more sophisticated techniques so as to obtain a "better" overbooking limit. FHA is evaluating two alternatives:

1. Setting the overbooking limit in a way that the percentage of customers denied boarding is less than 1%.
2. Setting the overbooking limit in a way that maximizes the expected profit. FHA estimates that a customer denied boarding costs $500 and that the average fare is 387 (this is approximately the weighted average of class 1 and class 2 fares in Table 19.3 with the weights being the mean demands for the two classes).

The booking limit required in (1) can be obtained from (19.13) as the largest value of L that satisfies

$$\frac{\sum_{k=44}^{L} (k-43) \binom{L}{k} (0.86)^k (0.14)^{L-k}}{0.86 L} \leq 0.01.$$

Searching over $L = 44, 45, \ldots$, it can be seen that $L^* = 48$ is the appropriate overbooking limit with a percentage of customers denied boarding of 0.7%.

---

* This result follows because the expected profit function in (19.14) is concave in L.

The booking limit required in (2) can be obtained from (19.16) as the largest value of L that satisfies

$$430 \sum_{k=43}^{L-1} \binom{L-1}{k} (0.86)^k (0.14)^{L-1-k} \leq 85.$$

Searching over $L = 44, 45, \ldots$, it can be seen that $L^* = 48$ is again the appropriate overbooking limit. In conclusion, it seems that a booking limit of 50 is a bit high given the survival probability of 0.86.

19.5   Case Study

You are the Manager of Revenue Optimization at FHA and your job is to improve the RM system (see Example 3 for the current system). FHA's current system calculates the bid-prices for each departure date only once, at the beginning of the first period (see Example 3). The CEO of FHA has taken a class in RM, and suggests the following options to improve revenue:

1. Upgrade FHA's bid-price system so that it produces updated bid-prices at the beginning of period 2.
2. Ignore the network effects, and simply use a flight-based (instead of a network-based) RM system (see Section 19.3.1). For products that consist of multiple flights, the stated fare will be used when determining the set of booking limits on each flight.
3. Modify FHA's bid-price system so that the optimal number of tickets (from Model 19.9) for each product is used to limit ticket sales, that is, if $x_{11} = 10$, then FHA will only sell up to 10 tickets for product 11. This will no longer be a bid-price system, but it is still based on Model 19.9.
4. Use a DAVN system (see Section 19.3.2.2).

Of course, these four options are mutually exclusive. It is up to you to decide which option is best. (Alternatively you can come up with another option.) To present your solution to the upper management, you need to prepare a detailed report. You should support your decision with a detailed quantitative analysis. In addition, you should include a discussion on the following points in your report: What are the drawbacks and advantages of each option? Can you think of a simple way to improve on any of the options? Is each option expected to perform better: (*i*) under high or low demand uncertainty? (*ii*) with high or low mean demands compared to capacity?

19.6   Challenges and Future Research Directions

In this chapter we present a brief overview of RM, specifically focusing on the use of OR techniques in this field. We believe that this chapter will provide the reader with a basic understanding of RM and serve as a good starting point for those new to RM. We refer the reader interested in a more detailed and a comprehensive material to the excellent text by Talluri and van Ryzin (2005) that we have consulted while preparing this chapter. The review article by Boyd and Bilegan (2003) is another highly useful reference.

Finally, we point out some of the current challenges and future directions of RM that we believe are the most important. We believe that a major challenge for RM is in applications in areas beyond the travel industry (e.g., airlines, hotels, and rental cars). Boyd and Bilegan (2003) identify the broadcasting industry and hospitals as two important areas for "nontraditional" RM applications. RM applications to trucking and manufacturing industries seem to be also promising.

Another challenge for RM is in coping with a changing, and a more competitive and uncertain, business environment. For example, in competing with low-cost carriers, major (legacy) airlines are bringing their fares down to low levels, which is jeopardizing profitability. This is prompting major airlines to come up with innovative techniques to benefit from their large fleets and networks.

Gallego and Phillips (2004) discuss such a novel approach. In particular, they consider a major airline flying multiple trips between two cities. In addition to the flight-specific products, the airline offers a cheaper, "flexible product," which guarantees the customer a flight between the two cities on a certain date and within a certain time window, but without specifying the exact time of departure. This provides the airline with some flexibility, and allows it to hedge against demand uncertainty by allocating the flexible product customers to the flights at a later time, when more demand information is obtained and uncertainty is reduced. Thus, the airline can better match its supply (capacity) with demand. This is just one example. We expect that such novel approaches that provide the firms with more flexibility, applied in conjunction with sophisticated RM techniques, will be the future of RM implementations.

Revenue management is a discipline that is spreading to more and more industries, each with its own challenges. When a firm embraces RM, it is usually a core function of the firm, which impacts many other units such as marketing, sales, pricing, and scheduling. As such, RM needs to be in tune with the market, industry, and the firm.

Practice Questions

1. What strategies can the manufacturing industry use to segment the market? Consider different types of manufacturing industries and discuss this question in the context of each industry. What types of manufacturing industries could benefit most from RM? Why?
2. What demand management decisions do retailers need to make? Answer this question in the context of different types of retailers.
3. How does the use of the internet facilitate RM implementation?
4. Explain why the equivalence between the newsvendor solution and that for the single-resource two-class problem discussed in Section 19.3.1.1 holds (see Remark 1).
5. Show that the expected profit function for the single-resource two-class problem, given in (19.4), is strictly concave in S_2.
6. Show that the expected profit function with overbooking, given in (19.14), is concave in L. Using this result, derive the optimality conditions in (19.15) and (19.16).

References

Bell, P. C., Revenue management for MBAs, *OR/MS Today*, Available at: http://www.lionhrtpub.com/orms/orms-8-04/frbell.html, August 2004.

Belobaba, P. P., Application of a probabilistic decision model to airline seat inventory control, *Opns Res.*, 37, 183, 1989.

Belobaba, P. P., Optimal versus heuristic methods for nested seat allocation. Presentation at the ORSA/TIMS Joint National Meeting, November 1992.

Boyd, E. A., Airline alliance revenue management. *OR/MS Today*, Available at: http://www.lionhrtpub.com/orms/orms-10-98/boyd.html, October 1998.

Boyd, E. A. and Bilegan, I. C., Revenue management and e-commerce, *Management Sci.*, 49, 1363, 2003.

Brumelle, S. L. and McGill, J. I., Airline seat allocation with multiple nested fare classes, *Opns. Res.*, 41, 127, 1993.

Carrol, W. J. and Grimes, R. C., Evolutionary change in product management: experiences in the car rental industry, *Interfaces*, 25, 84, 1995.

Cook, T.M., SABRE soars, *OR/MS Today*, Available at: http://www.lionhrtpub.com/orms/orms-6-98/sabre.html, June 1998.

Gallego, G. and Phillips, R., Revenue management of flexible products, *M&SOM*, 6, 321, 2004.

Geraghty, M. K. and Johnson, E., Revenue management saves National car rental, *Interfaces*, 27, 107, 1997.

Littlewood, K., Forecasting and control of passenger bookings, in *Proc. AGIFORS Annual Sympos.*, 12, 95, 1972.

Rothstein, M., OR and the airline overbooking problem, *Opns. Res.*, 33, 238, 1984.

Simon, J., An almost practical solution to airline overbooking, *J. Trans. Econ. Policy*, 5, 201, 1968.

Smith, B. C., Leimkuhler, J. F., and Darrow, R. M., Yield management at American Airlines, *Interfaces*, 22, 8, 1992.

Talluri, K. T. and van Ryzin, G. J., *The Theory and Practice of Revenue Management*, Springer, New York, 2005.

Thompson, H. R., Statistical problems in airline reservation control, *Opnl. Res. Quart.*, 12, 167, 1961.

Weatherford, L. R. and Bodily, S. E., Taxonomy and research overview of perishable asset revenue management: Yield management, overbooking, and pricing, *Opns. Res.*, 40, 831, 1992.

IV

Enabling Technologies

20

Ubiquitous Communication: Tracking Technologies within the Supply Chain

M. Eric Johnson
Dartmouth College

Information integration and total supply chain visibility are viewed as integral parts of supply chain excellence. Real time and accurate information on the status of goods in a supply chain requires the integration of several evolving technologies that enable tracking of items, cartons, totes, containers, trucks, ships, rails, and other conveyances continuously. In this article, we examine tracking technologies in the context of a case study of an integration project at a major retailer, focusing on the business case for investment. The case examines how technologies like Radio Frequency Identification (RFID) and Global Positioning System (GPS) can be used to improve supply chain performance and aid in reducing supply chain shrinkage. Based on the results of that case and others, we discuss some of the key lessons for engineers and managers interested in implementing tracking technologies. Finally, we discuss the benefits of automated identification and tracking as compared with traditional legacy systems like bar codes.

20.1 Introduction

For decades, the physical operating layer in logistics lived in disconnected isolation from the information layer of supply chain management. The movement of products within a manufacturing or distribution facility was nearly invisible. Of course, the information systems could show that they were somewhere in the facility, and possibly the designated storage location, but little beyond that—particularly if the items were in-transit. The same was true outside facilities. Goods that were shipped to a warehouse were "on the road, boat, or air," but little more was known other than possibly when they were received at their destinations. Today, all that is changing. The race to connect the physical logistics layer and the information layer is accelerating. Many technologies are emerging to close the gap including wireless devices (e.g., RFID tags, 802.11 and bluetooth-enabled devices, pagers, cellular), GPS, and legacy tracking, including EDI links and bar coding, all linked to the massive information backhaul capabilities of the internet. When the connection is complete, the ubiquitous communication capability will make physical items visible throughout the supply chain.

However, while there is much excitement about the technologies for tracking, implementation in real supply chains has been inhibited by costs, lack of uniform standards, and the inability of many firms to develop the compelling business cases required to justify the sizeable investments. In this article, we will examine some of the most popular tracking technologies and consider their impact on supply chains. While we will focus on RFID, we will also examine other technologies and their integration to create tracking solutions. After looking at some of the technologies, we will present a case study of a tracking project at the U.K. retailer Woolworths. The case examines how technologies like RFID and GPS can be used to improve supply chain performance and aid in reducing supply chain shrinkage. Using the case study, we will discuss many of the benefits of tracking and the barriers of implementing new technologies.

While Woolworths began in the United States, and has since vanished, the once U.S. subsidiary operations in the United Kingdom and Australia have continued to thrive in those countries by evolving their business models. In the United Kingdom, Woolworths competed in a range of retail formats from traditional general merchandise to large-scale Big W stores that offered a huge spectrum of merchandise. Woolworths managed an extensive distribution network that suffered from many supply chain problems such as accurate forecasts and reliable inventory information that would facilitate effective asset management. Like all retailers, Woolworths also faced significant product losses across its supply chain from theft and mislocation.

Shrinkage impacts all retailers, from direct merchants like L. L. Bean to large box retailers like Staples, and the problem is global. Total retail losses are estimated at €30 billion/year across Europe. Wal-Mart alone was estimated to lose nearly $1 billion to shrinkage each year. In a 2005 study conducted by the Tuck Business School in cooperation with the Merchant Risk Council, we found that supply chain losses within the U.S. retail supply chain (not including store theft) total nearly 1% of sales revenue. Product leakage occurs across the supply chain, from inbound freight to warehousing and outbound distribution. Beyond the losses, shrinkage also contributes to inventory inaccuracy—both in stores and in warehouses. This inaccuracy often leads to customer service defects, lost sales, and customer dissatisfaction (Raman et al. 2001, DeHoratius and Raman 2004). The Woolworths case shows how a novel integration of RFID and GPS technologies can help reduced shrink and improve inventory accuracy.

After discussing the lessons from the case, we will examine the barriers to implementing new tracking solutions, including costs, standards, and the ability to financially justify evolving technology. Then we will discuss legacy tracking solutions, such as bar codes, comparing them to new automated approaches. Finally, we will conclude with a look to the future evolution of RFID and related technologies.

20.2   Technology of Tracking

While there are many technologies that enable wireless, automated tracking—including active and passive RFID, 802.11, bluetooth, pagers, and cellular—by far the most attention has been focused on RFID. RFID is a means of storing and retrieving data through electromagnetic transmission to an RF compatible integrated circuit. The technology uses small radio transponders, called "tags," that are attached to the objects being tracked. The tags communicate with a reader (or antenna) when a tag is within range of the reader. The reader then passes information about the object to a host computer that processes the information and, in turn, passes the information over internal networks and the internet. Thus, as the tagged objects move in the supply chain, the movements can become visible through a web-interface.

Currently, RFID tags are available in many different configurations, employing different technologies that have cost and performance trade-offs. Tags are often broadly segregated into two major classifications: passive and active. Pure passive, or "reflective," tags do not contain an internal power source and are less expensive to manufacture. These tags typically have a short range (2–3 m) and rely on the energy radiated by the reader to power the circuit. For example, to track merchandise leaving a warehouse, readers could be positioned at the dock doors. As tagged merchandise comes within the range of the reader, the readers send signals to the tag and it would respond by transmitting its unique identification number. That number could be associated with the merchandise, so the system could quickly identify the merchandise and record its movement. Until recently, the costs of these tags (typically $0.20 or more) have prohibited wide-scale adoption for disposable packaging. Many industry analysts and researchers have predicted that a sub-$0.05 tag will represent a tipping point in mass implementation (Bartels 2005) for item-level tagging. However, to achieve such costs requires large chip manufacturing volumes creating a chicken-and-egg problem—low costs are required for high adoption, yet high adoption is required for low cost (Yates 2005). Of course, there are many other passive tags that have been employed in applications where the tags could be attached to a more permanent conveyance such as a pallet or tote (Johnson and Lee 2002). These tags cost anywhere from $0.50 to $10 or more depending on the technology, data storage capability, and operating range of the tag. Readers, on the other hand, typically cost $1000 to $2500 depending on their connection requirements. Wireless readers used in outside applications are more expensive while ones that could be connected by cable inside a building are at the lower end of the cost range.

Active tags contain both a radio transceiver and battery. They have a substantially larger range (100+ meters), and are considerably more expensive to manufacture, and require periodic battery replacement. Active tags have the ability to transmit their location and other information intermittently with the signals being monitored by readers in the vicinity. Active tags typically can store far more information that could also be updated through interaction with the reader. Simple active tags cost as little as a few dollars or hundreds of dollars, again depending on the technology, range, and capabilities. Readers also range in cost from $1000 to $10,000 or more for tower readers in outside applications. For example, the U.S. military has installed thousands of active tags on assets (e.g., truck and containers). These tags can transmit over long distances and operate on long-life batteries that last for years without interruption. The tag can be programmed to hold a substantial amount of information describing the contents of the container, its shipment origin, destination, etc. They can also be used to detect tampering or other security breaches (Machalaba and Pasztor 2004).

While RFID enables tracking at each discrete point in a network where a reader has been installed, many supply chain managers have also begun focusing on higher resolution systems that allow truly ubiquitous tracking. Such systems typically employ longer range wireless communication systems such as off-the-shelf pager or traditional cellular communications along with GPS location systems. The costs and supply chain capabilities of these technologies have all greatly benefited from their widespread consumer use. These maturing technologies have become far more accessible and cost-effective in the past five years. For less than $100, pocket-sized GPS devices allow items to be tracked exactly anywhere on Earth at any moment. With no more than a clear view of the sky, satellite-based GPS enables location visibility ensuring that products are never lost in the supply chain. GPS itself is enabled by a constellation of 27 Earth-orbiting satellites (24 in operation and three extras in case one fails). The U.S. military developed and implemented the satellite network as navigation system, latter opening it for commercial use. A GPS receiver locates four or more of these satellites, calculates the distance to each, and uses that information to deduce its own location based on the principle of trilateration. With an accurate location reading passed over an existing wireless pager/voice network (e.g., AT&T, T-Mobile and Cingular GSM/GPRS digital wireless networks) to a server, items can be tracked over the internet anywhere in the world.

Companies competing in the supply chain visibility space fall into one of four categories:

1. Hardware providers: Companies focused on developing a specific technology like bar code readers, RFID devices and readers (e.g., Texas Instruments, Alien Technology, Symbol, Intermec, Philips), or GPS hardware (e.g., Global Tracking Communications, Advanced Tracking Technologies).
2. Focused application providers: Companies who deliver solutions for specific tracking needs. Examples include Savi Technology, which focuses on active RFID-enabled networks for transportation tracking and security; WhereNet, which provides RFID tracking solutions operating in confined spaces like factories or warehouse; and @Road, which provides GPS-tracking solutions for trucking companies.
3. Visibility dashboard providers: Firms that capture and present tracking data using visualization software, typically in a web-interface. Examples here include companies such as Blue Sky Logistics and SeeWhy that provide logistics tracking dashboards. These firms focus on reporting and metrics, with the underlying tracking information gathered by others.
4. Integration service providers, who work to pull all the pieces of technology and systems together to provide a complete solution. Many large technology consulting firms are competing in this area (e.g., Accenture, IBM, and HP) along with smaller specialty providers like RedPrairie or Savi Technology.

Of course, there are many other firms who offer applications that leverage the tracking data for supply chain planning, forecasting, or execution. For example, TrueDemand Inc. offers forecasting tools based on warehouse movement data that aids in replenishment planning and RedPrairie offers RFID supply chain execution solutions that facilitate order processing. However, in the end, few firms have successfully integrated all of the elements required for a large-scale supply chain tracking system. Integrating all the players and the technologies has proved exceedingly difficult because integration of tracking technology is messy and cumbersome. There are companies who have developed tags and bar codes that can be attached to assets. In addition, several companies have figured out how to provide event and logistics management from the software end. Unfortunately, the tags and bar codes are useless without a system to read them. And most of the software applications are linear in nature and tend to only focus on one or two internal business processes. The entire supply chain network for a customer is more complex than that. It is a challenge to integrate all the levels of the chain, especially from RFID tag to reader to software to enterprise, so that all levels can experience real-time visibility simultaneously.

20.3 Case Company Background

F. W. Woolworths, a subsidiary of its U.S. parent, was founded in the United Kingdom in 1909 as part of its parent company's global expansion plan. The first store opened in Liverpool, beginning a rapid roll-out throughout the United Kingdom. While Woolworths may have begun in the United States, it quickly became one of the U.K.'s most loved retailers. Focused on product lines for the home, family, and entertainment, Woolworths always offered its customers excellent values on a wide range of products. F. W. Woolworths was subsequently listed on the London Stock Exchange with its U.S. parent retaining a majority shareholder. In 1982, Woolworths was acquired by Kingfisher, Europe's largest home improvement retailer. Following the acquisition, the new management implemented a strategy to focus the product offering, centralize accounting, invest in new information systems, rationalize the store base, reduce costs, and centralize distribution. Products were rationalized into clearly defined categories: entertainment, home, kids (toys and clothing), and confectionery. This enabled further development of the individual product ranges through the use of branded, own-brand and exclusive merchandise such as Ladybird Clothing and Chad Valley Toys.

In the late 1990s, the management extended the Woolworths brand into other retail formats and alternative channels to accelerate growth by taking advantage of changing retail trends. This resulted in the opening of the first "Big W" store in 1999 and Woolworths General Store in 2000.

Woolworths was divested from Kingfisher plc in 2001 and began trading on the London Stock Exchange. The divestiture enabled the Woolworths Group plc management to pursue (independently of Kingfisher) the recovery and growth strategies that best met its long-term objectives. By 2006, Woolworths maintained a portfolio of approximately 900 stores. Over 800 Woolworths, Woolworths General Stores, and Big W superstores offered housewares, toys, sweets, apparel, home electronics, and seasonal fare with sales of over £2.8 billion. The group's other retail outlets included MVC home entertainment and electronics boutiques (about 85 shops), EUK, United Kingdom's largest distributor of home entertainment products, and music and video publisher VCI.

Woolworths faced increased competition from all sides. Traditional U.K. grocery retailers such as Sainsbury and Tesco had aggressively expanded their offerings beyond traditional food items. Pharmacy chains such as Wilkinsons and Boots the Chemist had expanded their general merchandising offerings. Finally, Woolworths' Big W supercenters faced competition from Wal-Mart, which established a U.K. presence through its purchase of ASDA. This increased competition placed a great deal of pressure on already thin general merchandise margins.

20.4 Tracking Project Description

Woolworths serviced the general merchandising needs of its 800+ stores through four distribution centers (DCs). Two primary DCs, located in Castleton and Swindon, were geographically focused, carried the same "general merchandise" items, and serviced approximately 400 stores each.

The two seasonal distribution centers, located in Rugby and Chester, carried a revolving inventory of seasonal merchandise including everything from patio furniture to Christmas decorations. Merchandise bound for the stores was typically transferred to the stores in either a large steel roll cage or a reusable plastic tote box. Large items were shipped in one of 100,000 roll cages while smaller items where shipped in totes (Fig. 20.1). Totes destined for the same store were stacked on one of 16,000 dollies (roll cages without sides). Distribution center employees wheeled these roll cages and dollies onto trucks for delivery to the stores.

Woolworths had first experimented with RFID in 1999 as a security system for tracking individual products. The project, which involved tagging clothing moving from a distribution center to a single store, was not a success. The tags were too expensive, too unreliable, and did not provide the read range the company needed.

So in 2003, when Woolworths began work on a second experiment, the goal was to define a manageable project scope in terms of the products, vehicles, stores, and distribution centers to be included. Woolworths distributed the dollies and their associated totes (up to 10 per dolly) only from its Swindon warehouse; all 800+ stores were covered from this site. Therefore, this closed loop was ideal for a "proof of concept" and did not require tagging all 100,000 roll cages. The system would track these dollies (and the associated totes) out of the warehouse, to the stores and back again.

From Woolworths' previous RFID initiative, they felt that item-level tracking, was not economically justifiable. The £4 average consumer "basket" price did not support a passive chip implementation. However, they believed that a unique "Russian Doll" strategy could achieve item-level visibility without item-level tags. The strategy combined a number of technologies focused on reducing both process losses and theft from the point of pick through the point of delivery to store, including:

- Bar codes on the individual items and on the tote boxes.
- Short- and long-range RFID devices to track movements within the DC.
- Portable RFID readers to track movements from the DC to stores.
- GPS to track vehicles on route.

FIGURE 20.1   Totes on dollies and roll cages waiting for shipment.

- A Wireless Wide Area Network (WAN) to transmit data back to the control system.
- 16,000 active RFID tags.
- Integration services with Woolworths' fulfillment and transport planning systems.

By using this unique combination of technologies, Woolworths had complete visibility of the movement of each tagged dolly, increasing security of both product and distribution assets [see Johnson et al. (2004) for detailed description of the technologies used].

The outbound distribution process began in Woolworths' national distribution center in Swindon via two technologies:

- Automated Storage and Retrieval System: A high density rack retrieval system that used automated robots to stock merchandise stored in plastic totes on the shelves and later extract merchandise bound for the stores. This process, which required no manual intervention, was used primarily for high-value items. Totes retrieved from the rack were sent to a picking area where items were sorted for each store order.
- Pick-to-Light System: Employees selected small items destined for each store using a pick-to-light system that resulted in better than 99% accuracy. In the pick-to-light system, workers were guided visually by lights to the exact bin locations where the required articles were stored. These items were picked and placed into plastic tote boxes. In cases where the tote held high-value items, the entire tote would be shrink-wrapped to discourage theft.

All of the plastic tote boxes had unique bar codes. Totes destined for the same store were scanned and then stacked on dollies using an automated stacking system. In the past, the dollies had no unique identifier. For the pilot, Woolworths attached a Savi EchoPointTM active (battery-powered) RFID tag on each dolly. The simple active tag was developed by Savi Technology to be low cost (about £5) and was disposable with a battery life of about four to five years.

The dollies were recorded as they moved toward the dispatch bay. A short-range device, called a SignPost, located under the track of the sorting system, emitted a low-frequency signal that activated the RFID tag. To conserve energy, the tag spent much of the time in a suspended state until it was activated by a sign post (or reader). When activated, the tag broadcast the information it contained, which was read by long-range readers installed in the rafters of the building (readers could recognize tags from up to 100 m away). The software system associated the bar codes that were scanned on the totes with the RFID tag on the dolly. So the system knew which items were put in specific totes and which specific totes were put on a specific dolly.

Woolworths also tied Savi's SmartChain real-time logistics platform in with its transport planning system. That way they could track which vehicle was in the dispatch bay at a given time and the destination of the vehicle. Medium-range SignPost readers, which could wake up a tag from about 20 ft, were installed over the dispatch bays. When the dispatch bay team loaded dollies onto a vehicle, the tags on the dollies were activated and read, and the system compared the ID numbers to the truck's delivery instructions. If the wrong dollies were being loaded, the system alerted staff with flashing lights (Fig. 20.2).

Once the vehicle had been loaded with the right dollies, the doors on the truck were closed and an encrypted seal (an electronic lock activated by a randomly generated code), was placed on the doors. The code had a four digit number that the driver punched into his handheld computer. The vehicle was then ready to make its first delivery. Each driver had his own portable kit that included a Symbol PDT8100 handheld scanner. The unit also had a GPS-enabled wireless communication system, so Woolworths could track the truck's movements between the distribution center and the store. The GPS transmitter was set up to send a signal at different intervals along the trip. Since the cost of monitoring was based on how often the transmitter broadcasted, the interval could be lengthened if the goods in transit were inexpensive, or set for every five or ten seconds if there was high-value merchandise in the truck.

When the driver arrived at the store, he keyed in a four-digit number used to identify the particular store and the system would confirm his exact location by "geo-fencing." If he said he was at store "1234," the system knew the location of that specific store and confirmed that he was at the correct location. He then entered the four-digit seal number, which was required to correspond with the number that was entered at the dispatch bay. If the code did correspond, the lock was released, and he was then given instructions on the handheld regarding which dollies and totes to unload. As he unloaded

FIGURE 20.2 Dock door with RFID reader and indicator lights.

each dolly, the driver would scan it with the Symbol unit. The system would then confirm that he moved the correct dolly or tote. The system would warn him if he had delivered too many, too few, or the wrong ones.

Once delivery was complete, the driver then received an electronic signature from the store manager on the screen of his PDT8100, indicating that the store had received the entire shipment. He would also accept any returns that might be going from the store back to the distribution center and then close the transaction by securing the vehicle door and entering the seal number. When he got back into the vehicle, he connected the PDT8100 to its base station, and the information from the transaction was transmitted via the Mobitex wireless network into Microlise's Transport Management Center. From there, the data was forwarded to the Savi SmartChain platform where the asset movement history was recorded. The driver would then move on to the next drop and the process was repeated. This completed the audit trail.

The system could track dollies going to any of Woolworths' stores in the United Kingdom. The pilot, however, only equipped 15 trucks with the Symbol handheld computers with GPS transmitters. Consequently, the company could only track shipments to 30–40 stores.

The SmartChain tracking platform formed the focal point of the tracking process. Hence the existing enterprise systems also fed data into the SmartChain tracking platform. For instance, the tote-pick control system told Savi SmartChain which picks and which tote boxes were going in each store order. The transport management system provided the SmartChain software with information about which store order went on which vehicles when the vehicles were in the dispatch bay. The SmartChain platform brought all the data together to keep track of which picked items went into which tote, which tote went onto which dolly and which dolly onto which vehicle (the Russian Doll).

Woolworths purchased Savi's EchoPoint tags for each of the 16,000 dollies in the system. Although the original goal was to track only high-risk merchandise, like expensive clothes and CDs, the company decided to extend the system to all items shipped using the dollies. They found that they could afford the same level of protection to lower value merchandise as mobile phones and electronics.

In addition to the 15 PortaPOD mobile units used to relay real-time data back to the Transport Management Center, Woolworths also equipped two stores with fixed RFID units to track dollies from the distribution center. The mobile PortaPods units, however offered greater flexibility and could be used to track vehicles (and therefore their contents) via GPS throughout the trip from distribution center to the store.

20.5 Project Benefits

The project demonstrated the ability to integrate a unique combination of technologies (bar codes, short and long range RFID, wireless wide area networks, GPS, and existing order fulfillment systems) and deliver useful information and visibility to an extended supply chain. This was done in such a way as to be almost invisible to the user unless there was an error (almost all information was gathered automatically and only when there had been loading or delivery errors did the system notify the user).

Although the project covered only a small proportion of the goods delivered to the stores, it demonstrated the capability to have complete visibility of all goods from the moment they were picked, in transit, and delivered to the store. The project eliminated incorrect deliveries of dollies to the participating stores (i.e., process errors and potentially criminal activities) and also provided useful information in the event of a criminal investigation. The system had also been designed in such a way that it could easily be extended to cover more stores and also include merchandise in roll cages.

Woolworths identified six categories of potential benefits:

1. Shrinkage: Through better visibility of inventory and its whereabouts, process/delivery errors were identified and corrected on a real-time basis. The new system also provided an automated audit trail in the event of losses.
2. Bookstock Accuracy: A real-time, automated update of book stock within stores made stock records more accurate. This in turn enabled higher availability from lower inventory levels thereby improving customer service.
3. Reduced Labor Costs: The automated inventory verification process reduced manual check-in and updating of stock records. The increased accuracy also reduced the effort required to investigate stock losses.
4. Asset Management: Dollies, tote boxes, and roll cages are valuable assets themselves. The system provided greater visibility into their whereabouts, pinpointing blockages, and loss points. This tracking capability improved asset utilization and reduced unnecessary capital expenditures.
5. Transport Efficiencies: Automated tracking of vehicles not only generated a security benefit, but also improved vehicle routing, driver performance and training, and vehicle availability.

6. Identification of future RFID applications: By involving warehouse workers and drivers early in the pilot, the employees quickly felt ownership of the system and incorporated it into their everyday operations. Their excitement about the project led to many other suggested applications of RFID.

Developing a leading RFID application provided Woolworths with a platform from which it could learn about its further application, develop new processes and gather previously unavailable information about its inventory and its movements, its processes, and its assets. Woolworths recognized that no single initiative would provide a complete solution to eliminate shrinkage. This project did provide clear visibility to one area of potential shrinkage (the supply chain), reduced the opportunity for loss, and brought significant operational efficiencies. In conjunction with other initiatives, Woolworths felt that the RFID pilot produced significant reductions in shrinkage throughout the supply chain. It was, however, difficult to attribute quantifiable benefits to any individual component of the strategy.

20.6 Building a Business Case

The team at Woolworths realized that an initial success with RFID was no guarantee of future funding. The future of RFID at Woolworths depended on a strong business case—one that could stand up against other requests for investment. With the Kingfisher divestiture, Woolworths had gone from a company with a net income of £800 million to a net income of £25 million, and as a result, an investment of £3 million would be scrutinized. Given this environment, everyone was forced to compete for scarce funding resources in a company that traditionally viewed new store construction as the surest way to growth. The team would have to demonstrate that the £2–3 million earmarked for a typical new store would be better spent on infrastructure upgrades with a much more attractive ROI. Clearly, senior management would only fund the projects with the best return for shareholders.

Given the initial success, the team believed that a return on investment of less than one year was a realistic objective for a full-scale implementation. But they knew they had to be clear on where the savings would come from. The initial benefits found during the pilot project proposal were:

- Reduced Supply Chain Theft/Loss: The "Russian Doll" concept maintained a detailed audit trail of the merchandise as it moved through the supply chain and assigned inventory accountability to each participant (i.e., loading dock employee, delivery driver, receivables clerk). This accountability should not only serve as a deterrent, but also provide important evidence for any criminal investigation.

- Improved Vehicle Utilization: A recent piece of British legislation would require all commercial carriers to install electronic recording systems in their vehicles to ensure driver compliance with regulations governing daily driving time Working Time Directive. They viewed this requirement as an opportunity to enhance the required functionality with a GPS-enabled vehicle telemetric program. A Vehicle Telemetrics System would track and measure fuel economy, brake usage, and vehicle abuse in real-time. Preventive maintenance and measures (driving skills courses, driver evaluations, etc.) would be implemented to prolong vehicle lives and reduce vehicle downtime. Some estimates showed that transportation costs could be reduced by 8–10% using the smart system. The smart truck could be outfitted with a single black box that handled everything needed for RFID, telemetrics, and the Working Time Directive.

- Improved Asset Utilization: Both roll cages and dollies were expensive distribution assets, costing £100 and £40, respectively. Woolworths had approximately 100,000 roll cages throughout its distribution system. Individual stores sometimes hoarded extra roll cages as a safety stock or for other tasks throughout the store. Often the cages were simply forgotten, misplaced, or stolen. Each year, central logistics planners were forced to buy additional roll cages to prepare for the holiday season rush—typically 2–3% of the fleet was lost each year. Better asset tracking would allow planners to recall outstanding assets or chargeback any lost roll cages directly against the individual stores.

- Reduced Paperwork: The electronic tracking and signature system would eliminate the need for paper-based manifests and proof of receipt documents. This did not include the expected savings from resolving errors with manual data entry.
- Inventory and Availability: An additional area of potential savings that was difficult to quantify was the impact on inventory and availability. Inventory levels followed seasonal cycles, typically rising in the late summer and fall in preparation for Christmas. Woolworths achieved about 4.5 turns/year. Inventory was also linked to item availability in the store. With a more accurate stock count, availability could be improved or safety stock lowered or both. Retail studies had shown that a 1 point increase in availability could translate into 0.25–0.5% increase in sales.

The team estimated the cost of a larger deployment to be about £2–3 million, including:

- £1,000,000 for system hardware, including tags for all 100,000 roll cages, readers for the other three distribution centers, and the additional portable units for delivery drivers. The team felt confident that the tag cost would drop from the £8 they paid for the pilot units to under £5. That would leave £500,000 to purchase 100 dispatch bay readers, the necessary signposts, and the handheld devices for the trucks.
- £400,000 for Software Integration: The team believed that the majority of software capability and compatibility was built into the pilot system, and therefore only minimal efforts would be required to extend the capability to incorporate roll cages.
- £1,000,000 for the Vehicle Telemetric System: This was the cost of enhancing the mandatory delivery truck system with GPS-enabled vehicle performance monitoring and reporting capability.

20.7 Lessons from Woolworths

The pilot RFID implementation was viewed as technical success and won the *Supply Chain Solution of the Year Award* at the *European Retail Solutions Award Conference*. Reporters from both the United States and Europe visited the distribution center to see the system in action and hear how they had brought many leading edge technologies together to build the first such commercial tracking system. Yet, despite accolades from the press and the program's initial success, developing a clear business case to move forward with the other distribution centers was more difficult than many expected. Several of the potential benefits, while real, were simply not large enough. For example, the asset utilization and paper work reduction were clear, but amounted to a few hundred pounds. The vehicle utilization telemetrics appeared to have large potential, but the RFID and GPS system are only a part of the whole on-board Telemetrics project. It was hard to attribute much of the benefit to the RFID tracking project alone. Equally important, transportation had been outsourced, so determining who would invest and how the benefits would be shared was challenging.

Possibly one of the largest benefits was improved availability yielding improved customer service. Inventory record inaccuracy is a significant problem (Raman et al. 2001, DeHoratius and Raman 2004) that impacts the execution abilities of even the best retailers. Clearly, shrinkage results in inaccurate inventory records, making replenishment more difficult. Beyond losses, Woolworths also suffered from replenishment mistakes such as inventory delivered to the wrong store. This led to inaccuracy at both stores—one store with more of a product than reported in the inventory system and the other with less. A recent study of Wal-Mart's RFID pilot (Hardgrave et al. 2005) examined store "out of stocks" over a six-month period at 12 Wal-Mart stores—six using carton-level RFID and six control stores who had not implement RFID.

Throughout the study, store shelves were audited each day, and the number of SKUs without any stock (out of stocks) were recorded. Over the six-month period, both groups of stores reduced their out of stocks—with the RFID stores outperforming the control group. Controlling for the improvement in all stores, the authors of the study attributed a 16% reduction in out of stocks to RFID alone. They argued that RFID delivered this benefit by improving store inventory accuracy and the backroom and shelf stocking processes. For example, the high quality and timely inventory information make it possible to better track inventory in the store and move it more quickly to the shelf when needed.

Inventory inaccuracy and poor operational processes erode the stores' ability to reach the efficient frontier in transforming inventory into service (Fig. 20.3). Thus, stores are not able to fully convert inventory into service (shelf availability). With improved information and operating processes, the stores could either improve their service (without increasing inventory) or reduce their inventory (without reducing service).

While it seems likely that Woolworths could improve it's out of stocks, creating a direct link between inventory inaccuracy and product availability was a little more challenging. Managers argued that it was realistic to believe that improved accuracy could (modestly) yield a one-point improvement in availability (without increasing inventory). This would yield about £6.75 million of revenue (0.25% of 2.7B). With 28.5% gross margins, this translated into roughly £1.9 million of gross profit—thus less than a two-year payback on the project. However, the availability argument had been used for many improvement projects (such as bar codes, shipment audits, improved totes) in the past—so some in the management were skeptical.

Likewise, shrink reduction appeared to be a significant benefit. While the RFID project at Woolworths would not have a big impact on store losses (e.g., shoplifting), it certainly would impact the 56% of the total losses attributed to supply chain (56% of £75 million is £42 million). Just a 10% reduction in supply chain losses alone (£4.2 million) would pay for the project. This appeared to be the most compelling. Yet, Woolworths realized that no technology can eliminate theft, and thieves will certainly find ways to circumvent the tracking system. Thus the potential savings were hotly debated within the firm.

Along the potential benefits, the project also had several important risks:

1. Timing: Given the rapidly changing RFID landscape, such as lack of standards, rapidly dropping costs, and increasing technical capabilities, there were many incentives to wait. This was a key concern of Woolworths' executives. There was significant concern that buying too early would result in an expensive system with limited capabilities.
2. Cost: The full-scale rollout could cost far more than projected. One key area of concern was the software integration. Some had argued that integration work from the pilot could be leveraged

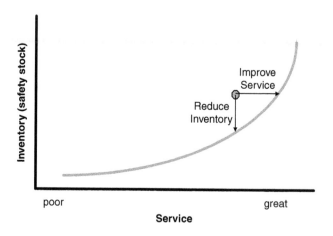

FIGURE 20.3   Using tracking systems to move to the efficient frontier.

resulting in only £400,000 requirement for full-scale integration. Many others argued that this was very optimistic given the rapidly evolving technologies. Moreover, there were some within Woolworths that argued that to fully capture the benefits of RFID, they needed to first replace their existing warehouse management systems and further redesign (or build a new) distribution center that was optimized for the tracking systems. This represented a roughly £50 million investment.

3. Feasibility: While the pilot was a technical success, there were still many uncertainties about the feasibility of a full-scale rollout—particularly reliable reads from the RFID devices in a wide range of operating environments.

4. Benefits: While the benefits looked compelling, the pilot could not fully validate the assumptions. Given the nature of the pilot, it was not possible to show (with hard data) that shrinkage would be reduced.

Faced with these risk, Woolworths decided to postpone a full-scale rollout for at least one to two years. The management felt that there were many other investment opportunities which appeared more compelling, including new stores and a new e-commerce operation—all multimillion pound investments. One executive commented, "Woolworths will invest in RFID when the time is right for us, but with trade as it is at present there are other things we must do first."

The Woolworths experience is not unique. While Wal-Mart appears to be showing progress in developing a compelling business case, many of their suppliers are still unclear if RFID and the tracking information will clearly benefit them (Overby 2005). Likewise, there are many firms who have conducted pilots, but have not yet felt confident enough in the business case to move to a full-scale rollout (Johnson et al. 2004). Given these difficulties in building a clear business case, one might wonder why RFID? Why not bar codes? In the following section, we will examine the benefits of RFID over bar codes.

20.8 Bar Codes and RFID

Radio Frequency Identification and other automated identification technologies have captured the imagination of engineers for years and held great promise in many applications. For example, many had dreamed of the RFID-enabled grocery store (Collins 2004), eliminating checkouts. However, the cost and capabilities have long made the concept unviable. Yet, RFID is steadily finding its way into everyday life, from automotive toll lanes that eliminate coins to building security systems that eliminate security guards. RFID offers many significant advantages over traditional bar code data collection. First, it does not require "line of sight" with the reader or even human interaction, so mis-reads are far less likely. Second, the tags can withstand harsh environments including rain, snow, and heat. Third, there are no moving parts that could be jarred loose. Fourth, the tags can hold vast amounts of information and be changed and reprogrammed. Fifth, RFID allows for simultaneous reading of multiple tags. And sixth, the data can be secured or locked. The key drawback in many applications remains its cost. For example, bar codes on consumer products could be implemented for less than $0.01 item. Many argue that the true costs are often hidden and that within a supply chain, bar coding for inventory and transportation management may cost as much as $0.10–$0.20 per read in direct labor and infrastructure cost.

For years, the common bar codes found on any product in a grocery or discount stores are the known as UPC A codes. It is simply a 12-digit number that has four parts. The first number denotes the type of bar code (e.g., "0" or "7" indicate the regular type found on most products, "2" indicates the product is a store weighted item like bulk food). The next five indicate the manufacturer and the following five identify the product code. Finally, the last number is called the checksum, which is a number used to do a simple test that determines if the code was read correctly (Fig. 20.4).

There are many other code types including UPC-E used on small products (takes less space) or EAN-13, which is the 13-digit European standard. EAN-13 is basically identical to UPC-A with the leading

UPC-A

1 (type) + 10 (mfg/prod) + 1 (checksum)

EAN-13

3 (country code) + 10 (mfg/prod) + 4 (supplements)

UPS
(MaxiCode)   93 characters

PDF-417   100's-1000's
                                              characters

FIGURE 20.4   A selection of bar codes.

two digits representing the country code. The United States began migrating to EAN-13 in 2005. Two-dimensional codes (like UPS's maxicode used on packages holding up to 93 characters) or PDF-417 codes hold more information—in some cases hundreds of characters.

The simplest RFID devices hold a little more information than a bar code but are equally static. Typical passive devices (write once read many) hold less than 2KB of data with so-called simple devices containing a 96-bit number. Of course there are advantages like the tag robustness (i.e., it can be encased so it is not easily damaged by water or trauma) and there is no need for direct line of sight. More expensive passive and active tags allow for all kinds of options like the ability to store new information, update that information, and broadcast that information (active tags). For example, a simple passive device that can be embedded into a shipping label is being piloted by many manufactures and retailers (including Wal-Mart). These labels can easily be applied to the outside of a carton or to a pallet of goods.

Radio Frequency Identification devices operate at several different frequencies with the most common being low-frequency (around 125 KHz), high-frequency (13.56 MHz), and ultra-high-frequency or UHF (860–960 MHz). The different frequencies have different operating characteristics that make them more appropriate for specific applications. For example, low-frequency tags conserve power and are good for penetrating objects with high-water content, such as fruit while high-frequency tags are more effective with objects made of metal. Although there is much talk of a $0.05 chip, in 2006 Gen 2 passives (simple, 96-bit EPC, short range) generally cost US $0.20 to US $0.40. However, many chip manufacturers offered discounts on large quantity purchases (10M or more). Tags embedded in a thermal transfer label typically cost $0.40 or more (RFID 2006).

Of course, there are many possibilities with more sophisticated passive and active tags, such as including sensors for temperature, humidity, etc. Many such capabilities could be very useful—for example, in monitoring the temperature of meat in transit or monitoring a container for security breaches. But these

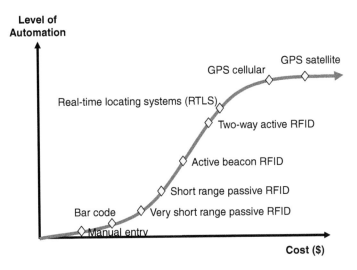

FIGURE 20.5 Cost and automation capabilities of tracking technologies.

capabilities all come with costs. The future of tracking supply chain technologies all hinge on the migration of the cost performance curve (Fig. 20.5). Clearly there are numerous applications as the costs are reduced. If technological developments and manufacturing volumes shift the curve back—making the adoption less expensive, the possibilities are nearly endless.

References

Bartels, N. (2005), RFID'S Tipping Point: Prices are Falling, but "Large Retailers" Worry about Impact on Vendor Viability, Manufacturing *Business Technology*, December.

Collins, J. (2004), Metro Readies RFID Rollout, *RFID Journal*, January.

DeHoratius, N. and A. Raman (2004), Inventory Record Inaccuracy: An Empirical Analysis, University of Chicago Working Paper.

Hardgrave, B., M. Waller, and R. Miller (2005), Does RFID Reduce Out of Stocks? A Preliminary Analysis, University of Arkansas Working Paper.

Johnson, M. E., M. Gozycki, and H. Lee (2004), Woolworths "Chips" Away at Inventory Shrinkage through RFID Initiative, Tuck School of Business Case Study #6-0020, Tuck School of Business, Dartmouth College.

Johnson, M. E. and H. Lee (2002), Quad Wants to be a Savi Player in Agribusiness, Tuck School of Business Case Study #6-0015, Tuck School of Business, Dartmouth College.

Machalaba, D. and A. Pasztor (2004), Thinking Inside the Box: Shipping Containers Get Smart, *Wall Street Journal*, January 15, B1.

Raman, A., N. DeHoratius, and Z. Ton (2001), Execution: The Missing Link in Retail Operations, *California Management Review*, 43, 136–154.

Overby, C. (2005), The State of Manufacturer and Retailer Collaboration, Forrester Research Report, November 21.

RFID (2006), The Cost of RFID Equipment, *RFID Journal*, available at: http://www.rfidjournal.com, accessed on 3/29/06.

Yates, S. (2005), Mitigating the Risks of Moving to Gen 2 RFID, Forrester Research Report, June 30.

21

Electronic Connectivity and Software

Darren M. Scott
McMaster University

21.1 Introduction

Two major functions comprise logistics: physical distribution and materials management (Hesse and Rodrigue, 2004). Physical distribution includes all activities involved in moving goods from points of production to final points of sale and consumption, such as transportation services, transshipment and warehousing services, to name a few. Materials management, on the other hand, consists of all activities related to the manufacturing of goods at all stages of production along a supply chain. Such activities include, among others, production planning, demand forecasting, inventory management, and purchasing.

All activities relating to logistics generate information, which is collected, stored, managed, manipulated, analyzed, and communicated in digital form via information technologies. This information may be spatial, such as customer addresses, or aspatial, such as customer orders. To remain competitive in a global marketplace, companies involved in logistics must deploy information systems and communication technologies appropriate for their business processes. These technologies include geographic information systems (GISs), enterprise resource planning (ERP) systems, warehouse management systems (WMSs), transportation management systems (TMSs), and electronic data interchange (EDI)—the topics of this chapter. Although these technologies may be costly to implement, these costs must be weighed against long-term savings in both time and money. Also, customer service is enhanced through these technologies.

The remainder of this chapter is organized as follows. Since GISs have received little attention in the literature concerning their logistics capabilities, they are discussed in detail in the following section. Data models and other data structures to support logistics applications are emphasized, along with advanced solutions for several logistics problems. Following this, the remaining four information technologies noted above are discussed briefly. A case study is then presented emphasizing the successful implementation of a WMS at Gentec International, a Canadian company located in Markham, Ontario. Concluding remarks are found in the final section.

21.2 Geographic Information Systems

21.2.1 What Is a GIS?

A GIS specializes in the collection, storage, analysis, and communication of georeferenced information—that is, information associated with specific locations on the earth's surface. The first GIS, the Canada Geographic Information System, was developed in the mid-1960s to inventory Canada's land resources and their existing and potential uses (DeMers, 2005; Longley et al., 2005). Since that time, the GIS industry has grown with worldwide revenue forecast at U.S. $3.63 billion in 2006 from sales of software, hardware, services, and data products (Daratech, Inc., 2006). Growth has been fueled largely by a seemingly endless variety of new application domains adopting GIS technology to manage assets and to solve real-world, geographic problems. Today, these domains include, among many others, crime, emergency management, health care, hydrology, military, sustainable development, urban planning, utilities, and *transportation*. In fact, transportation applications, which include those pertaining to logistics, have become so commonplace in recent years that they are known as GIS-T (GIS for transportation) applications (Miller and Shaw, 2001; Thill, 2000; Waters, 1999). Further, an increasing number of such applications reflect two of the latest frontiers in GIS—real-time navigation and tracking (e.g., Bowman and Lewis, 2006) and web-based deployment (e.g., Hamilton Street Railway Company, 2006).

Such frontiers illustrate two important features of a contemporary, comprehensive GIS—one, it integrates technology and two, it is extendable via customized programming. Technological integration is of particular concern to GIS-T given that often-used algorithms, such as those for solving the traveling salesman problem (TSP), the transportation problem, and the problem of facility location, have been developed outside of a GIS. As shown in Table 21.1, some software vendors have integrated such

TABLE 21.1 Logistics Capabilities of GIS Software Developed by Selected Commercial Vendors

Vendor	Website	Software	Version	Logistics Capabilities[a]						
				LE	TE	BR	VR	AR	NFM	FLM
Caliper Corp	http://www.caliper.com/	TransCAD	4.0	*	*	*	*	*	*	*
		Maptitude	4.8	*	*	*				
ESRI, Inc.	http://www.esri.com/	ArcGIS	9.1	*	*	*				*
		ArcLogistics Route	3.0	*		*	*			
Intergraph Corp.	http://www.intergraph.com/	GeoMedia Suite	6.0	*	*	*				
MapInfo Corp.	http://www.mapinfo.com/	MapInfo Professional	8.5	*	*					
		MapInfo Routing J Server	3.0	*		*				

[a] LE = locating events, TE = tracking events, BR = basic routing, VR = vehicle routing, AR = arc routing, NFM = network flow modeling, FLM = facility location modeling.

* Specific logistics capabilities associated with the software package.

Source: Compiled by author from first-hand knowledge of software and, in some instances, information found in the websites given in the table.

algorithms into their products. Among the software choices available today, TransCAD offers the greatest off-the-shelf algorithm functionality for GIS-T in general and logistics in particular.

Another technology important for some GIS-T applications is the global positioning system (GPS), which is a satellite-based, radio-navigation system. This technology is necessary for real-time navigation and for tracking events in real-time across the earth's surface. Fortunately, many GIS software vendors have integrated into their products technology for exploiting real-time GPS data (e.g., xy coordinates, time). ArcGIS Tracking Analyst, an ArcGIS extension, is one such example.

Extendibility, technological integration, and tools for the effective management and analysis of vast quantities of geographic information, not to mention the widespread availability of digital information from government agencies (usually at no cost) and commercial vendors (e.g., CanMap RouteLogistics enhanced street files for Canada, which are developed and sold by DMTI Spatial),* suggest that GISs can play an important role in logistics management. In fact, their use in logistics is already growing (Sutton and Visser, 2004). Although there are many comprehensive texts written on GIS (e.g., DeMers, 2005; Lo and Yeung, 2007; Longley et al., 2005, to name a few), there are very few dealing with logistics applications.[†] This chapter rectifies this shortcoming by first discussing how data are represented digitally within a GIS for logistics applications. Following this, advanced solutions for several logistics problems are reviewed briefly. However, as shown in Table 21.1, the extent to which commercial software vendors have integrated these capabilities into their current software varies considerably from one GIS package to another.

21.2.2 Data Models and Other Data Structures for Logistics Applications

21.2.2.1 Vector Data Model

Distribution centers, warehouses, terminals, customers, fleets and networks are all examples of real-world phenomena that must be represented digitally within a GIS to support logistics applications. Conceptually, such phenomena fall under the discrete object view of the world as opposed to the continuous field view, which describes phenomena that vary continuously across space (e.g., elevation, precipitation). Discrete objects are identifiable through their well-recognized boundaries (Lo and Yeung, 2007; Longley et al., 2005), and are best represented digitally by the vector data model as opposed to the raster data model,[‡] which is better suited to the continuous field view of the world.

Although software vendors have developed many proprietary variants of the vector data model over the years,[§] they all share a common conceptual foundation wherein each real-world object corresponds to a specific type of geometric object—namely, a point, line or area. Like the paper map, the purpose of the vector data model is to specify precisely the locations of these simple objects by one or more georeferenced xy coordinate pairs. This implies that a point is encoded by one coordinate pair, a line by a series of ordered coordinate pairs and an area by a closed loop of ordered coordinate pairs. In turn, the coordinate pairs forming lines and areas are connected by straight lines—hence lines and areas are often referred to as polylines and polygons, respectively.

Real-world objects encoded as points, lines, and areas are typically referred to as features—a convention that is adopted here. Features of the same geometric and thematic type are grouped together to form a layer, which is sometimes referred to as a feature class. As illustrated in Figure 21.1, layers are the

---

* Website is <http://www.dmtispatial.com/index.htm> (verified September 2006).

† Notable exceptions include Chapter 11 of Miller and Shaw's (2001) book on GIS-T, and the work by Sutton and Visser (2004).

‡ Treatments of the raster data model can be found in comprehensive texts written on GIS (e.g., DeMers, 2005; Lo and Yeung, 2007; Longley et al., 2005, to name a few).

§ ESRI, Inc., for example, has introduced a new vector data model with each of its software packages: the coverage with Arc/Info, the shapefile with ArcView, and the geodatabase with ArcGIS.

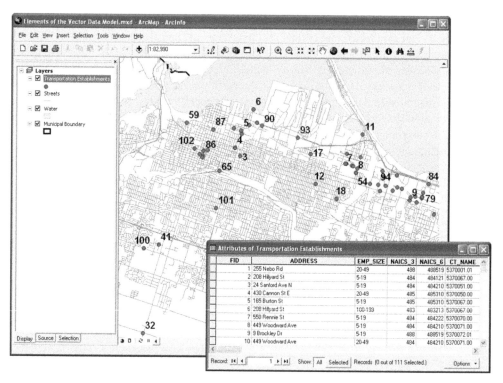

FIGURE 21.1 Example illustrating various elements of the vector data model. Each transportation establishment is labeled according to its unique feature identifier—that is, its FID value as shown in the attribute table.

basic organizational units for the vector data model. Besides geographic location, each feature in a layer is associated with one or more attributes, which describe characteristics of the feature. For instance, the transportation establishments shown in Figure 21.1 are characterized by several attributes including, among others, number of employees, street address, and 6-digit level North American Industry Classification System (NAICS) code.

The vector data model, as described earlier, has evolved since the early 1980s in accordance with advances in computer technology—notably, advances in hardware and a new approach to software engineering (i.e., object-oriented analysis and design, otherwise known as object orientation). This has given rise to several generic variants of the vector data model, which underlie the proprietary models developed by software vendors (e.g., ESRI's coverage, shapefile and geodatabase). These variants include the georelational data model and the object-based data model plus their derivatives.

The georelational data model has been used extensively in GIS for almost three decades. This model stores feature geometry, and therefore, locational information, separately from feature attributes. To be more precise, feature geometry is stored digitally in graphic files while associated attributes are stored in one or more tables of a relational database. The link between the two is maintained through a unique feature identifier. This linkage, which underpins the georelational data model, is illustrated by means of the transportation establishments shown in Figure 21.1. Specifically, the attributes found in each row of the attribute table are related to a specific point feature via a unique feature identifier, which is the value shown for FID in this case.

The georelational data model has two derivatives. On the one hand, features can be encoded as simple features—that is, spatial relationships among features forming a layer are not specified. This data model is known as the spaghetti or nontopological data model. On the other hand, features can be encoded as topological features—that is, spatial relationships among features forming a layer are specified explicitly. Obviously, this data model makes use of topology, which is a branch of mathematics that investigates qualitative properties of geometric objects that remain unchanged when the shape of an object is distorted in some way such as through twisting, stretching, or shrinking (Lawson, 2003). Specifically, three such properties are incorporated in the topological data model: (*i*) *adjacency*, which is information about neighboring features; (*ii*) *containment*, which is information about the inclusion of one feature within another; and (*iii*) *connectivity*, which is information about linkages among features (Lo and Yeung, 2007).

Adoption of the topological data model by software vendors predates adoption of its nontopological counterpart by almost a decade (Chang, 2008). However, despite its popularity among users, software vendors and data vendors due primarily to its interoperability across different GIS software packages, the nontopological data model, and more generally, the georelational data model are now being challenged by the object-based data model, which is a recent innovation. The object-based data model treats all features as objects. It differs from the georelational data model in two important ways. First, feature geometry is stored as an attribute along with other attributes in one file. In other words, there is no artificial separation between the spatial and attribute components of a feature. Second, features are associated with both attributes and methods. For example, in addition to attributes such as length and capacity, a link in a road network can have methods such as delete and compute congestion index (based on traffic flow and link capacity attributes). Like the georelational data model, the object-based data model has two derivatives based, once again, on the presence or absence of topology. In this case, however, topology, if implemented by the user, consists of relationship rules some of which apply to features in the same layer, while others apply to features in two or more participating layers.

21.2.2.2 Network Data Model

Networks are at the heart of many logistics applications including, among others, those pertaining to problems of routing and facility location. In a GIS, networks are modeled via the network data model, which is topological. Inclusion of topology among network features is necessary to facilitate the rapid processing of network algorithms, and thus, the timely solution to logistics, and more generally, GIS-T problems. Today, there are two variants of the network data model—one is georelational and the other, object-based. Despite this fact, the basic components of the data model remain the same—namely, links, nodes, and turns.

A link is simply a line feature defined by two end points or nodes. In a road network, for example, a link is a road segment between two intersecting roads. Each link forming a network has a unique feature identifier plus any number of other attributes. For most logistics applications, these attributes must include direction (i.e., one-way and two-way streets) and at least one measure of link impedance (i.e., free-flow travel time, congested travel time and/or physical length of the link), which may be directional. Other attributes may be necessary depending on the problem to be solved. Common attributes include link capacity, number of lanes, link type, speed and elevation (to model overpasses and underpasses), some of which may be directional.

A node is a point feature created where links intersect. Again, referring to the example of the road network, a node would occur where two roads intersect. Each node in a network is associated with a unique feature identifier. Other attributes can be included if they are necessary to solve a particular problem.

Unlike links and nodes, turns are not actual features in the network data model. Instead, turns occur at nodes and represent transitions between two links. The time it takes to make a turn is called turn impedance. Furthermore, turn impedance is directional. For example, in most road networks around the world, the time it takes to make a right-hand turn is typically less than the time it takes to complete a left-hand turn. Turns, although optional, are included in the network data model to achieve a greater correspondence with actual conditions affecting movement. In the case of road networks, for example, turn impedance can be significant during congested periods of the day, which suggests that the exclusion of such information may affect solutions to some logistics problems, especially those pertaining to routing. However, if included, turn impedance values, which are also known as turn penalties, can be assigned to the network in three ways: (*i*) as global penalties, which apply to the entire network; (*ii*) as link penalties, which apply to different types of links; and (*iii*) as specific penalties, which apply to specific pairs of links. The latter is accomplished via a turn table. Each row of the table contains information for a specific turn—namely, the pair of links involved in the turn and the turn penalty.

Networks for use in logistics and other GIS-T applications are derived from line layers such as street layers, rail layers, transit layers, and so on. The features in these layers are structured topologically according to the network data model supported by the GIS software being used. While proprietary object-based network data models are now emerging, such as the geodatabase network dataset of ArcGIS, most software packages still rely on the georelational network data model. In both cases, however, topology defines how links comprising a specific network are connected to one another at nodes. For links in the georelational network data model, these nodes are designated as *from nodes* and *to nodes* based on topological direction (see Fig. 21.2). An important difference between the geodatabase network dataset and the georelational network data model is that the former also allows for connectivity, via topological rules (specifically, connectivity groups), between different types of networks—for example, road networks and transit networks. In other words, the geodatabase network dataset can support intermodal applications whereas the latter cannot.

21.2.2.3 Routes

As mentioned at the outset of this chapter, one of the two major functions of logistics is physical distribution—that is, the movement of goods from points of production to final points of sale and consumption (Hesse and Rodrigue, 2004). Obviously, this implies that vehicles must be routed optimally through a network. Fortunately, many GIS software packages support such functionality (see Table 21.1), although the sophistication of problems handled varies considerably across packages. Recognizing the importance of routing to logistics and many other application domains, a few software vendors have even

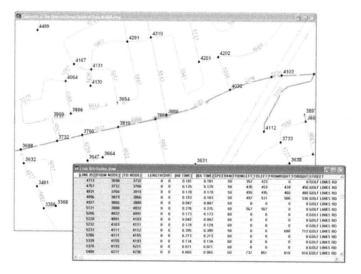

FIGURE 21.2 Example illustrating various elements of the georelational network data model: links (shown in gray), nodes (shown in black) and topology (i.e., connectivity of links at nodes as indicated by the FROM NODE and TO NODE fields in the link attribute table.

gone so far as to develop targeted software for tackling such problems only. Two examples of such software are ESRI's ArcLogistics Route and MapInfo's MapInfo Routing J Server.

In a GIS, a route is modeled as a composite feature—that is, it is composed of line features found in a line layer. However, unlike line features, such as street segments in a street layer, a route is associated with a linear measurement system that is stored with its geometry. This measurement system is used to locate linearly referenced data or events along a route. Linear referencing is a form of georeferencing whereby locations are specified using a distance measurement from a fixed reference point along a route, such as an intersection. Each route in a route system (i.e., a collection of routes) is identified via a unique feature identifier. Although routes can be created manually by the user, they are generated automatically by routing procedures, such as those discussed in Section 21.2.3. A necessary input to all such procedures is a network that has been structured topologically according to the network data model.

21.2.2.4 Matrices

A final data structure that is employed in many logistics applications including, among others, vehicle routing problems (VRP) and network flow problems (see Section 21.2.3), is the matrix. A matrix consists of rows and columns each of which is identified by a unique identifier that corresponds to a point or area feature in a layer. Each cell of a matrix (i.e., row and column intersection) contains a value measuring something about the row and column. For logistics applications, these values typically correspond to flows, distances, and travel times between places. Furthermore, values of impedance can be approximated via straight-line distances between locations or derived realistically via a network.

21.2.3 Logistics Applications

Geographic information systems can be used to solve a wide variety of logistics problems confronting businesses and other organizations. This section reviews briefly six advanced applications of said technology in the realm of logistics. However, as shown in Table 21.1, the extent to which commercial software vendors have integrated such capabilities into their current software varies considerably from one GIS package to another.

21.2.3.1 Locating Events

Geographic information systems offer several means for locating events in space, thus turning events into point features. An event defines the occurrence of some phenomenon at a particular location in space. This location can be specified by xy coordinates in any coordinate system (e.g., UTM coordinate system, latitude/longitude), as a distance measurement along a route from a fixed reference point (see Section 21.2.2.3) or as a street address. The important thing to note is that events are stored in event tables—that is, they are not associated with geometric point features. However, such features can be created easily from events. Moreover, such features are required in many logistics applications. For example, point layers containing depots and stops are a necessary input to VRP.

A powerful technique for creating such features is geocoding, which is arguably the most commercialized GIS-related operation across all application domains including logistics (Chang, 2008). The most common form of geocoding is address geocoding, also known as address matching, which is the process whereby new point features are created from addresses found in an event table. In logistics, for example, such event tables may correspond to suppliers, customers, warehouses, and distribution centers.

Address geocoding has three requirements: (*i*) an event table containing addresses; (*ii*) a reference layer, which is simply a street layer containing several necessary attributes—namely, street name, address ranges for both sides of the street defined according to topological direction, and typically, postal/ZIP codes defined in the same manner as address ranges; and (*iii*) a geocoding service, which specifies the event table and its relevant attributes, the reference layer and its relevant attributes, and various geocoding rules and tolerances. The geocoding engine matches each address against the street layer and uses a linear interpolation to locate the address within a specific address range. A point feature is then created for the event.

Figure 21.2 can be used to illustrate how address geocoding works. Assume that the address to be geocoded is 507 Golf Links Rd. The geocoding engine first matches the address to the line feature with a LINK ID of 4896. The topological direction of this feature is from node 3819 to node 3866 (as shown by the direction of the arrow), which means that the address range for the right-hand side of the street is from 506 to 536 (i.e., even numbers) and for the left-hand side, from 497 to 531 (i.e., odd numbers). 507 is an odd number and falls within the address range specified for the left-hand side of the street. The geocoding engine locates the point feature for the address on the correct side of the street (i.e., left) about 30% of the segment's length from node 3819.

21.2.3.2 Tracking Events

Real-time navigation and tracking events in real-time across the earth's surface have become increasingly important activities in logistics, not to mention other GIS application domains (Hallmark, 2004; Rose, 2004). These activities are possible because of the GPS, which is a satellite-based, radio-navigation system originally developed by the United States Department of Defense. Fortunately, many GIS software vendors have integrated into their products technology for exploiting real-time GPS data (e.g., *xy* coordinates, time). ArcGIS Tracking Analyst, an ArcGIS extension, is one such example.

21.2.3.3 Vehicle Routing

As can be seen in Table 21.1, virtually all software packages shown can solve basic routing problems, in particular the shortest path problem and, to a lesser extent, the traveling salesman problem (TSP). However, only two, TransCAD and ArcLogistics Route, can solve the more complicated vehicle routing problem (VRP), which is essentially a fleet version of the TSP. Solving the classic VRP involves determining how many vehicles, along with their routes and schedules, are required to service a set of demand locations or stops (e.g., customers) from a supply location or depot (e.g., warehouse, distribution center). The routes obtained through the VRP minimize the total travel time or distance traveled by the entire fleet of vehicles. Many variations of the classic VRP can be handled by TransCAD and ArcLogistics Route including the dispatching of vehicles from multiple depots, restrictions on the timing of stops, restrictions on the timing of vehicle dispatches, fixed and variable service times at stops, restrictions on route length or route duration, backhaul stops, mixed pick up and delivery, and open-ended routes (i.e., a vehicle does not return to the depot from which it was dispatched).

21.2.3.4 Arc Routing

Of the commercial GIS software packages reviewed for their logistics capabilities, only TransCAD is able to solve the arc routing problem (ARP), which involves dispatching people or vehicles from one or more depots to traverse a set of service links in a network. The solution to an ARP is a set of one or more routes that cover all such links with a minimal amount of deadheading (i.e., parts of a route that do not require service), which can be measured in terms of distance or time. Only one route is created per depot. If necessary, these routes can be split into shifts or parts by the arc routing procedure. Each shift can then be assigned to a specific person or vehicle. Arc routing can be applied to many logistics problems including garbage collection, school bus routing, mail delivery, and other door-to-door operations (e.g., reading water or electric meters).

21.2.3.5 Network Flow Modeling

Again, TransCAD is the only commercial GIS software package reviewed that is capable of solving network flow problems pertaining to logistics. The objective of all such problems is to identify the most efficient way to ship goods from origins (e.g., warehouses, distribution centers) to destinations (e.g., retail outlets). The transportation problem, also known as the Hitchcock problem, seeks to distribute goods in such a way that the demand at all destinations is met, the supply at each of the origins is not exceeded and the overall cost of distributing the good is minimized. An important input to the problem is a cost matrix, which can be generated quite easily in a GIS as travel times via a network or as straight-line distances. The solution to the problem is a matrix, showing the quantity of a good to be shipped from each origin to each destination. This optimal solution will always consist of $m + n - 1$ nonzero flows, where m is the number of origins and n is the number of destinations. If a network-based cost matrix is used to solve the transportation problem then the volume of good that flows over each network link can be computed and displayed.

Another network flow problem that can be solved using TransCAD is the minimum cost flow problem, which is essentially a general version of the previous problem that accounts for some form of link capacity. Examples of such capacities include height and weight restrictions on bridges, or congested network links that increase travel time (Sutton and Visser, 2004). The solution to the problem is a link flow table containing the flow of a good in each direction along every link in a network. These flows can then be joined to a street layer and displayed.

21.2.3.6 Facility Location Modeling

Both ArcGIS (ArcInfo license) and TransCAD can be used to solve a variety of facility location problems, which vary in terms of their objectives. For example, problems encountered in the public sector are often of two basic types: (*i*) minimizing emergency response times, as is the case when locating fire or police stations in an urban area; and (*ii*) maximizing public welfare, as is the case when locating libraries or schools to serve a population. The objective of most private sector problems, on the other hand, is to either minimize costs or maximize profits. Indeed, this was the case for Proctor & Gamble, which used a proprietary facility location model (specifically, a *p*-median problem) interfaced with a GIS (i.e., MapInfo) to identify inefficient distribution centers. The subsequent closure of these centers contributed to U.S. $250 million in cost savings for the company (Camm et al., 1997).

While objectives may differ, all facility location problems involve siting a number of facilities, such as warehouses or distribution centers, among a set of candidate sites (e.g., available land parcels for local problems, cities for regional or national problems) to serve a set of demand locations, such as retail establishments or other customers. According to Larson and Odoni (1981), the objectives give rise to three general types of facility location problems: (*i*) *median* problems, which identify locations that minimize the total or average cost between facilities and demand locations; (*ii*) *center* or *minimax* problems, which minimize the maximum travel cost incurred based on travel to/from a facility; and (*iii*) *requirements* problems, which locate facilities according to some prespecified performance standard such as an emergency response time.

Although implementation may vary across commercial GIS software packages, the basic inputs to a facility location problem consist of a point or area layer containing existing (if applicable to the problem being solved) and potential facility locations, a point or area layer containing demand locations and a cost or profit matrix. More often than not, point locations are determined via geocoding and cost matrices are approximated by network travel times or distances, although straight-line travel times and distances can be used. The solution to the problem consists of the new facility locations (as determined from the candidate sites), the assignment of each demand location to an existing or new facility and the cost or profit associated with each assignment.

21.3 Other Information Technologies

As mentioned at the beginning of this chapter, GISs are but one type of information technology used in logistics management. Other technologies include ERP systems, WMSs, TMSs, and EDI. These technologies are described briefly in the following sections.

21.3.1   ERP Systems

The ERP system is a recent innovation, having evolved over the past decade from material requirements planning (MRP) and manufacturing resource planning (MRPII) systems—the traditional production and scheduling tools used by manufacturers (Koh and Saad, 2006). Today, ERP systems are deployed throughout an enterprise to manage and coordinate digital information from various internal sources such as finance, accounting, sales, planning, production, purchasing, human resources, and logistics and distribution. As described by Dilworth (2000), an ERP system "can record transactions throughout a company to enter an order, produce the order, ship the order and account for the payment." In other words, an ERP system integrates information from essential, if not all, company processes.

The deployment of an ERP system throughout an enterprise can be very expensive—even exceeding U.S. $100 million (Koh and Saad, 2006; Miller and Shaw, 2001). However, customer expectations for shorter delivery lead times, improved quality and reduced costs have made the adoption of such a system a must for many manufacturing enterprises (Koh and Saad, 2006). While such systems are common across large companies, their uptake by small- and medium-sized enterprises (SMEs) remains in its infancy as the implementation costs are difficult to justify despite the need to compete in business-to-business (B2B) and business-to-consumer (B2C) markets (Muscatello et al., 2003). Recognizing this issue, commercial software vendors have responded with mid-range, less complex systems such as All-in-One by SAP.

In recent years, the ERP software industry has undergone considerable consolidation as some vendors have sought to expand revenues by acquiring rivals. According to Bragg (2005), the emerging ERP landscape is characterized by three groups of competitors: SAP, the *Acquirers* and the *Independents*. In 2004, the Acquirers included such software vendors as Oracle, Microsoft Business Solutions, SSA Global and Infor—all of which had grown through acquisitions. In 2004, for example, Oracle acquired PeopleSoft, which had purchased JD Edwards one year prior for U.S. $10.3 billion (Bragg, 2005). In 2004, the Independents included such software vendors as Intentia, IFS, IBS, QAD, and Epicor.

21.3.2   Warehouse Management Systems

Over the past two decades, largely as a result of global competition and supply-chain concepts, warehousing has evolved from an unavoidable and costly goods-storage activity to become a critical activity in the supply chain (Faber et al., 2002). Today, warehouses are designed to facilitate the timely movement of goods from one part of the supply chain to another, thereby reducing order processing costs. The high performance of warehousing operations demanded in today's marketplace is achieved in large part through the implementation of a WMS, along with other complementary information technologies such as barcodes, radio frequency identification (RFID) tags and, in some instances, GPS (Faber et al., 2002; Stough, 2001).

A WMS is a computer-based information system designed to manage a warehouse's resources (i.e., space, labor, and equipment) and operations. The digital information collected by a WMS enables the efficient flow of goods within a warehouse from time of receipt to time of shipping. This is achieved by tracking goods as they enter, are handled and removed from storage. This tracking is made possible by digital encoders located on vehicles or hand-held by workers, which read barcodes or RFID tags placed on items. This information is then transmitted to the WMS, updating information on available goods. Faber et al. (2002) note that implementation of a WMS can increase productivity, reduce inventories, improve space utilization within a warehouse, and reduce errors.

Up to about the early 1990s, nearly all WMSs in use were tailor-made—that is, they were designed to meet the needs of a particular warehousing operation (Faber et al., 2002). Today, however, this situation is the exception rather than the norm due to the changing role of the warehouse in the supply chain. This role has facilitated the development of standard WMS solutions by commercial software vendors. These vendors either specialize in WMSs only, such as Radio Beacon, or have acquired or developed such software as an add-on to other software packages (ARC Advisory Group, 2006). The latter is especially true in the case of ERP vendors, which now offer WMS solutions as part of their comprehensive solutions to information management. Such vendors include SAP and Oracle.

21.3.3 Transportation Management Systems

Transportation management systems have evolved considerably since they first appeared commercially in the late 1980s. While their primary task remains unchanged, that is, managing the physical flow of goods to customers, it is characterized today by an unprecedented level of sophistication aimed at improving customer service while reducing shipping costs. For example, many contemporary TMSs are capable of managing shipping activities as a whole rather than responding to one order at a time. Consequently, after considering customer constraints, orders for multiple customers are consolidated automatically into fewer shipments bound for specific destination zones, thereby realizing significant time and cost savings.

Today, TMSs are no longer engineered as stand-alone software applications residing on single computers within companies. Instead, they are designed for the enterprise using client–server technology. In other words, a single TMS can be deployed to manage multiple shipping locations for the enterprise. Also, such systems can facilitate customer service via the Internet through websites for placing and tracking orders in real time.

Over the past decade, the TMS vendor landscape has changed significantly largely due to the fact that transportation is no longer seen as a stand-alone activity in the supply chain. Instead, transportation is viewed, rightfully so, as the "glue" that links multiple parties and business processes together. According to Gonzalez (2005), this change is responsible, at least in part, for the demise of traditional stand-alone TMS software vendors. In most cases, such vendors have been acquired by other software companies seeking to enter the TMS market, notably those specializing in ERP software solutions. For example, SSA Global acquired Arzoon in 2004 and Oracle acquired G-Log in 2005.

21.3.4 Electronic Data Interchange

Electronic data interchange has facilitated B2B commerce since the 1970s (Humphreys et al., 2006). EDI involves the electronic transmission of information and documents, such as invoices or purchase orders, between computer systems in different organizations based on a standard, structured, machine-retrievable format (Sánchez and Pérez, 2003). EDI standards, called *X.12*, are set by the American National Standards Institute (ANSI). These standards are specialized within industries. By automating the exchange of standardized documents, EDI eliminates repetitive data entry and mistakes due to human error. Other benefits of EDI include reduced costs, increased productivity, minimal paper use, and improved business relationships.

For the most part, EDI has been limited to larger companies due to the high costs associated with implementation (Humphreys et al., 2006). In 1999, for example, only 10% of warehouses used EDI to facilitate communications between trading partners (Brockmann, 1999). EDI has, however, seen widespread use in several industries, especially the automotive and retail sectors (Bhatt, 2001; Weber and Kantamneni, 2002).

EDI exchanges information between trading partners either through direct telecommunication links between computer systems (i.e., point-to-point) or though a value-added network (VAN). A VAN is a third-party communication service that sorts information into electronic partner mailboxes for eventual download. Obviously, this implies that real-time communication is not possible through such a network.

Despite its benefits, the adoption of EDI by companies has failed to meet expectations (Chau and Jim, 2002; Zacharia, 2001). This is especially true today as companies are increasingly embracing the Internet for B2B transactions.

21.4   Case Study: Gentec International and Radio Beacon WMS

Gentec International (Gentec),[*] located in Markham, Ontario, is the only Canadian-owned and operated importer and distributor of brand name optical, photographic imaging, digital imaging, wireless telecom, and electronic accessory products. The company supplies over 3000 such products to many retailers operating in Canada including, among others, Best Buy, Future Shop, Henry's, Sears, Wal-Mart, and Zellers. Since its founding in 1990, Gentec's sales have grown over 425%, making it the leading accessory supplier in Canada. This growth has been fueled in large part by Gentec's regular investments in seamless information technologies to facilitate customer service and to manage its daily operations. Today, these technologies include an ERP system (i.e., PointForce WinSol by PointForce,[†] an operating division of TECSYS),[‡] EDI (i.e., ProEDI by Encomium Data International),[§] a TMS (i.e., Clippership by Kewill),[‖] and a WMS (Radio Beacon by Radio Beacon Inc.).[#]

Although Gentec is a small company, boasting only 64 employees in 2004 (Mark, 2004), it embraced the Radio Beacon WMS soon after its founding. In fact, the initial system, which was a beta version at the time, was installed in 1993 to manage operations at Gentec's high-cube, state-of-the-art, 5300 square meter warehouse, also located in Markham. The Radio Beacon solution was chosen because it was affordable, windows-based and easily integrated with Gentec's ERP software. Furthermore, Radio Beacon was willing to incorporate suggestions by Gentec and other users in subsequent upgrades of its product.

In 2003, Gentec updated its WMS with Radio Beacon for Microsoft SQL Server. At the same time, the company also deployed the Radio Beacon Demand Forecasting Module, an add-on to the Radio Beacon WMS, in an effort to improve its ability to meet the cyclical demand of its customers. In essence, the module analyzes customer order histories and uses basic replenishment and safety stock calculations to forecast product quantities necessary to meet demand over a given period. This forecasting module has helped Gentec meet its goal of filling and shipping orders within 24 h. In fact, Gentec guarantees that an order will leave its warehouse no later than 48 h after it is placed.

21.5   Conclusion

This chapter has reviewed five information technologies (i.e., GISs, ERP systems, WMSs, TMSs, and EDI) that are deployed by companies to facilitate their logistics-related activities. These technologies are used to collect, store, manage, manipulate, analyze and communicate digital information, which can be either spatial or aspatial. Moreover, many of these technologies can now be integrated seamlessly to support information exchange among business processes within a company. Also, EDI enables information exchange between companies (i.e., trading partners) through directly-linked computers or VANs.

The move toward seamless integration appears to be the primary reason for consolidation in the software industry. Specifically, commercial ERP software vendors are entering the WMS and TMS markets either by developing their own products or by acquiring companies specializing in such software solutions. Similar consolidation is not the case for the GIS market, which continues to be dominated by a few large vendors. Instead, commercial GIS vendors have responded to the need for logistics software by developing specialized products, such as TransCAD and ArcLogistics Route.

---

[*] Website is <http://www.gentec-intl.com/> (verified September 2006).
[†] Website is <http://www.pointforce.com/index.asp> (verified September 2006).
[‡] Website is <http://www.tecsys.com/index.shtml> (verified September 2006).
[§] Website is <http://www.proedi.com/default.asp> (verified September 2006).
[‖] Website is <http://www.kewill.com/shipping/index.asp>(verified September 2006).
[#] Website is <http://www.radiobeacon.com/>(verified September 2006).

It is clear that the information technologies reviewed in this chapter support the competitiveness of companies involved in logistics activities by enhancing customer service and reducing costs. Gentec International is a case in point. At the same time, however, information technology changes rapidly. In turn, this means that the most successful companies will be those that can adapt to such changes by investing regularly in information systems and communication technologies to support their business processes.

References

ARC Advisory Group, Warehouse management worldwide outlook: five year market analysis and technology forecast through 2010, available from <http://www.arcweb.com/StudyPDFs/Study_wms.pdf>, 2006 (verified September 2006).

Bhatt, G., Business process improvement through electronic data interchange systems: an empirical study, *Supply Chain Management: An International Journal*, 6, 60, 2001.

Bowman, D. and Lewis, D., Sears Holdings Corporation deploys GIS navigation and mapping system, available from <http://www.esri.com/news/arcnews/winter0506articles/sears-holdings.html>, 2006 (verified September 2006).

Bragg, S., ERP: state of the industry, available from <http://www.arcweb.com/Featured%20Research/ERP-The%20State%20of%20the%20Industry.pdf>, 2005 (verified September 2006).

Brockmann, T., 21 warehousing trends in the 21st century, *IIE Solutions*, 31, 36, 1999.

Camm, J.D., Chorman, T.E., Dill, F.A., Evans, J.R., Sweeny, D.J., and Wegryn, G.W., Blending OR/MS, judgment, and GIS: restructuring P&G's supply chain, *Interfaces*, 27, 128, 1997.

Chang, K.-T., *Introduction to Geographic Information Systems*, 4th ed., McGraw Hill, New York, 2008.

Chau, P.Y.K. and Jim, C.C.F., Adoption of electronic data interchange in small and medium-sized enterprises, *Journal of Global Information Management*, 10, 61, 2002.

Daratech, Inc., GIS/geospatial market grew 17% in 2005 to top $3.3 billion; sales led by growth in data products, available from <http://www.daratech.com/press/releases/2006/060706.html>, 2006 (verified September 2006).

DeMers, M.N., *Fundamentals of Geographic Information Systems*, 3rd ed., John Wiley & Sons Inc., New York, 2005.

Dilworth, J.B., *Operations Management: Providing Value in Goods and Services*, 3rd ed., Dryden Press, Fort Worth, 2000.

Faber, N., de Koster, R.B.M., and van de Velde, S.L., Linking warehouse complexity to warehouse planning and control structure: an exploratory study of the use of warehouse management information systems, *International Journal of Physical Distribution and Logistics Management*, 32, 381, 2002.

Gonzalez, A., Trends and predictions in the transportation management systems market, available from <http://www.arcweb.com/Featured%20Research/Trends%20and%20Predictions%20in%20the%20Transportation%20Management%20Systems%20Market.pdf>, 2005 (verified September 2006).

Hallmark, S.L., Other transportation applications of GPS, in *Handbook of Transport Geography and Spatial Systems*, Hensher, D.A., Button, K.J., Haynes, K.E., and Stopher, P.R., Eds., Elsevier, Oxford, 2004, 489.

Hamilton Street Railway Company, HSR trip planning, available from <http://www.busweb.hamilton.ca:8008/>, 2006 (verified September 2006).

Hesse, M. and Rodrigue, J.-P., The transport geography of logistics and freight distribution, *Journal of Transport Geography*, 12, 171, 2004.

Humphreys, P., McIvor, R., and Cadden, T., B2B commerce and its implications for the buyer-supplier interface, *Supply Chain Management: An International Journal*, 11, 131, 2006.

Koh, S.C.L. and Saad, S.M., Managing uncertainty in ERP-controlled manufacturing environments in SMEs, *International Journal of Production Economics*, 101, 109, 2006.

Larson, R.C. and Odoni, A.R., *Urban Operations Research*, Prentice Hall, Englewood Cliffs, 1981.

Lawson, T., *Topology: A Geometric Approach*, Oxford University Press, New York, 2003.

Lo, C.P. and Yeung, K.W., *Concepts and Techniques of Geographic Information Systems*, 2nd ed., Upper Saddle River, Pearson Prentice Hall, 2007.

Longley, P.A., Goodchild, M.F., Maguire, D.J., and Rhind, D.W., *Geographic Information Systems and Science*, 2nd ed., John Wiley & Sons Ltd., Chichester, 2005.

Mark, K., Making radio waves: Canada's Gentec future-proofs, available from <http://www.bizlink.com/purchasingfiles/issue-2-4.html>, 2004 (verified September 2006).

Miller, H.J. and Shaw, S.-L., *Geographic Information Systems for Transportation: Principles and Applications*, Oxford University Press, New York, 2001.

Muscatello, J.R., Small, M.J., and Chen, I.J., Implementing ERP in small and midsize manufacturing firms, *International Journal of Operations and Production Management*, 28, 850, 2003.

Rose, G., GPS and vehicular travel, in *Handbook of Transport Geography and Spatial Systems*, Hensher, D.A., Button, K.J., Haynes, K.E., and Stopher, P.R., Eds., Elsevier, Oxford, 2004, 451.

Sánchez, A.M. and Pérez, M.P., The use of EDI for interorganizational co-operation and co-ordination in the supply chain, *Integrated Manufacturing Systems*, 14, 642, 2003.

Stough, R.R., New technologies in logistics management, in *Handbook of Logistics and Supply Chain Management*, Brewer, A.M., Button, K.J., and Hensher, D.A., Eds., Pergamon, Oxford, 2001, 513.

Sutton, J.C. and Visser, J., The role of GIS in routing and logistics, in *Handbook of Transport Geography and Spatial Systems*, Hensher, D.A., Button, K.J., Haynes, K.E., and Stopher, P.R., Eds., Elsevier, Oxford, 2004, 357.

Thill, J.C., Ed., *Geographic Information Systems in Transportation Research*, Pergamon, New York, 2000.

Waters, N.J., Transportation GIS: GIS-T, in *Geographical Information Systems, Volume 2: Management Issues and Applications*, 2nd ed., Longley, P.A., Goodchild, M.F., Maguire, D.J., and Rhind, D.W., Eds., John Wiley & Sons Inc., New York, 1999, 827.

Weber, M. and Kantamneni, S., POS and EDI in retailing: an examination of underlying benefits and barriers, *Supply Chain Management: An International Journal*, 7, 311, 2002.

Zacharia, Z.G., The evolution and growth of information systems in supply chain management, in *Supply Chain Management*, Mentzer, J.T., Ed., Sage Publications, London, 2001, 289.

22

Reliability, Maintainability, and Supportability in Logistics

C. Richard Cassady
Edward A. Pohl
University of Arkansas

Thomas G. Yeung
Ecole des Mines de Nantes/IRCCyN

22.1 Introduction

The effective operation of any logistics system requires the utilization of many types of equipment, including trucks, aircraft, forklifts, computers, etc. Therefore, equipment integrity is fundamental to success relative to logistic system performance. Equipment integrity refers to equipment performance as it relates to reliability, maintainability, and supportability.

22.2 Basic Reliability Concepts

Although the word reliability has many colloquial interpretations, equipment reliability has a very precise, technical definition. Reliability is the probability that a unit of equipment correctly performs its intended function over a specified period of time when operated in its design environment. There are four important features of this definition: intended function, design environment, probability, and time. In the study of reliability, it is most always assumed that the unit of equipment is being used for its intended function and in its design environment. Therefore, our focus in this chapter is on the probabilistic and time-dependent aspects of the definition.

Although reliability is typically discussed at the equipment level, most units of equipment are comprised of hundreds of thousands of components. Reliability problems tend to occur at the component level. Therefore, reliability analysts typically study component reliability and aggregate component reliability methods at the equipment level.

Consider a unit of equipment that is comprised of n components, and consider some component $i \in \{1, 2, \ldots, n\}$. Let S_i denote the event that component i functions properly over the specified period of time, and let

$$R_i = P_r(S_i = 1) \qquad (22.1)$$

denote the equipment reliability of component i. Clearly, equipment reliability is a function of the individual component reliabilities. The nature of this function depends upon the functional relationships between the components. In most fundamental studies of equipment reliability, it is assumed that the individual components are probabilistically independent. We make this assumption, and we consider three possible physical arrangements of the components that comprise a unit of equipment: series, parallel, and everything else.

A series system is one in which all components must function properly in order for the unit of equipment to function properly. We represent this and all other system structures graphically using reliability block diagrams. The reliability block diagram for a three-component series system is given in Figure 22.1. Note that the connectors between components in a reliability block diagram do not necessarily imply physical connections.

For a series system, all components must function properly in order for the unit of equipment to function properly. Therefore,

$$S = \bigcap_{i=1}^{n} S_i \qquad (22.2)$$

and

$$R = \Pr(S) = \Pr\left(\bigcap_{i=1}^{n} S_i\right). \qquad (22.3)$$

Since the components are independent,

$$R = \prod_{i=1}^{n} \Pr(S_i) = \prod_{i=1}^{n} R_i. \qquad (22.4)$$

Consider a three-component series system having component reliabilities $R_1 = 0.98$, $R_2 = 0.93$, and $R_3 = 0.97$. For this system,

$$R = (0.98)(0.93)(0.97) = 0.884058. \qquad (22.5)$$

A parallel, or redundant system is one in which the proper function of at least one component implies proper function of the unit of equipment. The reliability block diagram for a three-component parallel system is given in Figure 22.2.

For a parallel system, the unit of equipment does not operate properly only if all components do not operate properly. Therefore,

$$S^c = \bigcap_{i=1}^{n} S_i^c \qquad (22.6)$$

FIGURE 22.1   Three-component series system.

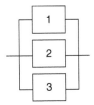

FIGURE 22.2 Three-component parallel system.

where the superscript c on an event denotes its complement. Thus,

$$\Pr(S^c) = \Pr\left(\bigcap_{i=1}^{n} S_i^c\right) = \prod_{i=1}^{n} \Pr(S_i^c) = \prod_{i=1}^{n}(1 - R_i) \tag{22.7}$$

and

$$R = \Pr(S) = 1 - \Pr(S^c) = 1 - \prod_{i=1}^{n}(1 - R_i). \tag{22.8}$$

For the somewhat common case in which the components are identical, then R_i denotes the reliability of any component and

$$R = 1 - (1 - R_i)^n. \tag{22.9}$$

Consider a three-component parallel system having component reliabilities $R_1 = 0.98$, $R_2 = 0.93$, and $R_3 = 0.97$. Then,

$$R = 1 - (1 - 0.98)(1 - 0.93)(1 - 0.97) = 0.999958. \tag{22.10}$$

It can be shown that any system configuration can be reduced to an equivalent configuration which combines the series and parallel system configurations. In some cases, this reduction results in a loss of component independence. However, we only consider here combined series-parallel configurations in which all components in the unit of equipment operate independently. Consider the system represented in Figure 22.3. Suppose $R_1 = 0.98$, $R_2 = R_3 = 0.91$, $R_4 = R_5 = 0.94$, $R_6 = 0.99$, $R_7 = R_8 = R_9 = 0.78$.

Let $R_{2,3}$ denote the reliability of the subsystem comprised of components 2 and 3. Then,

$$R_{2,3} = 1 - (1 - 0.91)^2 = 0.9919. \tag{22.11}$$

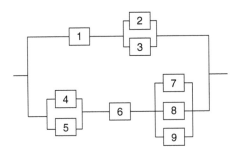

FIGURE 22.3 Example combined series-parallel system.

Likewise,

$$R_{4,5} = 1 - (1 - 0.94)^2 = 0.9964 \tag{22.12}$$

and

$$R_{7,8,9} = 1 - (1 - 0.78)^3 = 0.989352. \tag{22.13}$$

Let $R_{1,2,3}$ denote the reliability of the subsystem comprised of components 1, 2, and 3. Then,

$$R_{1,2,3} = (0.98)(0.9919) = 0.972062. \tag{22.14}$$

Likewise,

$$R_{4,5,6,7,8,9} = (0.9964)(0.99)(0.989352) = 0.975932. \tag{22.15}$$

Finally,

$$R = 1 - (1 - 0.972062)(1 - 0.975932) = 0.999328. \tag{22.16}$$

22.3   Design for Reliability

Designing for reliability involves a process known as reliability allocation. Upon specification of the overall reliability goal of the system, the determination must then be made as to how to allocate reliability to the various components and possibly subcomponents of the system to ensure the specified reliability goals are met. Let R^* denote the desired reliability of the system. Then reliability allocation should occur such that the following inequality holds

$$R^* \leq h(R_1, R_2, \ldots, R_n) \tag{22.17}$$

where h represents a function that aggregates the individual component reliabilities into the total system reliability.

Reliability allocation is an optimization problem in which some objective is optimized (generally minimizing cost) such that the minimum reliability constraint is not violated. If the components of a system are assumed to be serially related, then solving the following mathematical program seeks to optimize the reliability allocation process.

minimize

$$\sum_{i=1}^{n} c_i x_i \tag{22.18}$$

subject to:

$$\prod_{i=1}^{n} (R_i + x_i) \geq R^* \tag{22.19}$$

$$0 < R_i + x_i \leq b_i < 1 \qquad i = 1, 2, \ldots, n \tag{22.20}$$

where R_i denotes the current reliability of component i, x_i denotes the increase in reliability of component i, c_i denotes the per-unit cost of obtaining this increase, and b_i denotes the upper bound on the increase in reliability that can be obtained from component i. This optimization is nonlinear due to the multiplication of decision variables in the first constraint. Sophisticated solution procedures such as Langrangian relaxation must be utilized to obtain only approximate solutions to large-scale problems. This highlights the difficulty in using optimization in allocating reliability.

Because of the difficulty in using optimization in reliability allocation, several methods have been developed that do not require optimization. Two of the most popular methods are the ARINC and AGREE methods. For a summary of these methods, see Ebeling (1997).

A subset of reliability allocation involves redundancy allocation. In the previous section it was assumed that components were serially related and that the reliability of component i could be increased by x_i in some manner up to a bound b_i. In practice, the reliability of a component may be fixed and cannot be increased through the component itself, but rather through the addition of redundant components that are related in parallel. Thus, the determination must be made as to how many types of each component are necessary to achieve desired component reliability. Consider a situation in which optimization was performed in reliability allocation and it was determined that component 4 should have a minimum reliability of 0.95. Furthermore, assume that component 4 currently only has a reliability of 0.70, and that the reliability of the component itself cannot be improved. However, by adding additional component 4s related in parallel, we can increase the reliability to the desired level of 0.95. Thus, from Equations 22.8 and 22.16 we get

$$0.95 \geq 1-(1-0.70)^n, \tag{22.21}$$

where n denotes the number of components necessary to achieve a reliability of 0.95. Solving for n, we find that 2.49 components are required. However, since fractional components are not available we take the ceiling of the answer. Thus, three components are required to achieve at least a 0.95 reliability for component 4.

22.4 Time-Dependent Reliability

To this point, our analysis of equipment reliability has focused on a specific interval of time. A more useful analysis of equipment integrity considers time to failure as a random variable. Let T denote the time to failure of the equipment (or component) of interest (note that $T \geq 0$). The reliability function for the system is given by

$$R(t) = \Pr(T > t) \cdot \tag{22.22}$$

This implies a direct relationship between the reliability function and the cumulative distribution function (CDF) of T, that is,

$$F(t) = \Pr(T \leq t) = 1 - R(t) \cdot \tag{22.23}$$

The probability density function (PDF) of T is thus given by

$$f(t) = \frac{dF(t)}{dt} = -\frac{dR(t)}{dt} \cdot \tag{22.24}$$

Given the PDF, we can construct the CDF and the reliability function using

$$F(t) = \int_0^t f(u) \, du \tag{22.25}$$

and

$$R(t) = \int_t^\infty f(u)du \cdot \tag{22.26}$$

These three functions permit three equivalent methods for computing failure probabilities:

$$\Pr(a \le T \le b) = F(b) - F(a) = R(a) - R(b) = \int_a^b f(t)dt \cdot \tag{22.27}$$

The mean time to failure (MTTF) is given by

$$\text{MTTF} = E(T) = \int_0^\infty t\, f(t)dt = \int_0^\infty R(t)\, dt \cdot \tag{22.28}$$

Consider a unit of equipment that has survived until time t. The probability that it fails in the next instant of time Δt is given by

$$\Pr(T \le t + \Delta t | T > t) = \frac{\Pr(T \le t + \Delta t, T > t)}{\Pr(T > t)} = \frac{\Pr(t < T \le t + \Delta t)}{\Pr(T > t)} = \frac{R(t) - R(t + \Delta t)}{R(t)}. \tag{22.29}$$

The hazard (failure) rate function provides the instantaneous failure rate at time t.

$$\lambda(t) = \lim_{\Delta t \to 0} \frac{\Pr(T \le t + \Delta t | T > t)}{\Delta t} = \lim_{\Delta t \to 0} \frac{R(t) - R(t + \Delta t)}{R(t)\Delta t} = \lim_{\Delta t \to 0} \frac{-[R(t + \Delta t) - R(t)]}{\Delta t} \frac{1}{R(t)} \tag{22.30}$$

$$\lambda(t) = -\frac{dR(t)}{dt} \frac{1}{R(t)} = \frac{f(t)}{R(t)} \tag{22.31}$$

The hazard rate function provides an alternative way of characterizing a failure distribution. Hazard rate functions are often classified as increasing failure rate (IFR), decreasing failure rate (DFR), or constant failure rate (CFR). Obviously, these classifications depend on the behavior of the hazard rate function over time. Also, note that

$$R(t) = e^{-\int_0^t \lambda(u)du} . \tag{22.32}$$

One form of the hazard rate function that is widely recognized is the bathtub curve (see Fig. 22.4). The bathtub curve is most often used as a conceptual model rather than a mathematical model, and the evolution of the bathtub curve is summarized as follows. Early in a unit of equipment's life, failures occur primarily due to manufacturing defects. Over time, these "defective" units are fewer in number so the overall hazard rate decreases. This portion of the bathtub curve is often referred to as the "infant mortality" period. Once the early failures have stopped, the system enters its useful life. During this time, failures are purely random, so the hazard rate is constant. At the end of its useful life, the system begins to wear out and the hazard rate increases.

Perhaps the most popular failure distribution assumed in reliability is the CFR model. This model assumes that failures are purely random, and have an equal probability of occurring at every instant in

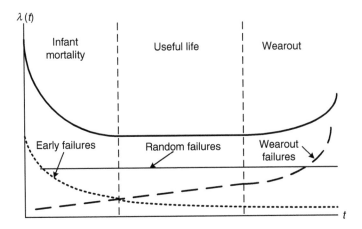

FIGURE 22.4 The bathtub curve. (Modified from Ebeling, C. E., *An Introduction to Reliability and Maintainability Engineering*, McGraw-Hill, Boston, 1997.)

time. Its popularity is due in large part to the ease in analyzing systems mathematically under the assumptions of this distribution. Suppose a unit of equipment's hazard rate function is CFR, that is,

$$\lambda(t) = \lambda \cdot \tag{22.33}$$

This implies

$$R(t) = \exp\left[-\int_0^t \lambda(u)du\right] = \exp\left[-\int_0^t \lambda \, du\right] = \exp(-\lambda u\big|_0^t) = e^{-\lambda t} \tag{22.34}$$

$$F(t) = 1 - R(t) = 1 - e^{-\lambda t} \tag{22.35}$$

$$f(t) = \frac{dF(t)}{dt} = \lambda e^{-\lambda t}. \tag{22.36}$$

Therefore, the CFR case corresponds to an exponential time to failure distribution. Note also that

$$\text{MTTF} = \int_0^\infty R(t)dt = \int_0^\infty e^{-\lambda t} \, dt = -\frac{1}{\lambda}e^{-\lambda t}\Big|_0^\infty = -\frac{1}{\lambda}(0-1) = \frac{1}{\lambda}. \tag{22.37}$$

The constant hazard rate function implies that the system does not become any more or less reliable over time.

Other than the exponential, the Weibull time to failure distribution is by far the most widely used and recognized time to failure distribution. In fact, research and industrial applications of the Weibull distribution are limited almost exclusively to reliability and maintainability (RAM) problems. The Weibull distribution is widely used because of its ability to capture IFR, DFR, and CFR (although the exponential already handles this case) behavior.

The hazard rate function of the Weibull failure distribution has the general form

$$\lambda(t) = \frac{\beta}{\eta}\left(\frac{t}{\eta}\right)^{\beta-1} \tag{22.38}$$

where $\eta > 0$, $\beta > 0$. Note that $\beta > 1$ implies IFR, $\beta = 1$ implies CFR and $\beta < 1$ implies DFR. Next,

$$R(t) = \exp\left[-\int_0^t \frac{\beta}{\eta}\left(\frac{u}{\eta}\right)^{\beta-1} du\right] = \exp\left[\left(\frac{u}{\eta}\right)^{\beta}\bigg|_0^t\right] = e^{-(t/\eta)^{\beta}}, \tag{22.39}$$

$$f(t) = -\frac{dR(t)}{dt} = \frac{\beta}{\eta}\left(\frac{t}{\eta}\right)^{\beta-1} e^{-(t/\eta)^{\beta}}, \tag{22.40}$$

and

$$\text{MTTF} = \eta\Gamma\left(1 + \frac{1}{\beta}\right) \tag{22.41}$$

where $\Gamma(x)$ is the gamma function evaluated at x. Since β affects the shape of the PDF, it is referred to as the shape parameter and is unit less. The scale parameter η is referred to as the characteristic life and has the same units as T, the time to failure.

22.5 Repairable Systems Modeling

Many types of equipment utilized in logistics systems can be repaired after failure. Constructing mathematical models of repairable system performance and using these models to optimize maintenance strategies require a basic understanding of several key reliability and maintainability concepts and mathematical modeling approaches. A *repairable system* (RS) is a system which, after failure, can be restored to a functioning condition by some maintenance action other than replacement of the entire system. Note that replacing the entire system may be an option, but it is not the only option.

Maintenance actions performed on a RS can be categorized into two groups: corrective maintenance (CM) actions and preventive maintenance (PM) actions. CM actions are performed in response to system failures, and they could correspond to either repair or replacement activities. PM actions are not performed in response to RS failure, but they are intended to delay or prevent system failures. Note that PM actions may or may not be cheaper and/or faster than CM actions. As with CM actions, PM actions can correspond to either repair or replacement activities. Finally, operational maintenance actions, for example, putting gas in a vehicle, are not considered to be PM actions.

Preventive maintenance actions can be divided into two subcategories. Scheduled maintenance (SM) actions are planned based on some measure of elapsed time. Condition-based maintenance (CBM) actions are initiated based on data obtained from sensors applied to the RS. Vibration data and chemical analysis data are two examples of the type of data used in CBM. While it is still a developing science, CBM provides the potential for just-in-time, cost-effective maintenance. However, SM is the only type of PM considered further here.

Repairable systems modeling refers to the application of operations research techniques (e.g., probability modeling, optimization, simulation) to problems related to equipment maintenance. RS models are typically used to evaluate the performance of one or more RSs and/or design maintenance policies for one or more RSs.

In our discussions, we assume that a RS is always in one of two states: functioning (up) or down. Note that a system may be down for CM or down for PM.

The performance of a RS can be measured in several ways. We consider three categories of RS performance measures: (*i*) number of failures, (*ii*) availability measures, and (*iii*) cost measures. Let $N(t)$ denote the number of RS failures in the first t time units of system operation. Because of the stochastic (random) nature of RS behavior, $N(t)$ is a random variable. Thus, we may focus our attention on the expected value, variance, and probability distribution of $N(t)$.

Availability can be loosely defined as the proportion of time that a RS is in a functioning condition. However, there are four specific measures of availability found in the RS literature (Barlow et al., 1965). All these measures are based on the RS status function:

$$X(t) = \begin{cases} 1 & \text{if system is functioning at time } t \\ 0 & \text{if system is down at time } t \end{cases}.$$ (22.42)

See Figure 22.5 for a graphical portrayal of $X(t)$.

The first of these availability measures is the availability function, $A(t)$, where

$$A(t) = \Pr[X(t) = 1].$$ (22.43)

Note that $A(t)$ is typically difficult to obtain and rarely used in practice. The second measure is *limiting availability*, A, where

$$A = \lim_{t \to \infty} A(t).$$ (22.44)

By far the most commonly used availability measure, limiting availability is often easy to obtain mathematically. However, there are some cases in which limiting availability does not exist. The third availability measure is the average availability function, $A_{avg}(T)$, where

$$A_{avg}(T) = \frac{1}{T} \int_0^T A(t)\, dt.$$ (22.45)

Average availability corresponds to the average proportion of "uptime" over the first T time units of system operation. Since it is based on $A(t)$, average availability is typically difficult to obtain and rarely used in practice. However, because it captures availability behavior over a finite period of time, it is a valuable measure of RS performance. The final availability measure is *limiting average availability*, A_{avg}, where

$$A_{avg} = \lim_{t \to \infty} A_{avg}(t).$$ (22.46)

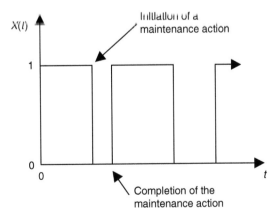

FIGURE 22.5 Graphical portrayal of $X(t)$.

When it exists, limiting average availability is almost always equivalent to limiting availability. To our knowledge, limiting average availability is almost never used in practice.

Cost functions are often used to evaluate the performance of a RS. The form of this function depends on the reliability and maintainability characteristics of the RS of interest. However, these functions typically include a subset of the following cost parameters.

c_f cost of a failure
c_d cost per time unit of "downtime"
c_r cost (per time unit) of CM
c_p cost (per time unit) of PM
c_a cost of RS replacement

Consider a RS that: (1) is modeled as a single component or a "black box"; (2) is intended to function 24 h per day, seven days per week; (3) has self-announcing (obvious) failures; (4) is binary-state—as in Equation 22.41; (5) is "as good as new" at time $t = 0$; (*vi*) is subjected to either SM or no PM. We utilize our own taxonomy to capture the essential elements of the models of such RSs. This taxonomy has six parts and can be summarized by 1/2/3/4/5/6 where 1 describes the probabilistic characteristics of the time to first failure of the RS, 2 describes the duration of CM, 3 describes the impact of CM on the age of the RS, 4 describes the type of PM policy (if any), 5 describes the duration of PM, and 6 describes the impact of PM on the age of the RS.

The first class of RS models that we address is based on concepts and results from renewal theory. We first consider a RS having the six characteristics related to our taxonomy. In the description G/G/P, the first G implies that the time to first failure of the RS is some type of random variable, that is, General distribution. The second G implies that the duration of CM is some type of random variable. The P implies that CM is perfect, that is, CM restores the RS to an "as good as new" condition. Note that the "as good as new" assumption is the key assumption and often the subject of criticism of the corresponding models (except when CM corresponds to RS replacement). The absence of the last three elements of the taxonomy implies that no PM is performed.

Let T_i denote the duration of the *i*th interval of RS function. Because of the "as good as new" assumption, $\{T_1, T_2, \ldots\}$ is a sequence of independent and identically distributed (iid) random variables. Let D_i denote the duration of the *i*th CM action, and note that $\{D_1, D_2, \ldots\}$ are assumed to be iid random variables. Therefore, each cycle (function, CM) has identical probabilistic behavior, and the completion of a CM action is a renewal point for the stochastic process $\{X(t), t \geq 0\}$.

Regardless of the probability distributions governing T_i and D_i, the limiting availability is easy to obtain (Barlow et al., 1965).

$$A = \frac{E(T_i)}{E(T_i) + E(D_i)} = \frac{\text{MTTF}}{\text{MTTF} + \text{MTTR}} \qquad (22.47)$$

Suppose T_i is a Weibull random variable having shape parameter $\beta = 2$ and scale parameter $\eta = 1000$ h. Then

$$R(t) = \Pr(T_i > t) = \exp\left[-\left(\frac{t}{\eta}\right)^{\beta}\right] \qquad (22.48)$$

and

$$MTTF = \eta\,\Gamma\left(1 + \frac{1}{\beta}\right) = 886.2 \text{ h.} \qquad (22.49)$$

Suppose D_i is a normal random variable having a mean (MTTR) of 25 h. Thus,

$$A = \frac{886.2}{886.2 + 25} = 0.9726. \tag{22.50}$$

For this example, availability and average availability values can be estimated using simulation.

In the description CFR/E/−, the CFR implies that T_i is an exponential random variable having failure rate λ, and the E implies that D_i is an exponential random variable having repair rate μ. Since the RS has a CFR, RS aging and the impact of CM are irrelevant. Note that the CFR/E/− model is a special case of the G/G/P model. For this RS (Barlow et al., 1965):

$$A(t) = \frac{\mu}{\lambda + \mu} + \frac{\lambda}{\lambda + \mu} e^{-(\lambda + \mu)t} \tag{22.51}$$

$$A = \frac{\mu}{\lambda + \mu} \tag{22.52}$$

$$A_{avg}(T) = \frac{\lambda\left[1 - e^{-(\lambda + \mu)T}\right] + \mu(\lambda + \mu)T}{(\lambda + \mu)^2 T}. \tag{22.53}$$

For example, suppose $\lambda = 0.001$ failures per hour (MTTF = 1000 h) and $\mu = 0.025$ repairs per hour (MTTR = 40 h). In this case,

$$A = \frac{0.025}{0.001 + 0.025} = \frac{1000}{1000 + 40} = 0.9615. \tag{22.54}$$

A plot of $A(t)$ can be found in Figure 22.6, and a plot of $A_{avg}(T)$ can be found in Figure 22.7.

In the description W/G/P/A/G/P, the W indicates that the time to first failure is a Weibull random variable with shape parameter β ($\beta > 1$) and scale parameter η. Note that since $\beta > 1$, the RS has an IFR. The A indicates that the RS is subjected to an age-based PM policy: If the RS functions without failure for τ time units, a PM action is initiated. Furthermore, the second G indicates that the duration of PM is a random variable, and the second P indicates that PM restores the RS to an "as good as new" condition. Note that PM may be worthwhile if PM is cheaper and/or faster than CM, since the RS has an IFR and PM reduces the age of the RS. Therefore, it may be of interest to derive an optimal PM policy for the RS. Specifically, we can modify our existing probability models to identify the value of τ that maximizes the limiting availability of the RS.

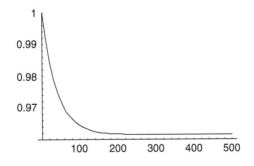

FIGURE 22.6 Example availability function—$A(t)$ vs. t.

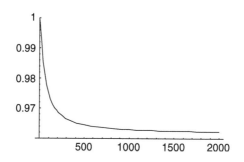

FIGURE 22.7   Example average availability function—$A(t)$ vs. t.

Let T denote the duration of an interval of RS function. Let $f(t)$ denote the PDF of T, and let $F(t)$ denote the CDF of T. Let D_{PM} denote the duration of a PM action, and let D_{CM} denote the duration of a CM action. Figure 22.8 contains a graphical portrayal of RS behavior under such a PM policy.

Because of our assumptions, the completion of any maintenance action corresponds to a renewal point, and

$$A = \frac{E(uptime)}{E(uptime) + E(downtime)} \tag{22.55}$$

We can derive $E(uptime)$ and $E(downtime)$ by conditioning on the value of T.

$$E(uptime) = \int_0^\tau tf(t)dt + \tau[1 - F(\tau)] \tag{22.56}$$

$$E(downtime) = E(D_{CM})F(\tau) + E(D_{PM})E[1 - F(\tau)] \tag{22.57}$$

Note that the integral in $E(uptime)$ typically must be evaluated numerically. Numerical analysis can then be used to compute limiting availability values for various values of τ. For example, suppose T is a Weibull random variable having $\beta = 2$ and $\eta = 80$ h. Suppose $E(D_{CM}) = 8$ h and $E(D_{PM}) = 2$ h. Figure 22.9 contains a plot (generated by Mathematica) of the limiting availability of the RS as a function of the age-based PM policy, τ. The optimal PM policy is $\tau^* = 47.5$ h with a corresponding limiting availability of 0.9182.

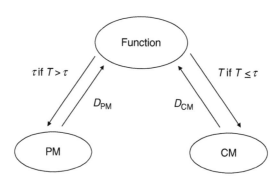

FIGURE 22.8   RS behavior under an age-based PM policy.

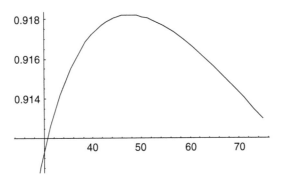

FIGURE 22.9 Example PM optimization—*A* vs. τ.

The second class of RS models that we address is based on the concept of minimal repair. In the description G/0/M, the G again implies that the time to first failure of the RS is a random variable. The 0 implies that CM is instantaneous, and the absence of the last three elements of the taxonomy implies that no PM is performed. The M implies that CM is minimal, that is, CM restores the RS to an "as bad as old" condition. Minimal CM, or minimal repair, implies that the RS functions after CM but its equivalent age is the same as it was at the time of failure. As with the "as good as new" assumption, the realism of the "as bad as old" assumption is often questioned.

Let *T* denote the duration of the first interval of RS function. Let *f*(*t*) denote the PDF of *T*, let *F*(*t*) denote the CDF of *T*, and let *z*(*t*) denote the hazard function of *T*. Then, $\{N(t), t \geq 0\}$ is a nonhomogeneous Poisson process (NHPP) having intensity function *z*(*t*) (Ross 1989). Since $\{N(t), t \geq 0\}$ is an NHPP having intensity function *z*(*t*), then *N*(*t*) is a Poisson random variable having mean *Z*(*t*), where *Z*(*t*) is the cumulative intensity function.

$$Z(t) = \int_0^t z(u)\, du \tag{22.58}$$

Furthermore, $N(t+s) - N(s)$ is a Poisson random variable having mean $Z(t+s) - Z(s)$ (Ross 1989).

$$N(t) \in \{0,1,\dots\} \tag{22.59}$$

$$E\big[N(t)\big] = Var\big[N(t)\big] = Z(t) \tag{22.60}$$

$$\Pr\big[N(t) = n\big] = \frac{e^{-Z(t)}\big[Z(t)\big]^n}{n!} \tag{22.61}$$

$$N(t+s) - N(s) \in \{0, 1, \dots\} \tag{22.62}$$

$$E\big[N(t+s) - N(s)\big] = Z(t+s) - Z(s) \tag{22.63}$$

$$Var\big[N(t+s) - N(s)\big] = Z(t+s) - Z(s) \tag{22.64}$$

$$\Pr\big[N(t+s) - N(s) = n\big] = \frac{e^{-[Z(t+s)-Z(s)]}\big[Z(t+s) - Z(s)\big]^n}{n!} \tag{22.65}$$

If T is an exponential random variable then $z(t)$ is a constant value λ and $\{N(t), t \geq 0\}$ is a Poisson process having rate λ. In this case, the impact of CM is irrelevant. Since CM is instantaneous, $\{N(t), t \geq 0\}$ is a Poisson process having rate λ and $N(t)$ is a Poisson random variable with mean λt (Ross 1989).

$$N(t) \in \{0, 1, \ldots\} \tag{22.66}$$

$$E[N(t)] = Var[N(t)] = \lambda t \tag{22.67}$$

$$\Pr[N(t) = n] = \frac{e^{-\lambda t} (\lambda t)^n}{n!} \tag{22.68}$$

Furthermore, $N(t + s) - N(s)$, the number of failures in the interval $(s, t + s]$, is also a Poisson random variable having mean λt (Ross 1989) . The implication of this result is that the number of failures in a given interval depends only on the length of the interval. Note that this is not true for an NHPP.

Suppose T is a Weibull random variable having shape parameter β and scale parameter η. Then,

$$z(t) = \frac{\beta}{\eta^\beta} t^{\beta-1} \tag{22.69}$$

$$Z(t) = \left(\frac{t}{\eta}\right)^\beta \tag{22.70}$$

and $\{N(t), t \geq 0\}$ is a power law process. If $\beta > 1$ $(\beta < 1)$, then the intensity function increases (decreases) and failures tend to occur more (less) frequently over time. Suppose $\beta = 1.75$ and $\eta = 1500$ h. Then,

$$E[N(1000)] = 0.4919 \tag{22.71}$$

$$E[N(2000) - N(1000)] = 1.1626 \tag{22.72}$$

$$E[N(3000) - N(2000)] = 1.7092 \tag{22.73}$$

$$\Pr[N(1000) > 2] = 0.0138 \tag{22.74}$$

$$\Pr[N(2000) - N(1000) > 2] = 0.1125 \tag{22.75}$$

$$\Pr[N(3000) - N(2000) > 2] = 0.2452 \tag{22.76}$$

Consider the W/0/M model with $\beta > 1$. Suppose the RS under consideration has an increasing intensity function. Over time, failures will tend to occur more frequently, and at some point, it will become economical to replace the system. Let τ denote the replacement time. Replacement of this type would be equivalent to perfect, instantaneous PM under a Block PM policy. The result is the W/0/M/B/0/P model. For such a RS, we can use a cost model to choose an optimal value of τ.

Let c_f denote the cost of a failure, let c_a denote the cost of replacing the RS, and let $C(\tau)$ denote the cost per unit time of RS ownership if the RS is replaced at time τ. Then:

$$E[C(\tau)] = \frac{1}{\tau}\{c_a + c_f E[N(\tau)]\} \tag{22.77}$$

$$E[C(\tau)] = \frac{1}{\tau}\left[c_a + c_f Z(\tau)\right] \tag{22.78}$$

$$E[C(t)] = \frac{c_a}{\tau} + \frac{c_f \tau^{\beta-1}}{\eta^\beta}. \tag{22.79}$$

Differentiation and algebraic manipulation yield (Ascher and Feingold, 1984):

$$\tau^* = \left[\frac{c_a \eta^\beta}{c_f(\beta-1)}\right]^{1/\beta}. \tag{22.80}$$

For example, if $\beta = 1.75$, $\eta = 1500$ h, $c_a = \$1000$, and $c_f = 75$, then the RS should be replaced after $\tau^* = 7768$ h.

22.6 Markov Models

In this section we focus our modeling efforts on using continuous time Markov chains (CTMC) to model RSs. A CTMC is a stochastic process that moves from state to state in accordance with a discrete time Markov chain (DTMC). It differs from a DTMC in that the amount of time it spends in each state before it transitions to another state is exponentially distributed (Ross 1989). Like a DTMC, it has the Markovian property whereby the "future is independent of the past, given the present." In this section, we assume that the CTMC has stationary (homogeneous) transition probabilities (i.e., P[$X(t + s) = j|X(s) = i$] is independent of s). Ross (1989) formally defines a CTMC as a stochastic process where each time it enters state i:

1. The amount of time it spends in state i before it transitions into a different state is exponentially distributed with a rate v_i.
2. When the process leaves state i, it will enter state j with some probability p_{ij}, where $\sum_{j \neq i} p_{ij} = 1$.

22.6.1 Kolmogorov Differential Equations

In the discrete time case, $p_{ij}^{(n)}$ represents the probability of going from state i to j in n transitions. In the continuous case we are interested in $p_{ij}(t)$ which represents the probability that a process currently in state i will be in state j in t time units from the present. Mathematically, we denote this by:

$$p_{ij}(t) = P\left[X(t + s) = j|X(s) = i\right]. \tag{22.81}$$

In the continuous time case we can define the intensity at which transitions occur by examining the infinitesimal transition rates:

$$-\frac{d}{dt}p_{ii}(0) = \lim_{t \to 0}\frac{1 - p_{ii}(t)}{t} = v_i \tag{22.82}$$

$$-\frac{d}{dt}p_{ij}(0) = \lim_{t \to 0}\frac{p_{ij}(t)}{t} = q_{ij} \tag{22.83}$$

where v_i represents the rate at which we leave state i and q_{ij} represents the rate at which we move from state i to state j. However, for Δt small, $q_{ij}\Delta t$ can be interpreted as the probability of going from state i to state j in some small increment of time Δt given we started in state i. Using the transition intensities, as well as making use of the Markovian property, one can derive the Kolmogorov differential equations for $p_{ij}(t)$. The backward and forward Kolmogorov equations are given by Equations 22.83 and 22.84. These equations can be used to derive the transient probabilities of a CTMC. This is best illustrated through the use of an example.

$$\frac{d}{dt}p_{ij}(t) = \sum_{k \neq i} q_{ik}p_{kj}(t) - v_i p_{ij}(t) \tag{22.84}$$

$$\frac{d}{dt}p_{ij}(t) = \sum_{k \neq i} q_{kj}p_{ik}(t) - v_i p_{ij}(t) \tag{22.85}$$

22.6.2 Transient Analysis

Consider a single component system that fails according to an exponential failure distribution with rate λ and whose repair time is exponentially distributed with rate μ. This system can be in one of two states. It can be working (state 0) or can fail and be undergoing repair (state 1). A state transition diagram for this system is given in Figure 22.10. This diagram shows the states and the associated transition rates between the states.

Using the state transition diagram and the Kolmogorov forward equation, Equation 22.84, we can derive the transition probabilities for the CTMC.

$$\frac{d}{dt}p_{ij}(t) = \sum_{k \neq j} q_{kj}p_{ik}(t) - v_i p_{ij}(t) \tag{22.86}$$

$$\frac{d}{dt}p_{00}(t) = q_{10}p_{01}(t) - v_0 p_{00}(t) \tag{22.87}$$

$$\frac{d}{dt}p_{00}(t) = \mu p_{01}(t) - \lambda p_{00}(t) \tag{22.88}$$

$$\frac{d}{dt}p_{00}(t) = \mu\left[1 - p_{00}(t)\right] - \lambda p_{00}(t) \tag{22.89}$$

$$\frac{d}{dt}p_{00}(t) = \mu - (\mu + \lambda)p_{00}(t) \tag{22.90}$$

$$\frac{d}{dt}p_{00}(t) + (\mu + \lambda)p_{00}(t) = \mu \tag{22.91}$$

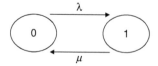

FIGURE 22.10   Single component state transition diagram.

Solving this differential equation, we obtain:

$$e^{(\lambda+\mu)t}\left[\frac{d}{dt}p_{00}(t)+(\mu+\lambda)p_{00}(t)\right]=\mu e^{(\lambda+\mu)t} \tag{22.92}$$

$$\frac{d}{dt}[e^{(\lambda+\mu)t}p_{00}(t)]=\mu e^{(\lambda+\mu)t} \tag{22.93}$$

$$e^{(\lambda+\mu)t}p_{00}(t)=\frac{\mu}{\lambda+\mu}e^{(\lambda+\mu)t}+c \tag{22.94}$$

Since
$$p_{00}(0)=1,\, c=\frac{\mu}{\lambda+\mu} \tag{22.95}$$

Therefore, $p_{ij}(t)$ for $i=j$ are given in the following:

$$p_{00}(t)=\frac{\lambda}{\lambda+\mu}e^{-(\lambda+\mu)t}+\frac{\mu}{\lambda+\mu} \tag{22.96}$$

$$p_{11}(t)=\frac{\mu}{\lambda+\mu}e^{-(\lambda+\mu)t}+\frac{\lambda}{\lambda+\mu}. \tag{22.97}$$

Note that $p_{00}(t)$ represents the probability that the system is operating at time t. This is also known as the system availability $A(t)$. If we take the limit of $p_{00}(t)$ as t goes to infinity we get the limiting or steady-state availability. The limiting availability is given in the following:

$$\lim_{t\to\infty}A(t)=\lim_{t\to\infty}p_{00}(t)=\frac{\mu}{\mu+\lambda} \tag{22.98}$$

In general, we can establish a set of N first-order differential equations which characterize the probability of being in each state in terms of the transition probabilities to and from each state. Mathematically, the set of N first-order differential equations is summarized in matrix form in Equation 22.99 and the general form of the solution to this set of differential equations is given by Equation 22.100.

$$\frac{dP(t)}{dt}-\lfloor T_R\rfloor P(t) \tag{22.99}$$

$$\underline{P}=Exp[T_R]t\cdot\underline{P}(0) \tag{22.100}$$

In Equation 22.100 T_R is the rate matrix. For our simple single system example, using Figure 22.10, we get the following rate matrix.

$$T_R=\begin{bmatrix}-\lambda & \mu\\ \lambda & -\mu\end{bmatrix} \tag{22.101}$$

In order to solve the set of differential equations one must compute the matrix exponential. There are several different approaches to computing the matrix exponential. Two such methods include the infinite series method and the eigenvalue/eigenvector approach. Such routines are readily available in many of the commercially available mathematical analysis packages (Maple, Mathematica, and MATLAB). In many instances, as the problem complexity increases, the Kolmogorov differential equations cannot be solved explicitly for the transition probabilities. In such cases, we will use numerical solution techniques; we might use simulation, or for a variety of reasons, focus our attention on the steady-state performance of the system.

22.6.3 Steady-State Analysis

For many systems, it is the limiting availability (a.k.a. steady-state availability), $A(\infty)$, which is of interest. Another common name for the steady-state availability is the uptime ratio. For example, the uptime ratio is of critical importance in a production facility. Similarly, for a communication system, the average message transfer rate will be the design transfer rate times the uptime ratio. Therefore, knowing the uptime ratio is essential to analyzing the performance of many systems.

We can compute the steady-state probabilities by making use of the following:

$$\text{let } \rho_j = \lim_{t \to \infty} p_{ij}(t) \tag{22.102}$$

We can then state the following:

$$v_j \rho_j = \sum_i p_i q_{ij} \quad \text{for all} \quad j = 0, 1, 2, \dots, N \tag{22.103}$$

$$\sum_j \rho_j = 1 \tag{22.104}$$

Expression 22.103 is called the "balance" equations. The balance equations state that the rate into each state must be equal to the rate out of each state for the system to be in equilibrium. Equation 22.104 states that we must be in some state, and the sum of the probabilities associated with each state must be equal to one. Using $N - 1$ of the balance Equations and Equation 22.104, we can easily derive the steady-state probabilities for each state.

22.6.4 CTMC Models of RSs

In this section we illustrate how to model and analyze a variety of RSs using continuous time Markov chains. We focus specifically on the single machine cases. Consider a single repairable machine. Let T_i denote the duration of the ith interval of machine function, and assume $\{T_1, T_2, \dots\}$ is a sequence of iid exponential random variables having failure rate λ. Upon failure, the machine is repaired. Let D_i denote the duration of the ith machine repair, and assume $\{D_1, D_2, \dots\}$ is a sequence of iid exponential random variables having repair rate μ. Assume no PM is performed on the machine.

Recall that $X(t)$ denotes the state of the machine at time t. Under these assumptions, $\{X(t), t \geq 0\}$ transitions among two states, and the time between transitions is exponentially distributed. Thus, $\{X(t), t \geq 0\}$ is a CTMC having the rate diagram shown in Figure 22.11.

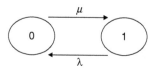

FIGURE 22.11 Single machine rate diagram.

We can easily analyze the "steady-state" behavior of the CTMC. Let ρ_j denote the long-run probability that the CTMC is in state j. We use balance equations to identify these probabilities. Each state of the CTMC has a balance equation that corresponds to the identity "rate in" = "rate out." For the rate diagram in Figure 22.11, the balance equations are:

$$\text{state 0: } \lambda \rho_1 = \mu \rho_0 \tag{22.105}$$

$$\text{state 1: } \mu \rho_0 = \lambda \rho_1 \tag{22.106}$$

These balance equations are equivalent, so we need an additional equation to solve for ρ_0 and ρ_1. We use the fact that the steady-state probabilities must sum to one.

$$\rho_0 + \rho_1 = 1 \tag{22.107}$$

We then use the two equations to solve for the two unknowns.

$$\rho_1 = \frac{\mu}{\lambda + \mu} \tag{22.108}$$

$$\rho_0 = \frac{\lambda}{\lambda + \mu} \tag{22.109}$$

Note that ρ_1 is equivalent to the steady-state availability found from taking the limit of the transient probabilities in Equation 22.98.

Let us consider another single machine example. Just like the first example, let T_i denote the duration of the ith interval of machine function, and assume $\{T_1, T_2, \dots\}$ is a sequence of iid exponential random variables having failure rate λ. Upon failure, the machine is repaired. But this time, each repair requires two distinct repair operations, A and B. Assume that the duration of repair is exponentially distributed with rate μ_j where $j = (A, B)$. For this example, assume there are enough resources available so that the repairs can be done concurrently.

This problem differs significantly from the first in that we now have four different states. State 0 is when the machine is operating, State 1 is when the machine is down and we are awaiting the completion of repair process A, State 2 is when the machine is down and we are awaiting the completion of repair process B, and State 3 is when the machine is down and we are awaiting the completion of both repair processes. The rate diagram for this model is shown in Figure 22.12. Using the rate diagram, the set of balance Equations 22.105 can be written as,

$$\mu_A \rho_1 + \mu_B \rho_2 = \lambda \rho_0 \tag{22.110}$$

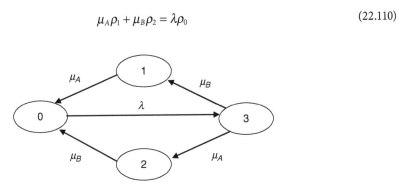

FIGURE 22.12 Rate diagram for multiple repair process.

$$\mu_B \rho_3 = \mu_A \rho_1 \tag{22.111}$$

$$\mu_A \rho_3 = \mu_B \rho_2 \tag{22.112}$$

$$\lambda \rho_0 = (\mu_A + \mu_B)\rho_3 \tag{22.113}$$

Using the balance equations in conjunction with the total probability equation we can solve for the individual steady-state values for each of the states. The state of interest is state 0, as it represents the system steady-state availability. Suppose the system has a mean time between failure of 100 h, and the mean repair time for process A is 5 h and for process B it is 2.5 h. Using the balance equations we can derive Equation 22.114 and determine that the system has a steady-state availability 0.96.

$$\rho_0 = \frac{\mu_A + \mu_B}{\lambda}\left(\frac{\mu_A + \mu_B}{\lambda} + \frac{\mu_B}{\mu_A} + \frac{\mu_A}{\mu_B}\right)^{-1} = \frac{0.2 + 0.4}{0.01}(62.5)^{-1} = 0.96 \tag{22.114}$$

References

Ebeling, C. E., *An Introduction to Reliability and Maintainability Engineering*, McGraw-Hill, Boston, 1997.

Barlow, R. E. and Proscan, F., *Mathematical Theory of Reliability*, John Wiley & Sons, Inc., New York, 1965.

Ross, S. M., *Introduction to Probability Models*, 7th ed., Harcourt Academic Press, San Diego, 1989.

Ascher, H. and Feingold, H., *Repairable Systems Reliability*, Marcel Dekker, Inc., New York, 1984.

23

Funding and Justifying Logistics

Ricki G. Ingalls
Yen-Ping Leow Sehwail
Oklahoma State University

Loay Sehwail
University of Wisconsin Oshkosh

23.1 Partnership for Success

Family Dollar Stores, Inc. is a company that operates a chain of self-service retail stores in low- to middle-income neighborhoods of the United States, focusing on rural and urban markets. Family Dollar was established in 1959 and the headquarters is located in Matthews, North Carolina, just outside of Charlotte. During the last 10 years, more than 3000 new stores have been added to the chain, with over 2000 added in the last five years. As of July 7, 2005, the company operated approximately 5732 stores in 44 states and in the District of Columbia. As the pace of store expansions accelerates, Family Dollar has become one of the fastest growing national chains of neighborhood convenience discount stores in the United States.

23.1.1 Establishing a Baseline

In 1998, Family Dollar operated 3600 locations in 39 states. New stores were opening at a rate of more than one per day. At that time, merchandise was moved from supply partners through three distribution centers that are located in Matthews, North Carolina; West Memphis, Arkansas; and Front Royal, Virginia. With the initiative to accommodate Western expansion and enhance supply chain efficiencies, Family Dollar decided to continue expanding their distribution facility network and was looking to open the fourth full-service distribution center in the Southwest region to service their stores in Texas, Oklahoma, Kansas, Nebraska, New Mexico, and Colorado. As soon as this decision was made, the company hired a site locator to start the search process for a site in the Southwest region that would eventually become a 907,000 square feet distribution center with 160 dock doors and over nine miles of conveyors, capable of delivering approximately 165 cartons per minute to the shipping dock.

23.1.2   Site Search

The first job of the site locator was to determine the geographical area of the new distribution center. This was done using a software that considered current stores and their demand, future stores and their demand, and the current distribution center locations. A key consideration was a site that could reach all its target stores within a 24-h period. With this information, the site locator determined that a site in North Texas would be best for the new distribution center.

With a geographic area in mind, the site locator considered 60 communities in the area as potential locations for the new distribution center. The site locator then approached the communities with information about the opening of a regional distribution center in the area. Some lead information such as the projected capital investment ($50–$60 million), the size of the location (85 acres), the size of employment (approximately 500 employees), and the project scope (distribution center) were released to all of the communities, but the name of the company was withheld. The site locator soon created the short-list of communities, with Grand Prairie, Texas being the leading candidate.

In June 1998, Duncan, Oklahoma, a community with a population of 23,000, was the last community to enter the race. Duncan has a strategic location with access to Interstate Highways I-35, I-44, and I-40 through U.S. Highway 81 and State Highway 7, which could be utilized to speed the deliveries of Family Dollar's merchandise to its stores. Wesley Devero, the president of Duncan Area Economic Development Foundation (DAEDF), was responsible for presenting Duncan and managing the project of trying to win Duncan as the location of Family Dollar's new regional distribution center. DAEDF is a nonprofit organization that was established in 1994. The objective of DAEDF is to promote the development of existing businesses, attract new businesses and diversify Duncan's economic base. DAEDF has been remarkably successful in a short period of time. With staff less than half the size of most comparable economic development foundations, DAEDF has become a model that others are trying to emulate. Although Duncan was the last community to enter the site selection process, it managed to garnet the attention of the Family Dollar executives because Duncan put together a very attractive incentive package that no other community could match. As a Duncan-based Family Dollar executive said, "Duncan is rolling out the red carpet."

23.1.3   Time Line

The whole process from beginning of the site selection process to the first day of operations in the Duncan distribution center was a just over a year. Table 23.1 shows the timeline that highlights the key events, starting from site selection to the operational of the new regional distribution center. The deal-making process with DAEDF was so fast and efficient that the development deal was closed in less than seven months.

The Duncan team led by Mr. Devero was well prepared for the Family Dollar questions and provided them with all the required information. With an 85-acre land in the Duncan South Industrial Park as a potential location, Duncan was ready to host the third largest building in the state at that time. After the site locator's initial visit to the location in Duncan, DAEDF started anticipating the site locator's next moves by taking several actions that the site locator came back and asked for, such as soil-boring tests, industrial site topography, and environment reports.

23.1.4   Incentive Package

Although officials with Family Dollar initially had their sights set on Grand Prairie, Texas for their newest distribution center, Duncan quickly proved to be just the location they were looking for. DAEDF managed to attract Family Dollar through an economic incentive package of approximately $13.85 million.

TABLE 23.1   Family Dollar Duncan Distribution Center Time Line

Timeline	Event Description
May 1998	Family Dollar hired consultant and site locator to initiate the search for possible locations
June 1998	Duncan entered the race as potential location for Family Dollar's new distribution center
July 1998	Site locator visited Duncan for the first time
August 1999	Duncan presented the incentive package to Family Dollar
September 1998	Family Dollar corporate executives' first visit to the Duncan site
October 2, 1998	Family Dollar officially announced Duncan as the location for their fourth distribution center
November 2, 1998	Ground breaking at Duncan site
May 18, 1999	First group associated hire date
June 1, 1999	Begin receiving merchandise
July 19, 1999	Begin shipping merchandise

In the competition to retain and attract employers, state and local governments are increasingly offering economic incentives that offset the high costs that businesses associate with locating in a new region or state. The incentives given by the City of Duncan and the State of Oklahoma are outlined in Table 23.2.

The economic incentive package offered by DAEDF could be largely classified into three categories:

Direct Financial Incentives

Direct financial incentives included an 85-acre tract of land located in the Duncan South Industrial Park, a $250,000 cash reward for employee training, and $4,380,000 from local and state quality jobs programs.

The innovative Oklahoma Quality Jobs Program ($3.88 million), administered by the Oklahoma Department of Commerce, provides quarterly cash payments of up to 5% of newly created payroll to qualifying companies. Generally, the program is applicable to manufacturing, research and development, central administrative offices, and selected service companies who achieve more than $2.5 million annualized payroll for new full-time employees within the first three years of the program. To date, more than 340 companies have claimed nearly $350 million in benefits from the Oklahoma Quality Jobs Program, creating more than 35,000 new jobs.

TABLE 23.2   Family Dollar Distribution Center Incentive Package

Incentive	Amount ($)
State quality jobs incentive	3,880,000
Sales and use tax exemption	2,500,000
Accelerated Federal Property Depreciation Schedule	1,870,000
Five year property tax exemption	1,500,000
State and local industrial access road (Hwy to industry park and within industry park)	1,270,000
85 Acre industrial site	510,000
Duncan cash incentive for jobs	500,000
Second water source and infrastructure built specially to Family Dollar property line	500,000
Utilities built specially to Family Dollar property line	420,000
Freeport (Inventory) tax exemption	320,000
Native American Federal Employment Tax Credit	300,000
Vocational technical school training for industry programs	250,000
Others	30,000
Total	13,850,000

The local quality jobs program is simply a $1000 cash payment by DAEDF for every job created up to a maximum of 500 jobs created. The incentive package also included a training fund of $500 per employee up to 500 employees.

Indirect Financial Incentives

The total indirect package is valued at more than $2 million, and included new state and local industrial access roads, a second water source built specifically for Family Dollar, infrastructure improvements to the site, paying for building permits, soil and boring tests, meter setting and tap fees, and other infrastructure and building fees.

The indirect financial incentives also included the costs associated with arranging a job fair. Duncan, represented through DAEDF, arranged for an initial job fair to assure Family Dollar of the available pool of quality labor in Duncan. DAEDF arranged for a career fair in September 1998 to show Family Dollar the quality of labor available. More than 1300 applicants showed up at the job fair, and after initial screening, it was decided by Family Dollar managers that 90% of these applicants have the required skills to work for the company.

Tax-Based Incentives

The total tax-based incentive package is valued at more than $6 million. The package included $2,500,000 sales and use tax exemption, $1,870,000 through accelerated federal property depreciation, $1.5 million dollars of property tax exemption over five years, and more than $300,000 in inventory tax exemption. The package also included the Native American Federal Employment Tax Credit which is a credit of up to $4000 per employee annually up to a total of $300,000 over a five-year period. Table 23.2 details all the incentives offered by Duncan to Family Dollar.

23.1.5 Summary

The Duncan financial incentive package was not the only factor in winning the Family Dollar distribution center. Family Dollar also cited the relationship with the community, the quality of life, availability of management staff, and the partnership with DAEDF.

With accelerating store expansion and a unique merchandising concept, Family Dollar is well positioned to continue providing consumers with convenience, low prices, and low overhead in a self-service retail environment.

In summary, Family Dollar went through the normal search process for building a new distribution center through starting by identifying the new geographical area for the new distribution center based on the forecasted demand for different stores and the expected growth. Family Dollar then reduced the number of communities to a short list and then negotiated economic incentive packages with the different communities on the short-list. After the negotiations were complete, Family Dollar made the final decision of Duncan, Oklahoma.

23.2 Introduction

Logistics management is an important area to analyze in Operations Management. Based on the Council of Logistics Management (CLM)—currently known as the Council of Supply Chain Management Professionals (CSCMP), one of the leading professional organizations for logistics personnel, the definition of logistics management is, "The process of planning, implementing, and controlling the efficient and effective flow and storage of goods, services and related information from point of origin to point of consumption for the purpose of conforming to customer requirements."

Ever since the CLM adopted this definition of logistics in 1986, the integration of transportation, procurement, inventory control, distribution management, and customer service has been a major thrust in many firms. Realizing the synergy that exists in these functions, many companies have extended the concept further upstream and downstream to include entities outside the company to include vendors and their vendors and also customers and their customers.

This chapter discusses the funding and justification of logistics activities. The remainder of the chapter is organized as follows:

1. The benchmarking phase in funding logistics activities.
2. Customer service as an important driver in funding logistics activities in order to achieve competitive advantage.
3. The importance on partnerships in logistics is presented in the next section. Two examples, Procter & Gamble (P&G) and Applied Industrial Technologies are used to illustrate the importance of logistics partnership and collaboration.
4. A discussion on the importance of logistics software is presented. A detailed discussion of supply chain network design software and warehouse and transport management systems is presented.
5. In the last section of this chapter, a detailed discussion about the advantages of teaming up with government, industry, and academia to achieve excellence in logistics.

In each section, referencing to the Family Dollar case study will be used to illustrate concepts and ideas.

23.3 Benchmarking

Benchmarking, which can be defined as comparing your performance to a baseline, is a good way to gauge your company's progress over time. The baseline can be a predefined internal objective(s) or a comparison to a competitor or an industry leader. Benchmarking may also be defined as the process of analyzing the best products or processes of leading competitors in the same industry or leading companies in other industries (Camp, 1995). The focal company then gains an understanding of the appropriate performance level and drivers behind the success (Zairi, 1996).

The benchmarking process provides areas for the company to identify for implementing the most effective solutions and realizing breakthroughs in performance. In this sense, benchmarking provides both motivation and learning in performance improvements. As the team in the company compares its internal practice with the best practice, benchmarking feedback reveals plenty of room for improvements and suggests how to imitate strategies, which have the potential to achieve better performance. Besides this motivational aspect, the team also becomes involved in the learning process of implementation. They engage in planning, controlling, and evaluating the life cycle of the improvement project (Simatupang and Sridharan, 2004).

There are three different types of benchmarking activities: benchmark internally within own company, benchmark competing firms, and benchmark companies outside industry. Benchmarking within a company can be done by benchmarking against previously stated objectives or by benchmarking against other entities within the same company.

In the Family Dollar case study, the company established the baseline for the guidelines for the new distribution center before engaging the site locator. The baseline information included conditions such as the size of the distribution center, the size and quality of labor needed, and the accessibility to markets that Family Dollar serves. Family Dollar also established a baseline for the completion and operation of the new distribution center.

23.4 Customer Service

Customer service is an important topic to practitioners and researchers alike. Practitioners have long recognized that exceptional customer service reaps the benefits of customer satisfaction, loyalty, and increased sales. In addition, as cross-functional cooperation increases within the firm, customer service serves as the overarching goal of the organization and is often included as such in corporate mission statements (Emerson and Grimm, 1998). Customer service can be defined as a customer-oriented philosophy, which integrates and manages all elements of the customer interface within a predetermined optimum cost-service mix.

Customer service can be viewed as an output of the logistics system. It involves getting the right product to the right customer at the right place, in the right condition, and at the right time, at the lowest possible cost. Good customer service leads to customer satisfaction, which is the one of the essential elements for the success of an organization. Coyle et al. (1996, p. 111) defined customer service as "an augmented product feature that adds value for the buyer."

Collins et al. (2001) stated that there has been an evolution over time in what is meant by customer service. Customer service in the 1970s and 1980s was reactive (to the customer complaint) and firm-oriented. Customer service moved to the concept of value addition in the late 1980s (Mantodt and Davis, 1993). Even then, these authors claim, the emphasis was on setting internally derived customer service standards based on what the company could do and not by what the individual customer wanted. Starting the 1990s, managers and researchers started recognizing that anticipating and exceeding the customer's expectations in a value-added way is what is required to retain and develop markets (Livingstone, 1992).

Regardless of how it is defined or perceived, customer service may be the best method of gaining competitive advantage for many firms (Lambert, 1993). Customer service can be used to differentiate a firm's products, keep customers loyal and increase sales and profits (Sharma and Lambert, 1994). The task for the logistics manager is to strike a balance between customer service levels, total logistics costs, and total benefits to the firm. It should be noted that some companies have discovered that customer service levels can be increased while total logistics costs are decreased (Coyle et al., 1996). Emerson and Grimm (1996, p. 29) described logistics customer service activities as providing "place, time and form utility, by ensuring that the product is at the right place, at the time the customer wants it, and in an undamaged condition."

When supply chain managers make strategic decisions, they are making decisions that trade-off three characteristics of the business: uncertainty, customer service, and cost. One of the most important survival and success factors for a company is the ability to deliver more value to the customer than their competition. Good customer value can be achieved only when product quality, service quality, and value-based prices exceed customer expectations. According to Earl Naumann's book *Creating Customer Value* (1995), the customer value triad consists of: product quality, service quality, and value-based prices, where product quality and service quality are the pillars that support value-based prices, if product or service quality is poor, value-based prices falls, but if price is too high sales suffer (Fig. 23.1).

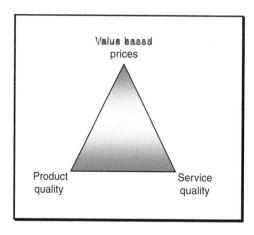

FIGURE 23.1 Customer value triad. (From Naumann, *Creating Customer Value: The Path to Sustainable Competitive Advantage.* Cincinnati, OH: Thomas Executive Press, South-Western College Publishing, 1995.)

Determining the location of the company's plant(s) and/or distribution center(s) is a strategic decision that affects not only the cost of transporting inbound raw materials and outbound finished products, but also customer service levels and order-to-delivery cycle times. Issues to consider include the location of customers, suppliers, available transportation services, the availability and wage rates of qualified employees, governmental cooperation, and other financial and operational considerations.

In the Family Dollar case study, the company focused on excellence in customer service by locating their new distribution center in a location that is within a one-day driving distance from the stores that will be serviced from the Duncan distribution center. Family Dollar realized that in order to survive and succeed in today's market environment, they needed to excel in serving their customers (retail stores). In 1998, when Family Dollar were looking to open their fourth distribution center, the objective of the Duncan distribution center was to service their stores in Texas, Oklahoma, Kansas, Nebraska, New Mexico, and Colorado. In 2005 (Family Dollar now has nine distribution centers in the United States), the Duncan distribution center supported approximately 775 stores in six states including Oklahoma, Kansas, southern Nebraska, western Missouri, western Louisiana, and the eastern half of Texas. With more than 500 stores in Texas alone and rapid growth throughout the rest of this territory, the Duncan DC supports Family Dollar's aggressive store growth.

23.5 Partnership to Permit Core Competency Focus

While the word partnership has been interpreted by some managers and educators to mean any business-to-business relationship, it is still the most descriptive term for closely integrated, mutually beneficial relationships that enhances supply chain performance. La Londe and Cooper (1989) defined a logistics partnership as "a relationship between two entities in a logistics channel that entails the sharing of benefits and burdens over some agreed upon time horizon." Ellram (1995) added the dimension of information sharing as "an agreement between a buyer and a supplier that involves a commitment over an extended time period, and includes the sharing of information along with a sharing of the risks and rewards of the relationship." Lambert et al. (1999) stated that although numerous other definitions include the key characteristics of shared risks/rewards, long-term focus, joint activities, and the concept of trust, they are incomplete and do not adequately emphasize the need for customization of the relationship. They defined a partnership as "a tailored business relationship based on mutual trust, openness, shared risk and shared rewards that results in business performance greater than would be achieved by the two firms working together in the absence of partnership."

A logistics partnership is a journey and not an event. A successful partnership is like a marriage. Neither just happens: both relationships require constant hard work from the parties involved. Both parties must understand each other's needs, and must be compatible, with shared values. Like a marriage, a successful logistics partnership requires open communications, mutual commitment to the partnership, fairness, and flexibility. Successful partnerships are cooperative and collaborative. They are long-term and built on trust (Tate, 1996). In order to have a real successful partnership, all parties should gain from this partnership.

Bowersox, in his book *Logistical Excellence* (1992), outlines five factors that are critical to the success of a logistics partnership. They include:

1. Selective matching—partners have compatible corporate cultures and values.
2. Information sharing—partners openly share strategic/operational information.
3. Role specification—each party in the partnership knows specifically what its role is.
4. Ground rules—procedures and policies are clearly spelled out.
5. Exit provisions—a method for terminating the partnership is defined.

Benefits from logistics partnership allow companies to manage the order fulfillment process effectively and efficiently, meet end customer requirements promptly, and face market demand variability with faster changes in production/resource planning. Today's level of interdependency among supply

TABLE 23.3 Logistics Collaboration Opportunities

Logistics Process	Collaboration Opportunities
Demand management—order processing	Integrated information and communication systems for order entry, status report, invoicing and documentation
Demand management—forecasting	Customer pattern identification, elimination/reduction of "bullwhip" effect, co-ordination of promotions and new product launchings, providing end customer demand information
Inventory planning—replenishment schemes	Vendor-managed inventory programs, scheduled replenishment programs, replenishment schemes, early alerts on fulfillment problems
Supply management	Production planning, especially useful in an increasing build-to-order process with shrinking intervals for production, product concept to design, suppliers on the product development side
Distribution management—transportation and warehousing	Packaging standards, distribution cost-reduction programs, picking and shipping schedules, shipment status reports

Source: Rey, M. Supply Chain Collaboration. *Global Purchasing and Supply Chain Strategies*, Available online at: http://www.bbriefings.com/pdf/976/6_rey.pdf#search='logistics%20collaboration', 2001, Last viewed (Sept., 7, 2005).

chain members is continually increasing. This interdependency requires companies to share forecasting and planning information as part of contractual terms to allow the whole community to perform according to market expectations. Competition forces companies to make alliances and partnerships to differentiate themselves in an increasingly homogeneous market. There are many different possible opportunities for partnership in logistics, Table 23.3 shows some examples.

Today's logistics models are powered by advanced software systems that allow companies to expand partnership through collaborative logistics networks on a large scale. Companies are forming web-based, as well as more traditional partnerships, to reduce the costs of transportation and inventory while raising the bas on customer service. Examples include companies such as Applied Industrial Technologies, General Mills, Georgia-Pacific, Procter and Gamble, Seneca Foods, Land O'Lakes, Kroger, Safeway, and DuPont.

Peter Strozniak (2003) from Frontline Solutions stated that analysts and executives believe this model is being adopted by more companies over time because of the increasing costs of transportation and the evolution of the supply chain from the conventional push system (in which companies push products to the marketplace), to a pull model system where the entire supply chain is reacting to the marketplace demands.

There are many business-to-business examples of logistics partnerships. Let us look at consumer packaged goods as an example of a partnership. If there are two companies that have made supply chain a household word, they are Wal Mart and Proctor & Gamble. Before these two companies started partnering back in the 1900s, retailers shared very little information with manufacturers. But these two giants built a software system that hooked P&G up to Wal-Mart's distribution centers. When P&G's products run low at a Wal-Mart distribution center, the system sends an automatic alert to P&G to ship more products. In some cases, the system goes all the way to the individual Wal-Mart store. It lets P&G monitor the shelves through real-time satellite link-ups that send messages to the factory whenever a P&G item swoops past a scanner at the register.

With this kind of minute-to-minute information, P&G knows when to make, ship, and display more products at the Wal-Mart stores. There is no need to keep products piled up in warehouses awaiting Wal-Mart's call. Invoicing and payments happen automatically as well. The system saves P&G so much in time, reduced inventory and lower order-processing costs that it can afford to give Wal-Mart "low, everyday prices" without putting itself out of business.

Another example is Applied Industrial Technologies (AIT), one of the leading distributors of industrial products in North America. It operates a chain of 450 service centers that sell maintenance, repair, and operational industrial products to large and small manufacturers. These service centers are supported by an extensive logistics and distribution network. Because of its growth during the last 10 years, AIT has repositioned and automated its distribution centers.

In the late 1990s, however, AIT saw a need to reduce its logistics costs and decided to form partnerships with other distributors or manufacturers to fill up truck capacity on dedicated routes that would reduce logistics costs. "We didn't know if it could work or not," says Jeff Ramras, vice president of marketing and supply chain solutions at AIT in Cleveland. "After talking to our transportation providers, they agreed to try it because we were pressuring them to bring their costs down. So we both started looking for partners."

Ramras says it was a difficult task because for the collaborative logistics model to work, AIT had to find companies that delivered goods regularly to service centers or stores. In addition, AIT had to identify companies that had distribution centers in the same locale.

"It was just a question, quite frankly, of who we could find," says Ramras. "Westco was our first partner. When they came on board, we started small and tried it to see if it would work. It did, so we started to expand it." Currently, about 70% of AIT's service locations are connected to the collaborative logistics network, and AIT has formed partnerships with companies such as John Deere, Lucent Technologies, Westco and Graybar.

Ramras says it was a difficult task because for the partnership logistics model to work, AIT had to find companies that delivered goods regularly to service centers or stores. In addition, AIT had to identify companies that had distribution centers in the same locale. By sharing truck space with its partners, AIT has seen its dedicated freight charges drop by nearly 30%. In addition, Ramras says customer service has improved and the need for the company's service centers to hold safety stock inventory has declined by 15% to 20%.

"Because these trucks are delivering products throughout the night, we have customers actually picking up merchandise at our service centers at 11:00 P.M. or midnight because they know the truck is coming," says Ramras. "Before, our customers would have to wait until the next day." Another benefit is that service centers don't have to pay for premium freight if they need products delivered the next day. Service center managers can order product as late as 5:00 P.M. because the trucks do not leave the distribution centers until much later. "It allows us to provide better service to our service centers, which allows them to service our customers better."

In the Family Dollar case study, the company looked at partnership from four dimensions: partnering with the customers, partnering with the suppliers, partnering with local shipping services, and partnering with the community. Family Dollar partnered with their customers to assure that stores carry lower inventories and get faster replenishment through building the Duncan distribution center within a day's drive from the stores that are to be serviced from Duncan. Family Dollar partnered with local shipping services through hiring and working with local companies on delivering products to their stores and customers. Family Dollar partnered with community in providing quality jobs to Duncan and the surrounding communities. Being one of the highest paying employers in the area, Family Dollar was successful in filling their employment needs.

23.6 Software

More companies are finding that logistics software can help them streamline many of their processes and can help their production cycle run more smoothly. Logistics software is designed to help businesses manage the steps in the production process, from the delivery of raw materials to the shipping of finished products to consumer outlets. Just as with other software related to supply chain management, the overall goals of logistics software are to boost profit margins and reduce cycle time in order to give the business a competitive advantage in the market. Logistics software focuses on transportation, which can be one of the most costly aspects of running a business, particularly with increases in shipping and gasoline prices. The software allows businesses to automate the management of mass quantities of transportation-related data so that it can be analyzed and so that the company can make informed decisions based on that analysis.

There is a wide range of benefits achieved simply by using logistics software. First, it can help companies get their products out to the public faster. By improving the delivery speed of necessary goods and by assisting in the selection of the most efficient shipping service, logistics software can cut days and even weeks off of the production cycle and delivery times. Additionally, logistics software greatly reduces human error. Completing the complicated calculations that were once necessary to analyze transportation data was not only time-consuming but was also all too frequently subject to human mistakes that often ended up costing the company a great deal of money. Since those calculations are imbedded in the software, those problems are no longer an issue. Furthermore, logistics software can help businesses save money. They can compare the rates of a variety of delivery agents and shipping services to help businesses locate the most cost-efficient based on more factors than price alone. Plus, the software helps businesses determine which method of transportation will be most efficient in terms of price and time for their products.

Logistics Software can generally be classified into one the following two groups: supply chain network design software and warehouse and transport management systems.

23.6.1   Supply Chain Network Design Software

Supply chain network design software is the planning software that has the potential to generate the most cost savings in supply chain costs. A major activity of network design software is to help companies decide the optimal location for a future facility, distribution center, or warehouse. The selection of a new site will have a major impact not only on a company's operating costs but also on its customer service levels.

New site selection software (also known as distribution-network optimization and modeling software) use optimization techniques to decide on the best new location a distribution center or a warehouse based on many input criteria such as transportation costs and capacity, distance to the customer, labor costs and availability, inventory costs and level, operation costs, and taxes and incentives.

Distribution-network optimization and modeling software typically asks modelers to input information about customer and supply location as well as desired service levels and lead times. The software then runs a series of "what if" analyses that calculate the effects of trade-offs between various costs and service factors and identifies optimal locations for warehouses and distribution facilities. For example, these tools use mathematical algorithms to conduct a "sensitivity analysis," which graphically represents how costs would increase the further away you move from a specified location. In short, the technology helps users make decisions based on facts instead of instinct.

In the Family Dollar case study, the company has used new site selection software to decide on the location of the new distribution center. Family Dollar input information such as customer and supply locations, customers projected demand, lead time, growth rates, inventory costs, labor and capacity to conclude that the new distribution center has to be located in Oklahoma or North Texas to achieve the company objectives.

There are several commercial as distribution-network optimization and modeling solutions, such as Insight's SAILS, Microanalytics' Optisite, LogicTools' LogicNet, and Supply Chain Designer from SSA Global (formerly CAPS Logistics).

23.6.2   Warehouse and Transport Management Systems

Warehouse and transport management systems (WMS/TMS) systems provide real-time views on material flows within the warehouse, that is, tracking and keeping note of the movement and storage of material within a warehouse facilitating the optimal use of space, labor, and equipments (ARC News, 2004; Piasecky, 2003). From the managers' point-of-view this means that a WMS enables to optimize

TABLE 23.4 Functionality and Benefits of TMS

Functionality	Claimed Benefits
Optimize delivery routes for retailers	Improve processes, drive saving and manage more business without increasing resources
Operational transportation control: booking, labeling and document printing, track and trace	Improve operational costs in collecting goods from suppliers and delivering to distribution center. Improved utilization of fleet
Transportation business control: load tendering	Reduced costs: improved invoicing and tendering system
Route planning	More precise scheduling: managing scale, constraints, and seasonal fluctuations of its operations
Real-time information	Streamlining reporting and analysis procedures: to achieve real-time inventory information

Source: Helo, P. and B. Szekely, *Industrial Management and Data Systems* 105 (1/2) 2005.

transactions to and from warehouse operators, recognize problem areas and major shifts in activity levels and patterns, while making it possible continuously to determine performance indicators, such as productivity, shipping and inventory accuracy, warehouse order cycle time, and storage density (Frazelle, 2002).

Typically WMS systems are well connected to material handling automation and transportation systems. Some WMS systems also include a route planning functionality that makes them related with the TMS systems. Some of the large suppliers of these software products are, amongst others, Marc Global Services, PeopleSoft, SSA Global, Microsoft Business Solutions, Oracle Corporation, JD Edwards, PULSE Logistics Systems (Helo and Szekely, 2005), and Global Concepts.

Transportation management systems are software applications that facilitate the procurement of transportation services, the short-term planning and optimization of transportation activities, and the execution of transportation plans with continuous analysis and collaboration (Rider, 2003, p. 62). TMS typically provide route planning, transportation control features, and advanced reporting. These software packages also automate the work of traffic controllers and provide a systematic way to generate documents and labels. Table 23.4 presents the functionality provided by typical TMS systems and the claimed benefits for business.

23.7 Teaming with Government, Industry, and Academia

In today's competitive environment, there are many opportunities to create partnerships with federal, local and state governments, other industrial companies, and universities. These partnerships take many forms including consortiums, government programs, and industrial coalitions. One of these partnerships is the Center for Engineering Logistics and Distribution (CELDi; visit site http://celdi.ineg. uark.edu. for information on CELDi), which is a National Science Foundation (NSF) Industry/University Cooperative Research Center (I/UCRC). The I/UCRC program at NSF encourages universities to work with companies and government agencies in the area of expertise that the center addresses.

Center for Engineering Logistics and Distribution is a multi-university, multi-disciplinary NSF sponsored I/UCRC. Research endeavors are driven and sponsored by representatives from a broad range of member organizations, including manufacturing, maintenance, distribution, transportation, information technology, and consulting. Industrial partners serve as the "thoughtleaders" with strong existing and ongoing financial commitment to logistics research. This partnership between academic institutions and industry represents the effective integration of private and public sectors to enhance the United States' competitive edge in the global market place. CELDi provides integrated solutions to logistics problems, through research related to modeling, analysis, and intelligent-systems technologies.

References

Bowersox, D.J. 1992. *Logistical Excellence.* Burlington, MA: Digital Press.

Camp, R.C. 1995. *Business Process Benchmarking: Finding and Implementing Best Practices.* Milwaukee, WI: ASQC Quality Press.

Collins, A., M. Henchion, and P. O'Reilly. 2001. Logistics customer service: performance of Irish food exporters. *International Journal of Retails and Distribution Management* 21 (1): 6–15.

Council of Supply Chain Management Professionals (CSCMP). http://cscmp.org/default.asp.

Coyle, J.J., E.J. Bardi, and J.C. Langley. 1996. *The Management of Business Logistics.* 6th ed., New York, NY: West Publishing Company.

Ellram, L. 1995. A managerial guideline for the development and implementation of purchasing partnerships. *International Journal of Purchasing and Materials Management* 31 (3): 10–16.

Emerson, C.J. and C.M. Grimm. 1996. Logistics and marketing components of customer service: an empirical test of the Mentzer, Gomes and Krapfel model. *International Journal of Physical Distribution and Logistics Management* 26 (8): 29–42.

Emerson, C.J. and C.M. Grimm. 1998. The relative importance of logistics and marketing customer service: A strategic perspective. *Journal of Business Logistics* 19 (1): 17–33.

Frazelle, H.E. 2002. *World-Class Warehousing and Material Handling.* New York, NY: McGraw-Hill.

Helo, P. and B. Szekely. 2005. Logistics information systems: an analysis of software solutions for supply chain co-ordination. *Industrial Management and Data Systems* 105 (1/2): 5–19.

Livingstone, G. 1992. Measuring customer service in distribution. *International Journal of Physical Distribution and Logistics Management* 22 (6): 4–6.

LaLonde, B. and M. Cooper. 1989. *Partnership in Providing Customer Service: A Third-Party Perspective.* Oak Brook, IL: Council of Logistics Management.

Lambert, D.M. 1993. Developing a customer-focused logistics strategy. *International Journal of Physical Distribution and Logistics Management* 23 (6): 12–19.

Lambert, D., M. Emmelhainz, and J. Gardner. 1999. Building successful logistics partnership. *Journal of Business Logistics* 20 (1): 165–181.

Mantodt, K.B. and F.W. Jr Davis. 1993. The evolution of service response logistics. *International Journal of Physical Distribution and Logistics Management* 23 (6): 56–64.

Naumann, E. 1995. *Creating Customer Value: The Path to Sustainable Competitive Advantage.* Cincinnati, OH: Thomas Executive Press: South-Western College Publishing.

Rey, M. 2001. Supply Chain Collaboration. *Global Purchasing and Supply Chain Strategies*, Last viewed (Sept., 7, 2005), Available online at: http://www.bbriefings.com/pdf/976/6_rey.pdf#search= 'logistics%20collaboration'.

Rider, S. 2003. Driving SCM leadership: The advantages of integrating TMS to your SAP suite. *World Trade* 10 (11): 62–3.

Sharma, A. and D.M. Lambert, 1994. Segmentation of markets based on customer service. *International Journal of Physical Distribution and Logistics Management* 24 (4): 50–56.

Simatupang, T. and R. Sridharan. 2004. Benchmarking supply chain collaboration: an empirical study. *Benchmarking* 11 (5): 484–504.

Strozniak, P. 2003. Collaborative logistics. *Frontline Solution*, Last viewed (Sept., 7, 2005). Available online at: http://www.frontlinetoday.com/frontline/article/articleDetail.jsp?id=65402.

Tate, K. 1996. The elements of a successful logistics partnership. *International Journal of Physical Distribution and Logistics Management.* 26 (3): 7–13.

Zairi, M. 1996. *Benchmarking for Best Practices.* Oxford, U.K.: Butterworth-Heinemann.

24

Logistics and the Internet

Teodor Gabriel Crainic
University of Québec in Montréal

M. Grazia Speranza
University of Brescia

24.1 Role of the Internet in Logistics

The beginning of electronic exchanges may be traced back to the 1970s. Exchanges were limited at that time to monetary transfers among a number of large firms and financial institutions. The 1980s saw the introduction of Electronic Data Interchange (EDI) systems that targeted the traffic of orders, waybills, invoices, and so on, with the promise of a significant reduction in paperwork and associated costs. EDI was embraced by large firms, particularly in the manufacturing and transportation sectors, and their suppliers, distributors and other partners did not really have another choice but to follow. These early efforts were marred by interoperability issues at the level of the communication hardware and software, which translated in high investment costs for firms that needed to exchange with partners using different systems.

A number of other systems based on electronic exchanges were also introduced during this period. Most were designed for utilization restricted to particular communities and quite independent of other electronic-exchange systems. Trading systems for traders on stock markets and airline-reservation systems for travel agents and airlines are two such examples. In the transportation sector, Intelligent Vehicle-Highway Systems (IVHS) were aimed at a larger audience, agencies monitoring and controlling traffic in major cities and intercity corridors and the people driving their automobiles on these roads, but were still ignoring large pans of the industry and the public, as well as the EDI systems motor carriers could have had. These systems evolved significantly during the 1990s and the beginning of the third millennium toward more open and integrated systems. More and more sophisticated and integrated information systems have impacted the organization of all companies, starting from the large ones and progressively involving the small and medium enterprises (SMEs). For example, reservation systems are now targeting the entire travel and leisure industry, bringing potential customers directly in contact with potential service providers. IVHS have similarly evolved into Intelligent Transportation Systems

(ITS) for all transportation modes and users, for passenger and freight transportation over all ranges of distances. This evolution and integration, far from being over, continue today, supported by the continuous evolution of electronic devices. It paralleled the evolution of the business community from the electronic commerce (e-commerce) concept to the electronic business (e-business) paradigm. The whole range of these systems makes up the so-called electronic society.

In this chapter we discuss the role and the impact that Internet has had on logistics (see also Chopra and Meindl, 2004; Shapiro, 2001; Simchi-Levi et al., 2003). In fact, we will refer as often to supply or value chain management as to logistics since the terms increasingly refer to the same concepts. Traditionally, and particularly in North America, the two terms designated two different points of view. Logistics referred to the planning and control of material flows and related information within a company, often restricted to the management of inventory and distribution activities. The concept of supply chain management has been developed from two different perspectives, the purchasing and supply management and the transportation and logistics management. It claimed a more comprehensive view of the firm activities and its linkages with partner organizations to ensure timely procurement and availability of materials for the firm as well as the production and delivery of services and goods to customers. The utilization of the term "value chain" is meant to emphasize the goal of this global management to produce value for all participants involved in the chain. The increased utilization of the Internet, electronic data exchanges, and electronic transactions has added new meaning and power to the value-chain concept and has brought logistics to adopt essentially the same holistic view of the relations of the firm with its upstream and downstream partners. This evolution is marked by the introduction of the term "logistics management" and the transformation of the U.S. Council of Logistics into the U.S. Council of Supply Chain Management Professionals.

The first major type of impact Internet has had on logistics is thus related to the possibility to integrate a company into a supply chain and, more generally, into the complex network of suppliers, partners, intermediaries, and customers the company belongs to. This has been made possible by means of typically Internet-based integrated information systems. Thanks to such systems, a more global view of logistics has become possible and, together with it, a more global optimization of the system has become possible. The second major type of impact is related to the possibility of using the Internet for e-business, a type of use that deeply modifies the way sellers and buyers interact.

While, at least in principle, the first type of impact has influenced the organization and the management of companies, it did not influence their suppliers and customers. It "simply" improved the efficiency and effectiveness of the supply chain. On the contrary, the second type of impact did modify the interactions between buyers and sellers, not only in terms of type and quality of the interactions but also in terms of selection of the suppliers and contacts with the customers.

It is also worthy to note that operations research-based methodologies empower electronic logistics and value chain management. Indeed, while information technologies provide the means to timely and accurately exchange information, operations research provides the means to transform this raw data into meaningful information and decisions. The following sections illustrate these concepts.

24.2   Internet-Based Information Systems

We briefly discuss here the evolution of information systems up to the advanced phase that includes Internet-based systems under the light of the impact that such systems have had on logistics and supply chain management. Companies have enhanced over the years their efficiency and effectiveness by streamlining and improving their internal operations, from procurement to warehousing, from production to distribution. Today's global competition has imposed not only a continuous re-assessment of internal processes but also a continuous enhancement of the efficiency and effectiveness of relations with external companies to ensure one's competitive place in the market. The objective of reducing costs while improving the service level can only be reached by a global optimization of the supply chain and a constant control of uncertainties.

The development of Inter-Organizational Information Systems (IOS) can be categorized in four phases (Shore, 2001; Williamson et al., 2004). In phase one, information technology did not contribute significantly to the information system. In phase two, the development of EDI had a tremendous impact on the automation of information flows. Many transactions, such as those related to ordering and invoicing, could be processed through EDI. The need of a value-added network (VAN), however, reduced the flexibility of suppliers that were connected to more than one customer because they needed to support different technologies. The number of EDI standards grew dramatically but currently universal standards allow companies to exchange data seamlessly between their computer systems. The phase 3 is characterized by more integration. The so-called Enterprise Resource Planning (ERP) systems are the result of the evolution of Manufacturing Resource Planning (MRPII) systems. Although ERP systems allow integration of companies with their suppliers and customers through an integrated database system, these systems were designed to integrate the various functions of an individual company. Moreover, their architecture is typically closed. Phase 4 is thus based on the development of web technologies, such as XML and Java, that enable the integration of the information system of a company with those of its strategic partners. Two-ways information flows improve quantity and quality of communication, reduce costs and times and increase satisfaction of all partners.

Obviously, moving from a phase to a successive one is costly for a company in financial terms but even more in terms of the enormous impact such a move has on the organization culture and operations as well as on the business processes and relations. A limited number of companies have adopted an integrated system. The variety of individual information systems represents a major obstacle to the foreseen integration. Technologies represent the enabling means to reach the proper level of communication in supply chain management, but are useless or may even have a negative impact on the business if not properly chosen, used and governed by managers.

The systems of phase 4 make possible to a company the accomplishment of a variety of important goals that can be summarized in the following (Simchi-Levy et al., 2003): (*i*) collect information and make it available to all parties involved in the supply chain; (*ii*) access any data from a single point of contact; (*iii*) plan activities based on information from the entire supply chain; and (*iv*) collaborate with partners.

Good logistics and supply chain management are made up of information and decision technologies. Information must display a number of fundamental characteristics to be useful when making decisions. Information must be accurate and accessible in due time. Delays may make information useless or even negative. Information must be of the right kind. Since information is costly, it is essential that valuable information be collected and useless information avoided. The first of these four characteristics allows partners to be aware of the situation through reliable and timely information. When delays are drastically cut to the minimum, the efficiency of decision-support technologies is enhanced and better and more timely decisions can be taken.

The second goal, the single point of contact, means that information is taken from a single source, what makes the information identical to everybody. It also means a coherent and timely fusion of electronic and more traditionally collected data (e.g., sale figures). While we devote the following section to the third goal, the fourth, the collaboration with partners, is probably the most evident. Retailer–supplier partnerships and relations with third-party logistics (3PL) providers are among the most important types of collaborations in the e-business environment.

24.3 Global Optimization through Global Information

The availability of information to all partners in the supply chain allows companies to take better decisions, decisions that are timely, based on reliable information and, most important, taken from a global and not local viewpoint. If a department takes decisions on the basis of its own goals, it may completely miss the company objectives, even work against them. The minimization of the costs of a company is not necessarily reached through the minimization, for instance, of the transportation costs alone. Such local optimization would mean always using the cheapest, and typically slowest, transportation mean and serving customers with direct routes at low frequency. Only a holistic view of the entire system allows companies to reach their objectives. Internet-based information systems are the most powerful enabling mean to move from a sequential optimization toward a global optimization.

If a supply chain would be isolated from other supply chains, a global optimization would really be possible. The fact that most supply chains interact with other supply chains may create conflicts among different supply chains, make the optimization of a specific supply chain more difficult to attain, and force companies to find a compromise. In any case, the compromise should be found at the highest possible level and not where it is unnecessary. Moreover, appropriate operations research-based methodologies are required to evaluate the many possible strategies and trade-offs and select the most appropriate ones for the company objectives.

A management practice that has pushed companies to move from a local and/or sequential to a global optimization is the vendor-managed inventory (VMI) policy. Consider a distribution system with a central warehouse from which products are shipped to retailers. With a VMI policy, the inventory of the retailers is controlled by the warehouse that organizes the deliveries in the globally most efficient way guaranteeing that no stock out will occur to any retailer. In a traditional distribution system, which we call retailer-managed inventory (RMI) system, the retailers decide when and how much to order, and in most cases will use an order-up-to level policy. Then, the warehouse has to organize the distribution. The decisions of the warehouse are constrained by the decisions of the retailers and for this reason less cost-effective. In Bertazzi et al. (2006), the savings that can be obtained with the VMI policy are quantified. Both the VMI and the RMI systems are organized at best, that is in such a way that the total cost of each system, that includes fixed and variable production costs, fixed and variable transportation costs and inventory costs, is minimized. Such optimization is obtained by means of operations research methodology. It turns out that a VMI policy allows savings that range from 8% to 50%, depending on the specific situation. The savings are higher if the production costs are small and the prevalent part of the expense is in transportation, because the major factor of saving is in the transportation activities that benefit the most from the coordination. The savings might be obtained even without the adoption of a VMI policy but through a policy of partnership with the retailers oriented toward the global optimization of the system performance.

A major issue to consider in order to move from local and/or sequential to global optimization is the motivation for the partners to collaborate, including the distribution among partners of the savings coming for the global optimization.

24.4 Electronic Business

The electronic commerce concept was introduced in the early 1990s and signaled the beginning of the commercialization of the Internet. E-commerce was essentially about transactions conducted between business partners to buy and sell goods and services. This definition is now considered as "old" and restrictive. The utilization of the e-business term signals a holistic view of business using electronic means: buy and sell, serve customers, share knowledge, set up intra- and inter-organization communications, collaborate and, eventually, integrate vertically (e.g., suppliers and distribution channels) or horizontally (e.g., community-of-interest marketplaces) with business partners, and so on, for competitive advantage. Logistics and transportation are significantly impacted by this evolution.

The evolution of e-business has been strongly affected by the speculative stock-market bubble of the turn of the millennium, the "new economy" hype and the fall of the so-called dot.com companies. This negative impact seems to have been overcome, however. There is significantly less hype now, other emerging economic sectors capturing the attention of the media and the stock-exchange traders (the huge success of the Google public offerings not withstanding). Yet, the pace of the penetration of electronic exchanges in business and society is accelerating and is becoming a fact of life.

24.4.1 Many Incarnations of Electronic Exchanges

There are several types and levels of electronic exchanges, often classified according to the nature of the transaction.

Business-to-business (B2B) is what most people think of when the e-business topic is mentioned. It refers to electronic exchanges and transactions among firms and organizations, irrespective of the amplitude or scope of the transactions involved. B2B is the direct descendant of e-commerce and earlier EDI efforts. EDI is now, in fact, one of the many facets and instruments of B2B e-business. Yet, it is the Internet that provides the enabling support for B2B e-business as for the other facets of the electronic society. We take a closer look at B2B exchanges in Section 24.4.2.

Business-to-customer (B2C) refers to relations between the firm and its customers. B2C e-business covers every type of activity the firms undertake to market and sell their products: passive websites describing the firm and its products, electronic broadcast of announcements and catalogs, interactive websites for direct sells to customers, individual e-messages to registered customers to maintain a high level of customer awareness relative to the firm, promote particular goods or services, or announce "new" and "spectacular" offers, and so on.

Firms at all levels of the digital scale, from "traditional" brick-and-mortar to "fully" virtual, engage in B2C activities. Almost all firms of the so-called developed world and many in the developing world implement web sites for promotion purposes. A continuously increasing number is also offering transaction capabilities that parallel or complement the traditional distribution channels, paper catalogs sent by slow mail, call centers for taking orders and customer support, phone solicitation, service counters, intermediaries (e.g., travel agencies), regular stores, etc. In this sense, one may qualify most brick-and-mortar firms as "click-and-mortar." Firms selling services or physical goods by electronic means exclusively have obviously developed their core business around the B2C concept. Firms such as Amazon.com that retail directly to customers from production or warehousing facilities illustrate the first category while virtual travel agencies (e.g., Expedia.com) belong to the second.

In all cases, however, the design and the operations of the physical logistic system are instrumental to the performance of the firm. With the exception of service-selling firms and of companies selling "virtual" products such as music, pictures, and software, all B2C activities involve the setup and operation of a distribution system. When regular or express carrier services are used to ship products to customers, the logistic system corresponds to a network of warehousing facilities, whereas the value-chain processes mainly address issues related to product procurement (selection of product providers, eventually), inventory management, and dispatch of product deliveries. In all other cases, routing issues have also to be addressed. As for regular operations of most firms, operations research methodology is at the core of the efficient and profitable logistics network setup and operation.

Electronic systems that provide services to individuals selling to organizations are identified as customer-to-business (C2B). Such services are not yet very much developed, however. Moreover, their impact on the logistics activities of the firm is limited. The impact on logistics of customer-to-customer (C2C) and government-to-citizens (G2C) electronic exchanges appears also rather limited but their role in the contemporary society is very important. C2C refers to all Internet-enabled exchanges between private citizens: personal web sites, music, and video exchanges, etc. In a broad sense, web sites that facilitate such exchanges belong to the C2C category, including for-profit organizations that, for example, manage auction sites dedicated to individual sellers and buyers (e.g., eBay). The G2C denomination encompasses all the electronic services that governments, at all levels, make available to the public, including commercial firms and not-for-profit organizations.

This very succinct overview of the various types of electronic exchange systems does not cover all aspects of the electronic society. The empowerment of the civil society, in particular through the proliferation of web sites representing nongovernmental organizations, special interest groups, and discussion forums (the so-called blogs) is only partially covered by the types defined earlier. This empowerment may and does impact the policies of both governments and industries. Thus, for example, a location for a new plant may be abandoned because of pressure from special interest groups through electronic mailing and discussion groups. Hence, a number of additional factors have to be included in the decision process, particularly at the strategic level. On the other hand, this does not change the nature of the logistics planning and management processes and, thus, given the limited space available, we do not address this topic in this chapter.

24.4.2   Business-to-Business Electronic Exchanges

Business-to-business e-business may further be analyzed according to the scope of the exchanges: either centered on a specific company or on a group of firms that collaborate, compete, or engage in both types of activities.

Most company-centric B2B activities concern the firm in relation with its suppliers or customers. The first case is identified as many-to-one B2B to reflect the flows of materials from many suppliers toward the firm and concerns the acquisition of materials and supplies. Traditionally, one differentiated between direct supplies for the core activities of the firm (e.g., production of the goods the firm sells) and other, so-called indirect, materials (e.g., supplies for the back-office), on the basis that the former requires long-term contracts directly and personally negotiated, while the latter may be decided on a case-by-case basis given current availabilities and prices. This differentiation persisted in the e-business era when the sourcing of direct materials was still deemed too strategic to be left to electronic negotiations and e-procurement, particularly on the spot market, was to be used for indirect materials only.

While not underestimating the value of direct negotiations and long-term contracts, one observes that this separation is fading away. Electronic markets are increasingly used to negotiate and acquire a wide gamut of supplies, materials, and services. Firms may negotiate on "open" electronic markets (e.g., the commodity markets) or implement private e-marketplaces where they usually control the negotiation processes and tools. The auctions by which many manufacturing and retail firms acquire long-term transportation services for their distribution routes illustrate the second case.

Business-to-business communities of interest are bringing together the main participants to the value chain within particular industrial sectors (e.g., steel, automobiles, chemicals, etc.) to facilitate communications and exchanges, help discover potential suppliers or customers and build partnerships, enforce quality standards, and implement efficient electronic markets. The e-business environment and the Internet are fostering the cooperation between "small" firms for group purchasing of items of common interest. They are also significantly enlarging the economic field, facilitating the discovery of potential partners in geographic zones not usually associated to one's own business practices.

On the sell side of B2B, one considers the firm in relation to its customers and the B2B is thus identified as one-to-many B2B. These relationships may take the form of B2C electronic marketing and selling activities, irrespective of the nature of the "customer," another firm, or organization, one's own distribution system or store, or an individual person. Such activities usually involve electronic markets as in the participation as seller to the communities of interest identified earlier.

One also identifies as B2B activities the coordination of the activities of the firm with those of its suppliers or customers. In a just-in-time environment, for example, one aims to coordinate its production, inventory, and distribution activities to the selling cycle and inventory management of its customers. These activities may also take place in a many-to-one environment where, for example, the arrival of supplies from various sources is made to match the production schedule. B2B coordination may also be enforced along the supply chain between a product provider, the transportation or distribution systems, and the final customer. Such e-logistics activities require advanced monitoring and information technologies and sophisticated operations research-based decision support systems.

Firms may also participate in many-to-many electronic exchanges. The earlier-mentioned communities of interest are an example of such activities; the participation to multi-lateral e-markets constitutes another. In multi-lateral electronic marketplaces, firms may be offering or acquiring goods or services. Some firms may do both. Many-to-many B2B may also be implemented within the organization, linking different departments, divisions, or participating firms. Private marketplaces to distribute tasks more efficiently constitute a prime example of such B2B activities.

Most B2B activities are performed as part of the value (supply) chain. They require an information exchange technology, as well as a number of modifications to the organizational structure and culture of the firm. They may also impact the actual procurement, production, and distribution activities and consequently the associated planning and operation processes. The development of models and methods to design and operate efficient e-logistic chains is still in its infancy but constitutes an exciting and challenging field for research and technological transfer.

Many B2B activities also involve participating in e-marketplaces and auctions. From the point of view of the market manager, be it the firm or a third party, the issue is what type of market to offer and how to design it. When one is only interested in buying or selling, the question is how to profitably participate in e-markets. We address these issues in the following section.

24.5 Electronic Auctions

Marketplaces emerged early on in human history as physical locations (e.g., the village square) where goods could be sold, bought, or exchanged. In time, marketplaces have grown in scope, size, and sophistication, but their main goals still consist in discovering "partners" as well as negotiating the quantities to exchange and the corresponding prices. Electronic marketplaces are no exception. This section presents a brief description of e-markets following Crainic and Gendreau (2003).

In unilateral markets, one seller negotiates with several buyers, one-to-many markets, or, alternatively, one buyer deals with several sellers: the many-to-one case often encountered in firm procurement processes. Several buyers and sellers meet in multi-lateral markets. In such many-to-many settings, participants may be either buyers, or sellers, or both. Markets may be dedicated to one product only or encompass negotiations on several commodities simultaneously (whatever the setting, several qualities or grades may be defined for each commodity). Moreover, products may be indivisible (e.g., a container) or divisible (e.g., telecommunication capacity) and may be traded one or several at a time. Markets aim to be efficient, either locally (i.e., maximize the benefit of the unique seller or buyer or the surplus of a many-to-many market) or socially (i.e., maximize the overall social welfare of the participants). In the case of freight transportation, the markets where regular distribution routes are auctioned off to carriers for a certain period of time (e.g., from one to three years) belong to the one-to-many case, where the commodities (routes) may be either considered indivisible (service for the whole volume on a route is sold to one carrier only) or divisible (several carriers may serve the same route). Freight exchanges where loads (e.g., containers or full truck loads) of different shippers are offered to several carriers may be described as many-to-many, multi-commodity, indivisible markets.

Auctions constitute a broad and important class of market organization that involves a formal design and negotiation process (Abrache et al., 2004; Caplice and Sheffi, 2003; Pekeč and Rothkopf, 2003; Rothkopf and Park, 2001; Sheffi, 2004). Participants to an auction declare bids, that is, they indicate the quantities they are ready to buy, sell, or exchange, as well as the corresponding prices (several other conditions, such as technological or product quality restrictions, on the objects or services traded are usually part of the bid as well). Several participants will generally declare bids on the same object or group of objects. The auctioneer receives the bids and ensures that the negotiations proceed efficiently and fairly, according to the rules of the market. In particular, it verifies the legality of the bids and determines, through an allocation mechanism, which one among the conflicting bids wins (who gets what and how much) and at what price. Markets are said to be optimized when an optimization formulation and method is used as allocation mechanism.

Assuming bids include the true valuations of participants for the objects or services on the market, a simple and direct market mechanism would be "description of items on the market; call for proposals; submission of bids by participants; winner determination through the allocation mechanism; announcement of winners and implementation of the deals." There are very few (if any) opportunities to implement such an ideal mechanism. The primary cause lies with the lack of information available to the auctioneer regarding the true intentions, possibilities, and limits of participants and, in particular, their true valuations of the items on the market. Participants are generally reluctant to disclose such information, even to a "neutral" agent under secrecy commitments, because it includes proprietary data on the economics and processes of the firm. Moreover, such data may not even be completely available to the firm. Thus, for example, the "true" valuation of an item often becomes clear only during the negotiation according to the other items that are on the market or have been acquired.

Approximate market mechanisms are therefore the norm. One of the most widely used market mechanisms, particularly when public institutions and governments are involved, are the so-called closed-envelope mechanisms where participants bid once and the auctioneer selects winners based on these proposals. "Lowest (or highest, as the situation commands) bid wins" is the usual selection criterion, with consideration given to characteristics such as product quality or participant reliability to perform the service. It is known, however, that this mechanism does not offer any guarantee of efficiency and that, in fact, it often yields inefficient allocations. Second-price criteria, where the highest bidder wins but pays the second highest price (the so-called Vickrey-Clarke-Groves auctions), have been extensively studied, since they present the advantage of inciting participants to bid truthfully, but little used because they are open to manipulations by participants (e.g., signaling and collusion).

Multi-round mechanisms are increasingly proposed to address these shortcomings. Multi-round markets attempt to bring participants to progressively "reveal" their true intentions and valuations and, thus, to achieve the best allocation possible. In such a setting, participants make initial bids and the auctioneer determines a temporary allocation and prices. The auctioneer returns this information to participants (according to predetermined privacy rules), which may then modify their bids (in quantity, value, composition, etc.) or submit new ones. Predetermined rules guide the definition and modification of bids, pace the auction, and determine its end.

To illustrate, consider a periodic multi-lateral market for geographically dispersed heterogeneous commodities (e.g., natural resources) where buyers express technological requirements on the mix of goods they purchase (Bourbeau et al., 2005). The market opens up periodically (e.g., once a day, once a week) and the agents in the corresponding economic sector negotiate using the centralized and optimized market. All possible multi-lateral trades are thus solved simultaneously. Commodities are classified by type or quality. Sellers may offer several commodity types separately or mixed up in lots. Buyers need to combine different grades and qualities, while technological constraints limit the quantities of each type of commodity they may acquire. Transportation costs are significant. The objective of this type of markets is to explicitly optimize both the production and transportation of resources in the industry.

Let K be the set of products. The definition of a product is domain specific. Generally speaking, however, a product is a generic classification reference, such as a quality of ore, a wood species, a type of grain, etc. It is a commodity differentiated by type and quality. Products are combined in lots that sellers and buyers may trade. To simplify the presentation (but with no loss of generality), a lot is sold by one and only one seller, is attached to a specific location and has its own idiosyncratic quality. Since a producer may sell more than one lot, the number of lots may exceed the number of producers. More importantly, a lot has its own composition of various products (e.g., oil or ore grade). Let $b(k,l)$ denote the proportion of product k in lot l, where $\sum_k b(l,k) = 1$. A maximum quantity Q^l is available for lot l. Lots are to be acquired by buyers grouped in set J. Buyers face technological constraints and use proprietary recipes and, thus, desire to acquire particular mixes of products. Given the diversity in product characteristics displayed by the lots on sale, buyers express preferences for the various lots by specifying two quality adjustment coefficients, one multiplicative, $r(j,l)$, and one additive, $s(j,l)$, which indicate that for buyer j one unit of lot l is equivalent to $r(j,l)$ units of a standard lot and is worth $s(j,l)$ monetary units more than a standard unit. Let $M(j,k)$ and $m(j,k)$ denote the maximum and minimum proportions, respectively, of product k that buyer j is ready to accept in the mix it purchases, while Q^j indicates the maximum total quantity of all products buyer j requires. Unit transportation costs between the seller of lot l and buyer j are denoted $t(j,l)$.

At the core of the market lies a market-clearing mechanism: a formal procedure that determines the "optimal" allocation, *who sells what (how much) to whom and at what price*, of goods given the participants and the market state. Assuming buyers and producers are willing to reveal information to the market, participants are asked to communicate their preferences, that is, their production cost or utility functions (their willingness to pay), together with all relevant technical information: transportation costs, technical constraints, etc. Denote by $U_j(\cdot)$ the utility function of buyer j and by $C_l(\cdot)$ the cost or production function of lot l. The decision variables of the optimized multi-lateral allocation mechanism, $q(j,l)$, indicate how much each buyer j buys of lot l. The corresponding optimization formulation is:

$$\text{Maximize } \sum_j U_j \left(\sum_l r(j, l) \, q(j, l) \right) - \sum_l C_l \left(\sum_j q(j,l) \right) - \sum_j \sum_l (t(j, l) - s(j, l)) \, q(j, l)$$

$$\text{Subject to } \sum_l q(j, l) \le Q^j \quad \text{for all buyers } j$$

$$\sum_j r(j, l) \, q(j, l) \le Q^l \quad \text{for all lots } l$$

$$\sum_l b(k, l) \, r(j, l) \, q(j, l) \le M(j, k) \sum_l r(j, l) \, q(j, l) \quad \text{for all buyers } j \text{ and products } k$$

$$\sum_l b(k, l) \, r(j, l) \, q(j, l) \ge m(j, k) \sum_l r(j,l) \, q(j, l) \quad \text{for all buyers } j \text{ and products } k$$

$$q(j, l) \ge 0 \quad \text{for all buyers } j \text{ and lots } l$$

The market maximizes buyer surplus minus the production and transportation costs subject to all technological constraints. Under classical, but strong, economic theory hypotheses, buyer utility functions are assumed concave (continuous, piece-wise linear, with strictly decreasing marginal buyer benefit), while costs are assumed convex (continuous, piece-wise linear, with strictly increasing marginal cost of producing a given lot). Using standard optimization methodology, the allocation mechanism yields prices and quantities that equilibrate supply and demand and that are the solution (dual and primal) of an optimization (maximization) formulation. It is unlikely, however, that such a mechanism will operate ever mainly due to the unwillingness of participants to disclose all their data, especially full supply and demand functions (assuming buyers and sellers do know these data) to even the most secure auctioning system. Multi-round auction mechanisms are therefore proposed in most such cases.

In many markets there is the need to negotiate items in bundles—allocation of airport take-off and landing time slots, wireless communications spectrum licenses, distribution routes, commodities for specific production recipes, loads to form closed (i.e., returning at the origin depot) multi-stop routes, assets in financial markets, supply chain formation and coordination, and so on. All these cases have one thing in common—they all trade items of different nature that are interrelated from the perspective of the participants. The value of one item to a participant depends on whether or not the participant managed to obtain (or sell) a number of other items. For example, the value of a load to a carrier will depend on whether or not one or several other loads may be secured to ensure a round trip may be constructed such that the vehicle is "always" moving loaded. Items may be complementary or substitutable. More precisely, if A and B are two items and $v(\cdot)$ denotes the evaluation function of the participant, A and B are said to be complementary if $v(A, B) > v(A) + v(B)$, and substitutable if $v(A, B) < v(A) + v(B)$. Loads in the previous example are complementary. Several loads that are available at about the same time, between the same pair of cities, are substitutable. Abrache et al., 2004; De Vries and Vohra, 2003; Pekeč and Rothkopf, 2003; Rothkopf and Park, 2001 review issues and contributions related to combinatorial markets.

Combinatorial auctions refer to marketplaces in which participants are allowed to bid on combinations, or bundles of items, and are increasingly considered as an alternative to simultaneous single-item auctions. Being able to bid on bundles clearly mitigates the exposure problem, since it gives the participants the option to bid their precise valuations for any collection of items they desire. On the other hand, combinatorial auctions often require the market maker and the participants to solve complex decision problems.

To illustrate, consider a simple combinatorial auction where a firm "sells" m different (indivisible) distribution routes, represented by set G, to n potential buyers (carriers), who are allowed to submit sealed bids on bundles of routes. A bid made by buyer j, $1 \leq j \leq n$, is defined as a tuple $(S, p(j, S))$ where S is a subset of routes in G and $p(j, S)$ is the amount of money buyer j is ready to pay to obtain bundle S. If route i is in bundle S, $\delta(i, S)$ is equal to 1 and is 0 otherwise. Define $x(j, S) = 1$ if S is allocated to buyer j, and 0 otherwise. The auctioneer must decide which bids win and which ones lose, under the condition that no single route is allocated to more than one carrier, and such that its revenue from the sale of the right to service the routes is maximized. The winner determination problem can be formulated as follows:

$$\text{Maximize} \sum_j \sum_{S \subseteq G} p(j,S)\, x(j,S)$$

$$\text{Subject to} \sum_j \sum_{S \subseteq G} \delta(i, S)\, x(j, S) \leq 1 \quad \text{for all routes } i \text{ in } G$$

$$\sum_{S \subseteq G} x(j, S) \leq 1 \quad \text{for all carriers } j, \, 1 \leq j \leq n$$

$$x(j,S) \in \{0,1\} \quad \text{for all bundles } S \subseteq G, \text{ or all carriers } j, \, 1 \leq j \leq n$$

The formulation is a weighted set packing problem (De Vries and Vohra, 2003). Similar to most combinatorial markets, this winner determination problem is NP-hard (Rothkopf et al., 1998), the number of possible bids growing exponentially with the number of routes on the market, and straightforward solution approaches do not work in most actual settings. Significant research is thus dedicated to combinatorial auction mechanism design issues, as well as to the associated operations research and combinatorial optimization methodologies. These efforts have already resulted in the successful utilization of combinatorial auctions to many applications. In the case of freight transportation, combinatorial auctions appear as powerful mechanisms to auction the right to service regular distribution routes and to design freight exchanges where many shippers and carriers meet to determine who will move the loads on the market and at what price.

Participants to combinatorial auctions face serious challenges, however, and research efforts start to be dedicated to this topic. The first and foremost challenge faced by participants in electronic auctions is clearly to identify which items are of interest to them and acceptable price ranges for these items. This is obviously further compounded in the case of combinatorial auctions by the need to build attractive bundles and to price them.

A major issue to be addressed by participants when there are significant value interactions is that of exposure risk. Exposure basically occurs when a participant is successful in obtaining only part of a bundle of complementary items it is interested in; more precisely, there is exposure when a participant obtains part of a bundle at a price that is higher than its value to it, but is not guaranteed to obtain the remainder of that bundle. Obviously, such a situation will be encountered when participants try to obtain items on different markets or when markets do not allow for combinatorial bidding. Thus, for example, a carrier may identify loads that would make a closed "full" route on the different markets set up in the regions visited. These loads are still complementary for the carrier, but it is impossible to negotiate them together. A similar situation can occur when participants bid simultaneously in parallel auctions for items that display significant substitution effects. An extreme case is the situation where a participant is interested in a single item among a collection of items (e.g., loads or routes) that are auctioned off simultaneously. Should the participant bid too aggressively, it may very well end up clinching two or more of these items. On the other hand, if the participant is too timid in its bidding, it may end up with none of the items.

The two previous situations highlight the need for participants to develop appropriate bidding strategies in relationship with their needs and their tolerance to risk. These strategies must also account for the presence of competitors on the markets: both the number of these competitors (or a reasonable estimate of it) and their bidding behavior are key elements in this assessment. It must be emphasized, once again, that the nature and number of markets a participant is involved in have a dramatic impact on the type of strategy a participant needs to develop. For instance, it is much easier for a participant to develop an effective strategy for obtaining a bundle of complementary items if all these items are traded on a single marketplace that allows combinatorial bidding.

Advisors may be defined as specialized decision-support software specifically designed to support participants in the complicated negotiation processes involved in the most sophisticated electronic markets, such as simultaneous auctions for several goods, sequential auctions or combinatorial auctions. Advisors may have several functions, according to the degree of sophistication of the firm in relation to Internet and the cyberspace: identify promising marketplaces and loads, assess the competition, build and price bids, determine a bidding strategy, develop contingency plans in case one ends up losing on loads that had been identified as attractive, conduct the negotiation, close the deal, etc. For most of these functions, the associated models and methods are encapsulated into software agents that help automate the negotiation process [see Crainic and Gendreau (2003) for a more in-depth discussion of this issue].

24.6   Case Example: FreightMatrix.com

FreightMatrix (FMX) is an e-marketplace that supports shippers and carriers with the objective of providing services and optimizing costs and profits.

The web site FreightMatrix.com has been created by i2 Technologies, a company leader in the production of software for logistics and supply chain management. FreightMatrix.com has two sections: FMX private marketplace and FMX public marketplace.

FMX private marketplace is designed for the customers who intend to create their own web site and are interested to have a secure access to a number of logistic services. Large and mid-size shippers, 3PLs, and exchanges use FMX private marketplace to manage on-line their own supply chain or the supply chain of their customers. The services offered by the private marketplace are of operational, tactical, and strategic nature. Among the operational and execution services are shipment optimization, shipment execution, freight matching, tariff management, parcel and small package management, intelligent messaging, and monitoring. For example, Freight matching provides producers, distributors, and 3PLs with a tool to quickly and simply buy transportation for loads on a spot basis. Moreover, it allows carriers to identify demands for transportation of loads and improve the vehicles utilization rate. Parcel management can be integrated with existing systems, gives access to information in real time, and reduces errors and costs. Among the tactical planning services we find, rate quotation, transportation planning, transportation bid request (more about this service will be discussed later), and export trade compliance and among the strategic planning services: transportation bid collaboration (TBC) (more about this service will be discussed later) and network planning and optimization. The latter is a tool that supports location decisions, allowing the decision makers to evaluate cost trade-offs.

The advantages for the customers of FMX private marketplace are the use of high-quality software; complete saving of the costs for the private hosting, software, and maintenance; a variety of services; and the possibility of connection in a secure environment, easily accessible to all partners of the supply chain. FMX public marketplace is mostly of interest to carriers and small shippers. Several of the tools made available through the private marketplace are available here too, sometimes in a simplified form.

Two possibilities are offered by the FMX marketplace to shippers to get transportation services: transportation bid request (TBR) and TBC. Both services are available in the private and in the public marketplace. TBR is a short-period tool aimed at providing a quick answer to a transportation demand. Demands are posted individually or in groups. Carriers access this service and pick up the demands that best fit into their plans.

TBC is a more complex auction mechanism to manage long-term contracts. The basic structure of the auction is as follows. Shippers auction all or a major part of their network of routes. Then, carriers provide their offers on the auctioned routes. TBC optimizes the system by finding the cheapest assignment of offers to transportation demands. In Figure 24.1 the auction process is depicted. Three basic steps are shown: shipper bid support (SBS), the tool to create the shipper network, carrier bid response (CBR), the tool used by carriers to submit their offers and carrier bid optimizer (CBO), the tool used to optimize the offers and conduct what-if analysis. After the data have been imported from the database into SBS, the shipper creates its sets of routes, possibly combining routes in groups, and posts the sets on the web from which carriers download them to later insert their offers. The carriers usually have one week to prepare their bids. Then, the shippers may update their demands and, finally, routes and offers are taken from SBS and optimized by CBO. After the optimization takes place, the shipper may negotiate the contracts with the carriers or conduct another round. In such a case the process is repeated.

Shipper bid support deals with one shipper at a time. The shipper auctions its routes to several carriers. The only way for carriers to bid on sets of routes of different shippers is that a 3PL or a consortium presents the sets. Each route is represented as an origin–destination pair. The shipper can define requirements on a route, for instance, that the route has a stop in between, that there are two drivers (for long routes), that the load be shipped within 24 h. Moreover, for each route the distribution of the demand over time must be provided. The shipper may invite any set of carriers to bid. CBR is the software used by carriers to create their offers. Three types of offers are possible: the single bids, the combo bids, and the reserve bids. If the shipper has chosen the "partial bidding" option, the carriers have the possibility to submit, for each route, an offer for a demand lower than that required by the shipper. CBO in this case may assign the route to more than one carrier. The single bids represent the traditional way to submit bids. The price offered by a carrier for a route is the price per mile multiplied by the distance and the number of trips plus a team charge, in case two drivers are required. A combo bid

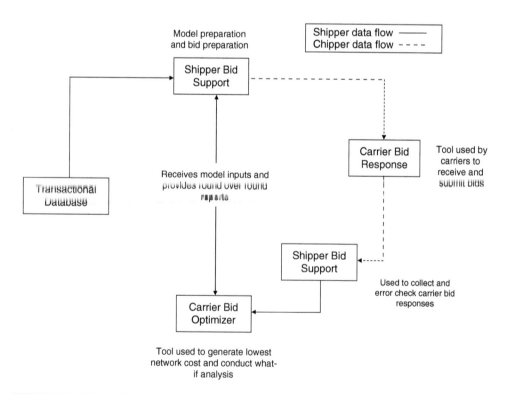

FIGURE 24.1 The auction process.

is a bid submitted on a set of routes. In this case, the price for the individual routes may be different from each other. To clarify the complexity of creating bids for carriers, consider the following example. A set of routes A, B, and C with demands 125, 99, 1200 (total demand 1424) is posted together with other sets of routes D and E with demands 500 and 750 (total demand 1250). If a carrier is interested to any of the two but not to both he can make a combo bid on the two sets and set a limit on the possibility to cover the demand equal to 1424. The reserve bid allows a carrier to set a price for ranges of demand. If a carrier offers a reserve bid for a certain amount of demand, the price holds for any lower demand. Take the following example. A route has demand 200. Carrier A submits a single bid with price $1.25 for a demand equal to 100. A carrier B submits a combo bid with price $1.21 for a demand equal to 175. In this case, the route is not covered by any bid. If a carrier F submits a reserve bid for $1.28, then the route will be assigned to carriers B (for a demand equal to 175) and F (for a demand equal to 25). CBO optimizes the offers by assigning the routes to the bids in such a way that the total cost is minimized. A route may happen not to be covered by any bid for any of the following reasons: no carrier submits a bid that includes the route, the price offered by the carriers is higher than the maximum set by the shipper, the carriers cover only part of the demand. Since more rounds are possible, it may happen that a route is not covered in a first round but is covered in a successive round. It is possible for a shipper to ensure that a reliable carrier gets a minimum amount of work or that an unreliable carrier is penalized.

Carrier bid optimizer can also be used to conduct what-if analyses. A typical use for the shipper is to see what changes if the expense of a carrier is limited to a lower bound or an upper bound or if the number of loads assigned to a carrier is limited. Changes in the prices can be tested as well. If more rounds are required the whole process would be repeated. The carriers can see the routes assigned to them and modify the bids. A carrier cannot see the routes assigned to other carriers. There is no limit in the number of rounds that a shipper can conduct. However, the average number of rounds is two and it has been seen that the best results are obtained with two to four rounds. A typical engagement lasts from few weeks to several months.

References

Abrache, J., T.G. Crainic, and M. Gendreau, Design Issues for Multi-object Combinatorial Auctions, 4OR 2(1):1–33, 2004.

Bertazzi, L., G. Paletta, and M.G. Speranza, Minimizing the Total Cost in an Integrated Vendor-Managed Inventory System, to appear in *Journal of Heuristics*, 11(5–6):393–419, 2005.

Bourbeau, B., T.G. Crainic, M. Gendreau, and J. Robert, Design for Optimized Multi-lateral Multi-commodity Markets, *European Journal of Operational Research* 163(2):503–529, 2005.

Caplice C. and Y. Sheffi, Optimization-Based Procurement for Transportation Services, *Journal of Business Logistics* 24: 109–128, 2003.

Chopra, S. and P. Meindl, *Supply Chain Management* 3/E, Prentice Hall, 2006.

De Vries, S. and S. Vohra, Combinatorial Auctions: A Survey, *INFORMS Journal on Computing* 15(3):184–309, 2003.

Crainic, T.G. and M. Gendreau, Advanced Fleet Management Systems and Advisors: Converging Technologies for ITS and e-Business, CD-ROM of the *10th World Congress and Exhibition on Intelligent Transport Systems and Services*, Madrid, Spain, 2003.

Pekeč, A. and M.H. Rothkopf, Combinatorial Market Design, *Management Science* 49(11):1485–1503, 2003.

Rothkopf, M.H. and S. Park, An Elementary Introduction to Auctions, *Interfaces* 31(6):83–97, 2001.

Rothkopf, M.H., A. Pekeč, and R.M. Harstad, Computationally Manageable Combinatorial Auctions, *Management Science* 44:1131–1147, 1998.

Shapiro, J., *Modeling the Supply Chain*, 2nd edition, Duxbury Press, 2001.

Sheffi, Y., Combinatorial Auctions in the Procurement of Transportation Services, *Interfaces* 34(4):245–252, 2004.

Shore, B., Information Sharing in Global Supply Chain Systems, *Journal of Global Information Technology Management* 4:27–50, 2001.

Simchi-Levi, D., P. Kaminski, and E. Simchi-Levi, *Designing & Managing the Supply Chain*, McGraw-Hill Irwin, 2003.

Williamson, E.A., D.K. Harrison, and M. Jordan, Information Systems Development within Supply Chain Management, *International Journal of Information Management* 24:375–385, 2004.

Emerging and Growing Trends

25

Reverse Logistics, Green Logistics, and Packaging

James R. Stock
University of South Florida

25.1 Introduction

Since early 1970s, logistics management has become a significant component of business strategy. Never has this been more evident than in the present era of supply chain management (SCM). However, in spite of the general increase in awareness of the impact of logistics on costs, customer service, and supply chain performance, the areas of reverse logistics, green logistics, and packaging have not garnered a lot of attention relative to most other activities of logistics and SCM. This chapter examines these three logistics-related areas, with specific emphasis on cost, revenue, service, and environmental issues.

25.1.1 Definitions

First, let us define the major terms or concepts that are discussed in this chapter. Reverse logistics can be defined as follows: "The role of logistics in product returns, source reduction, recycling, materials substitution, reuse of materials, waste disposal, and refurbishing, repair and remanufacturing; from an engineering logistics perspective, it is referred to as reverse logistics management and is a systematic business model that applies across the enterprise in order to profitably close the loop on the supply chain" [1].

Green logistics has not been defined specifically, but adapting a definition of green marketing proposed by Stock [1], the term can be defined as: "The practice of incorporating environmental topics such as recyclability, product labeling, biodegradable packaging, reusable containers, nonpolluting products, and other 'environmentally friendly' issues, into the logistics efforts of the enterprise" [1].

Finally, packaging can be defined as follows: "A coordinated system of preparing goods for safe, efficient and effective handling, transport, distribution, storage, retailing, consumption and recovery, reuse or disposal combined with maximizing consumer value, sales and hence profit" [2].

With those definitions established, let us examine these elements within the broader contexts of logistics and SCM.

25.1.2 Logistics and SCM Impacts on the Economy and the Environment—Macro Perspective

25.1.2.1 Economic Issues

Economic conditions throughout the globe have resulted in rising levels of affluence of consumers, not only in industrialized nations, but in developing nations as well. In turn, this has led to expanding national and international markets for goods and services. Thousands of new products and services have been introduced in this century and are sold and distributed to customers in every corner of the world. Business firms have increased in size and complexity to meet the challenges of expanded markets and the proliferation of new products and services. Multiple-plant operations have replaced single-plant production. The distribution of products from point-of-origin to point-of-consumption has become an enormously important component of the gross domestic product (GDP) of industrialized nations.

In the United States, logistics in 2006 accounted for 9.9% of GDP [3]. U.S. firms spent an estimated $801 billion on freight transportation; about $446 billion on warehousing, storage, and inventory carrying costs; and approximately $58 billion to administer, communicate, and manage the logistics process—a total of $1305 billion. Investment in transportation and distribution facilities, not including public sources, is estimated to be in the hundreds of billions of dollars. Considering its consumption of land, labor, and capital, and its impact on the standard of living, logistics is clearly big business!

As a component of the GDP of every country, logistics affects the rate of inflation, interest rates, productivity rates, energy availability and costs, and other aspects of the economy. Improvements in a nation's productivity have positive effects on many factors including prices paid for goods and services, the national balance of payments, currency values and exchange rates, the ability to compete more effectively in global markets, industry profit levels (higher productivity implies lower costs of operation to produce and distribute an equivalent amount of product), the availability of investment capital, and economic growth that leads to higher employment.

Perhaps the best way to illustrate the role of logistics in an economy is to compare logistics expenditures with other societal activities. In an industrialized economy, business logistics costs can be 10 times that of advertising, twice the amount spent on national defense, and equal to the cost of medical care [4]. Thus, by improving the efficiency of logistics operations, logistics makes an important contribution to the economy as a whole.

Additionally, logistics supports economic transactions. It is an important activity in facilitating the sale of virtually all goods and services. To understand this role from a systems perspective, consider that if goods do not arrive on time, customers cannot buy them. If goods do not arrive in the proper place, or in the proper condition, no sale can be made. Thus, all economic activity throughout the supply chain will suffer.

25.1.2.2 Environmental Issues

Green logistics deals with environmentally related logistics and supply chain management issues. The environmental aspects of logistics have gained increased business awareness in the past two decades, especially throughout Europe and Asia. The transportation and disposal of hazardous materials are frequently regulated and controlled. Organizations are increasingly required to remove and dispose of packaging, packaging materials, and related items used in the manufacture, storage or movement of their products. These issues, if not addressed correctly, complicate forward and reverse logistics tasks by potentially increasing costs and having negative customer service implications.

Key elements of green logistics include source reduction/conservation (use less), recycling (reuse what we do use), substitution (use environmentally friendly items) and disposal (dispose of what we can't reuse). These elements relate to products, packaging, and the facilities, equipment and materials used to carry out logistics and supply chain activities. For example, with respect to products, items can be remanufactured, repaired and refurbished to "as new" condition and resold. Evidence of this can be seen in the large amount of refurbished computers, office copiers, and other products offered for resale in retail and Internet stores.

Modules and components of defective products can be recovered and used as spare parts for other products needing these items for their repair or refurbishing. Remanufactured or refurbished products can be used to provide replacements for items returned to sellers by customers under warranty programs. For example, Maxtor, the computer hard drive manufacturer, refurbishes hard drives that are returned as defective and uses them to replace hard drives that are covered under consumer warranties. Remanufacturing a hard drive is much more cost efficient than producing a new hard drive, yet it performs to the same specifications [5].

In the packaging area, one third-party reverse logistics service provider recovers the packaging materials (e.g., Styrofoam peanuts, shredded paper, bubble wrap) and reuses them to repackage items that are returned to manufacturers for credit, and sometimes, can sell the used packaging material to others who need it for their own product packaging. Additionally, many firms utilize various kinds of reusable containers. Plastic totes can be used multiple times, often recovering the higher initial cost of these containers after several uses of them. Pallet recycling has become big business, with firms such as CHEP International, a pallet recycling and rental company, providing pallets that can be reused over and over. Also, package sizes and package configurations can be modified to allow more products to be included in a container, utilizing the same amount of packaging materials and thus reducing the packaging costs per unit or item.

With respect to facilities, equipment and the materials used to carry out logistics and supply chain activities, many illustrations abound. Better computerized algorithms for routing and scheduling transportation vehicles result in less emissions (because of less vehicle travel) and more efficient use of resources (vehicles, energy). Utilizing standard forklift trucks in materials handling that are powered by electricity, natural gas, gasoline, or alternative fuels, can have significant impacts on fuel usage levels and environmental pollution. Use of rail and piggyback movements can provide more environmentally friendly transportation than air freight or truck transport, assuming that customer service levels can be met using rail rather than air or motor.

25.2 Reverse Logistics

25.2.1 Reverse Logistics Costs

As mentioned earlier, logistics is an important component of GDP. While the largest share of that component occurs from forward logistics activities, a portion comes from reverse logistics and product returns. While estimates vary, many writers have reported that reverse logistics accounts for 5–6% of total logistics costs [6]. Using the United States as an example, that percentage would equate to U.S. $60–70 billion, a very large sum indeed!

From an individual firm perspective, Figure 25.1 provides some general financial implications of reverse logistics and product returns activities based on the DuPont Company's Strategic Profit Model [5,7]. The figure shows that there are a number of financial, service, and competitive benefits that can result from performing reverse logistics well. With each benefit, there will be some impact on either income or asset accounts, or both. From an income perspective (the top one-half of Fig. 25.1), implementation of optimal reverse logistics processes can have a positive impact on sales. By spending full-time effort on reverse

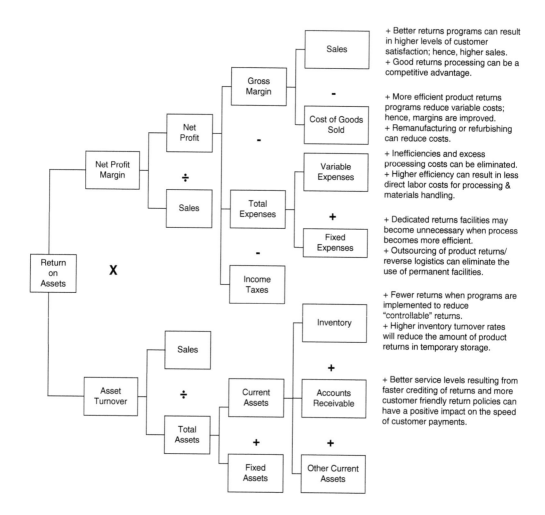

FIGURE 25.1 Product returns and the strategic profit model. (From Stock, J. R., *Product Returns/Reverse Logistics in Warehousing: Strategies, Policies and Programs*, Warehousing Education & Research Council, Oak Brook, IL, 2004. With permission.)

logistics, costs can be reduced and/or output increased (through handling more products in less time). This will tend to reduce variable expenses. In combination, net profits can be increased, ultimately resulting in improvements in return on assets, net worth, and other financial measures of performance.

In the lower one-half of Figure 25.1, which represents items from the balance sheet, improving reverse logistics will usually result in lowered levels of inventory as product returns are processed more quickly. With improved service levels resulting from faster handling of product returns, sales can improve. The effect will be an increase in asset turnover as sales improve and total assets decrease. When combined with improvements in net profit margin on the income side of the financial equation, the organization can have a significant increase in return on assets and return on net worth.

25.2.2 Reverse Logistics Process

Figure 25.2 presents a general overview of the various activities that occur in reverse logistics [1]. The figure includes a number of supply chain entities including manufacturers, third-party service providers, retailers, customers, and suppliers. Activities include those related to both products and packaging.

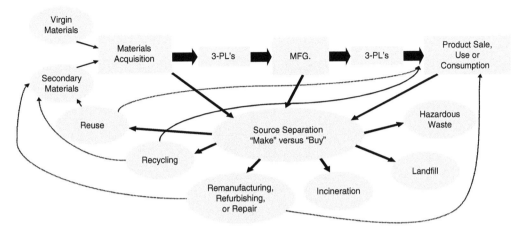

FIGURE 25.2 Reverse logistics within the supply chain. (From Stock, J. R., *Development and Implementation of Reverse Logistics Programs*, Council of Supply Chain Management Professionals, p.22, Oak Brook, IL, 1998; Council of Supply Chain Management Professionals. With permission.)

The handling of return goods, as well as salvage and scrap disposal, are parts of the larger process referred to as reverse logistics, and are important components of logistics. Buyers may return items to the seller due to product defects, overages, incorrect items received, trade-ins, or other reasons. Return goods handling has been likened to going the wrong way on a one-way street because the great majority of product shipments flow in one direction. Many logistics systems are ill-equipped to handle product movement in a reverse channel.

In many industries where customers return products for warranty repair, replacement, remanufacturing, or recycling, reverse logistics costs are higher relative to forward logistics costs. The cost of moving a product back through the system from the consumer to producer may be as much as nine times the cost of moving the same product from producer to consumer. Often the returned goods cannot be transported, stored, and/or handled as easily, resulting in higher costs per item/unit processed.

Logistics is also involved in removal and disposal of waste materials from the production, distribution, or packaging processes. If waste materials cannot be used to produce other products, they must be disposed of in some manner. Whatever the by-product, the logistics process must effectively and efficiently handle, transport, and store it. If the by-products are reusable or recyclable, logistics manages their transportation to remanufacturing or reprocessing locations. Often, these activities are outsourced by the company to various third parties [1].

25.2.3 Product Returns

"No one likes product returns, but this black sheep of the supply chain is gaining new respect as companies better understand the impact of returns management on customer relationships, brand loyalty and the bottom line" [8]. While there seems to be an endless number and variety of products that flow through reverse logistics channels, the actual process used to handle product returns and perform other reverse logistics tasks is quite similar. Figure 25.3 shows the typical six-step process for product returns.

Before product returns arrive at a firm's facility where they are processed and dispositioned, several tasks must occur in the "prereceipt" phase. Products can be returned to the point of sale or directly to the seller, as would be the case for most Internet or mail-order catalog purchases. Some products returned to the point of sale can be disposed immediately (e.g., resold as new, marked down in price, scrapped), but those requiring more sophisticated processing typically go to a processing facility located geographically

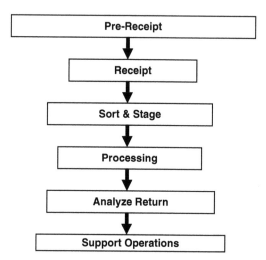

FIGURE 25.3   Stages in the product returns process. (From Stock, J. R. *Stages in the Product Return Process*, Unpublished Working Paper, 2006. With permission.)

between the buyer and seller. In a business-to-business environment, customers will contact the seller and usually obtain authorization to return the items.

Once the decision has been made by the customer, retailer, or other entity to return the product backwards through the logistics or supply chain channel, products enter the next phase of the cycle—the "receipt" phase.

Returns are received at a processing facility, usually a warehouse or distribution center. Delivery occurs by a variety of transportation carriers, such as FedEx, UPS, DHL, postal service, LTL trucking companies, or a firm's private fleet that has picked up returns when making deliveries to retail stores or customer locations.

When the product returns arrive, they often do not look the same as they did when they were initially shipped via forward logistics. About 30% of all returns come back on pallets containing the same or mixed items, but most arrive as loose packages, boxes, totes and/or cartons [5]. Full pallets or cases of returns are the easiest and fastest to process because they are much like the original shipments sent to customers. They mirror the initial handling of finished goods being shipped to customers. However, most items returned do not closely resemble the original shipments. Items are of a wide assortment and variety and some are in their original packaging while others are not. Many product returns received directly from consumers or through retail stores come back in a seemingly infinite number of packages or containers that must be handled individually. Their final disposition will usually be somewhat different, with store returns more likely than customer returns to go directly back into inventory for resale.

When items are received, they are often marked with the date of arrival. Often, this may be as simplistic as a piece of colored paper representing the date of arrival in the returns processing facility. The paper may also contain information on how many pallets came from the same customer; or how many cartons are on each pallet. The way the items are returned and received will greatly influence how products are sorted and staged in the next step—"sort and stage."

In this stage of the returns process, items are subjected to a general sort, that is, they are separated by how the products have been returned (e.g., pallets, cartons, packets) or the type of return (which could be identified from the address, color of the label, size and/or number of items being returned). For example, if a combination of individual small packages, cartons and pallets of products are returned, the pallets might be separated for quick processing, while cartons and individual items are processed normally.

In the next stage—"processing"—products are initially examined and basic data entered into the organization's data system. Physical products are typically sorted by stock keeping unit (SKU). If there are products that can be returned to vendors for full credit, those items are often handled during this stage.

At this stage, customer credits can be given since specific product information is entered into the firm's data system. Most customer credits do not occur, however, until later in the process because a determination must be made as to whether the product should have been returned and whether the product met the firm's conditions for accepting returns.

In one consumer products company, returns move to processing stations where diverters direct them to each station so that each processing line has an equal number of return items in their queue. Having the appropriate information (i.e., customer ID and return authorization numbers) on the return label allows for items from the same customer to be diverted to the same conveyor line for simultaneous processing. The success of this procedure depends on having the right information on the item being processed. Having a return authorization (RA) attached to the product allows this process to occur more quickly and efficiently.

Much of the information about the returned product will be captured during this stage, except for the method of disposition which will be determined during a subsequent stage of the process. Computer terminals and bar code scanners are often used to collect data on the following: company name; date and time; SKU number; description of the item; number of items processed (if multiple items in one order); location of item in inventory; package code (if product is available in multiple packages); condition description; and reason for the return [5].

In the "analyze return" stage of the returns process, the critical decisions relating to product disposition are made. The recoverable value of the returned products will depend to a large extent on the ability of employees to determine the optimal disposition strategy for each item. Hence, these employees must be highly trained and knowledgeable about remanufacturing, repair or refurbishing options, allowable versus nonallowable returns, and the economic outcomes possible with each disposition option. For example, products that can simply be repackaged for resale will return greater economic benefit than items that must be remanufactured, repaired, or refurbished before they can be resold. Similarly, being able to resell an item after some type of processing is a much better alternative than disposing of items as scrap or salvage, since these latter strategies offer the lowest recovery rates for returned products.

At this stage of the returns process, almost all data about the product, customer, and disposition options should have been collected. Additionally, firms should also have data on various aspects of the returns process so that its efficiency and effectiveness can be measured.

In the final stage of the returns process—"support operations"—products are either returned directly to inventory for resale, repaired, or refurbished for resale, returned to vendors for credit, sent to outlet or discount stores, donated to charities, sold as scrap or salvage, or destroyed. If remanufacturing, repair, or refurbishment are required, appropriate diagnostics, repairs and assembly/disassembly operations can be performed in the returns processing in order to put the items into a saleable condition. The degree of remanufacturing, repair, or refurbishing that occurs should be examined relative to the potential value of the product once it has been improved.

If products are going to be returned to vendors, employees determine the appropriate quantities and/or time windows acceptable to vendors and ship them back accordingly. In some instances, processed items will be resold in outlet stores, serve as replacements for warranty repairs, sold to wholesalers, off-price retailers, or offshore buyers. Returned items can also be donated to charities or relief organizations. If none of the preceding options are viable, products will be sold as scrap or salvage. The least desirable option would be to destroy product returns either by sending them to landfills or destroying them through incineration. In those instances, no value is obtained for the product.

25.2.4 Best Practices in Reverse Logistics and Product Returns

When enterprises perform reverse logistics activities well, they can gain a competitive advantage over other organizations that do not view these activities as a priority. Competitive advantage can be in the form of lower costs to process product returns, higher recovery rates for products that are returned, and improved levels of customer satisfaction resulting from more efficient and effective returns management. Specifically, organizations can do a number of things that can potentially result in competitive advantage in reverse logistics and product returns processing [5].

For example, enterprises that make reverse logistics and product returns processing a priority are likely to have full-time personnel responsible for managing these activities. Part-time effort usually results in suboptimal performance in processing product returns quickly and inexpensively. Full-time effort usually results in better overall returns management and lower costs. When managed on a full-time basis, higher recovery rates for returned products are achieved. These rates of recovery, often 80% or higher, are substantially greater than the 60% achieved by organizations that manage reverse logistics and product returns part-time. When large amounts of product returns are involved, these higher recovery rates can mean significant increases in revenues and profits. Additionally, when optimal returns management occurs, service levels are higher because customers get their refunds or replacement products quicker and more efficiently (i.e., at lower cost).

For processing product returns, "leading edge" organizations are more likely to require customers to use RAs. RAs are either requested from the seller via telephone or directly using electronic methods (Internet, fax, e-mail). By having customers notify the organization as to what items are being returned, they are better able to forecast when and how many products will be received by their returns processing facilities, thus allowing them to optimize personnel and facility utilization.

When the firm knows what customers want to send back, they can determine if the item is worth returning and if so, the best way of returning it. These firms can evaluate the value of the item to be returned given its most likely disposition strategy and subtract the costs of returns processing (including transportation, labor, and materials) to determine if the payback is there for the item. If yes, the customer sends the item back using a RA. If not, the customer can be instructed as to how to disposition the product at their location.

Another vital aspect of successful product returns and reverse logistics operations is the ongoing measurement and evaluation of overall order cycle time for items being returned. Many organizations utilize customized software for monitoring and evaluating product return flows. This software is either developed in-house or acquired from a software vendor and specifically modified for the firm. Detailed process maps exist that include narratives of each component of the product returns and reverse logistics processes.

Once the process has been mapped and software has been implemented to manage product returns, "leading edge" firms invest heavily in the formal training of employees who are involved in performing and/or administering product returns and reverse logistics activities. Training methods often used include mentoring programs, shadowing existing employees, and extensive written and/or visual training materials for individual or group education. As part of this effort, workers are cross-trained to perform multiple forward and reverse logistics and product return activities since most returns are handled in a facility that performs both forward and reverse logistics.

Regular audits are conducted on the product returns and reverse logistics processes to ensure that optimal strategies and programs are in place, including audits of safety procedures, administrative policies, security protocols, operational activities, workforce efficiency and productivity, and overall systems accuracy.

25.3 Environmental or Green Logistics

25.3.1 Logistics Activities

Reverse logistics and product returns are both economic and environmental in scope. When viewed from an environmental perspective, we refer to it as "green logistics." Figure 25.4 provides a visual picture of the relationships between reverse logistics and green logistics [9].

Product returns, marketing returns, and identifying secondary markets for returns are reverse logistics activities which have limited environmental aspects, while packaging reduction, air and noise emissions,

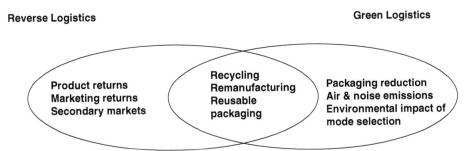

FIGURE 25.4 Comparison of reverse logistics and green logistics. (From Rogers, Dale S. and Tibben-Lembke, R., *Journal of Business Logistics*, 22, 2, 2001. With permission.)

and transportation mode selection have significant environmental impacts. Activities relating to recycling, remanufacturing, and reusable packaging are examples of elements that have a combination of both reverse and green logistics aspects.

Each aspect of reverse and green logistics that have environmental impacts will be jointly managed and coordinated within organizations that have adopted and implemented ISO 14000 standards.

25.3.1.1 ISO 14000 Series of Standards

The International Organization for Standardization (ISO) developed the ISO 14000 series of international standards for environmental management. Similar to the ISO 9000 standards dealing with quality management, ISO 14000 deals with various environmental issues such as designing products for the environment, environmental labeling, communication of environmental performance, environmental performance evaluation, and organizations having written environmental policies.

ISO 14000 is a series of standards for environmental management issued by the ISO [10]. Of most importance to firms involved in reverse and green logistics are ISO 14001 and ISO 14004. "ISO 14001 is an environmental management standard; that is, it is concerned with the formulation of environmental policy and objectives, including environmental impacts. ISO 14004 provides guidelines for introducing an eco-management system" [1]. In combination with ISO 9000 which relates to quality management, more than 750,000 organizations in 154 countries have implemented these standards [10].

25.3.1.2 ISO 14000 Components

Within the ISO 14000 series, several elements are important to logisticians and supply chain managers, specifically:

- ISO 14040—Life cycle assessment (description of environmental performance of products).
- ISO 14062—Design for environment (improvement of environmental performance of products).
- ISO 14020—Environmental labels and declarations (information about environmental aspects of products).
- ISO 14063—Environmental communication (communication on environmental performance).
- ISO 14030—Environmental performance evaluation (description of environmental performance of organizations) [11].

25.3.1.3 ISO 14000 Implementation

From a strategic perspective, the effect of ISO 14000 implementation includes the following:

- Policy stating a commitment to a specified level of environmental performance
- Planning process and strategy to meet stated performance commitment

- An organizational structure to execute the strategy
- Specific objectives and targets
- Specific implementation programs
- Communications and training programs
- Measurement and review processes [12]

There are no other standards as widely recognized and accepted as ISO 14000. Therefore, gaining certification provides evidence to the outside world that a firm considers environmental management programs in its strategy and operations. This is advantageous because organizations do not have to establish their own set of standards or certification [1].

25.4 Packaging

25.4.1 Packaging and Other Logistics and Supply Chain Elements

Packaging serves a marketing function and a logistics function. From a marketing perspective, packaging identifies the product, provides information about the product, and promotes the product to customers. From a logistics perspective, the package organizes, protects and identifies the product so that it can move efficiently and effectively through the supply chain [13]. Environmental aspects of packaging are important because of the issue of reverse logistics [14].

The package should be designed to provide for the most efficient storage and handling, and to minimize damage. Good packaging goes hand-in-hand with having the right materials handling equipment and allows efficient utilization of both storage space, transportation cube, and weight constraints. From an environmental perspective, the package and packaging materials should be "environmentally or eco-friendly."

Twede and Parsons [15] have stated that "logistical packaging affects the cost of every logistical activity, and has a significant impact on the productivity of logistical systems. Transport and storage costs are directly related to the size and density of packages. Handling cost depends on unit loading techniques. Inventory control depends on the accuracy of manual or automatic identification systems. Customer service depends on the protection afforded to products as well as the cost to unpack and discard packing materials. And the packaging postponement/speculation decision affects the cost of the entire logistical system. Furthermore, the characteristics of the logistics system determine the requirements and costs for packaging" [15].

25.4.1.1 Trade-Offs in Packaging

Often, packaging trade-offs are not given a lot of emphasis in logistics and supply chain management decision-making. However, packaging can have significant impacts on costs and customer service. For example, assume that a product carton is made of corrugated cardboard and measures $12'' \times 12'' \times 8''$, while another carton is $12'' \times 12'' \times 16''$. Assume that the smaller option costs \$.30 less per carton. Assume also that the smaller option requires less packing material which can save an additional \$.50. This would represent a savings of \$.80 for each small carton substituting for a larger carton. Considering that hundreds, thousands, or millions of packages may be distributed each year by an organization, the savings can be quite significant [13,16].

To illustrate, NKL, a vertically integrated Norwegian food cooperative, redesigned the packaging of just one of its products, Maggi (potato flakes manufactured by Nestlé), and realized 500,000 NOK savings during the first year after all costs were paid [1]. "This total was comprised of savings to retailers of 375,000 NOK (included less deliveries and fewer trucks for the same amount of product), warehousing cost savings of 66,000 NOK (required less handling and warehousing at supplier and retailer), expediting/transportation cost savings of 240,000 NOK (less transport costs due to 25% less EU-pallets)," [1] and various packaging design savings of 118,000 NOK (required less packaging). Given the costs of the program of 300,000 NOK, the trade-off was very positive.

Examples such as NKL are not as common as one might expect. In order for NKL to achieve such cost savings (and also customer service improvements), cooperation and collaboration across the supply chain was necessary. Suppliers, vendors, NKL, transportation carriers, warehouses, packaging companies, and the manufacturer had to get together and come up with a joint solution. Typically, such cooperation across supply chain members does not occur. However, there are a number of potential benefits that can accrue to organizations if they develop joint packaging solutions. Some general examples include the following:

- Lighter packaging can save transportation costs.
- Careful planning of packaging size/cube can allow better warehousing and transportation space utilization.
- More protective packaging can reduce damage and requirements for special handling.
- More environmentally conscious packaging can save disposal costs and improve the company's image.
- Use of returnable containers provides cost savings as well as environmental benefits through the reduction of waste products [17,18].

25.4.1.2 Types of Packaging

In the area of packaging, there are multiple aspects of the package that have direct relevance to environmental issues. First, there is the primary package. This is the package that comes in direct contact with the product. Second, there is the secondary package. This packaging often wraps around the primary package or joins multiple primary packages together, such as six-pack rings for canned beverages. Third, there is the tertiary package, which is the shipping box or container that is used to transport the product [19].

Table 25.1 identifies some of the most common forms of packaging, with the most widespread being new or used corrugated cardboard. While corrugated has many fine qualities, (initial cost, weight), plastic is very durable and recyclable.

Most forms of packaging can be recycled. Throughout the world, most government and commercial sectors are involved in various kinds of recycling programs, and packages and packaging materials are a significant part of the majority of those programs. Sometimes packages and packaging materials can even be reused. For example, when some products are returned to Avon (cosmetics and other consumer products), some of the same cardboard boxes used to ship products back to the company from distributors are reused for temporary storage of other returned items (usually small items), reducing the need for new containers to be used—resulting in environmental and cost reduction benefits.

TABLE 25.1 Comparison of Distribution Containers by Construction Material*

Material	One-Way Corrugated	Reusable Corrugated	Reusable Fiberboard	Plastic
Initial cost	$.95	$2.00	$6.33	$14.81
Estimated life	1 trip	5 trips	50 trips	250 trips
Average cost per trip	$.95	$.40	$.127	$.059
Weight	1.5 lbs	2.4 lbs	3.9 lbs	6.6 lbs
Durability	Poor	Fair	Good	Excellent
Additional costs	1. Setup	1. Setup	1. Return	1. Return
	2. Disposal	2. Break down		
		3. Return		
		4. Setup		
		5. Disposal		

*Based on 2 cubic feet size and order quantity of 500.
Source: Selection of Distribution Containers, Orbis Corporation, Oconomowoc, WI, December 1992. With permission.

Additionally, many firms utilize plastic totes for moving products around within their warehouses or distribution centers, or, they may use them for transporting small items to retail stores, pharmacies, etc. While the initial cost of these totes is higher, they are more durable, resulting in a payback period that is relatively sort. In some cases, the payback period can be as short as a few months or as long as two to three years.

25.4.1.3 Reusable Containers

As an alternative to disposal and recycling, reuse is a very environmental friendly option. It can also make financial sense to utilize reusable containers as opposed to "one-way" packaging. Numerous companies such as John Deere, Herman Miller, IBM, Ford, General Motors, Chrysler, and Toyota have invested in reusable containers to reduce costs and provide environmental benefits [20].

The use of reusable containers, which includes both packages and the pallets used to transport many products, is impacted by a variety of factors. With such containers, organizations must be able to coordinate activities in order to effectively and efficiently bring back these items to the source. Extra handling will be required. When firms have their own transportation (i.e., private carriage), it is relatively easy to return the reusable containers using backhauls occurring after product deliveries. When a third-party performs the transportation service, it is a bit more difficult taking back reusable containers because someone other than the firm is performing the service and there may be other company's products on the delivery vehicles [21].

Utilizing third-party providers of reusable containers is very popular for shipping pallets. CHEP USA is an international company specializing in pallet rental. The company rents reusable pallets to customers and manages their distribution throughout the supply chain. Industries that are heavy users of shipping pallets include grocery stores, drugstores, mass merchandisers, and warehouse clubs. More than 80% of all shipments in these sectors take place on pallets and many companies in these sectors utilize CHEP pallets [21]. Internationally, CHEP and firms such as Deutsche Bahn have established pallet pools, where pallets can be shared by different companies that belong to the pool [22]. Generally, reverse logistics and product returns activities are outsourced more frequently in Western Europe and Asia-Pacific than in North America or Latin America [23].

There is also an investment in the containers themselves. As shown in the examples in Table 25.1, the initial cost of plastic containers is more than 14 times that of one-way corrugated. In fact, each reusable option (reusable corrugated, reusable fiberboard) is a multiple of 2 to 6 times the initial cost of one-way corrugated. However, there are significant differences in the estimated life or number of trips that reusable plastic containers can make before they "wear out." With such a difference in estimated life of the container, even when the additional costs due to shipping weight and storage are considered, the potential benefits of reusable containers can be sizable.

The final decision to utilize reusable containers must be based on a variety of factors, not just estimated life of the containers, shipping costs and storage costs. Other factors to consider include stackability of the containers, closure style and security of the containers, container space efficiency, ability of the reusable container to be handled manually or by automated equipment, and the return ratio or number of reusable containers that actually are returned to the company and reused.

25.4.1.4 Container/Packaging Waste

As previously mentioned, packaging promotes and protects the contents of the package. Usually, when the package has arrived at its final destination, that is, at the point where the customer consumes or uses the product contained in the package, it is no longer needed, at least for its original purposes. At that point, containers are destroyed, recycled or reused. If they are not reused, they can be disposed by sending them to landfills, or they can be recycled by the company or a third-party. From an environmental perspective, disposal is the least attractive option. In theory, all packaging materials can be recycled, although some more easily than others. Sometimes, less packaging can be used if the product can be sufficiently protected (see Table 25.2).

TABLE 25.2 Source Reduction Measures for Packaging Applications

Source Reduction Activity	How Measured	Examples
Lightweighting of package using the same material.	Weight reduction per unit of package.	Aluminum glass, and plastic beverage containers. Corrugated boxes.
Lightweighting via material substitution or partial substitution.	Weight reduction per unit of package.	Glass bottles w/polystyrene "sleeve." Plastic in place of glass, steel, or paper. Aluminum in place of steel.
Reconfiguration of package to create more efficient design and package.	Quantity of product packaged per volume or weight of material.	Large size or bulk packaging of products. Shrink wrap in place of boxes.
Elimination of primary, secondary, or tertiary packaging.	Weight and/or volume of material eliminated.	Elimination of a paper or plastic wrap—over a primary package. Shipping in bulk instead of containerboard.
Bulk packaging.	Weight and/or volume of material compared to conventional practice.	Shipment of fresh produce in large containers. "Generic" products in bins, not packaged.
Use of composite materials to create more efficient package.	Lower volume or weight compared to alternatives.	Orange juice concentrate in multi-layer cans. Liquid products in aseptic packages, e.g., fruit juice. Multi-layer bottles and containers.
Use of consumable package.	Weight and/or volume of discard compared to conventional practice.	Dissolvable package for detergents.
Extended package life via reuse/refilling/reconditioning/durability.	Quantity of package per unit of product delivered.	Reconditioned drums and pails. Refillable milk, water, beverage containers. Refillable condiments and household chemical containers. (Note: may require major restructuring of product dist. system.)
Product design for reducing package requirements.	Quantity of package per unit of product delivered.	Concentrates, powders, reshape product

Source: Franklin, William E. and Warren A. Bird, *Source Reduction: A Working Definition*, prepared for Council on Plastics and Packaging in the Environment by Franklin Associates, LTD Prairie Villages, KS, December 29, 1989, p. 20, With permission.

To illustrate, Anheuser-Busch Companies, a brewer and theme park operator, reduced the amount of aluminum used in its 24-ounce beer cans, thereby saving 5.1 million pounds of materials. It saved 7.5 million pounds of paperboard by reducing the thickness of its 12-pack bottle packaging [24]. International Truck and Engine Corporation, a North American manufacturer of medium and heavy duty trucks, allows suppliers to ship parts to the company using returnable containers. This reduced packaging waste by one-half. Additionally, through recycling efforts, the company recycles more than 1600 tons of corrugated boxes, 3400 tons of wooden pallets, and 22,000 tons of metals each year [24].

25.5 Case Example

"Reverse logistics in the automotive sector has its own unique challenges given the over-sized nature of the freight and the cost of moving parts back upstream in the supply chain. For Hyundai Motor America, ... one of its greatest pains has been properly managing returns from dealers" [25].

The company, which sells automobiles and sports utility vehicles, has more than 640 dealerships in the United States. One of the products being returned by dealers were automatic transmissions destined for remanufacturing. Historically, Hyundai dealers sent returns to brokers or directly to the company. Hyundai would give the dealers credits, usually without much inspection or evaluation of the transmissions that were received. In order to better manage the process, the company outsourced the process to Roadway Reverse Logistics, a third-party reverse logistics provider.

The new process is much different than the "old way of doing things" and the efficiency and effectiveness of the returns system are much improved. "Now, once a transmission has been removed from a vehicle at the dealership and a new or remanufactured one put in, Roadway picks up the core or defective unit, inspects it at its facility, and issues credit to the dealer for sending the core back. Roadway then batches the cores on pallets, ships them via rail to San Diego, then on to Tijuana, Mexico, where they are remanufactured" [25]. Besides improving customer service levels to dealerships by providing them with return credits sooner and more accurately, Hyundai has also realized cost savings of US $250,000 per year from reduced order processing costs and transportation savings.

25.6 Summary

In an industrialized economy, business logistics costs are a significant amount of GDP. By improving the efficiency of logistics operations, logistics can make an important contribution to the economy. As part of the overall logistics process, reverse logistics and green logistics are vital components. Reverse logistics, which accounts for 5–6% of total logistics costs, relates to the entire process of bringing items back into the organization and includes product returns, recycling, remanufacturing, repair, and refurbishing.

There are a number of financial, service and competitive benefits resulting from performing reverse logistics well. To achieve these potential benefits, it is important that firms optimize the product returns process. Once customers, retailers, or other entities return products backwards through the logistics or supply chain channel, those products enter a multi-phase process which includes receipt, sort and stage, processing, analyze return, and support operations.

When enterprises perform reverse logistics activities well, they can gain a competitive advantage over other organizations that do not view these activities as a priority. Competitive advantage can be in the form of lower costs from processing product returns more efficiently, obtaining higher recovery rates for products that are returned, and improving the levels of customer satisfaction that result from more effective product returns management. Reverse logistics and product returns are both economic and environmental in scope. When viewed from an environmental perspective, we refer to it as "green logistics."

Green logistics deals with environmental aspects of logistics and supply chain management. The environmental aspects of logistics have become more important in recent years, especially in Europe and Asia. Key elements of green logistics include source reduction/conservation (use less), recycling (reuse what we do use), substitution (use environmental friendly items) and disposal (dispose of what we can not reuse). These elements relate to products, packaging, and the facilities, equipment and materials used to carry out logistics and supply chain activities. The package should be designed to provide for the most efficient storage and handling, and to minimize damage. Good packaging goes hand-in-hand with materials handling equipment and allows efficient utilization of both storage space, transportation cube and weight constraints. From an environmental perspective, the package and packaging materials should be "environmentally or eco-friendly."

In the area of packaging, there are several issues relating to the package that have direct relevance to the environment: (*i*) the primary package (the package that comes in direct contact with the product); (*ii*) the secondary package (this packaging often wraps around the primary package or joins multiple primary packages together); and (*iii*) the tertiary package (the shipping box or container that is used to transport the product). Packaging promotes and protects the contents of the package. Typically, when the package arrives at the point where the customer consumes or uses the product contained in the package, it is no longer needed, at least for its original purposes. At that point, containers are destroyed, recycled, or reused.

In sum, all organizations can benefit from optimal planning, implementation, and control of reverse logistics, green logistics, and packaging. "With increasing management attention being given to reverse logistics and environmental programs, these areas will see greater development. From a corporate perspective, it makes economic sense to develop optimal reverse logistics processes. From an environmental perspective, eco-friendliness translates into societal benefits. Such a win–win situation will be attractive to many organizations" [1].

References

1. Stock, J. R., *Development and Implementation of Reverse Logistics Programs*, Council of Logistics Management, Oak Brook, IL, 1998. Used with permission of the Council of Supply Chain Management Professionals.
2. Saghir, M., *Packaging Logistics Evaluation in the Swedish Retail Supply Chain*, licentiate thesis, Department of Design Sciences, Packaging Logistics, Lund University, Lund Sweden, 2004.
3. Wilson, R., *The New Face of Logistics, 18th Annual State of Logistics Report*, Council of Supply Chain Management Professionals, presented in Washington, DC, June 6, 2007.
4. Delaney, R. V., *CLI's 'State of Logistics' Annual Report*, press conference remarks to the National Press Club, Washington, DC, June 15, 1990.
5. Stock, J. R., *Product Returns/Reverse Logistics in Warehousing: Strategies, Policies and Programs*, Warehousing Education & Research Council, Oak Brook, IL, 2004.
6. Raimer, G., In reverse, *Materials Management and Distribution*, 12, 12, 1997.
7. E. I. DuPont de Nemours and Company, *Executive Committee Control Charts*, 1959.
8. Murphy, J. V., New solutions help companies find the return in returns management, *Global Logistics and Supply Chain Strategies*, 8, 44, 2004.
9. Rogers, D. S. and Tibben-Lembke, R., An examination of reverse logistics practices, *Journal of Business Logistics*, 22, 129, 2001.
10. International Organization for Standardization, http://www.iso.org
11. *Environmental Management: The ISO 14000 Family of International Standards 2002*; http://www.iso.org/iso/en/prods-services/otherpubs/iso14000/index.html. Terms and definitions are taken from ISO standards which can be obtained from ISO national member bodies (listed with contact details on the ISO web site www.iso.org and from ISO Central Secretariat sales@iso.org). Used with permission.
12. Nestel, G. An introduction to ISO 14000, in *The ISO 9000 Handbook*, 3rd ed., Peach, R. W., Ed., McGraw-Hill, New York, 1997, 485.
13. Stock, J. R. and Lambert, D. M., *Strategic Logistics Management*, 4th ed., McGraw-Hill/Irwin, New York, 2001, Chapter 11.
14. Klostermann, J. E. M. and Tukker, A., eds., *Product Innovation and Eco-efficiency*, Kluwer Academic Publishers, Dordrecht, The Netherlands, 1998.
15. Twede, D. and Parsons, B., *Distribution Packaging for Logistical Systems: A Literature Review*, Pira International, Surrey (UK), 1997, 1. Used with permission of Pira International.
16. Gooley, T. B., Is there hidden treasure in your packaging? *Logistics Management*, 35, 23, December 1996.
17. Andel, T., Conversion to returnables wins believers, *Transportation & Distribution*, 36, 94, September 1995.
18. Kroon, L. and Vrijens, G., Returnable containers: an example of reverse logistics, *International Journal of Physical Distribution and Logistics Management*, 25, 56, 1995.
19. Coalition of Northeastern Governors (CONEG), *Final Report of the Source Reduction Task Force*, Policy Research Center, Washington, DC, September 8, 1989, 18.
20. Mollenkopf, D., Closs, D., Lee, S. and Burgess, G. Assessing the viability of reusable packaging: a relative cost approach, *Journal of Business Logistics*, 26, 169, 2005.
21. Kopicki R., Berg, M. J. and Legg, L., *Reuse and Recycling—Reverse Logistics Opportunities*, Oak Brook, IL, Council of Logistics Management, 1993, Chapters 5 and 7. Used with permission of the Council of Supply Chain Management Professionals.

22. de Brito, M. P., *Managing Reverse Logistics or Reversing Logistics Management?* Ph.D. thesis, Erasmus University, Rotterdam, 2004, Chapter 4.
23. Langley, Jr., C. J. van Dort, E., Topp, U., Allen, G.R. and Sykes, S.R., 2006. *Third-Party Logistics: Results and Findings of the 11th Annual Survey*, Capgemini US, LLC, Cambridge, MA, 2006, Exhibit 6.
24. US Environmental Protection Agency, *WasteWise 2004 Annual Report*, US Government Printing Office, Washington, DC, October 2004.
25. O'Reilly, J., Rethinking reverse logistics, *Inbound Logistics*, 25, 114, July 2005.

26

Global Logistics Concerns

David Bennett
Aston University

26.1 Globalization of Operations and Logistics

In recent years there has been rapid growth in the globalization of goods production and service provision, which is expected to continue well into the future. An indicator of the extent of this growth in globalization is the consumption of imported goods in the United States, which is expected to increase from about US$1 trillion in 2002 to close to US$2.5 trillion by 2013. Of this, close to 50% is expected to originate from low-cost, developing, countries.

Future projections for the growth of globalization are based on changes in industry factors among which there are five that require special mention, these are:

1. The growth in demand for industrial and consumer products in emerging markets (e.g., the demand for machine tools and automobiles in China).
2. The development of supply bases in low-cost countries (e.g., supply of ferrous castings from China and India).
3. Factor cost advantages (US$1 per hour in some parts of China compared with US$22 per hour in the United States).

A. Domestic operations

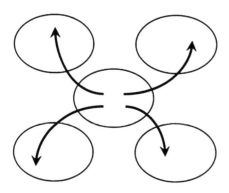

C. Global operations

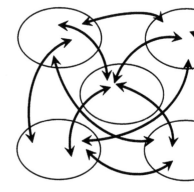

B. Regional operations

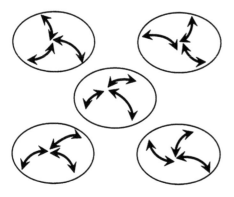

D. Linked global / regional operations

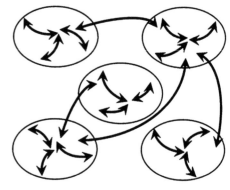

FIGURE 26.1 Four broad types of international operations network cofigurations. (From Slack, N., et al., *Operations Management*, London, Pearson, 2004.)

  4. Regulatory changes (e.g., tariff barriers and quotas).
  5. The free flow of capital across the globe (in the form of foreign direct investment).

Initially, more labor-intensive industries (e.g., toys, textiles, and footwear) and the first-stage skill-intensive industries (electronics) saw their supply chains being globalized. The next phase then saw the start of globalization of the supply chains of further skill-intensive industries such as pharmaceuticals and automotive components.

Slack et al. (2004) show how the internationalization of operations has created increasingly complex network configurations, especially when they involve a combination of regional and global transfers of products, parts, and materials (see Fig. 26.1).

A particular effect of the globalization of logistics is that shocks in one part of the globe (such as SARS, avian flu, or terrorist strikes) have a much wider impact than was the case in the past. In the short term, such shocks result in a sudden and urgent demand for certain products (medical kits, rescue equipment, etc.), while in the longer term risk management decisions will necessitate production facilities being geographically more widespread.

26.2 Role of Logistics Different in Organizations

All logistics operations involve the planning and coordination of the movement of materials. However, the globalization of operations described in the previous section has greatly extended the nature and

TABLE 26.1   Alternative Logistics Philosophies

Philosophy	Description	Possible Rationale	Sample Logistics Roles
Insourcing	Maintain logistics capabilities in house	Logistics core to the business More effective/efficient to maintain in-house	Provider of core execution services Preferred partner for third party logistics (3PL) services in regions outside of core coverage
Transition	Undecided whether to maintain in-house or outsource Choices made on a business unit basis	Logistics knowledge exists within the organization Concern about becoming over-reliant on outsource providers Challenging business needs for certain business units	Provider of core execution services Preferred partner for 3PL services in regions outside of core coverage 3PL services (i.e., order management, transportation management) Consulting services for supply chain optimization
Outsourcing	Outsource logistics capabilities	Logistics not considered core to the business Lacking the skill and knowledge to do so	Provider of core execution services Preferred partner for 3PL services in regions outside of core coverage 3PL services (i.e., order management, transportation management) Consulting services for supply chain optimization Lead Logistics Provider services Potential acquisition of assets and people (preferably with shared facilities)

scale of logistics activities. A particular question for companies today is whether to retain control of these logistics activities themselves or to rely on the services of specialist logistics services providers. Table 26.1 shows some of the alternative logistics philosophies that lie behind this decision for companies. The three main philosophies are described (insourcing, transition, and outsourcing), together with the possible rationale for their adoption and some sample logistics roles for each.

26.3   Global Outsourcing of Operations

For around the last 20 years global competition has emerged as a major driving force in shaping business strategies. The continued growth in global markets and supply has placed an increasing demand on the logistics function (Cooper 1993; Fawcett et al. 1993; Razzaque 1997). In order to reduce cost and handle their operations more effectively and efficiently, firms are increasingly subcontracting, or outsourcing, many of their core operations as well as peripheral activities, often from low-cost countries. Both value-adding and non-value-adding manufacturing and logistics activities can be effectively outsourced (McIvor 2005; Razzaque and Chang 1998). Consequently, international, or global, outsourcing is expected to drive rapid growth in international trade. Of the US$2.2 trillion in global spending, approximately 20% is currently in the form of outsourced products, and this is expected to grow up to 60% over the next 10 to 15 years resulting in an extra $750 billion per annum in trade. Table 26.2 shows

TABLE 26.2   Estimated Scope for Outsourcing and Untapped Potential

Sector	Total Spending in 2003 (US$ billions)	Percentage That Could be Outsourced	Potential Tapped (%)	Size of Untapped Potential (US$ Billions)
Consumer	250	75	<20	150
Semi-conductors	180	50	<25	70
Telecoms	300	60	<30	130
Automotive	600	50	17	250
Retail	850	50	<25	320

Source: Bloomberg, Expert interviews, McKinsey analysis.

the estimated potential for outsourcing in a number of sectors and the amount of that potential that is currently tapped. It can be seen that there is still huge untapped potential, which provides an indication of the large possibilities for further growth in this area.

The Asian transport market is projected to grow from US$700 billion in 2005, to over US$1.3 trillion by 2020. Most of this growth will be captured by the Northeast Asia "mega market" of China, Japan, Taiwan, and South Korea, which is expected to grow from US$550 billion in 2005 to US$900 billion by 2020, accounting for 70% of market share. This region's market growth of 4% is forecast to be driven by the increase in domestic consumption, which is expected to grow by 4% to 5% per annum, and the continuing trend of global outsourcing to China, with a projected compound annual growth rate (CAGR) of about 6–8%. Northeast Asia appears to be the main source of opportunity for logistics providers because of their economic power and geographical distribution.

The size of economic power wielded by these Northeast Asian economies is reflected in their performance against all macro-economic indicators, that is,

- 22% of worldwide gross domestic product (GDP), and 77% of all Asian GDP
- 24% of worldwide trade (growing at 9.4% CAGR)
- 9 out of the top 10 intra-Asian air trade lanes
- 70% of Asia's transport and storage market

The region's geography, with key centers well linked and almost 200 medium- to large-sized cities in close proximity make Northeast Asia the key center of commerce for the wider Asian area.

It can therefore be seen that Northeast Asia is a huge market, but how does it compare with other parts of Asia that are set to emerge over the next 15 years? Looking at the two next biggest economies, India and the Association of South East Asian Nations (ASEAN), these both require a huge annual growth rate of 12% to catch up with China by 2020. This is almost double their forecast growth during this period of 5–6% per annum, so the chance of them catching up during this period is slight. Even if we look at other macro-indicators, such as the value of exports, India will need to grow at 25% a year and Southeast Asia at 13% a year to catch up with China's exports by 2020.

A further aspect of manufacturing that has a profound effect on global logistics activities is the demand for greater product variety and differentiation. This trend towards "mass customization" means that many firms are using the concept of postponement to produce end-product variety, while at the same time enabling production efficiency and economies of scale in the upstream processes. Dornier et al. (1998) show how outbound postponement and inbound outsourcing are influenced by the relative demand for customized or standardized products. The four resulting supply chain structures are shown in Figure 26.2, with the simplest being "rigid," where products are highly standardized and inbound outsourcing is low, while the most complex is "flexible," where products are extensively customized and inbound outsourcing is high.

26.4   High Value Industry Trends and Implications for Logistics Services

Industries with high-value goods and substantial spending on air freight are more likely to offer the most interesting opportunities for logistics players to offer value-added and premium services. The highest value industries are:

- High-tech (semi-conductors, advanced video technologies, telecommunications equipment etc.)
- Fashion
- Life science
- Fast moving consumer goods (FMCG)
- Automotive

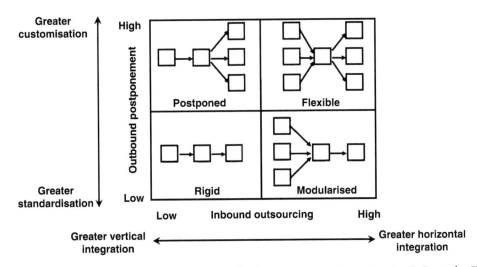

FIGURE 26.2 Global logistics—framework for supply chain structures. (From Dornier, P.-P., et al., *Global Operations and Logistics: Text and Cases*, New York, John Wiley & Sons, 1998.)

High-tech will continue to be a dominant "high value" sector for Northeast Asia, growing to over 70% of the share in trade with the rest of the world and around 80% of private consumption. There are several key trends and potential implications associated with this phenomenon, which are as listed in the following.

China becoming the center of the high-tech world, characterized by:

- A massive move of manufacturing and supply bases to China.
- The Chinese market overtaking the United States for many high-tech products.
- High-tech original equipment manufacturers (OEMs) increasingly taking control of distribution.
- Chinese high-tech companies mostly still being local and subscale.

Cost pressure, with:

- Value flowing to both ends of value chain (branding, channeling).
- Many segments commoditizing, with fast price erosion and consolidation.
- Many players scrambling for new business models such as original design manufacturers (ODMs) etc.
- Increasing focus on core competencies, readiness to outsource.

Supply chain velocity increasing, meaning:

- Shorter product life cycles, "fashion" trends in consumer demand.
- Product variety and supply chain complexity increasing.
- Product price erosion drives shorter lead time requirements.

Higher flexibility requirements, meaning:

- Most supply chains are globalizing and changing continuously.
- Unsophisticated supply chain management induces volatility.
- Component supply constraints demand continuous changes.
- Channel power overrides value chain pains.

The emergence of new business models, for example:

- Direct distribution models, which are winning in key segments such as personal computers.
- The direct-to-retail/operator approach by Asian ODMs/OEMs.

- Direct sourcing organizations.
- "No-touch" outsourcing models.

There is much evidence to corroborate the belief about China's increasing, and perhaps soon to be unrivalled, position in the global high-tech industry by 2012. By this time, China is expected to have increased its share of global production in segments such as notebook computers and DVD players to over 80%. Consequently, multinational corporations (MNCs) are increasingly taking control of distribution channels within China and the wider Northeast Asia region. For example, companies such as Electrolux and Dell have built unique brands and started to penetrate the local Chinese market. However, the market for high-tech products in China is still fragmented and small in relation to its population, as well as being served by many Taiwanese companies. Looking at the value-creation chain, it can be seen that most value today is being captured by "brands." Meanwhile the more central positions in the value chain, namely design and manufacturing, are tough places to be. Prices are falling rapidly, especially of good made in China, having reduced by between 10% and 20% each year between 1999 and 2003 across most high-tech consumer goods. The faster introduction of new designs is also putting greater pressure on the product life cycle. For mobile telephones, the average life has reduced by over 60% in the last seven years, while for personal computers the reduction in average life has been almost as steep at 50%. Consequently, there has been a growing need to identify new, more profitable, business models in segments that are "commoditizing." For example, considering the case of mobile telephone handsets, the rise of "direct-to-operator" models has been tremendous. These essentially connect the design provided by technology providers to the telecoms operators in a seamless fashion. Direct sourcing is also increasing, especially among retailers in Asia and China, with major international retailers such as Wal-Mart, Carrefour, and Best Buy having all exhibited a trend toward increased direct purchasing from the region in the past two years.

Essentially, the needs of a customer for high-tech goods vary along three principal dimensions:

1. Customer type (i.e., whether they are distributors, ODMs, OEMs, contract manufacturers, etc.).
2. Type of offering (i.e., whether the needs are for inbound or outbound logistics).
3. Customer philosophy (i.e., whether the services required are complex, and integrated or basic outsourcing).

Each of these three dimensions can be examined in more detail, by looking first at how concerns about the supply chain differ between four customer groups, that is,

1. For component/semi-conductor manufacturers:
   - For major silicon chip manufacturers, flexibility of delivery.
   - Lack of global networks for other suppliers.
2. For distributors:
   - Threat of breakdowns in intermediation.
   - Pressure to expand the logistics role.
   - Lack of scale and skill in managing logistics.
3. For Electronic Manufacturing Services (EMSs) and ODMs:
   - Leading players requiring third party logistics (3PL) support to close their logistics gaps (i.e., geography, lack of scale).
   - Tier 2 or 3 EMS/ODM players sub-scale to offer global logistics management and supply chain visibility.
4. For OEMs:
   - Drive in Europe and the Americas to outsource logistics management.
   - Intent to reduce the number of suppliers.

There are also common concerns among these four customer types including:

- High indirect logistics cost to cope with volatility and rapid obsolescence.
- Flexibility and responsiveness required to address stringent delivery constraints.
- Consistency of capacity and visibility required by OEM/EMS to guarantee customer service level.

26.5   Logistics Relationships

Relationships between buyers and sellers can range from arm's length transactions to close strategic alliances (Lambert et al. 1999). The ability to manage buyer–seller logistics relationships within a cross-cultural context has become a crucial success factor (Luo et al. 2001).

West (1989) suggests that the greatest weakness of the United States when dealing with Asia is probably a lack of basic knowledge and understanding of the Asian context. In his view, the task that lies ahead for U.S. companies may be construed as one of cross-cultural literacy in order to understand Asian values better. Asians appear to look and act alike to most Americans. However, the cultural diversity among Asian countries in fact may often be greater than the differences between any one of them and the United States.

Trust has often been cited by researchers as an antecedent to the type of relationship that exists between buyers and sellers (Golicic et al. 2003). In turn, trust can allow lower transaction costs in more uncertain environments and thereby provide firms with greater competitive advantage (Barney and Hansen 1994). Trust also facilitates long-term relationships between firms (Ganesan 1994; Ring and van de Ven 1994), which is an important component in the success of strategic alliances (Morgan and Hunt 1994; Browning et al. 1995; Gulati 1995).

Trust is a quality that will characterize a good relationship. The element of trust is an indispensable component of a healthy, growing, relationship between buyers and sellers, but it is not an absolute requirement for any relationship. A buyer may remain in a relationship with a seller either because it needs the relationship or simply because it perceives no alternative. The buyer's level of commitment to its relationship with a seller will depend on the extent to which the relationship derives from dedication, rather than from constraint. Trust-based relationships foster dedication (Leonard 1999). Peters points out that, "In our world gone mad, trust is, paradoxically, more important than ever." In a world of increasingly commodity like products and services, a relationship founded on trust is the only genuinely sustainable competitive edge (Peppers and Rogers 2004).

Trust in logistics relationships is not homogeneous and should be treated differently according to the relevant cultural dimensions (Aquilon 1997). Cultural differences are perhaps the most prominent obstacles which defy solutions to problems in inter-organizational relationships. Payne (1995) discusses the cultural barriers that affect international negotiations. Internal values, attitudes, beliefs, and the feelings of a particular society inevitably help to shape the nature of relationship. Payne also insists that many alternatives are foreclosed and many opportunities for resolution of conflicts are squandered or overlooked due to cultural perceptions and misperceptions. He believes it is almost impossible not to project one side's cultural values onto the other side, even if each is culturally sensitive. For example, rational behavior in the United States is often viewed as irrational by other countries, and vice versa.

26.6   Supply of Transportation Services

When viewed at a high level, the contemporary landscapes of transportation services can be characterized by distinct market segments, generally served by one natural owner. "Carriers" (those that physically transport goods) have a leading position in the markets for deferred, low-price type services for full-load shipments, mainly by surface modes. At the other end, "integrators" (those who arrange a full-load for door-to-door transportation) control the express markets, as well as day-defined parcels by air and ground, while "forwarders" (those who organize movement of goods on behalf of an exporter, an importer, or another company or person) have a leading position in nonexpress consolidated freight across all modes.

However, what might this picture look like in the next 15 years? Could the emergence of new modes such as high-speed ocean ships or improved ground infrastructure allow conversion of airfreight into surface modes? Will forwarders enter the express markets by upgrading their business system to enable time-defined express services for "consolidation freight" on a large scale? Will integrators gain significant share in the "consolidation freight" market by building platforms to handle large volumes of freight at competitive cost through their integrated networks?

26.6.1   Air to Ground Transportation

In recent years, ground transportation has overtaken overnight air freight in terms of growth leadership from, indicating a relative shift from "air" to "ground." In the United States the growth of different transportation service classes has followed the cycles of the economy. During times when the economy has been strong all classes have grown, but expensive overnight services have grown fastest. During times when the economy has been weak the slower, cheaper, deferred transportation services have gained in importance, while overnight services have slowed down. Since 2001, the overnight market has declined, while ground transportation has accelerated and deferred services have remained flat. Surveys of customer preferences conducted by the industry indicate that the relative shift to ground transportation will be sustained. Part of the reason may be structural changes and a shift in customers' methods of communication. For example, the emergence of e-mail has particularly caused the decline in overnight letters and documents. Customers have also moved away from overnight-dependent supply chains and have adjusted to reliable, deferred, and ground products as a viable alternative.

Since the 1970s the European market has also seen a shift from rail to road transportation, and now has a similar percentage of items shipped by road as the U.S. shippers have come to demand more flexibility and speed, and they see road transport as more efficient, more reliable, and faster than rail. In the established developed economies air freight has also never gained importance; particularly in the express document and parcel market, since most major metropolitan centers are within overnight trucking distance of each other. Trucking is consequently a significantly cheaper alternative for these routes.

In Asia scheduled infrastructure initiatives could allow the market to also shift from air to ground. When completed, the major infrastructural initiatives currently planned are likely to result in a significant increase in the share of international ground transportation across the continent. Some of these initiatives aim to improve connections between the existing road and rail networks within Asia and to Western Europe. However, implementation of seamless international infrastructure is slow, so rail in particular is not expected to be a major option in the near future.

Some of the infrastructure initiatives in Asia are as follows:

The Asian Highway Network: 32 member countries of Economic and Social Commission of Asia Pacific Countries (ESCAP) agreed in principle for an Asian Highway Network. The aim was to link major national roads and to maximize the use of existing infrastructure. The construction of new highways will be avoided except where deemed necessary to complete "missing links." By 2010, around 140,000 km of roads at a cost of US$16 billion are planned to be available.

The Trans-Asian Railway (TAR) was initiated in the 1960s with the objective of providing a 14,000 km rail link between Singapore and Istanbul, with possible onward connections to other parts of Europe and Africa. However, due to the huge differences in national standards and levels of technical development a step-by-step approach had to be adopted by ESCAP. A formal agreement on the railway network is expected to be signed by 2006, while completion could take up to 25 years.

26.6.2   Air to Sea Transportation

The service gap between intercontinental air and sea freight is between 10 and 25 days on most routes, but the cost differential is a factor of about 10 in favor of sea. With the aim of narrowing the service gap, while keeping costs low, a number of new concepts for sea freight have emerged.

Development of intercontinental fast ships: Several attempts have been made to develop high-speed ocean freight (with twice the speed of current vessels) to compete with air freight in terms of reliability and reduce the time gap. One project called FastShip Atlantic aims to soon provide a seven-day, door-to-door service between the United States and Europe at half the cost of air freight.

Introduction of inter-regional fast ships: Inter-regional fast ships on short haul routes are already in operation, albeit mainly focused on passenger services. They are similar in concept to intercontinental fast ships, except that they have a shorter reach. They can be positioned either as a premium ocean freight service or a low-cost alternative to standard airfreight or even express air freight, depending on the distance of the route.

26.6.3 Improving the Competitiveness of Carriers

Historically, an estimated two-third of all air freight shipments have been handled through forwarders and one-quarter through integrators, with less than 10% going directly through carriers.

Given the poor long-term performance of the air cargo and combination carrier industry, it seemed an obvious strategy that they would look to establish direct relationships with end-customers as a means of trying to improve margins. Many have tried this in the past, but without any real success beyond serving selected niches. One major reason for the problems facing air freight is the heavy dependence on forwarders' volumes, making it difficult to compete with this important customer segment for the same shippers. Various attempts have been abandoned after adverse reactions from forwarding customers. Limited network reach, lack of peak capacity, and lack of value-added services (such as customs clearance and warehousing) also make it difficult to sell the entire capacity exclusively to end users.

26.6.4 Challenge for Forwarders

One of the biggest challenges for a forwarder trying to match the value proposition of an integrator is to gain the ability to offer a predefined service at any time, as opposed to a service being made available and priced upon request. Given the brokerage nature of the business, prices cannot be quoted instantly based on a fixed schedule of rates. Capacity must be booked and transport time specified. Then confirmation generally depends on availability.

It is unlikely that forwarders will be able to establish close enough relationships, and operational integration, with a number of different carriers to overcome these and other issues entirely. Even if they succeed and narrow, or close, the service gap, this would most likely come at the price of taking over a much larger share of the overall capacity risk and higher operations costs, with a fundamental impact on the economics for forwarders. It remains to be seen how the major global forwarders can enhance their core business systems with the help of new technology, but an immediate step change in their ability to provide express services is not expected. The area of competition with integrators is much more likely to emerge in the field of supply chain solutions, where coordination is an important aspect of the logistics function, rather than in basic transportation services.

26.7 Technology Shifts

Investments in information technology (IT) for managing supply chains have grown rapidly in recent years. These investments include the costs of licenses, hosting, hardware, and maintenance of supply-chain related information systems.

Supply-chain integration and optimization requires visibility of information. This drives the demand for IT solutions and fuel investments in new logistics technologies. The adoption of radio frequency identification (RFID) will further increase reliance on IT as large amounts of data would have to be compiled and integrated with the existing enterprise management systems.

The market for IT for supply chain management comprises the following segments (with percentage shares and CAGR):

- Application software license: share = 32% (CAGR = 12%)
- Implementation services: share = 39% (CAGR = 7%)
- Application software maintenance: share = 19% (CAGR = 4%)
- Application hosting: share = 3% (CAGR = 14%)
- Others: share = 7%

The market is highly fragmented with the top four suppliers currently being SAP, "i2," Manugistics, and Manhattan Associates (in order of market share) holding 25% of the market.

Use of RFID can greatly improve efficiency and reduce cost, but requires major investments. Possible applications of RFID in supporting chain management include:

- Dynamic tracking of objects and information transparency relating to:
  - Goods in transit
  - Distribution fleet management
  - Warehouse inventories
- Security and authentication systems for:
  - Maintenance management
  - Theft prevention

The implications for supply chain management of RFID can be divided into opportunities and challenges. Among the opportunities are:

- Improvements in efficiency and productivity, especially in warehousing and inventory management.
- Reductions in warehouse labor cost by 4% through effective handling.
- Savings of 20% to 40% in inventory and stockout cost through better stock replenishments.

While the challenges include:

- Implementation costs of necessary IT systems.
- Information "overflow."

Radio frequency identification has to be carefully assessed since the cost for developing the necessary IT systems are massive and early experiences have shown that the technology still has a relatively high failure rate (more than 20%). However, as RFID technology becomes established its lower cost has the potential of lowering the entry barrier for small- and medium-sized logistics players wishing to serve first-tier MNCs.

26.8 Final Comments about Industry Trends and Their Implications for Logistics

In summary, the implications for future global logistics systems of the aforementioned discussion can be highlighted as relating to the following factors:

- Globalization of production and the trend towards outsourcing.
- The demand for flexibility of supply and logistics services among customers.
- Transparency with regard to tracking inventories and goods in transit.
- IT integration with existing logistics infrastructures.
- Potential cost savings leading to greater competitiveness among current and aspiring logistics providers.

These implications result in demands on integrated logistics and supply chain networks that leverage strengths across:

- Modes of transport (especially short-term shifts).
- Global and domestic networks (in particular China becoming a dominant consumer nation).
- Systems [for business to consumer (B2C) distribution].
- Services (integration along the entire supply chain).

Many customers will value one-stop solutions, supported by supply chain networks, for helping them to buy time to react to changes in customer needs and to mitigate risks. The trends will create a demand for logistics service providers that can provide multi-modes of transport, offer combinations of global and domestic networks, integrate systems, and develop complete supply chain solutions.

26.9 Case Study: DHL's Emergence as a Global Logistics Provider

26.9.1 DHL's Origins and Initial Growth

DHL was founded in San Francisco in 1969. Its founders were Adrian Dalsey, Larry Hillblom, and Robert Lynn. Hence the company name, which was simply created from the initial letters of their own last names. DHL initially provided a service flying "bill of lading" papers between San Francisco and Honolulu so that shipped cargo could obtain faster customs clearance. The success of this initial operation enabled the company to grow rapidly and during the next three years it expanded across the Pacific to the Philippines, Japan, Hong Kong, Singapore, and Australia. In 1974, DHL expanded into Europe, opening an office in London, and between 1976 and 1978 it launched services in the Middle East, Latin America, and Africa.

As an emerging global parcel and freight shipping company DHL's natural international competition was from the established players in the industry. These included U.S. companies such as United Parcel Service (UPS), established in 1907, and FedEx, founded in 1913, which between them had a virtual duopoly in the American market. The other major international competitor was TNT, which was founded in Australia in 1946 and was acquired by the Dutch KPN postal company in 1996.

DHL, therefore, continued its growth by moving into new markets that had not been exploited by these established companies. In 1983, it was the first air express forwarder to serve Eastern European countries. Then in 1986, it became the first express company active in China through the establishment of a joint venture there. In 1991, DHL was the first international express service to restart service to Kuwait after the Gulf war and in 1993 it invested US$60 million in a new hub facility in Bahrain.

26.9.2 "New" DHL

In 1998, the German company Deutsche Post World Net acquired 25% of the shares in DHL. Deutsche Post's origins were with the established Deutsche Reichspost in 1924, Germany's public postal system, which subsequently became Deutsche Bundespost in 1959. Then in 1989, under Germany's postal reforms, it was separated into its postal, banking, and telecoms businesses with the former East German postal service being incorporated in 1990. Two further rounds of reform saw the creation of Deutsche Post, the sale of private shares, liberalization of postal services and the eventual dissolution of the Federal Ministry of Post and Telecommunications in 1997. In 1998, the German Government sold its remaining shares and this opened the way for Deutsche Post to become a forerunner in Europe for providing cross-border postal services with the formation of Deutsche Post Express, which was set up in 1998 as a parcel and distribution network for Germany and Europe. It also placed Deutsche Post in a position where it could consider acquisitions as a means of achieving growth.

In 1999 DHL, while still independent, purchased the Dutch shipping company Van Gend & Loos EuroExpress and merged it with its existing operations in the Netherlands. In 2001, Deutsche Post World Net, which had become a completely public company since 2000, then acquired a majority (51%) of DHL shares, and the remaining 49% in 2002. The "new" DHL was launched by merging the old DHL with EuroExpress and Danzas, a Swiss freight company it had acquired in 2000. This opened the way for DHL's subsequent rapid expansion through natural growth and further acquisitions. The most important recent developments have been the acquisition in 2003 of Airborne Express in the United States and the purchase in 2005 of the U.K. contract logistics company Exel, which had only just itself acquired another company, Tibbett & Britten, and employed 111,000 people. These acquisitions created a global DHL workforce of 285,000 people and around US$65 billion in annual sales.

DHL's growth since its acquisition by Deutsche Post has been significant. Its global outlook has enabled the company to outstrip its U.S. rival FedEx, which has 138,000 employees and US$32 billion in sales, and TNT, which has 164,000 employees and US$12 billion in sales. It is also in a position to challenge the world leader, UPS, which has 407,200 employees and US$42.6 billion in sales.

Other key figures for DHL are:

- Number of offices: around 6500
- Number of hubs, warehouses and terminals: more than 450
- Number of gateways: 240
- Number of aircraft: 420
- Number of road vehicles: 76,200
- Number of countries and territories: more than 220
- Shipments per year: more than 1.5 billion
- Destinations covered: 120,000

An important aspect of DHL's ability to provide a global logistics service is its use of air transport capacity to support its commercial activities. There are four DHL owned airlines:

1. European Air Transport, based in Brussels, which provides capacity for DHL's European network as well as longhaul services to the Middle East and Africa, using Boeing 757SF/PF and Airbus A300B4 aircraft.
2. DHL Air UK, based at East Midlands airport in the United Kingdom, which provides services on DHL's European Network using Boeing 757SF aircraft.
3. DHL's Middle East airline, based at Bahrain International Airport, which serves a wide variety of Middle East destinations including Afghanistan and Iraq, using a variety of regional aircraft.
4. DHL's Latin American airline, based in Panama City, which services a wide range of destinations in Central and South America using Boeing 727 aircraft.

26.9.3   DHL's Role in Logistics

DHL's role as a company providing services for transporting parcels and freight has been extended into that of being a 3PL company. Its expertise in this area is mainly derived from its acquisition of Danzas and Exel, both of which had a special reputation as logistics providers. DHL's services therefore now include customized logistics solutions for the entire supply chain. These services are supplied through two subsidiaries, DHL Exel Supply Chain for procurement logistics, warehousing, sales operations and value-added services, and DHL Global Forwarding for worldwide project logistics services based on air and sea freight.

Among the supply chain services provided are:

- Supply Chain Management—analysis, design, and engineering. Lead logistics provider (LLP) and lead service provider (LSP) services.
- Warehousing—engineering, design, and management. Vendor management, just in time.

- Value-added services—order management, quality control, outbound fulfillment—reverse and return logistics and other services.
- Distribution—industry-specific, local, and transnational. Network and fleet planning and optimization.
- Outsourcing—takeover and management of in-house logistics including distribution, transport, back-office, supply chain, and after sales.

Among the project logistics services provided are:

- Transport and logistics design—packaging design (industrial). Multimodal transport design, feasibility studies and consulting.
- Project logistics management—risk management planning. Project services for freight forwarding, document process management, expediting, including order management, tracking and tracing.
- Project cargo logistics monitoring—packaging control. Port/airport handling supervision.
- Heavy load installation—job-site delivery up to final positioning at factory, work site, or project destination.

DHL offers these logistics services to a number of key industries, as described below:

- Electronics and telecommunications—by helping to reduce inventory and cycle times while providing control and visibility to final delivery. Enables focus on product availability, optimization of the product flow and supply chain costs as well as services for the after-sales market.
- Automotive—faclitating the design and management of optimal tailor-made solutions. Handling flows of components and service parts from suppliers through inbound logistics. Optimizing the material flow prior to arriving at the assembly plant through component logistics. Deals with the flow of service parts to dealers.
- Healthcare and pharmaceuticals—managing the variety and complexity of different healthcare supply chains from primary, secondary, and hospital to medical devices. Total control of product inventory, source and status. Security, via compliance with strict healthcare distribution requirements such as extreme temperature control. Facilitating direct trade with end-users through tailor-made services and web technologies. Support for global marketing activities.
- FMCG—optimizing logistics in food, beverages, wines and spirits, personal and home care, tobacco and domestic appliances. Bring more value into flow via warehouse and value-added services including cross-docking at shared distribution centers which increases supply chain responsiveness and efficiency. Supporting efficient consumer response (ECR) with enabling technologies such as European Article Number (EAN) coding to reduce lead times and costs.
- Fashion—solutions for getting new fashion garments to markets fast, from sourcing textiles to managing swatches or samples, to shipping hanging or flat-pack garments to retail outlets around the world. Reducing purchase order lead time and responding faster to market demands with a variety of transport options, warehousing services, and retailer fashion networks.
- Chemicals—specialized transport and storage of bulk and packed chemicals. Integrated end-to-end supply chain services as well as environmental and security standards using dedicated, solutions for commodities, specialties and the life sciences. Implementation of logistics with strict attention to quality, from core transport (inbound, inter-facility, and outbound) to warehousing and inventory management. Supply chain design, re-engineering and complex IT services.

26.9.4   3PL Logistics Market

Third party logistics is a major growth area for companies involved with transport. In 2005, its value was US\$179 billion and it represented the largest area of outsourcing by businesses; bigger even than contract manufacturing. Yet, there is some uncertainty about how the market will develop. There are differences among the large global corporations about what they want from logistics providers. Some require a "one-stop" 3PL that can provide all the services and geographic coverage that their customers need. This has helped to encourage mergers, so that larger logistics providers can provide more services and cover the globe. DHL and UPS are pursuing this strategy. However, many large customers are not willing to put all of their eggs in one basket and risk their entire supply chains by putting them in the hands of on one or two partners. Also they do not want to limit their negotiating potential.

This situation has meant that there is space for a number of smaller, specialist, logistics providers that are not part of very large, global, parcel and freight shipping companies. One of the most important of these is the Swiss company Kuehne and Nagel, with 40,000 employees and US$10.7 billion in sales. Others have a more regional focus such as SembCorp Logistics of Singapore, with 2700 employees and US$713 million in sales, covering Asia, and Ryder Systems of Miami, with 15,625 employees and $2.1 billion, covering North America.

26.9.5 Future of 3PL

With the rapid growth in outsourcing, greater internationalization of manufacturing, and the drive toward lean operations there seems little doubt that the demand for 3PL will continue to rise. China's emergence onto the world stage also means that it has become an important player, with so many products being sourced and manufactured there. This means that companies doing business with China are looking for a logistics partner with experience and operations in the country. This in turn is forcing many logistics companies to establish a presence in China either through acquisitions, start-ups, or joint ventures. DHL-Sinotrans Ltd is DHL's joint venture based in Beijing. In 2003, it embarked on a five-year investment plan worth US$200 million aimed at significantly increasing its capacity.

Technology is another area of development for 3PL providers. As well as investing in new and more modern equipment for transporting goods they are also showing interest in technologies for identifying, tracking, and expediting. For example, RFID is increasingly being used, especially for the food and beverage industries where delivery schedules are critical and perishability is a major consideration. Early in 2006, DHL opened an innovation center in Germany in partnership with IBM, Intel, Philips, and SAP. Development work in the center focuses on RFID technology, geodata technology for optimizing travel routes and networks, as well as logistics-related Global Positioning System (GPS) applications. The center is part of an innovation initiative aimed at making supply chains more efficient by uniting the flow of information and physical goods through increased automation, visibility, and improved collaboration among trading partners.

Acknowledgment

The author would like to acknowledge the contribution of Desmond Gay, DHL Singapore, for the material he provided in preparing this chapter.

References

Aquilon, M. (1997). Cultural dimensions in logistics management: a case study from the European automotive industry. *Supply Chain Management* Vol. 2(No. 2): pp. 76–87.

Barney, J. B., and Hansen, M.H. (1994). Trustworthness as a source of competitive advantage. *Strategic Management Journal* Vol. 15: pp. 175–190.

Browning, L. D., Beyer J. M., and Shetler, J. C. (1995). Building cooperation in a competitive industry: SEMATECH and the semiconductor industry. *Academy of Management Journal* Vol. 38: pp. 113–151.

Cooper, J. C. (1993). Logistics strategies for global businesses. *International Journal of Physical Distribution and Logistics Management* Vol. 23(No. 4): pp. 12–23.

Dornier P.-P., Ernst, R. Fender, M. and Kouvelis, P. (1998) *Global Operations and Logistics: Text and Cases*, New York, John Wiley & Sons.

Fawcett, S. E., Birou, L., and Taylor, B. C. et al. (1993). Supporting global operations through logistics and purchasing. *International Journal of Physical Distribution & Logistics Management* Vol. 23 (No. 4): pp. 3–11.

Ganesan, S. (1994). Determinants of long-term orientation in buyer–seller relationships. *Journal of Marketing* Vol. 58: pp. 1–19.

Golicic, S. L., Foggin, J. H., and Mentzer, J. T. (2003). "Relationship magnitude and its role in interorganizational relationship structure." *Journal of Business Logistics* 24(1): 57–75.

Gulati, R. (1995). Does familiarity breed trust? The implications of repeated ties for contractual choice in alliances. *Academy of Management Journal* Vol. 38: pp. 85–112.

Lambert, D. M., M. A. Emmelhainz, et al. (1999). Building successful logistics partnerships. *Journal of Business Logistics* Vol. 20(No. 1): pp. 165–181.

Leonard, B. (1999). *Discovering the Soul of Service*. New York, Free Press.

Luo, W., R. I. Van Hoek, et al. (2001). Cross-cultural logistics research: a literature review and propositions. *International Journal of Logistics: Research and Applications* Vol. 4(No. 1): pp. 57–78.

McIvor, R.T. (2005) *The Outsourcing Process: Strategies for Evaluation and Management*, Cambridge, Cambridge University Press.

Morgan, R. M. and Hunt, S. D. (1994). The commitment-trust theory of relationships marketing. *Journal of Marketing* Vol. 58: pp. 20–38.

Payne, R. J. (1995). *The Clash with Distant Cultures: Values, Interests, and Force in American Foreign Policy*. New York, State University of New York Press.

Peppers, D. and Rogers, M. (2004). *Managing Customer Relationships*, John Wiley & Sons, Inc.

Razzaque, M. A. (1997). Challenges to logistics development: the case of a third world country—Bangladesh. *International Journal of Physical Distribution and Logistics Management* Vol. 27(No. 1): pp. 18–38.

Razzaque, M. A. and Chang, C. S. (1998). Outsourcing of logistics functions: a literature survey. *International Journal of Physical and Logistics Management* Vol. 28(No. 2): pp. 89–107.

Ring, P. S. and van de Ven A. H. (1994). Developmental processes of cooperative interorganizational relationships. *Academy of Management Review* Vol. 19(No. 1): pp. 90–118.

Slack, N., Chambers, S. and Johnston, R. (2004). *Operations Management* , London, Pearson.

West, P. (1989). "Cross-cultural literacy and the Pacific rim, *Business Horizons*, Vol. 32(No.2): pp. 3–13.

27

Outsourcing and Third-Party Logistics

Xiubin Wang
University of Wisconsin at Madison

Qiang Meng
National University of Singapore

27.1 Introduction

This chapter is organized into three sections. In Section 27.2, we introduce outsourcing practices in a general way. In Section 27.3, we present several models for different outsourcing problems that have appeared in the recent literature. Section 27.3 serves as a starting point to model realistic problems in outsourcing. In Section 27.4, we present a successful case study followed by a summary.

27.2 Outsourcing Practices

The practice of outsourcing has, by now, been weaved into the fabric of business. Eighty-five percent of all North American and European companies have outsourced at least one function (Logan et al., 2004). Sixty percent of Fortune 500 companies surveyed have at least one logistics outsourcing contract (Vaidyanathan, 2005). Outsourcing is a practice that creates opportunities for positive synergy by bringing together the core competencies of two companies.

Outsourcing is an agreement between a business and a third-party service provider for ongoing management and improvement of activities related to part of, or entire, business functions, infrastructure and operating processes. Outsourcing often means finding new suppliers and new ways to secure the delivery of raw goods, materials, components, and services. It aligns a company and its supply base across each link in the supply chain to minimize the cost for purchased materials and services.

Third-party logistics (3PL) is a special form of outsourcing that relates to logistical processes. Logistical activities typically include transportation, warehousing, inventory control, distribution, and materials procurement. A contractor who provides services directly to the outsourcer is often called a "third-party logistics service provider." When the number of 3PL providers is beyond the normal range of management, because of the complexity of the outsourced processes, the fourth-party logistics (4PL) provider is introduced to interface between the outsourcer and the 3PL service providers. In this way, the 3PL service providers are responsible to the fourth-party provider, and the latter is responsible to the company for final delivery.

Outsourcing takes different forms. It may be based on a single transaction, on a continuous relationship over a certain period of time, or on a combination of the two. A general trend is toward a stable relationship over a relatively long period of time, which is called "relational outsourcing." A combination of the two sometimes is adopted. For example, a firm could have outsourcing relationship with several service providers on a long-term basis. The specific amount of business outsourced to each provider, however, may be decided on a daily basis or on the transactional basis.

27.2.1 Most Frequent Reasons for Outsourcing

Clearly, the logistics service providers have advantages in providing their assigned services. We present two examples of these advantages.

Example 1: Exploit Economies of Scale (Density)

Outsourcing helps formation of a value network on which each company thrives by exploiting the economies of scale (density). The economies of scale take place if the per-unit cost decreases as the output increases. Economies of density happen if the per-unit cost decreases when the operational density increases. For example, computer companies reduce their costs by outsourcing keyboard manufacturing to specialized manufacturing firms in Taiwan or South Korea. The keyboard manufacturing firms decrease the unit cost of the keyboard by increasing their volume of outputs. Part of this decrease may be passed back to the computer companies.

Example 2: Role of Freight Density and Technology

Transportation is another typical service that often is outsourced. With a higher freight density, the transportation service providers may offer services with lower cost than when provided by the outsourcers themselves. The advantage of freight density is illustrated in the following example: A trucking firm has a regular operation on three freight lanes sequentially on a line—from A to B, C to D, and E to F. Additional services of freight lanes from B to C and from D to E enjoy the advantage of freight density as the empty travel distance is reduced. Synergies are generated by reducing each other's empty travels. Obviously, specialized trucking firms have the advantage of operating an economy of density.

Technology plays an important role in utilizing freight density. In the transportation area, routing and scheduling technologies have seen increasingly important applications. It might not be economically feasible for small firms to maintain such technological capabilities of their own. Large trucking firms, such as Schneider National and JB Hunt, have sophisticated software to make routing and scheduling assignments such that the number of drivers may be reduced to the minimum, the driving distances shortened, and a large cost savings achieved.

27.2.2 Advantages of Outsourcing

There are many advantages that promote outsourcing. In the following pages, we make a brief list of general ones.

27.2.2.1 Utilizing Core Competency Fully

The core competency is an area such as production, operation, or management in which the company shows superiority of competition to its competitors in the market. It can be a process, such as assembling or manufacturing. It also can be an aspect of general management, such as coordination of diverse production skills and integration of multiple technologies. As an example, Dell Computer showed one of its core competencies in the 1990s in managing its supply chain and organizing an efficient distribution system. The core competency is a building block for lean organization. It must be nurtured in-house to retain the company's long-term competitiveness. In addition, the core competency represents a unique competitive edge that is difficult to be imitated by competitors. Core competencies typically have the following characteristics:

- Having architectural support
- Being continuously reviewed through performance measures, and being enhanced through training and development
- Being embodied into company ethics and demonstrated in multiple products and services
- Enduring and well-recognized in the market

Outsourcing enables the company to strategically direct its resources to its core competencies. For instance, General Motors Saturn outsourced part of its logistics activities to Ryder Dedicated Logistics, and General Motors was able to focus on its manufacturing activities. As another example, Nike Inc. concentrates on design, development, sourcing, and marketing, while outsourcing many other operations, such as manufacturing, to outside contractors.

27.2.2.2 Cost Reduction and Control

Outsourcing reduces the overall operational cost of the outsourcer. In an early study performed by Ernst & Young and the University of Tennessee, organizations using 3PL companies reduced logistics costs by an average of 7.8%, achieved 21.6% cutback of logistics assets, and a reduction of order cycle from 6.3 to 3.5 days. The reason for this is that the outsourcing company benefits from the savings passed back by the contractors because of the realized economies of scale.

In outsourcing, the service provider usually is chosen from several competing companies in the market. One might say that outsourcing places a cap on a company's cost of the process and operation outsourced, and that this cap is determined by the dominating market. On the other hand, while part of the benefit from economies of scale is passed back to their clients, logistics providers are the first beneficiaries of the economy of scale for providing services to a large number of clients.

27.2.2.3 Better Dealing with Variable Demand

An advantage of outsourcing is its flexibility. The company remains flexible by not locking itself into a long-term financial commitment. This advantage becomes obvious when the market is uncertain. Otherwise, if the demand turns out to be much lower than expected, the operation will have to be stopped. It could be very costly to terminate an unprofitable operation. In addition to the financial cost, early termination may lead to legal lawsuits and union strikes. Outsourcing is also preferable when demand has periodic swings.

27.2.2.4 Service and Technology Improvement

Outsourcing can improve the company's level of service, credibility and even market image. As an example, some advanced technology might be beyond the capability of development by itself due to the company's current financial or technical constraints. Therefore, the outsourcer may seek service from specialized providers who have the technological expertise. Furthermore, the contractor can take care of technology upgrade, training, and maintenance. Often times, outsourcing causes the company to raise its internal standards, leverage the use of new technology, and adopt new business ethics.

27.2.2.5 Easy Business Expansion

Outsourcing can help business expansion through increased market access. The service providers can be highly specialized and very strong in marketing or distribution. Numerous third-party service users indicate a broader geographical coverage of their products through using service providers.

27.2.2.6 Risk Sharing

Outsourcing helps share and reduce associated risks. The company could transfer, reduce, or eliminate risks by outsourcing to highly specialized service providers. First, the binding contract may make the provider responsible for part or all of the cost. Second, providers could be very experienced and skilled in assessing and dealing with related uncertainties. The providers could use various risk management tools to mitigate exposure to loss. Therefore, they may help insulate their clients from various risks, such as a price spike.

We have enumerated a list of reasons why companies choose outsourcing. The primary reason is to refocus on their core competencies.

Be aware that outsourcing is not a stand-alone function. It may be considered part of the business re-engineering process. Re-engineering is the fundamental rethinking and radical redesign of business processes to achieve dramatic improvements in critical, contemporary measures of performance, such as cost, quality, service, and speed.

27.2.3 Reasons for Not Outsourcing

It appears that outsourcing is a necessity, based on our earlier analysis. Why are there businesses that still overlook these advantages? In the following points, we take a closer look at the disadvantages of outsourcing.

27.2.3.1 Unclear Relevant Costs

It is often difficult to estimate the cost of outsourcing. Relatively, it is easy to identify the direct cost, such as that for sourcing providers, drafting contract, and managing outsourcing process by estimating the labor hours needed and the relevant financial support necessary. However, there are indirect, hidden costs and costs hard to allocate among business processes. An example: There is no clear cut to divide the company administrative cost among different business processes and activities. In addition, the cost of potential outsourcing failure is often beyond calculation.

27.2.3.2 Loss of Control and Flexibility

Signing a contract means going into a commitment, and it therefore may reduce flexibility. The outsourcing company may risk being locked into a long-term obligation, and it risks overdependence on the service provider. In the event of a failure of the outsourced system, the outsourcing company is under the mercy of the service provider's capability of, and commitment to, quick recovery. The concern about over-reliance on the provider—and, therefore, loss of control over the process—seems legitimate. Some outsourcings may have long-term impact on costs and on the company's latitudes of making changes to affected processes. For example, once the core information system is outsourced, the equipment, operating system, programming languages, and interfaces will be largely in the hand of the IT service provider. A switch to a different provider later could be accompanied by a significant cost.

27.2.3.3 Lack of Improvement and Innovation

Outsourcing may reduce the company's learning capability. Also, it might impair the company's ability to integrate with internal processes. During outsourcing, the provider may not make necessary innovations due to their lack of incentives. At least, the innovation may not be as efficient and timely as when the process is done internally.

27.2.3.4 Risks Involved

There are various potential risks associated with outsourcing. One of them is that confidential information may be lost. Therefore, businesses might have proprietary concerns about data confidentiality, especially about marketing and financial data. Other risks may include a loss of key markets. Be mindful that outsourcing intellectual or other skills of a core competency may be a bad strategy (Lonsdale and

Cox, 2000). In an example, Schwinn outsourced its bicycle frame manufacturing to Giant Manufacturing. After a few years, Giant entered the bicycle market and greatly cut back Schwinn's business.

27.2.3.5 Labor Issues

Outsourcing could lead to layoffs. Early termination may lead to union strikes and legal lawsuits. Proprietary information such as client lists and other marketing data may leave, along with the furloughed employees. Outsourcing could, therefore, be disruptive to continuous operations of the company. There also can be a negative impact on company morale. In turn, the morale affects productivity and rates of employee turnover.

To summarize, the primary reasons why a company does not choose to outsource can be classified as follows:

1. Costs might not be reduced. Anticipated savings may not materialize.
2. Logistics is too important to outsource.
3. Enough (logistics) expertise is available in-house.
4. Control would diminish, and time/effort spent on logistics would not decrease.
5. Service providers might not cooperate in the future in updating the technology and improving the process as needed.

In fact, outsourcing is an alignment of core competencies of two companies. However, the two are different business entities. They could have different objectives, each maximizing its own benefit. It is often the case that one's objective is achieved at the price of the other. Finding common ground between competing objectives largely determines the success and failure of this practice, and it may seriously affect the firm's long-term profitability.

27.2.4 Types of Outsourced Services

Operations kept in-house usually are core competencies of the company. Those outsourced typically are ones either without long-term stable demand or without shared skills or resources with other currently in-house processes. In particular, the outsourced services typically include the following:

- Operations that are resource-intensive, in relatively discrete areas, or those that provide support services
- Operations with fluctuating working patterns
- Operations subject to quickly changing markets and technology, and operations for training, recruitment, and staff retaining

If we look at the outsourced services in terms of their specialty areas, services outsourced may belong to one of the following areas:

27.2.4.1 Information Systems

This includes data warehousing, logistics information systems, enterprise resource planning systems, EDI, and decision support systems typically for demand forecasting, and network planning, etc.

27.2.4.2 Logistics

Examples include outsourcing the transportation function to one or more carriers, and using professional warehousing management firms to manage the warehousing operations, inventories, and distributions. Outsourcing warehousing operations often is seen in public warehouses where owners of the warehouses conduct warehousing operations. The owner operators decide space allocation to SKU, order picking and assembly, cross docking, and so on. As another example, logistics network design is a task critical to a firm's long-term profitability, and is often outsourced to highly specialized firms.

27.2.4.3 Manufacturing

This means outsourcing part of a firm's manufacturing processes. Take a look at the computer companies such as Dell and Compaq. They do not make keyboards, mice, monitors, or hard drives. Nor do they make central processors. All they do is order the semi-products/parts and have them assembled before they are marketed. In this way, the computer firms may focus just on marketing, management of the supply chain system, and other core competencies. Nike is another example that outsources most of its manufacturing processes to vendors all over the world from Korea to China.

27.2.4.4 Freight Payment and Audit Services

Carriers must be paid within a specific number of days. For convenience, traffic managers participate in bill-paying services. A variety of firms offer this service, including banks. The payment service provides summaries of traffic activity that are useful to shippers when planning future freight consolidation. Computerized programs help detect duplicate billings. The rate charged is audited and is ensured to be correct. It is claimed that an average overcharge is 4–5% of the total dues.

27.2.4.5 Customer Service

As an example, outsourcing warranty repairs has seen enormous growth in recent years. In particular, this practice is popular in the PC industry. Under a service warranty, the manufacturer incurs a repair cost each time an item needs repair and a goodwill cost while an item is awaiting and undergoing repair. In outsourcing the warranty repairs, the manufacturer often can improve turnaround times by using geographically distributed vendors, and also can decrease cost by not having to maintain an in-house repair facility (Opp et al., 2005).

Logistics outsourcing has grown rapidly during the past few years. Seventy-five percent of all the freight moved and 65% of all the freight lifted in the United Kingdom in 2000 was done by third-party operators. The most frequently outsourced logistics functions in 2003 were freight payment services, shipment consolidation, direct transportation services, customs brokerage, warehouse management, and freight forwarding (Lieb and Brooks, 2004). TNT and UPS Supply Chain Solutions are examples of companies that offer these services.

27.2.5 Reasons for Failed Outsourcing Partnerships

Unfortunately, not all outsourcing companies achieve their established objectives. Companies that fail in outsourcing experience cost increases for bringing the processes back in-house, or for repeating the process of outsourcing. Reasons for failure vary. The following are some common ones:

27.2.5.1 Lack of Shared Objectives

In an outsourcing relationship, one entity may reduce its cost at the expense of the other. A shared objective is one that saves cost for both parties. As an example, the contract may specify rewards to the supplier for on-time delivery of products or services above a certain percentage (say, 98%), or a penalty for that below a certain point (such as 92%).

27.2.5.2 Poor Communication

Unexpected circumstances may arise so that modifications to the agreement or contract are needed. The relationship easily can fail if the outsourcing company and its supplier do not communicate in time about differences between what they each expect and what they each possess, as well as the changes in their expectations.

27.2.5.3 Failure to Manage Relationship

Companies that have made a complex outsourcing arrangement may fail in building or preparing an internal infrastructure for relationship management. In an extreme case, there may be no contact person designated at all.

27.2.6 Outsourcing Process

The following steps generally are followed:

Step 1 Making a Decision: To Outsource or Not?

The primary decision should be made between outsourcing and keeping operations in-house. The company must have its clearly defined core competencies and evaluate its own competitive position in the market. The trade-off between the benefit and cost of outsourcing must be calculated carefully. Since the outsourcing decision is sometimes too complicated to make, the company can seek help from external consultants. External consultants could be a good complement to the company's expertise.

Step 2 Making a List of Potential Providers

After the decision has been made to outsource, the outsourcing company has to analyze and articulate the specific requirements of the service or process to be outsourced. The company then makes a list of potential service providers. This can be done based on the provider's production capacity, as well as its technical strength. A formal request for proposal (RFP) is sent to these companies. The RFP must include specific criteria for selection of candidate service providers.

Step 3 Shortening the List and Signing the Contract

Once the applications for service provision have been received, a short list of potentially promising service providers is generated based on the criteria specified in the FRP earlier. These selected providers are to be examined further through phone calls, or to be requested for additional materials. Site visits to these providers' facilities also may be arranged. Two or three of the most preferred providers are prioritized in the end to negotiate for a contract.

A contract is a commitment. It needs to be carefully prepared. There are many factors to be considered in making the contract, such as the following:

- Control over the process outsourced
- Confidentiality of proprietary information
- Risk of cost escalation
- Provisions of termination
- Performance measure and minimum performance requirement
- Incentives

The incentive usually includes reward terms for good performance.

Other Steps

As said earlier, it is necessary to specify the requirement of regular communications between the parties involved in outsourcing. Periodic meetings can be held, and a liaison officer helps maintain a permanent line of communication between them.

Outsourcing is an art-of-practice, in many ways. Many practitioners are in favor of consolidating service providers because it usually brings down the service cost. In some situations, however, companies choose to split the outsourced service among multiple service providers. Multi-sourcing prevents providers from becoming complacent and gives them incentives to compete based on their performances. An example is chemical giant DuPont, which outsourced IT services to Accenture and Computer Sciences Corp. (CSC) in 1997. In a 10-year contract, CSC handles the infrastructure and Accenture handles SAP AG software deployment (Gibson, 2005).

27.3 Outsourcing Modeling

Modeling outsourcing problems is interdisciplinary. There is no single theory or technique universal to all situations. Theories in linear programming, stochastic processes, and statistics, as well as game theory, often are used, as will be seen clearly in the examples introduced in the following. In what follows, several representative models are presented. The models introduced shall not be taken as the only choices available. Rather, they should serve just as examples to show the true interdisciplinary nature of the modeling. Engineers need to know that modeling is just one aspect of the complex outsourcing practice. And business managers shall be conscious that good modeling contributes to significant cost savings.

Example 1: Warrantee Service Outsourcing

A manufacturer outsources its after-sale repair services of K identical items to V service vendors. The decision is to preassign the items to the vendors. Vendor i ($i = 1, 2, ..., V$) has s_i identical servers, each with exponential service times with rate μ_i. The time between failures for a single item is distributed exponentially with rate λ. We assume that the information about s_i, μ_i, and λ is known to the manufacturer. For each repair performed by vendor i, the manufacturer must pay the vendor a fixed amount c_i. The manufacturer also must consider loss of goodwill associated with long waits for repair. To account for this, the manufacturer incurs a goodwill cost at the rate of h_i per unit time that an item spends in queue and service at vendor i. The objective of this problem is to allocate the repairs among the vendors such that the total cost incurred to the manufacturer is minimized (Opp et al., 2005).

We introduce a simple static model for an approximate solution to this problem. First, we use a single server with service capacity of $s_i\mu_i$ to approximate the total service capacity at vendor i. We assume that items preallocated to vendor i will always go to vendor i for repair services. Denote by k_i allocation of items to vendor i. Of course, we assume that the total service capacity of the providers exceeds the service needs on the average. The average stay time of each customer allocated to vendor i is about equal to $1/s_i\mu_i - \lambda k_i$ based on queuing theory.

Then the outsourcing allocation problem may be modeled as a nonlinear optimization problem as follows:

OBJ:
$$\min \sum_{i=1}^{V} \left(\lambda c_i k_i + \frac{h_i \lambda k_i}{s_i \mu_i - \lambda k_i} \right) \tag{27.1}$$

Subject to:

$$\sum_{i=1}^{V} k_i = K \tag{27.2}$$

$$k_i \geq 0 \text{ and integer } i = 1, ..., V \tag{27.3}$$

Here, the cost associated with vendor i is $f_i(k_i) = \lambda c_i k_i + h_i \lambda K_i / s_i \mu_i - \lambda k_i$, in which the first term accounts for the repair cost and the second for queuing cost. It is easy to verify the convexity of the objective function by deriving the Jacobian matrix, as we have assumed that $s_i \mu_i - \lambda k_i > 0$. Furthermore, this static allocation problem is a separable convex allocation problem. A greedy algorithm may be obtained readily.

Greedy Algorithm for Program 1

Step 1: Set $= 0$, for all $i = 1, ..., V$.
Step 2: Choose a vendor $j \in \arg\min_{i=1,...,V} \{f_i(k_i + 1) - f_i(k_i)\}$.
Step 3: Set $k_j = k_j + 1$.

Step 4: If $\sum_{i=1}^{V} k_i < K$, repeat from Step 2; Else, stop: (k_i) is optimal.

We have presented probably the simplest algorithm used in this problem. Interested readers may continue to think about replacing the duration cost at each vendor with its exact expression, assuming that the arrival rate for repair is λk_i, and that there exist multiple identical servers at each vendor. The objective function of the new formulation is still convex and separable. Furthermore, if one thinks that the algorithm is "too" static, exploration of dynamic models is available, as in Opp et al. (2005).

Example 2: Combinatorial Auction for Transportation Service Outsourcing

It has been increasingly popular to use combinatorial auction for transportation service procurement in recent years. A bid with freight lanes bundled and priced is called combinatorial bid. An auction that allows combinatorial bid is called a combinatorial auction. Most shippers use annual auctions to procure transportation services, leading to annual contracts (Sheffi, 2004). Successful practice at Sears Logistics Services reduced the annual freight cost from \$190 million to about \$165 million (Ledyard et al., 2002). By using combinatorial auctions, they can reduce their operating costs while protecting carriers from winning lanes that do not fit their networks, thereby improving carriers' operations, as well.

Let us consider the shipper's problem first. A shipper, every year, has a set of freight lanes for service. Each lane could have several hundred truckloads of movement. Instead of calling for bids for each of the lanes, the shipper first may bundle them to avoid empty backhauls so that carriers may incur lower service cost and, therefore, may bid less for the service. How to generate the bundles is a theoretical problem for shippers.

Although there has been a lack of theoretical discussion about the optimality of combinatorial bundling, researchers have been developing heuristic methods for generation of bundles. A recent example is presented as follows (Song and Regan, 2005):

OBJ:
$$\min \sum_{j=1}^{|J|} e_j y_j \tag{27.4}$$

Subject to:

$$\sum \delta_{ij} y_j = u_i, \qquad \forall i \tag{27.5}$$

$$y_j \text{ is binary and } j \in J \tag{27.6}$$

The above set-covering formulation generates candidate bundles for carriers to bid on. In this formulation, e_j is the cost of bundle j. J is the set of all candidate bundles, assuming a sufficiently large set of bundles are available. δ_{ij} is the number of truckload movements on lane i covered by bundle j. u_i is the number of truckload movements demanded on lane i. y_j is one when bundle j is selected, and zero otherwise.

There are many assumptions underlying the set-covering formulation for candidate bundle generation. The complexity involved in obtaining the cost e_j from bundle y_j is immense. The cost e_j is hard to evaluate, as it depends on potential fleet condition and availability of other lanes. This complexity arises out of the nature of combinatorial problems. Change of a single vehicle status or commitment leads to change of assignments to all vehicles potentially. One may ask a legitimate question of whether it is worthwhile to pursue the optimal solution to such a complex problem.

Next, we turn to the carriers' problem. The carriers may have different valuation of the lanes, depending on how the lanes in auction fit their service networks. If carriers (bidders) are given the freedom to generate bundles of their own interest, how can carriers generate bundles to form a bid? There are several theoretical issues involved in this problem. These issues are interrelated. First, what bidding language to use? Basic bidding languages include OR and XOR. In an OR bid, the structure is expressed in the form of A OR B OR C. Each element in a bid can be won independent of the winning status of other elements. In this case, the bidders may be subject to the exposure problem. The exposure problem takes place if the bidder wins both A and B when they do not want both. In an XOR bid, the bid is expressed as A XOR B XOR C, and only one of the elements (A, B, or C) can be won. Secondly, how to generate the elements in a given bidding language? In addition, consider OR or XOR, respectively. Bundle generation is definitely a function of resulting revenue and cost. Therefore, the third question arises: How to price the bundles? All three questions are determinants of a good combinatorial bid. Furthermore, evaluation of a bid definitely requires knowledge of the outcomes. There must be a certain description of the outcomes in terms of their associated probabilities. The probability of an outcome may depend on pricing as well. Obviously, all of the key issues are intertwined. Of course, the carriers need to take into account their service capacity available and current contracts as well.

To understand the complexity of the problem, we take a look at the synergy yielded. The following definition helps understand the synergy of two bundles (Song and Regan, 2005):

Definition: Denoted by *empty(S)* a carrier's true cost of serving a set of new lanes *S* if and only if these lanes are awarded. We say two disjoint sets of lanes S_i and S_j are:

- Complementary: if $empty(S_i) + empty(S_j) > empty(S_i \cup S_j)$;
- Substitutable: if $empty(S_i) + empty(S_j) < empty(S_i \cup S_j)$;
- Additive: if $empty(S_i) + empty(S_j) = empty(S_i \cup S_j)$.

The degree to which bundles are complementary or substitutable varies with availability of other lanes. This fact complicates the combinatorial bid generation problem.

We further look at how to evaluate a combinatorial bid just to show the complexity of this problem.

Assuming that there is a price associated with each bundle in a combinatorial bid, a combinatorial bid may be evaluated in the following way (Wang and Xia, 2005):

$$\sum_{w \in \omega} \overline{p}(w)\left\{r(w) - emp(w + S_g)\right\}$$

(27.7)

Here, *w* represents a particular outcome. ω is the set of all outcomes. $\overline{p}(\cdot)$ is the probability of a certain outcome. $r(\cdot)$ is the revenue of a certain outcome. $emp(\cdot)$ is the travel cost in terms of empty distance to serve a certain set of lanes. S_g represents the set of lanes in which the carrier already has been engaged.

It is shown that even evaluation of a combinatorial bid is an *NP*-hard problem in a strong sense.

A heuristic bundling method for bidders (carriers) is proposed as follows:

Step 1: Use the nearest insertion method to make assignments to all of the drivers.

1. Each driver starts with a null assignment (i.e., from depot back to depot).
2. Corresponding to each assignment (route), an unassigned lane has an insertion cost at its best insertion point. Consequently, each unassigned lane has the least insertion cost among all the routes.
3. Choose the unassigned lane with the smallest value of least insertion cost, and insert it into the corresponding route. This lane has been assigned.
4. Repeat from (2) until all of the lanes have been assigned, or until no feasible assignment is possible.

Step 2: Include the lanes assigned to each driver in a bundle.
Step 3: Encompass all of the bundles in an OR bid.

For large vehicle routing problems, this heuristic method shows its computational advantage over an integer-programming-based method designed for vehicle routing and scheduling. Note that even sophisticated vehicle routing and scheduling methods are still heuristics in bid generation.

Example 3: Linear Programming Model for Production Outsourcing

Production outsourcing sometimes can be defined as a linear programming problem. It identifies which products to manufacture in-house and which to outsource. The following example and model are from Coman and Ronen (2000). Consider a production facility consisting of N resources $(1, 2, ..., j ..., N)$ and manufacturing M different products. The decision is how much of each product should be outsourced so that the total profit may be maximized.

We first introduce the definitions as follows:

b_j Resource j's capacity (in working minutes) per planning period.
a_{ij} Number of minutes required by resource j to process product i.
RM_i Cost of raw material i.
MP_i Market price of product i.
DQ_i Market demand quantity for product i.
OE Operating expenses.
CP_i Price of purchasing their own raw materials and deliver products by contractors.

Because of capacity constraints, the company can manufacture only X_i units of products i, where $X_i \leq DQ_i$. The throughput from manufacturing one unit of product i is $MP_i - RM_i$. Production of X_i units generates a throughput of $X_i(MP_i - RM_i)$. To satisfy market demands, the company orders $(DQ_i - X_i)$ from outside contractors. This generates a throughput of $(MP_i - CP_i)$ per contracted unit. Thus, total profits from product $i = X_i(MP_i - RM_i) + (DQ_i - X_i)(MP_i - CP_i)$. Total profit from all M products equals $\sum_{i=1}^{M}[X_i(MP_i - RM_i) + DQ_i - X_i)MP_i - CP_i)] - OE = \sum_{i=1}^{M}(CP_i - RM_i)X_i + Const$. The total throughput of the plant is restricted by its production capacity of resources as well as market demands.

The LP formulation of this problem is presented as follows:

$$Max \sum_{i=1}^{M} X_i(CP_i - RM_i) \tag{27.8}$$

$$\sum_{i=1}^{M} a_{ij}X_i \leq b_j; \ j \in \{1, ..., N\} \tag{27.9}$$

$$X_i \leq DQ_i; i \in \{1, ..., M\} \tag{27.10}$$

As indicated, only two variables are relevant. These are contractor markup per product and time per product at the bottleneck resource. The order of priority of products for production can be determined by the ratio of contractor markup per bottleneck minutes, $(CP_i - RM_i)/a_{ij}$ = (contractor price – raw materials price)/constraint minutes, where j = constrained resource (bottleneck).

From the discussion, it can be seen that the less greedy the contractor is, the higher the incentive to outsource to it. This model was demonstrated to be analytically robust and, at the same time, simple to implement. Thus, the LP-generated ratio requires only two variables per product: contractor markup and work time at the bottleneck.

27.4   Sample Case and Summary

27.4.1   Sample Case

A national retail chain store received the following benefits from a comprehensive logistics service provider, Transplace.com. The company was focused on rapid expansion of its chain stores, and preferred to focus its internal efforts on display, marketing, and rapid store growth. Transplace.com was engaged to design and manage the inbound transportation supply chain. The service package included

TABLE 27.1   Performance Comparison before and after Outsourcing

Before	After
$1.8 billion revenue and 1150 stores	$4.5 billion revenue and over 3000 stores
85% less-than-truckload inbound shipments	Less than 2% less-than-truckload inbound shipments
Seven-day transit time	1.5-day transit time
No visibility to shipment	Total shipment visibility through Internet/EDI
Rampant freight damage	Freight damage eliminated
Seven DCs constraining growth	Eight DCs now support more than double business
High inventory level	– 50% reduction of the inventory
High transportation cost	– 20% transportation cost reduction
Poor on-time performance	– 98% on-time performance
Poor utilization of outbound fleet	– 25% increase in outbound fleet utilization and avoidance of original fleet expansion budget

Source: www.transplace.com.

conversion of inbound vendor shipments to Transplace.com management, consolidation of less-than-truckload shipments, optimal mode selection, cross docking, carrier management, optimization of the retailer's private fleet and networking of the retailer's freight with additional Transplace.com freight. The supply chain objectives are maximum inventory velocity and lowest total transportation cost. Table 27.1 provides a comparison between the retail store's performances before and after outsourcing.

27.4.2   Summary

Outsourcing is a strategic means for cost reduction by bringing together two or more companies with complementary competitive advantages. 3PL is the outsourcing related to logistics activities. The increasing importance of focusing on core competencies has fueled the development of 3PL. 4PL is a supply-chain integrator that delivers a comprehensive supply chain solution to its client when 3PL does not meet the needs. Nowadays, new technologies have been creating opportunities for service outsourcing.

Outsourcing may be considered as an engineering process. One may view it as having a life cycle that includes start-up and normal operation, as well as salvaging. It could help manage the process with systems approach and control theory. In this chapter, we have introduced briefly the outsourcing process, its advantages, disadvantages, and outsourcing failures.

Outsourcing modeling is a relatively new front in the logistics engineering area. It has a set of tools from probability theory, linear programming, to dynamic control. In this chapter, we have presented several representative examples.

References

Coman, A. and Ronen, B., Production outsourcing: a linear programming model for the theory of constraints, *International Journal of Production Research*, 38, 1631, 2000.

Gibson, S., Get the right balance; picking outsourcing partner takes skill, *eWeek*, 22, 18, 2005.

Ledyard, J.O., Olson, M., Porter, D., Swanson, J.A., and Torma, D.P. The first use of a combined value auction for transportation services, *Interfaces*, 32, 4, 2000.

Lieb, R.C. and Brooks, B., Lock haven, *Transportation Journal*, 43, 24, 2004.

Logan, S.M., Faught, K., and Ganster, D. Outsourcing a satisfied and committed workforce: a trucking industry case study, *International Journal of Human Resource Management*, 15, 147, 2004.

Lonsdale, C. and Cox, A., The historical development of outsourcing: the latest fad? *Industrial Management and Data Systems*, 100, 444, 2000.

Opp, M., Glazebrook, K., and Kulkarni, V.G.., Outsourcing warranty repairs: dynamic allocation, *Naval Research Logistics*, 52, 381, 2005.

Sheffi, Y., Combinatorial auctions in the procurement of transportation services, *Interfaces*, 34, 245, 2004.

Song, J. and Regan, A.C., Approximation algorithms for the bid construction problem in combinatorial auctions for the procurement of freight transportation contracts, *Transportation Research B*, 39, 914, 2005.

Vaidyanathan, G. A framework for evaluating third-party logistics, *Association for Computing Machinery. Communications of the ACM*, 48, 89, 2005.

Wang, X. and Xia, M., Combinatorial bid generation for transportation service procurement, *Transportation Research Record, Journal of Transportation Research Board*, 1923, 1989, 2005.

28

Brief Overview of Intermodal Transportation

Tolga Bektaş
Teodor Gabriel Crainic
University of Québec in Montréal

This chapter focuses on Intermodal Freight Transportation broadly defined as a chain made up of several transportation modes that are more or less coordinated and interact in intermodal terminals to ensure door-to-door service. The goal of the chapter is to present intermodal transportation from both the supplier and the carrier perspectives, and identify important issues and challenges in designing, planning, and operating intermodal transportation networks, focusing on modeling and the contributions of operations research to the field.

28.1 Introduction

In today's world, intermodal transportation forms the backbone of world trade. Contrary to conventional transportation systems in which different modes of transportation operate in an independent manner, intermodal transportation aims at integrating various modes and services of transportation to improve the efficiency of the whole distribution process. Parallel to the growth in the amount of transported freight and the changing requirements of integrated value (supply) chains, intermodal transportation exhibits significant growth. According to the U.S. Department of Transportation (2006), the value of the multimodal shipments, including parcel, postal service, courier, truck-and-rail, truck-and-water, and rail-and-water increased from about $662 billion to about $1.1 trillion in a period of nine years (1993–2003).

Major players in intermodal transportation networks are shippers, who generate the demand for transportation, carriers, who supply the transportation services for moving the demand, and the intermodal network itself composed of multimodal services and terminals. The interactions of these players and their individual behavior, expectations, and often conflicting requirements determine the performance of intermodal transportation systems. The goal of this chapter is therefore to be informative on intermodal transportation, from both the supplier and the carrier perspective, identify important issues and challenges in designing and operating intermodal transportation networks, and point out major operations research contributions to the field. A more in-depth discussion of these topics may be found in, for example, Crainic and Kim (2007), Macharis and Bontekoning (2004), and Sussman (2000).

The chapter is structured as follows. Section 28.2 presents the basics on intermodal transportation, with an emphasis on its foremost components: containers, carriers, and shippers. We then discuss, in Sections 28.3 and 28.4, respectively, the major issues and challenges of intermodalism from the shippers' and carriers' perspective. Section 28.5 provides a brief description of intermodal terminals and the operations performed therein. Section 28.6 is dedicated to the case of rail intermodal transportation, as an illustration of the main discussion.

28.2   Intermodal Transportation

Many transportation systems are multimodal, that is, the infrastructure supports various transportation modes, such as truck, rail, air, and ocean/river navigation, and carriers operating and offering transportation services on these modes. Then, broadly defined, intermodal transportation refers to the transportation of people or freight from their origin to their destination by a sequence of at least two transportation modes. Transfers from one mode to the other are performed at intermodal terminals, which may be a sea port or an in-land terminal, for example, rail yards, river ports, airports, etc. Although both people and freight can be transported using an intermodal chain, in this chapter, we concentrate on the latter.

The fundamental idea of intermodal transportation is to consolidate loads for efficient long-haul transportation (e.g., by rail or large ocean vessels), while taking advantage of the efficiency of local pick up and delivery operations by truck. This explains the importance of container-based transportation. Freight intermodal transportation is indeed often equated to moving containers over long distances through multimodal chains. Intermodal transportation is not restricted, however, to containers and intercontinental exchanges. For instance, the transportation of express and regular mail is intermodal, involving air and land long-haul transportation by rail or truck, as well as local pick up and delivery operations by truck (Crainic and Kim 2007). In this chapter, we focus on container-based transportation.

An intermodal transportation chain is illustrated in Figure 28.1. In this example, loaded containers leave the shipper's facilities by truck to a rail yard, where they are consolidated into a train and sent to another rail yard. Trucks are again used to transport the containers from this rail yard to the sea container terminal. This last operation may not be necessary if the sea container terminal has an interface to the rail network, in which case freight is transferred directly from one mode to the other. Containers are then transported to a port on another continent by ocean shipping, from where they leave by either trucking or rail (or both) to their destinations.

28.2.1   Containers

A container, as defined by the European Conference of Ministers of Transport (2001), is a "generic term for a box to carry freight, strong enough for repeated use, usually stackable and fitted with devices for

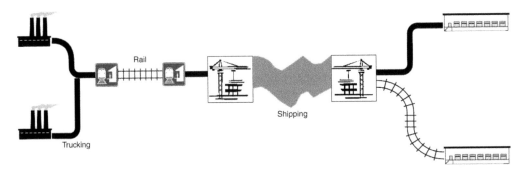

FIGURE 28.1 An intermodal transportation network.

transfer between modes." The fact that the standards on container dimensions were established very early also explains its popularity. A standard container is the 20-foot box, which is 20 feet long, 8'6" ft high and 8 ft wide. This is refereed to as a twenty-foot equivalent unit (TEU). However, the widely used container size is the 40-foot box (a number of longer boxes are sometimes used for internal transport in North America). Containers are either made of steel or aluminum, the former being used for maritime transport and the latter for domestic transport.

Intermodal transportation relies heavily on containerization due to its numerous advantages. First, containerization offers safety by significantly reducing loss and damage, since the contents of a container cannot easily be modified except at origin or destination. It is worth mentioning in this respect that the safety level of container transportation is currently being significantly increased by electronic sealing and monitoring to address preoccupations with terrorist treats, illegal immigration, and smuggling. Second, due to its standard structure, transfer operations at terminals are fast and performed with a minimal amount of effort. This results in reduced cargo handling, and thus a speed-up of operations not only at the terminals, but through the whole transport chain. Third, containers are flexible enough to enable the transport of products of various types and dimensions. Fourth, containerization enables a better management of the transported goods. Due to these reasons, the use of containers significantly decreases transport costs.

Containerization has had a noteworthy impact on both land transportation and the way terminals are structured. An example for the former can be seen in rail transportation, where special services have been established by North-American railways, enabling container transportation by long, double-stack trains. As for the latter, ports and container terminals have either been built or undergone major revisions to accommodate continuously larger container ships and efficiently perform the loading, unloading, and transfer operations. Container terminal equipment and operating procedures are continuously enhanced to improve productivity and compete, in terms of cost and time, with the other ports in attracting ocean shipping lines.

28.2.2 Carriers

In an intermodal chain, carriers may either provide a customized service, where the vehicle (or convoy) is dedicated exclusively to a particular customer, or operate on the basis of consolidation, where each vehicle moves freight for different customers with possibly different origins and destinations.

Full-load trucking is a classic example of customized transportation. Upon the call of a customer, the truck is assigned to the task by the dispatcher. The truck then travels to the customer location, is loaded, and then moves to the destination, where it is unloaded. Following this, the driver is either assigned a new task by the dispatcher, kept waiting until a new demand appears in the near future, or repositioned to a location where a load exists or is expected to be available about the arrival time. The advantages of full-load trucking come from its flexibility in adapting to a highly dynamic environment and uncertain future demands, offering reliability in service and low tariffs compared to other modes of transportation. The full efficiency of full-load trucking is achieved. Customized services are also offered, for example, by charted sea or river vessels and planes.

In many cases, however, trade-offs between volume and frequency of shipping, along with the cost of transportation, render customized services impractical. Consolidation, in such situations, turns out to be an attractive alternative. Freight consolidation transportation is performed by less-than-truckload (LTL) motor carriers, railways, ocean shipping lines, regular and express postal services, etc. A consolidation transportation system is structured as a hub-and-spoke network, where shipments for a number of origin-destination points may be transferred via intermediate consolidation facilities, or hubs, such as airports, seaport container terminals, rail yards, truck break-bulk terminals, and intermodal platforms. An example of such a network with three hubs and seven regional terminals (origin and destination points for demand) is illustrated in Figure 28.2. In hub-and-spoke networks, low-volume demands are first moved from their origins to a hub where the traffic is sorted (classified) and grouped (consolidated). The aggregated traffic is then moved in between hubs by high frequency and high capacity services. Loads are then transferred to their destination points from the hubs by lower frequency services often utilizing smaller vehicles. When the level of demand is sufficiently high, direct services may be run between a hub and a regional terminal. Although a hub-and-spoke network structure results in a more efficient utilization of resources and lower costs for shippers, it also incurs a higher amount of delays and a lower reliability due to longer routes and the additional operations performed at terminals. The planning methodologies evoked in Section 28.4 aim to address these issues.

Land consolidation transportation services are offered by LTL motor carriers and railways. The flexibility, high frequency, and low cost of trucking transportation resulted in a high market share of freight transportation being captured by this mode. This situation, which may be observed world-wide, resulted in very large truck flows and road congestion, and contributes significantly to the high level of pollutant emissions attributed to the transportation sector. The trend is slowly changing, however. On the one hand, recent policy measures, particularly in the European community, target the mode change from

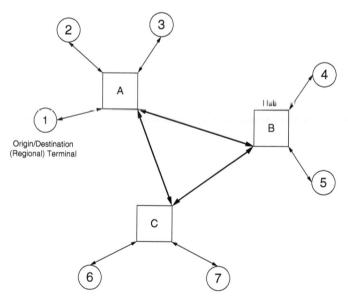

FIGURE 28.2   A hub-and-spoke network.

road-based to intermodal. On the other hand, the continuous and significant increase in container-based international traffic, which generates large flows of containers that need to be moved over long distances, favors rail (and, in a smaller measure, river) based transportation.

Railways have risen to the challenge by proposing new types of services and enhanced performances. Thus, North-American railways have created intermodal subdivisions which operate so-called "land-bridges" providing efficient container transportation by long, double-stack trains between the East and the West coasts and between these ports and the industrial core of the continent (so-called "mini" land-bridges). New container and trailer-dedicated shuttle-train networks are being created within the European Community. These initiatives have succeeded. Thus, intermodal transportation is the fastest growing part of railroad traffic. Thus, for example, in a period of only two years, from 2003 to 2005, the amount of rail intermodal traffic in the United States grew from 7.33 million to 8.71 million containers (Association of American Railroads 2006). Note that the 8.71 million containers transported by rail in the United States in 2005 represent about 75% of the total intermodal units moved in that year, the remaining 25% being trailers. In Canada, there is a similar trend in the number of intermodal carloads, which grew by 336,000 or 77.5% during the past decade. This number represents 26.3% of the industry's overall growth in originated carloads during this period (The Railway Association of Canada 2006). Section 28.6 discusses rail intermodal transportation in some more depth.

Maritime and air-based modes are used for the intercontinental legs of intermodal transportation. Heavily container-based, the former provide the backbone of nonbulk international trade. Efficiency reasons result in continuously larger container vessels being commissioned for inter continental movements. The operation of such ships, which cannot pass through the Panama Canal, should not stop too frequently, and cannot even berth in most ports, has had a number of important consequences. In particular, maritime and land transportation routes have been modified through, for example, the creation of the North-American land-bridges and the introduction of a "new" link into the intermodal chain: super-ships stop at a small number of major seaports and containers are transferred to smaller ships for distribution to various smaller ports. For maritime transportation, variations in travel times are larger, and travel and loading/unloading times are longer, compared to that of most land-based trips. As for the air mode, although being increasingly used for intermodal transport, its relatively higher costs still make it interesting mostly for high-value or urgent deliveries.

All consolidation-based transportation modes involved in intermodal transportation must provide efficient, reliable, and cost-effective services. Comparatively, railways face the biggest challenge, in competing with trucking to provide shippers the level and quality of service they require for their land long-haul transportation needs. Sections 28.4 and 28.6 discuss these issues.

28.2.3 Shippers

Shippers generate the demand for transportation. Defining its logistics strategy represents a complex decision process, and the choice of the transport mode is only a part of this whole strategy. This process is generally assumed to have a three-level decision structure, composed of long-, medium-, and short-term decisions (Bolis and Maggi 2003). In the long run, shippers define their logistics strategies in terms of their customer network and production. Medium-term plans include decisions as to inventories at production, warehousing, and distribution facilities, frequency and amount of shipping, flexibility of service, etc. Finally, the shipper decides, at the short-term level, the attributes of the services required for its shipments, such as maximum rates, transport time, reliability, and safety.

When such decisions are made, shippers consider the availability and the characteristics of the services offered on the market by carriers and intermediaries such as freight brokers and third-party logistics providers. These decisions are based on several factors detailed in the following section. It is worth mentioning, however, that in many cases shipper decisions have a greater importance for the outcome of the service rather than the way it is delivered.

28.3   Shipper Perspectives on Intermodal Transportation

Although intermodal networks are formed as combinations of individual transportation modes and transfer facilities, a shipper perceives it as a single integrated service. Shippers, therefore, expect intermodal services to function similar to unimodal services, especially in terms of speed, reliability, and availability.

The shipper's decision to use a particular transportation mode is generally based on several criteria. A number of studies mostly based on surveys and data analyses have been conducted to identify the specific service characteristics often deemed important in the shipper decision process. McGinnis (1990) identified six factors that affect the shipper's decision in choosing a specific transportation mode, namely (1) freight rates (including cost and charges), (2) reliability (delivery time), (3) transit times (time-in-transit, speed, delivery time), (4) over, short, and damaged shipments (loss, damage, claims processing, and tracing), (5) shipper market considerations (customer service, user satisfaction, market competitiveness, market influences), and (6) carrier considerations (availability, capability, reputation, special equipment). Amongst these, shippers were observed to place more emphasis on overall service than cost, although freight rates still had a significant importance. Five years later, in an update to McGinnis' study, Murphy and Hall (1995) observed that reliability, as opposed to cost or any other factor, had more influence on the U.S. shipper decisions.

As important as the shipper's choice of transportation modes may be, shipper perception of modes and services is believed to have a higher impact on the overall decision process. To quote Evers et al. (1996), "shippers decide on the mode of transportation and specific carrier only after they have formed perceptions of the alternative services. They compare their perceptions, and possibly other information, with criteria they have developed. Then, they use a decision-making process to choose the transportation method that best meets their established criteria." Hence, shipper perception is a central input component to the decision-making process in mode selection.

In a study conducted to identify the determinants of shipper perceptions of modes, Evers et al. (1996) proposed six factors: (*i*) timeliness, (*ii*) availability, (*iii*) firm contract, (*iv*) suitability, (*v*) restitution, and (*vi*) cost, the first two being observed to have a greater effect than the others. Thus, the more a carrier focuses on improving shipper's perceptions of these six factors, the more likely it is for it to be used. Extending their analysis to an individual carrier level, Evers and Johnson (2000) found that the future collaboration of a shipper with an intermodal service is affected by the shipper's satisfaction with, and ability to replace, the carrier. They identified that determinants of shipper's perceptions at the carrier level were communication, quality of customer service, consistent delivery, transit times, and competitive rates, with the first two factors being the most important drivers of the overall perception. Although these results do confirm what is expected of a carrier, it is interesting to see that the perception of modes may be different from that of carriers. More specifically, the two chief factors influencing the overall perception for modes, namely timeliness and availability, are replaced by communication and customer service for the individual carriers.

Studies performed in the early 1990s revealed that shippers have varying perceptions of alternative transportation modes. In general, shippers were found to perceive truck transportation best, followed by intermodal (rail-truck) transportation, and rail transportation. Prior research indicates that perceptions of railroad service have improved since deregulation in 1980, both in terms of rates and service. More recent research shows that price, time, and reliability are important factors in the decision process, but, frequency and flexibility also emerge as significant decision factors when firms operate in a just-in-time (JIT) context. Empirical evidence exists, on the other hand, as to the fact that, in Europe, rail has no acceptance problems but that of service quality (Bolis and Maggi 2003).

Shipper perceptions, influenced by past experiences, expectations, common knowledge, carrier advertising, modal image, and misinformation (Evers et al. 1996), may not always reflect the real situation. After all, a shipper's perception of a mode or an individual carrier is a tentative view. This, however, may be altered through marketing efforts by the carrier. Carriers thus should strive to ensure that the shipper's perception is in line with reality as much as possible as such efforts can be beneficial, in terms of increased usage of their service. Failure to put forth such efforts may easily place the carrier at a disadvantage with respect to its competitors.

28.4 Carrier Perspective on Intermodal Transportation

Carriers face a number of issues and challenges in providing an efficient and cost-effective service to the customer, which may be examined according to the classical categorization of planning decisions, namely strategic (long term), tactical (medium term), and operational (short term) level of planning and management of operations. We briefly address these issues in this section by focusing on consolidation-based carrier cases. This choice is motivated by the complexity of the planning issues in this context and on the fact that intermodal transportation forms a consolidation-based system. For a more in-depth treatment, one may consult the reviews of Christiansen et al. (2004, 2007), Cordeau et al. (1998), Crainic (2003), Crainic and Kim (2007), Crainic and Laporte (1997).

28.4.1 System Design

At the strategic level, the carrier is concerned with the design of the physical infrastructure network involving decisions as to the number and location of terminals (e.g., consolidation terminals, rail yards, intermodal platforms), the type and quantity of equipment (e.g., cranes) that will be installed at each facility, the type of lines or capacity to add or abandon, the customer zones to serve directly, and so on. The term system design encompasses issues pertinent to strategic level decisions. It is often the case that such decisions are made by evaluating alternatives using network models for tactical or operational planning of transportation activities. When specific models are developed for strategic planning issues, these usually take the form of static and deterministic location formulations addressing issues related to the location of consolidation or hub terminals and the routing of demand from its origin to its destination terminals.

To illustrate the methodological approaches proposed to address strategic planning issues for consolidation-based carriers, we examine the issue of determining the sub-structure of such a system. This problem consists in determining the locations of the hubs on a given network, the assignment of local terminals to the hubs that are established, and the routing of loads of each demand through the resulting network. In its most basic form, the problem is addressed by a multi-commodity hub-location formulation, assuming that all traffic passes through two hubs on its route from its origin to its destination. In this simple formulation, it is assumed that there can be no direct transport between non-hub terminals, based on the hypothesis that inter-hub transportation is more efficient due to consolidation. Furthermore, one also assumes that there are neither capacity restrictions on hubs, nor fixed costs associated with establishing a link between a regional and a consolidation terminal.

The problem is modeled on a directed network $G = (N, A)$, with N as the set of nodes (or vertices) and A as the set of arcs (or links). Nodes are identified as origins (set O), destinations (set D), and hubs or consolidation nodes (set H; sets O, N, and H are not necessarily disjoint). The set of commodities (types of containers) that move through the network are represented by the set P. The amount of commodity $p \in P$ to be transported from origin terminal $i \in O$ to destination terminal $k \in D$ is denoted by d_{ik}^p. There are three decision variables: y_j is a binary variable equal to 1 if a consolidation terminal is located at site j and 0, otherwise. y_{ij} is a binary variable equal to 1 if terminal i is linked to hub j and 0, otherwise. Finally, variable x_{ijlk}^p denotes the amount of flow for commodity p with origin i, destination k, passing through terminals j and l in the given order. The following formulation then solves the aforementioned network design problem with exactly M hubs installed in the network, where c_{ij}^p, c_{lk}^p, and c_{jl}^p stand for the unit transportation costs between origin terminals and hubs, hubs and destination terminals, and inter-hubs, respectively.

$$\text{Minimize} \sum_{p \in P} \left\{ \sum_{i \in O} \sum_{j \in H} c_{ij}^p y_{ij} \left(\sum_{l \in H} \sum_{k \in D} x_{ijlk}^p \right) + \sum_{l \in H} \sum_{k \in D} c_{lk}^p y_{lk} \left(\sum_{i \in O} \sum_{j \in H} x_{ijlk}^p \right) + \sum_{j \in H} \sum_{l \in H} c_{jl}^p y_{ij} y_{kl} \left(\sum_{i \in O} \sum_{k \in D} x_{ijlk}^p \right) \right\}$$

subject to

$$\sum_{j \in H} y_j = M,$$

(28.2)

$$\sum_{j \in H} \sum_{l \in H} x_{ijlk}^P = d_{ik}^p \qquad i \in O, \; k \in D, \; p \in P$$

$$y_{ij} \le y_j \qquad i \in O, \; j \in H, \tag{28.3}$$

$$y_{kl} \le y_l \qquad k \in D, \; l \in H, \tag{28.4}$$

$$x_{ijlk}^p \le d_{ik}^p y_j \qquad i \in O, \; k \in D, \; j, \; l \in H, \; p \in P, \tag{28.5}$$

$$x_{ijlk}^p \le d_{ik}^p y_l \qquad i \in O, \; k \in D, \; j, \; l \in H, \; p \in P, \tag{28.6}$$

$$y_j \in \{0,1\} \qquad j \in H, \tag{28.7}$$

$$y_{ij} \in \{0,1\} \qquad i \in O, \; j \in H, \tag{28.8}$$

$$x_{ijlk}^p \ge 0 \qquad i \in O, \; k \in D, \; j, \; l \in H, \; p \in P. \tag{28.9}$$

The objective function of the formulation minimizes the total transportation cost of the system. Constraints 28.1 state that exactly M hubs should be located in the network. To ensure that the demands are satisfied, constraints 28.2 are used. Constraints 28.3 and 28.4 guarantee that a terminal is assigned to a hub only if the hub is established. Constraints 28.6 and 28.7 serve a similar purpose in terms of routing the flows using only selected hubs.

The previous quadratic formulation was first introduced by O'Kelly (1987), while, Campbell (1994) proposed its first linearization. Both formulations are difficult, as are more complex models that include capacities, fixed costs to open facilities, link terminals to hubs, or establish transportation connections, more complex load routing patterns, etc. Consequently, although several contributions have been made relative to the analysis of hub location problems and the development of solution procedures, this is still an active and interesting field of research (see, e.g., the surveys of Campbell et al. 2002 and Ebery et al. 2000).

28.4.2 Service Network Design

Designing the service network of a consolidation-based carrier refers to constructing the transportation (or load) plan to serve the demand, while at the same time operating the system in an efficient and profitable manner. These plans are built given an existing physical infrastructure and a fixed amount of resources, as determined during the system design phase.

Service network design is concerned with the planning of operations related to the selection, routing, and scheduling of services, the consolidation of activities at terminals, and the routing of freight of each particular demand through the physical and service network of the company. These activities are a part of the tactical planning at a system-wide level. The two main types of decisions that are considered in service network design are to determine the service network and the routing of demand. The former refers to selecting the routes, characterized by origin-destination nodes, intermediate stops and the physical route, and attributes, such as the frequency or the schedule, of each service. The latter is concerned with the itineraries that specify how to move the flow of each demand, including the services and terminals used, the operations performed in these terminals, and so on. Although minimization of total operating cost is the main criterion of the service network design objective, improving the quality of service measured by its speed, flexibility, and reliability is increasingly being considered as an additional component of this goal. Service performance measures modeled, in most cases, by delays incurred by freight and vehicles or by the respect of predefined performance targets, are then added to the objective function of the network optimization formulation. The resulting generalized-cost function thus captures the trade-offs between operating costs and service quality. For further details, the reader is referred to Crainic (2000) and Crainic and Kim (2007).

Formulations for service network design either assume that the demand does not vary during the planning period (static formulations) or explicitly consider the distribution of demand as well as the service departures and the movements of services and loads in time (time-dependent formulations). In both cases, however, modeling efforts take the form of deterministic, fixed cost, capacitated, multicommodity network design formulations. To illustrate such approaches, we provide here the multimodal multicommodity path-flow service network design modeling framework proposed by Crainic and Rousseau (1986) (see Powell and Sheffi 1989 for a complementary formulation) for the static case.

In this formulation, the service network, defined on a graph $G = (N, A)$ representing the physical infrastructure of the system, specifies the transportation services that could be offered. Each service $s \in S$ is characterized by its (*i*) mode, which may represent either a specific transportation mode (e.g., rail and truck services may belong to the same service network), or a particular combination of traction and service type; (*ii*) route, defined as a path in A, from its origin terminal to its destination terminal, with intermediary terminals where the service stops and work may be performed; (*iii*) capacity, which may be measured in load weight or volume, number of containers, number of vehicles (when convoys are used to move several vehicles simultaneously), or a combination thereof; (*iv*) service class that indicates characteristics such as preferred traffic or restrictions, speed and priority, and so on. To design the service network thus means to decide what service to include in the transportation plan such that the demands and the objectives of the carrier are satisfied. When a service is operated repeatedly during the planning period, the design must also determine the frequency of each service.

A commodity $p \in P$ is defined as a triplet (origin, destination, type of product or vehicle) and traffic moves according to itineraries. An itinerary $l \in L_p$ for commodity p specifies the service path used to move (part of) the corresponding demand: the origin and destination terminals, the intermediary terminals where operations (e.g., consolidation and transfer) are to be performed, and the sequence of services between each pair of consecutive terminals where work is performed. The demand for product p is denoted by d_p. Flow routing decisions are then represented by decision variables h_l^p indicating the volume of product p moved by using its itinerary $l \in L_p$. Service frequency decision variables y_s, $s \in S$, define the level of service offered, that is, how often each service is run during the planning period. Let $F_s(y)$ denote the total cost of operating service s, and $C_l^p(y, h)$ denote the total cost of moving (part of) product p demand by using its itinerary l. Further, a penalty term $\theta(y, h)$ is included in the objective function capturing various relations and restrictions, such as the limited service or infrastructure capacity. The following formulation can then be used for the service network design problem:

Minimize
$$\sum_{s \in S} F_s(y) + \sum_{p \in P} \sum_{l \in L_p} C_l^p(y, h) + \theta(y, h)$$

subject to

$$\sum_{l \in L_p} h_l^p = d_p \qquad\qquad p \in P,$$

$$(y_s, x_l^p) \in X \qquad\qquad s \in S, l \in L_p, p \in P,$$

$$y_s \geq 0 \text{ and integer} \qquad s \in S,$$

$$h_l^p \geq 0 \qquad\qquad\qquad l \in L_p, p \in P,$$

where $(y_s, x_l^p) \in X$ stand for the classical linking constraints (i.e., no flow may use an unselected service) as well as additional constraints reflecting particular characteristics, requirements, and policies of the firm (e.g., particular routing or load-to-service assignment rules). The objective function of this formulation describes a generic cost structure, flexible enough to accommodate various productivity measures related to terminal and transportation operations. As an example, one may consider service capacity restrictions as utilization targets, which may be allowed to be violated at the expense of additional penalty costs. The last component of the objective function, albeit in a nonlinear form, may be used to model such a situation.

The network design, in general, and service network design, in particular, problems are difficult and transportation applications tend to be of large dimensions with complicating additional constraints. A number of important contributions to both methodological developments and applications have been proposed, and are reviewed in the references indicated at the beginning of this section. Many challenges still exist, however, and make for a rich research and development field.

28.4.3 Operational Planning

The purpose of operational level planning is to ensure that the system operates according to plan, demand is satisfied, and the resources of the carrier are efficiently used. Most methodologies aimed at carrier operational-planning issues explicitly consider the time dimension and account for the dynamics and stochasticity inherent in the system and its environment, some having to be solved in real or near-real time (e.g., dynamic resource allocation).

Main operational-level planning issues relate to empty vehicle distribution and repositioning, also sometimes called fleet management, crew scheduling, including the assignment of crews to vehicles and convoys, and allocation of resources, such as the dynamic allocation empty vehicles to terminals, motive power to services, crews to movements or services, loads to driver-truck combinations, routing of vehicles for pick up and delivery activities, and the real-time adjustment of services, routes, and plans following modifications in demands, infrastructure conditions (e.g., breakdowns, accidents or congestion), weather conditions, and so on.

We will not attempt here to review this field of research, which has been studied extensively for various modes of transport. We refer the reader to the surveys by Christiansen et al. (2004, 2007), Cordeau et al. (1998), Crainic and Laporte (1997), Crainic and Kim (2007), Powell et al. (1995, 2007), Powell and Topaloglu (2005), Toth and Vigo (2002) and the references therein, for details on these planning issues. It is worth mentioning, however, that few efforts were dedicated to container fleet management issues: Crainic et al. (1993) proposed a series of deterministic and stochastic models for the allocation and management of a heterogeneous fleet of containers where loaded movements are exogenously accepted; Cheung and Chen (1998) focused on the single-commodity container allocation problem for operators of regular ocean navigation lines and proposed a two-stage stochastic model; while Powell and Carvalho (1998; see also Powell et al., 2007) and (Powell and Topaloglu, 2005) addressed the problem of the combined optimization of containers and flatcars for rail intermodal operations using an adaptive dynamic stochastic programming approach. Significant more research is thus required in this field.

28.5   Intermodal Terminals

Intermodal terminals may belong to a given carrier, rail yards, for example, or be operated independently on behalf of public or private firms (e.g., air and sea and river ports). The main role of these facilities is to provide the space and equipment to load and unload (and, eventually, store) vehicles of various modes for a seamless transfer of loads between modes. When containerized traffic is of concern, the operations performed are restricted to the handling of the containers and not the cargo they contain. Terminal operations may also include cargo and vehicle sorting and consolidation, convoy make up and break down, and vehicle transfer between services. Some terminals, sea ports and airports, in particular, also provide the first line of Customs, security, and immigration control for a country. Thus, for example, North American sea port terminals are being equipped with container scanning facilities for enhanced control and security. Terminals thus form perhaps the most critical components of the entire intermodal transportation chain, as the efficiency of the latter highly depends on the speed and reliability of the operations performed in the former. Avoiding unplanned delays and the formation of load or vehicle bottlenecks is one of the major goals in operating intermodal terminals. Given the space limitations of this chapter, we only indicate the major classes of operations and issues related to intermodal terminals. For a more in-depth study of these issues, see the reviews of Crainic and Kim (2007), Günther and Kim (2005), and Steenken et al. (2004).

The intermodal transfer of containers between truck and rail takes place at rail yards. When containers arrive at a rail yard by truck, they are either directly transferred to a rail car or, more frequently, are stacked in a waiting area. Containers are then picked up from the waiting area and loaded unto rail cars that will be grouped into blocks and trains. When containers arrive by train to the terminal, they are transferred to trucks using the reverse operations.

Major operations in rail yards are: classification (sorting of rail cars), blocking (consolidation of rail cars into blocks), and train make-up (forming of blocks to trains). A significant amount of research exists on planning these operations. In most cases, there are no differences between intermodal and regular rail traffic with respect to blocking and train make up planning and operations, even when particular terminals are dedicated to handling intermodal traffic. Bostel and Dejax (1998) deal with the models that target planned terminals dedicated to the transfer of containers among intermodal shuttle trains.

A container port terminal provides transfer facilities for containers between sea vessels and land transportation modes, in particular, truck and rail. Such terminals are composed of three areas. The sea-side area includes the quays where ships berth and the quay-cranes that facilitate the loading and unloading of containers into and from ships. The truck and train receiving gates are located on the land-side area, which constitutes an interface between the land and sea transportation systems. Rail cars are loaded and unloaded in this area. Finally, there is the yard area, reserved mostly for stacking loaded and empty containers, and for loading and unloading the trucks.

Operations at a container port terminal can be partitioned into three classes: The first class consists of operations that deal with berthing, loading, and unloading of container ships. Following the arrival of a ship at the terminal, it is assigned a berth and a number of quay cranes. A number of planning problems arise here, such as determining the berthing time and position of a container ship at a given quay (berth scheduling), deciding on the vessel that each quay crane will serve and the associated service time (quay-crane allocation), and establishing the sequence of unloading and loading containers, as well as the precise position of each container that is to be loaded into the ship (stowage sequencing). The operations belonging to the second class are associated with receiving/delivery trucks and trains from/to the land-side. Containers arrive at the gate of the terminal, either by truck or train. Following inspection, the trucks are directed to the yard area where containers are unloaded and stacked. Trucks then either leave empty or pick up a new container. Empty trucks also call at the terminal to pick up containers. When containers arrive by rail, they are transferred, via a gantry crane, onto a transporter, which moves them to the designated area. The same transporter-gantry crane combination is used to load containers on departing trains. (Several variations exist according to the layout and operation mode of the terminal; the fundamental planning issues are still the same, however.) The last class of operations is concerned with container handling and storage operations in the yard. Determining the storage locations of the containers in the yard, either individually or as a group, is referred to as the space-allocation problem. This issue is a critical planning component as the way containers are located in the yard greatly affects the turn-around time of ships and land vehicles. Decisions regarding alloca-tion and dispatching of yard cranes and transporters, often performed in real or quasi-real time, complete the yard-related set of issues. The references indicated at the beginning of this section detail these issues, present the main methodologies proposed to address them, and identify interesting research perspectives and challenges.

28.6 Case Study: New Rail Intermodal Services

The performance of intermodal transportation directly depends on the performance of the key individ-ual elements of the chain, navigation companies, rail and motor carriers, ports, etc., as well as on the quality of interactions between them regarding operations, information, and decisions. The Intelligent Transportation Systems and Internet-fueled electronic business technologies provide the framework to address the latter challenges. Regarding the former, carriers and terminals, on their own or in collabora-tion, strive to continuously improve their performance. The rail industry is no exception. Indeed, com-pared to the other modes, railways face significant challenges in being a part of intermodal networks and competing intensively with trucking in offering customers timely, flexible, and long-haul transpor-tation services.

The traditional operational policies of railroads were based on long-term contracts, providing "sure" high volumes of (very often bulk) freight to move. Cost per ton/mile (or km) was the main performance measure, with rather little attention paid to delivery times. Consequently, rail services in North America, and mostly everywhere else in the world, were organized around loose schedules, indicative cut-off times for customers, "go-when-full" operating policies, and significant marshalling (classification and consoli-dation of cars) activities in yards. This resulted in rather long and unreliable trip times that generated both inefficient asset utilization and loss of market share. This was not appropriate for the requirements of inter-modal transportation and the North American rail industry responded through (Crainic et al., 2006):

1. A significant restructuring of the industry through a series of mergers, acquisitions, and alliances which, although far from being over, has already drastically reduced the number of companies resulting in a restricted number of major players.
2. The creation of separate divisions to address the needs of intermodal traffic, operating dedicated fleets of cars and engines, and marshalling facilities (even when located within regular yards). Double-stack convoys have created the land bridges that ensure an efficient container movement across North America.
3. An evolution toward planned and scheduled modes of operation and the introduction of booking systems and full-asset-utilization operating policies.

Most Western Europe railways have for a long time now operated their freight trains according to strict schedules (similarly to their passenger trains). This facilitated both the interaction of passenger and freight trains and the quality of service offered to customers. Particular infrastructure (e.g., low over-passes) and territory (short inter-station distances) make for shorter trains than in North America and forbid double-stack trains. Booking systems are, however, being implemented and full-asset-utilization and revenue management operating policies are being contemplated. Moreover, shuttle-service networks are being implemented in several regions of the European Union to address the requirements of intermo-dal traffic (e.g., Andersen and Christiansen 2006; Pedersen and Crainic 2007).

Booking systems bring intermodal rail freight services closer to the usual mode of operation of pas-senger services by any regular mode of transportation, train, bus, or air. In this context, each class of customers or origin-destination market has a certain space allocated on the train and customers are required to call in advance and reserve the space they require. The process may be phone or Internet based but is generally automatic, even though some negotiations may occur when the train requested by the customer is no longer available. This new approach to operating intermodal rail services brings advantages for the carrier, in terms of operating costs and asset utilization, and the customers (once they get used to the new operating mode) in terms of increased reliability, regular and predictable service and, eventually, better price.

A full-asset-utilization operation policy generally corresponds to operating regular and cyclically scheduled services with fixed composition. In other words, given a specific frequency (daily or every x days), each service occurrence operates a train of the same capacity (length, number of cars, tonnage) and the same number and definition, that is, origin, destination, length, of blocks (groups of cars traveling together as a unit from the origin to the destination of the block; blocks result from classification operations at yards). Assets, engines, rail cars and, even, crews, assigned to a system based on full-asset-utilization operation policies can then "turn" continuously following circular routes and schedules (which include maintenance for vehicles and rest periods for crews) in the time-space service network, as illustrated in Figure 28.3 for a system with three yards and six time periods (Andersen et al., 2006). The solid lines in Figure 28.3a represent services. There is one service from node 1 to node 3 (black arcs) and one service from node 3 to node 2 (gray arcs), both with daily frequency. Dotted arcs indicate repositioning moves (between different nodes) and holding arcs (between different time representations of same node). One feasible vehicle circuit in the time-space service network is illustrated in Figure 28.3b. The vehicle operates the service from node 3 to node 2, starting in time period 1 and arriving in time period 3. Then from period 3 to period 4 the vehicle is repositioned to node 1, where it is held for two time periods. In period 6 the vehicle operates the service from node 1 to node 3, arriving at time 1 where the same pattern of movements starts all over again.

Freight carrier systems operating according to such policies require the same type of planning methods as when full-asset-utilization policies are not enforced. Yet, their particular characteristics lead to significant differences that require revisiting models, methods, and practices. The field is very new and, consequently, very little work has yet been done, particularly with respect to these new intermodal rail systems, which may be observed both in North America and Europe.

To illustrate differences and the corresponding challenges, consider that to adequately plan services according to a full-asset-utilization operating policy requires the asset circulation issue to be integrated into the service network design model. The requirement may be achieved by enforcing the condition that at each node of the network representation, that is, at each yard and time period, the

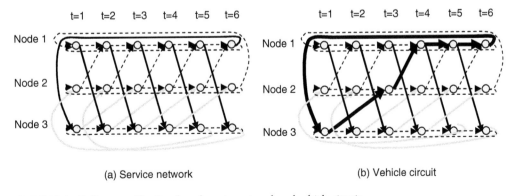

FIGURE 28.3   Full-asset-utilization-based service network and vehicle circuit.

(integer) design flow must be balanced. Schematically, the service network design model of Section 28.2 becomes

Minimize
$$\sum_{s \in S} F_s(y) + \sum_{p \in P} \sum_{l \in L_p} C_l^p(y, h) + \theta(y, h)$$

subject to

$$\sum_{l \in L_p} h_l^p = d_p \qquad\qquad p \in P,$$

$$(y_s, x_l^p) \in X \qquad\qquad s \in S, l \in L_p, p \in P,$$

$$\sum_{s \in S} y_{si^+} - \sum_{s \in S} y_{si^-} = 0 \qquad s \in S, i \in N,$$

$$y_s \geq 0 \text{ and integer} \qquad s \in S,$$

$$h_l^p \geq 0 \qquad\qquad l \in L_p, p \in P,$$

where si^+ and si^- indicate, respectively, that service $s \in S$ arrives and stops or terminates at, and that it initiates or stops and departs from node (yard) $i \in N$ in the appropriate period, while the third set of constraints enforces the balance of the total number of services arriving and departing at each yard and time period. Such requirements increase the difficulty of the planning problem. Pedersen, Crainic, and Madsen (2006) and Andersen et al. (2006, 2007) present formulations and propose solution methods, but significant research work is still needed.

Many other issues have to be addressed and offer an exiting research perspective. Consider, for example, that although bookings tend to "smooth" out demand and decrease its variability, the stochasticity of the system is not altogether eliminated. Regular operations tend to be disrupted by a number of phenomena, including the fact that arrival of ships in container port terminals are not regularly distributed and Customs and security verification may significantly delay the release of containers. When this occurs, rail operations out of the corresponding port are severely strained: there might be several days without arrivals, followed by a large turnout of arriving containers. Optimization approaches (e.g., Crainic et al., 2006) may be used to adjust service over a medium-term horizon in such a way that a full-asset-utilization policy is still enforced, but a certain amount of flexibility is added to services to better fit service and demand. Such approaches may become even more effective when appropriate information sharing and container-release time mechanisms are implemented.

Acknowledgments

Partial funding for this project has been provided by the Natural Sciences and Engineering Research Council of Canada (NSERC) through its Discovery Grants and Chairs and Faculty Support programs.

References

Andersen, J. and Christiansen, M., Optimal Design of New Rail Freight Services, Department of Industrial Economics and Technology Management, Norwegian University of Science and Technology, 2006.

Andersen, J., Crainic, T.G., and Christiansen, M., Service Network Design with Asset Management: Formulations and Comparative Analyzes, Publication, CIRRELT, Montreal, Canada, 2007.

Andersen, J., Crainic, T.G., and Christiansen, M., Service network design with management and coordination of multiple fleets, *European Journal of Operational Research*, 2006 (forthcoming).

Association of American Railroads, Class I Railroad Statistics, 2006, available on-line at http://www.aar.org/PubCommon/Documents/AboutTheIndustry/Statistics.pdf.

Bolis, S. and Maggi, R., Logistics Strategy and Transport Service Choices: An Adaptive Stated Preference Experiment, *Growth and Change*, 34(4): 490–504, 2003.

Bostel, N. and Dejax, P., Models and Algorithms for Container Allocation Problems on Trains in a Rapid Transshipment Shunting Yard, *Transportation Science* 32(4): 370–379, 1998.

Campbell, J.F., Integer programming formulations of discrete hub location problems, *European Journal of Operational Research*, 72(2): 387–405, 1994.

Campbell, J.F., Ernst, A.T., and Krishnamoorthy, M., Hub Location Problems in Facility Location: Application and Theory, Drezner, Z. and Hamacher, H. (Eds.), Springer Verlag, Berlin, 373–407, 2002.

Christiansen, M., Fagerholt, K., and Ronen, D., Ship Routing and Scheduling: Status and Perspectives, *Transportation Science* 38(1): 1–18, 2004.

Christiansen, M., Fagerholt, K., Nygreen, B., and Ronen, D., Maritime Transportation, Chapter 4 in *Transportation, Handbooks in Operations Research and Management Science*, Barnhart and G. Laporte (Eds.), North-Holland, Amsterdam, 189–284, 2007.

Cheung, R.K. and Chen, C.-Y., A Two-stage stochastic network model and solution methods for the dynamic empty container allocation problem, *Transportation Science* 32(2): 142–162, 1998.

Cordeau, J.-F., Toth, P., and Vigo, D., A survey of optimization models for train routing and scheduling, *Transportation Science* 32(4): 380–404, 1998.

Crainic, T.G., Service network design in freight transportation, *European Journal of Operational Research*, 122(2): 272–288, 2000.

Crainic, T.G., Long-Haul Freight Transportation, in *Handbook of Transportation Science*, Hall, R.W., (Ed.), Kluwer Academic Publishers, New York, 2nd Edition, 451–516, 2003.

Crainic, T.G. and Kim, K.H., Intermodal Transportation, Chapter 8 in *Transportation, Handbooks in Operations Research and Management Science*, C. Barnhart and G. Laporte (Eds.), North-Holland, Amsterdam, 467–537, 2007.

Crainic, T.G. and Laporte, G. Planning models For freight transportation, *European Journal of Operational Research*, 97(3): 409–438, 1997.

Crainic, T.G. and Rousseau, J.-M., Multicommodity, multimode freight transportation: a general modeling and algorithmic framework for the service network design problem, *Transportation Research Part B: Methodological*, 20(3): 225–242, 1986.

Crainic, T.G., Bilegan, I.-C., and Gendreau, M., Fleet Management for Advanced Intermodal Services—Final Report, Report for Transport Canada, Publication CRT-2006-13, Centre de recherche sur les transports, Université de Montréal, 2006.

Crainic, T.G., Gendreau, M., and Dejax, P.J., Dynamic and stochastic models for the allocation of empty containers, *Operations Research* 41(1): 102–126, 1993.

Ebery, J., Krishnamoorthy, M., Ernst, A., and Boland, N., The capacitated multiple allocation hub location problem: formulations and algorithms, *European Journal of Operational Research* 120(3): 614–631, 2000.

European Conference of Ministers of Transport, Terminology on Combined Transport, page 44, United Nations, New York and Geneva, 2001, available online at http://www.cemt.org/online/glossaries/termcomb.pdf

Evers, P.T. and Johnson, C.J., Performance perceptions, satisfaction and intention: the intermodal shipper's perspective, *Transportation Journal*, 40(2): 27–39, 2000.

Evers, P.T., Harper, D.V., and Needham, P.M., The determinants of shipper perceptions of modes, *Transportation Journal*, 36(2): 13–25, 1996.

Günther, H.-O. and Kim, K.H., *Container Terminals and Automated Transport Systems*, Springer-Verlag, Berlin, 2005.

Macharis, C. and Bontekoning, Y.M., Opportunities for OR in intermodal freight transport research: a review, *European Journal of Operational Research* 153(2): 400–416, 2004.

McGinnis, M.A., The relative importance of cost and service in freight distribution choice: before and after deregulation, *Transportation Journal*, 30(1): 12–19, 1990.

Murphy, P.R. and Hall, P.K., The relative importance of cost and service in freight distribution choice before and after deregulation: an update, *Transportation Journal*, 35(1): 30–38, 1995.

O'Kelly, M.E., A quadratic integer program for the location of interacting hub facilities, *European Journal of Operational Research*, 32(3): 393–404, 1987.

Pedersen, M.B. and Crainic, T.G., Optimization of Intermodal Freight Service Schedules on Train Canals, Publication, CIRRELT Montréal, 2007.

Pedersen, M.B., Crainic, T.G., and Madsen, O.B.G., Models and Tabu Search Meta-heuristics for Service Network Design with Vehicle Balance Requirements, Publication CRT-2006-22, Centre for Research on Transportation, Montréal, 2006.

Powell, W.B. and Carvalho, T.A., Real-time optimization of containers and flatcars for intermodal operations, *Transportation Science* 32(2): 110–126, 1998.

Powell, W.B. and Sheffi, Y., Design and implementation of an interactive optimization system for the network design in the motor carrier industry, *Operations Research* 37(1): 12–29, 1989.

Powell, W.B. and Topaloglu, H., Fleet Management, in Applications of Stochastic Programming, Math Programming Society—SIAM Series on Optimization, S. Wallace and W. Ziemba (Eds), SIAM, Philadelphia, PA, 185–216, 2005.

Powell, W.B., Bouzaïene-Ayari, B., and Simaõ, H.P., Dynamic Models for Freight Transportation, Chapter 5 in Transportation, Handbooks in Operations Research and Management Science, C. Barnhart and G. Laporte (Eds), North-Holland, Amsterdam, 285–365, 2007.

Powell, W.B., Jaillet, P., and Odoni, A., Stochastic and Dynamic Networks and Routing, in Network Routing, Volume 8 of Handbooks in Operations Research and Management Science, Ball, M., Magnanti, T.L., Monma, C.L., and Nemhauser, G.L., (Eds), North-Holland, Amsterdam, 141–295, 1995.

Steenken, D., Voß, S., and Stahlbock, R., Container terminal operation and operations research—a classification and literature review, *OR Spectrum* 26(1): 3–49, 2004.

Sussman, J. Introduction to Transportation Systems. Artech House. Boston, 2000.

Toth, P. and Vigo, D., The Vehicle Routing Problem, Volume 9 of SIAM Monographs on Discrete Mathematics and Applications, SIAM, 2002.

The Railway Association of Canada, Railway Trends 2006, Ontario, 2006, available on-line at http://www.railcan.ca/documents/publications/1349/2006_10_24_RAC_Trends_en.pdf.

U.S. Department of Transportation, Freight in America, Washington D.C., 2006, available on-line at http://www.bts.gov/publications/freight_in_america/), page 24.

29

Logistics in Service Industries

Manuel D. Rossetti
University of Arkansas

29.1 Introduction

According to the U.S. Department of Commerce the service sector has been the fastest growing section of the U.S. economy during the last 50 years. In fact, as of 1999, the service sector accounted for up to 80% of the U.S. economy. This remarkable statistic is related to structural shifts first from agriculture, to manufacturing, and now to services within the United States over the last century. To understand the application of logistics to the service sector, we must first examine the types of firms that constitute the service sector and the types of services they provide. Cook et al. (1999) provide a comprehensive review of the ways in which the service industry has been classified over the last 50 years. The U.S. Census Bureau conducts a survey of the service sector to understand revenues, growth, and the effect of the service sector on the U.S. economy. The primary classification used for the service sector is based on the National American Industry Classification System (NAICS). The primary categories for the service sector include:

- Transportation and warehousing
- Information
- Finance and insurance
- Real estate and rental leasing
- Professional, scientific, and technical services
- Administrative and support, and waste management and remediation services
- Health care and social assistance
- Arts, entertainment, and recreation
- Other services (except for Public Administration)

From this classification, it is easy to see why the service sector constitutes such a large part of the U.S. Gross Domestic Product. This handbook concentrates on many aspects of the transportation and warehousing category. This chapter will discuss issues related to logistics applied to other service sector areas.

Murdick et al. (1990) offer one of the more useful definitions of a service: "Services can be defined as economic activities that produce time, place, form, or psychological utilities." This definition allows us to conceptualize services as nontangible deliverables; however, services are often inseparable from actual physical products. For example, when a doctor performs an operation to replace a hip in a patient, the doctor is performing a service; however, the service cannot be provided without the artificial hip. This connection to the physical delivery of a product has allowed many firms that were previously purely manufacturing oriented to move into the delivery of services; services that add value or utility to their customer base. A classic example of this is International Business Machines. While still a leading manufacturer, IBM is now arguably one of the most competitive providers of service (maintenance, repair, software, training, consulting, etc.).

While all business activity is customer focused, the service industry's primary focus is on the delivery of service directly to the customer. That is, within a service transaction, the customer is often involved directly in the experience. For example, in the entertainment industry (e.g., theme parks), it is the customer's direct interaction that provides the "entertainment" service. Service sector firms have a special need to address the following questions:

1. Who are my customers? How do they demand service?
2. What are the elements of the service provided to the customer? In other words, define the service content for the customer. In addition, how will we measure customer satisfaction with the services?
3. What operating strategies are important to providing service to the customer?
4. How should the service be delivered to the customer? What should constitute the delivery system and what are the capacity characteristics of the system?

The first question must be answered so that the firm can begin to address the latter questions. In answering question 1, the firm must define the characteristics of the customer, for example, what are their attributes, demographics, requirements, etc. In addition, the firm must understand how customer demand will be realized. The demand for services may be highly variable and stochastic in nature. Characterizing the behavior of customers and their resulting demand over time through forecasting is important in any industry, but may be especially difficult in service industries since the delivered product is often intangible. The second question forces firms to try to make the intangible, tangible. That is, the more the service content can be described and measured, the easier it will be to decide on how to deliver the service, which is the key to questions 3 and 4. In question 4, the firm needs to organize its operating strategies around the customer and the service content, and in question 5, the firm begins to answer how the service will be delivered. It is this latter question, especially the issue of capacity planning, which is critically important to logistics planners.

All these questions imply that it is critically **important for service** industries to know their customers and to design and implement their logistics delivery structure based on customer requirements that may change over time. Within service industries is not always possible to build up inventory and it may take a long time to create increased capacity through new facilities. For example, the airline industry has widely varying demand patterns, but can react to demand changes only by adding flights, crews, planes, etc., all of which are discrete capacity changes. These increments are only possible in a timely manner if excess capacity already exists. The hotel and car rental industries also experience similar demand and capacity requirements. In addition, the time needed to react is longer, especially if it involves the movement or location of a facility to meet customer demand. For example, in the financial services industry it takes time to build a banking infrastructure in order to serve a growing population area and there is the risk that the customer demand may not materialize.

As a simple example for the four questions, we might consider the theme park industry. The first question requires an understanding of the customer. The customers of a theme park are primarily families or groups of people in the younger age demographic who are willing to spend dollars on a leisure activity that they can do together. An operative concept is doing the activity together. Families or small groups will want to move together, ride together, eat together, etc. The service delivery must be designed to facilitate the delivery of service to these groups. Customer demand is time varying and stochastic within this industry both at a seasonal level but also on an hourly basis. To address question 2, we need to understand the total service content. While in the park, the customers not only want to be entertained, but they also have many other needs that must be met (e.g. food, water, transportation, rest areas, medical response, etc.). The last two questions begin to involve questions concerning logistics. For example, the decision of where to locate rides, rest rooms, food services, etc. are all predicated on understanding the demand for these items and the customer's trade-off in walking to and competing for such facilities. In addition, if the park decides to have a light rail system for moving customers within the park, the operating characteristics of the transport system must be designed. Moving customers from point to point within the park has direct customer contact; however, there will also be all the other logistical support activities to get the food, supplies, costumes, etc. into the park so that the customer can have a "fun" experience. Mielke et al. (1998) discusses the application of simulation to theme park management as well as some of the issues mentioned. In summary, within the service industry we start with customer needs and then develop logistics delivery mechanisms to meet those needs.

While service industries are unique in many respects, they also have similar characteristics to standard manufacturing and distribution systems, especially in the scope of the delivery processes. In the following sections, we discuss how the design of service logistics can be considered at two levels: within the facility logistics and between the facilities logistics. We provide examples of applications at each of these two levels.

29.2 Within-the-Facility Logistics

Within-the-facility logistics involves the design and operation of the physical plant with respect to logistical goals and objectives. The techniques and issues of within-the-facility logistics are discussed elsewhere in this volume. In Chapters 11 and 12 we present an overview of some of these issues as they relate to service industries. In general, the issues involved in within-the-facility logistics may include:

1. Sizing and planning the capacity of the logistical functions
2. Designing the layout and flow of the logistics
3. Material handling system selection and operation
4. Efficiently operating the logistics system

While the flow of items (food, medicine, inventory, etc.) is a critical aspect of service logistics, systems involving service have additional requirements involving customers. When considering the customer within intra-facility logistics, there are two main issues to consider: (1) how the logistics system may indirectly affect the delivery of service to the customer, and (2) how the logistics system directly affects the customer. Keeping with our theme park example, we know the rides must be maintained and repaired. If the service parts logistics is not designed properly then the guest may not get their rides. The guest's service is affected indirectly. Even though their service is affected, the guest is not directly interacting with the logistics system, that is, the service parts supply chain. In the second issue, often the major purpose of the logistics system is to move the customers. For example, within an amusement park, our service is entertainment; however, a critical logistical issue is how to get the "guests" to and from the attractions. This involves the design and operation of the guest handling systems, such as elevators, people movers, shuttle busses, mono-rails, etc. In this case, the logistics system is directly interacting with the customer and directly affecting their service.

The rental car industry is a service industry that requires careful layout of the facilities with which the customers directly and indirectly interact. For example, in Johnson (1999) the layout of the check-out areas, parking and washing areas, and check-in areas were examined for the impact on customer waiting and service efficiency. Simulation was used to examine multiple layout scenarios and to determine the impact on customer service. In other situations, such as public bus transport and air-travel, the service is the transport of people. In these instances, the waiting time of the customers using the logistics system becomes even more important. In Takakuwa and Oyama (2003), airport terminal design is examined via simulation to understand the waiting time of passengers using the service. Such models require the detailed analysis of internal flows as customers utilize the service. This often necessitates an analysis of the capacity of the system and the scheduling of the availability of the staff and or handling systems. Thus, the design and layout questions go together with tactical planning issues such as staffing and scheduling of service. For example, Rossetti and Turitto (1998) examined the use of dynamic hold points within a transit system to prevent the phenomenon of "bunching" within bus schedules. That is, busses catching up with each other on a route and then traveling together. This is not only highly detrimental to operating efficiency, but also causes poor service because the scheduled arrival times are not met.

While many service systems directly involve the movement of people, many others such as retail stores, hospitals, and banks rely on people, but their service is not the movement of people *per se*. In these situations, the logistics engineer must consider how the logistics system may indirectly affect the customer. For example, within a retail system, the main purpose of the logistics system is to move the goods to the store for sale to the customer. While there are many logistical decisions to get the items to the store, we must also consider the affect on the customer in the store. The "back room" logistics, in retail stores becomes an important consideration, especially how and when to move the items to the shelf so as to minimize the disruption of customer shopping. This consideration of the disruption to customer shopping causes many (if not most) retailers to have goods delivered in the late evening, with shelf stocking occurring in the over night hours. This customer service decision drives the replenishment processes, the truck delivery schedule, and ultimately distribution center operations. To further illustrate how the customer affects logistics system design and operation, we will examine the implications of using mobile robots to perform delivery functions within a hospital with a case study. As we will see, in considering service systems it is important to consider the entire system, especially the customer.

29.2.1 Case Study: Mobile Robot Delivery Systems in Hospitals

In previous research, the author was asked to analyze the delivery functions within a hospital, and in particular, examine whether or not autonomous mobile robotic carriers could be utilized in such an environment. Portions of this discussion are based on Rossetti et al. (2000)[*] and from Rossetti and Seldanari (2001). The hospital selected for analysis was the University of Virginia Medical Center (UVA-MC) located in Charlottesville, Virginia. At the time of the case study, the UVA-MC was a 591-bed, eight-floor complex and represents a medium to large size hospital facility that, at the time, handled about a 454 daily bed census with close to $420 million in annual operating expenses. The hospital has two elevator banks. One elevator bank is located on the west side of the hospital while the other is located on the east side of the hospital. Each bank of elevators consists of two rows of three elevators each. For each elevator bank, one row of three is reserved for visitors and the other row is reserved for hospital personnel. Figure 29.1 illustrates the basic layout of the floors of the hospital.

As illustrated in Figure 29.2, the hospital has many delivery components. This case study focuses on the use of mobile robots for pharmacy and clinical laboratory deliveries. The mobile robot examined, see Figure 29.3, in this case study is manufactured and sold by Cardinal Health Inc. (www.cardinal. com/pyxis). The Pyxis HelpMate robotic courier is a fully autonomous robot capable of carrying out delivery missions between hospital departments and nursing stations. The Pyxis HelpMate robotic system uses a specific world model for both mission planning and local navigation. The world is represented as a network of links (hallways) and an elemental move for the robot is navigating in a single

---

[*] With kind permission of Springer Science and Business Media.

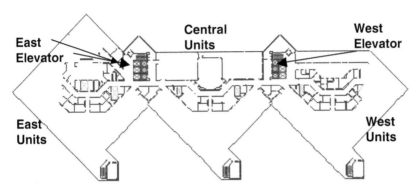

FIGURE 29.1 Generic hospital floor layout.

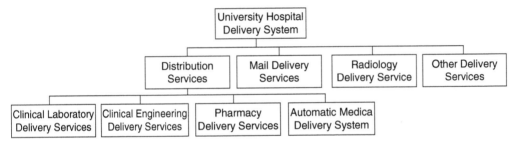

FIGURE 29.2 Components of a hospital delivery system.

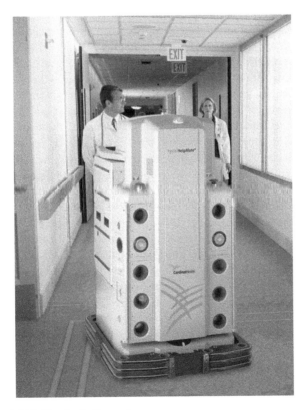

FIGURE 29.3 The Pyxis HelpMate® robotic courier.

hallway, avoiding people and other obstacles. In situations where more than one robot is present, a computerized supervisor properly spaces the robots along the hallways, since the robots compete for space and for the elevators.

The pharmacy and clinical laboratory delivery processes utilized human workers to complete the deliveries. Each hospital floor consists of a number of hospital units. Each hospital unit collects specimens during the course of its operation. The specimens require transport to the clinical laboratory located on the second floor of the hospital where they are tested. The results of the tests are reported back to the hospital units via the hospital's laboratory information network.

The clinical laboratory process collects specimens that are placed on floors 3 to 8 from the 29 medical units of the hospital. For routine pick-ups and deliveries, the human courier follows a predefined route. Each courier is assigned two floors: one person for the third and fourth floors, a second person for the fifth and the sixth floors, and a third person for the seventh and eighth floors. Couriers wait in the personnel lounge until it is time to start the shift. At the beginning of the shift, couriers make their way to the top floor of their route and visit each unit assigned to their route on their way to the clinical laboratory. If they have picked up items during the route, they deliver the items to the clinical laboratory; otherwise, they repeat their route. During the shift, there are three breaks that are scheduled for couriers: two breaks of 15 min each and 1 break of 30 min. When a specimen requires STAT delivery, the courier picks up the specimen and then takes the best direct route to the clinical laboratory for delivery. Any items that have already been picked up along the route are also dropped off at the laboratory. The courier then travels back to the unit that was next on the route before they responded to the STAT delivery. The determination of whether or not a specimen is STAT is dependent on the nurses or the doctors and their determination of the patient's medical needs.

Courier delivery for pharmaceuticals is broken into two distinct delivery processes. These are the delivery of routine pharmacy medicines and the delivery of STAT pharmacy medicines. Three couriers are assigned to deliver medicines to the appropriate units. Couriers performing routine deliveries are each assigned three floors: (3, 4, 5) for one courier and (6, 7, 8) for another courier. One courier performs STAT delivery to all the floors. The delivery process for routine pharmacy is similar to the clinical laboratory delivery process. The courier picks up the medicines at the central pharmacy located in the basement of the hospital and destined for units along their route. The courier uses the elevator to travel to the top floor of the route and then visits each unit on the route. At the nursing station within the unit, a box is kept for pick-ups and deliveries. The courier drops off the medicines at their destinations and picks up any unused medicine for return to the pharmacy. After completing all the floors on the route, the courier returns to the pharmacy to drop off unused medicines and to pick up a new batch of medicines for delivery. Figure 29.4 illustrates that the demand for pharmacy delivery services varies significantly by hospital unit and by time of day. The clinical laboratory demand characteristics were similar, but with less variability. These sorts of demand processes are often characteristic of service systems and complicate the scheduling of the staffing requirements. In addition, for a hospital, we must take into account the fact that the facility must provide service 24 hours per day, 7 days a week, 365 days per year.

As indicated, the human courier process is labor intensive and requires low-skilled labor. This sort of process is a prime candidate for automation. The four key issues of within-the-facility logistics must be addressed. The solution approach for this case study involved material handling design and selection. Because of the service nature of this system, special care was taken to ensure that capacity and performance can meet customer requirements. In order to understand whether or not mobile robots would be beneficial for this situation, we used simulation modeling and cost analysis to compare alternatives involving robots to the current operating situation. The simulation models were built using the Arena simulation environment. Both the robotic couriers and the human couriers were modeled using the guided transporter modeling constructs available within Arena. To analyze the delivery processes, four models were developed. The first model described the current system with human couriers. The second and third models described the operation of the system with mobile robots serving as the primary delivery mechanism with independent operation of clinical laboratory and pharmaceutical deliveries.

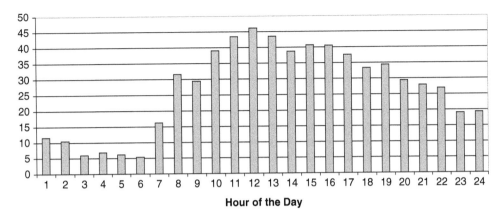

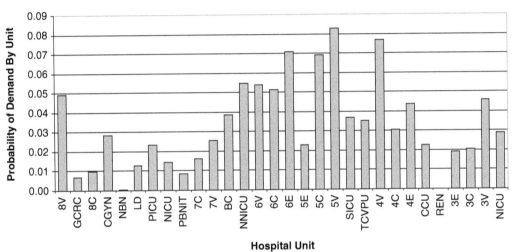

FIGURE 29.4 Pharmacy time demand process. (From Rossetti, M.D., Kumar, A., and Felder, R. *Health Care Management Service*, Vol. 3, pp. 201–213, 2000; Springer Sciences and Business Media. With permission.)

The mobile robot models acted essentially the same as the human courier model except for minor changes in accommodate the speed of the robots, the amount of time they wait at the hospital units to be loaded by nurses, and elevator interactions. The fourth model combined the processes associated with clinical laboratory and pharmaceutical deliveries and utilized mobile robots as the delivery mechanism.

Within this simulation modeling paradigm, each floor consists of a network of links and nodes. Links and nodes have a capacity, limiting the maximum number of robots allowed to occupy them at any time. A careful choice of link and node capacity makes it possible to avoid deadlocks in floors where more than one robot can travel simultaneously; moreover, it enables the modeling of a "space cushion" between two consecutive robots in the same hallway. The length of the space cushion can be set by choosing the appropriate minimum length for each link. Only three relevant differences distinguish human couriers from robots: human couriers do not compete for space while they are walking along the hospital hallways. During an 8-h shift each courier is allowed to take three breaks: two breaks are short (15 min), one break is long (30 min), and robots must use the elevators to move from one floor to another. One elevator in each bank is retrofitted for use by the robots. If a robot needs to use the elevator, the elevator's calling mechanism ensures that the elevator does not respond to any other floor calls.

In addition, the elevator has a weight sensor to indicate if all riders have disembarked. Essentially, these rules ensure that no people are in the elevator when the robot is using it. If more than one robot is requesting the elevator, the robots are served according to FIFO logic, regardless of their current floor. Human couriers use the elevators as any other human passenger since they do not have a special priority. All the other human courier behaviors are modeled the same as the robot's behaviors except for traveling speed.

Robot speed modeling deserves particular mention: it was modeled as a triangular random variable to account for possible interference with humans in the hospital hallways (if a robot is blocked it automatically stops and computes a path around the obstruction) and to account for velocity changes because of curves and long straight hallways. A new value of the velocity for a link is assigned every time the robot traverses the link. The parameters for the triangular distribution (min, mode, max) were: (0.274 m/s, 0.508 m/s, 0.63 m/s) based on vendor recommendations. Human courier velocity was modeled in the same way to account for courier unscheduled breaks or fatigue. The parameters for the triangular distribution (min, mode, max) for humans were: (0.381 m/s, 0.762 m/s, 0.875 m/s) based on standardized data and observation of the human couriers. The model data and the model were verified and validated using standard statistical simulation techniques, see for example Chapter 11 of Banks et al. (1996). Further details of the simulation models are given in Rossetti et al. (2000) and from Rossetti and Seldanari (2001).

In comparing the system with human carriers to the systems involving robotic carriers, a number of key issues needed to be examined. The first issue was to determine whether or not the robotic carriers were competitive in terms of cost, both in operating cost and in terms of any capital investment. The second issue was that the robotic carriers be competitive in terms of performance. The key performance measures included (but were not limited to) turn around time, delivery time variability, and cycle time, and utilization. Turn around time refers to the time from when the delivery is requested to when it is completed. Delivery time variability is the standard deviation of the turn around time. Cycle time is the time it takes a courier to complete one cycle of its assigned route, and utilization refers to the percentage of total time spent carrying items for delivery.

The simulation models were run and the performance measures collected. As indicated in Tables 29.1 and 29.2, robotic couriers are extremely cost-effective. The cost of the robots included the cost of the equipment (robots, batteries, carrying compartments, robot communication system, door actuators, etc.), cost of installation, and cost of operation and maintenance (service contract, monitoring personnel, energy, etc.). A net present value calculation was performed over a five-year planning horizon. The cost of the courier system was based on a loaded hourly rate of $10.26/h for 24 h/day and 365 days/year. In order to obtain full yearly coverage over sick days, vacations, etc. one person is considered equivalent to 1.4 FTE.

The two-robot alternative has lower cost but it has difficulty matching the performance of the three-courier model. A one-for-one replacement of the couriers with robots reduces the cost by roughly 74% with only an approximate 20% increase in turn-around time. The six-robot alternative dominates the other alternatives by maintaining low cost and significantly improving the turn-around time and the delivery variability. For the combined model in Table 29.3, robots perform both pharmacy and clinical

TABLE 29.1   Clinical Laboratory Summary of Performance Measures

	Two Robots	Three Robots	Six Robots	Courier
COST	$81,110	$107,605	$178,027	$407,614
TAT (min)	47.28 (1.97)	33.54 (1.07)	18.9 (0.44)	28.08 (2.16)
DV (min)	24.77 (1.87)	16.67 (0.82)	8.63 (0.04)	20.72 (2.83)
CT (min)	67.03 (2.01)	42.25 (0.87)	20.72 (0.33)	26.3 (1.57)
UTIL	92.50% (0.44)	91.90% (0.63)	81.70% (1.52)	88.33% (0.68)

Source: Rossetti, M.D., Kumar, A., and Felder, R. *Health Care Management Service*, Vol. 3, pp. 201–213, 2000; Springer Sciences and Business Media. With permission.

TABLE 29.2 Pharmacy Model Summary Results

Performance Index	Alternatives		
	Two Robots	Three Robots	Courier
Cost	$86,141.00	$104,579	$281,742
Turn-around time	102.25 (15.06)	71.16 min (13.25)	55.87 (9.21)
Delivery variability	86.88 (22.97)	57.87 min (19.724)	49.22 (13.86)
Average cycle time	57.37 (2.11)	42.35 min (1.255)	30.86 (1.11)
Utilization	13.28% (3.22)	56.87% (7.323)	11.69% (1.97)

Source: Rossetti, M.D., Kumar, A., and Felder, R. *Health Care Management Service*, Vol. 3, pp. 201–213, 2000; Springer Sciences and Business Media. With permission.

laboratory delivery. The combined delivery had a 75% decrease in cost, a 34% decrease in turn around time, a 38% decrease in delivery variability while virtually matching the cycle time and utilization performance of the courier-based system.

From these results, we can see that mobile robots are a highly competitive alternative to human couriers. To further explore the indirect affects of such a system, an extensive sensitivity analysis involving the trade-offs between both quantitative and qualitative factors using the analytical hierarchy process (AHP) was performed. The AHP [see Saaty (1977, 1994, 1997)] is a technique that can be applied to address problems that have multiple conflicting performance measures. Several examples of the application of AHP to transportation system planning and automation introduction in manufacturing can be found in the literature [see e.g., Albayrakoglu (1986); Khasnabis (1994); and Mouette and Fernandes (1997)]. AHP is based on the analysis of a hierarchy structure. Decision analysis techniques like AHP are especially relevant within service industries because these techniques attempt to incorporate nonquantitative measure of performance into the decision process.

Figure 29.5 presents the complete AHP tree used in the analysis. As can be seen in the figure, both quantitative performance measures are captured as well as qualitative performance measures. In particular, there are other important considerations such as safety, noise, and technical innovation that are important to both the users of the delivery system (doctors/nurses/administrators) and to the customers of the service system (patients). For example, this analysis incorporates the effect of additional elevator delay on the patients and their families caused by the use of the elevator by the robots. In addition, the robots make noise as they actuate the hospital unit doors. These issues are important from a patient's point of view.

The stability of the system response after modifications in the decision-maker preference structure affecting the AHP Global Priorities was checked through a sensitivity analysis. The analysis investigated whether or not preference structure modifications could benefit the human-based solution. The analysis showed that the robotic delivery system is preferable with respect to the human-based system with an overall confidence level of 99.986% based on the preference structure of the decision-makers. In addition,

TABLE 29.3 Summary for Combined Delivery

Performance Index	Results in Absolute Terms	
	Six Robots	Courier
Cost	$178,076	$689,356
Turn around time	28.14 (1.461)	42.69 (5.055)
Delivery variability	12.30 (1.404)	20.01 (2.963)
Average cycle time	28.97 (0.722)	28.70 (0.937)
Utilization	89.69% (0.508)	86.72% (1.031)

Source: Rossetti, M.D., Kumar, A., and Felder, R. *Health Care Management Service*, Vol. 3, pp. 201–213, 2000; Springer Sciences and Business Media. With permission.

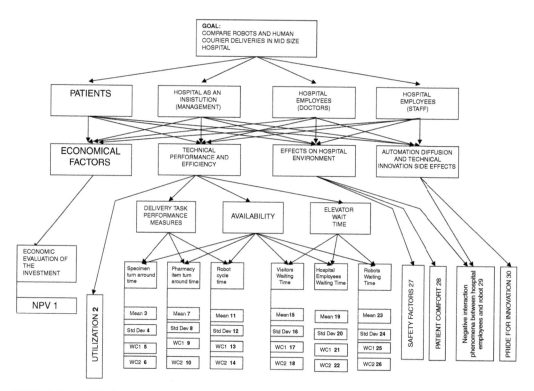

FIGURE 29.5 AHP hierarchy structure. (From Rossetti, M.D. and Seldanari, F., Computers and Industrial Engineering, 41, 301–333, 2001. With permission from Elsevier.)

while some changes to the priorities can allow the human-based system to be preferable, the priority weights were not realistic for practical situations. A complete discussion of the sensitivity analysis is found in Rossetti and Seldanari (2001).

In this case study, we illustrated that when analyzing service systems there is a critical need to consider all relevant factors that may directly or indirectly affect the customer. In this case study, we saw that through a systems perspective, considering the current system, alternatives, and a complete understanding of the socio-technical factors, we can make strong decisions involving logistics systems that support service systems. Examining the affect of logistics alternatives in this manner is especially important in within-the-facility logistics because of the proximity of the customer to the logistics solution. In the following section, we briefly discuss a case study involving between-the-facility logistics or supply chain management. In these types of situations, the end customer often drives the overall requirements for the logistics solution, but the actual solution involves rather typical supply chain decision-making concepts.

29.3   Between-the-Facility Logistics

Between the facility logistics refers to the structure and processes required to move materials between facilities in order to meet customer requirements. The structure and processes constitute the logistical network. Logistical network design involves determining the number, location, type, and capacity of the facilities. In addition, the network connectivity (who supplies what to whom) and the inventory requirements become important issues in creating time and place utility for customers. Supply chain management constitutes the key paradigm by which logistics network designs will impact customer service. A large amount of research has been conducted on supply chain modeling, inventory policy determination, and operations research techniques applied to logistics. This volume highlights many of these supply chain issues.

Within the area of logistical network design, the specialized area of facility location (discussed elsewhere in this volume) has been able to make significant contributions to how service facilities are located. As an example, extensive research has been performed on questions of network design for which service coverage is the main issue. The p-center, p-median, and other covering problems all are specific examples of models that are motivated from service coverage. For example, the p-center problem attempts to locate p facilities on a network in such a way as to minimize the maximum travel time from a user to the closest facility. This type of model is important in locating emergency response teams, hospital facilities, schools, etc. A review of network design problems can be found in Daskin (1995). In addition, a comprehensive review of the application of operations research techniques to emergency services planning can be found in Goldberg (2004). Min and Melachrinoudis (2001) examine the use of location-allocation models within the banking industry. In location-allocation problems, we jointly solve the location of the facilities as well as the allocation of the services that the facilities provide. An interesting aspect of many service systems is the hierarchical nature of the systems. For example, banking customers can receive different services from full service banks, satellite banks, automatic teller machines, and drive-through facilities. Min and Melachrinoudis (2001) developed an optimization model for deciding which services to offer at which levels of the hierarchy while maximizing the profitability and customer response to services while minimizing the risks associated with offering the services. These types of problems often require the dynamic and stochastic modeling of customer behavior [see e.g., Wang et al. (2002) who incorporate queuing analysis into the location of automatic teller machines within a service network].

Location analysis provides the network structure within which logistic processes must be allocated and then operated. It is the operation of these processes within the supply chain that motivates service industries to consider two main alternatives: in-house logistics versus outsourced logistics. For many service industries like banking and health care, logistics is simply a support activity. Because of this, two key questions that firms in the service industry face are (*i*) whether or not logistics strategy is a key to their success with their customers, and (*ii*) whether or not they should execute the logistic processes internally or whether they should rely on an external provider of logistics.

We will illustrate these concepts with a discussion of logistics within the healthcare industry. According to Burns and Wharton School Colleagues (2002), the health care value chain consists of the producers of medical products such as pharmaceuticals and medical devices, the purchasers of these products such as wholesalers and group purchasing organizations, the providers including hospitals, physicians, and pharmacies, the fiscal intermediaries (insurers, HMOs, etc.), and finally the payers (patients, employers, government, etc.). Within the healthcare value chain, see Figure 29.6, physical products (drugs, devices, supplies, etc.) are transported, stored, and eventually transformed into health-care services for the patient. Burns and Wharton School Colleagues (2002) discusses a number of configurations of healthcare value chains that are used to manage the inventory supply process from producer to payer. Each configuration will result in its own performance in terms of cost and reliability of service in delivering the medical services.

Hospital executives must make decisions regarding whether to maintain inventory and distribution functions in-house, to outsource them to external firms, or to engage in collaborative ventures with such external firms. Figure 29.6 illustrates the healthcare value chain with the key decision of in-house logistics and outsourced logistics contrasted. This decision is typically a major issue in corporate strategy for many service industries. For example, in recent years, hospitals have formed large "integrated delivery networks" (IDNs) that combine multiple hospitals and often, large physician groups. A chief intent of these strategies has been to achieve economies of scale to reduce rising healthcare costs, although such economies have proved elusive (Burns and Pauly, 2002). Initially, IDNs sought these economies by consolidating finance and planning functions, yielding little cost savings. IDNs have only recently begun to pursue these economies through integration of their supply chain activities. Because such activities (products, services, and handling) comprise up to 30% of a hospital's cost structure, the potential for cost savings through consolidation and scale economies seems more promising.

Based on these recent trends, it is clear that hospital service providers have recognized that efficient logistics is a key to their success in holding down costs. Even if a firm decides that logistics is a key

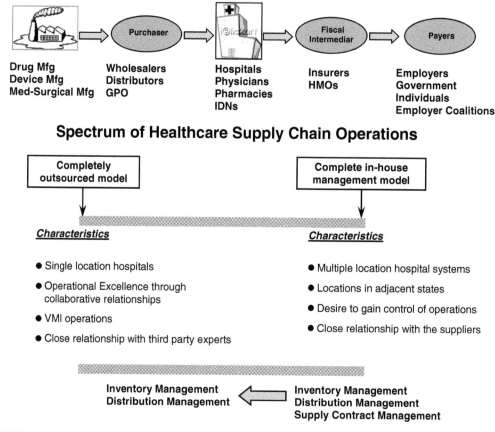

FIGURE 29.6 The health care value chain.

component of their service delivery strategy, it does not mean that the firm should perform the logistics functions internally. They may instead consider their service their key competency and perform that function exceptionally well, while delegating logistics to a firm that excels in logistics. Thus, the question of whether or not these functions should be internalized or externalized remains open and depends on many factors

To illustrate some of these factors we will discuss two contrasting healthcare providers: Mercy Health Systems and The Nebraska Medical Center. Sisters of Mercy Health System is a hospital system based in St. Louis, Missouri, that operates facilities and services in a seven-state area encompassing Arkansas, Kansas, Louisiana, Mississippi, Missouri, Oklahoma, and Texas (see Fig. 29.7). With a total of 26,000 coworkers, 815 integrated physicians, 3100 medical staff members, and 3600 volunteers Sisters of Mercy Health Systems (hereforth referred as Mercy) can be classified as a medium- to large-scale hospital system. Its members include 18 acute care hospitals providing more than 4000 licensed beds, a heart hospital, a managed care subsidiary (Mercy Health Plans), physician practices, outpatient care facilities, home health programs, skilled nursing services and long-term care facilities. Mercy is also the ninth largest Catholic healthcare system in the United States, based on net patient service revenue. Established in 1986, the Health System is operated through regional "Strategic Service Units" (SSUs) which enjoy mutual benefits of local management and system strength. Mercy programs and services are driven by the specific needs of each SSU's community, and local operating autonomy is valued. A key component of Mercy's service is its focus on five quality factors: (*i*) information about programs that educate patients,

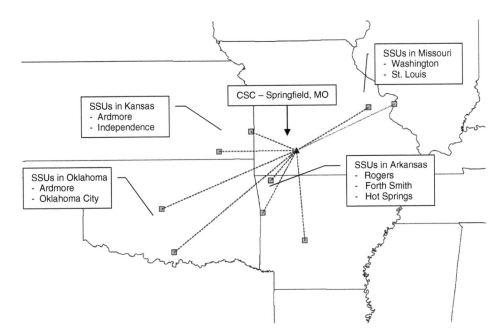

FIGURE 29.7 Mercy health systems strategic service units.

technology that enables and expands service, customized patient service experience, doctor quality, and nurse quality. As we can see, the key competency for Mercy is patient service, not logistics; however, Mercy recognized that logistics can be a key enabler in improving their quality factors.

As part of their quality improvement efforts Mercy launched a program called Mercy Meds. Mercy Meds entailed a comprehensive transformation of medication administration processes that incorporated technology, supply chain management, strategic partnerships and improved work processes to enhance safety and efficiency in the delivery of medications to patients. Supply chain management activities are a key part of Mercy Meds and play an important role in the provision of high-quality service to patients. The newly designed process begins at a consolidated services center (CSC) which serves as a centralized warehouse and distribution center for the entire Mercy organization. As a part of the Mercy Meds scheme, Mercy took a unique step of becoming its own pharmaceutical distributor. Through a partnership with the nation's largest pharmaceutical wholesaler, AmerisourceBergen, the CSC purchases, stores, repackages, bar-codes, and distributes pharmaceuticals used across Mercy.

Through the Mercy Med effort, Mercy developed in house expertise that eventually led Mercy to establish an in-house organization, Resource Optimization and Innovation (ROi) to manage its internal and external supply chain activities. ROi operates as a "for profit" internal entity which is now responsible for group purchasing functions, logistics, distribution, and other supply chain activities for the entire Mercy system. For example, Mercy now operates a private fleet in order to meet its SSU's specific delivery requirements. Mercy's success in implementing their in-house model can be attributed to their size (8 SSUs, over 3600 beds), location (relatively close geographically), specific logistics requirements (high fill rate requirements, specific delivery requirement), and the recognition by top management that logistics can be a key enabler.

To contrast the in-house model, we will briefly discuss how The Nebraska Medical Center (NMC) handles its inventory and logistics processes, and why it chose to outsource its logistics functions. The NMC formed in October 1997 with the merger of Clarkson Hospital, Nebraska's first hospital, and University Hospital, the teaching hospital for the University of Nebraska Medical Center (UNMC), located in Omaha. The NMC is a single location medium-sized hospital with 689 beds and over 950 physicians. Prior to outsourcing their inventory functions, the NMC operated as many other similar size hospitals. They had their own warehouse, located at a site close to the hospital, where all pharmaceutical, medical surgical, and other supplies were received and stored. Demand for various items from the hospital was fulfilled through the warehouse. The NMC purchased most of its pharmaceutical and medical surgical requirements from a large distributor of national name-brand medical/surgical supplies to hospitals.

Like many hospitals of its size, the NMC managed its own inventory and supply chain operations and placed orders with the distributor as required. Typically, hospitals do not have strong and matured inventory and supply chain management functions, as those are not core competencies of the hospitals. But these functions may have a significant impact on the overall customer service offered by the hospitals and also on the cost of operations as almost 30% of a hospital's total operating cost are materials costs which include pharmaceutical and medical surgical supplies. The executives at NMC started thinking about improving the operations and recognized that there was a significant opportunity in inventory management and supply chain functions. The NMC wanted to focus on its core competency, healthcare services, and wanted to offer differentiation in healthcare services at a lower cost and better returns on investment. After benchmarking and studying the vendor managed inventory program between P&G and Wal-Mart, NMC decided to look for similar opportunities in NMC operations. Following over two years of studying the intricacies of it operations, NMC identified a third party partner (Cardinal Health) who could run their inventory and supply programs. Cardinal Health was picked because of its capabilities of offering inventory management and supply chain services by integrating its information systems with NMC's systems. Cardinal Health already had inventory hardware and software systems in place via their Pyxis inventory systems and thus already understood many of NMC's requirements.

Under the new program Cardinal would not only manage inventory of all pharmaceutical and medical surgical supplies at NMC but also own the entire inventory at NMC for a monthly fee and a share in the yearly savings. One of the main drivers for this initiative was the freeing up of capital for NMC. Typically, capital is scarce at smaller sized hospitals and the new initiative freed over $80 million dollars for NMC. This capital, which was investment in inventory, could then be reinvested in other technological improvements and improvements in service offerings. Now Cardinal manages the inventory levels at NMC and is penalized for any out-of-stock situations. Inventory locations in departments are replenished four times a week on Monday, Tuesday, Thursday, and Friday from its distribution center in Omaha, NE which is just two miles from NMC. The replenishment lead time from this distribution center is around 9 h and orders can be placed at the piece level. Any emergency requirements are fulfilled in a shorter time if necessary. NMC also receives some supplies from Cardinal Health's distribution centers in Chicago, IL, and Kansas City, KS, which typically have a lead time of three days but orders can be placed only in case packs. NMC's in-stock performance has improved significantly after implementation of this new program and is currently over 99.5%.

The NMC is an example of a hospital that neither has the demand levels necessary to bypass the Group Purchasing Organization (GPO) nor a matured supply chain function within the hospital. Thus, outsourcing the supply chain operations and inventory management functions to a third party expert who could use its expertise has proven to provide significant benefits. The dynamics of supply chain operations in the healthcare industry are constantly changing as the roles of distributors, GPOs, and manufacturers change. Hospitals need to identify the best supply chain strategy based on size, in-house management capabilities and the presence of trustworthy and capable third party experts in the vicinity that can reduce the operating costs and improve healthcare services by focusing on their core competencies. Both the examples, Mercy Health Systems and The Nebraska Medical Center, illustrate that no matter the size of the service system, logistics can become a key enabler and a strategy for success in providing improved customer service.

29.4   Summary

Service industries differ from traditional manufacturing industries in many important ways. One of the key differences is the cost structure associated with service firms. The study performed for the Council of Logistics Management [Arthur D. Little, Inc. and Penn State University (1991)] indicates that up to 75% of the total operating cost for service-oriented companies is labor and capital. That is, the majority of costs are fixed. In contrast, manufacturing firms have a different cost structure with much more cost tied up in inventory. The cost structure of service firms presents a unique problem in that an extra customer adds little marginal cost, but may add significant revenue. In other words, the customer is truly "king." Because so much cost is tied up in labor and capital, it is critically important that service industries design their service delivery systems from a customer and cost-efficient standpoint. The optimal design of logistics functions is and will continue to be a key enabler for service companies looking for competitive advantage in the market place.

Acknowledgments

I would like to thank Amit Bhonsle for his excellent work on the Mercy Health System and The Nebraska Medical Center examples and for allowing me to utilize some of his work. In addition, I would like to thank Marty Comody at The Nebraska Medical Center and Gerald Ledlow at Mercy Health Systems for allowing us to perform our case study work. I would also like to thank Francesco Selandari for all his efforts on the original Pyxis Helpmate robotic system research that led to the case study discussed in this chapter. Finally, I would like to thank Sannathi Chaitanya for assisting in finding some of the references associated with this chapter.

References

Albayrakoglu, M.M. (1986). Justification of new manufacturing technology: a strategic approach using the analytic hierarchy process, *Production and Inventory Management Journal*, vol. 37, no.1, pp. 71–76.

Banks, J., Carson, J.S. II, and Nelson, B.L. (1996). *Discrete-Event System Simulation*, Englewood Cliffs, NJ: Prentice-Hall International.

Burns, L.R. and Wharton School Colleagues (2002). *The Health Care Value Chain: Producers, Purchasers, and Providers*, San Francisco: Jossey-Bass.

Burns, L.R. and Pauly, M.V. (2002). Integrated delivery networks: a detour on the road to integrated healthcare?, *Health Affairs*, July–August, 128–143.

Cook, D.P., Goh, C-H., and Chung, C.H. (1999). Service typologies: a state of the art survey, *Production and Operations Management*, vol. 8, no. 3, pp. 318–338.

Daskin, M.S. (1995). *Network and Discrete Location: Models, Algorithms and Applications*, John Wiley and Sons, Inc., New York.

Goldberg, J.B. (2004). Operations research models for the deployment of emergency services vehicles, *EMS Management Journal*, vol. 1, no. 1, January–March, pp. 20–39.

Johnson, T. (1999). Using Simulation to Choose Between Rental Car Lot Layouts, in the *Proceedings of the 1999 Winter Simulation Conference*, P.A. Farrington, H.B. Nembhard, D.T. Sturrock, and G.W. Evans, eds. IEEE Piscataway New Jersey.

Khasnabis, S. (1994). Prioritizing transit markets using analytic hierarchy process, *Journal of Transportation Engineering*, vol. 120, 74–93.

Arthur, D. Little, Inc. and Penn State University (1991). *Logistics in Service Industries*, The Council for Logistics Management, Oak Brook, IL.

Mielke, R., Zahralddin, A., Padam, D., Mastaglio, T. (1998). Simulation Applied to Theme Park Management, in the *Proceedings of the 1998 Winter Simulation Conference*, D.J. Medeiros, E.F. Watson, J.S. Carson, and M.S. Manivannan, eds. IEEE Piscataway New Jersey.

Min, H. and Melachrinoudis, E. (2001). The three-hierarchical location-allocation of banking facilities with risk and uncertainty, *International Transactions in Operational Research*, vol. 8, 381–401.

Mouette, D. and Fernandes, J.F.R. (1997). Evaluating goals and impacts of two metro alternatives by the AHP, *Journal of Advanced Transportation*, vol. 30, 23–35.

Murdick, R.G., Render, B., and Russell, R.S. (1990). *Service Operations Management*, Allyn and Bacon, Boston, MA.

Rossetti, M.D. and Turitto, T. (1998). Comparing static and dynamic threshold based control strategies," *Transportation Research Part A*, vol. 32, no. 8, pp. 607–620.

Rossetti, M.D. and Seldanari, F. (2001). Multi-objective analysis of hospital delivery systems, *Computers and Industrial Engineering*, vol. 41, 309–333.

Rossetti, M.D., Kumar, A., and Felder, R. (2000). Simulation of mid-size hospital delivery processes, *Health Care Management Science*, vol. 3, 201–213.

Saaty, T.L. (1977). A scaling for priorities in hierarchy structures, *Journal of Mathematical Psychology*, vol. 15, 234–281.

Saaty, T.L. (1994). How to make a decision: the analytic hierarchy process, *Interfaces*, vol. 24, no. 6, 19–43.

Saaty, T.L. (1997). Transport planning with multiple criteria: the analytic hierarchy process applications and progress review, *Journal of Advanced Transportation*, vol. 29, no. 1, 81–126.

Takakuwa, S. and Oyama, T. (2003). Simulation Analysis of International Departure Passenger Flows in an Airport Terminal, in the *Proceedings of the 2003 Winter Simulation Conference*, S. Chick, P.J. Sanchez, D. Ferrin, and D.J. Morrice, eds. IEEE Piscataway New Jersey.

Wang, Q., Batta, R., and Rump, C.M. (2002). Algorithms for a facility location problem with stochastic customer demand and immobile servers, *Annals of Operations Research*, vol. 111, no. 1–4, 17–34.

30

Securing the Supply Chain

Luke Ritter
Trident Global Partners

30.1 Introduction

A compelling requirement exists to continue enhancing security throughout the global trade community. This effort is often euphemistically and universally referred to as supply chain security. In the modern world of global trade, however, it is important to make two key observations about this term. First, is that the processes, procedures, and essential elements that comprise a business entity's logistics solutions these days are much more like a network than a chain. And second, that so much of the average corporations value and related ability to provide products and services can be directly related to these activities that it may be more appropriate to think of them as value chains versus supply chains.[1] The resultant edits would logically bring us to the term: value network security—certainly not as elegant or widely accepted a term as supply chain security, but arguably a better descriptor for the issues at hand. That said, in deference to the more commonly accepted terminology, and in order to stay in step with conventional wisdom, we will use the term supply chain security throughout this chapter.

As a fairly universally accepted rule of thumb, the ultimate goal for any supply chain security initiative is to be able to continuously and effectively operate a business while facing increasingly higher threat levels. Stakeholders in the global trade community are continually being forced to address and respond to greater threats to continuity of operations than any time in recent history. Crimes perpetrated by terrorists, smugglers, stowaways, and others with illegal intent all have the potential to significantly disrupt the core business being conducted throughout the global transportation network. In most cases, exploiting weaknesses in a businesses security strategy is a critical success factor that can be directly attributed to the success or failure of these events. Ultimately, the crimes perpetrated by these individuals can serve to disrupt the free flow of commerce, and subsequently reduce the competitive posture of the businesses that are impacted by their effects. Global trade activity is the undeniable lifeblood of the world's economy, and the "mission critical" linkages that are provided by the global supply chain must be protected.

The potential economic impact of disruptive events in the supply chain is far from trivial. In January 2007, a Mediterranean Shipping Company container vessel was damaged in a severe storm while transiting between Antwerp and Durban. The ship was subsequently beached on a sandbar in an effort to keep her from sinking. Insurance professionals have estimate that this ship will likely be a total loss, and the economic impact of this event has conservatively been estimated at over $300 million dollars. This incident prompted industry experts to ponder what the economic impact would be if one of the largest container ships afloat—three times the size of this ship—were to be lost at sea. The answer is daunting. Some have speculated that the impact could ultimately rise as high as $5 billion dollars—for one ship and its cargo.[2]

Stakeholders throughout the global trade community continue to be subject to significant man-made, as well as natural hazards. The potential benefit that exists related to an organization's ability to develop and implement an aggressive security strategy have never been greater, primarily because the threats to the vital functions that drive the global supply chain have never been greater.[3] Entities throughout the global trade community who are endeavoring to secure their critical supply chain infrastructure cannot be expected to protect against all hazards, at all times—but one universal truth seems to hold true— as the authors of *Securing Global Transportation Networks* have pointed out, there is significant business value to be derived by "developing strategies that serve to avoid, minimize, or at least survive the effects of a major disruptions" in the supply chain.[4]

Just in time supply chain initiatives, increasingly more sophisticated approaches to the movement of goods throughout the world, and the steady increase in global trade capacity related to the effects of globalization and developing economies, have combined to increase congestion at ports, highways, airports, and rail yards. Terminal operators, railroads, steamship lines, retailers, transportation intermediaries, and others can expect to face increasing pressure to find a way to keep global freight moving. At the same time, threats from potential disruptions continue to evolve and appear to be growing increasingly more sophisticated, and unfortunately, potentially more catastrophic. It is this intersection between business process improvement and security imperatives that continues to underscore the importance of solutions that represent both a business and a security benefit to an organization.

This chapter explores the forces that are driving the requirements for supply chain security solutions, highlights the most fundamental elements of supply chain security, investigates the concept of value creation associated with managing security as a core business function, explains some of the public and private sector security initiatives that have been initiated, and finally takes a look at the road ahead for supply chain security.

30.2   Security Imperative

Supply chain executives are faced with a difficult dilemma: how to find the right balance between effective security measures and efficient movement of freight. Balance is critical in this instance. A supply chain solution that enables freight to flow totally unencumbered but lacks the associated protection to ensure on-time and in-tact delivery, is no more attractive to the industry than a solution that includes inefficient and restrictive security measures and chokepoints. The ability to facilitate the secure and efficient movement of freight is a perquisite for daily operations as well as for economic growth in virtually any organization that has a logistics component.

The quality revolution taught us that business process improvement is directly linked to an organization's willingness to commit to continual, incremental improvement—security is no different. Enhancing the supply chain security posture of an organization, as a complementary component to efficiency in the supply chain, requires the same commitment. The definitive answer to the security imperative must undoubtedly include the process of adopting an enterprise approach to security that cultivates and leverages multiple core security competencies to create value.

Effective supply chain security requires attention to a unique set of capabilities, and often demands expertise in an array of specific disciplines from freight operations to infrastructure protection to transportation technology. Sensitivity to the nuanced nature of the individual missions, capabilities, and cultures of many federal, state, and local law enforcement agencies can be critical. The ability to build bridges among supply chain stakeholders and to remain current in relevant best practices—public, private, domestic and global, are all critical components of an effective supply chain security solution. Finally, reliable analyses of emerging issues and industry trends, and the ability to adapt and respond rapidly to complex threats and challenges to business continuity are the characteristics that differentiate organizations with successful supply chain security solutions from the rest of the market place.

30.3 Primary Threats to Supply Chain Security

The Introduction presented the idea that the genesis of tangible threats to supply chain solutions around the globe is a combination of both natural and man-made events. This does not narrow down the list very much. In fact, there is a wide variety of these events that all have the potential to cause significant disruptions: labor unrest, sabotage, illicit drug activity, governmental instability, illegal immigration, smuggling, hurricanes, power outages, and of course, violence as a political tool—terrorism. All of these sources have, and will continue to threaten the delicate balance that exists throughout the world of global trade. Labor unrest resulted in a 10-day closure of all of the ports on the West Coast of the United States in 2002, and resulted in serious economic impact.[5] Hurricane Wilma crippled railroad infrastructure in the state of Florida for weeks in 2005. And terrorist attacks in recent years on subway systems in England, Spain, and India have all resulted in deaths, destruction of vital transportation infrastructure, and significant economic losses in the global supply chain.

In an important recent work on supply chain security called *The Resilient Enterprise*,[6] the author and MIT professor argues that it is not just major corporations that are dependent upon global supply chains. He points out that virtually all modern products enterprises are "interwoven networks of companies involved in getting goods to market." He goes on to explain that in addition to the normal competitive pressures that have developed as a result of "just in time" manufacturing practices and the associated customer service standards, businesses must be able to address vulnerabilities associated with "high-impact/low-probability events." Because of the networked nature of modern supply chains, companies can be impacted by disruptive events that happen outside of the discreet confines of their business operations. Responding to disruptions that affect second- and third-tier suppliers, businesses that a firm may not routinely come in direct contact with, can be just as important as dealing with other security events.

With the understanding that any and all of these threats pose the potential for significant disruptions, we may now focus on three specific threats that have both a high probability of occurrence in the global supply chain, as well as a high potential for significant impact on businesses:

30.3.1 Meteorological Events

Natural meteorological disturbances (hurricanes, earthquakes, tsunamis, etc.) can, (a) be counted on to occur, (b) happen with some frequency throughout the world, and (c) frequently result in significant or even catastrophic supply chain disruptions and associated economic losses. Even if the event itself does not cause significant damage, disruptions related to the temporary or permanent loss of human capital can have the same result on business operations.

30.3.2 Smuggling

For perhaps as long as there have been supply chain solutions, they have been used as a conduit for smuggling. These activities can consist of relatively innocuous (yet still illegal and threatening) activity related to gray market goods or drugs, to the more serious issues related to smuggling weapons, people, and explosives. The greatest potential for catastrophic disruptions related to smuggling exists in the unauthorized transportation of weapons of mass destruction, or the materials and components used to build these weapons.

30.3.3 Terrorist Attack

Conventional wisdom in this area dictates that terrorists favor attacks on targets that have the potential to result in significant media exposure—normally with the goal of providing exposure to their associated political message. This source also tends to favor attacks that result in extensive loss of life or economic disruption. Supply chain infrastructure represents an attractive target as it satisfies both of these prerequisites.

Following is a list of threat scenarios that have commonly been used to assess vulnerability in the supply chain:

- Hijacking of a vessel/airplane/train/bus.
- Tampering with cargo in transit.
- Smuggling dangerous substances into a facility.
- Blockage of key transportation conduits.
- Use of the supply chain to transfer weapons.

All of these scenarios represent the potential for catastrophic disruptions in the supply chain. Unless supply chain decision-makers establish mitigating strategies to address these and other disruptions, the potential for these events to disrupt normal operations will remain high.

30.4 Fundamental Elements of Supply Chain Security

The fundamental building blocks of any supply chain security solution are the same. These essential elements include: access control; assessments and plans; asset tracking and accountability; cargo screening; command and control; communications and IT; and surveillance and monitoring.[7]

30.4.1 Access Control

In order to establish and maintain a secure operational environment, facilities involved in global trade must possess the fundamental ability to control access. The ability to determine if an individual is authorized to enter/exit, has a reason for being in a facility, has the required clearance to be in a sensitive or controlled area, and has valid credentials can all contribute significantly to a comprehensive supply chain security solution. Access control systems currently being employed throughout the global supply chain community range from very simple, administrative systems to more sophisticated, technologically advanced systems that can automatically validate, record, and permit or restrict access to a location.

Some of the high-level objectives that have been presented for the U.S. Government's Transportation Worker Identification Credential (TWIC) program are a good indication of the type of capabilities that should be an integral part of a sound access control strategy:

- Allow for a positive match through the use of a secure reference biometric
- Allow for the centralized ability to interface with other federal agencies and databases
- Allow for centralized record control

- Reduce the risk of fraudulent/altered credentials through use of state-of-the-art antitamper and anticounterfeit technologies
- Minimize the requirement for redundant credentials and background investigations

Finally, effective access control solutions tend to be those that are designed and implemented with reliability and maintainability in mind. When properly implemented, effective access control systems can support continuity of operations throughout the global supply chain.

30.4.2 Assessments and Plans

Many supply chain security solutions have recently been developed to comply with established governmental security requirements. While these solutions may serve the purpose of meeting a minimum regulatory standard, they do not necessarily account for the predatory and adaptive nature of the modern terrorist threat, nor do they attempt to provide continuity and resilience in the face of catastrophic natural events.

Effective management of assessments and planning, by each individual business entity, can contribute significantly to the goal of enhancing business continuity throughout the global supply chain. Careful planning can enable supply chain professionals to establish flexible, responsive policies and procedures that address a wide array of threats. By continually monitoring, tracking changes, and adjusting policies and procedures that are aligned with current assessments, the impact of disruptive events can be mitigated.

30.4.3 Asset Tracking

Asset tracking in a competitive supply chain environment is an integral part of every-day operations. The requirement to identify movement in the supply chain (visibility) is driven by both business and security requirements. Technology has facilitated asset tracking by providing increased access to location data within the confines of a designated operational area. The advent of radio frequency identification (RFID) tags, optical character recognition (OCR) systems, other global positioning system (GPS) enabled tracking systems, as well as the associated software, continues to offer operators the capability to enhance both their business and security profiles.

30.4.4 Cargo Screening

Within the last several years, the federal government has established aggressive cargo screening requirements for freight entering United States ports. The U.S. Department of Homeland Security's strategy is based on 100% screening of cargo shipment data for all containerized cargo destined for American ports. Between 2004 and 2006, U.S. Customs and Border Protection increased the percentage of containers processed through the Container Security Initiative from 48% to 82%.[8]

The Container Security Initiative continues to build bi-lateral cooperative relationships in which United States and foreign customs authorities work with their international counterparts to build working relationships to enhance global supply chain security. Radiation scanning equipment is being installed at all major U.S container ports, as well as other key international load centers. Supply chain professionals throughout the world have also been affected by the "24 Hour Rule." Implemented in 2003, this law requires carriers to provide cargo manifest information to U.S. law enforcement entities at least 24 h before a container is loaded onto the vessel destined for the United States. All of these initiatives have been designed and implemented in an attempt to have better visibility throughout the global supply chain.

Effective cargo screening programs throughout the global trade community may share some common goals:

- Designated primary and secondary inspection stations that are at a safe distance from critical operational assets

- Screening solutions that quickly and thoroughly scan for suspect elements
- Archiving capability to validate and record screening data
- Open architecture technology capable of sharing data
- Universally deployable, reliable solutions

By properly employing cargo screening tools, supply chain professionals have the ability to reduce data errors, reduce transaction processing, reduce risk, reduce waiting time, reduce manpower costs, increase throughput, and mitigate risk. Effective cargo screening has the potential to enhance existing business processes and decrease the likelihood that a disruptive event will translate into a significant negative operational impact and associated economic loss.

30.4.5   Surveillance and Monitoring

Supply chain security solutions are not complete without the ability to monitor, detect, alert, and record suspicious events. By providing advanced warning related to potential disruptive events, supply chain stakeholders can enhance their overall security posture, and potentially create multiple opportunities to mitigate threats.

The fundamental building blocks of an effective surveillance and monitoring capability should include:

- Surveillance—continuous monitoring to detect abnormal conditions.
- Detection—identification of abnormal conditions.
- Alarms—notification sent to the appropriate authority for action.
- Tracking—abnormal condition monitored until resolved.
- Archiving—data storage to preserve a complete record of events.

Current state-of-the-art surveillance and monitoring systems offer all-weather, day and night, wide-area detection and tracking capabilities. Some of these systems employ high-performance camera technologies and radars to increase the probability of detection and enhance tracking capabilities. Many of these systems come bundled with software packages that provide additional value-added features such as remote operating capability, digital recording and playback, and mapping.

30.4.6   Command and Control

It is hard to argue that the ability to effectively direct, coordinate and control supply chain security assets is not a valuable resource. Realizing this potential, however, can require significant investments in technology, training, and human capital. Many supply chain entities have implemented discreet technology solutions that provide security data, but have stopped short of investing in the decision support tools that can transform that data into knowledge.

Interfacing with local, state, federal, and even international law enforcement agencies has become a normal and customary component of supply chain security. Where various stakeholders throughout the global supply chain tend to differentiate themselves, in this case, is in their ability to manage situational awareness, including creating a "common operational picture" that incorporates security data from multiple sources. The best of these systems provide near real-time display of security information and facilitate decision making, particularly under adverse conditions. Effective command and control applications can significantly enhance resilience and enhance the supply chain security posture of a business.

30.4.7   Staffing, Training, and Exercises

Proper employment of a security training and exercises regime can serve to reinforce core competencies and increase awareness. To be effective, this training must be linked to practical and realistic threat and vulnerability scenarios which have the potential to significantly disrupt supply chain operations. Exercises should be designed to test the full spectrum of response capabilities within an organization, and should emphasize the impact that lack of resilience can have. Many progressive organizations within the global supply chain community have discovered that staffing, training, and exercises should reflect an "effects-based approach" in order to have maximum impact.

A comprehensive approach to supply chain security staffing, training, and exercises can require expertise from a broad spectrum of functional areas, including security and law enforcement, transportation operations, command, control and communications, systems integration, and business continuity planning.

30.5 Supply Chain Security and Total Security Management

A new concept has recently emerged in supply chain security that is worth noting. Using the paradigm of "Total Quality Management" as a benchmark, and comparing some of the fundamental challenges and lesson learned from the quality revolution, the authors of *Securing Global Transportation Networks: A Total Security Management Approach*[9] introduced the concept of Total Security Management (TSM). By going back to the most fundamental question that supply chain professionals must ponder, in this case: "does security matter?" Ritter et al.[9] argue that not only does it matter, but that security matters enough to be managed as a core business function to create value—much the same way that quality is now managed throughout the global business community.

In the beginnings of the quality revolution, many serious, thinking professionals decided that quality did not matter enough to be managed as a core business function, and they rejected the notion that consumers would pay for higher quality—they all turned out to be mistaken. Perhaps the same is true today concerning the heated debates that continue to develop regarding the value and return on investment related to security initiatives, which only time will tell.

The definition of TMS is as follows: "The business practice of developing and implementing comprehensive risk management and security best practices for a firm's entire Value Chain, including an evaluation of suppliers, distribution channels, and internal policies and procedures in terms of preparedness for disruptive events such as terrorism, political upheaval, natural disasters, and accidents."[10]

As the definition demonstrates, this approach to supply chain security ultimately includes all of the business partners in a firm's supply chain, including second- and third-tier carriers and suppliers, and any business entity that could reasonable be considered part of the overall business enterprise. This is an important point. Simply engaging with the first-tier stakeholders in a supply chain solution does not provide the depth or breadth required to optimize the return on a security investment. It is also the metaphoric equivalent of locking the front door and leaving the back door open. Unless all relevant parties in a firm's total supply chain solution are considered in the solution set, the resulting initiative runs the risk of being incomplete, and perhaps inadequate.

Five "Strategic Pillars" form the foundation of this approach, and represent the baseline framework for supply chain security solutions that are developed using this methodology. The strategic pillars are:

- Total security practices must be based on creating value that can be measured.
- Total security involves everyone throughout your value chain.
- Total security implies continual improvement.
- Total security helps firms avoid, minimize, or survive disruptive events.
- Total security requires resiliency and business continuity planning as essential business functions.[11]

At the end of the day, it will be entirely up to the private sector to determine whether or not there is enough merit to this methodology to pursue it as a primary driver for the security solution set. This will undoubtedly be an evolutionary process, as the quality revolution was, and could take many years to evolve to the point where it becomes conventional wisdom.

What remains reasonably obvious today is that significant security vulnerabilities remain in the global supply chain. Both the public and private sector leaders essentially have two choices: (a) accept the status quo and hope that these vulnerabilities are not exploited in a way that leads to major disruptions and economic damage, or (b) take action to close the gaps that exist and transfer the global supply chain from an attractive target into a secure and resilient mainstay for the global business community.

30.6 Public–Private Partnerships

The ultimate solution to the challenges posed by threats to the global supply chain must be, by default, the result of concerted effort from both the public and the private sectors. It is essentially a universal truth that governments of the world bear the responsibility for protecting their citizens. In the case of the United States, this responsibility extends to include "all enemies, foreign and domestic." We also now know that the critical transportation infrastructure that the global supply chain depends upon for productive, efficient business operations is a popular target for those foreign and domestic enemies. This means that governmental responsibility to provide security now frequently crosses over into the business domain. Bridges, tunnels, rail yards, ports, fuel farms, power plants, and airports all must be protected in the context of providing a common defense for a nation's citizenry.

At the same time, there is usually a major disconnect in this process that exists. While governments are charged with providing security, they rarely own or operate the critical transportation infrastructure that needs to be protected. Experts have estimated that perhaps as much as a full 90% of transportation and logistics infrastructure is owned and operated by the global private sector. How then do we reconcile a situation where a government is expected to protect assets that are almost entirely outside of their direct control?

The answer to this question is still evolving as this book is being written. But it is safe to say that part of the definitive answer has to do with establishing and maintaining effective public–private partnerships.

The *Journal of Commerce* conducted a survey of supply chain and logistics professionals in 2006 (Fig. 30.1). One of the questions that respondents were asked to answer was this: "How effective do you feel the government (United States) is in preventing an ocean container attack in the United States?" The answers were interesting. Only 5% of the survey sample responded with "Very Effective." Exactly half of the respondents responded "Fairly Effective." But a full third of the professionals sampled answered "Fairly Ineffective." So to what can we attribute this lack of confidence in the government's ability to provide security in the supply chain? It is a combination of two things: (a) all of the governments in the world could not possibly spend enough money to solve all of the supply chain security challenges that exist throughout the global supply chain, and (b) effective public–private partnerships have not fully evolved to address these issues.

To understand this situation, it is helpful to outline a few of the factors that can contribute to the success or failure of a public–private partnership for security. Effective public–private partnership in this case is not a situation where the government is underwriting all security initiatives. Too often in the past, business entities have attempted to avoid responsibility for security by looking back to the

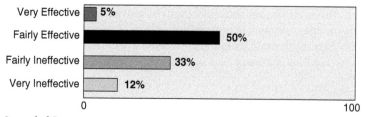

FIGURE 30.1   *Journal of Commerce* survey, 2006.

government as the sole source for this protection. If private sector businesses own and operate 90% of this infrastructure that needs to be protected, it follows then that they should at least share the burden of ensuring security. Public–private partnerships should not represent an excuse for the private sector to avoid implementing a substantial, enterprise security program. The operative word in this case is partnership, which demands a real commitment from both parties. Similarly, for these programs to succeed, they should not be small scale and strictly voluntary. There is only a limited degree of protection that can reasonably be expected to be derived from this kind of structure. Finally, public–private partnerships should not be events that happen only once. Much like any legitimate process improvement initiative, they should be oriented toward continual, incremental improvement. Once established, these programs need to be monitored, evaluated, and cultivated to ensure success.

Additional essential elements for successful public–private security partnerships also include collaborative and globally oriented approaches. The supply chain profession is global in nature, and it follows then that security partnerships should be developed using the same focus point. Collaboration among all of the pertinent stakeholders involved in the partnership is essential. In most cases, these programs will only be as good as the effort that is made to include key players in the process. Finally, in order to be effective, public–private partnerships for supply chain security must be mutually beneficial, and should be designed to reinforce traditional roles.

Governments of the world are not typically responsible for solving business challenges—businesses are. Supply chain security is a business challenge. By effectively performing its traditional role of legislation and regulation, governments have the potential to establish and enforce significant incentives that encourage and reward effective security initiatives. Similarly, when an appropriate and tangible reward mechanism exists for pursuing enterprise security, the global business community has both the incentive and the ability to focus an impressive amount of resources on the challenges presented by supply chain security.

30.7 Road Ahead

"*Ex Scientia Tridens*" (from knowledge, seapower), the U.S. Naval Academy motto, reminds us that knowledge and information can and should be at the root of security initiatives. In the case of global supply chain security, this is certainly true as well. Mitigating threats in the supply chain posed by natural and man-made hazards, using information technology solutions, is becoming increasingly more important.

In general, the security focus in the supply chain industry over the last five years has been on developing solutions that establish some baseline of compliance-oriented protection. But the real value in implementing strategic security initiatives has to do with solutions that go beyond minimal compliance, and ensure business resilience. Resilience translates to business value over time. And a businesses ability to mitigate all-hazards threats should ultimately become a key differentiator in the global marketplace.

At the SecurePort 2007 conference in Houston, Texas, an industry professional event focused on the delivery and exchange of innovative ideas, several current and former U.S. Coast Guard Admirals made reference to the term "unity of results." If these Admirals have their way, the trend in the United States will be a transition to a "unity of results" approach from the more commonly accepted "unity of effort" mentality. This is a very important differentiation in the way that public and private enterprises can address security concerns. By focusing on an end result that is designed to create both business and security benefits, end users can do more than simply check boxes toward compliance. On the information management side of this equation, for example, using this approach can convert security data into "actionable intelligence," or knowledge that provides an advantage in the effort to mitigate risk.[12] A "knowledge-based approach" to command, control and communications in a security environment is critically important in order to ensure that security information does not become just another cumbersome stockpile of data.

TABLE 30.1 Gap between Perceived Security and Actual Security

Security Perception	Security Reality
The Port of _____ is not an attractive target for terrorists.	The United States' 129 highest volume ports have been subjected to a risk-based evaluation and the Port of _____ was placed on the list of the 66 highest risk ports in the United States.
2002 Maritime Transportation Security Act (MTSA) has provided the solution to port security requirements at the Port of _____.	ILWU Port Security Director Michael Mitre highlighted the "… problem of system-wide noncompliance with existing port security regulations"—speaking before the Senate HLS Committee.
Port administration authorities are confident that the Port of _____ has adequate security in place to protect against primary threats.	Interviews with port police officers, security audits, and state documents point to discrepancies such as gaps in fences, unattended gates, inoperable alarm and camera systems, and insufficient command and control capability.
Security investments made by the Port of _____ fund "best of breed" security solutions.	The "low bidder" requirements driven by the Port of _____ procurement laws practically preclude best of breed solutions by default.
Security expenditures only represent a net cost with no potential for associated business return on investment.	Security initiatives at Port of _____ can provide a real return on investment if designed with that goal in mind.
Federal and state government should be responsible for all port security and are capable of providing adequate security for all transportation infrastructure at the Port of _____.	A large percentage of the assets at the Port of _____ are owned and operated by the private sector and public–private partnerships are critical to secure commerce.

Unlocking the true latent power that exists in the private sector, to invest in security initiatives that strengthen the global supply chain networks, is an important part of the future of supply chain security. This must be done cooperatively, with the public sector, in an appropriate and mutually beneficial way. Some experts are still skeptical about the industry's ability to collaborate in this way, and to elevate the importance of security and resilience to the point where we collectively achieve a balance between risk and reward. Technology is required in many cases to execute these strategies, and technology initiatives can be costly. In most cases, however, the cost of implementation for effective security initiatives can become insignificant in the face of the related return on investment—particularly when major disruptions are avoided or mitigated.

The global supply chain community has an opportunity to invest in security and resilience in an effort to ensure a secure operational environment, and long-term economic viability. Because the global trade community is, in actuality, a diverse and distributed network, these initiatives stand to benefit the local, national, and even the global business community. In the end, business and security objectives can be complimentary: increased productivity; enhanced competitive posture, acceleration of business development; increased customer satisfaction; enhanced resilience; and ultimately, continued growth and prosperity for the world's supply chain economy.

30.8 Case Example: Security at a Major U.S. Port

Table 30.1 illustrates the gap that can exist between common perception of security and the actual security posture of a major node in the global supply chain. The information presented in the Security Reality column is factual, but the name of the port has been redacted due to the sensitive nature of the topic.

References

1. Ritter, Barrett, and Wilson, *Securing Global Transportation Networks: A Total Security Management Approach* (New York: McGraw Hill, 2006).

2. Peter Leach, Sunk costs, *Journal of Commerce*, (05 February 2007), p. 11.

3. Luke Ritter, The Elephant in the Room is ROI, *Cargo Security International*, (06 December 2006), pp. 20–23.

4. Ritter, Barrett, and Wilson, *Securing Global Transportation Networks: A Total Security Management Approach* (New York: McGraw Hill, 2006).

5. Chris Isidore, Bush to Seek to Re-open Ports, CNN Money, http://money.cnn.com/2002/10/08/news/ports/index.htm (October 2002).

6. Yossi Sheffi, *The Resilient Enterprise*, (Massachusettes: The MIT Press, 2005).

7. Reference refers to a White Paper: Essential Elements of Supply Chain Security, developed by Trident Global Partners, (June 2005).

8. U.S. Department of Homeland Security, <www.dhs.gov> (April 2007).

9. Ritter, Barrett, and Wilson, *Securing Global Transportation Networks: A Total Security Management Approach* (New York: McGraw Hill, 2006).

10. Ibid.

11. Ibid.

12. SecurePort 2007, Houston, TX, <http://www.secureportusa.com>, (January 2007).

Index

D

F

J

K

S

Milton Keynes UK
Ingram Content Group UK Ltd.
UKHW051903071024
449327UK00025B/2068